EARTH'S CLIMATE
PAST and FUTURE

EARTH'S CLIMATE

PAST and FUTURE

William F. Ruddiman

University of Virginia
Lamont–Doherty Earth Observatory

W. H. Freeman and Company
New York

Acquisitions Editors: *Melissa Wallerstein and Nicole Folchetti*

Development Editor: *Barbara Brooks*

Director of Marketing: *John Britch*

Media/Supplements Editor: *Bridget O'Lavin*

Project Editor: *Diane C. Davis*

Copy Editor: *Barbara Salazar*

Indexer: *Louise Ketz*

Cover Designer: *Paula Jo Smith*

Text Designer: *Victoria Tomaselli*

Illustration: *Robert Leo Smith Jr.*

Illustration Coordinator: *Bill Page*

Photo Researcher: *Vikii Wong*

Production Coordinator: *Paul W. Rohloff*

Composition: *Black Dot Graphics*

Cover and Title Page photographs: Top: *Icebergs. William Manning/The Stock Market.* Bottom: *Hall of mosses, Hoh Rain Forest, Olympia National Park, Washington. Galen Rowell.*

Library of Congress Cataloging-in-Publication Data

Ruddiman, W. F. (William F.), 1943–
    Earth's climate : past and future / William F. Ruddiman.
        p. cm.
Includes bibliographical references and index.
    ISBN 0-7167-3741-8 (pbk.) (EAN: 9780716737414)
    1. Climatology. 2. Climatic changes. I. Title.
    QC981 .R76 2000
    551.6—dc21   00-012165

Printed in the United States of America

Third printing

W. H. Freeman and Company
41 Madison Avenue, New York, NY 10010
Houndmills, Basingstoke RG21 6XS, England

To five colleagues who headed the effort to make the study of Earth's climate a science: John Imbrie, John Kutzbach, Wally Broecker, Nick Shackleton, and Murray Mitchell.

William F. Ruddiman is a member of the Department of Environmental Sciences of the University of Virginia. He holds a 1964 undergraduate degree in Geology from Williams College and a 1969 Ph.D. in marine geology from Columbia University. His first job was Senior Oceanographer at the U.S. Naval Oceanographic Office from 1969 until 1976. He then returned to Lamont-Doherty Observatory as a Senior Research Associate, a position he held until 1991. He took a year's leave from Lamont in 1981 to work at the National Science Foundation in the Ocean Science Division as part of the Marine Geology and Geophysics Program. After returning to Lamont, he was Associate Director of the Oceans and Climate Division from 1982 to 1986 and Adjunct Professor in the Department of Geology. He joined the University of Virginia in 1991 as Professor in the Department of Environmental Sciences and served as department chair from 1993 to 1996. At Virginia he has taught courses in climate change, physical geology, and marine geology.

Professor Ruddiman's research interests center on changes in climate. His early research focused on orbital-scale climate change, particularly in and around the north and equatorial Atlantic Ocean, much of it published with his colleague Andrew McIntyre. These efforts included investigations of marine plankton assemblages, ice-rafted debris, oxygen isotopes, volcanic ash, windblown particles, sediment mixing, and bottom currents. He was a member of the CLIMAP project from 1978 to 1984 and served as Project Director from 1980 until 1981. He was also a member of the COHMAP project from 1980 until 1989, and served on the steering committee throughout that time. His co-edited volumes resulting from this work include *North America and Adjacent Oceans During the Last Deglaciation* (with Herbert Wright, vol. K-3 of the DNAG series produced by the Geological Society of America, 1987), and *Global Climates Since the Last Glacial Maximum* (with Herbert Wright et al., University of Minnesota Press, 1993).

His more recent work has focused on the longer-term (tectonic-scale) physical and geochemical effects of the uplift of the Tibetan Plateau and other high topography on regional and global climate. This work compares climate model simulations against paleobotanic and geologic evidence of past climate change.

In 1997 he edited a book titled *Tectonic Uplift and Climate Change*, published by Plenum Press. His work on plateau uplift with his colleagues Maureen Raymo, John Kutzbach, and Warren Prell has been featured in BBC and NOVA television documentaries.

Professor Ruddiman is a Fellow of the Geological Society of America and a member of the American Geophysical Union. His recent service has included chairing review committees for the 1997 five-year renewal phase of the Ocean Drilling Project and for the AGU journal *Paleoceanography*, as well as serving on the initial steering committee for the ESH (Earth System History) Program at the National Science Foundation and as a member of the IGBP PAGES (Past Global Changes) Committee. He has participated in fifteen oceanographic cruises and was co-chief on leg 94 of the Deep-Sea Drilling Project and on leg 108 of the Ocean Drilling Project.

He lives on the side of a large hill (Pisgah Hill) in the Shenandoah Valley with his childhood sweetheart and wife, Ginger, and several dogs and cats. His hobbies include rock-wall building, gardening, and walking the Appalachian Mountains, remnants of the continental collisions that produced the supercontinent Pangaea.

Brief Contents

Contents

Preface

Over the last half century, scientists have pieced together a basic outline of how Earth's climate has varied in the past, and they have also uncovered many of the explanations behind these changes. As a result, some parts of the multifaceted story of Earth's climate history have begun to capture the attention of the public at large. Nonscientist friends frequently ask me questions about climate change, and new acquaintances are also curious about my efforts in this field. People are naturally curious about climate, probably in part because it is a subject that the average person can easily comprehend. This curiosity is spurred by the realization that human beings are now beginning to alter Earth's climate.

The Central Goal of *Earth's Climate:* Accessibility

Many college and university students share this natural curiosity, yet my sense is that schools are reaching only a tiny fraction of the students who might be attracted to the subject of climate. The central goal of this book is to make the story of Earth's climate accessible to many more people, teachers as well as students.

What has stopped us from reaching more students? Part of the problem is the enormous breadth of the subject: it is difficult to know enough about the many facets of climate to teach it well. Even active researchers concentrate on one or two of the many disciplines used to study climate—geology, ecology, paleobotany, glaciology, oceanography, meteorology, biogeochemistry, climate modeling, atmospheric chemistry, and hydrology, among others. In addition, the efforts of most research scientists tend to focus on just one or two of the many time scales of climate change (tectonic, orbital, deglacial and millennial, historical, and future).

It is nearly impossible then for most scientists in the field to keep abreast of emerging developments within all these disciplines and at all these time scales. So the first need is to help potential teachers broaden their base of knowledge enough to make them comfortable teaching the whole subject. This project requires a textbook that starts from a ground-zero knowledge base and moves step by step through logically developed explanations of the critical material. Such a text needs to be accompanied by graphics that succinctly summarize key points and by lists of follow-up references that give instructors access to background material in the literature.

Another part of the accessibility problem consists of the needs and expectations of the students. To interest a broad array of students in a scientific subject today requires at a minimum lively color graphics and a text written in plain English. The color graphics make the subject more interesting to students put off by dull or technical treatments. A clearly written text makes the subject more accessible to students and helps teachers convey a deeper understanding of the subject.

This book attempts to address these needs. Each part of the text has been rewritten in response to outside reviews by instructors and climate experts, and under the capable and thoughtful guidance of a content editor who is not a scientist. The result is a book written in plain English and containing over 400 new color graphics either created explicitly for *Earth's Climate* or adapted—greatly simplified—from the literature, plus numerous photographs.

I have worked to avoid the "lists and tables" structure that can deaden the teaching of science by reducing it to a series of disconnected facts. Instead, I use a combined storytelling and hypothesis-testing approach. In many chapters, evidence of past climate change serves to pose a basic question: What caused the observed changes? Most chapters are devoted to testing hypotheses that have been put forward by scientists. Some of this hypothesis-testing ends in a resolution of the mystery; in other cases, the causes become better constrained but remain unspecified. In either case, the student is invited to play detective in the problem-solving process by assessing the hypotheses posed against the accumulating evidence.

The Audience for *Earth's Climate*

Earth's Climate should prove useful at three levels in departments of earth and environmental sciences, as well as in related science departments such as geography, botany, ecology, and oceanic and atmospheric sciences:

1. The book can be used at upper undergraduate and graduate levels for students already specializing in

climate-related studies. At this level, it should help teachers broaden the scope of such courses toward more comprehensive coverage of additional time scales.

2. *Earth's Climate* will also prove useful for many other departmental majors who can benefit from taking at least one course that gives them some sense of Earth history. Many earth and environmental science departments have seen their enrollments decline in courses covering historical geology, stratigraphy, and paleontology, and many of these courses have simply disappeared from college and university curricula. A course taught from this book will inevitably take a historical perspective and will also give students baseline knowledge in many aspects of stratigraphy and some aspects of paleontology, using the unifying context of climate to convey information in an engaging and accessible way.

3. The potentially largest benefit of this book will be for instructors who want to create lower-level undergraduate courses accessible to many students, whether future scientists or not. This is the main body of students we are currently failing to reach with the story of Earth's climate. In my experience, most students who avoid science courses are fully capable of mastering the minimal amount of hard science included in this text. Rather than presenting numerous equations and calculations, this book draws on the intuitive knowledge we all possess as its main means of instructing. Yet *Earth's Climate* is not a dumbed-down insult to the intelligence of students: it requires them to put in a serious and thoughtful effort to learn the subject.

Each year I ask for a show of hands to find out which of my lower-level undergraduate students are English majors. I tell them I have targeted the course specifically for them, I invite them to let me know how they are coping during the term, and I track their progress. Most end up doing very well, basically just as well as the potential science majors. The possible benefits of a broadly appealing course that can satisfy a science requirement are obvious: greater numbers of students in classes and more majors.

The Organization of *Earth's Climate*

Earth's Climate falls under the general heading of Earth system history. It explores the basic climatic responses of Earth's major systems and subsystems (ice, water, air, vegetation, and land), and it traces their interactions through Earth's history. This book is intended to be self-sufficient, with no supplementary text required.

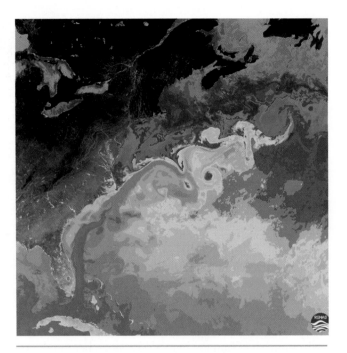

Ocean temperatures in the North Atlantic: Science or art? Although seemingly the product of the creative hand of an artist, these splashes of color are in fact a map of the surface temperatures of the western North Atlantic Ocean east of North America. The narrow band of bright red color marks the warm (27°C, 80°F) Gulf Stream, swirling through the cooler ocean temperatures and colors to the north and east. This image, created from infrared measurements made by NOAA's AVHHR (Advanced Very High Resolution Radiometer) satellite in June of 1984, forms the borders for each part and chapter introduction in *Earth's Climate* and also for resource material at the end of each chapter. (O. Brown, R. Evans, and M. Carle, University of Miami Rosenstiel School of Marine and Atmospheric Science, Miami, Florida.)

Part I surveys Earth's present-day climate system and the approaches used by climate scientists to unravel Earth's climatic history. Chapters 2 and 3 serve both as an introduction to the study of climate and a reference base of background knowledge to be used repeatedly through the course.

Parts II through V are organized by time scale. Because we know relatively little about the positions of continents, tectonic features, and climatic responses during the first 90% (4 billion years) of Earth's history, only Chapter 4 and part of Chapter 5 are devoted to this long interval. The rest of the book deals with the remaining 10% of Earth's history, moving from tectonic changes (Part II) to orbital changes (Part III) to deglacial and millennial changes (Part IV) and finally to historical and future changes (Part V). Because each succeeding part covers intervals that are one or two orders of magnitude shorter than the one preceding it, more recent climate changes receive progressively more detailed treatment at increasingly high resolution.

Necessary background material is embedded in a few chapters. The first part of Chapter 5 contains a review of those parts of plate tectonic theory necessary for students to understand tectonic-scale changes in climate without having taken a course in physical geology. Chapter 8 reviews aspects of Earth's orbit in sufficient detail to enable students to understand orbital-scale climate changes.

I chose the basic progression from older to younger intervals and from slower to faster changes for several reasons. The main reason is simply that this is the way the history of Earth's climate actually developed, so it seems the most natural way to tell the story. In addition, shorter-term climate changes tend to ride on the back of longer-term changes, and it is necessary to understand the former to give a context to the latter.

Another important advantage of organizing a book by time scale is that all the interconnected issues critical to each time scale can be presented coherently. This strategy allows students to see all at once the evidence of climate change, the hypotheses posed to explain the evidence, and the extent to which some hypotheses have survived years of critical testing by other scientists or have been rejected. These interconnections among data, theory, and theory testing are the real dynamic of science, and telling these stories coherently makes science come alive.

The Approach of *Earth's Climate*

Although distinguished by time scale, the parts of the book are actually linked together by the recurrence of critical themes through Earth's history. Several themes reappear in every part of this book, including:

- The causes (forcing) of climate change

- The natural response times of the many components of Earth's climate system

- Interactions and feedbacks among these numerous components

- The role of carbon as it moves within the climate system at each time scale.

In my experience, repeated revisiting of these themes helps students really learn them. These themes also bring history to life by connecting the deep past to the more recent past, to the present, and to the future.

In undertaking this book, I faced important choices about depth versus breadth in the coverage of each topic. I focused on climate changes that are large in amplitude and extensive in geographic reach. This choice reflects my judgment that larger-scale changes are of greater fundamental importance to Earth's climate than smaller ones. In addition, large-amplitude climate changes are generally recorded with greater clarity than small ones, because they stand out from the uncertainties inherent in our measurements of past climate. In other words, they have a good signal-to-noise ratio.

My bias toward large-scale responses results in repeated emphasis on a few key indices of global climate:

- The presence of ice sheets at high latitudes

- Changes in the strength of low-latitude monsoon circulations

- The reaction of ice-free land surfaces outside monsoon regions

- The flow of deep water

- The movement of carbon among reservoirs in the climate system.

These indices appear and reappear in every part of the book. Other indices of the response of the climate system appear in individual case studies throughout the book, but they are inevitably tied to a specific local context.

I also focus on the response times, interactions, and feedbacks among these major components of the climate system:

- Ice (ice sheets, mountain glaciers, and sea ice)

- Air (mainly the lower layer of the atmosphere)

- Land surfaces

- Water (the surface and deep ocean, as well as lakes and rivers)

- Vegetation.

Climate changes involve multiple interactions among these components of the climate system. One way to make sense of this complex system is to understand the inherent rate at which each of its components responds both to the primary causes of climate change and as part of a web of interactions within the system. In taking this approach, I deemphasize chaotic or unpredictable climate changes in favor of those we can understand and to varying degrees can predict.

Another recurrent focus is the testing of hypotheses by means of climate models. Although models are imperfect representations of reality, the best of them are preferable to unsupported arm-waving arguments. I have tried to make clear the cases in which models can help to resolve issues and the cases in which they are not yet capable of doing so.

A particularly critical focus of the book is the flow of carbon through the climate system. Greenhouse gases have played a major role in the climatic changes of the past, and they will play a comparably important role in the climate changes of the next several centuries. Different kinds of exchanges of carbon among Earth's major carbon reservoirs have determined the level of CO_2 in the atmosphere through time, with direct consequences for the climate system.

Inevitably, these choices of focus have resulted in some narrowing of the book's scope. Many newly emerging geochemical techniques have received no mention, although instructors should have no problem incorporating them into the larger structure of this book. Also, many interesting climate phenomena have been omitted, including entire phases of Earth history before 400 million years ago and shorter-term oscillations during more recent tectonic-scale intervals. Again, it should prove simple to fit these periods of study into the larger context provided by this book.

In writing *Earth's Climate*, I intentionally avoided the artificial distinction usually made between older (paleo) and younger (modern/instrumental) parts of Earth's climate history. Studies of climate on time scales ranging from hundreds of millions of years ago up to the last several decades all face similar limitations: in the subset of climate phenomena that have actually been measured, in the signal-to-noise ratios of the measurements, in the spatial coverage of measurement sites, and in the temporal duration and resolution of the measured signals. No climate change is perfectly known or understood, whether it occurred 200 million years ago or 200 years ago. As a result, I simply treat the flow of time as continuous, which of course it is.

For this reason, I generally avoid using the prefix "paleo" (paleoclimate, paleoceanography, paleobotany) to describe older intervals. And I minimize the use of geologic nomenclature to subdivide time: the only such terms used in this book are Cretaceous and Cenozoic. Otherwise, time is referred to in units of years: billions (Byr), millions (Myr), thousands, and hundreds.

Teaching from *Earth's Climate*

Any course taught from this book will at least in a general way follow time's arrow, mainly because that is the basic structure of the book. Within this structure, the book continuously adds to its base of knowledge, as later chapters increasingly draw on information conveyed in earlier chapters. This makes it easy for instructors to follow the book's basic structure, but harder to begin teaching from later chapters.

The most direct option to teaching from *Earth's Climate* is to present a one-semester course covering as much as possible of *all* the time scales. I have taught one-semester courses based on this material each autumn for the last several years, both to 20 to 30 upper-level environmental science majors and to 100 or so lower-level nonscientists. The lecture breakdown in the accompanying table has worked well for both courses. The outline is based on a standard of forty lectures per term, each lasting 50 minutes.

It is not necessary to try to teach this entire book from front to back. Other possible options include these:

- Move quickly through the foundation chapters in Part I in a minimal number of lectures, using Chapters 2 and 3 primarily as a reading reference source throughout the rest of the term. Chapter 1 is critical both as a general introduction and because it contains essential material on the operation of components of the climate system, including response times and feedbacks. Chapters 2 and 3, in contrast, are dense with new information that most students will not fully absorb on first exposure. But the material in these two chapters provides important background for case studies explored later in the course, and students will best learn this material as they revisit it later in specific storytelling, hypothesis-testing contexts within the assigned reading.

- Skip most of Part II (tectonic-scale climate change). Some teachers may not want to emphasize tectonic-scale changes and may prefer to begin with shorter time scales (orbital, deglacial-millennial, and historical). It should be possible to teach a coherent course by omitting several topics in Part II: the climate of Pangaea (Sections 5-4 and 5-5), the Cretaceous/Tertiary impact (the last section of Chapter 6), and tectonic-scale changes in sea level (Sections 6-3 and 6-4). Those who wish to skip the tectonic section entirely will need to make sure that students read Box 7-1, covering oxygen isotopes, because this technique is used repeatedly in Parts III–V. In addition, Chapter 6 contains important material on the Cretaceous greenhouse (Sections 6-1 and 6-2) that reappears in Chapters 18 and 19 because of its relevance to Earth's sensitivity to carbon dioxide and to future climate change.

Instructional Features

In addition to a text written in plain English with hundreds of new color illustrations, *Earth's Climate* contains a range of features designed to aid teachers and students.

Boxes set off four categories of special topics from the main text. "Tools of Climate Science" describe key

Lectures Based on *Earth's Climate*

Lecture	Topic	Section and Chapter
1	Overview of the climate system	Part I, Chapter 1
2	Response times and feedbacks	
3	Radiation and the atmosphere	Chapter 2
4	Heat transfer in the atmosphere and ocean	
5	Deep water, ice, and vegetation	
6	Climate data and models	Chapter 3
7	Faint young Sun	Part II, Chapter 4
8	Basics of plate tectonics	Chapter 5
9	The BLAG theory: CO_2 input	
10	The uplift weathering theory: CO_2 removal	
11	Climate on the supercontinent Pangaea	
12	The Cretaceous greenhouse climate	Chapter 6
13	Tectonic-scale changes in sea level	
14	Oxygen isotopes	Chapter 7
15	Causes of cooling during the last 55 Myr	
16	Orbital variations	Part III, Chapter 8
17	Changes in insolation received on Earth	
18	Orbital changes in monsoons	Chapter 9
19	Orbital changes in ice sheets (1)	Chapter 10
20	Orbital changes in ice sheets (2)	
21	Ice core records: CO_2, CH_4, dust	Chapter 11
22	Carbon isotopes and orbital changes in deep water	
23	Orbital-scale interactions in the climate system	Chapter 12
24	Survey of the last glacial maximum	Part IV, Chapter 13
25	The tropical cooling debate	
26	The last deglaciation	Chapter 14
27	Climate changes in the last 7000 years	
28	Millennial changes: $\delta^{18}O$ in ice sheets	Chapter 15
29	Millennial changes in other regions	
30	Cause of millennial changes	
31	Historical climate: the Little Ice Age	Part V, Chapter 16
32	Historical climate: tree rings and corals (El Niño)	
33	Historical climate: instrument records	
34	Impacts of climate on early humans and civilizations	Chapter 17
35	Anthropogenic inputs of gases	
36	The greenhouse debate: natural changes	Chapter 18
37	The greenhouse debate: CO_2 sensitivity	
38	Future climate: industrial emissions	Chapter 19
39	Future climate: CO_2 levels and the fate of CO_2	
40	Future climate: $2xCO_2$ and $4xCO_2$ worlds	

scientific techniques and methods in depth. "Climate Interactions and Feedbacks" highlight and develop important ideas or methods. "Climate Debates" describe contentious scientific issues and points of view. The last group of boxes, "Looking Deeper into Climate Science," was created specifically to bracket and isolate more quantitative material. These boxes are appropriate for use in upper-level courses for majors, but should probably be omitted in general-interest, lower-level undergraduate courses for nonmajors and nonscientists.

Key Terms appear in boldface in the text and are listed at the end of each chapter and defined in the Glossary.

Review Questions provide a recap of major issues at the end of the chapter.

Additional Resources listed at the end of each chapter provide instructors and students with access to materials that can deepen their insight into the subject material. Most are books and journal articles that are broad surveys of a subject area or that were the original source of key findings cited in the text. Those listed as "Basic Reading" are written in a more accessible style than those listed as "Advanced Reading."

Media Resources Web-based resources, including helpful Web links and study aids, are available on the W. H. Freeman and Company Web site: http://www. whfreeman.com/ruddiman.

Instructor Media All text art is available on the Instructor's Section Web site.

Acknowledgments

W. H. Freeman and Company is one of the last publishing companies willing to put significant resources into book development, the long process that turns an author's rough first draft into a finished product. The first part of this process includes choosing outside reviewers, distributing copies of the manuscripts, reminding reviewers to do their work on time, and returning reviewed copies to the author for revision. These and other chores were very capably overseen by the acquisitions editors, Melissa Wallerstein and Nicole Folchetti.

Among the many helpful people, my old friend and colleague Tom Webb put in an enormous critical editorial effort on most of the manuscript and I warmly thank him for his work. I am also grateful to the persons who reviewed chapters or parts of the book:

Paul Baker
Duke University
Subir Banerjee
University of Minnesota
Jay Banner
University of Texas, Austin
William B. N. Berry
University of California, Berkeley
Julia Cole
University of Arizona
Tom Crowley
Texas A&M University
P. Thompson Davis
Bentley College
John Gosse
University of Kansas
Andrew Ingersoll
California Institute of Technology
Emily Ito
University of Minnesota
George Jacobsen
University of Maine
David Kemp
Lakehead University
Zhuangjie Li
University of Illinois, Champaign–Urbana
R. Timothy Patterson
Carleton University
Maureen Raymo
Massachusetts Institute of Technology
Mike Retelle
Bates College
John Ridge
Tufts University
David Rind
Goddard Institute of Space Sciences
Lisa Sloan
University of California, Santa Cruz
Howard Spero
University of California, Davis
Aondover Tarhule
University of Oklahoma
Lonnie Thompson
Byrd Polar Research Center
Thompson Webb
Brown University
Al Werner
Mount Holyoke College
Herb Wright
University of Minnesota

My development editor, Barbara Brooks, guided improvements in the original book manuscript. She worked effectively and with unflagging cheerfulness for many months toward our common goal of a book explaining climate science in a simple logical way. She also kept my energy and spirits from sagging during the long, demanding rewrite process.

The project editor, Diane Davis, then guided the book toward and into production, and also helped turn my graphics sketches into useful figures. Thanks also go to Barbara Salazar for careful copyediting in the "old-school" style. The text designer, Vicki Tomaselli, is the main reason for the visual attractiveness of the book, and she also helped with the figures.

I also gratefully acknowledge the work of Bob Smith, the graphics artist who produced over 400 color figures based on my rough sketches and on figures adapted from the literature. His work was funded by the National Science Foundation, through the farsightedness of Herm Zimmerman, then Atmospheric Sciences Program Manager. Without the help he courageously provided, the book would not have been rendered in full color.

Finally, I lovingly acknowledge the quiet sacrifice made by my wife, Ginger, who let me plug away on the endless effort needed to produce a book like this, in effect giving up hundreds of hours we could have spent together (and now will).

Recent fluctuations of mountain glaciers   An engraving made in 1850 shows the Argentière glacier extending far down the valley in the French Alps (top). A century later the glacier had retreated far up the mountainside (bottom). (From E. L. Ladurie, *Times of Feast, Times of Famine: A History of Climate Since the Year 1000* [New York: Doubleday, 1971] in J. Imbrie and K. P. Imbrie, *Ice Ages* [Cambridge, Mass., and Springfield, N.J.: Harvard University Press and Enslow Publishers, 1979].)

Framework of Climate Science

Climate change is a topic that stirs our curiosity. Although most people may actually know little about climate, they correctly sense that it has something to do with long-term changes in temperature, rainfall, vegetation, and other features measurable on Earth.

Some climate changes in the past are more familiar than others. In Earth's relatively recent past, the hot, dry interval that produced the Dust Bowl of the 1930s drove thousands of farmers from the Great Plains west to California. A century earlier, slightly cooler air temperatures caused glaciers to expand down the sides of mountains beyond the positions to which they have retreated today.

Considerably further back in time, 21,000 years ago, the climate was so cold that enormous ice sheets covered Canada and northern Europe, and sea level was some 110 meters (360 feet) lower than it is today because so much water was stored in ice on land. Even further back, 100 million years ago, much warmer conditions eliminated all ice from the face of the Earth, even at the South Pole. Today most people are aware that Earth is projected to warm in the future because humans are beginning to alter climate on a global scale, but scientists are still debating how large the warming will be.

Clearly Earth's climate has changed many times over the vast span of its existence, and it will do so again. The chapters in Part I provide a general framework for understanding climate change by addressing several questions:

- **What are the components of Earth's climate system?**
- **How does climate change differ from day-to-day weather?**
- **What factors drive changes in Earth's climate?**
- **How do the many parts of Earth's climate system react to these driving forces and interact?**
- **How do scientists study past climates and project changes that lie in our future?**

CHAPTER 1

Overview of Climate Science

Life exists nearly everywhere on Earth because its climate is favorable. We live within, upon, and surrounded by the climate system: the air, land surfaces, oceans, ice, and vegetation.

People are naturally curious about climate change, a vital thread in the tapestry of Earth history, along with the evolution of life and of the physical form of this planet.

But the study of climate also matters for a practical reason: its relevance to climatic changes we face in the near future. We are leaving an era when natural changes governed Earth's climate and are are about to enter a time when changes caused by human activity will gradually come to predominate climate change even on a global scale.

In this chapter we will survey the factors that cause Earth's climate to change. We will also see how the field of climate science came into being in recent decades, how scientists study climate, and why a knowledge of the history of climate change will help to inform us about changes looming in our near future.

Climate and Climate Change

Even from distant space, it is obvious that Earth is the only habitable planet in our solar system (Figure 1-1). More than 70% of its surface is a welcoming blue, the area covered by life-sustaining oceans. The remaining 30% of Earth's surface, the land, is partly blanketed in green, darker in forested regions and lighter in regions where grass or shrubs predominate. Even the brown deserts and some of the white ice contain life.

Earth's favorable climate enabled life to evolve on our planet. **Climate** is a broad composite of the average condition of a region, measured in terms of such things as temperature, amount of rainfall or snowfall, snow and ice cover, and winds. Climate specifically applies to longer-term changes (years and longer), rather than to the shorter fluctuations that last hours, days, or weeks and are referred to as **weather**, or the even shorter changes between day and night. "Climate" can be also used to refer to the mean state of an entire planet or to specific regions, such as particular continents or oceans.

Earth's climate is highly favorable to life both in an overall, planet-wide sense and at more regional scales. Its surface temperature averages a comfortable 15°C (59°F) and much of its surface ranges between 0° and 30°C (32° and 86°F) and can support life (Box 1-1).

Although we take Earth's habitability for granted, climate can change over time, and with it the degree to which life is possible, especially in regions vulnerable to change. During the several hundred years in which humans have been making scientific observations of climate, actual changes have been relatively small. Even so, climatic changes significant to human life have occurred. One striking example, the retreat of ice from a mountain valley in the European Alps between the middle of the nineteenth century and the middle of the twentieth, appears in the introduction to this part. The immediately preceding centuries had been colder: advances of Alpine glaciers had overrun mountain farms and even some small villages.

Scientific studies carried out a century or more ago reveal that these historical changes in climate are tiny in comparison with the much larger changes that happened earlier in Earth's history. For example, early studies show that at times in the distant past ice covered much of the region that is now the Sahara Desert, and trees flourished in what are now Antarctica and Greenland.

1.1 Geologic Time

Understanding these climatic changes of the past begins with a difficult challenge: coming to terms with the enormous span of time over which Earth's climatic history has developed. Human life spans are generally measured in decades. Many of the phases of our lives, such as childhood, adolescence, and education at college, come and go in a few years, and most of the focus of our daily lives tends to be on needs and goals that we

FIGURE 1-1 **The habitable planet** Even seen from distant space, most of Earth's surface looks inviting to life, especially its blue oceans and green forests, but also its brown deserts and white ice. All these areas are prominent parts of Earth's climate system. (NASA.)

BOX 1-1 TOOLS OF CLIMATE SCIENCE

Temperature Scales

Three temperature scales are in common use in the world today. For day-to-day nonscientific purposes, most people in the United States use the Fahrenheit scale, developed by the German physicist Gabriel Fahrenheit. It measures temperature in degrees **Fahrenheit** (°F), with the freezing point of water at sea level set at 32°F and the boiling point at 212°F.

Most other countries in the world, and most scientists as well, routinely use the **Celsius** (or centigrade) scale developed by the Swedish astronomer Anders Celsius. It measures temperature in degrees Celsius (°C), with the scale set so that the freezing point of water is 0°C and the boiling point of water is 100°C.

These equations convert temperature values between the two scales:

$$T_C = \tfrac{5}{9}(T_F - 32) \qquad T_F = \tfrac{9}{5}T_C + 32$$

where T_F is the temperature in degrees Fahrenheit and T_c is the temperature in degrees Celsius.

Many scientific calculations make use of a third temperature scale developed by the British physicist Lord Kelvin (William Thomson) and known as the **Kelvin** scale. This scale is divided into units of Kelvins, rather than degrees Kelvin. The lowest point on the Kelvin scale (absolute zero, or 0 K) is the coldest temperature possible, the temperature at which motions of atomic particles effectively cease. The Kelvin scale does not have negative temperatures, because no temperature colder than 0 K is possible.

Temperatures above absolute zero on the Kelvin scale increase at the same rate as the Celsius scale, but with a constant offset. Absolute zero (0 K) is equivalent to −273°C, and each 1K increase on the

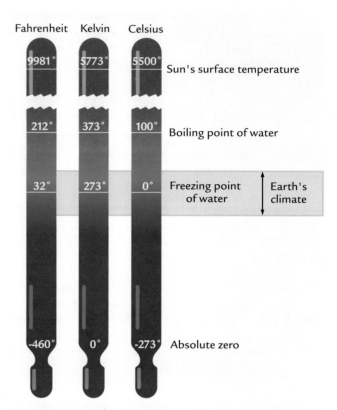

Temperature scales Scientists use the Celsius and the Kelvin temperature scales to measure climate changes. Temperatures at Earth's surface vary mainly within a small range of −50°C to +30°C, just below and above the freezing point of water. (Adapted from W. F. Kaufman III and N. F. Comins, *Discovering the Universe,* 4th ed., © 1996 by W. H. Freeman and Company.)

Kelvin scale above absolute zero is equivalent to a 1°C increase on the Celsius scale. As a result, 0°C is equivalent to 273K.

hope to satisfy within days or weeks. We feel fortunate if we can trace our family history as far back as a century, to the "olden days" of our grandparents or great-grandparents. From this natural human perspective, events recorded in Egypt, China, Greece, and Rome a few thousand years ago are considered "antiquity."

Almost all of Earth's long history lies immensely far beyond this human perspective. Earth formed 4.55 billion years (Byr) ago, or 4,550,000,000 years if we show all the zeros! Most of the earliest part of Earth's history

is either unknown or known only in the most sketchy way. One reason for the gaps in our knowledge is the climate system itself: the relentless action of air and water on Earth's surface has eroded away many of the early deposits that might have helped us reconstruct and understand more of this history.

This book focuses mainly on the last several hundred million years of Earth's history, equivalent to less than 10% of its total age. Our focus must be limited because many aspects of Earth's history grow increasingly vague as they recede in the past, so that scientists are forced to speculate more and more about fewer and fewer hard facts. But as more information becomes available from the younger part of the climatic record, our chances of measuring and understanding climate change increase.

Even the last 10% of Earth's history covers time spans beyond imagining. The climate scientists who daily study records spanning thousands to millions or tens of millions of years probably understand time only in a technical way, primarily in the sense of a shorthand technique for cataloging and filing information. Geologists often refer to these unimaginably old and long intervals as "deep time", hinting at their remoteness to our senses. Like the scientists who study climate change, you will learn in this book to catalog deep time, even if not to comprehend it in a literal sense.

The plot of time to the left in Figure 1-2 shows that most of the focus of this book fits into a fraction of Earth's history too small even to be shown. There are two ways to overcome this problem and convey a better sense of the portion of deep time covered by this book. One is by again starting with a plot of Earth's full age, but then progressively expanding out and magnifying shorter and shorter intervals to show how they fit into the whole (Figure 1-2, center). The other method is to plot time on a logarithmic scale that increases by successive jumps of a factor of 10 (Figure 1-2, right). This kind of plot compresses the longer parts of the time scale and expands the shorter ones so that they all fit onto one plot.

1-2 How This Book Is Organized

Within its focus on the most recent 10% of Earth's age, this book is organized by time scale (see Figure 1-2). Part II covers climatic changes that have occurred during the last several hundred million years, an interval during which mammals evolved from primitive to more distinctive forms. Part III looks at the last 3 million years, a time span when our species was rapidly evolving toward its present form. Part IV explores changes over the last 50,000 years, an interval during which humans initially lived a primitive hunting-and-gathering life, then developed and practiced agriculture, and later created the first recorded human civilizations. Part V zeroes in on the last 1000 years, spanning the most recent historical era.

This progression from longer to shorter time scales is a natural pathway because faster changes in climate at the shorter time scales are embedded in and superimposed upon the slower changes at the longer time scales (Figure 1-3). The longest time scale shows a slow warming between 300 and 100 million years (Myr) ago, followed by a gradual cooling in the last 100 million years (Figure 1-3A). This gradual cooling eventually led to the appearance of massive ice sheets that advanced and retreated many times during the last 3 million years at cycles of tens of thousands of years (Figure 1-3B). Superimposed on these climatic cycles were still shorter oscillations that lasted a few thousand years and that were distinctly larger during times when climate was colder (Figure 1-3C). The last 1000 years has been a

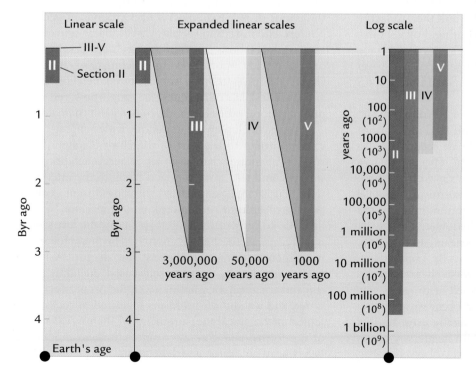

FIGURE 1-2 Earth history Earth's age is 4.55 billion years. Most of the focus of this book fits into a very small fraction of this immense interval and can be represented only by a series of magnifications or by plotting time on a log scale that increases by factors of 10.

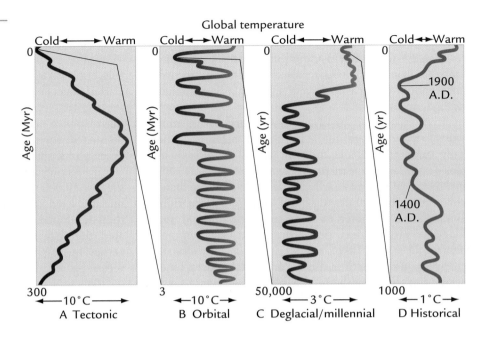

FIGURE 1-3 Time scales of climate change Changes in Earth's climate span several time scales, arrayed from longer to shorter: (A) the last 300 million years, (B) the last 3 million years, (C) the last 50,000 years, and (D) the last 1000 years. Here progressively smaller changes in climate at successively shorter time scales are magnified out from the larger changes at longer time scales.

time of relatively warm and stable climate, with only small oscillations toward slightly cooler conditions (Figure 1-3D).

Each of these successively more detailed time scales reveals shorter oscillations embedded within longer ones, just as cycles of daily heating and nighttime cooling are embedded within longer seasonal cycles of summer heating and winter cooling. To understand the extreme heat reached during a specific afternoon in July in the northern hemisphere, it first makes sense to realize that such a July afternoon occurs in the larger context of the hottest season of the year, and then to factor in the additional effect of daytime heating. For a similar reason, it makes sense to follow time's arrow and trace climate changes from older to younger eras, and from the larger cycles to the smaller ones superimposed on them.

As the book progresses from older to younger time scales, you will notice a change in the kind of information scientists have extracted from past records of climate change. In part this development reflects a change in the amount of detail that can be retrieved from these records, called their degree of **resolution**. Because older records tend to have less resolution, much of the focus of Part II of the book is on the longer-term average climatic states over millions of years, and the way they differ from our climate today. By comparison, younger records tend to have progressively greater resolution, and Parts III through V look at progressively shorter-term changes in climate that occur within intervals of thousands, hundreds, and finally even tens of years. We will examine the resolution issue more closely in Chapter 3.

Development of Climate Science

As scientists began to discover striking examples of major climatic changes earlier in Earth's history, their curiosity naturally grew about what had happened and why. In the middle and late twentieth century, and especially in its last two decades, interest in the study of climate change and its causes grew enormously. The few amateur scientists and university professors who studied climate in relative isolation during the nineteenth century and the early twentieth have now been replaced by thousands of researchers with backgrounds in geology, physics, chemistry, and biology working at universities and national laboratories and research centers throughout the world (Figure 1-4). Today climate scientists use aircraft, ships, satellites, sophisticated new biological and chemical lab techniques, and high-powered computers, among other methods, to carry out their studies.

Studies of climate are incredibly wide-ranging. They vary according to the part of the climate system being studied, including changes in air, water, vegetation, land surfaces, and ice. They vary according to the interval of time that is emphasized, from the long-distant geological past (deep time) going back tens of millions of years or more to historical changes spanning the last several hundred years to changes examined with sophisticated instruments during just the last few decades. Climate studies also vary by the techniques used, including physical and chemical measurements of the properties of air, water, and ice and of life-forms fossilized in rocks; biological or botanical measurements of the myriad shapes of life-forms; and computer simulations to model the behavior of air, water, and vegetation.

FIGURE 1-4 **National research centers** The National Center for Atmospheric Research (NCAR) in Boulder, Colorado, is one of several national laboratories and university centers at which Earth's climate is studied. (NCAR.)

This huge diversity of studies also covers a broad array of scientific disciplines. Some studies are directed solely at improving our understanding of the modern-day climate system: *meteorologists* study the circulation of the atmosphere, *oceanographers* explore the circulation or chemistry of the ocean, *chemists* investigate the composition of the atmosphere, *glaciologists* measure the behavior of ice, and *ecologists* analyze life-forms on land or in the water. Chapter 2 gives an overview of many features of the climate system now under study.

Other scientists study changes in climate or climate-related phenomena in Earth's recent or more distant past: *geologists* explore the broader aspects of Earth's history; *geophysicists* investigate past changes in Earth's physical configuration (continents, oceans, mountains); *geochemists* analyze past chemical changes in the ocean, air, or rocks; *paleoecologists* study past changes in vegetation and their role in the climate system; *climate modelers* evaluate possible causes of climate change; and *historians* comb archives for writings that will enable them to reconstruct past climates. Much of this book focuses on these studies of past climates, because the biggest changes have been those in the past, and because these changes help us to understand those that may lie ahead.

In recent decades, studies of Earth's climatic history have begun to cross the traditional disciplinary boundaries and merge into an interdisciplinary approach referred to as "Earth system science" or "Earth system history." Such efforts recognize that the many parts of Earth's climate system are interconnected, so that investigators of climate must look at all the parts in order to understand the whole. This entire book is an example of this **Earth system** approach.

Beyond varying in their goals and disciplines, scientists until recently have also tended to be distinguished by the time scales they investigate. Those who study Earth's past history have been split into groups working on the very distant past, the not-so-distant past, and the more recent historical climatic record. And to an even greater extent, those who work on aspects of Earth's history have tended to work almost entirely apart from those who study the modern climate system.

These distinctions based on time scale are beginning to disappear as scientists draw on one another's work for insights. Lessons learned about how the climate system operates today can be directly applied to the past, and in many cases the opposite is also true.

This book further blurs artificial distinctions based on time. Because it is organized by time scale, it implicitly shows climatic change for what it is: a continuum from the distant past right up to the present day and on into the near future. This is also the way Earth's climatic history has developed and will continue to develop, with no breaks in the flow of time. As used in this book, the term **climate science** refers to this vast field of research and its many components by capturing the inherent unity of this *multidisciplinary* and *interdisciplinary* new field of research, as well as its linkage of the past, the present, and the future.

1.3 How Scientists Study Climate Change

Like any other field of scientific exploration, climate science moves forward by an interactive combination of observations and theory. Most climate scientists spend much of their time gathering and analyzing data drawn from the kinds of climatic archives reviewed in Chapter 3 and using a variety of techniques introduced in subsequent chapters. The results of this research are written up and published, in part for the practical reason of demonstrating the progress necessary to help scientists obtain additional funding for further research. At a more basic level, progress in science depends on the free exchange of ideas, and climate researchers publish to tell the scientific community what they have discovered.

Scientists invariably need to explain and interpret their research results, and occasionally they devise a new idea called a **hypothesis**. Hypotheses are informal ideas that have not been widely tested by the larger community of scientists doing similar research. Most hypotheses are discarded, either because they are found to disagree with some basic scientific principle or because they make predictions that other observations contradict.

A hypothesis may occasionally reach a higher status because it is capable of explaining a wide array of observations. It then becomes a **theory**. Scientists further test successful theories by making additional observations, by developing new techniques to analyze data in new ways, and by devising models to simulate the workings of the climate system.

Models are useful because they give climate scientists an independent way to test whether a particular theory can explain the data they have collected. Experiments run on climate models are often based directly on geologic data that define various configurations of Earth's surface at key times in the past. The results that emerge from modeling experiments based on these configurations are compared and tested against climates that actually existed in the past and can be determined from climatic data.

This ongoing work may eventually disprove the predictions of a current theory. Science moves forward in part by disproving and discarding the less worthy among the existing hypotheses and theories. In some cases, new work may not only support an existing theory but refine and improve it, giving it even greater power to explain an even wider range of basic scientific observations. Only a few theories survive years and decades of repeated testing by energetic and imaginative scientists. Those that do are sometimes called "unifying theories" and are generally regarded as close approximations to the "truth," although it is impossible to prove that a theory is true, only that it is untrue.

These expanding efforts to understand climate change have led to a scientific revolution that has accelerated through the late twentieth century and into a new century. The mystery of climate change yields its secrets slowly, and many important questions still remain unanswered, but the revolution in knowledge has been immense, as this book will show.

This revolution has reached the point where it has begun to take its place alongside the two great earlier revolutions in knowledge of Earth history. The first was the development by Charles Darwin and others in the nineteenth century of the theory of **evolution**, which led to an understanding of the origin of the long sequence of life-forms that have appeared and disappeared during the history of this planet. The second was the grand geological synthesis during the 1960s and 1970s of the theory of **plate tectonics**, which has given us an understanding of the physical basis for the slow motions of continents and changes in ocean basins across Earth's surface through time, as well as for associated phenomena such as volcanoes, earthquakes, and mountain ranges. Key elements of the plate tectonic theory are covered in Part II.

On a practical level, improvements in knowledge of the climate changes of the past can also offer us guidance about future changes. Earth has performed many large-scale, natural "experiments" during its long history, and some of them have aspects in common with the kinds of changes that climate scientists expect to occur in the near future. For example, the level of **greenhouse gases** in the air (gases such as carbon dioxide, which trap and retain heat and warm Earth's surface, as explained in Chapter 2) has varied greatly through time, sometimes rising to values far higher than today's, sometimes sinking far below the present level. Changes in the concentrations of greenhouse gases contribute to the appearance and disappearance of ice sheets on Earth's surface through time. We will see in the final chapter of this book that we can look to these earlier times in Earth's history for lessons about what may happen in the future as human activity continues to increase the concentrations of greenhouse gases in the atmosphere.

Overview of the Climate System

In this section we take a first look at Earth's **climate system**, consisting of air, water, ice, land, and vegetation. At the most basic level, changes in these components through time are analyzed in terms of *cause* and *effect*, or, in the words used by climate scientists, **forcing** and **response**. The term "forcing" refers to those factors that drive or cause change; "responses" are the effects—the climatic changes that occur.

1.4 Components of the Climate System

Figure 1-5 provides an initial impression of the vast array of factors involved in studies of Earth's climate. It shows the air, water, ice, land, and vegetation that are the major components of the climate system, as well as processes at work within the climate system, such as precipitation, evaporation, and winds. The processes at

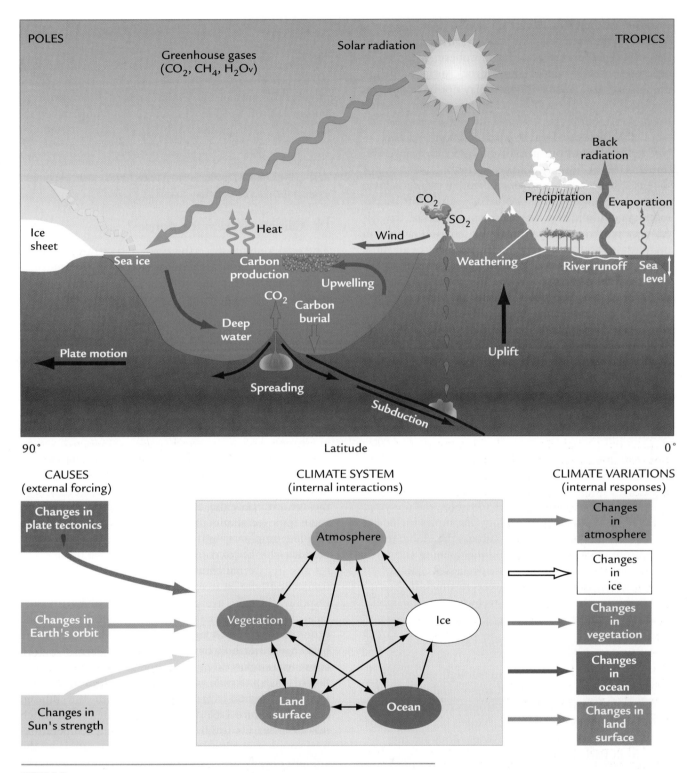

FIGURE 1-5 **Earth's climate system and interactions of its components** Studies of Earth's climate cover a wide range of processes, indicated at the top. Climate scientists organize and simplify this complexity, as shown at the bottom. A small number of factors drive, or "force," climate change. These factors cause interactions among the internal components of the climate system (air, water, ice, land surfaces, and vegetation). The results are the measurable variations known as climate responses.

work in creating Earth's climate extend from the warm tropics to the cold polar regions and from the Sun in outer space down into Earth's atmosphere, deep into its oceans, and even beneath its bedrock surface. All the processes shown, many of which may not be familiar to you at this point, will be explored as we proceed through this book.

The complexity of the top part of Figure 1-5 is reorganized in the bottom part of the figure to provide an initial sense of how the climate system actually works. The relatively small number of external factors shown on the bottom left force (or drive) changes in the climate system, and the internal components of the climate system respond by changing and interacting in many ways (bottom center). The end result of all these interactions is a number of observed variations in climate that can be measured (bottom right). If we look at this process the way an engineer would, the climate system runs like a machine: the factors that drive climate change are the input to that machine, the climate system is the machine, and the resulting variations in climate are the output from the machine.

1-5 Climate Forcing

Three fundamental kinds of climate forcing exist in the natural world:

- *Tectonic processes* generated by Earth's internal heat affect its surface by means of processes that alter the basic geography of Earth's surface. These processes are part of the theory of plate tectonics, the unifying theory of the science of geology. Examples include the movements of continents across the globe, the uplift of mountain ranges, and the opening and closing of ocean basins. These processes change very slowly over millions of years or much longer. The basic processes of plate tectonics are explained in Part II of this book.

- *Earth-orbital changes* result from variations in Earth's orbit around the Sun. These orbital changes alter the amount of solar **radiation** (sunlight and other energy) received on Earth by season and by latitude (from the warm low-latitude tropics to the cold high-latitude poles). Orbital changes occur over tens to hundreds of thousands of years. They are the focus of Parts III and IV.

- *Changes in the strength of the Sun* also affect the amount of solar radiation arriving on Earth. One example appears in Chapter 5: the strength of the Sun has slowly increased throughout the 4.55 Byr that Earth has existed. In addition, shorter-term variations that occur over decades, centuries, and millennia may be partially responsible for climatic

changes at the shorter time scales that are one part of the focus of Part V of this book.

A fourth factor capable of influencing climate, but not in a strict sense part of the natural climate system, is the effect of humans on climate, referred to as **anthropogenic forcing**. This forcing is an unintended byproduct of agricultural, industrial, and other human activities, and it occurs by way of additions to the atmosphere of materials such as carbon dioxide (CO_2) and other gases, sulfate particles, and soot. We will examine these effects in Part V.

1-6 Climate System Responses

The components of Earth's climate system vary widely in their physical and chemical characteristics, and climate scientists can measure or estimate the climatic responses of major parts of the climate system. These responses include changes in global mean and regional temperatures, in the extent of ice of various kinds, in the amounts of rainfall and snowfall, in the strength and direction of the wind, in the circulation of water at the ocean's surface and in its depths, and in the types and amounts of vegetation. We will examine these components of the climate system in more detail in Chapter 2 and review the methods used for studies of past climates in Chapter 3.

A useful way of thinking about how the climate system responds to the factors that drive climate change is in terms of a characteristic **response time**, a measure of the time it takes the climate system to react fully to some imposed change in the forcing. We start with the simple example in Figure 1-6, showing a Bunsen burner beneath a beaker of cool water. The Bunsen burner represents an external climate forcing (like the Sun's radiation), and the water temperature is the climatic response (like the average temperature of Earth's surface). When we light the burner, it begins to heat the water. If the flame is far enough away, the water in the beaker will not boil but will gradually warm toward a constant temperature, and after a long interval will finally reach and maintain that **equilibrium** value.

The response time (shown beneath the Bunsen burner in Figure 1-6) is a measure of the rate at which the water warms toward this equilibrium temperature. The simplest way to define this response rate is by the amount of time it takes for half the remaining warming needed to achieve equilibrium to occur. This is most easily seen in the early phase of the heating response, when the water temperature moves the first 50% of the way toward equilibrium.

The form of the temperature increase is inverse and exponential: that is, the rate of warming is very rapid at first but progressively slows as time passes. This seems intuitively reasonable: a response should naturally be faster when the ultimate equilibrium state is more distant and should slow as the response nears the final equilibrium state.

This concept of response time is a powerful and yet concise way of characterizing how the climate system responds to external forcing. Although the heating curve is exponential, the mathematically defined response time of the beaker of water remains exactly the same throughout the entire experiment. Regardless of position along the response curve, the passage of one response time always moves the system half the remaining way toward equilibrium

The overall sequence shows the response of the water temperature moving from 50% ($\frac{1}{2}$) to 75% ($\frac{3}{4}$) to 87.5% ($\frac{7}{8}$) to 93.75 ($\frac{15}{16}$) of the way toward equilibrium.

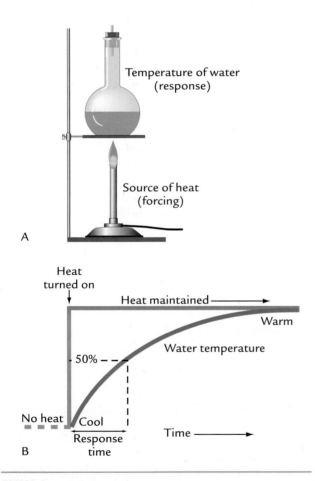

FIGURE 1-6 Response time Earth's climate system has a response time, suggested conceptually by the reaction of a beaker of water to heating by a Bunsen burner. The response time is the rate at which water in the beaker warms toward an equilibrium temperature. (Adapted from J. Imbrie, "A Theoretical Framework for the Ice Ages," *Journal of the Geological Society* (London) 142 [1985]: 417–32.)

After seven response times have elapsed, almost all (99.2%, or $\frac{127}{128}$) of the eventual equilibrium has been achieved. Note that the *absolute amount* of change continually decreases through time, but that the *proportional* change toward the ultimate goal stays the same: half the remaining response within each successive response time that passes.

Each part of the climate system has its own characteristic response time (Table 1-1), ranging from hours or days up to thousands or even tens of thousands of years. The atmosphere has a very fast response time: significant changes are possible in hours to days, as we know from daily cycles of heating and cooling. The land surface reacts somewhat more slowly but it still shows large heating and cooling changes on time scales of hours to days to weeks. Beach sand can become too hot to walk on during just a single summer afternoon, but it takes somewhat longer to chill the upper layer of soil in winter to the point where it can freeze.

Liquid water has an inherently slower response time than air or land. The response of shallow lakes or of the wind-stirred upper 100 meters of the ocean is measured in weeks to months. This is evident in the way lakes cool off seasonally, but not as fast as the land does. For the deeper ocean layers that lie remote from interactions with the atmosphere, response times can range from decades to hundreds or even a thousand years or more.

Ice is the slowest-responding part of the system. Although the thin layer of sea ice on a polar ocean can grow and melt over time scales of just months to years, thicker mountain glaciers react over time spans of decades to centuries, and the massive kilometers-thick ice sheet that covers the entire continent of Antarctica has a response time of 10,000 years or more. This slow reaction is well captured in the common use of the word "glacial" to mean very slow.

The response times discussed so far all pertain to physical, or inorganic, parts of the climate system. Vegetation, the organic component, is a somewhat different case, but the concept still applies. Unseasonable frosts can kill leaves and grass overnight, and abnormally hard freezes can do the same to the woody tissue of trees, in both cases producing a vegetation response measured in hours. Still, seasonal spring greening of the landscape and autumn loss of this leafy green material take weeks or months to complete. In addition, pioneering vegetation that occupies newly exposed ground (for example, bare ground left behind by a melting glacier) may take tens to hundreds of years or more to come to full development because of the slow dispersal of seeds and the time needed for them to germinate and produce mature trees. Each of these various kinds of changes in vegetation involves a distinctive response time tied to the specific context.

TABLE 1.1　Response Times of Various Climate System Components

Component	Response time (range)	Example
Fast responses		
Atmosphere	Hours to weeks	Daily heating and cooling Gradual buildup of heat wave
Land surface	Hours to months	Daily heating of upper ground surface Midwinter freezing and thawing
Ocean surface	Days to months	Afternoon heating of upper few feet Warmest beach temperatures late in summer
Vegetation	Hours to decades/centuries	Sudden leaf kill by frost Slow growth of trees to maturity
Sea ice	Weeks to years	Late-winter maximum extent Historical changes near Iceland
Slow responses		
Mountain glaciers	10–100 years	Widespread glacier retreat in 20th century
Deep ocean	100–1500 years	Time to replace world's deep water
Ice sheets	100–10,000 years	Advances/retreats of ice sheet margins Growth/decay of entire ice sheet

1-7　Time Scales of Forcing versus Response

As we move through this book, you may notice an underlying change in emphasis concerning the relationship between the forces that drive climate change and the responses of the climate system. This shift in emphasis reflects the fact that the relative size of the time scale over which the forcing is applied versus that at which the climate system can naturally respond determines much about what actually happens in the climate system. Several hypothetical examples shown in Figure 1-7 can help to explain this idea:

• *Forcing is slow in comparison with the climate system's response.* If the changes in climate forcing are very slow in comparison with the response times in the climate system, the system will simply passively track the forcing, one for one (Figure 1-7A). This is equivalent to increasing the flame of the Bunsen burner in Figure 1-6 so slowly that the water temperature can easily keep pace with the gradual application of more heat.

　The situation in Figure 1-7A is typical of many climate changes that occur over the long tectonic

time scales discussed in Part II. For example, continents can be slowly carried by plate tectonic processes toward higher or lower latitudes at rates averaging about 1 degree of latitude (100 km or 60 miles) per million years. As the landmasses move toward lower latitudes, where incoming solar radiation is stronger, or toward higher latitudes, where it is weaker, temperatures over the continents will react to these very slow changes in solar heating with an imperceptibly subtle year-by-year response. Because the response time of air over land is short (hours to weeks; see Table 1-1), the average temperature over the continent easily keeps pace with the very slow change in average overhead solar radiation spread over millions of years. Over tectonic time scales, much of the key climate forcing changes very slowly, and the climate responses track right along with it.

• *Forcing is fast in comparison with the climate system's response.* At the other extreme of the hypothetical range of cases, the response time of the climate system may be much slower than the time scale of the change in forcing (Figure 1-7B). In this case,

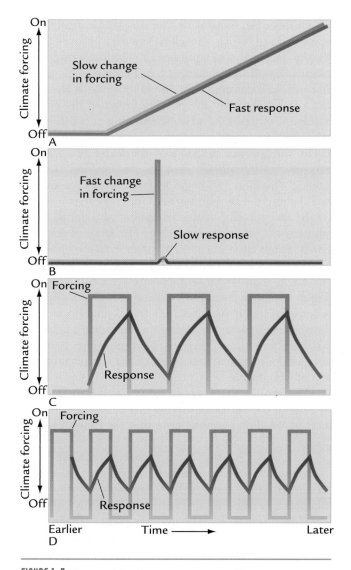

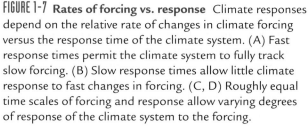

FIGURE 1-7 **Rates of forcing vs. response** Climate responses depend on the relative rate of changes in climate forcing versus the response time of the climate system. (A) Fast response times permit the climate system to fully track slow forcing. (B) Slow response times allow little climate response to fast changes in forcing. (C, D) Roughly equal time scales of forcing and response allow varying degrees of response of the climate system to the forcing.

there is little or no response to the climate forcing. This is equivalent to turning the Bunsen burner on and then off so quickly that the temperature of the water in the beaker has no time to react. Because this example is trivial (it produces no climate response on Earth), no examples of it are included in this book.

There are, however, cases of a climate forcing applied so briefly (over a few years or less) that it does produce a small climate response on Earth.

One example is the eruption of Mount Pinatubo in the Philippine Islands in the summer of 1991. The fine particles emitted by the volcano blocked part of the Sun's radiation and caused Earth's average temperature to fall by 0.5°C within less than a year. Changes from much larger eruptions might even produce enough cooling to increase the snow cover at high latitudes for a year or two. If this cooling persisted for decades or centuries, it might even be enough to cause ice sheets to begin to grow. But most fine volcanic particles stay in the upper layers of the atmosphere for only a few years, and one or two cold years can have no permanent effect in building ice sheets. The ice sheet response time (tens of thousands of years; see Table 1-1) exceeds the one year of forcing by too much.

- *The time scales of forcing and the climate system's response are similar.* The most interesting situation lies between the two extremes: cases in which the time scale of the climate forcing and that of the climate system's responses are more nearly equal. This situation produces a more dynamic response of the climate system, and one that is typical of much of what actually happens in the real world.

    Consider a different experiment with the Bunsen burner and the beaker of water. This time, the Bunsen burner (again the source of climate forcing) is abruptly turned on, left on awhile, turned off, left off awhile, turned on again, and so on (Figure 1-7C). These changes cause the water to heat up, cool off, heat up again, and so on. The water temperature responds by trying to cycle back and forth between two different equilibrium values, one at the cold extreme with the flame off and one at the warm extreme with the flame on.

    But these intervals of heating and cooling may not be long enough to allow the water enough time to reach either equilibrium value. If the flame is turned on and off too quickly, the water temperature cycles back and forth but never reaches the full equilibrium shown in Figure 1-6.

    If we compare the two cases shown in Figures 1-7C and 1-7D, we find that the frequency with which the flame is turned on and off has a direct effect on the size of the response attained by the temperature of the water in the beaker. The example in Figure 1-7D assumes the same equilibrium values for the water temperature and the same position of the Bunsen burner as in Figure 1-7C; the only difference is the length of time the flame under the beaker is left on or off. If the flame is switched on and off much more rapidly than the response time of the water, the water temperature has too little time to reach the equilibrium temperatures (hot or cold) and the amount of response is muted (Figure 1-7D). But if the flame stays on or off for longer time intervals between changes, the temperature of the water can approach much closer to the full equilibrium states (Figure 1-7C).

In the real world, climate forcing rarely acts in the on-or-off way implied by the example of turning the burner flame on and then off. Instead, it more commonly cycles back and forth in smoother oscillations. If we again use the Bunsen burner concept, this situation is analogous to keeping the burner (the climate forcing) on at all times, but slowly varying the intensity of the flame in repeated cycles (Figure 1-8). This produces cycles of warming and cooling of the water, with these cycles lagging behind the shifts of heat applied from the burner.

Even though the smooth cycles of forcing and response in Figure 1-8 look quite different from the cases examined in Figure 1-7, the underlying physical response of the beaker of water (or, by extension, of the climate system) remains exactly the same. The temperature of the water in the beaker continues to react at all times with the characteristic response time defined earlier (see Figure 1-6). The difference now is that the climate forcing (the Bunsen burner) is acting as a moving target. In effect, the climate system response (the water temperature) keeps chasing after this moving target but can never catch up to it.

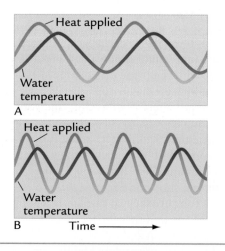

FIGURE 1-8 **Cycles of forcing and response** Many kinds of climate forcing vary in a cyclical way and produce cyclic climate responses. The amplitude of climate responses is related to the time allowed to attain equilibrium: (A) Climate changes are larger when the climate system has ample time to respond. (B) The same amplitude of forcing produces smaller climate changes if the climate system has less time to respond. (Adapted from J. Imbrie, "A Theoretical Framework for the Ice Ages," *Journal of the Geological Society* (London) 142 [1985]: 417–32.)

In this case, there is neither one single equilibrium value toward which the climate system moves (as in Figure 1-6) nor two equilibrium values working in an alternating off-on sequence (as in Figures 1-7C and D), but instead a continuously changing series of equilibrium values set up by the moving target of the climate forcing. For example, when the forcing curve (the heat applied) first starts to move upward to begin a new cycle, the delayed reaction of the water temperature response is initially to slow its descent and then to reverse direction and begin to rise, but always lagging well behind the Bunsen burner's forcing curve.

It also remains true throughout this more complicated case that the *rate of response of the climate system is fastest when the climate system is farthest from the equilibrium it seeks.* This was clearly the case for the simple changes in water temperature response shown in Figure 1-6, with faster rates of heating occurring initially when the system was furthest from equilibrium, and slower heating later on during the subsequent approach toward equilibrium. The same concept applies here, but it is harder to visualize because of the complications of the moving target.

You should try to verify for yourself that this idea applies to the cycles of forcing and response shown in Figure 1-8. For example, note that the fastest changes in water temperature occur at times when the position of the forcing curve is furthest away from that of the response; these are times when the forcing curve reaches maximum or minimum values. Also note that the water temperature response in Figure 1-8 is hardly changing at all during times when the two curves (forcing and response) are close together or crossing. These are times of maximum or minimum values of water temperature, but little or no net change.

As was the case for the on-off changes shown in Figures 1-7C and D, the frequency of occurrence of the smooth cycles of forcing has an effect on the amplitude of the responses. This is apparent in the differences between the two cases shown in Figures 1-8A and B. Slower cycling of the forcing produces a larger response (stronger maxima and minima), because the climate system has more time to react before the forcing cycles back in the opposite direction (see Figure 1-8A). In contrast, faster cycling of the forcing (even though it moves back and forth between the same maximum and minimum values) produces a smaller response (see Figure 1-8B), because the climate system has less time to react before the forcing turns in the other direction.

The relationships between forcing and response shown in Figure 1-8 are particularly useful for understanding the orbital-scale climatic changes explored in Parts III and IV of this book. Changes in incoming solar radiation due to changes in Earth's orbit occur over tens of thousands of years, and this also happens to be the climate response time characteristic of large ice sheets that grow and melt over these orbital time scales. This approximate match of the time scales of forcing and response sets up cyclic interactions very much like those shown in Figure 1-8.

Even at much shorter time scales, Earth's climate system provides us with examples of the kind of forcing and response shown in Figure 1-8. On a seasonal scale in the northern hemisphere, the summer Sun is most nearly overhead and therefore strongest on June 21 (at the summer solstice), but the hottest air temperatures are not reached until late July over land and late August or later over the ocean. A similar lag of the coldest winter days (occurring in January, February, or later) behind the weakest Sun occurs at the winter solstice, December 21. At the even shorter time scale of a single day, the strongest solar heating occurs near noon, but the warmest temperatures are generally not reached until hours later, just before evening. These examples show that the response time concept is an integral part of the entire climate system.

1-8 Response Rates and Interactions within the Climate System

The water-beaker examples shown so far summarize the response of the climate system by a single curve, as if it were capable only of a single response. But we have already seen in Table 1-1 that the system has many components with different response times. Each responds to the same forcing at its own tempo.

An easy way to grasp these different responses is to imagine that some change is abruptly imposed on the climate system from the outside (for example, a sudden strengthening of the Sun's radiation). Each part of the climate system will respond to this sudden increase in heating in a way analogous to the beaker of water sitting over the Bunsen burner in Figure 1-6, except that each will tend to respond at the tempo dictated by its own unique response time (Figure 1-9). The fast-response parts of the climate system will warm up more quickly, and the slow-response parts will do so more slowly, producing an array of different responses through the system.

We can also apply this idea of different response times to the case in which the factor causing climate change varies in smooth cycles (Figure 1-10). Here again, each part of the climate system will tend to respond at its own rate, and this will again produce several different patterns of response. In the example shown in Figure 1-10, some fast-response parts of the climate system respond so quickly to the climate forcing that they track right along it. In contrast, other slow-response parts of the climate system lag well behind the forcing. Seasonal changes in tropical monsoons are an

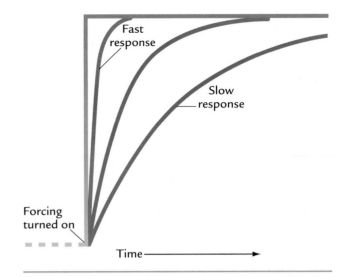

FIGURE 1-9 Variations in response times An abrupt change in climate forcing will produce climate responses ranging from slow to fast within different components of the climate system, depending on their inherent response times.

example of fast responses to climate forcing, whereas ice sheets are an example of slow responses that lag thousands of years behind.

In addition, various kinds of interactions can occur within the climate system. Consider again the example in Figure 1-10. Let us assume that the climate forcing curve shown in this figure represents changes in the amount of the Sun's heat that reaches a particular region over long intervals of time (tens of thousands of years), and that the slow response curve represents cycles in the size of ice sheets that lag thousands of years behind because of their long response time.

Now we focus on a single point in time shown by the asterisk in Figure 1-10. We consider a time when a huge ice sheet has gradually built up over Canada and the northern United States (this has actually happened

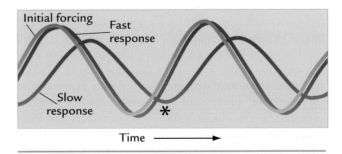

FIGURE 1-10 Variations in cycles of response If the climate forcing occurs in cycles, it will produce differing cyclic responses in the climate system, with the fast responses tracking right along with the forcing cycles while the slower responses lag well behind.

many times in the past, most recently 20,000 years ago). We further assume that this ice sheet has reached the maximum size in its normal cycles of growth and decay, but that it has not yet begun to retreat. This explains the location of the asterisk on the blue (cold) part of the slow-response curve. And we further imagine that the heating from the Sun has begun a slow, long-term increase, but that this increase has not yet begun to melt any of the ice.

What would be the response of air temperatures over the land just south of the ice limits at this particular point in time? Would the air start to warm with the slow strengthening of the overhead Sun? If so, its climate response would track right along with the initial forcing curve in Figure 1-10, with little or no delay.

Or would the air temperatures still be influenced by the huge mass of ice sitting nearby just to the north? If so, its response might instead follow the slower, delayed response pattern of the ice. In this case, the ice sheet would in effect be acting as a significant player within the climate system by exerting an influence of its own on Earth's climate. In an ultimate sense, the ice is a climate response that is driven by slow changes in the Sun, but in a here-and-now sense the ice also has an identifiable effect as a form of climate forcing separate from that of the Sun.

Because both these explanations sound and are plausible, another possibility exists: the response of the air temperatures could be influenced by *both* the overhead Sun and the nearby ice. In this case, the air-temperature response might fall somewhere in between the two, faster than the response of the ice but lagging behind the forcing from the Sun.

This example tells us something important about how interactive the climate system really is. Although we can think of the components of the climate system in one sense as just responding passively to the forces that drive climate change, in actuality they are interactive players in a dynamic system. This interactive role blurs the distinction between forcing and response, and it makes the effort to figure out what is really going on in the climate system more challenging (and interesting) than it otherwise would be.

The response-time concept is also relevant to projections of climate change in the near future. Part V addresses the effects of humans on climate in the recent past and the near future, especially the buildup of greenhouse gases (mainly CO_2 from burning fossil fuels such as oil, coal, and natural gas). The changes in Earth's climate in the next few centuries will be unusual in that this large climate forcing from humans will be applied with unusual speed to the climate system. After a few centuries, the fossil fuel that generates excess CO_2 in the atmosphere will be used up, and Earth's climate may then begin to return to its previous state.

But before that happens, Earth will face a century or more of substantial change in its climate. The major question humanity faces is this: How large a disruption will this spike of excess CO_2 cause? The answer will ultimately be determined by the size of the spike (the anthropogenic forcing that humans apply) and the response times of several major components in the climate system. Some parts of the system will respond right away to the greenhouse-gas forcing with large changes, while others will respond sluggishly, if at all. Part of the challenge to climate scientists is to sort out all these different responses, and the response-time concept is a helpful way to do this.

1-9 Feedbacks in the Climate System

Another important kind of interaction in the climate system is the operation of **feedbacks**, processes that alter climate changes already under way, either by amplifying them (*positive feedbacks*) or by suppressing

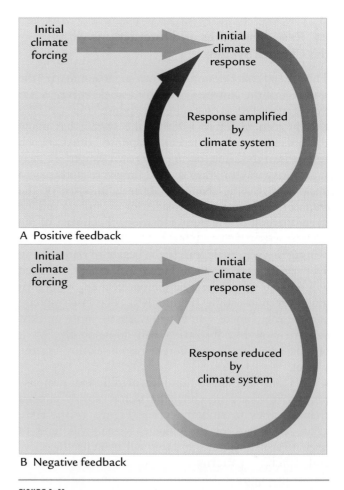

A Positive feedback

B Negative feedback

FIGURE 1-11 **Climate feedbacks** (A) Positive feedbacks within the climate system amplify climate changes initially caused by external factors. (B) Negative feedbacks mute or suppress the initial changes.

them (*negative feedbacks*). Figure 1-11 shows the basic way climate feedbacks operate.

Assume that some external factor (for example, a change in the strength of radiation from the Sun) causes Earth's climate to change. That change will consist of many different responses among the various internal components of the climate system. The changes in some of these components will then further perturb climate through the action of feedbacks.

Positive feedbacks produce additional climate change beyond that triggered by the initial factor (Figure 1-11A). For example, a decrease in the amount of heat energy sent to Earth by the Sun could allow snow and ice to spread across regions at high latitudes that had not been covered by them before. Because snow and ice reflect far more sunlight (heat energy) than bare ground, an increase in the area they cover should decrease the amount of heat taken up by Earth's surface and further cool the climate. Because positive feedbacks amplify changes in both directions, an initial warming caused by increasing energy from the Sun would be amplified by the climate system into a larger warming.

Negative feedbacks work in the opposite sense, by muting climate changes (Figure 1-11B). When an initial climate change is triggered, some components of Earth's climate system respond in such a way as to reduce the initial change. For example, one of the major debates in the field of climate science today concerns the extent to which clouds act as a negative feedback and counter part of the warming effect caused by increasing concentrations of CO_2 in the atmosphere.

Some feedbacks work almost immediately in the climate system, amplifying or suppressing climate changes with no delay. Other feedbacks depend on the delayed response times in the climate system for their effects. Several feedbacks are described in the overview of the climate system in Chapter 2; others are introduced in later chapters.

Key Terms

climate (p. 3)
weather (p. 3)
Fahrenheit (p. 4)
Celsius (p. 4)
Kelvin (p. 4)
resolution (p. 6)
Earth system (p. 7)
climate science (p. 7)
hypothesis (p. 8)
theory (p. 8)
evolution (p. 8)
plate tectonics (p. 8)
greenhouse gases (p. 8)
climate system (p. 8)
forcing (p. 8)
response (p. 8)
radiation (p. 10)
anthropogenic
   forcing (p. 10)
response time (p. 10)
equilibrium (p. 10)
feedbacks (p. 16)

Review Questions

1. How does climate differ from weather?

2. In what sense does climate science differ from traditional sciences such as chemistry and biology?

3. How does climate forcing differ from climate response?

4. In the example in which the Bunsen burner is lit and the beaker of water at first warms quickly and then more slowly, does the response time of the water change through time?

5. Explain, in a general way, how climate change depends on the rate of application of climate forcing and the natural response time of the climate system.

6. The climate system consists of many components with different response times. What is the total range over which these responses vary?

7. Do positive feedbacks always make the climate warmer? How could they make it colder?

Additional Resources

Basic Reading

Understanding Climate Change. 1975. Washington, D.C.: U.S. National Academy of Science.

Advanced Reading

Imbrie, J. 1985. "A Theoretical Framework for the Ice Ages." *Journal of the Geological Society* 142:417–32.

CHAPTER 2

Earth's Climate System Today

In this chapter we examine Earth's climate system, which basically functions as a heat engine driven by the Sun. Earth receives and absorbs more heat in the tropics than at the poles, and the climate system works to compensate for this imbalance by transferring energy from low to high latitudes. The combination of solar heating and movement of heat energy determines the basic distributions of temperature, precipitation, ice, and vegetation on Earth.

We first look at Earth from the viewpoint of outer space as a sphere warmed by the Sun and characterized by global average properties. Then we explore how solar radiation is absorbed by the ocean and land at lower latitudes, transformed into several kinds of heat energy, and transported to higher altitudes and latitudes by wind and ocean currents. Next we examine the kinds of ice that accumulate at the high latitudes and high altitudes where little of this redistributed heat reaches. The chapter ends by reviewing characteristics of life that are relevant to climate.

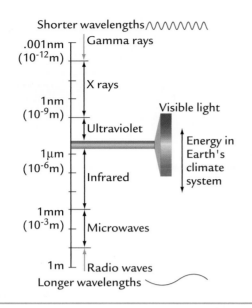

Shorter wavelengths

.001nm (10⁻¹²m) — Gamma rays

1nm (10⁻⁹m) — X rays

Ultraviolet

Visible light

1μm (10⁻⁶m) — Infrared — Energy in Earth's climate system

1mm (10⁻³m) — Microwaves

1m — Radio waves

Longer wavelengths

FIGURE 2-1 **The electromagnetic spectrum** Energy moves through space in a wide range of wave forms that vary by wavelength. Energy from the Sun that heats Earth arrives mainly in the visible part of the spectrum. Energy radiated back from Earth's surface moves in the longer-wavelength infrared part of the spectrum. (Modified from W. J. Kaufman III and N. F. Comins, *Discovering the Universe*, 4th ed., © 1996 by W. H. Freeman and Company.)

Heating Earth

Earth's climate system is driven primarily by heat energy arriving from the Sun. Energy travels through space in the form of waves called **electromagnetic radiation**. These waves span many orders of magnitude in size, or wavelength, and this entire range of wave sizes is known as the **electromagnetic spectrum** (Figure 2-1).

The energy that drives Earth's climate system occupies only a narrow part of this spectrum. Much of the incoming radiation energy from the Sun consists of visible light at wavelengths between 0.4 and 0.7 μm (1 μm, or micrometer = 1 millionth of a meter), sometimes referred to as **shortwave radiation**. Some ultraviolet radiation from the Sun also enters Earth's atmosphere, but radiation at still shorter wavelengths (X rays and gamma rays, measured in nanometers, or billionths of a meter) does not affect climate.

2-1 Incoming Solar Radiation

The most basic way to look at Earth's climate system is to consider its average properties as a sphere. In this way we reduce its three-dimensional complexities to a single global average value typical of the entire planet, as if we were space travelers looking at Earth from a great distance.

Radiation from the Sun arrives at the top of Earth's atmosphere with an average energy of 1368 watts per square meter (W/m²). These watts are the same units of energy used to measure the brightness (or more accurately the power) of a household light bulb. If Earth were a flat, one-sided disk directly facing the Sun, and if it had no atmosphere, 1368 W/m² of solar radiation would fall evenly across its entire surface (Figure 2-2, top).

But Earth is a three-dimensional sphere, not a flat disk. A sphere has a surface area of $4\pi r^2$ (r being its radius) that is exactly four times larger than the surface area of a flat one-sided disk (πr^2). Because the same amount of incoming radiation must be distributed across this larger surface area, the average radiation received per unit of surface area on a sphere is only one-quarter as strong (1368/4 = 342 W/m²). Another way of looking at this is to see that half of Earth's rotating surface is dark at any time because it faces away from the Sun at night, while the daytime side, warmed by the Sun, receives radiation at indirect angles except at the one latitude in the tropics where the Sun is directly overhead in any given season (Figure 2-2, bottom).

The 342 W/m² of solar energy arrives at the top of the atmosphere, mainly in the form of visible radiation. About 70% of this shortwave radiation passes through Earth's atmosphere and enters the climate system (Figure 2-3). The other 30% is sent directly back out into space after reflection (or scattering) by clouds,

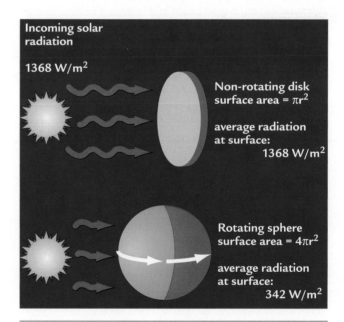

FIGURE 2-2 **Average solar radiation on a disk and a sphere** The surface of a flat nonrotating disk that faces the Sun (top) receives exactly four times as much solar radiation per unit of area as the surface of a rotating sphere, such as Earth (bottom).

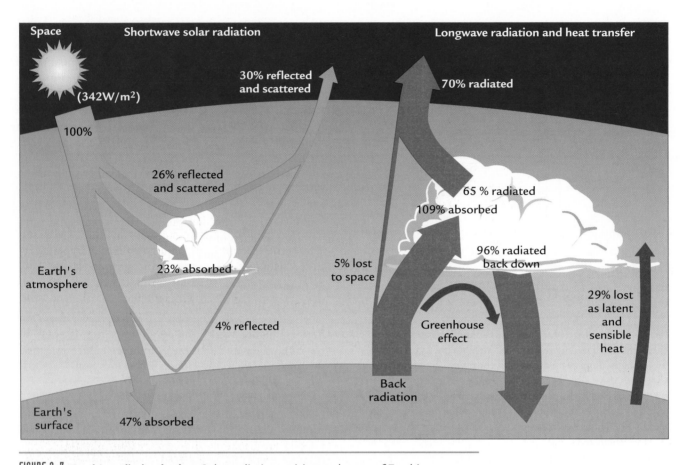

FIGURE 2-3 Earth's radiation budget Solar radiation arriving at the top of Earth's atmosphere averages 342 W/m², indicated here as 100% (upper left). About 30% of the incoming radiation is reflected and scattered back to space, and the other 240 W/m² (70%) enters the climate system. Some of this entering radiation warms Earth's surface and causes it to radiate heat upward (right). The greenhouse effect (lower right) retains 95% of the heat radiated back from Earth's heated surface and warms Earth by 31 °C. (Adapted from T. E. Graedel and P. J. Crutzen, *Atmosphere, Climate, and Change*, Scientific American Library, © 1997 by Lucent Technologies, after S. H. Schneider and R. Londer, *Co-evolution of Climate and Life* [San Francisco: Sierra Club Books, 1984], and National Research Council, *Understanding Climate Change: A Program for Action* [Washington, D.C.: National Academy of Sciences, 1975].)

dust, and the more reflective regions at Earth's surface. As a result, the average amount of solar energy retained by Earth is 240 W/m² (0.7 × 342 W/m²).

Of the 70% of solar radiation that is retained within the climate system, about two-thirds is absorbed at Earth's surface and about one-third by clouds and water vapor in the atmosphere (see Figure 2-3). This absorbed radiation heats Earth and its lower atmosphere and provides energy that drives the climate system.

2-2 Receipt and Storage of Solar Heat

Because Earth is continually receiving heat from the Sun but is also maintaining a constant (or very nearly constant) temperature through time, it must be losing

an equal amount of heat (240 W/m²) back to space. This heat loss, called **back radiation**, occurs at wavelengths lying in the infrared part of the electromagnetic spectrum (see Figure 2-1). Because it occurs at longer wavelengths (5–20 μm) than the incoming shortwave solar radiation, back radiation is also called **longwave radiation**.

Any object with a temperature above absolute zero (−273°C, or 0K) contains some amount of heat that is constantly being radiated away toward cooler regions. Radiated heat can come from such objects as red-hot burners on stovetops, but it is also emitted from objects not warm enough to glow, such as the asphalt pavement that emits shimmering ripples of heat on a summer day.

Although it may seem counterintuitive, even objects with temperatures *far* below the freezing point of water are radiators; that is, they emit some heat. The amount of heat radiated by an object increases with its temperature. The radiation emitted is proportional to T^4, where T is the absolute temperature of the object in Kelvins. Objects with temperatures of $-272°C$ (1K) emit at least a tiny bit of heat energy and so can technically be considered radiators!

What is the average radiating temperature of our own planet? One way to try to answer this question is to average the countless measurements of Earth's surface temperature made over many decades to derive its mean surface temperature. If we do this, we find that the average surface temperature of Earth as a whole is $+15°C$ (288K, or 59°F). This value sounds like a reasonable middle ground between the very large area of hot tropics, averaging 25°–30°C, and the much smaller polar regions, which average well below freezing (0°C). It seems reasonable that the *surface* of planet Earth radiates heat with this average temperature.

But there is a problem with this simple way of looking at radiator Earth. If we now take all available measurements from orbiting satellites and space stations high above Earth, they tell us our planet is sending heat out to space as if it had an average temperature of $-16°C$ (257K or 3°F). This value is 31°C colder than the $+15°C$ average temperature we are certain is correct for Earth's surface. If both sets of measurements are accurate, why are these values so different?

The reason for this discrepancy is the **greenhouse effect** (see Figure 2-3). Earth's atmosphere contains greenhouse gases that absorb 95% of the longwave back radiation emitted from the surface, thus making it impossible for most heat to escape into space. The trapped radiation is retained within the climate system and reradiated down to Earth's surface. This extra heat retained by the greenhouse effect makes Earth's surface temperature 31°C warmer than it would otherwise be.

In effect, measurements made by satellites and space stations in outer space cannot detect the radiation emitted directly from the warmer surface of the Earth because of the muffling effect of the blanket of greenhouse gases and clouds. Instead, most of the heat actually radiated back to space is emitted from an average elevation of 5 kilometers, equivalent to the tops of many clouds—still well within the lowest layer of Earth's atmosphere (Box 2-1). These cold ($-16°C$) cloud tops emit radiation at an average value of about 240 W/m², exactly the level needed to offset the amount of solar radiation retained within Earth's climate system and keep it in balance.

The two main gases in Earth's atmosphere are N_2 (nitrogen) at 78% of the total and O_2 (oxygen) at 21%, but neither is a greenhouse gas because neither traps outgoing radiation. In contrast, the three most important greenhouse gases form very small fractions of the atmosphere. Water vapor (H_2O_v) averages less than 1% of a dry atmosphere, but it can range to above 3% in the moist tropics. Carbon dioxide (CO_2) and methane (CH_4) occur in much smaller concentrations of 0.035% and 0.00018%, but they are also important greenhouse gases.

Clouds also contribute to the retention of heat within the climate system by trapping outgoing radiation from Earth's surface. This role in warming Earth's climate works exactly opposite to the impact of clouds in reflecting incoming solar radiation and cooling our climate. The relative strength of these two competing roles varies with region and season.

Many important characteristics of Earth's climate, such as the amount of incoming sunlight, vary with latitude. Incoming solar radiation is stronger at low latitudes, where sunlight is concentrated more nearly overhead, than at high latitudes, where the Sun's rays strike Earth at a more indirect angle and cover a wider area (Figure 2-4). As a result, larger amounts of solar radiation reach the same unit area of Earth's surface in the tropics than near the poles (Figure 2-5).

This unequal distribution of incoming solar radiation is aggravated by unequal absorption and reflection

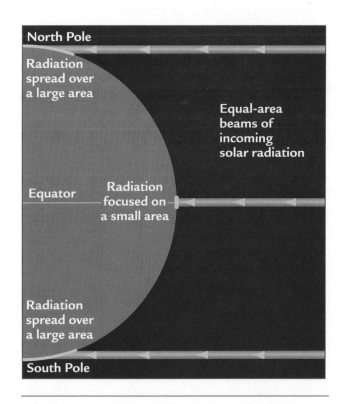

FIGURE 2-4 **Unequal radiation on a sphere** More solar radiation falls on a unit area of Earth's surface near the equator than at the poles because of the more direct angle of incoming radiation. (Adapted from L. J. Battan, *Fundamentals of Meteorology* [Englewood Cliffs, N.J.: Prentice-Hall, 1979].)

BOX 2-1 LOOKING DEEPER INTO CLIMATE SCIENCE

The Structure of Earth's Atmosphere

Earth's atmosphere is divided into four layers that extend far above its surface: the **troposphere,** which extends from the surface to between 8 and 18 kilometers; the **stratosphere,** extending up to 50 kilometers; and the overlying mesosphere (50–80 km) and thermosphere (above 80 km). Only the lowest two are critical in climatic changes.

The troposphere is both the layer within which we live and the layer within which most of Earth's weather happens. Storm systems that produce clouds and rainfall or snowfall are almost entirely confined to this layer. Dust or soot particles that are lifted by strong winds from Earth's surface into the lower parts of this layer are quickly removed by precipitation every few days or weeks. The troposphere is also the main layer within which we measure Earth's climate and its changes, particularly those at Earth's surface. As we will see later, about 80% of the gases that form Earth's atmosphere are contained within the troposphere.

Above the troposphere lies the stratosphere, a much more stable layer almost completely separated from the turbulent storms and other processes so common in the troposphere. Only the largest storms penetrate the stratosphere, and only its lowermost layer. Large volcanic eruptions occasionally throw small particles up out of the troposphere and into the stratosphere. Because no rain or snow falls in most of this layer, it may take years for gravity to pull these particles back to Earth's surface. The stratosphere forms 19.9% of Earth's atmosphere; the troposphere and stratosphere together account for 99.9% of its mass.

The stratosphere is also important to Earth's climate because it contains small amounts of oxygen (O_2) and ozone (O_3), which block ultraviolet radiation arriving from the Sun. This shielding effect accounts for a small fraction of the 30% reduction in incoming heat energy from the Sun. It also greatly

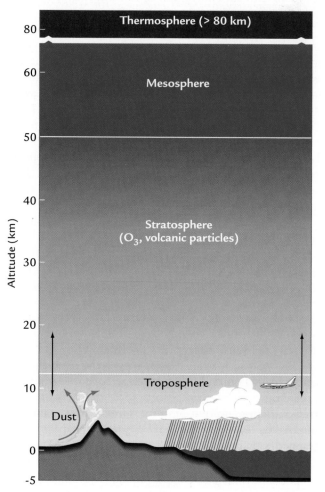

Structure of Earth's atmosphere Most day-to-day weather, as well as important changes in Earth's climate, occur in the lowermost layer of the atmosphere, called the troposphere, which varies in height from 8 to more than 18 km.

reduces the exposure of life-forms on Earth to the harmful effects of ultraviolet radiation, which can cause skin cancers and genetic mutations.

---

by Earth's surface at different latitudes. A smaller fraction of the incoming radiation is absorbed at higher latitudes than in the tropics mainly because (1) solar radiation arrives at a less direct angle (see Figure 2-4) and (2) snow and ice surfaces at high latitudes reflect more radiation (see Figure 2-5).

The percentage of incoming radiation that is reflected rather than absorbed by a particular surface is referred to as its **albedo** (Table 2-1). Snow and ice surfaces at high latitudes have albedos ranging from 60% to 90%, with larger values typical of freshly formed snow and ice, and somewhat lower values for snow or

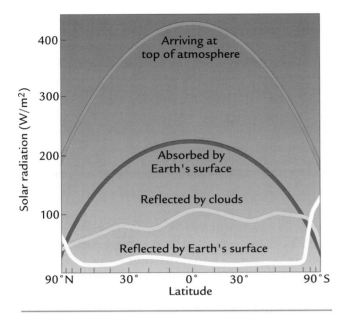

FIGURE 2-5 **Pole-to-equator heating imbalances** Incoming radiation is higher in the tropics than at the poles. Less reflective surfaces at low latitudes absorb a larger fraction of incoming radiation, while highly reflective snow and ice surfaces at the poles reflect more, aggravating the pole-to-equator imbalance in absorbed radiation. (Adapted from R. G. Barry and R. J. Chorley, *Atmosphere, Weather, and Climate,* 4th ed. [New York: Methuen, 1982].)

TABLE 2-1 Average Albedo Range of Earth's Surfaces

Surface	Albedo range (percent)
Fresh snow or ice	60–90%
Old, melting snow	40–70
Clouds	40–90
Desert sand	30–50
Soil	5–30
Tundra	15–35
Grasslands	18–25
Forest	5–20
Water	5–10

Adapted from W. D. Sellers, Physical Climatology *(Chicago: University of Chicago Press, 1965), and from R. G. Barry and R. J. Chorley,* Atmosphere, Weather, and Climate, *4th ed. (New York: Methuen, 1982).*

ice that contain dirt or are partly covered by pools of melted water. In contrast, snow-free land surfaces have much lower albedos (15–30%) and ice-free water reflects even less of the incoming radiation (below 5% when the Sun is overhead).

The albedo of any surface also varies with the angle at which incoming solar radiation arrives. For example, water reflects less than 5% of the radiation it receives from an overhead Sun, but a far higher fraction of the radiation from a Sun lying low in the sky (Figure 2-6). This same tendency holds true for the other surfaces. Because 70% of Earth's surface is low-albedo water, Earth's surface has an average albedo near 10%.

These factors combine to make Earth's surface more reflective near the poles than in the tropics. The Antarctic ice sheet and extensive areas of nearby sea ice in the southern hemisphere have very high albedos, as do the Arctic sea ice cover and the extensive winter

FIGURE 2-6 **Sun angle controls heat absorption** All of Earth's surfaces (here water) absorb more solar radiation from an overhead Sun (A) than from a Sun lying low in the sky (B). (Adapted from L. J. Battan, *Fundamentals of Meteorology* [Englewood Cliffs, N.J.: Prentice-Hall, 1979].)

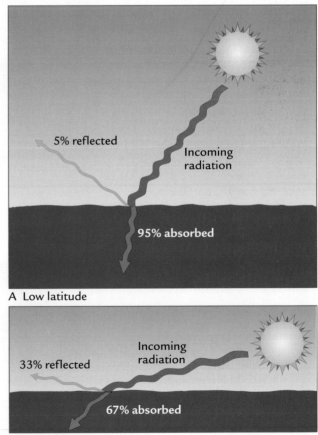

snow cover in Eurasia and North America in the northern hemisphere. Clouds counteract some of this imbalance in surface albedo by reflecting somewhat more solar radiation at low latitudes than at high latitudes.

The net result is that both polar regions absorb even less of the already low amounts of incoming solar radiation, while the tropics absorb even more of the already-high radiation. The overall effect of these albedo differences is to increase the solar heating imbalance between the poles and the tropics (see Figure 2-5).

Radiation and albedo also vary seasonally. The 23.5° tilt of Earth's axis as it revolves around the Sun causes the northern and southern hemispheres to tilt alternately toward and away from the Sun, and this motion causes seasonal changes in solar radiation received in each hemisphere (Figure 2-7A). From our Earthbound perspective, we experience this orbital motion as a shift of the overhead Sun through the tropics from a latitude of 23.5°N on June 21 to 23.5°S on December 21. This change in the Sun's angle results in large seasonal changes in the amounts of solar radiation (W/m²) received on Earth (Figure 2-7B).

Accompanying these seasonal changes in solar heating are large seasonal changes in the albedo of Earth's surface with latitude (Figure 2-8). In the south polar region, the ice sheet over Antarctica remains intact through the year, but an extensive ring of sea ice surrounding Antarctica expands and contracts every year across an area of 16 million square kilometers. In contrast, the Arctic Ocean has a multiyear cover of sea ice that fluctuates much less in extent through the year, and the main seasonal albedo change in the northern hemisphere comes from the winter expansion and summer retreat of snow cover on Asia, Europe, and North America. Both of these changes in albedo play important roles in long-term climate change (Box 2-2).

FIGURE 2-7 **Earth's tilt and seasonal radiation** (A) The tilt of Earth's axis in its annual orbit around the Sun causes the northern and southern hemispheres to lean directly toward and then away from the Sun at different times of the year. (B) This change in relative position causes seasonal shifts between the hemispheres in the amount of solar radiation received at Earth's surface. (A: adapted from F. K. Lutgens and E. J. Tarbuck, *The Atmosphere* [Englewood Cliffs, N.J.: Prentice-Hall, 1992; B: adapted from A. L. Berger, "Milankovitch Theory and Climate," *Reviews of Geophysics* 26 [1988]: 624–57.)

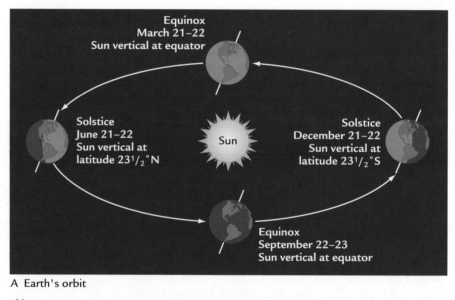

A Earth's orbit

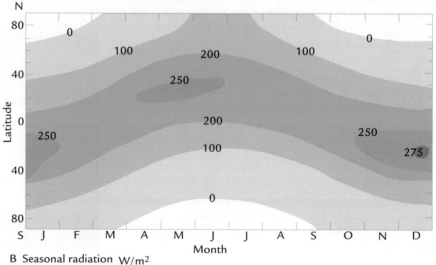

B Seasonal radiation W/m²

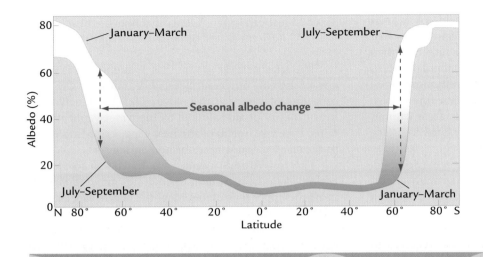

FIGURE 2-8 Albedo changes with season Average albedo increases in the northern hemisphere in winter (January–March) mainly because of increased snow cover over land and also because of more extensive sea ice. Albedo increases in the southern hemisphere's winter (July–September) because of more extensive sea ice. (Adapted from G. Kukla and D. Robinson, "Annual Cycle of Surface Albedo," *Monthly Weather Review* 108 [1980]: 56–68.)

BOX 2-2   CLIMATE INTERACTIONS AND FEEDBACKS

Albedo/Temperature

The large difference in albedo between highly reflective snow or ice and land or water surfaces that absorb radiation is important in climate change. Surface albedos can increase by 75% (from 15% to 90%) when snow-free land areas become covered with snow, and over oceans that become covered by sea ice. As a result, a surface that had previously absorbed most incoming radiation will now reflect it away, with significant implications for climate change.

Assume that climate abruptly cools, for any reason (perhaps a decrease in the output of the Sun). Part of the climate system's natural response to a cooling is an increase in the land area covered by snow and in the ocean area covered by sea ice. The expansion of these light, high-albedo surfaces will cause an increase in the percentage of incoming radiation reflected back out to space, and a decrease in the amount of heat absorbed at the surface.

The loss of absorbed heat in these regions will in turn cause the local climate to cool by an additional amount beyond the initial cooling. This is an example of the concept of positive feedback, introduced in Chapter 1. The positive feedback process also works in the opposite direction: an initial warming will reduce the cover of snow and ice, increase the amount of heat absorbed by exposed land or water, and further warm climate. Climate scientists estimate that any initial climate change will be amplified by about 40% by this positive feedback effect. An initial cooling of 1°C would be amplified to a total cooling of 1.4°C by this process.

Climate scientists call this positive feedback **albedo–temperature feedback.** In a larger sense, its net effect is to increase Earth's overall sensitivity to climate changes. The greater the area on Earth covered by snow and ice, the more sensitive the planet as a whole becomes to imposed changes in climate. Because most of the regional albedo contrast on Earth is localized at the equatorward limit of snow and sea ice, the albedo–temperature feedback most strongly affects climate at higher latitudes near these limits.

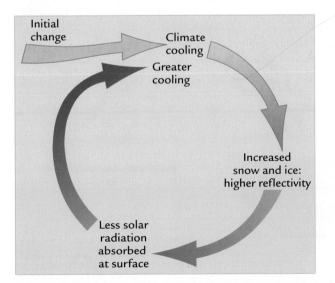

Albedo–temperature feedback When climate cools, the increased extent of reflective snow and ice increases the albedo of Earth's surface in high-latitude regions, causing further cooling by positive feedback. The same feedback process amplifies climate warming.

Clouds also affect the regional receipt of heat at Earth's surface. Areas such as subtropical deserts that are free of clouds receive far more solar radiation than do most parts of the oceans, especially subpolar oceans where frequent storms produce nearly continuous cloud cover. The cloud cover over tropical rain forests also reduces the receipt of radiation there. But areas without clouds also back-radiate more heat from Earth's surface than do areas with heavy cloud cover.

Water is the key to Earth's climate system (Box 2-3). Absorption and storage of solar heat are strongly affected by the presence of liquid water because of its high **heat capacity,** a measure of the ability of a material to absorb heat. Heat energy is measured in units of **calories** (one calorie is the amount of heat required to raise the temperature of 1 gram of water by 1°C). Heat capacity is the product of the density (in g/cm³) of a heat-absorbing material and its **specific heat,** the number of calories absorbed as the temperature of 1 gram of this material increases by 1°C:

$$\text{Heat capacity} = \text{Density} \times \text{Specific heat}$$
$$(\text{cal/cm}^3) \qquad (\text{g/cm}^3) \qquad (\text{cal/g})$$

The specific heat of water is 1, higher by far than any of Earth's other surfaces. The ratios of the heat capacities of water:ice:air:land are 60:5:2:1. Much of the heat capacity of air is linked to the water vapor it contains. Likewise, much of the heat capacity of land is due to the small amount of water held in the soil.

The low-latitude oceans are Earth's main storage tanks of solar heat. Sunlight penetrates into and directly heats the upper tens of meters of the ocean, especially in the tropics, where the radiation arrives from a Sun high in the sky. Equally important, winds blowing across the ocean's surface stir the upper layers and rapidly mix solar heat as deep as 100 meters (Figure 2-9). In contrast, even though tropical and subtropical landmasses generally become very hot under the strong Sun, they are not capable of storing much heat because heat is conducted down into soil or rock at very slow rates (see

FIGURE 2-9 **Difference in heating of land and oceans** During the seasonal cycle of solar radiation (top), ocean surfaces heat and cool slowly and only by small amounts because temperature changes are mixed through a layer 100 m thick (lower left). In contrast, land surfaces heat and cool quickly and strongly because of their low capacity to conduct and store heat (lower right). (Adapted from J. E. Kutzbach and T. Webb III, "Late Quaternary Climatic and Vegetational Change in Eastern North America: Concepts, Models, and Data," in *Quaternary Landscapes,* ed. L. C. K. Shane and E. J. Cushing [Minneapolis: University of Minnesota Press, 1991].)

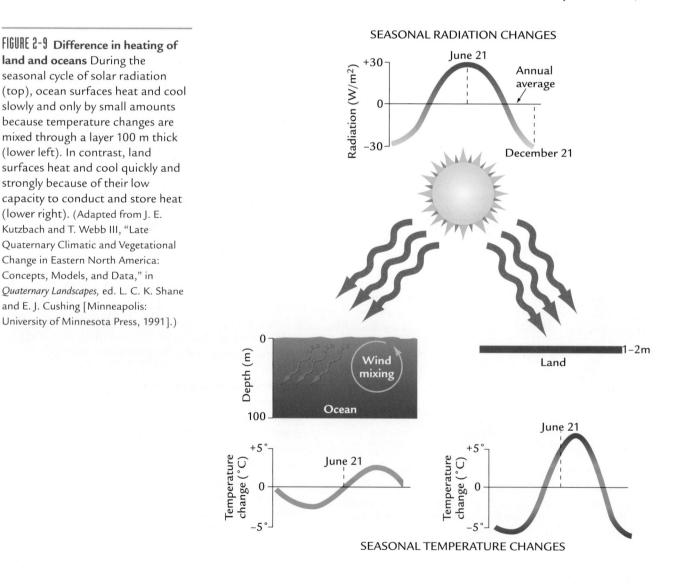

BOX 2-3 CLIMATE INTERACTIONS AND FEEDBACKS

Water in the Climate System

Water is a critical part of the miracle of life on Earth. Of all the planets in the solar system, only here on Earth can water exist in three forms: as a liquid (in oceans and lakes), as a frozen solid (ice), and as a gas (water vapor). The essence of our good fortune comes from the fact that water not only accounts for the largest fraction of our bodies and those of most other organisms, but also is the medium we drink for survival, the substance we swim in and skate or ski on for amusement, and part of the air we breathe from day to day without ill effect.

Water is also a vital component of the climate cycle. The oceans contain over 97% of Earth's water, with just 2% held in glacial ice and less than 1% in all the rest of Earth's reservoirs. The **hydrologic cycle,** or continual recycling of water among all these reservoirs, including the much smaller amounts held in the atmosphere, in plants, in lakes, in rivers, and in soil, is vital to the operation of the climate system, as we will see in this and future chapters.

Water and the climate system (A) Only on planet Earth can water exist in the form of a solid (ice), a gas (water vapor), or ordinary liquid water, depending on the prevailing temperature and pressure. (B). Water moves through the climate system in different forms as part of the global water cycle. (D. Merritts et al., *Environmental Geology,* © 1997 by W. H. Freeman and Company.)

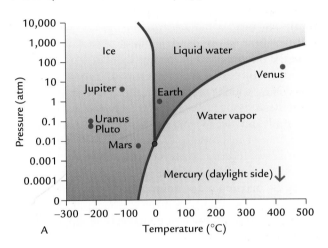

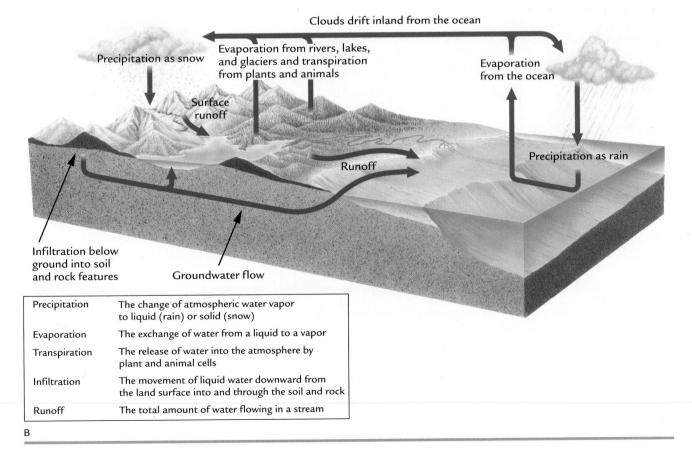

Precipitation	The change of atmospheric water vapor to liquid (rain) or solid (snow)
Evaporation	The exchange of water from a liquid to a vapor
Transpiration	The release of water into the atmosphere by plant and animal cells
Infiltration	The movement of liquid water downward from the land surface into and through the soil and rock
Runoff	The total amount of water flowing in a stream

Figure 2-9, bottom). As a result, the large amount of heat stored in the low-latitude ocean provides most of the fuel that runs Earth's climate system.

One consequence of the large difference in heat capacity between land and water is evident in Earth's response to seasonal changes in solar heating: a map showing only the seasonal *range* of temperatures at Earth's surface succeeds in outlining the shapes of most of its continents and oceans (Figure 2-10). The low heat capacity of landmasses allows them to respond strongly to seasonal changes in heating, while the high heat capacity and strong wind mixing of the ocean's upper layers limit its seasonal response (see Figure 2-9, bottom). The largest temperature changes occur over the largest, most continental landmass, Asia, with lesser changes on smaller continents such as Australia (see Figure 2-10). Small seasonal changes also characterize the tropics, which receive consistently strong solar radiation throughout the year and remain warm.

Land and ocean surfaces also differ markedly in their *rates* of response to seasonal heating changes (see Figure 2-9). Interior regions of large continents heat up and cool off quickly because of their low heat capacity, reaching maximum (or minimum) seasonal temperatures about one month after the seasonal radiation maximum (or minimum). In contrast, the upper ocean responds much more slowly because of its high heat capacity, reaching its extreme temperature responses only two or more months after the June 21 and December 21 solar radiation peaks in each hemisphere.

These differences in amplitude and timing of response between land surfaces and the upper ocean layers are referred to as differences in **thermal inertia**. The fast-responding land has a low thermal inertia; the slower-responding upper layers of the ocean have a high thermal inertia.

2-3 Heat Transformation

The heat energy received and stored in the climate system is exchanged among water, land, and air through several processes. As we saw earlier, some of the absorbed heat is lost from Earth's warm surface by long-wave back radiation, but most back radiation is trapped by greenhouse gases and radiated back down toward Earth's surface (see Figure 2-3).

Two other important kinds of heat transfer occur within the climate system. One process involves the transfer of **sensible heat** by moving air. Sensible heat is the product of the temperature of the air and its specific heat. It is also the heat that a person directly senses as it is carried along in moving air masses. Surfaces heated

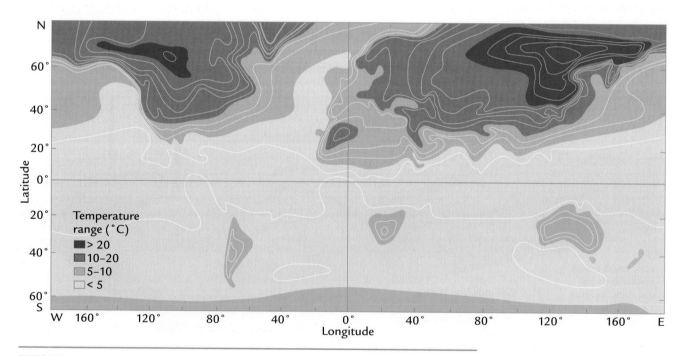

FIGURE 2-10 Sensitivity of solar heating to land vs. ocean The total change in mean daily surface temperature between summer and winter is greatest over large landmasses and much smaller over oceans and small continents. The locations of most continents and oceans can be detected on this map from their temperature responses alone. (Adapted from A. S. Monin, "Role of Oceans in Climate Models," in *Physical Basis of Climate and Climate Modeling,* Report no. 16, GARP Publication Series [Geneva: World Meteorological Organization, 1975].)

by the Sun warm the lowermost layer of the overlying atmosphere. The heated air expands in volume like any heated gas, becomes lighter (less dense), rises higher in the atmosphere, and carries sensible heat along with it in a large-scale process known as **convection**.

Convection of sensible heat by air is analogous to what occurs in a pot of water heated from below on a stovetop: the stovetop heating warms the lower layers of water and causes them to expand, rise, and transfer heat upward (Figure 2-11, top). The same process happens when air is heated and rises. Sensible heating at low latitudes is largest over land surfaces in summer, especially in dry regions, because the low water content and small heat conductance of land surfaces allow them to warm to much higher temperatures than the oceans and they transfer this heat to the overlying air (Figure 2-11, bot-

tom). As air and water move horizontally across Earth's surface, they transport sensible heat.

The second form of heat transfer within the climate system involves the movement of **latent heat.** This more powerful process of heat transfer also depends on the convective movement of air, but in this case the heat carried by the air is temporarily hidden, latent in the water vapor. Transfer of this latent heat occurs in two steps: (1) initial evaporation of water and storage of heat in water vapor, and (2) later release of stored heat during condensation and precipitation, usually far from the site of initial evaporation.

To understand this process, we take a closer look at the behavior of water (Figure 2-12). We learned earlier that it takes 1 calorie of heat energy to raise the temperature of 1 gram of water by 1°C within the range of 0°–100°C. These stored calories are all available to be returned from the water to the air for each 1°C that the water subsequently cools.

But additional changes happen when water changes state, either to ice (a solid) or to water vapor (a gas). Large amounts of latent heat are stored during the warming process that transforms ice to water or water to water vapor, and this stored heat is available for later release during the cooling process that transforms water vapor back to water or water back to ice (see Figure 2-12).

When ice warms to a temperature of 0°C, any addition of heat (usually from the atmosphere) causes it to melt. Melting of solid ice to form liquid water requires a large input of energy (80 calories per gram of ice). During the melting process, the temperature of the ice holds at 0°C, because all the additional energy is being used for melting.

The opposite happens when water freezes: caloric energy (heat) is liberated. Water chilled to 0°C by loss of heat to the atmosphere begins to freeze if still more heat is extracted. The freezing process liberates the same 80 calories of heat energy per gram of water that had been stored in the water as the ice melted. Because the heat energy liberated during the freezing process was in effect hidden in the water, this released energy is called the **latent heat of melting**.

A similar process governs transitions between liquid water and water vapor (see Figure 2-12). Turning liquid water heated to 100°C into water vapor (gas) requires the input of 540 calories per gram of water, again with no change in temperature during the change of state. Condensation of water vapor back to liquid water in clouds or fog liberates the same amount of energy, called the **latent heat of vaporization**.

Despite what Figure 2-12 implies, water does not have to boil to be turned into vapor. Evaporation occurs across the entire range of temperatures at which liquid water exists, including all the temperatures of the surface ocean and the lower atmosphere. As with boiling,

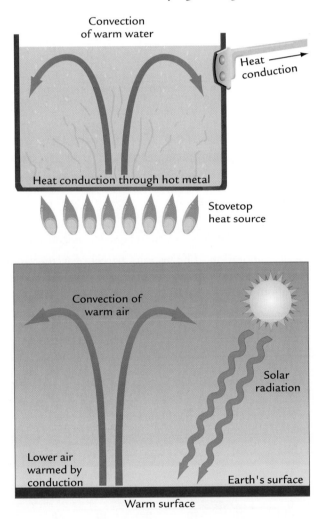

FIGURE 2-11 **Convection of heat** A kettle heated on a stovetop conducts heat to the lower layers of water, which then rise and convect heat upward (top). Similarly, a land surface warmed by the Sun transfers heat by conduction to the lower layers of the atmosphere, which then rise and convect heat upward (bottom).

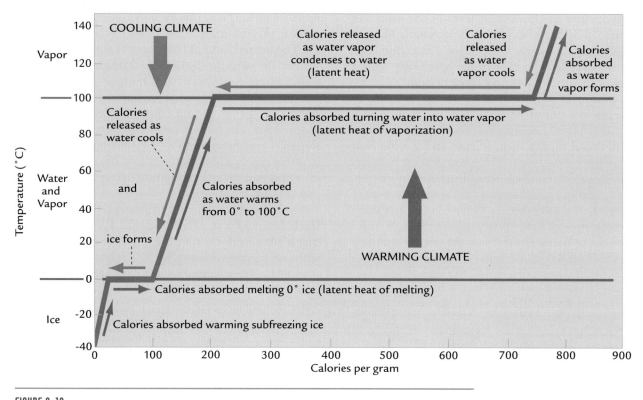

FIGURE 2-12 **Heat transformations** Heat calories are absorbed from the atmosphere when ice, water, or water vapor is warmed. This heat is released back to the atmosphere during cooling. Heat is also absorbed and stored during changes of state from ice to liquid water or from liquid water to water vapor and this stored latent heat is available for release when water vapor condenses to water or when water freezes to form ice. (Adapted from H. V. Thurman, *Introduction to Oceanography*, 6th ed. [New York: Macmillan, 1991].)

the change of state from liquid to vapor at lower temperatures requires the addition of a large amount of heat energy, and this energy is stored in the water vapor in latent form for later release.

The amount of water vapor that can be held in air is limited, much like the amount of sugar that can be dissolved in coffee. Attempts to add more water vapor to fully saturated air will cause condensation (the formation of dew or raindrop nuclei at the so-called **dew point** temperature). This limit of full saturation, measured in grams of water per cubic meter of air and called the **saturation vapor density**, roughly doubles for each 10°C increase in air temperature (Figure 2-13). Warm tropical air at 30°C can hold almost ten times as much water vapor as cold polar air masses near or below 0°C. As a result of this relationship, water vapor is an important positive feedback in the climate system (Box 2-4).

Evaporation of water from Earth's surface in warmer regions stores excess heat energy in the warm atmosphere. This energy stored in water vapor is carried along with the moving air, both vertically and horizontally. When condensation and precipitation occur, the stored latent energy is released as heat, far from the site of evaporation. The average parcel of water vapor stays in the air for 11 days and travels over 1000 kilometers, about the distance from Los Angeles to Denver.

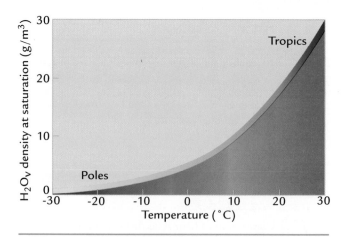

FIGURE 2-13 **Water vapor content of air** Warm air is capable of holding almost ten times as much water vapor (H_2O_v) as cold air. (Adapted from L. J. Battan, *Fundamentals of Meteorology* [Englewood Cliffs, N.J.: Prentice-Hall, 1979].)

BOX 2-4 CLIMATE INTERACTIONS AND FEEDBACKS

Water Vapor

Water vapor, Earth's major greenhouse gas, varies in concentration from 0.2% in very dry air to over 3% in humid tropical air. This strong dependence on temperature produces an important positive feedback in the climate system called **water vapor feedback.**

Assume that climate warms for any reason. A warmer atmosphere can hold more water vapor, and the increased greenhouse gas traps more heat. This large greenhouse effect in turn further warms Earth, amplifying the initial warming through a positive feedback loop. The same positive feedback process works when the climate cools: initial cooling reduces the amount of water vapor held in the atmosphere and produces additional cooling because of the reduced greenhouse effect. It is estimated that direct positive feedback from water vapor can triple the size of an initial climate change. This estimate is based on the action of water vapor in a clear sky; it ignores the more complicated effects that occur when water vapor condenses and forms clouds.

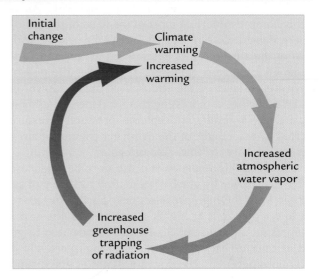

Water vapor feedback When climate warms, the atmosphere is able to hold more water vapor (the major greenhouse gas in the atmosphere), and the increase in water vapor leads to further warming by means of a positive feedback. This feedback works in reverse during cooling.

Heat Transfer in Earth's Atmosphere

Tropical and subtropical latitudes below 35° have a net excess of incoming solar radiation over outgoing back radiation, while at latitudes higher than 35° the opposite is true (Figure 2-14). Most of this excess heat is stored within a thin upper layer of the tropical ocean, and it is this heating imbalance that drives the general circulation of the Earth's atmosphere and oceans. Roughly two-thirds of the net transport of heat from the Earth's equator toward its poles occurs in the lower atmosphere.

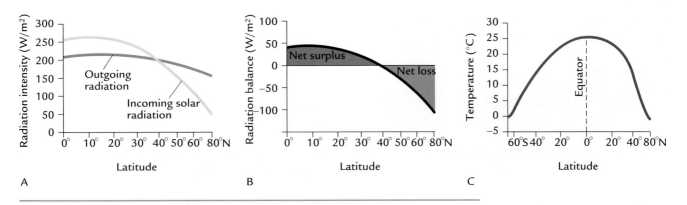

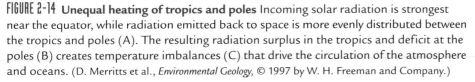

FIGURE 2-14 **Unequal heating of tropics and poles** Incoming solar radiation is strongest near the equator, while radiation emitted back to space is more evenly distributed between the tropics and poles (A). The resulting radiation surplus in the tropics and deficit at the poles (B) creates temperature imbalances (C) that drive the circulation of the atmosphere and oceans. (D. Merritts et al., *Environmental Geology*, © 1997 by W. H. Freeman and Company.)

2-4 Overcoming Stable Layering in the Atmosphere

Both sensible and latent heating are associated with upward motion or convection in the atmosphere. But how high does this convected air go, and why? Air is highly compressible, and most of the mass of the atmosphere is held close to Earth's surface by gravity, with 50% of the air molecules below 7 kilometers and 75% below 10 kilometers (Figure 2-15). Under the pull of gravity, each layer of the atmosphere presses down on the layers below, compressing them and increasing their density. As a result, atmospheric pressure—pressure exerted by the weight of the overlying column of air—increases toward the lower elevations.

Opposing this tendency of dense air to be held close to Earth's surface by gravity is a natural tendency of air to flow from the high pressures near Earth's surface to the lower pressures at higher elevations. In effect, the compressed air at lower elevations has nowhere to go and so it pushes back against the overlying layers. These two opposing forces, the downward pull of gravity and the resistance directed upward, tend to remain in a stable but delicate balance that limits the amount of vertical air motion.

Over limited areas, however, parcels of air will rise if they become less dense than the surrounding air. The major way this process occurs is by warming of Earth's surface and its lower atmosphere. Heating causes the lower layers of air to expand, become lower in density, and rise buoyantly in the atmosphere (like a balloon full of hot air).

As these parcels of heated air rise to higher elevations, processes come into play that involve changes in temperature and in the density of the moving air due to changes in pressure. Climate scientists refer to these changes as **adiabatic** processes. In the case of rising parcels of air, the lower pressures encountered at higher elevations result in an additional expansion of the air beyond the amount originally caused by heating at Earth's surface. But expansion requires the expenditure, or loss, of heat energy, and no source of heat exists to replace the lost heat. As a result, the rising air parcels begin to cool and increase in density, especially in comparison with the surrounding air that is not rising. Eventually the parcel stops rising at the level where its density matches that of the surrounding air. This upward loss of heat is a dry process that is independent of the amount of water vapor in the air.

Latent heating is the second process that can destabilize the atmosphere, and it occurs as a wet process driven by water vapor, which weighs roughly a third less than the mixture of gases that form Earth's atmosphere. Evaporation adds water vapor to the atmosphere at low elevations and causes a net decrease in the density of the air, which can then rise to higher elevations. As before, the moist air rises, expands, and at some point cools to the temperature at which it becomes fully saturated with water vapor. Then condensation begins. Condensation releases latent heat, which partially opposes the cooling of the rising air parcel due to its expansion. The release of this heat in the air parcels also makes them more buoyant (less dense) and allows them to rise much higher in the troposphere (sometimes to 10 to 15 km). Eventually the cumulative loss of most of the water vapor in the parcels stops the release of latent heat that had kept them rising.

The rate at which Earth's atmosphere cools with elevation is called the **lapse rate**. This rate ranges from 5°C/km to as high as 9.8°C/km, but typically averages 6.5°C/km both at middle latitudes and for the planet as a whole. Localized lapse rates are highest in very dry regions where the only factor at work on rising air parcels is adiabatic expansion and cooling. For the areas of Earth's surface where rising air carries water vapor, the release of latent heat at higher elevations warms the air at upper levels and reduces the lapse rate to values near the lower end of this range.

2-5 Tropical–Subtropical Atmospheric Circulation

The large-scale circulation in Earth's atmosphere that transports heat from low to high latitudes is summarized in Figure 2-16. The red arrows on the left show

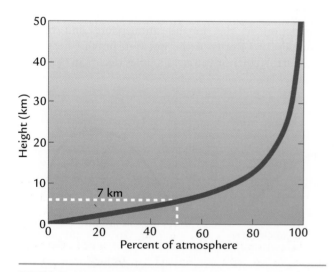

FIGURE 2-15 Distribution of air with elevation Each layer of air in Earth's atmosphere presses down on the underlying layers, increasing the pressure on the lower layers and at the surface. Most of the mass of Earth's atmosphere lies at lower elevations. (Modified from R. G. Barry and R. J. Chorley, *Atmosphere, Weather, and Climate*, 4th ed. [New York: Methuen, 1982].)

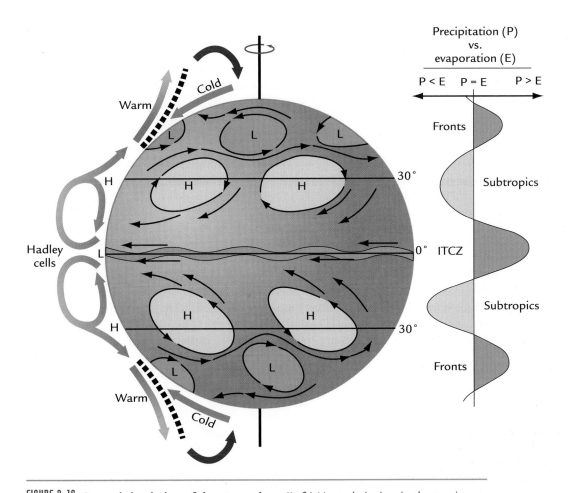

FIGURE 2-16 **General circulation of the atmosphere** (Left) Heated air rises in the tropics at the intertropical convergence zone (ITCZ) and sinks in the subtropics as part of the large-scale Hadley cell flow, which transports heat away from the equator. Additional poleward heat transfer occurs along moving weather systems (fronts) at middle and higher latitudes, with warm air rising and moving poleward and cold air sinking and moving equatorward. (Right) Rising air in the tropics causes a net excess of precipitation over evaporation, while dry air sinking in the subtropics produces more evaporation than precipitation. Higher latitudes tend to have small excesses of precipitation over evaporation. (Adapted from S. H. Schneider and R. Londer, *Co-evolution of Climate and Life* [San Francisco: Sierra Club Books, 1984]; E. Bryant, *Climate Process and Change* [Cambridge: Cambridge University Press, 1998]; and J. P. Peixoto and M. A. Kettani, "The Control of the Water Cycle," *Scientific American*, April 1973.)

heat being transported by warm air, with blue arrows indicating movement of cold air. The large-scale patterns of precipitation and evaporation resulting from these air motions are summarized on the right. This summary figure is important to the discussion that follows, so refer to it often as we go along.

Tropical heating drives a giant tropical circulation pattern called the **Hadley cell** (see Figure 2-16, left). Warm air rises in giant columns marked by towering, puffy (cumulonimbus) clouds created by evaporation of water vapor from tropical oceans and subsequent condensation at high altitudes. Condensation produces a narrow zone of high rainfall in the rising part of the

Hadley cell near the equator. The rising motion in the tropical part of the Hadley cell represents an enormous transfer of heat through the atmosphere from low to high altitudes.

Air parcels that rise and lose water vapor in the tropics move toward the subtropics in both hemispheres, transporting sensible heat and other energy from lower to higher latitudes. In the subtropics, this air sinks toward the surface near 30° latitude. The sinking air is then warmed by the increasing pressure of the atmosphere at lower elevations (another adiabatic process), and it gradually becomes even drier and able to hold still more water vapor. This Hadley cell flow prevents condensation from occurring in much of the subtropics and makes these latitudes a zone of low average precipitation and high evaporation, in regions such as the Sahara Desert.

The Hadley cell circulation is completed at Earth's surface, where trade winds from both hemispheres blow from the subtropics toward the tropics and replace the rising air. As warm dry air carried by the trade winds passes over the tropical ocean, it continually extracts water vapor from the sea surface. The region near the equator where the northern and southern trade winds meet is called the **intertropical convergence zone (ITCZ)**. Water vapor carried by the trade winds contributes to the rising air motion and abundant rainfall along the ITCZ.

Viewed on a daily or weekly basis, the actual circulation at low latitudes consists of small-scale cloud systems that develop explosively in limited regions. But when the circulation is averaged over seasons or years, it takes the form shown by the schematic Hadley cell in

Figure 2-16. This flow is an important part of the poleward transfer of heat, with a net upward movement and release of latent heat in the tropics followed by a net horizontal transfer of sensible heat to the subtropics at high elevations.

These large-scale movements of large masses of air alter the pressure (weight of air) at Earth's surface. Air movement upward and away from the tropics reduces the weight of the column of overlying air and produces low surface pressures near the ITCZ. Air moving into the subtropics and then down toward Earth's surface increases the pressure there.

Because solar heating is the basic driving force behind the Hadley cell circulation, the seasonal shifts of the Sun between hemispheres also affect the location of the ITCZ. It moves northward during the northern hemisphere's summer (June to September) and southward during the southern hemisphere's summer (December to March). The slow thermal response of the land and oceans causes the seasonal shifts of the

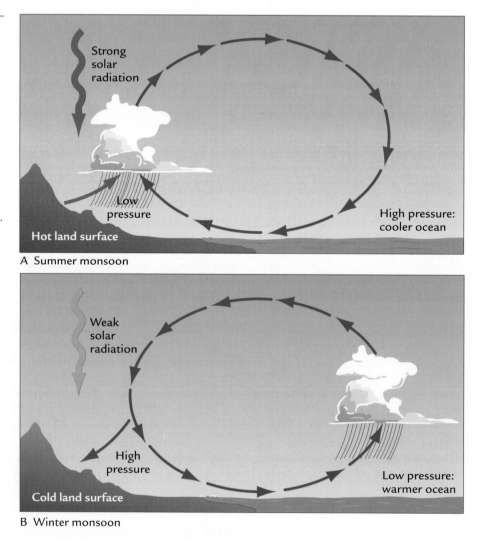

FIGURE 2-17 Monsoonal circulations
(A) In summer, more rapid heating of land surfaces than of the ocean produces rising motion over the continents and draws moist air in from the ocean, producing precipitation over land. (B) In winter, more rapid cooling of the land surfaces than of the ocean produces sinking motion over the continents and sends cold dry air out over the warmer ocean, shifting most winter precipitation out to sea.

A Summer monsoon

B Winter monsoon

ITCZ to lag more than a month behind those of the Sun.

Important seasonal transfers of heat between the tropical ocean and land, called **monsoons**, arise from the fact that water responds more slowly than land to these seasonal changes in solar heating because of its larger heat capacity and high thermal inertia. The *summer monsoon* circulation is basically an in-and-up flow of moist air that produces precipitation. The strong, direct solar radiation in summer at low and middle latitudes heats Earth's surface (Figure 2-17A). Because soil contains relatively little water, land surfaces have low thermal inertia and heat up quickly. The ocean, with its much higher thermal inertia, absorbs the heat, mixes it through a layer up to 100 meters thick, and warms up far more slowly and to a smaller extent than the land.

These different responses to solar heating set in motion a large-scale land-sea circulation, the monsoon. Initially, dry air heated rapidly over the continental interior rises, and the upward movement of this mass of air produces a region of low pressure over the land. Air is then drawn in toward the low-pressure region from the cooler oceans. The moist air coming in from the oceans is slowly heated and joins in the prevailing upward motion.

As the moist air rises, it cools and its water vapor condenses. Condensation produces heavy precipitation and releases substantial amounts of latent heat, which fuels an even more powerful upward motion (see Figure 2-17A). This net in-and-up circulation in summer monsoons is a two-stage process: initially a dry process due to the rising of sensible heat, and later a wet process linked to ocean moisture and release of latent heat.

The strongest summer monsoon circulations on Earth today occur over India. Heating of the large high landmass of southern Asia focuses a strong wet summer monsoon against the Himalaya Mountains (see Chapter 7).

The *winter monsoon* circulation is the reverse of the summer monsoon. The basic flow is a down-and-out motion of cold, dry air from land to sea (Figure 2-17B). In winter the Sun's radiation is weaker, and land surfaces cool by back radiation. Because of differences in thermal inertia, land surfaces cool faster and more intensely than the oceans. Air cooled over the land sinks toward the surface and creates a region of high pressure where the extra mass of air piles up. Air flows outward from this cell toward the oceans at lower levels. Because the sinking air holds little moisture, the outflow from the continents to the oceans is cold and dry.

These near-surface monsoon circulations are part of a much larger circulation that links the subtropical continents and oceans (Figure 2-18). In summer the upward motion prevailing over land produces an excess of atmospheric mass (and high pressures) near 10 kilometers of altitude. This high pressure results in a high-

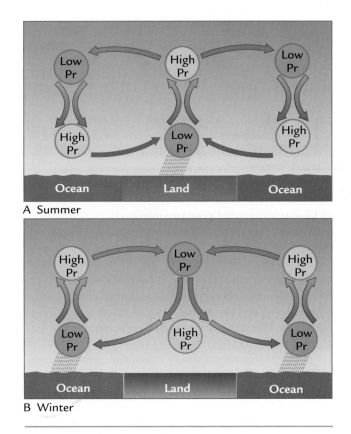

A Summer

B Winter

FIGURE 2-18 **Large-scale monsoon circulations** Air motion associated with monsoon circulations at larger scales is upward over the land and downward over the ocean in summer, but the exact reverse in winter. Precipitation is heaviest in regions of low pressure (Pr) and upward motion. (Adapted from J. E. Kutzbach and T. Webb III, "Late Quaternary Climatic and Vegetational Change in Eastern North America: Concepts, Models, and Data," in *Quaternary Landscapes*, ed. L. C. K. Shane and E. J. Cushing [Minneapolis: University of Minnesota Press, 1991].)

level flow of air out toward the adjacent oceans, and this flow produces regions of high pressure in the lower atmosphere over the subtropical ocean in summer (Figure 2-18A).

In winter the slow thermal response of the oceans keeps them relatively warm, and they provide heat to the cold air flowing out from the land. The heat gained by the atmosphere results in rising motion and increased precipitation, and the upward movement of air produces strong low-pressure cells over the oceans at higher latitudes (Figure 2-18B).

2-6 Atmospheric Circulation at Middle and High Latitudes

The giant Hadley cells are a simple and convenient summary of basic atmospheric circulation across that half of Earth's surface area lying between 35°S and

BOX 2-5 LOOKING DEEPER INTO CLIMATE SCIENCE

The Coriolis Effect

The direction of movement of fluids (air and water) is complicated by Earth's rotation. One way to visualize this effect is to contrast the actual motion of a person during a single day at either pole in comparison with that of a person on the equator. A person standing exactly at one of the poles spins around in place once each day, without ever moving from that point. In contrast, a person standing on the equator and facing east (the direction of Earth's rotation) zooms through space at 500 meters per second around the 40,000 km of Earth's circumference once each day, all the while facing east and not spinning at all. Earth's rotation accounts for this shift from a spinning motion at the poles to a one-way trip through space at the equator.

To appreciate the effect Earth's rotation has on moving objects, we can track the movement of a small airplane across the northern hemisphere. If the plane sets out from a point P on Earth's surface and flies in a straight line in relation to the stars, it will appear to an Earthbound observer located under the plane at point P_1 to be moving toward the northeast. Several hours later, with the plane still moving in exactly the same direction *in relation to the stars,* Earth's coordinate system (its reference directions of north, south, east, and west) will have rotated out from under the airplane, and the airplane will now appear from an Earthbound perspective at point P_2 to be moving to the southeast rather than the northeast. In effect, the rotation of Earth's coordinates out from under the plane makes the plane appear to have turned to the right, *although it has not really turned at all.*

Earth's rotation has the same effect on moving air and water. Air naturally tends to flow from regions of high to low pressure, but rotation causes its motion to appear to be deflected to the right (again, from an Earthbound perspective). This *apparent* deflection is called the **Coriolis effect,** after the French engineer who first described it.

In the northern hemisphere, air moving from a zone of high pressure to low pressure always appears to be deflected to the right. This deflection causes air moving outward from a high-pressure cell to acquire a net clockwise spin, and air moving in toward a low-pressure region acquires a counterclockwise spin for the same reason.

Both the direction of deflection and the spinning of air around the highs and lows are exactly reversed in the southern hemisphere because the direction of Earth's rotation viewed from the perspective of the South Pole is exactly opposite that of the view from the North Pole. At the equator, where Earth's rotation has no net spinning motion, the Coriolis effect drops to zero.

35°N latitude (see Figure 2-16). The circulation at latitudes above 35° is more difficult to summarize.

To understand the transfer of heat from middle to high latitudes, we begin with the regions of high surface pressure in the subtropics. The air that sinks to Earth's surface in the subtropical branch of the Hadley cell converges or piles up there, creating a nearly continuous band of high surface pressures in the subtropics (Figure 2-19). Superimposed on this basic pattern is a tendency for the monsoonal flow of air from land to sea in summer to produce oval-shaped cells of high pressure over the subtropical oceans (see Figure 2-17A).

Because air naturally flows away from regions of higher pressures toward areas where pressure is lower, the subtropical high-pressure cells send air moving outward near Earth's surface in all directions. Ultimately, however, the path taken by this air is not directly from regions of high to low pressure. Earth's rotation deflects the path of this air in a *relative* sense (Box 2-5), and this

deflection produces a clockwise spin of air around subtropical highs in the northern hemisphere and a counterclockwise spin in the southern hemisphere.

Returning to the summary map of Earth's circulation in Figure 2-16, notice that the trade winds that flow out from the subtropical high-pressure zones toward the equator in the northern hemisphere are deflected to the west by Earth's rotation and given a net easterly trajectory: from northeast to southwest. These are the trade winds that move toward the tropics.

The same Coriolis deflection turns air flowing poleward from the subtropical highs toward the east, resulting in a net southwest-to-northeast ("westerly") flow. This surface flow of warm air out of the subtropics transports heat into high latitudes where the radiation balance is negative (see Figure 2-14B). Much of this heat transport in the atmosphere is ultimately tied to the large transfer of latent heat from the ocean to the atmosphere in the tropics.

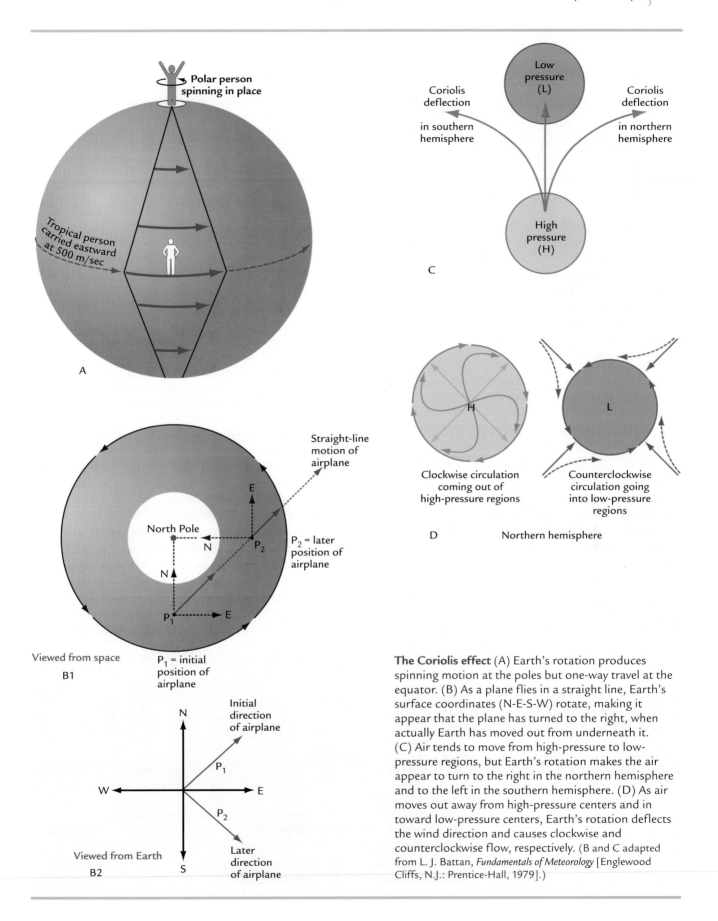

The Coriolis effect (A) Earth's rotation produces spinning motion at the poles but one-way travel at the equator. (B) As a plane flies in a straight line, Earth's surface coordinates (N-E-S-W) rotate, making it appear that the plane has turned to the right, when actually Earth has moved out from underneath it. (C) Air tends to move from high-pressure to low-pressure regions, but Earth's rotation makes the air appear to turn to the right in the northern hemisphere and to the left in the southern hemisphere. (D) As air moves out away from high-pressure centers and in toward low-pressure centers, Earth's rotation deflects the wind direction and causes clockwise and counterclockwise flow, respectively. (B and C adapted from L. J. Battan, *Fundamentals of Meteorology* [Englewood Cliffs, N.J.: Prentice-Hall, 1979].)

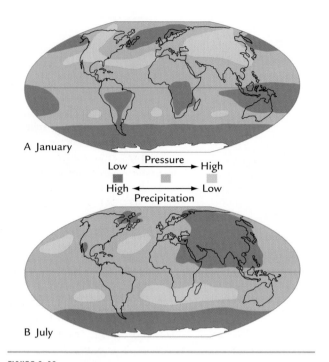

A January

Low ← Pressure → High

High ← → Low

Precipitation

B July

FIGURE 2-19 **Seasonal pressure patterns** A band of high surface pressure occurs in the subtropics in both hemispheres in both seasons. During summer, this zone is interrupted over land (especially Asia) by areas of low pressure produced by summer monsoon circulations. (Adapted from E. Bryant, *Climate Process and Change* [Cambridge: Cambridge University Press, 1998].)

Toward higher middle latitudes in both hemispheres, the circulation of the lower atmosphere is a complex zone of transition between the prevailing warm flow coming out of the subtropics and a much colder equatorward flow from higher latitudes (see Figure 2-16). The surface flow at these latitudes is dominated by an ever-changing procession of high- and low-pressure cells moving from west to east, separated by *frontal zones*, or regions near Earth's surface where large changes in temperature occur over short distances in association with fast-moving air. Both the poleward movement of warm air and the equatorward movement of cold air along these frontal systems have the effect of warming the higher latitudes in a net sense, adding a major contribution to heat redistribution on Earth.

Precipitation generally exceeds evaporation in the temperate middle latitudes, for two reasons: the cooler air temperatures reduce the rates of evaporation, and precipitation associated with moving low-pressure cells is heavy.

Because the low-pressure cells move rapidly from west to east across the middle latitudes, they encounter and interact strongly with the topography of the land. Air flowing in from the ocean tends to carry large amounts of water vapor. As this air encounters mountain ranges that block its flow, it is forced to rise to higher elevations, and it cools. Water vapor condenses from the cooling air and produces heavy precipitation on the sides of mountains that face upwind toward warm oceans, such as the Olympic Mountains of Washington State. This is referred to as **orographic precipitation** (Figure 2-20).

Air that has been stripped of much of its water vapor then sinks on the downwind side of the mountains. As it moves to lower elevations, it is compressed and warmed, and it gains in capacity to store even more water vapor without condensation occurring. As a result, the lee or *rain shadow* sides of mountain ranges are areas of lower precipitation. This process also reinforces the natural tendency of mid-continental regions far from the oceanic sources of moisture, such as the Great Plains of the United States, to be dry.

At higher elevations in the mid-latitude atmosphere, winds flow more steadily from west to east. Narrow rib-

FIGURE 2-20 **Orographic precipitation** As moist air masses driven up against mountains rise and cool, water vapor condenses and produces precipitation. The air masses subsiding on the downwind side of the high topography warm, retain water vapor, and suppress precipitation. (Adapted from F. Press and R. Siever, *Understanding Earth*, 2nd ed., © 1998 by W. H. Freeman and Company.)

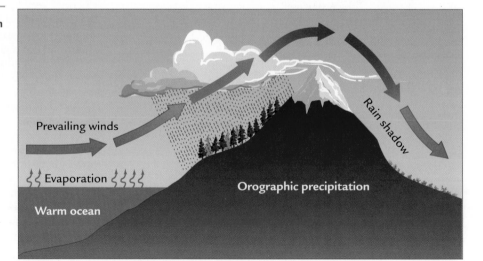

Prevailing winds

Evaporation

Warm ocean

Rain shadow

Orographic precipitation

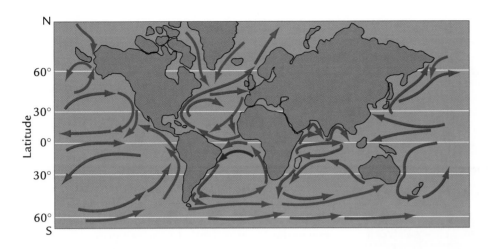

FIGURE 2-21 **Surface ocean circulation** The surface flow of the oceans is organized into strong wind-driven currents. These currents encircle large spinning gyres in the subtropical oceans. Currents moving out of the tropics carry heat poleward, while currents moving away from the poles carry cold water equatorward. (Modified from S. Stanley, *Earth Systems History*, © 1999 by W. H. Freeman and Company.)

bons of faster flow called **jet streams** occur at altitudes of 5 to 10 kilometers in two regions: a persistent but weaker jet near 30° latitude in the subtropics, near the sinking branch of the Hadley cell; and a more mobile jet that wanders between latitudes 30° and 60° above the moving high- and low-pressure cells. The jet at middle latitudes is especially strong in winter. Almost hidden by these prevailing west-to-east motions and the meandering paths of these jet streams is a net transport of heat and water vapor from low to high latitudes. Poleward meanders in the jet stream carry warm air to the north, and equatorward meanders carry cold air south.

Heat Transfer in Earth's Oceans

The uppermost layer of the ocean is heated by solar radiation. Like air, water expands as it warms and becomes less dense, but in this case the warmest layers are already at the top of the ocean, so they simply float on top of the colder, denser deep ocean. Winds mix the stored solar heat to maximum depths of 100 meters, a small fraction of the 4000-meter average depth of the oceans. Some of this warm water is transported from the tropics toward the poles, and this poleward flow carries about half as much heat as is transported by the atmosphere.

2-7 The Surface Ocean

Most of the surface circulation of the oceans is driven by winds, and one of the most prominent results is huge **gyres** of water at subtropical latitudes (Figure 2-21). These spinning gyres are mainly the result of an initial push (or drag) of the winds on the ocean surface, and of the Coriolis deflection of the moving water (see Box 2-5).

Blowing wind exerts a force on the upper layer of the ocean and sets it in motion in the same direction as the wind. The Coriolis effect turns this surface flow of

water to the right in the northern hemisphere (and to the left in the southern hemisphere). The top layer of water in turn pushes (or drags) the underlying layers, which are deflected a little farther to the right than the surface layer and are also slowed by friction. This process continues down into the water column to a depth of about 100 meters, creating a downward spiral of water gradually deflected farther and farther to the right (Figure 2-22). The net transport of water in this

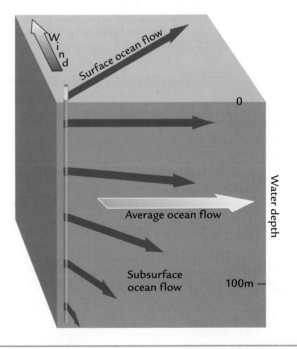

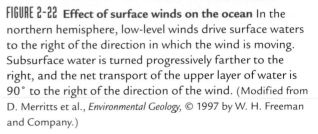

FIGURE 2-22 **Effect of surface winds on the ocean** In the northern hemisphere, low-level winds drive surface waters to the right of the direction in which the wind is moving. Subsurface water is turned progressively farther to the right, and the net transport of the upper layer of water is 90° to the right of the direction of the wind. (Modified from D. Merritts et al., *Environmental Geology*, © 1997 by W. H. Freeman and Company.)

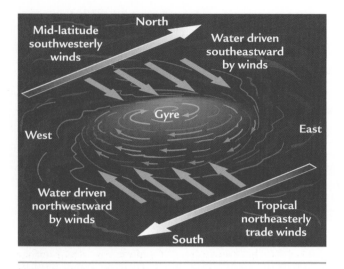

FIGURE 2-23 **Subtropical gyres** In the northern hemisphere, mid-latitude southwesterly winds and tropical northeasterly trade winds drive warm water toward the centers of subtropical gyres, forming a thick lens of warm water that circulates in a clockwise gyre.

entire 100-meter layer of ocean water is 90° to the right of the wind in the northern hemisphere (and to the left south of the equator).

In the North Atlantic Ocean, the prevailing low-altitude winds are the tropical trade winds and mid-latitude westerlies. Trade winds blowing toward the southwest push shallow waters toward the northwest, and mid-latitude westerlies blowing toward the northeast push surface water to the southeast (Figure 2-23). Together these winds drive the uppermost layer of water into the centers of the subtropical gyres and pile up a lens of warm water.

Sea level in the center of this lens sits 2 meters higher than the surrounding ocean. Water that flows away from this lens is turned to the right by the Coriolis deflection, and this creates a huge subtropical ocean gyre spinning in a clockwise direction (counterclockwise for gyres in the southern hemisphere). The edges of the continents also play a role in forming these gyres by acting as boundaries that contain the flow within individual ocean basins.

Subtropical gyres extend all the way to depths of 600 to 1000 meters. In the North Atlantic, most of the water moving into the deeper parts of the gyre comes from its northern margin, where the prevailing westerly winds are strong enough to push large volumes of water toward the south just underneath the lens of warm surface water.

Viewed over long intervals of time, most of the flow in subtropical gyres consists of water moving around in giant spirals. The volume of water circulating is enormous, about 100 times the transport of all Earth's rivers flowing into the oceans. Almost hidden in this recirculation is a much smaller amount moving *through* the gyres and carrying heat toward high latitudes.

The prevailing flow toward the equator in the deeper parts of the gyres must be balanced by a return flow toward the poles, and this flow is concentrated in narrow regions along the western gyre margins. In the North Atlantic, the poleward transport occurs in the **Gulf Stream** and its continuation, the **North Atlantic Drift** (see Figure 2-21). As the Gulf Stream emerges from the Gulf of Mexico, it forms a narrowly concentrated outflow of warm salty water headed north.

Another factor that affects poleward heat transport only in the North Atlantic Ocean is a giant vertical circulation cell linked to the deeper circulation of the ocean (Figure 2-24). A large volume of surface water sinks to depths below 2 kilometers in the higher latitudes of the North Atlantic, and this sinking water must be balanced by a compensating inflow of surface water from the south.

The effects of this circulation cell are felt even beyond the North Atlantic. The surface circulation typical of most oceans carries heat from the warm equator to the cold poles, as partial compensation for the imbalances set up by uneven solar heating and heat absorption at the surface. This normal equator-to-pole flow is

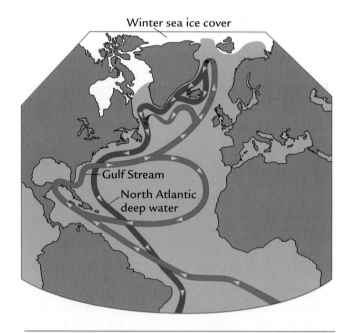

FIGURE 2-24 **Sinking of surface water** Warm salty water flowing northward in the North Atlantic Ocean chills and sinks north of Iceland and in the Labrador Sea, between North America and Greenland. This cold deep water flows south out of the Atlantic at depths of 2 to 4 km. (Modified from D. Merritts et al., *Environmental Geology,* © 1997 by W. H. Freeman and Company.)

reversed in the South Atlantic, where the net direction of heat transport is from south to north, *against* the planet's temperature gradient (see Figure 2-21). Temperate water flows into the South Atlantic from the middle latitudes of the Indian and Pacific oceans, moves northward across the equator, and sinks in the high-latitude North Atlantic.

The net northward transport of heat in the North Atlantic is often referred to as a "conveyor belt" (see Figure 2-24), but the overall flow has also been compared to an airport baggage carousel. Most of the luggage (heat) spins around and around the carousel (warm water recirculating in the subtropical gyre), while only a small amount enters the carousel (comes across the equator from the South Atlantic) or is removed (heads farther north in the Gulf Stream and its continuation).

This warm, northward-moving water in the Atlantic transfers a huge amount of heat to the atmosphere. At latitudes above 50°N, the large temperature contrast between the warm North Atlantic waters and the cold overlying air produces a loss of sensible heat from the ocean to the atmosphere that is comparable to the amount of heat delivered locally by incoming solar radiation.

The fundamental circulation and heat transport of the oceans are less well understood than those of the atmosphere, mainly because of the difficulty of maintaining long-term monitoring stations at sea. The closer oceanographers look at the surface flow, the more complicated it turns out to be, with smaller gyres of water recirculating within larger gyres. At even smaller scales, spinning cells of water 100 kilometers wide move erratically across the ocean and transport large amounts of heat. These are analogous to moving low-pressure and high-pressure cells in the atmosphere.

2-8 Deep-Ocean Circulation

The poleward flow of warm water that counters some of Earth's heat imbalance occurs above the **thermocline**, a zone of rapid temperature change between warm upper layers and cold water filling the deeper ocean basins. Actually, two thermoclines exist: (1) a deeper permanent portion that is maintained throughout the year, and (2) a shallower portion that changes as a result of seasonal heating by the Sun (Figure 2-25).

We've just seen how the warm poleward flow above the thermocline in the North Atlantic is balanced by sinking of cold water at high latitudes and movement of this cold deep water toward the equator. This overturning circulation is called the **thermohaline flow**. This term refers to the two main processes that control formation of deep water: temperature ("thermo-") and salinity ("-haline," from the same root as "halite," a synonym for rock salt).

Deep waters form and sink because they become more dense than the underlying water, as a result of any of several mechanisms. Seawater contains dissolved salt (on average near 35 parts per thousand, or ‰, by mass), and this salt content, or **salinity**, makes it 3.5% more dense than freshwater. The density of ocean water can be increased at lower latitudes when the atmosphere evaporates freshwater as water vapor, leaving the remaining water saltier (denser). The density of seawater can also can be increased at high latitudes by formation of sea ice during **salt rejection**, a process that stores freshwater in sea ice and leaves the salt behind.

Another way to increase the density of ocean water is by cooling it. Saltwater is slightly compressible, which means it loses volume and gains density when it is cooled by the atmosphere. Cooling can occur either because warm ocean water is carried poleward into cooler regions or because colder air masses move to lower latitudes.

Salinity and temperature often work together to raise the density of water. Initially, evaporation or formation of sea ice increases the salinity and density of

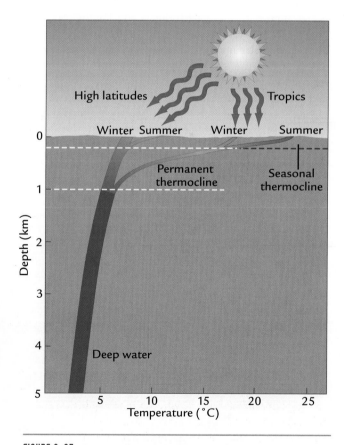

FIGURE 2-25 **Thermoclines** The permanent thermocline (100–1000 m) separates cold deep water from shallower layers affected by changes in Earth's surface temperature. Shallow seasonal thermoclines (0–100 m) vary in response to seasonal solar heating of the upper ocean layers.

surface waters as a kind of preconditioning process. Cold air then cools the water, further increases its density, and causes it to sink.

The large-scale thermohaline flow in the deep ocean is closely linked to temperature and its effect on density. The very dense waters filling the deepest ocean basins today form at higher, colder latitudes. The progressively less dense waters that fill successively shallower depths of the ocean form at less frigid latitudes farther from the poles.

Most of the deepest ocean of the world is filled by water delivered from just two regions, the high-latitude North Atlantic Ocean and the Southern Ocean, near Antarctica (Figure 2-26). No deep water forms in the high latitudes of the Pacific Ocean today because the salinity of the surface water is too low and the surface waters there are not sufficiently dense.

The surface waters that sink and form deep water in the North Atlantic initially acquire high salinity in the dry subtropics as a result of strong evaporation. Some of the water vapor taken from the tropical Atlantic Ocean is exported westward over the low mountains of Central America into the Pacific Ocean, leaving the Atlantic saltier than equivalent latitudes in the Pacific. Some of the salty water left in the Atlantic is carried northward by the Gulf Stream and North Atlantic Drift. Frigid air masses from the surrounding continents then extract sensible heat from the water in winter and further increase its density to the point where it sinks.

This water mass is called **North Atlantic deep water**. Sinking occurs in two regions in the North Atlantic, one north of Iceland, the other east of Labrador (see Figure 2-24). Together these two sources of deep water fill the Atlantic Ocean between depths of 2 and 4 kilometers. This flow moves southward with a total volume 15 times the combined flow of all the world's rivers. Eventually, much of this flow rises toward the sea surface in the Southern Ocean and joins the waters circling eastward around Antarctica.

An even colder and denser water mass forms in the Antarctic region and flows northward in the Atlantic below 4 kilometers (see Figure 2-26). This water mass, called **Antarctic bottom water**, fills the deep Pacific and Indian oceans. Some forms near the Antarctic coast when seawater is chilled by very cold air masses and as a result of salt rejection as sea ice forms. Some also forms by intense cooling in gaps in the extensive sea-ice cover well away from Antarctica.

Two smaller water masses are prominent at intermediate depths of the North Atlantic. **Antarctic intermediate water** forms far north of Antarctica at latitudes 45°–50°S. This water is warmer and less dense than North Atlantic deep water and flows northward above it at depths above 1.75 kilometers. **Mediterranean overflow water** forms in the subtropical Mediterranean Sea as a result of winter chilling of surface waters with a very high salt content caused by strong evaporation. Water vapor extracted by subtropical evaporation leaves Mediterranean waters much saltier than normal ocean water.

Deep ocean water is one of the slowest-responding parts of the climate system. On average, it takes more than 1000 years for a parcel of water that leaves the surface and sinks into the deep ocean to emerge back at the ocean's surface. The oldest and slowest-moving deep

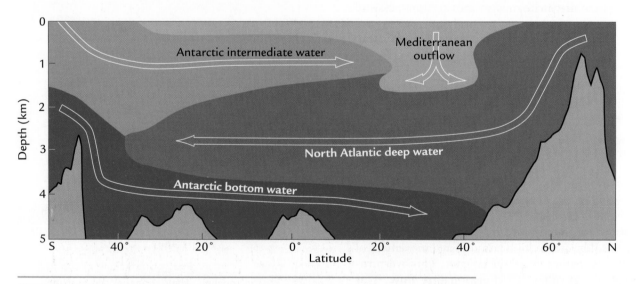

FIGURE 2-26 **Deep Atlantic circulation** Water filling the North Atlantic basin comes from sources in the high-latitude North Atlantic, the Southern Ocean near Antarctica, and (at shallower depths) the Mediterranean Sea. (Adapted from E. Berner and R. Berner, *Global Environment* [Englewood Cliffs, N.J.: Prentice-Hall, 1996].)

water is found in the Pacific Ocean, while the Atlantic has younger, faster-moving water. The long journey of water through the deep ocean keeps much of it out of touch with the climatic changes that affect the atmosphere and surface ocean.

With all this water sinking into the deep ocean, how does it get back to the surface? It turns out that climate scientists don't really know the answer to this question very well. In the Atlantic, some of the southward-flowing North Atlantic deep water rises to the surface near Antarctica, and very strong winds mix it into the upper layers of the Southern Ocean. But what happens to the large volume of Antarctic water moving north in the deep Pacific and Indian Oceans, where no deep water forms?

For decades a widely accepted explanation has been that deep water injected into the ocean in specific regions gradually mixes into the central ocean basins and slowly moves upward across the thermocline and

into the warm surface waters. But this highly diffuse return flow has been difficult to detect because it is spread across such a large area. Recent measurements show that this upward diffusion of water is too slow to account for much of the return flow.

In addition, a process called **upwelling**, rapid upward movement of subsurface water from intermediate depths, occurs in two other kinds of ocean regions. Both upwelling processes are initiated by surface winds and aided by the Coriolis effect:

- When surface winds in the northern hemisphere blow parallel to coastlines along the path shown in Figure 2-27A, they push water away from the land. To replace surface water pushed offshore, water rises from below. The upwelling water is cooler than the nearby surface water that has remained at the surface and has been warmed by the Sun.

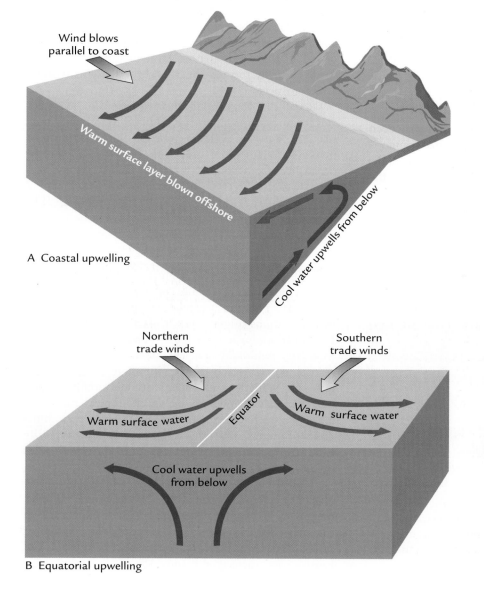

FIGURE 2-27 **Upwelling** Cool subsurface water rises along coastal margins (A), where winds drive warm water offshore, and near the equator (B), where winds drive surface waters away from the equator. (Modified from D. Merritts et al., *Environmental Geology*, © 1997 by W. H. Freeman and Company.)

• A second kind of upwelling occurs along the equator, especially in the eastern end of ocean basins (Figure 2-27B). Trade winds push surface waters away from the equator. Warm surface water is driven northward in the northern hemisphere and southward in the southern hemisphere by the opposing Coriolis deflections north and south of the equator. The movement of warm water away from the equator causes upwelling of cooler water from below.

Ice on Earth

Despite the redistribution of heat by air and water, temperatures cold enough to keep water frozen solid occur at high altitudes and latitudes. Ice is one of the most important components of the climate system, because its properties are so different from those of air, water, and land.

2-9 Sea Ice

Although freshwater freezes at 0°C, typical seawater resists freezing until it is cooled to −1.9°C. As sea ice forms, it rejects almost all the salt in the seawater. Because sea ice is less dense than seawater, it floats on top of the salty ocean.

When sea ice forms, it seals off the underlying ocean from interaction with the atmosphere. This change is vital to regional climates. Without an ice cover, high-latitude oceans transfer large amounts of heat to the atmosphere, especially in winter, when air temperatures are low (Figure 2-28A). This heat transfer keeps temperatures in the lower atmosphere close to those of the ocean surface (near 0°C).

But if an ice cover is present, this heat release stops, and the reflective ice surface absorbs little incoming solar radiation. Because of these changes, winter air

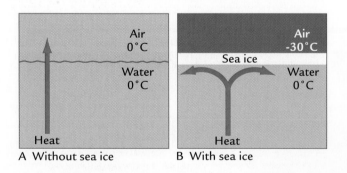

FIGURE 2-28 Effect of sea ice on climate Whereas heat can escape from an unfrozen ocean surface (A), a cover of sea ice (B) stops the release of heat from the ocean to the atmosphere in winter and causes air temperatures to chill by as much as 30°C.

temperatures can cool by 30°C or more in regions that develop a sea-ice cover (Figure 2-28B). In effect, an ice-covered ocean behaves like a snow-covered continent. This change forms a prominent part of the albedo–temperature feedback process examined in Box 2-2.

Many ocean surfaces are only partially ice-covered. Gaps ("leads") produced in the ice by changing winds allow some heat exchange with the atmosphere and moderate the climate effects of a full sea-ice cover. Also, in summer, meltwater pools may form on the ice surface, and this water, along with a gradual darkening of the melting ice, may absorb more solar radiation.

The formation and melting of sea ice are driven mainly by seasonal changes in solar heating. In the Southern Ocean, most of the sea ice melts and forms again every year, over an area comparable in size to the entire Antarctic continent it surrounds. This annual ice cover averages 1 meter in thickness, except where strong winds cause the ice to buckle and pile up in ridges. In contrast, the landmasses surrounding the Arctic Ocean constrain the movement of sea ice and allow it to persist for 4 or 5 years. Older sea ice in the center of the Arctic may reach 4 meters in thickness, while annually formed ice around the margins is about 1 meter thick.

Recall that large inputs and extractions of heat calories from the atmosphere are required to form and melt sea ice, and the cycle of freezing and melting also involves exchanges of heat with the slow-responding ocean because of its high heat capacity. For these reasons, seasonal extremes in sea-ice cover lag well behind the seasonal extremes of heating by solar radiation. The maximum extent of sea ice is usually reached in the spring, the minimum extent in the autumn.

2-10 Glacial Ice

Glacier ice occurs mainly on land, in two forms. **Mountain glaciers** are found in mountain valleys at high elevations (Figure 2-29, top). Because glaciers can exist only where mean annual temperatures are below freezing, mountain glaciers near the equator are restricted to elevations above 5 kilometers (Figure 2-29, bottom). At higher and colder latitudes, mountain glaciers may reach down to sea level. Typical mountain glaciers are a few kilometers in length and tens to hundreds of meters in width and thickness. They typically flow down mountain valleys, constrained on both sides by rock walls.

Continental ice sheets are a much larger form of glacier ice, typically hundreds to thousands of kilometers in horizontal extent and 1 to 4 kilometers in thickness. The two existing ice sheets, which cover most of Antarctica and Greenland, represent roughly 3% of Earth's total surface area and 11% of its land surface.

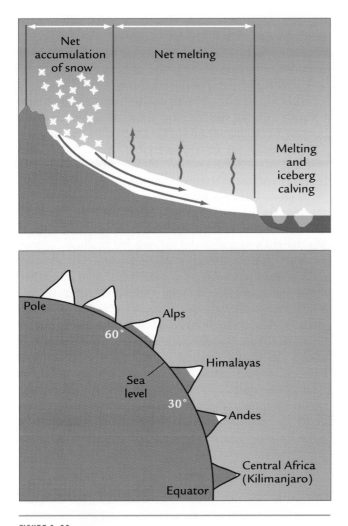

These great masses of ice depress the underlying bedrock below the elevation it would have had if no ice were present (Figure 2-30, bottom). About 30% of the total ice thickness sits below the original (undepressed) bedrock level; the other 70% protrudes into the atmosphere as a broad smooth ice plateau.

Snow that falls on the higher parts of mountain glaciers and ice sheets gradually recrystallizes into ice. The ice then moves toward lower elevations under the force of gravity. Ice in mountain glaciers is affected by gravity because of the steep slopes of mountain valleys. Continental ice sheets are high plateaus, and gravity moves ice from higher to lower elevations.

Ice deforms and moves in the upper 50 meters of a glacier in a brittle way: fracturing and forming crevasses, emitting loud cracking noises in the process. Below about 50 meters, ice deforms more gradually, by slow plastic flow. Parcels of ice may take hundreds to thousands of years to travel through mountain glaciers. For the central domes of ice sheets, where the flow is directed deep into the center of the ice mass, the trip may take tens of thousands of years.

As ice flows, its layers are stretched and thinned. If the ice is very cold (below −30°C), it behaves in a stiff manner, and its sloping edges can be relatively steep. Impurities such as dust also tend to stiffen the ice. If the ice is somewhat warmer (close to the freezing point), it

FIGURE 2-29 **Mountain glaciers** Mountain glaciers accumulate snow at colder, higher elevations. (Top) The snow turns to ice, flows to lower elevations where temperatures are warmer, and melts. (Bottom) Mountain glaciers can exist near sea level at high latitudes, but survive only at elevations above several kilometers in the warm tropics. (Modified from F. Press and R. Siever, *Understanding Earth*, 2nd ed., © 1998 by W. H. Freeman and Company.)

These ice sheets contain some 32 million cubic kilometers of ice, equivalent to about 70 meters of sea level change.

Continental ice sheets indeed have dimensions comparable to those of sheets—usually more than 1000 times as wide as they are thick—but their surfaces do have structure (Figure 2-30, top). The highest regions on the ice sheets are rounded **ice domes**, with the elevations sloping gently away in all directions. Domes may be connected by high broad ridges with gentle sags called **ice saddles**. On the sides of ice sheets, ice flows in fast-moving **ice streams** from which **ice lobes** protrude beyond the general ice margins.

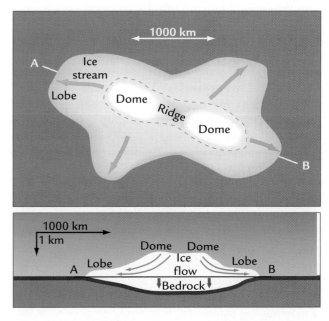

FIGURE 2-30 **Continental ice sheets** The central portions of large continent-sized ice sheets have high central domes connected by ridges. Ice streams on the flanks carry ice to lobes protruding from the ice margins (top). In cross section, snow accumulates on the high part of an ice sheet, turns to ice, and flows to the lower margins (bottom).

is more plastic—softer and easier to deform—and it will tend to relax into gentler slopes.

Ice also moves by mechanisms favorable to sliding on its basal layers. Large amounts of water can accumulate at the base of mountain glaciers, causing them to **surge** down valleys at rates far in excess of their normal movement. Lobes of continental ice sheets also move by sliding along their bases. In regions where ice streams occur, the ice may move several meters per day, or 100 to 1000 times faster than the rest of the ice mass.

Ice streams occur for two reasons. First, the pressure from the weight of the overlying ice may cause some of the ice at the bottom to melt and create a thin layer of water on which the ice can slide. Second, this water may percolate into and saturate soft unconsolidated sediments lying beneath the outer margins of the ice sheet, causing them to lose their mechanical strength or cohesiveness. These water-lubricated sediments provide a slippery *deformable bed* on which the overlying ice can easily slide.

Under certain conditions, **ice shelves** may form over shallow ocean embayments, and several shelves exist today on the margins of Antarctica. In these regions, gravity pulls ice out of the interior of the continent to the embayments, where it spreads out in shelves tens to hundreds of meters thick. Bedrock surrounding these embayments and at the shallow depths below it provides friction that keeps the ice from sliding away into the ocean. Immense **tabular icebergs** occasionally break off from these shelves and float away. An iceberg the size of the state of Connecticut broke off from Antarctica a few years ago.

The bottom of the western part of the Antarctic ice sheet lies below sea level, and this portion is called the West Antarctic **marine ice sheet**. Because marine ice sheets have bases lying below sea level, they are highly vulnerable to sea level changes and respond to changes in climate much more quickly than ice sheets that sit higher on the land.

Mountain glaciers and continental ice sheets ultimately exist for the same reason: the overall rate of snow falling across the entire ice mass equals or exceeds the overall rate at which ice is lost by melting and other means. Climate scientists analyze the conditions over present-day glaciers and ice sheets in terms of their **mass balance**, the average rate at which ice either grows or shrinks every year. The concept of mass balance can also be applied to different portions of glaciers: mass balances are positive at upper elevations, where **accumulation** of snow and ice dominates, but negative at lower elevations, where rapid **ablation** (loss of ice) occurs.

Ice accumulation occurs in regions where temperatures are cold enough both to cause precipitation to freeze and also to allow new-fallen snow to persist through the warm summer season. For mountain glaci-

ers, subfreezing temperatures occur on the highest parts of mountains, where the air is coldest. For continent-sized ice sheets, which often exist at sea level, the cold required to sustain ice is found at high polar latitudes and on the high parts of the ice sheets.

Ablation of glacial ice by melting occurs when temperatures exceed the freezing point. Melting can occur because of absorption of solar radiation or by uptake of sensible or latent heat delivered by warm air masses (and by rain) moving across the ice. Ablation can also occur by **calving**, the shedding of icebergs to the ocean or to lakes. Calving differs from the other processes of ablation in that icebergs leave the main ice mass and move elsewhere to melt, often in an environment much warmer than that near the ice sheet.

The boundary between the high-elevation region of positive ice mass balance and the lower area of net loss of ice mass occurs at a mean annual temperature near −10°C for ice sheets but closer to 0°C for mountain glaciers. The mass balance at high elevations on the Greenland and Antarctic ice sheets is positive because of the absence of melting to offset the slow accumulation of snow (Figure 2-31). At elevations above 1 to 2 kilometers, air is so cold that it contains little water vapor. Although the precipitation that falls on these higher parts of the ice sheets is all snow, accumulation rates are low. This is especially the case for the frigid

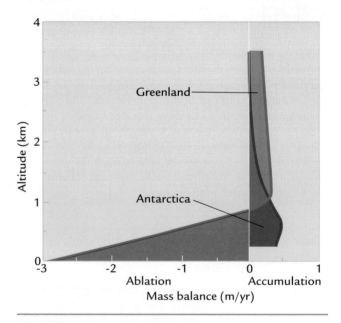

FIGURE 2-31 Ice mass balance Snow accumulates on the upper parts of ice sheets where melting does not occur. At lower elevations, accumulation is overwhelmed by net loss of ice by ablation due to melting (in Greenland) and to calving of icebergs (in Antarctica). The units shown convert snow to equivalent thicknesses of ice (in meters). (Adapted from J. Oerlemans, "The Role of Ice Sheets in the Pleistocene Climate," *Norsk Geologisk Tidsskrift* 71 [1991]: 155–161.)

Antarctic continent, centered on the South Pole and surrounded by an ice-covered ocean. The mass balance is more positive on the sides of these ice sheets, where air masses carry more moisture and cause more snowfall, yet ablation is not strong.

The mass balance on ice sheets is negative at lower elevations, usually because mean annual temperatures above 0°C accelerate the rate of melting. In Greenland, some of the low-elevation precipitation also falls as rain rather than snow, which further promotes ablation. In Antarctica, freezing conditions persist at sea level even in summer, and no melting occurs. The Antarctic ice sheet loses mass mainly by calving icebergs into the ocean.

Earth's Biosphere

To this point, we have examined only the physical side of the climate system, expressed mainly by variations in temperature, precipitation, winds, and pressure, but these physical parts of the climate system also interact with its organic parts (Earth's **biosphere**). Many of these interactions result from the movement of carbon (C) through the climate system and in turn affect the distribution of heat on Earth.

Carbon moves among and resides in several major reservoirs. The amount of carbon in each reservoir is typically quantified in gigatons (or 10^{15} grams) of carbon. Relatively small amounts of carbon reside in the atmosphere, the surface ocean, and vegetation; a slightly larger reservoir resides in soils, a much larger reservoir in the deep ocean, and a huge reservoir in rocks and sediments (Figure 2-32A).

Carbon takes different chemical forms in these different reservoirs. In the atmosphere, it is a gas (CO_2). Carbon in land vegetation is organic, as is most carbon in soils, while that in the ocean is mostly inorganic, occurring as dissolved ions (atoms carrying positive or

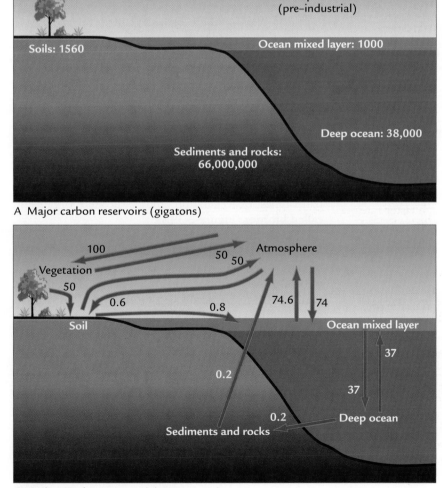

A Major carbon reservoirs (gigatons)

B Carbon exchange rates (gigatons/year)

FIGURE 2-32 **The carbon cycle** The major carbon reservoirs on Earth vary widely in size (A) and exchange carbon at differing rates (B). Larger reservoirs (rocks, the deep ocean) exchange carbon much more slowly than smaller reservoirs (air, vegetation, the surface ocean). (Adapted from J. Horel and J. Geisler, *Global Environmental Change* [New York: John Wiley, 1997], and from National Research Council Board on Atmospheric Sciences and Climate, *Changing Climate*, Report of the Carbon Dioxide Assessment Committee [Washington, D.C.: National Academy Press, 1993].)

negative charges). Despite these differences in form, carbon is exchanged freely among all the reservoirs, changing back and forth between organic and inorganic forms as it moves.

Rates of carbon exchange among reservoirs vary widely (Figure 2-32B). In general, the sizes of the reservoirs are inversely related to their rates of carbon exchange. The small surface reservoirs (the atmosphere, surface ocean, and vegetation) exchange all their carbon with one another within just a few years. The much larger deep-ocean reservoir is partly isolated from the surface reservoirs by the thermocline and exchanges carbon with the surface ocean and the atmosphere over hundreds of years. The carbon buried in sediments and rocks interacts very slowly, moving in and out of the surface reservoirs only over hundreds of thousands of years or longer.

Plants grow on land if the conditions necessary for **photosynthesis** (the production of plant matter) are met: sunlight is needed to provide energy, and nutrients (mainly phosphorus and nitrogen) provide food for plant growth (Figure 2-33). With these conditions satisfied, plants draw CO_2 from the air and water from the soil to create new organic matter, while oxygen is liberated to the atmosphere:

$$6CO_2 + 6H_2O \xrightarrow[\text{Oxidation}]{\text{Photosynthesis}} \underset{\text{Plants (organic C)}}{C_6H_{12}O_6 + 6O_2}$$

During the time that plants grow by photosynthesis, and also during the time they are mature but no longer growing, plants take the water they need from the soil and give it back to the atmosphere (see Figure 2-33). This process, known as **transpiration** (and also as "respiration"), is a highly efficient way to return water vapor to the atmosphere, and it can occur at much faster rates than ordinary evaporation from vegetation-free ground.

After plants die (either by seasonal die-back or by reaching the end of their natural lifetime), oxygen is consumed in destroying their organic matter during **oxidation**. Oxidation can occur either through rapid burning (in fires) or by slow decomposition in the presence of oxygen, with the same ultimate result. Through either process, oxidation converts organic carbon back to inorganic form, as shown by the equation above.

2-11 Response of the Biosphere to the Physical Climate System

Trees, shrubs, and other plants accomplish most photosynthesis on land. CO_2 and sunlight are usually available over the continents, but the distributions of temperatures and rainfall critical to photosynthesis and

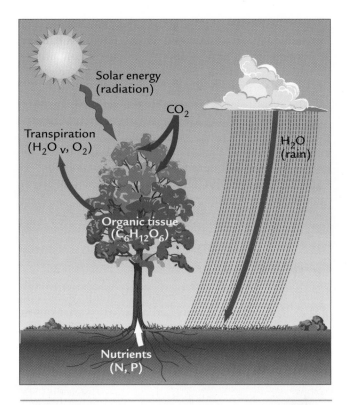

FIGURE 2-33 Photosynthesis on land Plant life on land uses sunlight, CO_2 and H_2O from the atmosphere, and soil nutrients in the process of photosynthesis. Plants also return water vapor (H_2O_v) and oxygen (O_2) to the atmosphere by transpiration. (Adapted from F. T. Mackenzie, *Our Changing Planet* [Englewood Cliffs, N.J.: Prentice-Hall, 1998].)

plant life vary widely. To a large extent, rainfall (Figure 2-34A) determines the total amount of organic (live) matter present, called the **biomass**, and the predominant types of vegetation and associated organisms, called **biomes** (Figure 2-34B).

The tendency for rainfall to be abundant along the ITCZ produces tropical rain forest biomes with dense biomasses. Toward the dry subtropics, rain forests grade into **savanna** (scattered trees in a grassland setting) and then to the sparse scrub vegetation typical of deserts. Total biomass decreases toward the subtropics along with precipitation.

Large-biomass **hardwood forest** (maple, oak, hickory, and other leaf-bearing trees) occurs in the wetter portions of the middle latitudes (eastern North America, Europe, Asia) and on the upwind side of mountain ranges facing the ocean, while low-biomass biomes such as grasslands and desert scrub are found in drier interior regions in the rain shadow of mountain ranges (see Figure 2-20). **Conifer forest** (spruce and other trees with needles) dominates toward the higher

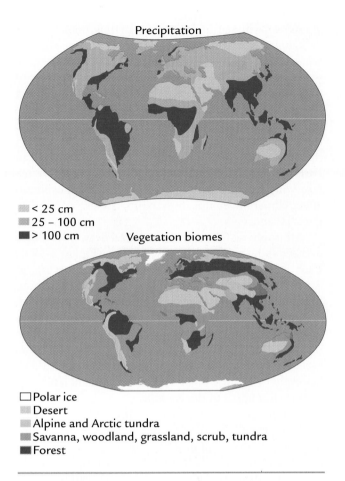

Precipitation

< 25 cm
25 – 100 cm
> 100 cm

Vegetation biomes

☐ Polar ice
Desert
Alpine and Arctic tundra
Savanna, woodland, grassland, scrub, tundra
Forest

FIGURE 2-34 **Precipitation and vegetation** Global precipitation (top) is highest in the tropics and along mountain slopes that receive moisture-bearing winds from the ocean, and lowest in subtropical deserts and over polar ice. Vegetation biomes (bottom) largely reflect the patterns of precipitation, with high-biomass forests in regions of high precipitation and low evaporation. (Top: adapted from L. J. Battan, *Fundamentals of Meteorology* [Englewood Cliffs, N.J.: Prentice-Hall, 1979]; bottom: adapted from E. Bryant, *Climate Process and Change* [Cambridge: Cambridge University Press, 1998].)

latitudes of the northern hemisphere, but the fringes of the Arctic Ocean are surrounded by a wide band of scrubby **tundra** vegetation with low biomass above ground and large amounts of carbon stored below ground. Ice sheets are free of life forms except cold-tolerant bacteria.

Life in the oceans depends on a different combination of the same factors as on land. Obviously, water is abundantly available in the oceans, and CO_2 is plentiful in surface waters that exchange CO_2 with the atmosphere. In addition, light from the Sun is widely available in the upper layers of the ocean into which it penetrates (Figure 2-35).

With all these conditions favorable to photosynthesis, why isn't the surface ocean an enormous photosynthesis machine? The answer is simple: a lack of the nutrients nitrogen (N) and phosphorus (P). Nutrient food sources are scarce in most parts of the surface ocean.

A floating form of microscopic plant life called **phytoplankton** lives in the surface layers of the ocean and uses sunlight for photosynthesis. These minute organisms extract nutrients and incorporate them in the soft organic tissues of their bodies. Phytoplankton have short life spans (days to weeks), and when they die, they sink to deeper waters, leaving the surface layer depleted of nutrients. Thus the rates of photosynthesis in these sunlit surface waters are limited.

Initially near the surface, and mainly later at depths well below the surface, the slow decay and oxidation of the soft tissues of these sinking organisms releases nitrogen and phosphorus back into ocean water. Because most of these nutrients are released into and stay in the deeper ocean, their scarcity in surface waters limits the amount of life that can exist across most ocean areas.

In the few parts of the surface ocean where upwelling occurs, nutrients are more plentiful, and they result in greater **productivity**, or rates of photosynthesis by

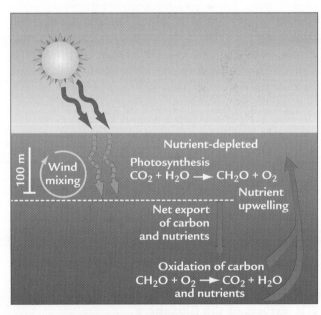

FIGURE 2-35 **Photosynthesis in the ocean** Sunlight penetrating the surface ocean causes photosynthesis by microscopic plants. As they die, their nutrient-bearing organic tissue descends to the seafloor. Oxidation of this tissue at depth returns nutrients and inorganic carbon to the surface ocean in regions of upwelling.

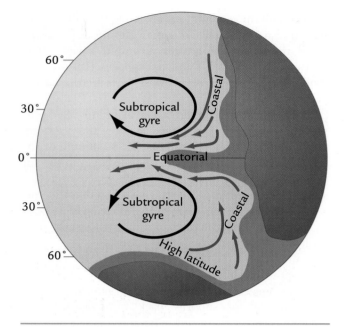

FIGURE 2-36 Ocean productivity The greatest amount of photosynthesis in the surface ocean occurs along shallow continental margins and in coastal, equatorial, and high-latitude regions where nutrients upwell from below.

phytoplankton (Figure 2-36). Wind-driven upwelling along some coastal margins returns nutrients to the surface from below and supplements nutrients delivered to continental shelves by rivers and resuspended during storms. As a result, surface waters near continental margins tend to be relatively productive. Upwelling in the eastern equatorial Pacific and Atlantic also increases rates of photosynthesis and productivity in those regions.

The Southern Ocean around Antarctica is another productive region (see Figure 2-36). Deep water from the Atlantic flows toward the surface in this area, bringing nutrients up from below. Strong winds mix these nutrients into the surface layers, producing the rich biomass of the Southern Ocean. Because sunlight is not plentiful, and because the long season of sea-ice cover limits the amount of time in which photosynthesis can occur, nutrients in the surface waters of the Southern Ocean are never depleted, even when productivity is highest.

2-12 Effects of the Biosphere on the Climate System

Life affects climate in many ways. One way is by providing positive feedback to physical processes that affect climate (Box 2-6). A second way is through changes in the amount of greenhouse gases in the atmosphere, especially carbon dioxide (CO_2) and methane (CH_4).

As we will see in later parts of this book, *all* the exchanges of carbon shown in Figure 2-32 have affected atmospheric CO_2 and climate, but at different time scales. The slow movement of carbon into and out of rock reservoirs affects atmospheric CO_2 and climate on the tectonic (million-year or longer) time scale, explored in Part II. Somewhat faster exchanges between the surface and deep ocean reservoirs affect climate on the orbital (ten-thousand-year) time scale, examined in Parts III and IV. Rapid exchanges among the surface reservoirs (vegetation, the ocean, and the atmosphere) affect climate over the shorter time scales reviewed in Parts IV and V.

Atmospheric CO_2 trends measured over the last four decades show two superimposed effects (Figure 2-37A). Each year a small drop in CO_2 values occurs in April–May and a comparable rise the following September–October. This oscillation reflects cycling of vegetation in the northern hemisphere: CO_2 is taken from the air by plant photosynthesis every spring and released by oxidation every autumn. The signal follows the tempo of the northern rather than the southern hemisphere because most of Earth's land (and land vegetation) lies north of the equator.

The second trend evident in the CO_2 curve is its gradual overall increase (see Figure 2-37A). This increase results mainly from burning of fossil-fuel carbon and secondarily from **deforestation** (clearing of vegetation from the land), which releases carbon to the atmosphere through burning and oxidation. The rapid increase in consumption of fossil fuels by humans over the last two centuries has tapped into huge reservoirs of coal, oil, and gas in rocks that naturally release their carbon at slow rates and has greatly accelerated these rates (Part V).

Methane (CH_4) is a second important atmospheric greenhouse gas, although far less plentiful than CO_2. It has many sources, including swampy lowland bogs, rice paddies, the stomachs and bowels of cows digesting vegetation, and termites. Common to all these CH_4 sources is the decay of organic matter in an oxygen-free environment. At the end of the twentieth century, methane concentrations in the atmosphere had risen by well over a factor of 2 above their natural (preindustrial) level to above 1700 parts per billion (Figure 2-37B). This recent increase in methane is the result of human activities (Part V).

As we noted earlier, both CO_2 and CH_4 trap part of Earth's back radiation, keep the heat in the atmosphere, and make Earth warmer than it would otherwise be. This warming in turn activates the positive feedback effect of water vapor (H_2O_v), the most important greenhouse gas. The combined effects of these three greenhouse gases in the recent past and near future are the focus of the last part of this book.

BOX 2-6 CLIMATE INTERACTIONS AND FEEDBACKS

Vegetation–Climate Feedbacks

The type of vegetation covering a land region can affect its average albedo. The two major types of vegetation in the Arctic, spruce forest and circumarctic tundra, interact in different ways with freshly fallen snow and produce surfaces with very different albedos. Snow that falls on tundra covers what little scrub vegetation exists and creates a high-albedo surface that reflects most incoming solar radiation. Snow that falls on spruce forests is blown from the trees and falls to the ground, allowing the dark-green surface of the exposed treetops to absorb most incoming solar radiation.

When climate cools, these contrasts in albedo produce an important positive feedback. Tundra gradually advances southward and replaces spruce forest, expanding Earth's high-albedo surface area. With more solar radiation reflected from this surface, the reduction in absorbed heat leads to further cooling. This process is called **vegetation–albedo feedback**. This same positive feedback works during times of climate warming: as forest replaces tundra, more solar heat is absorbed, and the climate warms even more.

A second type of vegetation feedback depends on the way vegetation recycles water. Land vegetation draws water needed for photosynthesis from the ground. Some of the water is handed off to the atmosphere as water vapor during times when the plants are actively extracting CO_2 from the air. Trees transpire much larger volumes of water vapor than grass or desert scrub, and this contrast is responsible for the second kind of positive feedback. When climate becomes wetter, forests gradually replace grasslands in some regions. The trees transpire more water vapor back to the atmosphere, thus increasing the amount available for rainfall. This positive feedback (called **vegetation–precipitation feedback**) works in the reverse sense when climate dries.

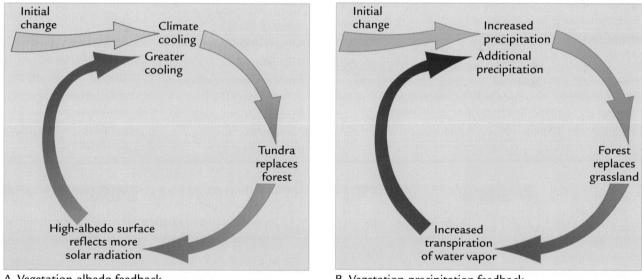

A Vegetation-albedo feedback

B Vegetation-precipitation feedback

Vegetation–climate feedbacks (A) When high-latitude climate cools, replacement of spruce forest by tundra raises the reflectivity (albedo) of the land in winter and causes additional cooling as a positive feedback. (B) When climate becomes wetter, replacement of grasslands by trees increases the release of water vapor back to the atmosphere and causes increases in local rainfall as a positive feedback.

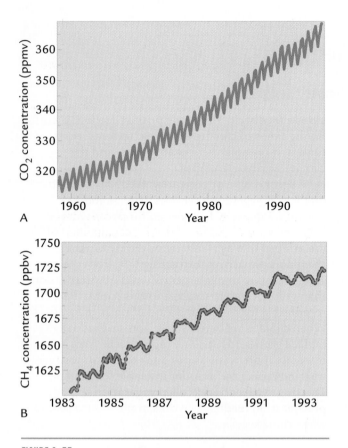

FIGURE 2-37 Recent increases in carbon dioxide and methane Instrument measurements record rapid rises of the greenhouse gases CO_2 (A) and CH_4 (B) in recent years. The gases are measured in parts per million by volume (ppmv) and parts per billion by volume (ppbv). (A: adapted from H. H. Friedli et al., "Ice Core Record of the $^{13}C/^{12}C$ Ratio of Atmospheric CO_2 in the Past Two Centuries," *Nature* 324 [1986]: 237–38; B: after M. A. K. Khalil and R. A. Rasmussen, "Atmospheric Methane: Trends over the Last 10,000 Years," *Atmospheric Environment* 21 [1987]: 2445–52.)

latent heat of vaporization (p. 29)
dew point (p. 30)
saturation vapor density (p. 30)
water vapor feedback (p. 31)
adiabatic (p. 32)
lapse rate (p. 32)
Hadley cell (p. 33)
intertropical convergence zone (ITCZ) (p. 34)
monsoon (p. 35)
Coriolis effect (p. 36)
orographic precipitation (p. 38)
jet streams (p. 39)
gyres (p. 39)
Gulf Stream (p. 40)
North Atlantic Drift (p. 40)
thermocline (p. 41)
thermohaline flow (p. 41)
salinity (p. 41)
salt rejection (p. 41)
North Atlantic deep water (p. 42)
Antarctic bottom water (p. 42)
Antarctic intermediate water (p. 42)
Mediterranean overflow water (p. 42)
upwelling (p. 43)

mountain glaciers (p. 44)
continental ice sheets (p. 44)
ice domes (p. 45)
ice saddles (p. 45)
ice streams (p. 45)
ice lobes (p. 45)
surge (p. 46)
ice shelves (p. 46)
tabular icebergs (p. 46)
marine ice sheet (p. 46)
mass balance (p. 46)
accumulation (p. 46)
ablation (p. 46)
calving (p. 46)
biosphere (p. 47)
photosynthesis (p. 48)
transpiration (p. 48)
oxidation (p. 48)
biomass (p. 48)
biomes (p. 48)
savanna (p. 48)
hardwood forest (p. 48)
conifer forest (p. 48)
tundra (p. 49)
phytoplankton (p. 49)
productivity (p. 49)
deforestation (p. 50)
vegetation–albedo feedback (p. 51)
vegetation–precipitation feedback (p. 51)

Key Terms

electromagnetic radiation (p. 19)
electromagnetic spectrum (p. 19)
shortwave radiation (p. 19)
back radiation (p. 20)
longwave radiation (p. 20)
greenhouse effect (p. 21)
troposphere (p. 22)
stratosphere (p. 22)
albedo (p. 22)

albedo–temperature feedback (p. 25)
heat capacity (p. 26)
calories (p. 26)
specific heat (p. 26)
hydrologic cycle (p. 27)
thermal inertia (p. 28)
sensible heat (p. 28)
convection (p. 29)
latent heat (p. 29)
latent heat of melting (p. 29)

Review Questions

1. How does solar radiation arriving on Earth differ from the back radiation emitted by Earth?

2. What kind of radiation is trapped by greenhouse gases? What is the effect on Earth's climate?

3. What different and opposing roles do clouds play in the climate system?

4. How does reflection of solar radiation from Earth's surface add to the effects of uneven solar heating in creating a pole-to-equator heat imbalance?

5. What processes cause air to rise from Earth's surface?

6. What causes the monsoon circulation to reverse from summer to winter?

7. Describe the main pathway by which heat in the atmosphere is transported toward the poles.

8. Why does rain fall on the sides of mountains in the path of winds from nearby oceans?

9. How do low-level winds create spinning gyres in the subtropical oceans?

10. Why does deep water form today at higher latitudes?

11. What effect does the formation of sea ice have on the overlying atmosphere?

12. What parts of ice sheets gain and lose mass? Why?

13. How closely does land vegetation (and total biomass) follow global precipitation trends?

14. What regions of the ocean are most productive? Why?

15. Describe two positive feedback processes discussed in this chapter.

ADDITIONAL RESOURCES

Barry, R. G., and R. J. Chorley. 1998. *Atmosphere, Weather, and Climate*. New York: Routledge.

Thurman, H. V. 1991. *Introductory Oceanography*. 6th ed. New York: Macmillan.

CHAPTER 3

Climate Archives, Data, and Models

Climate scientists use a wide range of techniques to extract, reconstruct, and interpret the history of Earth's climate. Much of this history is recorded in four climate archives: sediments, ice, corals, and trees. In this chapter we first examine the major climate archives. Then we explore how their climate records are dated, how much of Earth's history each archive spans, and the resolution of climate history yielded by each. We also examine some of the methods scientists use to study climate, but leave more specific information for later chapters.

The process of analysis and interpretation of climate data is aided by the use of climate models to test hypotheses of climate change in a quantitative way (Chapter 1). In this chapter we investigate physical models that simulate the circulation of Earth's atmosphere and ocean. Then we examine geochemical models used for a different purpose—tracking mass movements of tracers through the climate system. Both types of models yield quantitative insights into the relative importance of the major processes of climate change.

Climate Archives

Like written chronicles of human history, climate archives hold stories of climate change for those who can read them. For all of Earth's history before the invention of instruments to measure climate in the seventeenth century, the major climate archives are sediments, ice, corals, and trees. The relative importance of these archives differs among the time intervals under investigation. In this section, we examine the major archives of climate, how climate scientists date the records in these archives, and the time resolution that can be achieved.

3-1 Types of Archives

Although relatively recent climate changes can be studied in archives such as ice cores, tree rings, and corals, sediments are the major climate archive on Earth for over 99% of geologic time, primarily as continuous sequences deposited by water.

Sediments Rainfall and the runoff it produces erode rocks exposed on the continents and transport the eroded sediments in streams and rivers in both physical (granular) and chemical (dissolved) forms. The sediments are eventually deposited in receptive environments, mainly quieter waters where layer upon layer of sediment can be laid down in undisturbed succession. Most sediment is carried to the ocean, either right after it is first eroded or later, after temporary deposition on land followed by one or more cycles of reerosion and redeposition. Sediment delivered to the seafloor may be quickly dragged down beneath the continental margin by plate tectonic processes and destroyed (see Chapter 5), or it may persist for tens of millions of years on the seafloor. The relentless action of these two processes, erosion and tectonic activity, decreases the likelihood that older sedimentary records will be preserved as time passes.

For intervals before the last 170 million years, all surviving sedimentary records come from continents. Under favorable conditions, sediments may be preserved there for a long time in the deposits shown in Figure 3-1: thick sequences in deep continental basins that contain large lakes; thinner sequences in shallow interior seas during times when the ocean floods low-lying land; and thick lens-shaped piles of sediment along **continental shelves** (the barely submerged coasts of continents) and on the steeper **continental slopes** leading down to the deep ocean.

Sediments are useful climate archives to the extent that their deposition is uninterrupted. Major disturbances during and just after deposition come from wave action reaching several meters below sea level and from occasional large storms that produce turbulent disturbances that may reach tens of meters deep in the water column and erode previously deposited layers. These problems affect shallow marine regions. In addition, sediments deposited on the steep continental slope margins are vulnerable to dislodgment down into the deeper ocean by disturbances such as earthquakes.

In the longer term, erosion tied to sea level change is a major factor in disturbing sediment sequences. Through time, the sea moves up and down along the continental margins over a total vertical range of several hundred meters. Sediments tend to be deposited high on the margins when sea level is high, but waves and storms may erode older deposits and carry them to the deeper ocean when sea level falls. If the sea withdraws completely, the sediments may be exposed to the kinds of erosion processes that occur on land.

All these factors ultimately determine the quality of climate records preserved in sediment archives (Figure 3-1). Sediments deposited on continental shelves when sea level is high form lens-shaped units separated

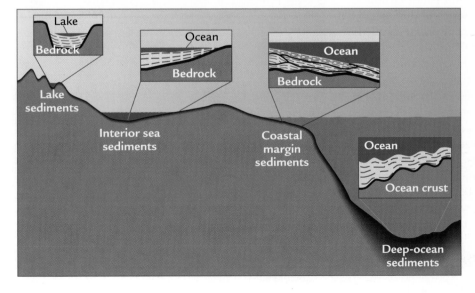

FIGURE 3-1 **Sediment archives** Layered sediments are major climate archives on all time scales. The insets show typical sediment layering in sediment archives from land and sea.

FIGURE 3-2 **Lake cores** Hundreds of cores have been taken from small lakes and analyzed for records of changes in pollen (vegetation) and lake level over the last several thousand years. (National Paleoclimate Data Center, NGDC, Boulder, Colo.)

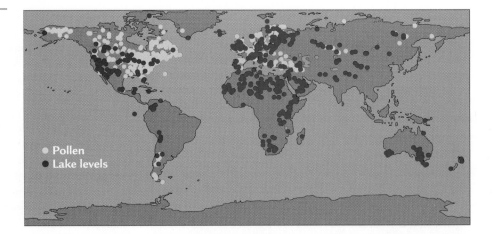

● Pollen
● Lake levels

by distinct surfaces where erosion has occurred. Deposition is often continuous within these sequences, but the rates vary widely and are highest in regions where rivers deliver sediments. Sediments deposited in interior seas during times when the ocean floods low-lying regions of the continents form thin sequences covering wide areas. Deposition rates vary widely with water depth but tend to be slower than on continental margins.

Sediments deposited in long-lived lakes in continental basins conform to the structural framework of the bedrock. Deposition tends to be most continuous in deeper parts of lakes. Sediments deposited in lakes that fill depressions left behind by melting glaciers are especially important climate archives for the last 20,000 years in several regions (Figure 3-2). Deposition rates in lakes vary with changing sediment sources and basin geometry.

Ice and wind are also powerful agents of sediment erosion and transport in some regions. Unfortunately, the deposits they leave in the geologic record are generally discontinuous and almost always difficult to date.

Ice sheets that reach maximum size and then begin to retreat leave behind long curving ridges called **moraines**. Moraines contain a jumbled mix of all the unsorted debris that ice can carry, ranging from boulders to clay. When an ice sheet advances again later, it usually destroys any moraines formed earlier and incorporates some of the older debris into newer deposits. Unraveling a climate history from this kind of record is like trying to decipher repeated episodes of writing and erasing on a blackboard. By contrast, coarse debris carried to the ocean and dropped by melting icebergs into the underlying ocean sediments enters a more continuous and easily dated archive.

Strong winds can also weather rocks and form fine sediment particles in regions with dry climates. Winds form sand dunes that slowly migrate across desert areas,

FIGURE 3-3 **Windblown loess** Strong winds have deposited thick layers of silt-sized grains in southeast China during the last 3 million years. The total thickness of these loess deposits can reach several hundred meters. In many regions people have created homes in the loess cliffs. (Courtesy of Steven Porter, University of Washington.)

but continuous reworking of the sand particles complicates efforts to use dunes as climate archives. Winds also pick up smaller silt-sized grains, lift them high in the air, and transport them far away from their original sources. In regions where the winds weaken, the silt may be deposited in thick sequences called **loess**. Loess deposits are excellent climate repositories of the last 3 million years, especially in China (Figure 3-3). Finer sediments carried far from the continents and deposited in ocean sediments are also useful indicators of climate.

For the portion of geologic time younger than 100 million years, climate scientists gain increasing access to an additional kind of climate archive: sediments preserved in ocean basins that have not yet been destroyed by tectonic recycling. Deep-sea sediments of younger age (the last several million years) cover almost two-thirds of Earth's surface (Figure 3-4A). Even older and more deeply buried sediments can be retrieved by the *JOIDES Resolution*, a ship capable of drilling into and recovering sediment sequences several kilometers thick (Figure 3-4B–D).

The deeper ocean is generally a quiet place with relatively continuous deposition, and it yields climate records of higher quality than most records from land, where water, ice, and wind are active agents of erosion.

Some deep-sea sediments are subject to disturbances, including dislodgment from steep slopes, physical erosion and reworking by deep currents, and gradual chemical dissolution by water in the deepest part of ocean basins. Despite these problems, many ocean basins have been sites of continuous sediment deposition over tens of millions of years.

Deposition of sediments is usually much slower in the ocean than on land. Deposition rates are higher in regions of the deep ocean that receive influxes of sediments eroded from nearby continents, in sediments beneath organically productive surface waters, and in regions standing well above the chemically corrosive bottom waters.

Glacial Ice At the very cold temperatures found at high latitudes and high altitudes, annual deposition of snow can pile up continuous sequences of ice that range in thickness from small mountain glaciers tens to hundreds of meters thick to large continent-sized ice sheets several kilometers thick (Figure 3-5). Ice-core archives contain many kinds of climatic information, although only in specific geographic regions (Figure 3-6).

The deep central portion of the Antarctic ice sheet has layers that extend back over 400,000 years, and the central Greenland ice sheet has ice dating back 100,000

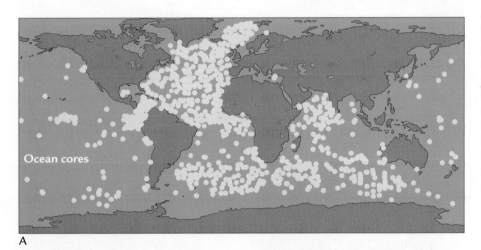

A

B

C

D

FIGURE 3-4 Ocean drilling
(A) Hundreds of ocean sediment cores are archives of past climatic changes. (B, C, D) The longest cores have been retrieved by drilling operations on the *JOIDES Resolution*, run by the international Ocean Drilling Program. (A: National Paleoclimate Data Center, NGDC, Boulder, Colo.; B, C, D: Ocean Drilling Program, Texas A&M University.)

FIGURE 3-5 **Ice archives** Ice is an important archive of many climate signals. Ice cores retrieve climate records extending back thousands of years in small mountain glaciers (A) to as much as hundreds of thousands of years in continent-sized ice sheets (B).

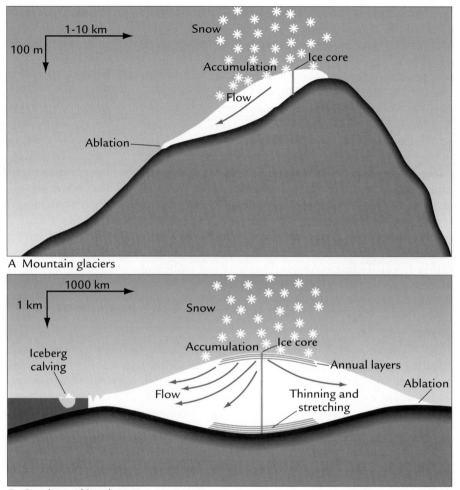

A Mountain glaciers

B Continental ice sheets

FIGURE 3-6 **Ice cores, corals, and tree rings** Ice cores, corals, and tree rings are archives of climate change in more recent Earth history. (National Paleoclimate Data Center, NGDC, Boulder, Colo.)

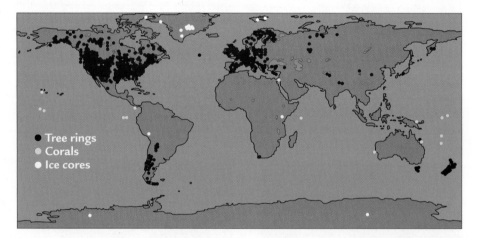

years or more. Many mountain glaciers have records of the last 10,000 years of climate change. Deposition rates range from a few centimeters a year in the coldest and driest areas to meters per year in less frigid, wetter regions.

Other Climate Archives In areas of sufficient rainfall, groundwater percolating through bedrock dissolves and redeposits limestone (calcite, or $CaCO_3$) layers in *caves*. These deposits, isolated from the active erosion occurring at or near Earth's surface, contain records of climate over time intervals ranging back several hundred thousand years.

Trees are climate archives for the interval of the last few tens, hundreds, or (in exceptional cases) thousands

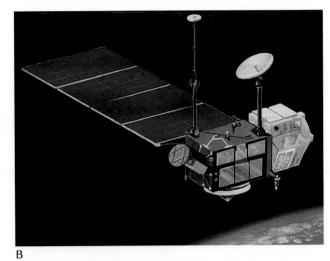

B

FIGURE 3-7 **Instrument measurements** Instruments that have been used to measure climate range from the primitive thermometers of the seventeenth century (A) to the multiple sensors flown aboard the TOPEX/Poseidon satellite (B). (A: The Granger Collection; B: NASA.)

A

of years. The outer softwood layers of many kinds of trees are deposited in millimeter-thick layers that turn into hardwood. These annual layers are best developed in mid-latitude and high-latitude regions that experience large seasonal climate changes (Figure 3-6).

In clear sunlit waters at tropical and subtropical latitudes, *corals* form annual bands of $CaCO_3$ that hold several kinds of geochemical information about climate (Figure 3-6). Individual corals may live for time spans of years to tens or hundreds of years.

Within the last few thousand years, humans began to keep **historical archives** of climate-related phenomena. Examples include the time of blooming of cherry trees in Japan, the success or failure of grape and grain harvests in Europe, and the number of days with extensive sea ice in such regions as Iceland and Hudson Bay in Canada. These records precede (and in some cases overlap) the **instrumental record** of the last 100 to 200 years. The first instruments invented to measure climate directly were thermometers in the eighteenth century, and recent human ingenuity has invented many instruments that can measure climate remotely from space (Figure 3-7).

3-2 Dating Climate Records

Climate records in older sedimentary archives are dated by a two-step process. First, scientists use the technique of **radiometric dating** to measure the decay of radioactive isotopes in rocks. (Isotopes are forms of a chemical element that have the same atomic number but differ in mass.) Dates are obtained on hard crystalline **igneous rocks** that once were molten and then cooled to solid form. In the second step, dates obtained from igneous rocks provide constraints on the ages of the sedimentary rocks that occur in layers between the igneous rocks and form the main archives of Earth's early climate history.

Radiometric Dating and Correlation Radiometric dating is based on the radioactive decay of a **parent isotope** to a **daughter isotope**. The parent is an unstable radioactive isotope of one element, and radioactive decay transforms it into the stable isotope of another element (the daughter). This decay occurs at a known rate, the *decay constant*, which is a measure of the likelihood of parent-to-daughter decay per amount of parent present per unit of time. This rate of decay in effect forms a clock with which we can measure age.

An event of some kind is required to start this clock ticking. The igneous rock most commonly used for dating is basalt, which cools very quickly from molten lava. The event that starts the clock ticking is the cooling of this molten material to the point where neither parent nor daughter can migrate in or out. At this point, the rock forms a **closed system**, and the only changes occurring are those caused by internal radioactive decay.

In the simplest example of a closed system, the decay of a parent to a daughter produces the changes shown in Figure 3-8: the parent decays away exponentially, while the daughter shows an exactly opposite (and compensating) exponential increase in abundance. The **half-life** is a convenient measure of the rate at which this process occurs: one half-life is the time needed for half the parent present to decay to the daughter. The first half-life reduces the parent to half its initial abundance, the second reduces it to half of that half, or one-quarter, and so on.

Because various radioactive parents have a wide range of half-lives, each is most useful over a different part of Earth's history (Table 3-1). Radioactive isotopes remain useful for at least the first five or six half-lives after the clock is set, after which too little parent is usu-

ally left to permit reliable dating. The long, slow decay series from uranium (U) to lead (Pb) is useful for rocks that are nearly as old as Earth itself. The decay from potassium (K) to argon (Ar) is used widely for dating much of Earth's history.

Several factors complicate radiometric dating. Unlike the simple case shown in Figure 3-8, the initial abundance of the daughter is rarely zero: usually some daughter isotope was present in the igneous rock when the decay clock was set. Other problems arise when the system is not fully closed to the migration of parent or daughter isotopes.

If both igneous and sedimentary rocks are present in a specific region (Figure 3-9), the igneous rocks constrain the ages of the sediment sequence by *cross-cutting relationships*. In the example shown, the age of each sedimentary layer can be obtained from the nearby igneous rocks by determining which is older or younger than the other. For example, a layer of igneous rock that spreads across the top of a layer of sediment or intrudes into it must postdate the time the sediment was deposited and thereby provides a minimum age for that layer of sediment.

In actual practice, it is rare to find enough igneous rocks in any one location to date sediments this way.

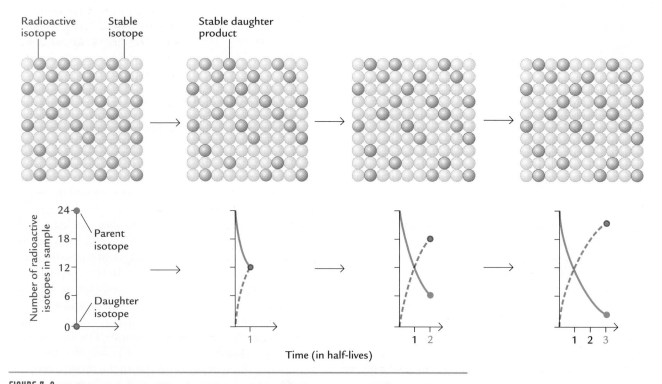

FIGURE 3-8 Radioactive decay Time is determined by measuring the gradual decay of a radioactive parent isotope to a daughter isotope. The half-life is the time needed for half the parent to decay. The relative abundances of parent and daughter isotopes follow the trends shown at the bottom. (D. Merritts et al., *Environmental Geology*, © 1997 by W. H. Freeman and Company.)

TABLE 3-1 Radioactive Decay Used to Date Climate Records

Parent isotope	Daughter isotope	Half-life	Useful for ages:	Useful for dating:
Rubidium-87 (^{87}Rb)	Strontium-87 (^{87}Sr)	47 Byr	100 Myr	Granites
Uranium-238 (^{238}U	Lead-206 (^{206}Pb)	4.5 Byr	>100 Myr	Many rocks
Uranium-235 (^{235}U)	Lead-207 (^{207}Pb)	0.7 Byr	>100 Myr	Many rocks
Potassium-40 (^{40}K)	Argon-40 (^{40}Ar)	1.3 Byr	>100,000 years	Basalts
Thorium 230 (^{230}Th)	Radon-226* (^{226}Ra)	75,000 years	<400,000 years	Corals
Carbon-14 (^{14}C)	Nitrogen-14* (^{14}N)	5,780 years	<50,000 years	Anything that contains carbon

*Daughter is a gas that has escaped and cannot be measured.

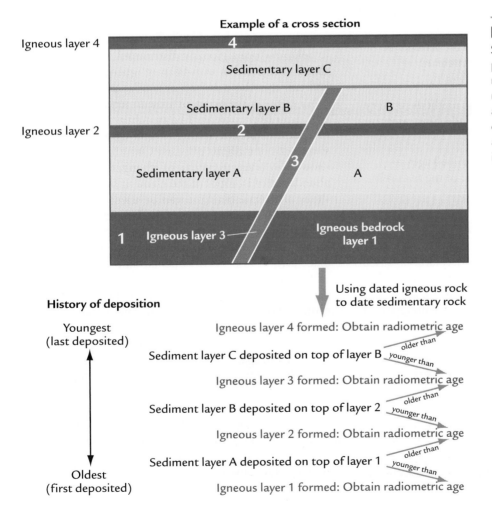

Example of a cross section

Igneous layer 4

4

Sedimentary layer C

Sedimentary layer B   B

Igneous layer 2

2

Sedimentary layer A   A

3

1   Igneous layer 3   Igneous bedrock layer 1

History of deposition

Youngest (last deposited)

Oldest (first deposited)

Using dated igneous rock to date sedimentary rock

Igneous layer 4 formed: Obtain radiometric age

Sediment layer C deposited on top of layer B — older than / younger than

Igneous layer 3 formed: Obtain radiometric age

Sediment layer B deposited on top of layer 2 — older than / younger than

Igneous layer 2 formed: Obtain radiometric age

Sediment layer A deposited on top of layer 1 — older than / younger than

Igneous layer 1 formed: Obtain radiometric age

FIGURE 3-9 Dating sediments Sediments can be dated by their position in relation to igneous rocks. First the igneous rocks are radiometrically dated. Then the ages of the sedimentary layers are constrained by their positions above, below, or between the igneous layers.

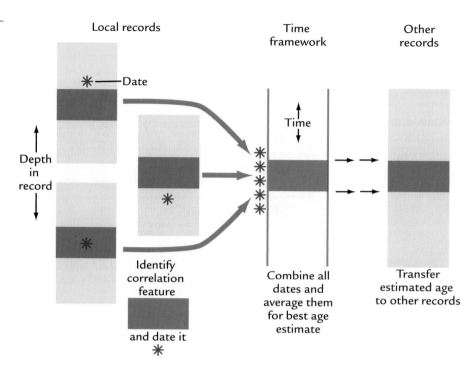

FIGURE 3-10 Dating and correlation
Actual dating of most sedimentary records relies on a stepwise process of dating and correlation. First, some kind of distinctive feature is dated in many locations by radiometric methods. Then the available dates are combined to derive the best possible estimate of the age of the feature. This age is then assumed to apply in other regions where the feature occurs.

Instead, sediment sequences are dated by a multistep process of dating and correlation (Figure 3-10). The key steps are:

1. Identifying a distinctive feature in the sediments that can be correlated among sedimentary sequences in many regions of the world.

2. Dating this feature by radiometric methods in every location where igneous rocks provide constraints on its age.

3. Combining all these locally determined ages to arrive at a best worldwide estimate of the feature's age.

4. Projecting this age estimate into any sedimentary sequence in which the feature can be found.

Distinctive fossils are the features commonly used for correlation in the older geologic record. This method depends on the fact that a *unique and unrepeated* sequence of organisms has appeared and disappeared through Earth's entire history, leaving fossilized remains in the geologic record. The most useful fossils are the shortest-lived but geographically most widespread forms. Extinctions of old forms and first appearances of new forms also provide distinct levels for radiometric dating and correlation.

Other information in sedimentary records can be dated and correlated by a process of *pattern matching and transfer*. This form of correlation is based on physical measurements whose distinctively varying patterns can be matched from region to region. As with the unique sequence of organisms in the fossil record, if a

pattern can be dated in one region and then found in another, the associated age scheme can be transferred wherever the pattern is recognized.

Chapter 5 explains how reversals of Earth's magnetic field produce a globally distributed magnetic pattern that is first dated in iron-bearing igneous rocks on land and can then be matched to similar patterns found in other rocks in the ocean or in sedimentary rocks with a high iron (Fe) content. The matching of these distinctive patterns permits dating of sediments deposited over many tens of millions of years. Chapters 10 through 12 show how measurements of changes in the relative abundance of isotopes of the element oxygen in $CaCO_3$ shells of marine plankton can be used to correlate and date deep-sea sediments over the last 3 million years.

Radiocarbon Dating In the younger geologic record, a different radiometric method can be used to date carbon-bearing sediments and other kinds of archives directly, rather than by extrapolation from initial dating of igneous rocks. Widely used since the late 1950s, **radiocarbon dating** is useful for dating lake sediments and other younger geologic deposits containing carbon.

Neutrons constantly streaming into Earth's atmosphere from space convert ^{14}N (nitrogen gas) to ^{14}C (an unstable isotope of carbon). Organisms living on Earth extract carbon from the atmosphere for photosynthesis, and a small part of the carbon used is radioactive ^{14}C. The death of the organism closes off carbon exchange with the atmosphere and starts the decay clock ticking. The ^{14}C parent decays to the ^{14}N daughter, a gas that escapes. The amount of ^{14}C lost is measured by comparing its abundance against that of a stable isotope of carbon (^{12}C) that is not subject to removal by radioactive decay. With half of the original ^{14}C lost every 5780 years, radiocarbon dating is most useful over five or six half-lives, or about 30,000 years, and in some cases as much as 50,000 years (Table 3-1).

Another important technique relies on the same uranium (U) decay series used to date igneous rocks (Table 3-1), but uses it in a different way to date corals. For longer-range dating of igneous rocks (Figure 3-8), the abundance of U is compared with its ultimate long-term daughter product, lead (Pb). In contrast, thorium/uranium ($^{230}Th/^{238}U$) dating focuses on just a small part of this long series of radioactive decays. Ocean corals incorporate a small amount of ^{234}U and ^{238}U from seawater into their shells (in place of calcium), but no ^{230}Th. When the corals die, the parent (^{238}U) slowly decays and produces ^{230}Th in the coral skeleton. But the ^{230}Th daughter is also radioactive and decays away with a half-life of 75,000 years. Gradually

the amount of ^{230}Th present in the coral increases to a level that reflects a balance between the slow decay of the parent U and the faster loss of the daughter ^{230}Th. The Th/U clock is useful for dating over the last few hundred thousand years.

Counting Annual Layers Some climate repositories contain annual layers that can be used to date archives by the simple approach of counting back in time from the present. These annual layers form because of seasonal changes in accumulation of climate-sensitive materials.

The most visible forms of annual layering in ice (mountain glaciers and ice sheets) are the alternations between darker layers, containing dust blown in from continental source regions during the dry windy season, and lighter layers from other seasons with little or no dust (Figure 3-11A). Together these dark/light couplets form annual layers that are easily visible in the upper

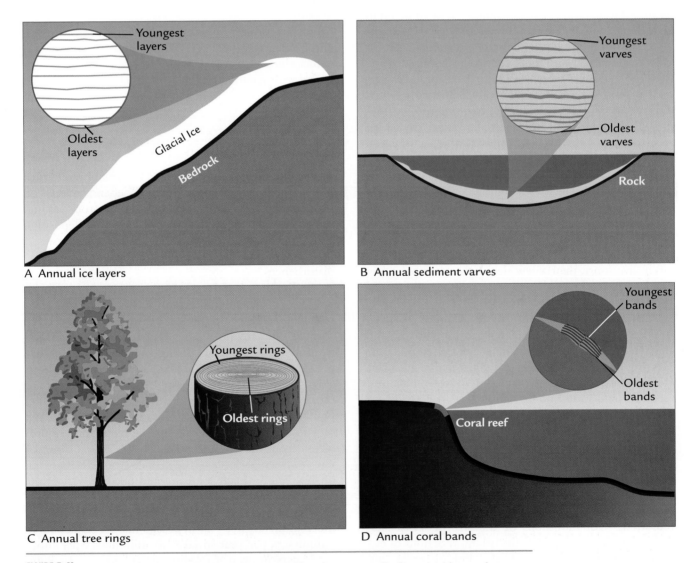

A Annual ice layers

B Annual sediment varves

C Annual tree rings

D Annual coral bands

FIGURE 3-11 **Annual layering** Four kinds of climate archives have annually deposited layers that can be used to date the climate records they contain: ice, varved lake sediments, trees, and corals.

parts of the glacial ice, but are gradually stretched and thinned deeper in the ice to the point where they cannot always be discerned. Ages of the deeper parts of the ice where annual layering no longer exists are usually estimated by methods based on models of how the ice flows.

Sediments in the deeper parts of some lakes contain annual-layer couplets called **varves** (Figure 3-11B). Varves are particularly common in the deeper parts of lakes containing little or no life-sustaining oxygen. The lack of oxygen suppresses or eliminates bottom-dwelling organisms that would otherwise obliterate annual layers by their physical activity. Varve couplets usually result from seasonal alternations between deposition of light-hued mineral-rich debris and darker sediment rich in organic material.

In regions of marked seasonal variations of climate, trees produce annual layers called **tree rings** (Figure 3-11C). These rings are alternations between layers of lighter, thicker wood tissue (cellulose) formed by rapid growth in spring and much thinner, darker layers marking cessation of growth in autumn and winter. As most individual trees live no more than a few hundred years, the time span over which this dating technique can be used is limited, but in some areas distinctive year-to-year variations in the thickness of tree rings can be used to splice records from longer-lived younger trees with records from trees that died much earlier and whose fossil trunks can still be found on the landscape.

In tropical oceans, corals record seasonal changes in the texture of the calcite ($CaCO_3$) incorporated in their skeletons (Figure 3-11D). The lighter parts of these **coral bands** are laid down in summer, during intervals of fast growth, and the darker layers during winter, when growth slows. Individual corals dated in this way rarely live more than a few decades or at most a few hundred years, but older records may be spliced into younger ones (as with tree rings).

Correlating Records with Orbital Cycles Another way of dating climate records is to use the characteristic imprint of climate cycles caused by variations in Earth's solar orbit. Changes in Earth's orbit around the Sun alter the amount of solar radiation received by season and by latitude (Part III). These orbital changes drive climate responses, including the strength of low-latitude monsoons and the growth and decay of ice sheets at middle and high latitudes.

Because the physical processes involved in these climate responses on Earth are understood reasonably well, it is possible to link the orbital-scale responses recorded in Earth's climate archives (such as changes in the strength of monsoons or the size of ice sheets) directly to the time scale of the orbital variations calculated from astronomy. With this link achieved, the record in the climate archive can be dated. Because vari-

ations in Earth's orbit back to at least 5 million years ago are well dated from astronomical calculations, this technique permits absolute dating of Earth history over this interval of time and even beyond.

Internal Chronometers The annual-layer-count and orbital-tuning techniques can also serve a similar purpose much farther back in time. Even in the absence of radiometric dates, some climate archives contain *internal chronometers* with which climate scientists can measure *elapsed time* (in number of years), rather than *absolute time* (in years before present).

For example, annual varves deposited in lake sediments millions of years ago still survive today in a few regions. Determining the actual age of the sequence by counting varves back in time from the present is impossible for such ancient deposits, because the varves were not continuously deposited up until the present day. However, the varves do supply an internal chronometer with which to count the years that elapsed during their deposition.

Orbital-scale responses can also be used as internal chronometers over longer time intervals. If we can detect climate-system responses to well-defined orbital cycles with lengths of 20,000 to 100,000 years, we can use these cycles as an internal chronometer and count elapsed time.

3-3 Climate Resolution

The extent to which the details of information recorded in climate archives can be resolved depends mainly on the interplay between two factors: (1) the processes that initially disturb the climate record soon after deposition and (2) the rate of accumulation of the record, which determines how fast it gains protection from these disturbances.

Sediment Archives Most sedimentary archives used for climate studies form in *low-energy* marine environments, those undisturbed by turbulent waves and storms. The only disturbance of particles raining down from above is physical stirring by organisms that live on the seafloor or within the upper sediment layers (Figure 3-12). Organisms living at the sediment surface thoroughly mix the uppermost layers. A much smaller number of animals burrow deep into the sediments, and only infrequently, so that subsurface sediments are protected from most disturbances as they are covered by newer layers. Eventually the deeper sediments pass out of the region of active mixing and become part of the permanent sedimentary record.

Typical rates of sediment deposition range from as much as meters per year in coastal marine sequences to millimeters per year in lakes to millimeters per thousand years in deep-sea sediments (Figure 3-13). Rates can easily vary around these average values by a factor

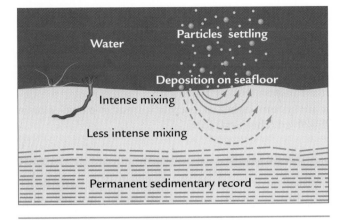

FIGURE 3-12 Sediment mixing Small animals crawl across and burrow into the upper layers of ocean sediments and mix particles that have fallen to the seafloor. Mixing blurs the degree of detail that can be resolved in the permanent climate record eventually buried deep below the surface.

of 10 because of local factors such as the degree of the sediment's exposure to turbulent water and the amount of sediment supplied locally by rivers or redistributed by currents.

The degree of disturbance by organisms that move across and burrow into the sediment surface also varies with environment. In highly productive coastal regions, large organisms burrow tens of centimeters or even meters down into the sediment. Relatively unproductive deep ocean basins have fewer and smaller bottom-dwelling organisms that typically burrow in no more than a few centimeters. Most lakes also have fewer and shallower burrowers. As a result, the resolution of sedi-

mentary records also varies with environment. Lakes usually have the best resolution and deep-ocean sediments the poorest, although locally rapid deposition can improve resolution in some ocean areas.

After particles pass through the upper layers, no further mixing occurs unless erosion reexposes the sequence back at the sediment–water interface. Increased pressure and loss of water caused by deep burial of sediments gradually compact the sediment layers but do not dramatically reduce the resolution they can provide if they are sampled carefully.

Ice Cores Annual layers of snow are visible at the surfaces of many mountain glaciers and rapidly deposited ice sheets (see Figure 3-11A). As the snow is buried and slowly recrystallized into ice, annual layers remain resolvable to a depth that depends on their initial thickness at the time of deposition. Below this level, the layering is lost. In cores from mid-latitude ice sheets such as the one on Greenland, where deposition of snow is rapid, the annual layering may remain visible tens of thousands of years into the past. In polar ice sheets such as the one that covers eastern Antarctica, where only a small amount of snow accumulates each year, annual layering may not occur even at the ice surface.

Tree Rings and Corals At middle and high latitudes where trees produce annual layers, tree rings become a permanent record of annual climate change, unless they are later disturbed by fire or by sporadic boring by insects or excavation by birds. Similarly, $CaCO_3$ bands in corals form a permanent record of seasonal to annual climate change.

The types of climate archives, the maximum time span of record they contain, and the highest resolution

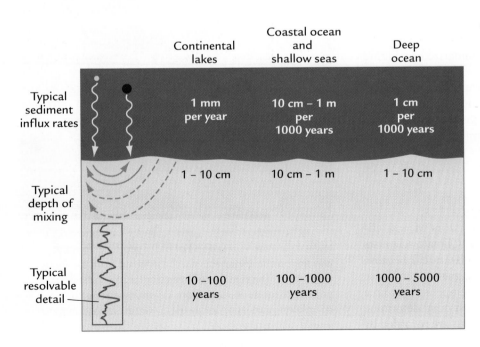

	Continental lakes	Coastal ocean and shallow seas	Deep ocean
Typical sediment influx rates	1 mm per year	10 cm – 1 m per 1000 years	1 cm per 1000 years
Typical depth of mixing	1 – 10 cm	10 cm – 1 m	1 – 10 cm
Typical resolvable detail	10 –100 years	100 –1000 years	1000 – 5000 years

FIGURE 3-13 Climate resolution The degree of resolution of climate records in sediment archives is related to the rate of deposition (and burial) of sediment and to the amount of activity of organisms burrowing into the sediments.

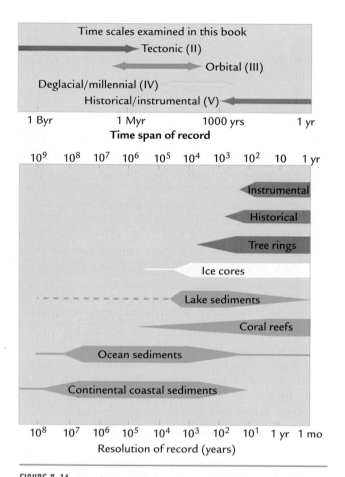

FIGURE 3-14 Resolution of climate records Climate archives vary widely in the length of the records they contain and in the degree of resolution they yield. A log scale (changing by powers of 10) is needed to show all geologic time in a single plot. (Adapted from J. C. Bernabo, *Proxy Data: Nature's Record of Past Climates* [Washington, D.C.: National Oceanic and Atmospheric Administration, 1978].)

achievable for each archive are summarized in Figure 3-14. Because Earth has existed for so long and because the amount of information on climate increases rapidly toward the younger part of the record, this information is displayed in a log time scale that changes by powers of 10. Also shown at the top are the time spans covered by the major parts of this book.

Climate Data

Climate archives contain many indicators of past climate referred to as **climate proxies**. Climate scientists use the term "proxy" (meaning "substitute") because the process of extracting climate signals from these indicators is not direct, like reading temperature from a thermometer. Instead, scientists must first determine the mechanism by which climate signals are recorded by

proxy indicators in order to decipher climate changes.

The two climate proxies that are most commonly used are (1) **biotic proxies**, which are based on changes in the composition of plant and animal groups, and (2) **geological-geochemical proxies**, which quantify mass movements of Earth's materials through the climate system, either as discrete (physical) particles or in dissolved (chemical) form.

3-4 Biotic Data

Because no seafloor older than 170 million years exists, broad-scale reconstructions of earlier oceanic environments are not possible. As a result, fossil remains from the continents are the main climate proxy for older tectonic-scale intervals.

Most of the organisms that have ever existed on Earth are now extinct, and the further back in time we

FIGURE 3-15 Past vegetation For older geologic intervals, climate on the continents can be inferred from distinctive vegetation. The remains of trees similar to modern-day palms are found in rocks from Wyoming dating to 45 million years ago. Today frigid winters in Wyoming would kill palm trees. (Chip Clark.)

look, the less recognizable the fossils appear. Using biotic proxies to reconstruct past climates over longer tectonic time scales often requires climate scientists to rely on the general resemblance of past forms to their modern-day counterparts, either in general appearance or in specific features that can be measured.

Because fossil remains of plants tend to be more numerous than those of animals in geologic records from continents, vegetation plays a central role in the reconstruction of ancient climates. Often the simple presence of a critical temperature-sensitive form is useful as a climate indicator in older deposits. For example, warmer climates tens of millions of years ago are inferred from the presence of palmlike trees at high northern latitudes (Figure 3-15).

For the younger continental record, climate scientists more commonly use the relative abundance of climate-sensitive vegetation indicated by pollen sequences deposited in sediments (Figure 3-16). Minute pollen grains are produced in vast numbers by vegetation, distributed mostly by wind, and deposited in lakes. Pollen grains are best preserved in oxygen-poor parts of lakes. Pollen can be identified initially by major vegetation type (trees, grass, and shrubs) and

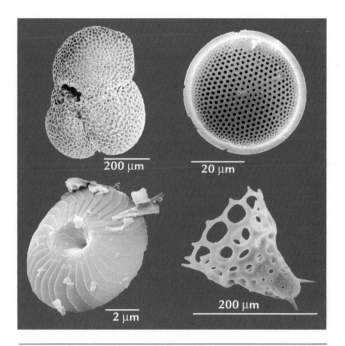

FIGURE 3-17 Plankton: a proxy indicator of climate in the ocean Four types of shelled remains of plankton are common in ocean sediments: $CaCO_3$ shells are represented by sand-sized planktic foraminifera (upper left) and small clay-sized coccoliths (lower left); SiO_2 shells include silt-sized diatoms (upper right) and sand-sized radiolaria (lower right). For scale, small grains of sand are 60 μm or larger in diameter. (Modified from W. F. Ruddiman, "Climate Studies in Ocean Cores," in *Paleoclimate Analysis and Modeling,* ed. A. D. Hecht [New York: John Wiley, 1977].)

then further subdivided (spruce trees indicate a cold climate, oak trees indicate warmth). Larger remains of vegetation that cannot have been carried far from their point of origin are also examined to make sure that the pollen in a lake sequence is representative of the nearby vegetation. These larger **macrofossils** include cones, seeds, and leaves.

In the oceans, four major groups of shell-forming animal and plant **plankton** are used for climate reconstructions (Figure 3-17). Two groups form shells made of calcite ($CaCO_3$). Globular sand-sized animals called **planktic foraminifera** (upper left) inhabit the upper layers of the ocean. Small spherical algae called Coccolithophoridae secrete tiny $CaCO_3$ plates called **coccoliths** (lower left) in these sunlit waters.

Two other groups of hard-shelled plankton secrete shells of opaline silica ($SiO_2 \cdot H_2O$) and tend to thrive in productive, nutrient-rich surface waters (Figure 3-17). **Diatoms** (upper right) are silt-sized plant plankton often shaped like pillboxes or needles. **Radiolaria** (lower right) are sand-sized animals with ornate shells often shaped like old military helmets.

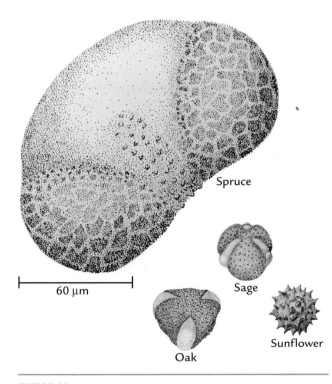

Spruce

Sage

Oak

Sunflower

60 μm

FIGURE 3-16 Pollen: a proxy indicator of climate on land For younger intervals, climate on land can be reconstructed from changes in the relative abundance of distinctive types of pollen. For scale, small grains of sand are 60 μm or larger in diameter. (Courtesy of Alan Solomon, Environmental Protection Agency, Corvallis, Ore.)

FIGURE 3-18 **Distribution of ocean sediments** The predominant type of sediment on the seafloor of the world ocean today varies regionally, with ice-rafted sediment in polar areas, SiO_2-rich sediment in productive areas, $CaCO_3$-rich sediment on higher rises and ridges, and windblown deep-sea silt and clay in basins far from continents. Coastal regions contain mainly debris from the land. (Modified from W. H. Berger, "Deep-Sea Sedimentation," in *The Geology of Continental Margins*, ed. C. A. Burke and C. L. Drake [New York: Springer-Verlag, 1974].)

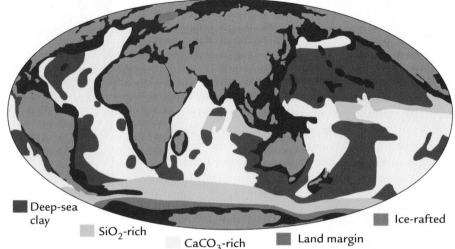

■ Deep-sea clay   ■ SiO_2-rich   ■ $CaCO_3$-rich   ■ Land margin   ■ Ice-rafted

Sediments rich in $CaCO_3$ fossils occur in open-ocean (noncoastal) waters and are most abundant at depths above 3500–4000 meters (Figure 3-18). Below that level, corrosive bottom waters dissolve calcite shells. SiO_2-shelled diatoms inhabit deltas and other coastal areas and extract silica from river water flowing off the land. Both radiolaria and diatoms are abundant in regions where highly productive waters upwell to the surface from below.

Plankton and pollen share traits that make them especially useful as biotic climate proxies. Both are widely distributed: plankton live in all oceans, and pollen are produced everywhere on continents except under ice sheets. Also, because fossil remains of these two groups are so abundant in sediments (usually thousands in a tablespoon-sized sample), their relative abundances can be determined with a much higher degree of accuracy than those fossil types that show up only sporadically in the climate record. In addition, populations of plankton and pollen in different areas tend to be dominated by a small number of species with well-defined climate preferences. The only other organisms on Earth with comparable ranges and abundance are insects, which rarely leave fossil remains.

Over the last few million years, the species of plankton and vegetation still in existence today gradually came to dominate biotic assemblages in the ocean and on land. This progressive trend opens up new avenues for quantitative reconstructions of past climates. Climate preferences of living species of plankton and vegetation can be determined by comparing their present-day distributions against measurements of current climate in the same areas. These modern-day climate preferences then become the basis for reconstructing past climates from past fossil assemblages. Because the same species have existed on Earth

for several million years, relationships defined between modern climate and organism abundances can be used to interpret similar assemblages in sediment archives as old as a few million years or more.

3-5 Geological and Geochemical Data

Mass movements of materials through the climate system are tied to processes of erosion, transport, and deposition, mainly by water but also by ice and wind. Most climate studies of the older tectonic-scale parts of Earth's history rely on physical debris deposited in sedimentary archives on the continents as the main proxy used for inferring past climates.

Sediment structures representative of distinctive depositional environments are shown in Figure 3-19. These sediment textures reveal erosional scraping and subsequent deposition of unsorted sediments by ancient ice sheets in cold environments, the existence of hyper-arid deserts covered by moving sand dunes under arid conditions, and deposition of sediment by water in moist environments. Sediments deposited by turbulent water and by downslope dislodgment from steep slopes also leave distinctive structures that can reveal past climates.

Although sediment structures are a useful basis for broad inferences about climate, poor dating control and the prevalence of erosion make detailed study of most older continental records difficult. Older deposits are also more likely to have been altered or deformed with the passage of time, diminishing their value for climate studies.

In contrast, the availability of ocean sediments from the last 170 million years brings new opportunities for climate research because of their relatively continuous deposition, better dating, and wider geographic coverage. As a result, the basic distribution of sediment types

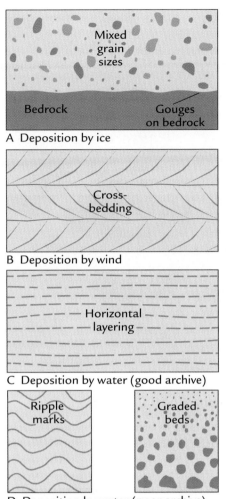

A Deposition by ice

B Deposition by wind

C Deposition by water (good archive)

D Deposition by water (poor archive)

FIGURE 3-19 **Sediment textures** Ice, wind, and water leave distinct textures in sediments and sedimentary rocks. Sediments deposited by ice are an unsorted jumble of debris scraped and eroded from bedrock (A). Sediments deposited by wind often show cross-bedding attributable to bouncing and rolling of sand grains (B). Sediments deposited by water are horizontally layered in the quiet environments most often used as climate archives (C), but they show such features as ripple marks and graded bedding in regions of more turbulent, high-energy conditions (D).

that carry distinctive information on climate can be mapped, and changes in their patterns of deposition can be quantified as **burial fluxes** (measures of the mass of sediment deposited per unit of area per unit of time). Local changes in sediment burial fluxes can be linked to variations in climate in the regions where the sediments originated and also to climate changes elsewhere on the globe.

Sediment is eroded from the land and deposited in ocean basins in two forms. One is debris eroded and

transported as discrete particles or grains as a result of **physical weathering**, the process by which water, wind, and ice physically detach pieces of bedrock and reduce them to smaller fragments. One example shown in Figure 3-20 is coarse **ice-rafted debris** (sand and gravel) eroded by ice sheets and delivered by icebergs that melt in ocean waters. Other examples include finer **eolian sediments** (silts and clays) lifted from the continents and blown to the ocean by winds, as well as the wide range of grain sizes and types in **fluvial sediments** carried by rivers to the ocean.

Geological and geochemical techniques can unravel the original sources of sediments formed by physical weathering. Microscope counts of sand-sized grains in marine sediments can distinguish different sources on the basis of distinct mineral types. In recent years, geochemical analyses of distinctive elements and isotopes have become the main method of tracing mineral grains back to specific source regions on the continents.

The second major way of removing sediments from the land is by **chemical weathering** and subsequent transport of dissolved ions (charged ions or compounds) to the oceans in rivers (Figure 3-21). Chemical weathering occurs mainly in two ways: (1) by **dissolution**, the dissolving of carbonate rocks (such as limestone, made of $CaCO_3$) and evaporite rocks (such as rock salt, made of $NaCl$) in water, and (2) by **hydrolysis**, the addition of water to continental silicate rocks, such as basalts and granites, during weathering.

Both processes depend on the fact that atmospheric CO_2 and rain (H_2O) combine in soils and rock crevices to form carbonic acid (H_2CO_3), a weak acid that attacks the rocks chemically. After weathering, rivers carry off

FIGURE 3-20 **Sediment particles** Deep-ocean sediments contain granular debris from land that reveals the climate of the source region. For example, sand-sized grains of quartz and other minerals rafted in from ice sheets by icebergs indicate cold climates. (Courtesy of Gerard Bond, Lamont-Doherty Earth Observatory of Columbia University.)

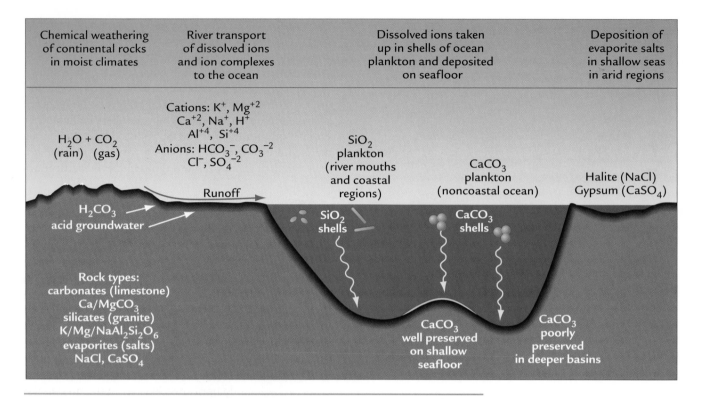

| Chemical weathering of continental rocks in moist climates | River transport of dissolved ions and ion complexes to the ocean | Dissolved ions taken up in shells of ocean plankton and deposited on seafloor | Deposition of evaporite salts in shallow seas in arid regions |

Cations: K^+, Mg^{+2} Ca^{+2}, Na^+, H^+ Al^{+4}, Si^{+4}

Anions: HCO_3^-, CO_3^{-2} Cl^-, SO_4^{-2}

$H_2O + CO_2$ (rain) (gas)

SiO_2 plankton (river mouths and coastal regions)

$CaCO_3$ plankton (noncoastal ocean)

Halite (NaCl) Gypsum ($CaSO_4$)

Runoff

H_2CO_3 acid groundwater

SiO_2 shells

$CaCO_3$ shells

Rock types: carbonates (limestone) $Ca/MgCO_3$ silicates (granite) $K/Mg/NaAl_2Si_2O_6$ evaporites (salts) $NaCl$, $CaSO_4$

$CaCO_3$ well preserved on shallow seafloor

$CaCO_3$ poorly preserved in deeper basins

FIGURE 3-21 Chemical weathering, transport, and deposition Chemical weathering slowly attacks rocks on land and sends dissolved ions into rivers for transport to the ocean. Ocean plankton incorporate some of the dissolved ions in their shells, which fall to the seafloor and form part of the geologic record. Some dissolved ions are also deposited in shallow evaporating pools on continental margins where the climate is dry.

many dissolved materials, including ions (Ca^{+2}, Mg^{+2}, Na^+, K^+, Sr^{2+}, Cd^{2+}, Al^{4+}, and Cl^-), and ion complexes (HCO_3^-, CO_3^{-2}, and $SiO(OH)_2$).

Some of these dissolved ions (Si^{+4}, Ca^{+2}, and CO_3^{-2}) are taken up in the shells of plankton (Figure 3-21). A small fraction ends up in the shells of **benthic foraminifera**, sand-sized animals that live on the seafloor and form calcite ($CaCO_3$) shells from Ca^{+2} and CO_3^{-2} ions in deep waters. Because all of these kinds of $CaCO_3$ and SiO_2 (opal) shells preserved in ocean sediments are the end products of chemical weathering on land and ion transport in rivers, they are useful for tracking changes in large-scale fluxes of calcium, silicon, carbon, and oxygen over time.

The rate of deposition of $CaCO_3$ in ocean basins responds to several factors: changes in rates of delivery of Ca^{+2} and HCO_3^- ions by rivers due to variations in weathering on land; changes in the relative amounts of $CaCO_3$ deposited on shallow continental shelves versus deeper ocean basins; and regional changes in the corrosiveness of bottom waters in deep ocean basins. Regional-scale changes in rates of deposition of SiO_2 mainly reflect local variations in the productivity of surface waters and to a lesser extent the rate of delivery of silicon from the continents to the oceans by rivers and the corrosion of SiO_2 shells settling through the water column.

Because it takes a long time in the lab to analyze the chemical properties of individual samples taken from thick sedimentary sequences, many recent studies have turned to logging techniques that quickly detect and record key physical or chemical properties of the sediments that correlate closely with their composition. Sediments are logged by moving the sediment cores through a detection unit that uses sound waves or some other nondestructive technique to sense the properties of the sediments without touching them. For example, a 16-meter section of equatorial Pacific sediments shows a good correlation between logged analyses of bulk density and $CaCO_3$ percentages determined by laboratory analyses (Figure 3-22). This pattern emerges because $CaCO_3$ has a higher density than the opal (SiO_2) fraction in this core.

A wide range of important climatic data is also stored in the isotopes of elements in the $CaCO_3$ shells of planktic organisms and benthic foraminifera. Cases

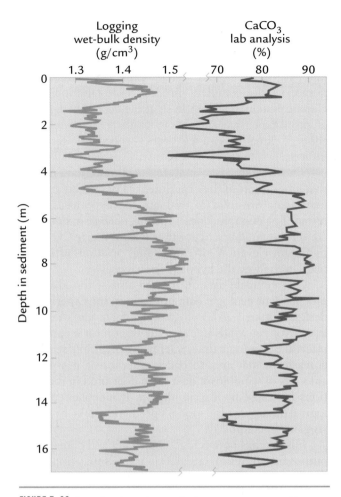

FIGURE 3-22 **Logging** Logging techniques can sense sediment properties such as bulk density without the need to remove and analyze actual samples. Logged properties often correlate closely with changes measured from individual samples in the laboratory, as in the case of sediment bulk density (left) and a record of the percentage of $CaCO_3$ (right) in a Pacific Ocean core. (Adapted from T. D. Herbert and L. A. Mayer, "Long Climatic Time Series from Sediment Physical Property Measurements," *Journal of Sedimentary Petrology* 61 [1991]: 1089–1108].)

that will be examined in detail in later chapters include isotopes of strontium (Sr), which measure changing river fluxes and exchanges of fluids with the seafloor (Chapter 7); isotopes of oxygen, which record changes in the global volume of ice and in local ocean temperatures (Chapters 7 and 10); and isotopes of carbon, which trace movements of organic material among reservoirs on the continents, in the air, and in the ocean, and also detect changes in deep-sea circulation (Chapter 11).

As we will see in later chapters, other geochemical proxies of climate gradually become available over more recent intervals of Earth's climate history. At orbital time scales, ice cores contain samples of air from past atmospheres, including past concentrations of CO_2 and CH_4 (Chapter 11). Other important ice-core proxies (Chapters 11–16) include changes in the thickness of snow deposited (related to the temperature and moisture content of the air); in the amount of dust blown in from various continents (which can be chemically separated into elements indicative of regional-scale sources); and in the amount of sea salt blown in from the ocean.

Cave deposits contain records of groundwater derived from atmospheric precipitation over intervals of several hundred thousand years. Changes in the chemical composition of this water reflect changes in the original source of the water vapor, in the atmospheric transport path to the site of precipitation, and in the groundwater environment. Sedimentary deposits in lakes record, in addition to changes in pollen, climatically driven fluctuations in lake levels (Chapters 13–15) and in many chemical tracers now under active investigation.

Trees record the amount of cellulose deposited in each annual layer (determined by the width and density of tree rings) as a function of climate, primarily changes in precipitation during the rainy seasons in dry regions and changes in summer temperatures in cold regions (Chapter 16). Annual coral bands contain a wide range of chemical information, including oxygen isotope ratios that record seasonal temperature and precipitation changes (Chapter 16).

Climate Models

Scientists who extract records from Earth's climate archives make discoveries that need explanations. At first scientists may just speculate somewhat casually about the causes of their findings, but often they turn to climate models to test their most provocative explanations. Because models *put numbers on ideas*, they can provide rigorous tests of the validity of proposed hypotheses. But models also simplify some aspects of reality, and their performance must be carefully and critically assessed.

In this section we examine two kinds of numerical models used by climate scientists. **Physical climate models** emphasize physical operations in the climate system, particularly the circulation of the atmosphere and ocean but also interactions with vegetation (biology) and with atmospheric trace gases (chemistry). In contrast, **geochemical models** track the movement of distinctive chemical tracers through the climate system.

3-6 Physical Climate Models

Most physical models are constructed to simulate the climate system as it exists *today*, and are judged by how well they do so. A simulation of modern-day climate is called the **control case**. Models must simulate modern climate reasonably well to be used for exploring past climates.

One reason climate simulations of the past are important is that they further test the performance of climate models under conditions very different from those today. If the models also succeed in reproducing past climates, scientists can better trust these models to simulate future climate change.

Simulations of past climates occur in a three-step process (Figure 3-23). The first step is to specify the model experiment to be run. Some aspect of the model's representation of the modern world is altered from its present form in order to reflect changes that have occurred on Earth in the past. For example, we could increase the level of CO_2 in the model atmosphere, drop the height of its mountains, remove or add ice sheets, or move the continents around on Earth's surface. The features that are altered to test hypotheses of climate change are called **boundary conditions**.

The second step is the actual operation of the model. Physical laws that drive the flow of heat energy through Earth's climate system are incorporated into the internal workings of the model. These laws are activated to perform a **climate simulation**. The amount of "climate time" simulated by the model, as well as the amount of actual running time the model requires, is related to its level of complexity.

Simpler models are less expensive to run and can simulate the evolution of climate over longer intervals of time (thousands of years), but they lack many important parts of the climate system. Complex models incorporate a more complete physical representation of the climate system, but they do so at the cost of being slower, more expensive, and able to simulate only brief snapshots of climate over a few years at specified intervals of interest.

The third step is to analyze the **climate data output** that emerges from the experiment. These simulated climatic data are used to evaluate the hypothesis being tested. For example, does a specific change in boundary conditions cited in a hypothesis (atmospheric CO_2 level, mountain elevation, or continental position) affect climate in the way the hypothesis proposes?

Often climate data output can also be tested against independent geologic data not used in the experiment's design (Figure 3-23). For example, if a model run simulates stronger winds in a specific region for a particular interval of geologic time, scientists can sample sediment cores from that area to check whether or not greater amounts of windblown dust were deposited in the locations indicated by the simulation. If independent geologic data support the model's predictions, the hypothesis gains in credibility.

Mismatches between geologic data and climate data output from physical circulation models may imply several possible problems: key boundary conditions were specified incorrectly or were omitted from the experiment; the model does not adequately simulate some part of the climate system; or the geologic data with which the model's output was compared were misinter-

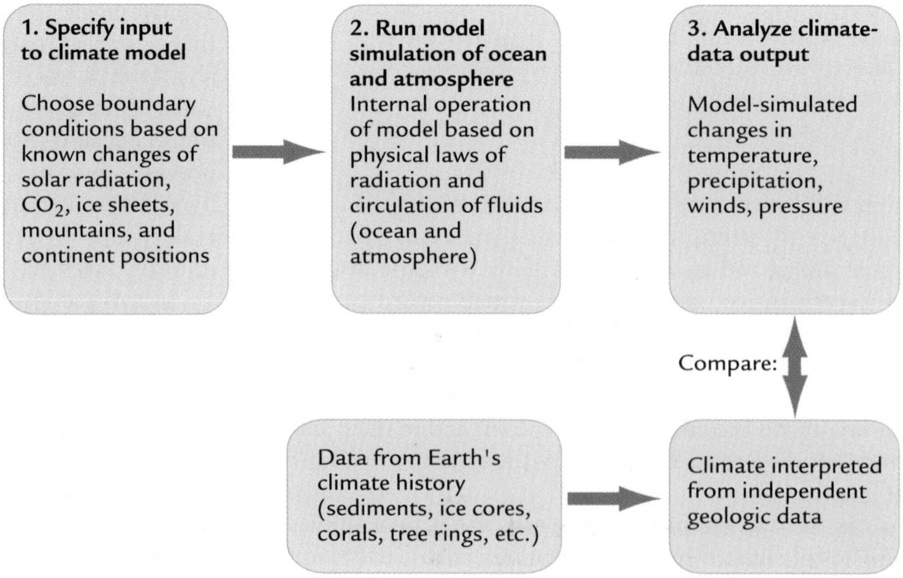

FIGURE 3-23 **Data-model comparisons** Models of Earth's climate are constructed to simulate present-day circulation. Then changes based on Earth's history (different CO_2 levels, ice sheet sizes, or mountain elevations) are inserted into the model, and simulations of past climates are run. The climate output is compared with independent geologic data to test the performance of the model.

1. Specify input to climate model

Choose boundary conditions based on known changes of solar radiation, CO_2, ice sheets, mountains, and continent positions

2. Run model simulation of ocean and atmosphere
Internal operation of model based on physical laws of radiation and circulation of fluids (ocean and atmosphere)

3. Analyze climate-data output

Model-simulated changes in temperature, precipitation, winds, pressure

Data from Earth's climate history (sediments, ice cores, corals, tree rings, etc.)

Compare:

Climate interpreted from independent geologic data

preted. Despite this large range of possible problems, the main cause of data-model mismatches is often obvious and can lead to useful refinements in boundary conditions, in data interpretation, or in model construction. The science of reconstructing past climates moves ahead best when the strengths and limitations of both the data and the models are constantly tested against each other.

As we review the physical models, we will also in effect be tracing their historical evolution through the last four decades. We start with models of atmospheric circulation, proceeding from simple to complex versions, then look at ocean models, and finally briefly review physical models that simulate changes in ice and vegetation.

One-Dimensional Atmospheric Models One-dimensional "column" models are the simplest kind of physical climate model of the atmosphere. They simulate a single vertical column of air that represents the average structure of the atmosphere of an entire planet (Figure 3-24). This air column is divided into layers

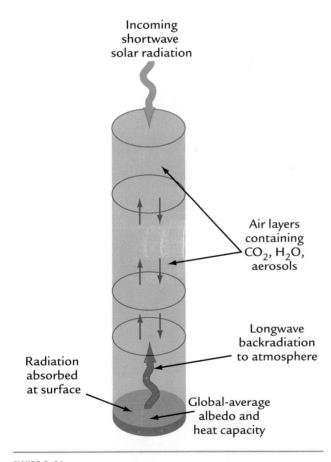

Incoming
shortwave
solar radiation

Air layers
containing
CO_2, H_2O,
aerosols

Longwave
backradiation
to atmosphere

Radiation
absorbed
at surface

Global-average
albedo and
heat capacity

FIGURE 3-24 **1-D models** One-dimensional column models are simplified representations of the response of an entire planet (its average land surface and the mean vertical structure of its atmosphere) to incoming solar radiation.

that are closely spaced near Earth's surface, where interactions are more complex, and are more widely spaced at higher elevations. Each layer contains climatically important constituents, such as greenhouse gases and dust particles. The model is driven by global-mean solar radiation, and a global-average representation of Earth's average surface albedo and heat capacity reflects some of the incoming radiation and absorbs and reradiates the rest.

Because column models are one-dimensional, the only possible directions in which air and the gases or particles it contains can be transported are up and down. Transport is simulated by the exchanges between adjoining layers shown by the thin arrows in Figure 3-24. Heat and water vapor also move between adjacent layers in the model.

One-dimensional (1-D) models are most appropriate for gaining an initial understanding of the way climate is affected by changes in concentrations of greenhouse gases and of airborne particles called **aerosols**, such as volcanic ash and dust. One example is the high CO_2 composition inferred for Earth's atmosphere in its earliest history (Chapter 4). Use of one-dimensional models to study climate is limited by their lack of such basic features as the pole-to-equator pattern of changes of temperature and moisture.

Two-Dimensional Atmospheric Models Two-dimensional (2-D) models move toward a more complete portrayal of the climate system. One type of 2-D model includes a multilayered atmosphere in one dimension and a second dimension representing Earth's physical properties averaged by latitude (Figure 3-25). Earth's modern-day surface at 80°–90°N in 2-D models has the properties of sea ice, because it is all ice-covered Arctic Ocean. Lower latitudes in the model are represented by the mixed properties typical of the oceans and landmasses found there today.

Adding a dimension that simulates processes that change with latitude (even in this simplified average way) makes it possible to use 2-D models to simulate several important processes that vary from pole to equator, such as the angle of incoming solar radiation, the albedo of Earth's surface, and the heat capacity of water versus that of land or ice.

Differences in radiation and albedo versus latitude set up a heat imbalance in 2-D models. To compensate for this imbalance, heat is transferred from warmer low-latitude regions toward the colder poles by a process called **diffusion**. In 2-D models, diffusion can be thought of as the rate of spreading of an ink blot across an area of absorbent paper, and it is expressed in square centimeters per second (cm²/sec). In this case. the spreading takes place within the plane of the 2-D model (Figure 3-25).

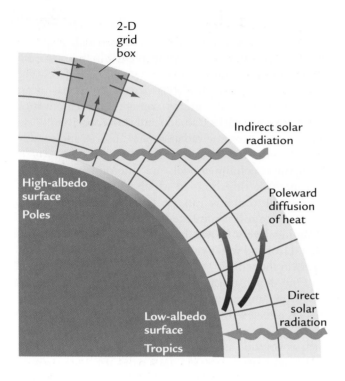

2-D
grid
box

Indirect solar
radiation

High-albedo
surface
Poles

Poleward
diffusion
of heat

Low-albedo
surface
Tropics

Direct
solar
radiation

FIGURE 3-25 2-D models The responses of latitude-averaged 2-D models to incoming solar radiation vary along lines of latitude because solar radiation arrives at different angles and because surface properties such as snow and ice cover vary with latitude.

Because 2-D models can simulate long intervals of time quickly and inexpensively, they are used to explore longer-term interactions among the surface ocean, sea ice, and land. They are also used in combination with models of slowly changing ice sheets (Chapter 10).

Missing from latitude-averaged 2-D models are important processes in the climate system that depend on the actual geographic positions of the continents and oceans. These processes include the geographic patterns of seasonal heating and heat storage (including the monsoon circulations) and the movement of low-pressure and high-pressure systems across the planet through time.

FIGURE 3-26 3-D GCMs General circulation models (GCMs) are full 3-D representations of Earth's surface and atmosphere, represented by individual grid boxes. Representations of Earth's surface within each grid box are entirely land, ocean, or ice. (Adapted from W. F. Ruddiman and J. E. Kutzbach, "Plateau Uplift and Climate Change," *Scientific American* 264 [1991]: 66–75.)

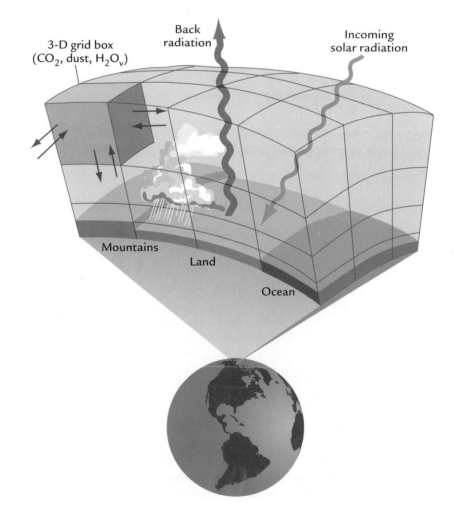

3-D grid box
(CO_2, dust, H_2O_v)

Back
radiation

Incoming
solar radiation

Mountains

Land

Ocean

Three-Dimensional Atmospheric Models Three-dimensional **general circulation models (GCMs)** provide the most complete numerical representations and simulations of the climate system. These 3-D models have the capacity to represent many key features: the spatial distribution of land, water, and ice, and the way their heat capacities, albedos, and other properties vary regionally; the elevation of mountains and ice sheets; the amount and vertical distribution of greenhouse gases in the atmosphere; and seasonal variations in solar radiation.

All these boundary conditions are specified for hundreds of model **grid boxes**, like those shown in Figure 3-26. The vertical boundaries of the grid boxes are laid out along lines of latitude and longitude at (and above) Earth's surface, and the box size shrinks near the poles because lines of longitude converge there. The horizontal boundaries of the grid boxes divide the atmosphere along lines of equal altitude above sea level. Models generally have 10 to 20 vertical layers, again more closely spaced near Earth's surface because the interactions with the land, water, and ice surfaces there are more complex than the flow higher in the atmosphere, which is considerably smoother.

The operation of GCMs incorporates the basic physical laws and equations that govern the circulation of Earth's atmosphere: the fluid motion of air; conservation of mass, energy, and other properties; and gas laws covering the expansion and contraction of air. All calculations in GCMs are carried out through the interaction of individual grid boxes with their immediate neighbors.

Model runs begin with the atmosphere in a state of rest. After solar heating causes air to begin to move, the model is run long enough to reach a state of equilibrium (Figure 3-27). Equilibrium occurs when the long-term drift in the simulated climate data disappears, and the only oscillations in climate remaining in the model are those analogous to short-term changes in weather over days and weeks. Running climate experiments on current-generation GCMs requires a simulation of about 20 years of climate. The first 15 years of the simulation are the "spin-up" interval, used to let the model attain a state of equilibrium, and the last 5 years of the simulation produce climate data that form the actual output of the model.

For the control-case simulation of modern-day climate, the regional climate-data output from the GCM are compared with regional instrumental measurements of temperature, precipitation, pressure, and winds in the present-day climate system averaged over the last several decades (Figure 3-28). Areas of major disagreement between the model and instrumental observations often become the focus of additional work to improve the model.

Improving Physical Models As we noted earlier, using GCMs to run experiments on past climates requires scientists to specify major changes in boundary conditions on the basis of geological changes from Earth's history. Two basic ways to do this are sensitivity tests and climatic reconstructions.

In a **sensitivity test**, just one boundary condition at a time is altered in relation to Earth's present-day configuration. When the output of such an experiment is compared with the output from the modern-day control case, the differences in climate between the two runs isolate and reveal the unique impact caused by a change in just that one boundary condition.

As we shall see, climate scientists have used GCM sensitivity tests to explore the effect of rearranging the continents into one giant supercontinent (Chapter 5), altering the heights of mountains (Chapter 7), adding large ice sheets (Chapters 10 and 13), and changing the amount of solar radiation (Chapter 14). It is also possible to run a series of sensitivity tests in which one new boundary condition is added for each successive new experiment in order to isolate the specific contribution of each new boundary condition (such as geography and CO_2 concentrations, in Chapter 6).

A climate **reconstruction** requires changing all known boundary conditions in a GCM at the same time in order to try to simulate the full state of the climate at some time in the past. This approach is more demanding than a sensitivity test, because all potentially critical boundary conditions are rarely known well enough to be used as input to a simulation. This method is used mainly to study glacial maximum and deglacial climates of the last 20,000 years, an interval for which climate

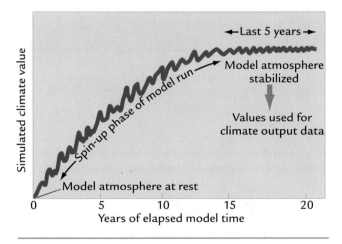

FIGURE 3-27 Model equilibrium GCMs require about 15 years of simulated climate change before they arrive at an equilibrium state. The final 5 years of the simulation are then averaged for use as the climate data output.

FIGURE 3-28 Control-case simulations
GCMs are developed by testing how well they reproduce modern-day climate (temperature, precipitation, and winds) based on present boundary conditions (CO$_2$, mountains, and land-sea distribution). This case compares observed January surface temperatures (A) with model-simulated values (B). (Adapted from J. Hansen et al., "Efficient Three-Dimensional Global Models for Climate Studies: Models I and II," *Monthly Weather Review* 111 [1983]: 609–62.)

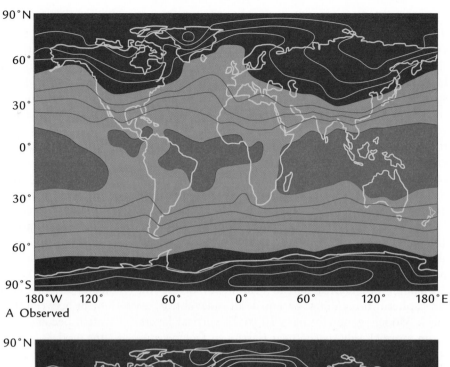

A Observed

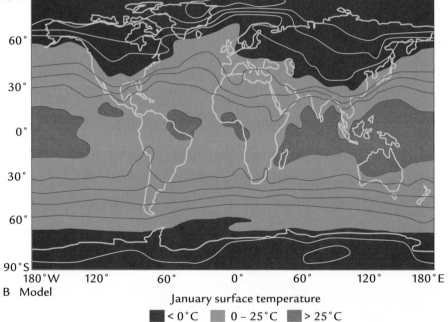

B Model

January surface temperature

■ < 0°C   ■ 0 – 25°C   ■ > 25°C

scientists have numerous records dated by ^{14}C methods (Part IV).

Every 1.5 years the power of the world's best computers increases by a factor of 10. Over time, this increase in computing power has gradually reduced the horizontal size of the grid boxes in GCMs. Typical grid boxes were once 8° of latitude by 10° of longitude, or as much as 1000 kilometers on a side. More recently GCM grid boxes have been reduced to 2° of latitude by 3° of longitude, or no more than 300 kilometers on a

side. The result has been improved resolution of coastal outlines of continents (including narrow isthmuses) and of small seas, larger ocean islands, and large lakes. For the first time, newer GCMs can "see" (resolve) New Zealand!

The shrinking of grid boxes has also improved the way elevation is represented in GCMs. Although low-resolution models captured the basic rounded shape of broad high plateaus and ice sheets, they smoothed high but narrow mountain ranges such as the Andes into low-

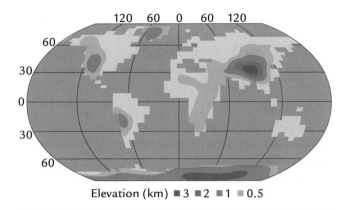

Elevation (km) ■3 ■2 ■1 ■0.5

FIGURE 3-29 **Model resolution** As computers gradually become more powerful, GCMs can be run with smaller grid boxes that give a more detailed portrayal of land-sea distributions and the elevations of mountains and ice sheets. (Adapted from S. Manabe and A. Broccoli, "The Influence of Continental Ice Sheets on the Climate of an Ice Age," *Journal of Geophysical Research* 90 [1985]: 2167–90].)

elevation blobs (Figure 3-29). Higher-resolution models increasingly distinguish these narrower features.

Increasing computer power has also allowed modelers to include more aspects of the climate system in GCMs. In older GCMs, features of the climate system such as soil moisture levels or vegetation types had to be fixed at modern values and were not allowed to interact with the model's atmosphere. In effect, these models were really only *atmospheric* GCMs (A-GCMs), far from full models of the climate system. With increasing computing power, such features are now included as interactive components, often as submodels that interact with the main model.

The modeling process is not a steady one-way march toward success. Initial attempts to include new

components in models are often so crude that they make the resulting climate simulations less realistic than those from models that had simply held these components fixed at modern values. Only with more refined treatments do the newly added components perform in a realistic way and make the resulting simulations clearly superior to the earlier versions.

Ocean GCMs Models of ocean circulation are at a more primitive stage of development than atmospheric GCMs. One reason is that climate researchers know much less about the modern circulation of the oceans, especially critical processes such as the brief but intense episodes of deep-water formation at high latitudes. As a result, scientists do not have as well defined a modern target for ocean models to attempt to reproduce in control-case simulations.

Three-dimensional ocean models (O-GCMs) are similar in construction to atmospheric GCMs (Figure 3-30). Their lower boundary is the seafloor, broken into flat stair steps marking boundaries between individual ocean grid boxes. The upper boundary of the ocean model is the air-sea boundary. The current size of the horizontal grid boxes that subdivide the ocean is 3° to 4° of latitude and longitude. The dozen or so vertical layers in the ocean are more closely spaced near the sea surface, where the flow is faster and interactions with the atmosphere are complex, than at greater depth, where ocean flow is generally slow. Typical climate-data output from ocean GCM experiments includes ocean temperature, salinity, and sea-ice extent.

Like atmospheric models, current ocean GCMs have important limitations imposed by the size of their grid boxes. They cannot capture the shape of very small openings, such as the modern-day mouth of the Mediterranean Sea at the Strait of Gibraltar. These narrow openings are important in the large-scale circulation of the ocean and critical to the success of ocean-

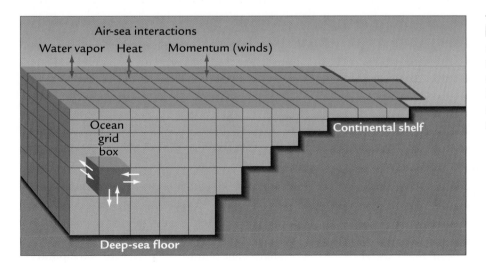

FIGURE 3-30 **Ocean GCMs** Ocean models use 3-D grid boxes that together represent the shapes of ocean basins. Exchanges of water, heat, and momentum between the ocean and the atmosphere occur at the sea surface.

model simulations. Most ocean models also cannot yet resolve details of flow in narrow swift currents such as the Gulf Stream.

Models that include the full structure of the ocean are not run directly coupled to (continuously interacting with) atmospheric models. Air and water respond to climate changes at different rates and consequently put different computing demands on models.

Ocean models can ignore interactions that occur on a daily cycle because these interactions have negligible effects on most of the slower-responding ocean circulation. As a result, O-GCMs need to calculate only changes in ocean circulation at time steps separated by a month or more. In contrast, daily changes are critical to models of the fast-responding atmosphere. Therefore, A-GCMs need to calculate changes in time steps that are just fractions of a day, and the model runs cover much shorter intervals of simulated time for the same amount of computer time and expense.

To overcome this basic incompatibility between the two kinds of models, exchanges between the atmosphere and ocean can be handled in several ways. One method is to drive an ocean model with an overlying atmosphere having properties (temperature, precipitation, and winds) that are kept fixed throughout the experiment. Often climate output data from an initial experiment with an atmospheric GCM are used to drive the ocean model in a second experiment.

Another method for linking ocean and atmosphere models is called *asynchronous coupling*, a procedure that involves an ongoing series of alternating runs, using first the atmosphere to drive the ocean, then the ocean to drive the atmosphere, and so on. The ocean and atmosphere exchange heat (via solar radiation and sensible heat), water and water vapor (via precipitation and evaporation), and momentum (via the winds that drive the surface currents). Going back and forth between the ocean and atmosphere models keeps the two systems from getting too far out of touch with each other, and it allows the overall simulation to progress much faster. When the atmospheric model is run only at selected intervals, the computer has no need to make short-term calculations of the atmospheric circulation throughout the entire simulation.

Ice Sheet Models Continent-sized ice sheets slowly grow and shrink on time scales of tens of thousands of years. Because GCMs reproduce only instantaneous snapshots of climate, they cannot simulate the slow evolution of ice sheets over such long intervals of time, but they can simulate the instantaneous effects that ice sheets have on the rest of the climate system.

Just as modern-day ice sheets are used as boundary-condition input to control-case simulations of modern climate, the estimated shapes of ice sheets that existed in the past can be used as boundary-condition input to simulations of past climates in GCMs. Such simulations are run to isolate the effects these huge, high-albedo masses of ice have on the circulation of the nearby atmosphere and ocean. The output from a GCM run spanning a few years of simulated time can also be examined to see if an ice sheet accumulated or lost mass during the brief simulation. The answer tells modelers whether such an ice sheet would have slowly melted or grown in the climate conditions simulated, or perhaps remained at constant size.

To learn about longer-term evolution of ice sheets, climate scientists use simpler physical ice sheet models. The most widely used ice sheet models have two dimensions, one vertical and the other varying with latitude but not longitude. These 2-D ice sheet models have been used to simulate the growth and decay of ice sheets in the northern hemisphere over tens of thousands of years in response to changes in solar radiation caused by changes in Earth's orbit. The models simulate several features discussed in more detail in Chapter 10, including changes of ice accumulation and melting with ice elevation, flow within the ice, and depression of underlying bedrock by the weight of the ice. The 2-D ice sheet models can also be linked to 2-D atmospheric circulation models to simulate interactions among the ice sheets, atmosphere, and land surface. The most complex ice sheet models are three-dimensional, with the ice accumulating on a specified land surface (such as North America) divided into grid boxes 50 to 100 kilometers on a side.

Vegetation Models Vegetation is important to climate modeling, both as a way to provide an independent assessment of the performance of climate models and in an entirely different sense as an active component within the climate system. The most complete evidence of the distribution of past vegetation comes from the last 20,000 years, because ^{14}C can be used to date pollen-bearing sediments in hundreds of lakes throughout the world. The distribution of pollen compiled at specific intervals of time can be compared against snapshot climate simulations from GCMs based on changes in ice sheets, solar radiation, and atmospheric CO_2 (Part IV). These comparisons provide a challenging test of the performance of GCMs.

The representation of vegetation within climate models has progressed through several stages. Early GCMs either ignored vegetation or included a representation of its physical effects on climate (such as albedo) in such a way that it did not interact with the climate system. More recent models have begun to incorporate the interactive climatic role of vegetation.

One such modeling approach works in two steps. First, climate data output (temperature, precipitation) from a GCM experiment are used as input to a vegetation model that simulates the resulting changes in vegetation. Then the simulated changes in vegetation are used as input to another GCM experiment that simulates the added climatic feedback effects caused by the *changes* in vegetation. Examples that incorporate the effects of the vegetation-temperature and vegetation–precipitation feedbacks discussed in Chapter 2 are found in Chapter 14.

3-7 Geochemical (Mass Balance) Models

Geochemical models are used to follow the mass movements of Earth's materials in physical or chemical form through the climate system. Unlike physical circulation models, most geochemical models do not reproduce the physical processes that govern the flow of air and water. Instead, geochemical models focus on the sources, rates of transfer, and ultimate depositional fate of the materials that are one product of processes active in Earth's climate system.

The geochemical materials examined include sediment particles that result from physical weathering (wind, water, and ice) and dissolved ions produced by chemical weathering (dissolution or hydrolysis). As we saw earlier in this chapter, rivers carry the products produced by both of these weathering processes to the ocean and deposit them in solid form as sediments. Geochemical models can also trace exchanges of biogeochemical materials such as carbon or oxygen isotopes that cycle back and forth among the atmosphere, ocean, ice, and vegetation.

Materials moving through the climate system are called **geochemical tracers**. The movements of conservative tracers (those that are neither created nor destroyed by radioactive decay or other transformations) can be evaluated by **mass balance models**. As we noted earlier, mass movements of geochemical tracers are quantified as flux rates (mass transfers per unit of time).

One-Way Mass Transfer Models The most basic kind of mass balance analysis simply traces permanent one-way transfers of some kind of material from its source or sources to a single site of deposition, such as debris eroded from the land and deposited in ocean sediments. If the material deposited has distinctive geochemical characteristics, it can be analyzed and its abundance quantified in terms of a flux rate—the rate of influx into that sedimentary archive. For example, climate scientists can quantify the rate of influx of ice-rafted debris to high-latitude polar oceans such as the North Atlantic by extracting all sediment that is sand-sized or larger and separating the mineral grains from the plankton shells. This analysis quantifies a climate-related process (the rafting of debris by icebergs).

The analysis can be carried a step further by counting the ice-rafted debris under a microscope to separate it into different grain types (such as volcanic grains, quartz, and limestone) indicative of distinctive source areas on the nearby continents. These analyses attempt to subdivide the total mass of ice-rafted debris into its smaller component fluxes. Further subdivisions can be made by analyzing the chemistry of individual ice-rafted grains for isotopes or other distinctive characteristics. This level of analysis might tell climate scientists which of several ice sheets has sent debris to the ocean during different intervals (see Chapter 15).

A more complicated situation arises if the material being examined is extremely small and has multiple sources. For example, fine silt and clay deposited in the North Atlantic Ocean could have been ice-rafted from North America or Europe, windblown from North Africa, or carried in by deep currents from other sources. Although it is easy to measure the total mass accumulation rate of fine sediment and then calculate its total rate of influx per unit of time, scientists may actually need to separate out the individual fluxes from these several sources.

Physically subdividing such fine material into its distinctive sources is impossible, but chemical analysis may offer an alternative. Subdividing the incoming material is possible if each source of fine sediment is tagged or marked with a distinctive chemical value (Figure 3-31). Typical chemical tags are the relative

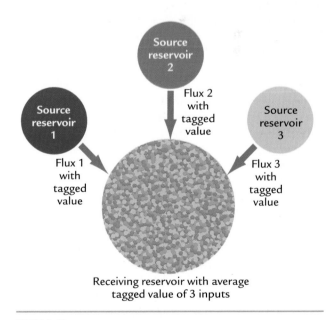

FIGURE 3-31 **One-way mass transfers** Geologists and geochemists often need to distinguish the separate contributions of several sources (usually linked to weathering of continental rocks) to a single depositional archive (such as ocean sediments).

abundances of elements or the ratios of isotopes of a single element. If distinctive chemical tags exist for tracers from each source, the contributions of the individual fluxes may be resolved by mass balance equations of this general form:

$$T_r = \frac{F_1 T_1 + F_2 T_2 + F_3 T_3}{F_1 + F_2 + F_3} \quad \left(\frac{\text{Sum of tagged fluxes}}{\text{Sum of fluxes}}\right)$$

where T_r is the mean value of tagged inputs in entire reservoir.

This equation assumes that the sediment reservoir that receives all three separately tagged fluxes blends them into a single average value typical of the entire fraction of fine sediment but made up of the sum of all the fluxes. When this kind of equation is used, the best-known terms can be used to solve for fluxes that are unknown or poorly known.

Chemical Reservoirs A different approach focuses on geochemical tracers that are delivered by rivers in dissolved form, pass through the ocean reservoir, and are deposited in ocean sediments. This type of mass balance model divides Earth into **reservoirs,** such as the atmosphere, ocean, ice, vegetation, and sediments. Here the ocean is the most important reservoir: it receives almost all erosional products from the continents, it interacts with all other reservoirs, and it deposits tracers in its well-preserved sedimentary archives.

The ocean reservoir is analogous to a bathtub (Figure 3-32). It receives geochemical-tracer inputs in a manner similar to the slow dripping of water from a faucet into a tub, and it loses geochemical-tracer outputs like water leaking slowly through the drain of the tub. In addition, the tracer stays in the ocean for a specific amount of time, the way water does in a drippy, leaky tub.

If the flux rates of a tracer into and out of a particular reservoir (the ocean) are equal, the system is said to be at steady state: no net gain or loss of the tracer occurs in that reservoir. By analogy, if the faucet input and the drain output are perfectly balanced (at steady state), the water level in the tub will stay the same, even while new water enters and leaves the tub.

The **residence time** is the time it takes for a geochemical tracer to pass through the reservoir, equivalent in our analogy to the time for the average molecule of water to pass from the faucet to the drain. For a reservoir at steady state, the residence time can be derived by this equation:

$$\text{Residence time} = \frac{\text{Reservoir size}}{\text{Flux rate in (or out)}}$$

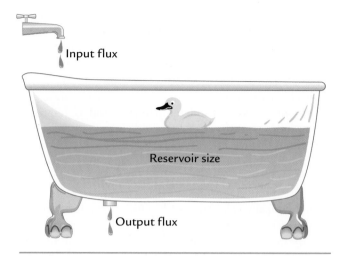

FIGURE 3-32 Geochemical reservoirs and fluxes Geochemical reservoirs are like bathtubs with the faucet and drain both left partly open. The faucet delivers the input flux, the drain takes away the output flux, and the balance between the input and output determines the water level in the tub (reservoir). At steady state, input and output are in balance, and the water level in the tub remains constant.

The reservoir concept is useful in geochemical studies of dissolved tracers with residence times in the ocean far in excess of the time it takes ocean water to mix thoroughly (1500 years). For this case, the distinctively tagged dissolved inputs from several sources are homogenized in ocean water into one average dissolved tracer value typical of the entire ocean. This global value is then recorded directly in the shells of marine organisms later deposited in ocean sediments.

The net result of this homogenization process is that shells from *any* sediment sample from *anywhere* in the ocean dating to a particular interval in time will reveal the chemical-dissolved value of this tracer in the entire ocean at that time. As a result, it is not necessary to measure large masses of sediment to determine the average ocean-reservoir value at a given time; just a few shells of that age will do. Climate scientists can then attempt to isolate the chemically tagged fluxes from specific types of sources, using the kind of mass balance equation provided above. For example, ratios of two isotopes of strontium are used to assess changes in river influxes through time (Chapter 7).

Reservoir-Exchange Models The methods discussed to this point have been based on one-way mass transfers in which geochemical tracers leave the interactive climate system by being buried in seafloor sediments out of touch with other reservoirs for many tens of millions of years. A different approach tracks a single geochemical tracer as it moves back and forth between two (or more) reservoirs (Figure 3-33). As before, this tracer carries a distinctive chemical tag as it moves, but in this case it never comes permanently to rest in either reservoir. The purpose of reservoir-exchange models is to monitor ongoing cycling of tracers between reservoirs through time.

The tracer moving back and forth between reservoirs is tagged with a distinctive value. The major ocean reservoir is much larger than the other reservoirs (ice sheets and vegetation). The tracer exchange is used to monitor changes in size of the smaller reservoir (the volume of ice or the amount of vegetation), but the history of exchanges is usually detected in sediments deposited by the larger (ocean) reservoir. The tracer exchanges cause both the volume and the mean tracer value of the ocean reservoir to change by small amounts that can be detected in the sediment archive and can be analyzed by means of another kind of mass balance equation:

$$\frac{\textit{Smaller reservoir}}{\begin{array}{c}(\Delta\ \text{mass})\ \times\\ (\text{tagged tracer value})\end{array}} = \frac{\textit{Ocean reservoir}}{\begin{array}{c}(\text{new mass})\ \times\\ (\Delta\ \text{mean tracer value})\end{array}}$$

where the Greek letter Δ (delta) means "change in."

One example is the transfer of water between the ocean and ice sheets on orbital time scales (discussed in Chapters 10 and 13). Exchanges of water between the small reservoir stored in fluctuating ice sheets and the much larger reservoir left behind in the ocean can be tracked by using the fact that the isotopic composition of oxygen in H_2O molecules in ice sheets is very different from that of the ocean from which the water comes. As a result, water moving back and forth is tagged with

distinctive oxygen-isotopic values. As water is removed from the ocean to form ice, it leaves seawater with a slightly different oxygen isotope composition. This value is recorded in the calcite shells of foraminifera living in the water. When the ice melts and water returns to the ocean, the average oxygen isotope composition of seawater returns to its previous value. Measurements of the past oxygen isotope composition of the ocean in calcite shells provide a way to estimate the volume of ice stored on land.

Another useful application of reservoir-exchange analysis examines fluxes of carbon among its many reservoirs (discussed in Parts III and IV). Fluxes of carbon between the small reservoir of carbon stored in vegetation on land and the much larger carbon reservoir in the ocean during glacial climate cycles can be tracked by using the fact that terrestrial carbon has a carbon isotope ratio different from that of marine carbon (Chapter 11). Net transfers of terrestrial carbon from land to sea can be detected by examining the average carbon isotope composition of the ocean recorded in the shells of calcite ($CaCO_3$) organisms buried in ocean sediments.

Time-Dependent Models The assumption that reservoirs remain at steady-state conditions may apply for the purpose of modeling the climate system over a brief span of time, but over longer intervals most geochemical transfer systems depart from a steady-state condition. Because climate changes continuously, rates of tracer input and removal also change for the atmosphere, land, ocean, ice, and vegetation, as do the sizes of the reservoirs.

Models that capture changing rates of tracer exchange over time are called **time-dependent models**. These models simulate reservoirs—tubs in our bathtub analogy (Figure 3-32)—in which the water level changes slowly through time because the faucet and drain are not in balance. For this group of models, the residence time of water in the tub is the time it would take for the tub to empty if the input at the faucet were suddenly cut off completely, or, similarly, the time it would take to double the amount of water in the tub if the drain were abruptly closed.

Time-dependent mass balance models are often used in two ways. One method assumes that abrupt changes occur in the rate of flux in or out (or in the size of the main reservoir) and then analyzes how long it takes the system to adjust to this new state and return to equilibrium. A more general use simulates the effects of tracer inputs or outputs that change continuously with time and explores the continuous record of their effects on the ocean reservoir through time.

As scientists attempt to gain progressively greater insights into changes in the climate system, geochemical

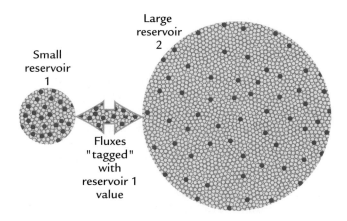

Large reservoir 2

Small reservoir 1

Fluxes "tagged" with reservoir 1 value

FIGURE 3-33 Reservoir exchange models Some geochemical models are designed to track reversible exchanges of important components such as water and carbon as they cycle between smaller reservoirs such as ice sheets and vegetation and the larger ocean reservoir.

models grow more complex in other ways. One example is the progressive subdivision of major reservoirs such as the ocean into smaller subunits, or "boxes." For example, the single ocean box used in the simplest models (Figure 3-33) may be divided into a surface layer that interacts both with the atmosphere and with a deep-ocean box. The surface layer may then be further divided into a warm, well-stratified low-latitude box that interacts slowly with the deep ocean and a cold high-latitude box that interacts more quickly. Further subdivisions might distinguish the Atlantic, Pacific, Indian, and Antarctic oceans and their separate boxes. In addition, while interactions *among* these boxes are generally modeled as simple exchanges of water, the more advanced models may also simulate the slow diffusion of a tracer *within* a large box.

At some point, the more complex geochemical box models may begin to incorporate so much of the actual physical processes at work in the ocean and atmosphere that they begin to resemble the physical simulations of ocean circulation run with the O-GCMs discussed earlier. Time-dependent box models may also make use of radioactive tracers with varying half-lives that change along circulation pathways. Radioactive tracers help climate scientists track ocean circulation patterns over the last few decades, and this information is useful for calibrating and testing ocean box models used to study earlier intervals.

KEY TERMS

continental shelves (p. 55)
continental slopes (p. 55)
moraines (p. 56)
loess (p. 57)
historical archives (p. 59)
instrumental record (p. 59)
radiometric dating (p. 59)
igneous rocks (p. 59)
parent isotope (p. 59)
daughter isotope (p. 59)
closed system (p. 60)
half-life (p. 60)
radiocarbon dating (p. 62)
varves (p. 64)
tree rings (p. 64)
coral bands (p. 64)

climate proxies (p. 66)
biotic proxies (p. 66)
geological-geochemical proxies (p. 66)
macrofossils (p. 67)
plankton (p. 67)
planktic foraminifera (p. 67)
coccoliths (p. 67)
diatoms (p. 67)
radiolaria (p. 67)
burial fluxes (p. 69)
physical weathering (p. 69)
ice-rafted debris (p. 69)
eolian sediments (p. 69)
fluvial sediments (p. 69)

chemical weathering (p. 69)
dissolution (p. 69)
hydrolysis (p. 69)
benthic foraminifera (p. 70)
physical climate models (p. 71)
geochemical models (p. 71)
control case (p. 72)
boundary conditions (p. 72)
climate simulation (p. 72)
climate data output (p. 72)

aerosols (p. 73)
diffusion (p. 73)
general circulation models (GCMs) (p. 75)
grid boxes (p. 75)
sensitivity test (p. 75)
reconstruction (p. 75)
geochemical tracers (p. 79)
mass balance models (p. 79)
reservoirs (p. 80)
residence time (p. 80)
time-dependent models (p. 81)

REVIEW QUESTIONS

1. How does the importance of each type of climate archive change according to the time scale being examined?

2. Why are ocean sediments and ice cores such important archives of climate?

3. How does the method of dating different climate records vary with the type of archive?

4. How does the resolution obtainable from sedimentary archives vary with the environment of deposition? Why?

5. Which two major groups of organisms that exist now and have existed for the past several million years are most important to broad-scale climate reconstructions? Why?

6. Describe how the products of physical and chemical weathering take different pathways before final deposition and provide different kinds of information about the climate system.

7. How are simulations of the past used to test the performance of climate models?

8. How do sensitivity tests run on climate models differ from climate reconstructions?

9. Why aren't models of the atmosphere and ocean coupled continuously to each other as they are run?

10. Describe two factors that make the ocean useful in geochemical mass balance models.

ADDITIONAL RESOURCES

Web Site

http://www.ngdc.noaa.gov/paleo/softlib.html.
Maintained by the World Data Center for
Paleoclimatology in Boulder, Colorado. Contains
climate data of all kinds, as well as the locations of
all sites that contain each type of data.

Advanced Reading

Bradley, R. S. 1998. *Paleoclimatology: Reconstructing
Climates of the Quaternary.* International Geophysics
Series, vol. 64. San Diego: Harcourt Academic
Press.

Hecht, A. D., ed. 1985. *Paleoclimate Analysis and
Modeling.* New York: John Wiley.

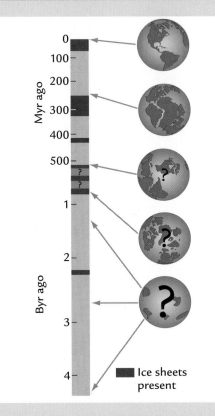

Past glaciations and continental positions. During Earth's 4.55-billion-year history, intervals when large continental ice sheets were present alternated with times when they were not (left). The earliest history of these changes is poorly defined because few ancient records are preserved. The movements of continents in relation to ocean basins are well known only for the last several hundred million years (right). (Globes adapted from D. Merritts et al., *Environmental Geology,* © 1997 by W. H. Freeman and Company.)

PART II

Tectonic-Scale Climate Change

In this part we examine the longest time scale of climate change on Earth, the *tectonic* scale, which encompasses most of Earth's 4.55-Byr history. Tectonic processes driven by Earth's internal heat alter Earth's geography and affect climate over intervals of millions of years. Before the last several hundred million years, the record of climate change is sparse, and even the basic configurations of continents and oceans are unknown. Yet we do know that Earth's climate remained moderate, neither freezing solid nor allowing oceans to boil away.

As the movements and locations of the continents become better known toward the present day, greater insights into cause-and-effect relationships of climate change emerge. We know that Earth's climate has oscillated between times when ice sheets were present somewhere on Earth (such as today) and times when no ice sheets were present. These oscillations over the last several hundred million years are the main focus of Part II.

For most of Earth's tectonic-scale history, evidence of climate change comes from sediments preserved on the continents and their margins. Uncertainties in dating and correlating these records make it difficult to explore sequences of shorter-term climate changes in the distant past. As a result, this part focuses mainly on gradual climate changes over many millions of years.

We explore these basic questions about Earth's tectonic-scale climate history:

- **Why has Earth remained habitable throughout its entire recorded history?**
- **What explains the changes in Earth's climate over the last several hundred million years?**
- **Why was Earth ice-free even at the poles 100 Myr ago?**
- **What are the causes and climatic effects of changes in sea level through time?**
- **How did the apocalyptic asteroid impact 65 Myr ago affect climate?**
- **What caused Earth's climate to cool over the last 55 Myr?**

CHAPTER 4

CO₂ and Long-Term Climate

hy is Earth habitable? With a mean temperature today of 15°C and only a small range of geographic variation around that average, life can flourish almost everywhere on this planet. The obvious first answer to this question is that our Sun is just the right distance from Earth to warm it without heating it too much. In addition, we have already seen in Chapter 2 that greenhouse gases warm Earth's climate by 31°C and contribute in an important way to its habitability.

The mystery deepens when we ask why Earth has remained habitable for most of the 4.55 Byr of its existence. Astronomers believe that over that immense interval our Sun has slowly increased in strength by 25% to 30%, yet somehow Earth's climate has varied only within narrow limits and has always remained habitable. Our planet's continuing habitability seems to require some kind of natural thermostat that allows its climate to warm up during **greenhouse eras** (times when no ice sheets are present) without boiling its oceans and lakes and to cool off during **icehouse eras** (times when ice sheets are present) without ever freezing solid. This chapter focuses on the search for Earth's thermostat.

Greenhouse Worlds

One hint that a factor other than distance to the Sun is involved in Earth's habitability comes from Venus, a "terrestrial" planet with an overall chemical composition similar to Earth's. Venus is a hot planet (its mean surface temperature is 460°C) that lies only 72% as far from the Sun as does Earth, and its closer position to the Sun should make it hotter (Figure 4-1).

Because the amount of solar radiation received by any planet varies *inversely* with the square of its distance from the Sun (d^2), Venus receives almost twice (1.93 times) as much solar radiation as Earth does. This relative amount comes from a calculation based on the relative distances:

$$\begin{array}{c} Earth \\ Venus \end{array} \quad \frac{(1)^2}{(0.72)^2} = \frac{1}{0.518} = 1.93$$

Although distance from the Sun seems a likely explanation for the much greater surface warmth of Venus, in fact this is not the answer. The upper atmosphere of Venus is shrouded in a thick cover of sulfuric acid clouds that reflect 80% of the incoming radiation and allow only 20% to reach the surface of the planet. In contrast, clouds on Earth reflect just 26% of the incoming radiation, allowing the other 74% to reach its surface. The large difference in average albedo (percent of radiation reflected; see Chapter 2) between the atmospheres of these two planets significantly alters the relative amount of solar energy reaching their surfaces. Even though Venus receives 1.93 times as much incoming solar energy, its higher albedo reduces the amount reaching its surface to just over half that of Earth:

$$1.93 \times \frac{0.20}{0.74} = 0.52$$

With just over half as much incoming solar radiation, how can Venus be so hot? The answer is that Venus has an atmosphere 90 times as dense as that of Earth, and 96% of that atmosphere is composed of carbon dioxide (CO_2), an important greenhouse gas. Enough sunlight manages to penetrate the thick atmosphere and reach the surface of Venus to heat it and cause it to emit long-wave radiation, as it does on Earth. But much of this outgoing back radiation is then trapped and retained as heat in the CO_2-rich atmosphere, making Venus much hotter than Earth.

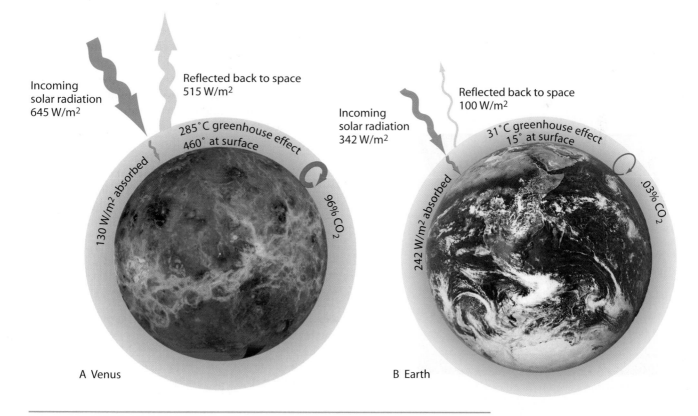

Incoming solar radiation 645 W/m²

Reflected back to space 515 W/m²

285°C greenhouse effect 460° at surface

130 W/m² absorbed

96% CO$_2$

A Venus

Incoming solar radiation 342 W/m²

Reflected back to space 100 W/m²

31°C greenhouse effect 15° at surface

242 W/m² absorbed

.03% CO$_2$

B Earth

FIGURE 4-1 **Why Venus is so hot** Venus (A) receives almost twice as much solar radiation as Earth (B), but its dense cloud cover permits less total radiation to penetrate to the surface. Yet Venus is much hotter than Earth because its CO$_2$-rich atmosphere creates a stronger greenhouse effect and traps much more heat. (NASA photos.)

In addition to being greenhouse planets, both Venus and Earth contain nearly equal amounts of carbon, if we total up the carbon residing in all the reservoirs on both planets. These similarities are not completely surprising, because both had a common origin in our solar system as rocky planets with similar compositions.

Yet the two planets differ in where they store their carbon. Most of Earth's carbon is tied up in its rocks, including its resources of coal, oil, and natural gas; much less carbon resides in Earth's atmosphere. The net greenhouse heating from atmospheric CO_2 on Earth is relatively small. The net effect, amplified by changes in water vapor—the major greenhouse gas—is 31°C (Chapter 2). In contrast, almost all the carbon on Venus resides in its atmosphere as CO_2 and produces an enormous net greenhouse warming (285°C) despite the near absence of water vapor as a greenhouse gas.

This comparison shows how vital greenhouse gases can be to the climate of planets. It also highlights the fact that Earth's comfortably small greenhouse effect is an important factor in its present-day habitability, in combination with its favorable distance from the Sun.

The Faint Young Sun Paradox

By studying the evolution of stars in the universe, astronomers can recreate the history of our own Sun over the 4.55 Byr existence of our solar system. Throughout this interval, the Sun's interior has been the site of an ongoing nuclear reaction that fuses nuclei

of hydrogen (H) together to form helium (He). Virtually every model developed by astronomers indicates that this nuclear reaction process has caused our Sun to expand and gradually become brighter. These models indicate that the earliest Sun shone 25% to 30% more faintly than today, and that its luminosity, or brightness, has slowly increased to its current strength. Stars similar to our Sun gradually grow larger through time and eventually burn out after many billions of years.

This insight from the field of astronomy creates an intriguing problem for climate scientists. A decrease of just a few percentage points in our Sun's present strength would cause all the water on Earth to freeze, despite the warming effect of our present-day greenhouse gases. If all our oceans and lakes were to freeze, the positive feedback caused by their high-albedo snow and ice surfaces reflecting solar radiation would make them difficult to melt. One-dimensional numerical climate models that simulate the mean climate of an entire planet (Chapter 3) suggest that an early Earth with so weak a Sun and with greenhouse gas levels at their present-day values would have remained completely frozen for the first 3 billion years of its existence (Figure 4-2).

This combined insight from astronomy and climate models collides with the fact that Earth's climate record shows that our planet has never frozen completely. Although the first half-billion years of Earth's existence left an incredibly sparse record, evidence of Earth's history becomes gradually more complete toward the present. Sedimentary rocks are a prominent part of this record, and most sedimentary rocks are made up of particles eroded from other rocks, reworked by running water, and eventually transported to their site of deposition. The prevalence of this evidence of running water early in Earth's history means it was not frozen solid.

The first evidence of ice-deposited sediments occurs in rocks dated to about 2.3 Byr ago, but these deposits were probably the result of glaciations localized in polar regions, similar to those on Earth today, and not an indication of a completely frozen planet. A debate is currently under way as to whether Earth reached a nearly frozen condition between 850 and 550 Myr ago (Box 4-1), but for most of Earth's history the sedimentary evidence leaves no doubt that most of the water on Earth remained unfrozen.

This conclusion is also supported by the continued presence of life on Earth. Primitive life-forms date back to at least 3.5 Byr ago, and their presence on Earth is incompatible with a completely frozen planet at that time. The succession of ever more complex life-forms that have continuously occupied Earth ever since is further proof against extreme cold (or heat).

So we are confronted with a mystery: With so weak a Sun, why wasn't Earth frozen for the first two-thirds

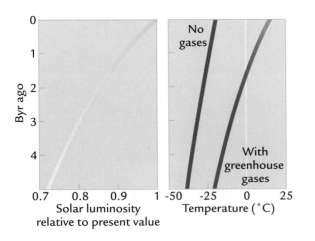

FIGURE 4-2 **The faint young Sun paradox** Astrophysical models of the Sun's evolution indicate it was 25% to 30% weaker early in Earth's history (left). Climate models show this situation would have produced a completely frozen Earth for well over half its early history if the atmosphere had the same composition as it has today (right).
(Adapted from D. Merritts et al., *Environmental Geology*, ©1997 by W. H. Freeman and Company.)

BOX 4-1 CLIMATE DEBATE

A Snowball Earth?

For obvious reasons, the occurrence of ice sheets at high latitudes does not mean that Earth must be completely frozen. Ice sheets occur at high latitudes today, yet the tropics remain hot because of the strong overhead Sun and the large amount of solar heat absorbed and redistributed by the tropical oceans. With the large pole-to-equator gradient in temperature that exists on Earth today, polar ice sheets can coexist with tropical heat.

Any continent-sized ice sheet in the tropics would indicate that all of Earth must be near the freezing point. The lower margins of continental ice sheets lie at or near sea level and are susceptible to melting. As a result, temperatures at sea level would have to be near or below freezing through most of the year for ice sheets to persist in the tropics. In effect, today's polar climates would have to invade the tropics to permit ice sheets to exist there.

Some climate scientists have suggested that Earth came very close to freezing totally during at least one interval in its long history. Rock types typical of glacial deposition found on several continents provide evidence that large ice sheets were present during the prolonged icehouse interval between 850 and 550 Myr ago. Most of the evidence comes from sedimentary rocks containing mixtures of coarse boulders and cobbles along with fine silts and clays. Because these ancient deposits

are difficult to date and correlate accurately, some scientists have inferred as few as two icehouse episodes and others as many as four within the interval between 850 and 550 Myr ago.

If we assume that these were episodes of glaciation, the key question is whether the ice occurred on continents located near the poles or at lower latitudes. Here scientists disagree. It becomes increasingly difficult further back in time to determine the latitudes at which ancient continents were located. Some evidence suggests that these glaciated continents were situated in the tropics, and this supposition has led to a hypothesis that Earth was nearly frozen at that time—the **snowball Earth hypothesis.**

Other evidence contradicts this interpretation and places these continents at higher latitudes, outside the tropics. If that was the case, Earth need not have approached a frozen condition. Just 20,000 years ago, ice sheets reached as far south as 40° latitude over North America, more than halfway to the equator, yet temperatures in the tropics remained within a few degrees of the warmth found there today. The alternative interpretation is that these older glaciations between 850 and 550 Myr ago occurred at times when the glaciated continents were outside the tropics, so that Earth need never have been close to a frozen state. The debate about a possible snowball Earth continues.

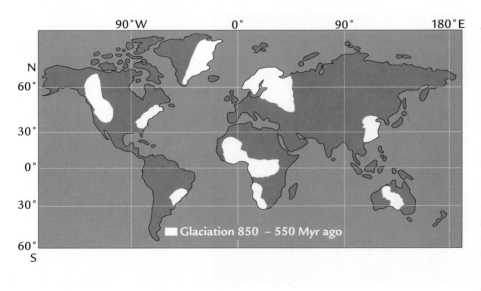

A snowball Earth? Evidence of several glaciations between 850 and 550 Myr ago is found on Earth's modern-day continents. If these glaciated regions were located at or near the polar regions, climate may have been little different from what it is today. But if they were located in the tropics, a snowball Earth may have existed. (Adapted from L. A. Frakes, *Climates Through Geologic Time* [Amsterdam: Elsevier, 1979], and from J. G. Meert and R. van der Voo, "Neoproterozoic (1000–540 Myr) Glacial Intervals: No More Snowball Earth," *Earth and Planetary Science Letters* 123 [1994]: 1–13.)

of its history? This mystery is known as the **faint young Sun paradox.** Even the possibility of a snowball Earth interval between 850 and 550 Myr ago only underscores this mystery. If Earth actually was almost frozen at a time when the Sun is estimated to have been only 5% weaker than it is today, then why would it not have been completely frozen for the preceding several billion years, an interval when the astronomical models indicate a far weaker Sun?

The most basic answer to the faint young Sun paradox is obvious: something must have kept the early Earth warm enough to offset the Sun's weakness. But this easy answer only confronts us with a more difficult problem: the process that warmed Earth early in its history, whatever it was, must no longer be doing so today, or at least must not be working as actively as it once did. If this same warming process had continued working at full strength right through the entire 4.55 Byr of Earth's history, it

would have combined with the steadily increasing warmth from the strengthening Sun (Figure 4-2) to overheat Earth and probably make it uninhabitable. That has not happened: somehow Earth has stayed within a moderate temperature range throughout the entire interval when the Sun's brightness was increasing.

The solution to the faint young Sun paradox appears to require a process that works the same way a **thermostat** (temperature regulator) works in a house. When outside temperatures fall in winter, the thermostat detects the cooling and turns on a heat source that keeps the house warm. When temperatures become too hot outside in summer, the thermostat activates a cooling source that keeps the house cool. The thermostat may allow the house temperatures to become a little warmer during the summer and a little cooler during the winter, but the interior does not experience the temperature extremes reached on the outside.

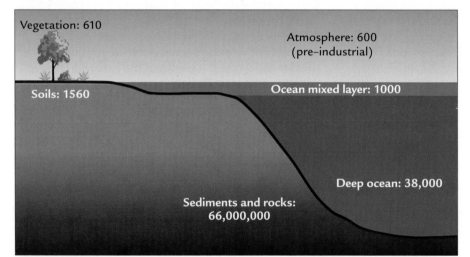

A Major carbon reservoirs (gigatons; 1 gigaton = 10^{15} grams)

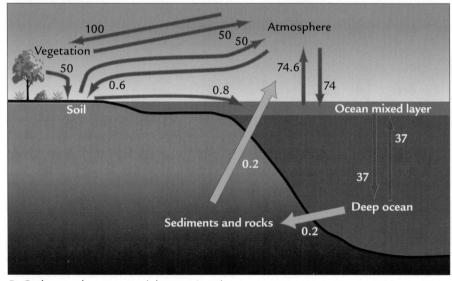

B Carbon exchange rates (gigatons/year)

FIGURE 4-3 Carbon exchanges with Earth's rocks (A) The largest reservoir of carbon on Earth lies in its rocks. (B) Over intervals of millions of years, slow exchanges among the rock and ocean/vegetation/soil/atmosphere reservoirs can cause large changes in atmospheric CO_2 levels. (Adapted from J. Horel and J. Geisler, *Global Environmental Change* [New York: John Wiley, 1997], and from National Research Council Board on Atmospheric Sciences and Climate, *Changing Climate,* Report of the Carbon Dioxide Assessment Committee [Washington, D.C.: National Academy Press, 1993].)

It seems as if such a thermostat may have been at work through Earth's history, warming it very early on when it would otherwise have frozen under a weak Sun, and later on detecting the increasing warmth from the strengthening Sun and cutting back on the heat provided.

One intriguing possibility is that greenhouse gases have been part of the mechanism that acts as Earth's thermostat. Our modern-day concentrations of greenhouse gases are not high enough to have counteracted the effects of a weak early Sun, but what if these gases were more abundant earlier in Earth's history and subsequently decreased in abundance?

This possible explanation seems more credible if we return to our comparison of Earth and Venus. These two planets have similar total amounts of carbon, yet they store it in entirely different reservoirs (in rocks on Earth, in the atmosphere on Venus). If the bulk of the carbon can reside in different reservoirs on different planets, why couldn't it move among reservoirs during the history of a single planet? More specifically, could the early Earth have held more carbon in its atmosphere (like Venus), and then transferred it to its rocks later in its history?

Carbon Exchanges between Rocks and the Atmosphere

To understand how carbon may have shifted among reservoirs, we need to take a closer look at the present-day carbon cycle on Earth. As we learned in Chapter 2, carbon on Earth today resides in many reservoirs (Figure 4-3A). Small amounts exist in the atmosphere, the surface ocean, and vegetation, a slightly larger reservoir in soils, a much larger reservoir in the deep ocean, and an immensely larger reservoir in rocks and sediments.

The rates of carbon exchange among these reservoirs vary widely (Figure 4-3B). In general, an inverse relationship exists between the size of a given reservoir and the rate at which it exchanges carbon. The small reservoirs (the atmosphere, surface ocean, and vegetation) all exchange carbon quickly, while the huge rock reservoir exchanges its carbon more slowly. As a result of the combined effects of reservoir size and exchange rate, carbon cycles quickly through the smaller reservoirs at the surface and more slowly through the deeper reservoir.

Because all reservoirs exchange carbon with the atmosphere, each has the potential to alter atmospheric CO$_2$ concentrations and affect Earth's climate. The relative importance of each carbon reservoir in Earth's climate history varies according to the time scale under consideration. We are concerned here with gradual climate changes over tectonic time scales, during which the effects of slow carbon exchanges between the rocks and the surface reservoirs persist over tens of millions of years. Over such long intervals, even these slow exchanges can produce large cumulative changes in the amount of CO$_2$ in the atmosphere.

4-1 Volcanic Input of Carbon from Rocks to the Atmosphere

Carbon cycles constantly between Earth's interior and its surface. It moves from the deep rock reservoir to the surface mainly as CO$_2$ gas produced during volcanic eruptions and the activity of hot springs (Figure 4-4). We will see shortly that CO$_2$ is also released to the atmosphere by oxidation of organic carbon in sedimentary rocks.

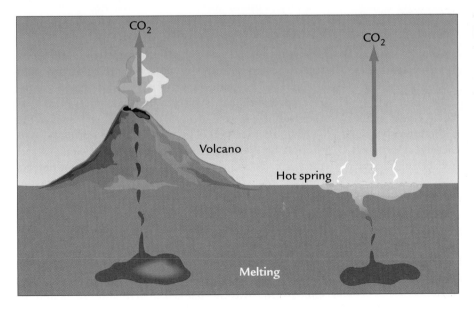

FIGURE 4-4 Input of CO$_2$ from volcanoes CO$_2$ enters Earth's atmosphere from deep in its interior through release of gases in volcanoes and at hot springs such as those found today at Yellowstone National Park in Wyoming.

The present-day rate of carbon input to the atmosphere from the rock reservoir is estimated at approximately 0.15 gigaton (or 0.15×10^{15} grams) of carbon per year (Figure 4-3). This value is probably known at best to within a factor of about 2, because volcanic explosions are irregular in time and because the amount of CO_2 released varies with each eruption. As we will see later, this rate of carbon input is roughly balanced by a similar rate of natural carbon removal (or at least a balance existed before the start of industrialization 250 years ago). This balance between natural carbon input and removal rates helped to keep the size of the "natural" (preindustrial) atmospheric carbon reservoir at 600 gigatons.

But how likely is it that this balance could have persisted over immensely long intervals of geologic time? We can evaluate this question by a simple thought experiment. Using the reservoir concept introduced in Chapter 3, we can calculate how long it would take for the atmospheric CO_2 level to fall to zero if all volcanic release of carbon from Earth's interior to the atmosphere abruptly ceased, but if carbon continued to be removed from the atmosphere at the same rate as before. We derive the answer by dividing the preindustrial atmospheric carbon reservoir size of 600 gigatons by the annual rate of carbon addition of 0.15 gigaton.

The answer is 4000 years. This number, although far longer than a human lifetime, is remarkably brief in the context of the several billion years of Earth's existence. It implies that the CO_2 content of our atmosphere could be highly vulnerable to changes in average rate of volcanic input that persist over that length of time.

In actuality, the atmosphere is not really this vulnerable because other rapid exchanges of carbon occur continuously between the atmospheric reservoir and several other reservoirs. These rapid exchanges have the effect of slowing and reducing the impact of the loss of carbon from Earth's interior.

In our hypothetical example of a sudden cessation of volcanic CO_2 input to the atmosphere, the actual scenario would probably develop more like this: As CO_2 levels in the atmosphere begin to fall, the other surface reservoirs (vegetation, surface ocean, soils) would begin to surrender some of their carbon to the atmosphere, slowing its rate of loss. For fast-reacting reservoirs, changes in one are soon felt by the others because of the rapid (and ongoing) exchange rates.

The combined size of all the near-surface reservoirs (atmosphere, vegetation, soil, and surface ocean) is 3700 gigatons, more than six times larger than the atmospheric reservoir alone. It would take roughly 24,700 years after volcanism ceased for these reservoirs to lose all their carbon (3700 gigatons divided by 0.15 gigaton/yr).

In addition, the large deep-ocean carbon reservoir would get into the act. If the surface reservoirs were all losing significant amounts of carbon, the deep ocean would feed carbon to the surface ocean at rates rapid enough to restore some of the loss, and the surface ocean would redistribute this carbon to the other surface reservoirs, including the atmosphere. If we take the large deep-ocean reservoir into account, the total size of these reservoirs amounts to 41,700 gigatons. It would take 278,000 years for a total shutdown of volcanic carbon input to deplete these combined reservoirs completely (41,700 gigatons divided by 0.15 gigaton/yr).

At this point we might be tempted to conclude that Earth's surface reservoirs, including its atmosphere, are *not* particularly vulnerable to changes in the amount of carbon coming out of (or going into) its rocks. But this conclusion would be incorrect. Even a time span as long as 278,000 years represents less than one–ten thousandth (0.01%) of Earth's 4.55-Byr age. Because Earth is so old, plenty of time is still available for the slow carbon exchanges with Earth's rock reservoirs to alter the amount of carbon in the surface reservoirs by large amounts. When we take Earth's great antiquity into account, it is still amazing that over this immense span of time Earth's volcanoes have somehow managed to keep delivering just enough carbon from Earth's interior to keep the atmosphere from running out of CO_2, but not so much as to overheat the planet. This achievement requires a delicate balance.

Even more amazing is the fact that this balancing act had to be maintained at the same time that the faint young Sun was slowly increasing in strength. A crude analogy for this long-term balancing act is that of a tightrope walker who has to keep his balance as he walks on a wire that is not only extremely narrow but also slopes uphill over a very long distance.

We noted earlier that this balancing act requires some kind of natural thermostat to moderate Earth's temperature, and that this thermostat could reside in the carbon system. An obvious question arises: Could the rate of volcanic input of CO_2 from Earth's interior have varied in such a way as to function as that thermostat?

The answer is no. The basic operating principle of a thermostat is that it first *reacts* to external changes and then *acts* to moderate their effects: a thermostat detects the chill of a cold night and sends a signal that turns on the furnace. Volcanic processes are not thought to operate in this way.

The volcanic activity that has occurred on Earth throughout its history has been driven mainly by heat sources located deep in its interior and generally far removed from contact with the climate system. Climatically driven changes in temperature penetrate only the outermost few meters or at most tens of meters of the land (or seafloor). As a result, climate changes confined to Earth's surface have no physical means of altering deep-seated processes in Earth's interior in such a way as to produce compensating thermostat-like changes in volcanic activity and CO_2 delivery to the surface.

Earth's thermostat lies elsewhere. It must be found in a process that responds directly to the climate conditions at Earth's surface.

4-2 Removal of CO₂ from the Atmosphere by Chemical Weathering

To avoid long-term buildup of CO_2 levels over time, CO_2 input to the atmosphere by volcanoes has to be countered by CO_2 removal. The major long-term process of CO_2 removal is tied to chemical weathering of continental rocks. We learned in Chapter 3 that two major types of chemical weathering occur on continents: *hydrolysis* and *dissolution*.

Hydrolysis Hydrolysis is the main mechanism for removing CO_2 from the atmosphere. The three key ingredients in the process of hydrolysis are minerals that make up typical continental rocks, water derived from rain, and CO_2 derived from the atmosphere (Figure 4-5).

Most of the continental crust consists of rocks such as granite, made of **silicate minerals** such as quartz (beach sand) and feldspar. Silicate minerals typically are made up of positively charged cations (Na^+, K^+, Fe^{+2}, Mg^{+2}, Al^{+3}, and Ca^{+2}) that are chemically bonded to negatively charged SiO_4 (silicate) structures. These sili-

cate minerals are slowly attacked by groundwater containing carbonic acid (H_2CO_3) formed when atmospheric CO_2 is combined with rainwater.

Part of the weathered rock is chemically converted to clay minerals (mainly compounds of Si, Al, O, and H) and left lying in soils on the continents. Chemical weathering also produces several types of dissolved ions and other complexes, including HCO_3^-, CO_3^{-2}, H_2SiO_4, and H^+. These ions are carried by rivers to the ocean, and some are incorporated in the shells of planktic organisms by biologic processes (Figure 4-5).

Dozens of chemical equations describe the process of chemical weathering—in fact, there is one equation for each of the many types of silicate minerals found on continents. The part of these processes that is most important to the carbon system can be represented by these reactions:

$$H_2O \quad + \quad CO_2$$

Rain   From atmosphere

↘   ↙

$$CaSiO_3 + H_2CO_3 \rightarrow CaCO_3 + SiO_2 + H_2O$$

Silicate rock   Carbonic acid   Shells of organisms
(continents)   (soil)

For simplicity, all the complexities of continental rocks are represented by just one of the many silicate minerals, $CaSiO_3$ (wollastonite). Carbon dioxide is taken from the atmosphere, incorporated in groundwater in soils to form carbonic acid, used in the chemical weathering reaction, and eventually deposited in the $CaCO_3$ shells of marine organisms. This reaction is a convenient shorthand summary for the way chemical weathering removes CO_2 from the atmosphere and

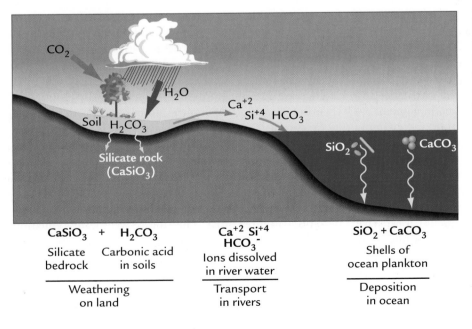

CaSiO₃ + H₂CO₃
Silicate   Carbonic acid
bedrock   in soils
_____
Weathering
on land

Ca⁺² Si⁺⁴
HCO₃⁻
Ions dissolved
in river water
_____
Transport
in rivers

SiO₂ + CaCO₃
Shells of
ocean plankton
_____
Deposition
in ocean

FIGURE 4-5 **Chemical weathering removes atmospheric CO₂** Chemical weathering of rocks on the continents removes CO_2 from the atmosphere, and the carbon is eventually stored in the shells of marine plankton.

buries it in ocean sediments, which in time turn into rocks. This process acts slowly but persistently over long intervals of geologic time and accounts for 80% of the 0.15 gigaton of carbon buried each year in ocean sediments and ultimately in rocks.

Dissolution It is important to distinguish weathering of silicates by hydrolysis from a second common type of weathering, dissolution. Dissolution is the familiar process that eats away at limestone bedrock and in some areas forms limestone caves. Rainwater and CO_2 again combine in soils to form carbonic acid (H_2CO_3) and attack limestone bedrock, and the dissolved ions created by dissolution again flow to the ocean in rivers. Dissolution can also be summarized by two simple reactions:

$$H_2O \quad + \quad CO_2$$
Rain   From atmosphere

$$CaCO_3 + H_2CO_3 \rightarrow CaCO_3 + H_2O + CO_2$$

Limestone   In soils   Shells of   Returned to
rock   organisms   atmosphere

Dissolution of limestone proceeds at rates averaging an order of magnitude faster than hydrolysis of silicates. Like the process of hydrolysis, dissolution extracts CO_2 from the atmosphere to attack rock. But unlike weathering of silicate rocks, limestone weathering returns all of the CO_2 to the atmosphere within the relatively short interval of time it takes for the dissolved HCO_3^- and CO_3^{-2} ions to reach the sea and become incorporated in the shells of organisms. As a result, no net removal of atmospheric CO_2 occurs during the overall process.

A CO_2 Balance In summary, slow weathering of granite and other silicate rocks on the continents by hydrolysis is the main way that CO_2 is pulled out of the atmosphere over very long time scales. In the context of Earth's delicate long-term balancing act, the rate of removal of carbon by chemical weathering must have very nearly balanced the rate of carbon input from volcanoes. If these rates had not been equal or very nearly equal, the system would have gotten out of balance and caused drastic changes in CO_2 levels and in climate.

The existence of this delicate balance does not imply that either the (volcanic) CO_2 input rate or (weathering) CO_2 removal rate remained constant through time. We will see in Chapter 5 that these two rates have almost certainly varied because of natural changes in tectonic processes on Earth. Yet Earth's long-term habitability requires that the rates of input and output must have always remained closely balanced even while they changed.

How has this near-perfect balance been possible? As we noted earlier, a thermostat can provide such a bal-

ance. In our search for Earth's thermostat within its carbon system, we have ruled out volcanic input of CO_2. The only other possibility is chemical weathering. If rates of chemical weathering are sensitive to climate, they may be able to act as Earth's thermostat.

Climate Factors That Control Chemical Weathering

Decades of laboratory experiments and many field studies show that rates of chemical weathering are influenced by three environmental factors. Temperature, precipitation, and vegetation all act in a mutually reinforcing way to affect chemical weathering.

Laboratory experiments show that higher temperatures cause more rapid weathering of individual silicate minerals. This kind of trend is consistent with other temperature-dependent chemical reactions in water or other aqueous solutions. These controlled experiments indicate that weathering rates roughly double for each 10°C increase in temperature.

Unfortunately, it is difficult to transfer these laboratory results to studies of the real Earth. So far, these experiments have examined only a few of the many silicate minerals that are common enough in Earth's crust to be important contributors to the overall rate of silicate weathering on a global scale. Natural chemical weathering rates are also difficult to determine in field studies because of the complicating effects from rapid carbonate dissolution. Because dissolution occurs many times faster than hydrolysis, the total amount of ions flowing down rivers can easily be dominated by ions derived from limestone dissolution, which does not control CO_2 levels in Earth's atmosphere, rather than from hydrolysis of silicates, which does control long-term CO_2 levels. Another problem with studying the real world is that humans have disturbed the natural chemistry of most of Earth's rivers by agricultural and industrial activities.

Still, we can apply the laboratory rule of thumb that says that silicate weathering rates double for each 10°C increase in temperature to the roughly 30°C range of mean annual temperatures found on Earth's surface (Figure 4-6A). It appears that silicate weathering rates should vary by a factor of at least 8 regionally from the equator to the poles as a result of temperature alone, with larger changes during extreme seasons on particular continents.

The second major control on weathering is precipitation (Figure 4-6B). The silicate weathering equation indicates that rainwater is central to the process of hydrolysis. Increased rainfall raises the level of groundwater held in soils, and combines with CO_2 to form carbonic acid and drive the weathering process.

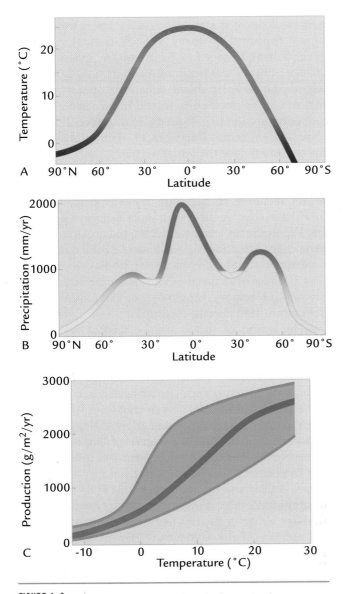

FIGURE 4-6 **Climate controls on chemical weathering**
Temperature (A) and precipitation (B) both show a
general trend from high values in warmer low latitudes to
low values in colder high latitudes. The total amount of
vegetation produced per year increases with temperature
(C), as well as with precipitation. (A and B: adapted from
R. G. Barry and R. J. Chorley, *Atmosphere, Weather, and Climate*, 4th
ed. [New York: Methuen, 1982]; C: adapted from R. L. Smith and
T. M. Smith, *Elements of Ecology* [Menlo Park: Addison Wesley
Longman, 1998].)

The effects of temperature and precipitation are so
closely linked in Earth's climate system that it is difficult
to measure their separate contributions to chemical
weathering. We learned in Chapter 2 that the amount
of water vapor that air can hold rises with temperature,
roughly doubling for a temperature increase of 10°C.
The heaviest rainfall on Earth occurs in the tropics

because warm tropical air holds more moisture than
cooler high-latitude air (Figure 4-6B). Polar regions
have much less precipitation because the atmosphere
holds so little water.

This relationship breaks down to some extent at
regional scales. One major complication is that warm
air obtains its large moisture content through evaporation, and extreme evaporation can dry soils and retard
silicate weathering. Water may evaporate in one region
but then fall as rain in another, creating regional wet-
dry patterns. As a result, temperature and precipitation
can change in opposite ways at smaller scales, with
opposing effects on chemical weathering. For example,
extreme minima in precipitation in some subtropical
regions greatly reduce chemical weathering, even
though relatively warm temperatures otherwise favor it.

Despite these smaller-scale complications, temperature and precipitation generally act together (Figure
4-6). A warmer Earth is likely to be a wetter Earth, and
both factors tend to act together to intensify chemical
weathering.

Vegetation also enhances chemical weathering.
Plants extract CO$_2$ from the atmosphere through
photosynthesis (Chapter 2) and deliver it to soils,
where it combines with groundwater to form carbonic acid. Although H$_2$CO$_3$ is a weak acid, it enhances
the rate of chemical breakdown of minerals. And
while firm numbers are difficult to determine from
experiments, scientists estimate that the presence of
vegetation on land can increase the rate of chemical
weathering by a factor of 2 to 10 over the rate on land
that lacks vegetation.

Vegetation is closely linked to precipitation and
temperature. We saw in Chapter 2 that the distribution
of vegetation on Earth's continents is closely linked to
precipitation (Figure 2-34). Dense rain forests are
found in regions with year-round rainfall, open forest
or savannas in areas with a short dry season, grasslands
in places with a long dry season, and deserts in areas
with little or no rainfall. Each step in the direction
of greater rainfall is a step toward more vegetation
and more total carbon biomass stored in vegetation and
soils.

In addition, the rate of photosynthetic production
of carbon across the planet is correlated with temperature, although with considerable scatter (Figure 4-6C).
Cold ice-covered regions produce little plant matter,
and seasonally or permanently frozen (but ice-free)
polar regions produce only sparse covers of tundra vegetation. Such regions store large amounts of carbon
belowground, but almost none of this deeper carbon
contributes to the annual cycle of productivity at the
surface. Production of carbon in warmer mid-latitude
and tropical regions is much greater (Figure 4-6C).

Chemical Weathering: Earth's Thermostat?

Now we have in hand the key components of a mechanism that can act as Earth's thermostat and regulate its long-term climate. The essence of this mechanism lies in two facts: the average global rate of chemical weathering depends on the state of Earth's climate, but weathering also has the capacity to alter that state by regulating the rate at which CO_2 is removed from the atmosphere.

The weathering thermostat works as a negative feedback that moderates long-term climate change. Consider what would happen if Earth's climate began to warm toward a greenhouse state (Figure 4-7A). Greenhouse climates imply a warm, moist, lushly vegetated Earth, the kind of world that climate models sug-gest would exist if atmospheric CO_2 levels were much higher than they are today. We will examine one such era in Earth's history in Chapter 6. Any initial climate change toward a warmer, moister, more heavily vegetated greenhouse Earth should enhance chemical weathering, but faster weathering should in turn speed up the rate of removal of CO_2 from the atmosphere. The result should be an increased rate of CO_2 removal that should oppose and reduce the initial warming.

The opposite sequence should happen if Earth's climate began to cool toward icehouse conditions (Figure 4-7B). Icehouse climates are typically cold, dry, and more sparsely vegetated, with more extensive snow and ice. We live in an icehouse climate today. An initial climate change (for any reason) toward a colder, drier, less vegetated Earth should reduce chemical weathering and slow the rate of removal of CO_2 from the atmosphere. Slower CO_2 removal should oppose and reduce the initial push toward climate cooling.

The action of these negative feedbacks does not mean that no climate change at all occurs. Any process that initially acts to warm Earth succeeds in doing so, but by an amount smaller than would have occurred without the action of the negative feedback. Conversely, any process that initially acts to cool Earth succeeds in doing so, but also to a reduced degree. Negative feedbacks simply moderate the degree of climate change. The existence of a climate-dependent negative feedback due to chemical weathering was proposed in 1981 by James Walker and his colleagues Paul Hays and James Kastings.

Now we can apply this negative feedback process to the mystery of the faint young Sun paradox and the Sun's evolution during Earth's history. Recall that we needed a global thermostat that could have made Earth warmer earlier in its history to counter the weakness of the early Sun, but that later throttled back on its warming effects as the strengthening Sun led to greater heating of Earth.

Earth's environment early in its history is poorly known, but it is widely thought to have included much more active volcanism that caused greater loss of volatile gases (including CO_2) from its interior. Many scientists believe that Earth's surface may even have been entirely molten for a few hundred million years after 4.55 Byr ago. In addition, ancient craters preserved on our moon and on other planets indicate that Earth was once under heavy bombardment by asteroids, meteors, and comets, and these collisions may have triggered greater volcanism as well. Radioactive elements deep in Earth's interior also released more heat that could have increased the amount of volcanism. Increased volcanic activity would have delivered more CO_2 to the atmosphere and may have helped to make Earth hot.

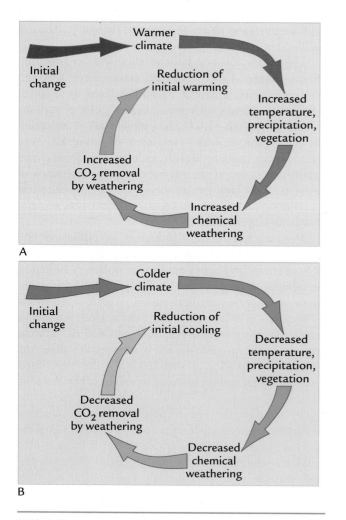

FIGURE 4-7 Negative feedback from chemical weathering Chemical weathering can act as a negative climate feedback that reduces the intensity of either an imposed climate warming (A) or cooling (B).

The interval beginning after 4 Byr ago is the time of the faint young Sun paradox. By no later than 4 Byr ago, evidence preserved in Earth's oldest rocks tells us, land surfaces existed across which water ran, eroding and depositing sediments we can sample and analyze today. Earth was not frozen solid, despite the weak Sun. By 3.5 Byr ago, we see the first indications of primitive life, another indication that Earth's climate was not extremely cold by this time. We have two basic choices to explain the fact that this early Earth was unfrozen.

One possibility is that volcanism was much higher early in Earth's history and produced enough CO$_2$ to counter the weak Sun, and then slowed down at exactly the right rate needed to add less CO$_2$ to the atmosphere as the Sun strengthened. The evidence for high rates of volcanism very early in Earth's history is consistent with this explanation, but it would be an amazing coincidence if the slowing of volcanism over the next 4 Byr or more of Earth's existence were paced at *exactly* the rate needed to offset the effects of the strengthening Sun.

The more plausible choice calls on changes in chemical weathering through the negative feedback just described. The weakness of the young Sun must have tended to make the early Earth cooler than it is today. But the rate of CO$_2$ removal by weathering was also probably much slower as well because of Earth's cooler temperature. In addition, early continents are thought to have covered a smaller area than they do today, and their compositions were probably less rich in granite-like silicate rocks. Slower CO$_2$ removal by weathering would also have been favored by the smaller area of continents (providing less rock surface to weather).

Slower rates of weathering would have left more CO$_2$ in the atmosphere over much of Earth's early history, perhaps 100 to 1000 times as much as today

(Figure 4-8A). The warmth produced by this high-CO$_2$ atmosphere could have countered most of the cooling caused by the reduced amount of incoming solar radiation.

Then, as Earth began to receive more solar radiation from the brightening Sun, its surface warmed and rates of chemical weathering gradually increased. Faster chemical weathering began to draw more CO$_2$ out of the atmosphere than previously, and the resulting drop in atmospheric CO$_2$ levels provided a cooling effect that counteracted the gradual increase in solar warming and kept Earth's temperatures moderate (Figure 4-8B). The centerpiece of this explanation is that the slow warming of Earth by the strengthening Sun would have directly caused the changes in weathering that moderated the final change in climate.

If chemical weathering is Earth's thermostat, we face still another question: What happened to all that CO$_2$ that once resided in the atmosphere and warmed Earth? The most likely answer is found by looking at the carbon reservoirs in Figure 4-3: the carbon removed from the atmosphere by weathering was gradually buried in sediments that turned into rocks.

In today's world, CO$_2$ removed from the atmosphere by weathering is deposited in ocean sediments that eventually become rocks (Figure 4-5). The same process would also have worked in the past, but over time it would have caused a slow but large net transfer from the atmosphere to the rocks. If this interpretation is correct, most of Earth's early greenhouse atmosphere lies buried in its rocks instead of concentrated in the atmosphere, as on Venus.

Some scientists have suggested that methane (CH$_4$) and ammonia (NH$_3$) outgassed from Earth's interior at a greater rate in the past also warmed the early Earth,

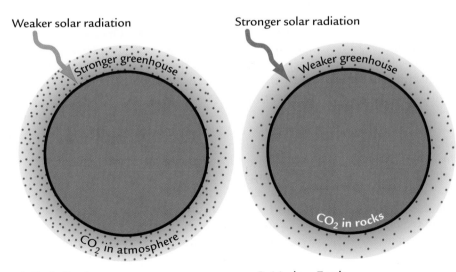

Weaker solar radiation   Stronger solar radiation

Stronger greenhouse   Weaker greenhouse

CO$_2$ in atmosphere   CO$_2$ in rocks

A Early Earth   B Modern Earth

FIGURE 4-8 Earth's thermostat? (A) The most plausible explanation of the faint young Sun paradox is that the weakness of the early Sun was compensated for by a stronger carbon greenhouse in the atmosphere. (B) When the Sun later strengthened, increased chemical weathering deposited the excess greenhouse carbon in rocks, and the greenhouse effect weakened enough to keep Earth's temperatures moderate. (Adapted from W. Broecker and T.-H. Peng, *Greenhouse Puzzles* [New York: Eldigio Press, 1993].)

although any such longer-term contribution is likely to have been smaller than that of CO_2. Because both of these gases tend to be broken down quickly in the atmosphere by chemical reactions, they would need to have been resupplied to the early atmosphere continually and in large amounts. The source of such an ongoing supply early in Earth's history is unclear.

In summary, chemical weathering is the most likely explanation for Earth's basic habitability over most of its 4.55-Byr existence (Figure 4-7). *Any* factor that heated Earth during any part of its history caused chemical weathering rates to increase, drew CO_2 out of the atmosphere at faster rates, and resulted in a partially opposing cooling. Conversely, any factor that cooled Earth set off the opposite sequence of events. The chemical weathering thermostat constantly acts to moderate long-term climate changes.

What about the role of water vapor in this process? We learned in Chapter 2 that water vapor is by far the most important greenhouse gas today, and the source of much of the 31°C greenhouse warming of our planet. What is the role of water vapor in these longer-term climate changes? Why wouldn't it play a larger role than CO_2?

CO_2 and water vapor work in a fundamentally different way within the climate system. Over tectonic time scales, CO_2 acts as a negative feedback that mutes climate changes (although we will see later that CO_2 may act in entirely different ways on time scales much shorter than a million years).

In contrast, the action of water vapor is the reverse of the effect of CO_2: it acts as a *positive* feedback that amplifies changes in climate (Chapter 2). As a result, water vapor cannot be invoked as the key first cause of the persistence of moderate climates during Earth's history. It amplifies, rather than moderates, climatic changes imposed by other factors.

The continuing presence of running water virtually throughout Earth's recorded history indicates that there

BOX 4-2 Looking Deeper into Climate Science

The Organic Carbon Subcycle

Approximately 20% of the carbon that cycles among Earth's carbon reservoirs today does so in organic form. An important part of this organic carbon subcycle is the process of photosynthesis, by means of which land plants extract CO_2 from the atmosphere and ocean plankton extract CO_2 from inorganic carbon dissolved in the surface ocean.

Most of the organic carbon fixed and temporarily stored in vegetation on land and in ocean plankton is recycled and quickly returned to the ocean atmosphere system. Recycling occurs by means of oxidation, which works like photosynthesis in reverse—it uses available oxygen in water or air to convert organic carbon back to inorganic form.

On land, oxidation consumes organic carbon just after the seasonal fall of leaves or die-back of green vegetation, and also after the death of such vegetation as the woody tissue of trees. In the oceans, it consumes organic debris slowly sinking out of the sunlit surface layers, where photosynthesis occurs.

Only a minute fraction of the organic carbon originally formed by photosynthesis ends up buried in the geologic record, and carbon from the land and the oceans is thought to contribute roughly equal amounts at present. Burial of organic carbon is favored in water-saturated environments (marine or terrestrial) characterized by (1) low oxygen levels, which minimize or stop the process of oxidation, and (2) high rates of production of organic matter, which consumes whatever oxygen is left in the water and lets the remaining organic debris escape oxidation. These conditions produce fine-grained carbon-rich muds that eventually turn into mudstones and then into harder rocks called shales.

This buried carbon represents a net loss of CO_2 from the atmosphere, or more accurately from the interactive carbon reservoirs in the ocean, atmosphere, soil, and vegetation. Once buried, the organic carbon stays in the rocks until tectonic processes return it to the surface. Organic carbon is returned from rocks to the atmosphere by two slow-acting processes: (1) weathering (and oxidation) of carbon-bearing rocks at Earth's surface, and (2) thermal breakdown of organic carbon in sediments and rocks deep in Earth's interior, with subsequent release of liberated CO_2 through volcanoes.

Because this subcycle of organic carbon carries one-fifth of the carbon moving between Earth's rocks and its surface reservoirs, it has the potential to have large-amplitude effects on the global carbon balance and on atmospheric CO_2 over tectonic-scale time intervals. Also, under certain conditions, the

has never been a shortage of liquid water on this planet. Unlike other planets in our solar system, Earth has an atmosphere that was never starved for a source of water vapor as a major greenhouse gas. More than enough liquid water was available to feed water vapor to the atmosphere and provide a positive feedback to temperature changes on this planet.

Is Life the Ultimate Control on Earth's Thermostat?

We have seen that chemical weathering provides a plausible—if not provable—thermostat mechanism to moderate Earth's climate, yet we have also seen that the processes involved in chemical weathering today are not strictly physical. Biological processes also participate in the carbon cycle. In addition, some of the carbon that moves through Earth's reservoirs does so in organic

form, as part of a separate (smaller) subcycle (Box 4-2). As we will see in later chapters, this organic carbon subcycle may also contribute to changes in atmospheric CO_2 and in Earth's climate. For these reasons, some scientists infer that life itself, rather than strictly physical-chemical factors, may be the thermostat that regulates Earth's climate.

4-3 The Gaia Hypothesis

The biologists James Lovelock and Lynn Margulis proposed in the 1980s that life itself has been responsible for regulating Earth's climate. They called their idea the **Gaia hypothesis,** after the ancient Greek Earth goddess. A crude analogy of how their hypothesis works is the way the fur on an animal fluffs out to create an insulated layer and keep the creature warm when the weather turns cold, but stays close to the animal's body when the air is warm. The animal in effect regulates its own

organic carbon subcycle has the potential to act more rapidly than the extremely slow inorganic cycle. Under the right conditions, such as times of abrupt onset of high productivity and burial of carbon in the ocean, large amounts of organic carbon can be quickly extracted from the

atmosphere, causing rapid reductions of CO_2 levels. Climate scientists have inferred that rapid burial of organic carbon caused large and rapid climatic coolings. In contrast, the rates of burial of inorganic carbon are constrained by the inherently slow process of chemical weathering of silicate rock.

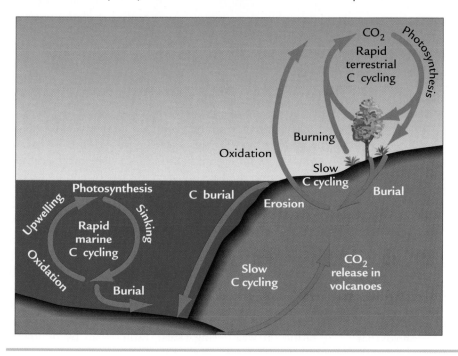

The organic carbon subcycle
About 20% of the carbon that moves between Earth's surface reservoirs (air, water, and vegetation) and its deep rock reservoirs does so in the organic carbon cycle. Photosynthesis on land and in the surface ocean turns inorganic carbon into organic carbon, most of which is quickly returned to the atmosphere or surface ocean. A small fraction of the organic carbon is buried in continental and oceanic sediments that slowly turn into rock. This carbon is eventually returned to the atmosphere as CO_2, either by erosion of continental rocks or by melting and volcanic emissions.

environment for its own good. The Gaia hypothesis holds that life regulates climate on Earth for its own good.

Supporters of this hypothesis cite several features of the chemical weathering thermostat concept that directly involve the action of life-forms: (1) the fact that carbon is at the center of the CO_2 cycle; (2) the action of land plants in contributing carbon dioxide to the soil to form carbonic acid and to enhance hydrolysis; and (3) the role of shell-bearing ocean plankton in extracting CO_2 from the ocean and storing it in their $CaCO_3$ shells. These modern-day biological processes clearly

are important parts of the processes of chemical weathering and carbon cycling. And by extension, they contribute to the thermostat that moderates Earth's climate today.

In its more extreme form, the Gaia hypothesis states that all evolution on Earth has occurred for the greater good of the planet such that it achieves the succession of life-forms needed to keep the planet habitable. This view is much more controversial: it goes far beyond Darwin's concept that evolution occurs to enhance the reproductive survival of each species.

Critics of the Gaia hypothesis point out that many of the most active roles played by organisms in the biosphere today are a relatively recent development in Earth's history, and that the role of life in the distant past was probably much smaller or even nonexistent. During Earth's long history, the life-forms that have existed differed considerably from those that exist today (Figure 4-9).

No known record of life exists before 3.5 Byr ago, although it is possible that primitive life-forms did exist and have escaped detection because the rock record is scarce and poorly preserved. By 3.5 Byr ago, primitive single-celled marine algae capable of photosynthesis had developed (Figure 4-10A). Over the next 3 billion years, slightly more complex organisms evolved, including moundlike clumps of marine algae called stromatolites that lived attached to the seafloor by 2.9 Byr ago, organisms that contained a cell nucleus by 2.5 Byr ago, and a variety of multicelled algae by 2.1 Byr ago.

Most of the more complex forms of life that we know today did not appear until relatively recent stages of Earth's history. Near 540 Myr ago, hard shells of many kinds of organisms abruptly began to appear in the fossil record. Before that time, the only fossilized records of life consisted of ghost impressions left imprinted on the surfaces of soft sediment layers. The first primitive land plants did not evolve until near 430 Myr ago (Figure 4-10B). These plants acquired the ability to survive because they had stems and roots to deliver water from the ground. The first treelike plants appeared by 400 Myr ago (Figure 4-10C). Trees and grasses are important in modern-day chemical weathering because they acidify groundwater by adding carbon to soils as litter and also help to produce organic acids.

Critics of the Gaia hypothesis claim that the life-forms that had previously existed on land for over 90% of Earth's history were too primitive to have had much effect on chemical weathering or to have played a central role in driving Earth's thermostat. In the view of these critics, the delicate climatic balance through most of Earth's history was achieved primarily by physical-chemical means (the effects of temperature and precipitation on weathering rates), rather than by biological means.

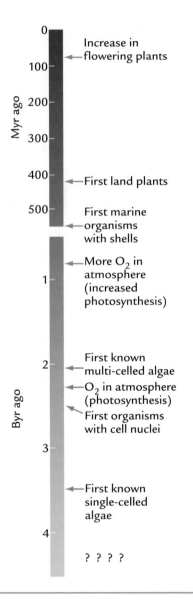

FIGURE 4-9 The Gaia hypothesis Over time, life-forms gradually developed in complexity and played a progressively greater role in chemical weathering and its control of Earth's climate. The Gaia hypothesis holds that life evolved in order to regulate Earth's climate.

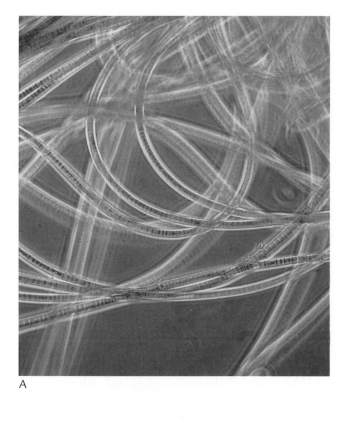

A

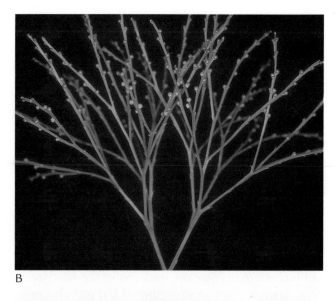

B

C

FIGURE 4-10 **Life-forms and weathering** Over Earth's 4.55-Byr history, plants evolved toward more complicated forms capable of a greater role in chemical weathering. Primitive organisms similar to the modern-day bacteria *Oscillatoria* (A) existed by 3.5 Byr ago. The first simple land plants with roots and stems similar to those of the modern-day plant *Psilotum* (B) appeared by 430 Myr ago. Increasingly complex treelike plants similar to modern tropical cycads (C) appeared by 400 Myr ago and led to the modern-day diversity of trees and shrubs. (A: Sinclair Stammers/Science Photo Library/Photo Researchers; B: William Ormoerod/Visuals Unlimited; C: Gerald Cubit.)

Similarly, critics note that the very late appearance of oceanic organisms with CaCO$_3$ shells near 540 Myr ago means that life had played no obvious role in transferring the products of chemical weathering on land to the seafloor for the preceding 4 billion years. Instead, most of the CaCO$_3$ in the oceans was deposited in warm shallow tropical seas where concentrations of dissolved ions increased to levels that permitted widespread chemical precipitation of CaCO$_3$, apparently with little or no biological intervention. Floating planktic plants capable of photosynthesis evolved still later, in the last 250 million years.

Supporters of the Gaia hypothesis respond to these criticisms with several counterarguments. First, they claim that critics underestimate the role played by primitive life-forms such as algae in the ocean and microbes on land early in Earth's history. They claim that modern-day bacteria that are very similar to the early primitive forms play a greater role in the weathering process today than is generally recognized, and that these organisms must also have been important early in Earth's history, when they were the only life-forms present on land.

One indication that early life-forms may have been important at a global scale is the first development of an oxygen-rich atmosphere near 2.3 Byr ago. Evidence for this important event includes the first appearance of rocks that show red staining (rusting) of iron (Fe) minerals.

The appearance of oxidized iron minerals nearly 2.3 Byr ago coincides roughly with the disappearance of minerals such as FeS ("fool's gold"), which can form only under reducing (oxygen-free) conditions. Minerals of this kind are commonly found in Earth's earliest rocks and persist through time until 2.3 Byr ago. The only conceivable source of the oxygen that caused this widespread change toward oxidized iron was photosynthesis by marine organisms, implying an active global-scale role for organisms far back in Earth's history.

Gaia supporters also point out that the general path of biological evolution matches Earth's need for progressively greater chemical weathering through time. The more primitive organisms played a smaller role in accelerating the process of chemical weathering during a time when it was to Earth's advantage to keep its CO_2 in its atmosphere to counter the weakness of the faint young Sun. Then, as the Sun strengthened and provided more heat to Earth, more advanced organisms capable of accelerating the weathering process appeared, accelerated the rates of weathering, and pulled CO_2 out of the atmosphere to keep the climate system in approximate balance.

The Gaia hypothesis is fascinating, but it is still unproved. To resolve this issue, scientists need far better quantitative measurements of the separate contributions of biological, chemical, and physical factors to rates of chemical weathering.

KEY TERMS

greenhouse era (p. 86)

icehouse era (p. 86)

snowball Earth hypothesis (p. 89)

faint young Sun paradox (p. 90)

thermostat (p. 90)

silicate minerals (p. 93)

Gaia hypothesis (p. 99)

REVIEW QUESTIONS

1. Why is Venus so much warmer than Earth today?

2. What factors explain why Earth is habitable today?

3. Explain the faint young Sun paradox.

4. What evidence suggests that Earth has always had some kind of long-term thermostat regulating its climate?

5. Why does volcanic input of CO_2 to Earth's atmosphere fail as a possible thermostat?

6. What climate factors affect the removal of CO_2 from the atmosphere by chemical weathering?

7. Why is chemical weathering a plausible thermostat for Earth's climate?

8. What kind of chemical weathering acts as a thermostat by affecting atmospheric CO_2?

9. Where did all the extra CO_2 from Earth's early atmosphere go?

10. What arguments support and oppose the Gaia hypothesis that life is Earth's real thermostat?

ADDITIONAL RESOURCES

Lovelock, J. 1995. *The Ages of Gaia: A Biography of Our Living Earth*. New York: W. W. Norton.

Kastings, J. F., O. B. Toon, and J. B. Pollack. 1988. "How Climate Evolved on the Terrestrial Planets." *Scientific American* (February), 90–97.

CHAPTER 5

Plate Tectonics and Climate

The last 550 million years of Earth history are far better known than the first 4 billion years (Figure 5-1). For the first time, scientists begin to know the locations of the continents and the shapes of the ocean basins. In addition, increasingly well preserved sedimentary rock archives hold abundant evidence of past climates, including a sequence of alternations between icehouse intervals, when ice sheets were present, and greenhouse intervals, when no ice existed. These fluctuations are the focus of this chapter.

First we examine how plate tectonic processes work. Next we explore the possibility that icehouse intervals simply occur when plate tectonic motions cause continents to drift across cold polar regions. Then we use climate models to investigate the range of factors that controlled climate 200 Myr ago, a time when all landmasses on Earth existed as a single giant continent. These investigations all reveal that changes in atmospheric CO_2 levels are needed to explain the sequence of icehouse/ greenhouse changes over the last 500 million years. Finally, we evaluate two hypotheses that link changes in plate tectonic processes to changes in CO_2 levels.

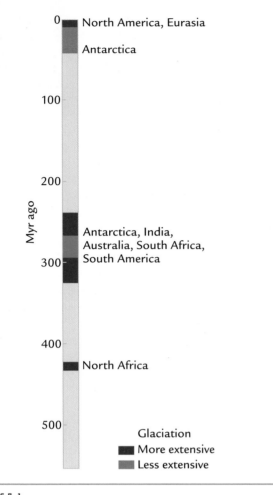

FIGURE 5-1 **Icehouse intervals** Three intervals of glaciation occurred during the last 550 million years.

Plate Tectonics

In 1914 the German meteorologist Alfred Wegener proposed that continents have slowly moved across Earth's surface for hundreds of millions of years. He based his hypothesis in part on the obvious fact that continental margins such as those of eastern South America and western Africa would fit together like pieces of a jigsaw puzzle if they were juxtaposed. Research in the last half of the twentieth century showed that Wegener was correct in deducing that continents can move, but that his hypothesis underestimated the mobility of Earth's outer surface. In fact, *all* of Earth's surface is on the move.

5-1 Structure and Composition of Tectonic Plates

Wegener's assumption that continents move in relation to ocean basins had a reasonable basis. The contrast between the elevated continents and the submerged ocean basins is the most obvious division on Earth's sur-

face. It also reflects a large difference in thickness and composition of the crustal layers of which continents and ocean basins are made (Figure 5-2).

Continental crust is 30-70 kilometers thick, has an average composition like that of granite, and is low in density (2.7 g/cm^3). The thick, low-density continental crust stands much higher than the floor of the ocean basins some 4000 meters below sea level. In contrast, **ocean crust** is 5-10 kilometers thick, has an average composition like that of basalt, and is higher in density (3.2 g/cm^3). Below each of these crustal layers lies the **mantle**, which is richer in the heavy elements iron (Fe) and magnesium (Mg) and has an even higher density (> 3.6 g/cm^3). The mantle extends 2890 kilometers into the Earth's interior, almost halfway to the center of the Earth at a depth of 6370 kilometers.

But differences in elevation, crustal thickness, and composition are not the primary causes of the movement of continents (and ocean basins). The critical factor is the way different layers of rock behave.

Two rock layers characterized by different long-term behavior exist well below Earth's surface (Figure 5-2). The outer layer, called the **lithosphere**, is

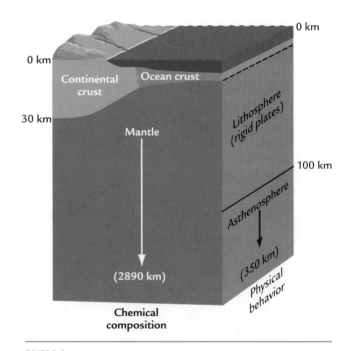

FIGURE 5-2 **Earth's structure** Earth's outer layers can be subdivided in two ways. One is by chemical composition: the basalts of the ocean crust and the granites in continental crust differ from each other and from the underlying mantle in chemical composition. The other division is by physical behavior: the lithosphere that forms the tectonic plates is a hard, rigid unit, whereas the underlying asthenosphere is softer and capable of flowing slowly.

100 kilometers thick and behaves in a hard, rigid, rock-like manner. The lithosphere encompasses not just the crustal layers (oceanic and continental) but also the upper part of the underlying mantle.

Below the lithosphere is a layer of partly molten but mostly solid rock called the **asthenosphere**, lying entirely within the upper section of Earth's mantle at depths between 100 and 350 kilometers. Over intervals of thousands to millions of years, the asthenosphere behaves like a soft, viscous fluid and flows easily. The fluidity and movement of this layer of hard rock over long intervals of time are the source of the movements of the overlying lithosphere, including Earth's surface.

The lithosphere is not a single rigid layer but consists of a dozen distinct **tectonic plates**, each a rigid unit drifting slowly across Earth's surface (Figure 5-3). Lithospheric plates move at rates ranging from less than 1 up to 10 centimeters per year, about the same as the rate of growth of a fingernail. Over a time span of 100 million years, 5 centimeters per year of plate motion adds up to 5000 kilometers, enough to create or destroy an entire ocean basin.

Most tectonic plates consist not of continents or ocean basins alone but of combinations of the two. The

South American plate, for example, consists of the continent of South America and the western half of the South Atlantic Ocean, all moving as one rigid unit.

These rigid tectonic plates have three basic types of edges, or margins. Almost all tectonic deformation on Earth (earthquakes, faulting, volcanoes) occurs at these plate margins (Figure 5-4).

Plates move apart at **divergent margins** at the crests of ocean ridges, such as the Mid-Atlantic Ridge. This motion allows new ocean crust to be created, and the new crust subsequently spreads away from the ridge. Plates diverging at ocean ridges carry not just the near-surface layer of ocean crust but also the much thicker layer of upper mantle below.

Plates come together at **convergent margins** (Figure 5-4, left). At these locations, the ocean crust plunges deep into Earth's interior at the sites of ocean trenches in a process called **subduction**. The subducting ocean crust rides on top of a much thicker layer of upper mantle.

Some convergent margins occur along continent-ocean boundaries, such as the western coast of South America, and narrow mountain chains such as the Andes form on the adjacent continents because of the

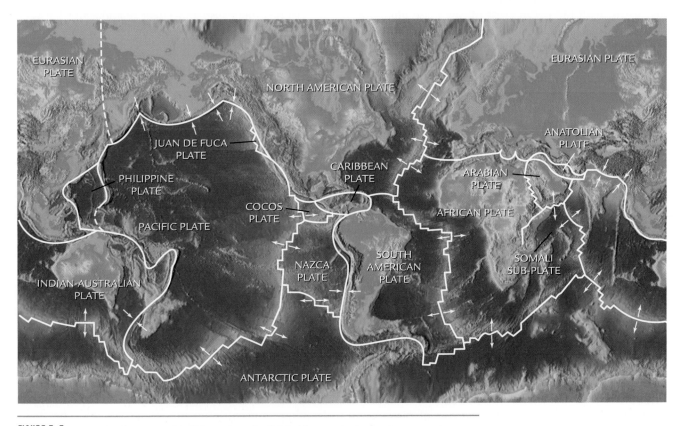

FIGURE 5-3 Tectonic plates Earth's lithosphere is divided into a dozen major tectonic plates and several smaller plates, which move as rigid units in relation to one another, as the arrows indicate. (F. Press and R. Siever, *Understanding Earth*, 2nd ed., © 1998 by W. H. Freeman and Company. False-color topography courtesy of Peter Schloss, NGDC, Boulder, Colo.)

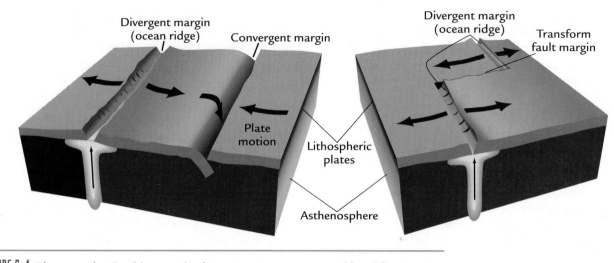

FIGURE 5-4 **Plate margins** Earth's tectonic plates move apart at ocean ridges (divergent margins), slide past each other at faults (transform fault margins), and push together at convergent margins, primarily during subduction at ocean trenches and less commonly during collisions of continents. (Modified from F. Press and R. Siever, *Understanding Earth*, 2nd ed., © 1998 by W. H. Freeman and Company.)

compressive (squeezing) forces produced when two plates move together. Subduction can also occur within the ocean, with the ocean crust of one plate plunging under the ocean crust of another plate, and forming volcanic ocean islands such as those of Japan. A much less common example of converging plates is the **continental collision** of landmasses such as India and Asia, which create massive high plateaus such as Tibet and the Himalaya mountain complex.

Plates also can slide past each other at **transform fault margins** (Figure 5-4, right), moving horizontally along faults such as the San Andreas Fault in western California. Sliding of plates at transform faults involves not just the 30 kilometers of continental crust but also the underlying 70 kilometers of upper mantle.

Geoscientists do not yet know the exact balance of forces that have produced the present (and past) distribution of plates and caused the direction and rate of movement of each plate. Ultimately plate movements are driven by the redistribution of Earth's interior heat. Part of this redistribution is upward movement of heat derived from Earth's molten outer core (below 2890 km) and from radioactive decay throughout the mantle. Although the upward motion of heat is the most obvious driving force behind plate tectonic processes, many scientists think that downward movement of cold lithosphere is at least equally important, if not more so.

Uncertainties about the mechanisms of plate tectonics are not critical here. What matters is that researchers can accurately measure the results of plate tectonic processes in changing Earth's surface geography and can compare these tectonic changes to changes in climate that occurred at the same times. These two lines of evidence can then point climate scientists toward possible cause-and-effect relationships between changes in Earth's tectonic system and its climate.

5-2 Evidence of Past Plate Motions

An enormous variety of evidence reveals the effects of plate tectonics in rearranging Earth's geography. The most important evidence is based on the fact that Earth has a **magnetic field**. Molten fluids circulating in Earth's liquid iron core today create a magnetic field analogous to that of a bar magnet (Figure 5-5). Compass needles today point to magnetic north, which is located a few degrees of latitude away from the geographic North Pole, which marks Earth's axis of rotation.

Some of Earth's once-molten rocks contain "fossil compasses" that record its magnetic field in the distant past. These natural magnetic compasses were frozen into the rock shortly after it cooled from a molten state. Now they give scientists working in the discipline of **paleomagnetism** a way to reconstruct past positions of continents and ocean basins.

The best rocks to use as ancient compasses are basalts, which are rich in highly magnetic iron. Basalts form the floors of all ocean basins and are also found on land in regions of frequent tectonic activity. They form from molten lavas, which cool quickly after being extruded onto Earth's surface. As the molten material cools, its iron-rich components align with Earth's magnetic field like a compass. After the lava turns into

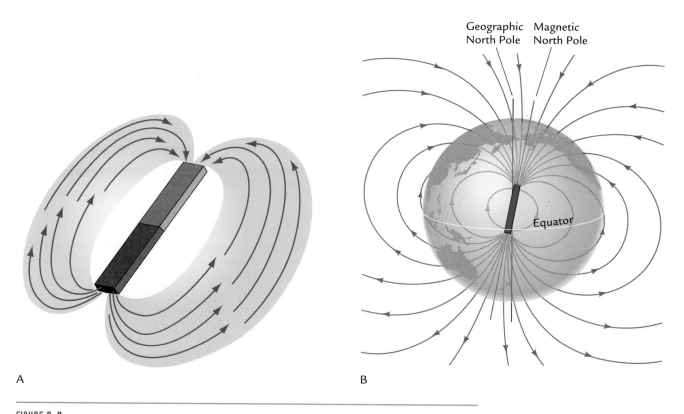

Geographic North Pole   Magnetic North Pole

Equator

A

B

FIGURE 5-5 **Earth's magnetic field** Like the magnetic field sensed by iron filings surrounding a bar magnet (A), Earth has a magnetic field that determines the alignment of compass needles (B). Basaltic rocks contain iron minerals that align with Earth's prevailing magnetic field shortly after the molten magma cools to solid rock. (B: F. Press and R. Siever, *Understanding Earth*, 2nd ed., © 1998 by W. H. Freeman and Company.)

basaltic rock (when its temperature drops below 1200°C), continued cooling to temperatures near 600°C allows the magnetic compasses to become permanently fixed in position in the rock.

Paleomagnetism is used to reconstruct changes in the configuration of Earth's surface in two ways. (1) Over the last 100 million years or more, paleomagnetic changes recorded in basaltic oceanic crust are used to reconstruct movements of plates and rates of spreading of the seafloor. Because no ocean crust older than 175 million years exists, this technique cannot be used farther back in time. (2) Back to about 500 million years ago, paleomagnetic compasses recorded in continental basalts can be reliably used to track movements of landmasses with respect to latitude (but not longitude).

Using paleomagnetism in this type of plate tectonic reconstruction requires establishment of a time framework for the magnetic signatures frozen into the rock. Basalts are well suited to this purpose because they cool so quickly. Little time elapses between the solidification of the molten magma into rock, which sets the age of the rock as determined by radioactive dating, and the later acquisition of a permanent magnetic signature. As a result, the radiometric age of the rock is effectively the same as the age of the magnetic signature. Radioactive decay of potassium (K) to argon (Ar) is a widely used means of dating basalts (Chapter 3), especially in regions repeatedly covered by lava flows, such as tectonically active parts of Iceland and East Africa.

Paleomagnetic Dating of Ocean Crust The usefulness of the paleomagnetic technique in tracing the movement of the seafloor during the last 175 Myr depends on an additional factor: Earth's magnetic field repeatedly reverses direction. Compasses that now point to magnetic north in the present-day "normal" magnetic field would have pointed to magnetic south (a position near the South Pole) during those times when the field was in a reversed orientation. Reversals happen at irregular intervals as large as several million years and as short as a few thousand years.

Samples are drilled out of individual layers of basaltic rock representing separate lava flows. The samples are measured for their magnetic signature (the direction toward which the Fe compasses embedded in

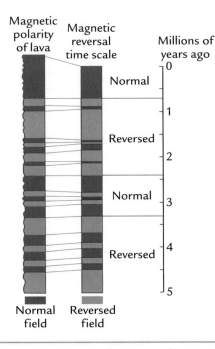

Magnetic polarity of lava

Magnetic reversal time scale

Millions of years ago

0

Normal

1

Reversed

2

Normal

3

Reversed

4

5

Normal field

Reversed field

FIGURE 5-6 Magnetic stratigraphy Magnetic signatures of individual lava flows from many regions (left) can be dated by radiometric methods and used to construct a time scale of past magnetic reversals (right). Intervals with present-day magnetic polarity (compasses pointing to the north magnetic pole) are said to have normal polarity; intervals when compasses would have pointed south were times of reversed polarity. (Modified from F. Press and R. Siever, *Understanding Earth,* 2nd ed., © 1998 by W. H. Freeman and Company.)

the rock point) and also for their K/Ar radiometric age. Each measured magnetic signature (normal or reversed) is plotted against a time axis to compile a **magnetic stratigraphy**, or time history of magnetic reversals (Figure 5-6). Because rocks from many regions on Earth have been found to yield the same sequence of magnetic normal and reversed intervals through time, this magnetic stratigraphy must be a worldwide phenomenon tied directly to an absolute time scale.

Soon after this magnetic reversal sequence was first established on land, marine geoscientists discovered stripelike magnetic patterns called **magnetic lineations** on the ocean floor (Figure 5-7). Ships surveying the ocean had for decades been towing instruments that measured Earth's regional magnetic field. The mapped pattern obtained from these shipboard measurements showed alternating strips of crust with relatively stronger and weaker magnetization. Near mid-ocean ridges, the magnetic lineations were found to be symmetrical around the ridge axis.

To the surprise of most scientists, this mapped pattern of highs and lows measured in the magnetic field at sea closely matched the pattern of normal and reversed

intervals defined by the magnetic stratigraphy developed from analysis of individual basalt layers on land (Figure 5-6). This unexpected match provided a direct way of dating the ocean crust anywhere ships have recorded the magnetic field.

The match of magnetic patterns proves that new (zero-age) crust is being formed at the crests of ocean ridges, and that the ocean crust and underlying lithosphere then slowly spread away in both directions. As a result, the age of the ocean crust steadily increases with distance away from the ridges (Figure 5-8). This important conclusion has been independently confirmed in many locations by drilling into the ocean crust and retrieving samples for K/Ar dating.

Scientists can use this information about the age of the ocean crust to evaluate causes of past climate changes in two ways. First, the dated magnetic lineations on the seafloor can be used to roll back the recent motions of the seafloor and restore the continents and oceans to their positions during the last 175 million years. Second, the lineations can be used to reconstruct the rate of **seafloor spreading**, the rate at which ocean crust forms and moves away from the ocean ridges. Changes in the rate of spreading define both the rate at which new ocean crust and lithosphere

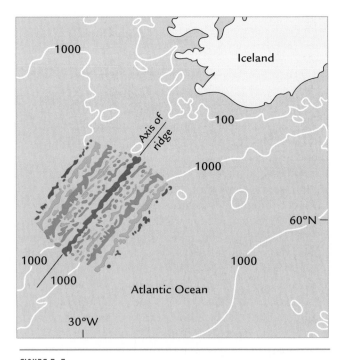

FIGURE 5-7 Magnetic stripes in the ocean The pattern of magnetic anomalies along a section of the Mid-Atlantic Ridge southwest of Iceland is symmetric around the ridge crest. Colored stripes mark regions with highly magnetized crust. (F. Press and R. Siever, *Understanding Earth,* 2nd ed., © 1998 by W. H. Freeman and Company.)

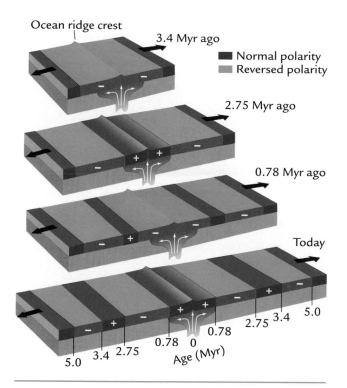

Ocean ridge crest

3.4 Myr ago

■ Normal polarity
■ Reversed polarity

2.75 Myr ago

0.78 Myr ago

Today

5.0

2.75 3.4

0.78

0.78 0

0.78 2.75 3.4 5.0

5.0 3.4 2.75

Age (Myr)

FIGURE 5-8 **Magnetization of ocean crust** Successive bands of ocean crust form as molten lava erupts at the seafloor, cools, and solidifies. The new crust is magnetized in the normal or reversed polarity prevailing at the time. As the plates move apart, equal amounts of magnetized crust are carried away from the ridge axis in both directions and can be used to date the seafloor. (Modified from F. Press and R. Siever, *Understanding Earth*, 2nd ed., © 1998 by W. H. Freeman and Company.)

are created at ocean ridges, as well as the rate at which older ocean crust and lithosphere are subducted at ocean trenches.

Paleomagnetic Determination of Past Locations of Continents Because no ocean crust older than 175 Myr survives for study, magnetic lineations on the seafloor cannot be used to roll back the spreading process and reconstruct the history of earlier plate motions. For earlier intervals, paleomagnetism must focus on basalts on the continents. But in this kind of study, the focus is not on detecting a distinctive sequence of magnetic reversals but rather on using the internal rock magnetic compasses to determine the past position of that rock (and of the portion of continental crust in which it is found) in relation to the magnetic poles.

The orientation of these fossil magnetic compasses is used to determine the latitude at which the rocks formed. For molten lavas that cool at high latitudes, the internal magnetic compasses point in a nearly vertical direction, because that is the direction of Earth's magnetic field at high latitudes (see Figure 5-5). In contrast,

lavas that cool near the equator have internal compasses oriented closer to horizontal, nearly parallel to Earth's surface. Subsequently, the basaltic rocks may be carried across Earth's surface by plate tectonic processes, yet their embedded magnetic compasses still carry a clear record of the latitude at which they first formed.

Because researchers can determine past latitudes but not longitudes of continents, this kind of reconstruction is more uncertain than those based on seafloor magnetic lineations. Fortunately, the magnetic data can be supplemented by other observations, such as the match of coastlines of eastern South America and western Africa, which provide clear evidence that these continents were joined before the present-day Atlantic Ocean formed.

Rocks older than 500 million years are less reliable for paleomagnetic work because of the increasing likelihood that their magnetic signatures have been reset to the magnetic field of a later time. Later overprinting is obvious for mountain-building episodes that grossly deform preexisting rocks and destroy the original magnetism. But it can also happen in a far more subtle way when tectonic activity produces hot fluids that circulate through adjacent rocks and reset their magnetism at low temperatures without deforming the rocks in any obvious way.

In summary, we can reconstruct the positions of continents on Earth's surface with good accuracy back to 300 Myr ago, and less accurately back to 500 million years ago. Subsequent to 175 million years ago, we can measure rates of seafloor spreading in several ocean basins, and subsequent to 100 million years ago, we can compile spreading rates over enough of the world's ocean to estimate the global mean rate of creation and destruction of ocean crust.

The Polar Position Hypothesis

An early hypothesis of long-term climate change focused on latitudinal position as a cause of glaciation of continents. The **polar position hypothesis** made two key predictions that could be tested: (1) ice sheets should appear on continents when they are located at polar or near-polar latitudes, but (2) no ice should appear anywhere on Earth if no continents exist anywhere near the poles. Note that to explain the occurrence of icehouse intervals this hypothesis calls not on worldwide climate changes but simply on the movements of continents on tectonic plates.

The fact that modern-day ice sheets occur on the polar continent of Antarctica and the near-polar landmass of Greenland makes this hypothesis seem plausible. Modern-day ice sheets exist at high latitudes for several obvious reasons (Chapter 2): cold temperatures caused by low angles of incident solar radiation, high

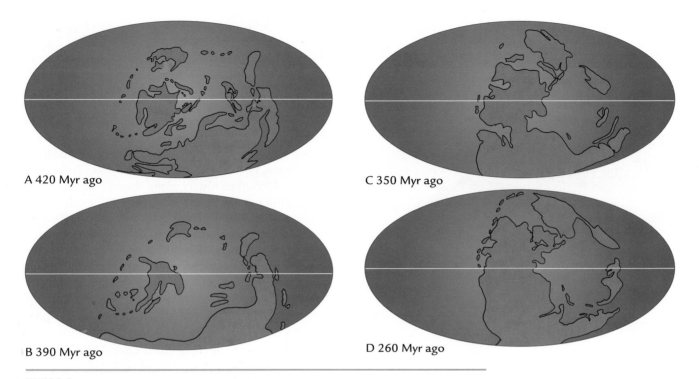

A 420 Myr ago

C 350 Myr ago

B 390 Myr ago

D 260 Myr ago

FIGURE 5-9 **Moving continents** (A–C) After 450 Myr ago, plate tectonic activity carried the southern continent of Gondwana across the South Pole on a path headed toward continents scattered across the northern hemisphere. Subsequent collisions formed the giant continent Pangaea (D). (Adapted from S. Stanley, *Earth System History,* © 1999 by W. H. Freeman and Company.)

albedos resulting from the prevalent cover of snow and sea ice, and sufficient moisture to maintain ice sheets despite melting along their lower margins.

5-3 Glaciations and Continental Positions since 500 Myr Ago

We can directly test the validity of the polar position hypothesis against evidence in the geologic record. Over the last 500 Myr, seafloor spreading has slowly moved continents across Earth's surface between the warmer climates of lower latitudes and the colder climates of higher latitudes. If latitudinal position alone controls climate, these movements should have produced predictable changes in glaciations over intervals of tens to hundreds of millions of years.

During the last 500 million years (Table 5-1), major continent-sized ice sheets existed on Earth during three icehouse eras: first a brief interval near 430 Myr ago, then a much longer interval from 325 to 240 Myr ago, and finally the current icehouse era of the last 35 million years. For the two long intervening intervals (425–325 Myr and 240–35 Myr), no evidence of large ice sheets exists anywhere on Earth.

Where were the continents located during these icehouse and greenhouse intervals? Did glaciation occur on continents positioned at high polar (or near-polar) latitudes? And did nonglaciated intervals coincide with times when all continents were located away from the poles?

Before 420 Myr ago, small fragments of land that were later to form parts of the modern continents of the northern hemisphere lay scattered across a wide range of latitudes (Figure 5-9A). These included landmasses roughly equivalent to modern-day northern Asia and North America.

The rest of the world's landmasses were combined in a southern supercontinent called **Gondwana**, equivalent to modern-day Africa, Arabia, Antarctica, Australia, South America, and India. This continent was located initially in the southern hemisphere on the opposite side of the globe, where the Pacific Ocean is now. Gondwana had begun a long trip across the South Pole, which would carry it northward until it collided with the other landmasses to form the giant supercontinent of **Pangaea**, a name meaning "All Earth" (Figure 5-9B–D).

An easy way of visualizing this motion is to plot the changing position of the magnetic south pole in relation to the land (Figure 5-10). Although this presentation makes it look as if the south magnetic pole were moving southward across Gondwana, in fact the Gondwana continent was moving northward across the pole.

TABLE 5-1   Evaluation of the Polar Position Hypothesis of Glaciation

Time (Myr ago)	Ice sheets present?	Continents in polar position?	Hypothesis supported?
430	Yes	Yes	Yes
425–325	No	Yes	No
325–240	Yes	Yes	Yes
240–125	No	No	Yes
125–35	No	Yes	No
35–0	Yes	Yes	Yes

How well does the pattern shown in Figure 5-10 explain the intervals of glaciation and nonglaciation listed in Table 5-1? The position of the south magnetic pole 430 Myr ago agrees with the occurrence of glaciation in the area of the modern-day Sahara Desert. Evidence of this glaciation survives as striations (grooves) cut into bedrock by the weight of the ice pressing down on the loose rubble carried in its base (Figure 5-11). The evidence of an ancient ice sheet in this area is in remarkable contrast to the modern desert, broiling under a hot tropical Sun.

At first this match seems to give us a positive confirmation of the polar position hypothesis, but on closer inspection problems emerge. One problem is that this glacial era was quite brief in terms of geologic time. Although its duration was once thought to be about 10 Myr, new evidence suggests an interval as short as 1 Myr. So brief a glaciation is not easily explained by the slow motion of Gondwana across the South Pole (Figure 5-9). We will revisit this problem later in the chapter.

A more perplexing problem is the lack of glaciations in the interval between 425 and 325 Myr ago, even though the Gondwana continent was still continuing its slow transit across the pole (Figure 5-10). Somehow land existed at the South Pole for almost 100 million years without any ice sheets forming. This observation argues against the hypothesis that a polar position was the *only* requirement for large-scale glaciation.

From 325 to 240 Myr ago, Gondwana continued its slow journey across the South Pole, and a huge region centered on the southern part of the continent was glaciated (Figure 5-10). The ice sheets were centered on modern-day Antarctica and South Africa, and they spread out into adjoining regions of South America, Australia, and India. Because of the correspondence

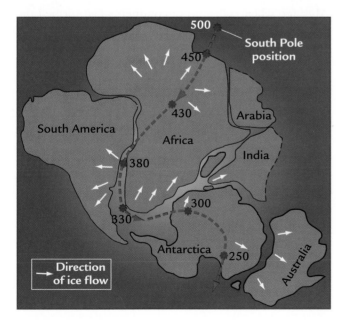

FIGURE 5-10 **Gondwana and the South Pole** Changes in the position of the south magnetic pole in relation to the Gondwana continent were caused by the slow movement of Gondwana across a stable pole. Glaciations occurred in the northern Sahara about 430 Myr ago and in southern Gondwana (South Africa, Antarctica, India, South America, and Australia) 325–240 Myr ago. The water shown between the modern-day continental outlines was land during Pangaean times. (Adapted from T. J. Crowley et al., "Gondwanaland's Seasonal Cycle," *Nature* 329 [1987]: 803-7, based on P. Morel and E. Irving, "Tentative Paleocontinental Maps for the Early Phanerozoic and Proterozoic," *Journal of Geology* 86 [1978]: 535-61].)

FIGURE 5-11 **Glacial striations** Sediment rubble carried in the bottom layers of ice sheets about 430 Myr ago gouged striations in North African bedrock similar to those in modern-day ice in Alaska, shown here. (Carr Clifton.)

between the area of Gondwana that was glaciated and its position in relation to the south magnetic pole, this long interval of glaciation is fully consistent with the polar position hypothesis.

By about 240 Myr ago, the southernmost part of Gondwana had moved northward past the pole, and glaciation of Gondwana ended. The lack of ice after that time agrees with the positioning of the land away from the pole. By that time, Gondwana had merged with the northern continents and formed the even larger supercontinent Pangaea.

After 180 Myr ago Pangaea began to break up. Its southernmost part, which included the modern-day continents of Antarctica, India, and Australia, moved back over the South Pole by 125 Myr ago, yet no ice sheet developed. Antarctica remained directly over the pole but free of ice from 125 Myr ago until near 35 Myr ago, when a small amount of ice appeared. Here again we face the mystery encountered earlier: How could a landmass centered right on the pole remain ice-free for almost 100 million years?

Clearly the polar position hypothesis cannot fully explain the sequence of glaciated and nonglaciated intervals over the last 500 Myr. The hypothesis *is* successful to this extent: ice sheets developed only on landmasses that were at polar or near-polar positions, consistent with the polar occurrences of ice sheets today. This correlation (Table 5-1) confirms that over the last 500 Myr continents had to occupy polar positions for large-scale glaciation to occur.

But the same record also tells us that the presence of continents in a polar position does not guarantee that ice sheets will form. Something is missing from the polar position hypothesis. Some other factor must be at work, a factor that controls climate in such a way as to allow ice sheets to form over polar continents during some intervals and prohibit them during others. Evidence we will examine in Chapter 6 confirms that climate at or near Earth's poles has not always been nearly as cold as it is today; at times it has been warm enough to keep ice from accumulating on polar continents. One likely cause is changes in concentrations of greenhouse gases.

Modeling Climate on the Supercontinent Pangaea

One fortunate aspect of studying the history of Earth's climate on tectonic time scales is the number of natural climate experiments Earth has run by altering its geography. Because the locations of continents are accurately known for the past 300 million years, climate scientists can use general circulation models (GCMs) to evaluate the impact on climate of geography as well as several other factors. Here we examine a time near 200 Myr ago when collisions of continents had formed the giant supercontinent Pangaea. Because this configuration differs from the more dispersed locations of continents today, Pangaea provides climate scientists with a very different and yet real Earth on which to test the performance of climate models.

Investigations of Pangaea raise two major climate-related questions. First, what level of atmospheric CO_2 is needed to explain Pangaean climate? And second, do GCM simulations of climate on this giant continent match independent geologic evidence preserved in the rock record?

5-4 Input to the Model Simulation of Pangaean Climate

Recall from Chapter 3 that GCM runs require the major physical aspects of a past world to be specified in advance as *boundary condition input* in order to initiate simulations of past climates. For experiments run on intervals lying in the more distant geologic past, the most basic physical constraint is the distribution of land and sea.

Earlier in this chapter we found that continental locations are difficult to determine for intervals before about 500 Myr ago, but that more and more information about past locations of the continents builds up for progressively younger intervals. By the time of the supercontinent Pangaea, we can be very sure of the continents' locations.

Pangaea was created by a sequence of continental collisions between 350 and 250 Myr ago. This landmass then remained intact until it began to break up after 180 Myr ago. The focus here is on the interval between 240 and 200 Myr ago, a long interval of relatively stable land-sea geometry that followed the collisions and preceded the breakup. The only tectonic change of significance during this interval was a very slow northward movement of Pangaea.

At 200 Myr ago, Pangaea stretched from high northern to high southern latitudes and was almost symmetrical around the equator (Figure 5-12A). Its southern half was the Gondwana landmasses: Antarctica, Australia, Africa, Arabia, South America, and India. Northern Pangaea consisted of the remaining landmasses, sometimes referred to as Laurasia: North America, Europe, and north-central Asia. A wedge-shaped tropical seaway indented far into Pangaea from the east, while the west coast had a smaller seaway in the northwest. This single landmass represented almost one-third of Earth's surface. It spanned 180° of longitude at its northern and southern limits, both near 70° latitude, and one-quarter of Earth's circumference (90°) at the equator.

Modelers have simplified this configuration for use as input to model simulations by making the land distribution exactly symmetrical around the equator (Figure 5-12B). This simplification requires only small changes in the way Pangaea is represented by the grid boxes in the model. The land-sea continental outlines are jagged because each grid box is specified as either all land or all ocean. One benefit of this simplification is that each seasonal model run in each hemisphere is the exact mirror image of the same season in the other hemisphere: the seasons simply switch back and forth between hemispheres. Using land-sea distributions symmetrical about the equator effectively doubles the number of years the model run simulates.

A second important decision on input to the model is global sea level. Evidence from Pangaean rocks indicates that global sea level 200 Myr ago was comparable to today's. Sea level was placed close to the structural edges of the continents, as it is today.

A third important decision is the uncertain distribution of elevated topography on the continents. One way to minimize problems caused by guessing wrong about the sizes of ancient mountains is to run a simulation based on an average elevation broadly distributed across all of Pangaea. In the simulation examined here, all land in the interior of Pangaea was represented as a low-elevation plateau at a uniform height of 1000 meters, with its edges sloping gradually down to sea level along the outer margins of the continents.

Another important boundary condition that needs to be specified, the CO_2 level in the atmosphere, is poorly known for 200 Myr ago. The CO_2 level chosen will obviously have a direct impact on Pangaea's simulated climate.

Fortunately, other considerations help climate modelers constrain the likely CO_2 level. Astrophysical models indicate that the Sun had not yet reached its present strength and was still about 1% weaker than it is today (Chapter 4). A weaker Sun should have made Pangaea significantly colder than the modern world, with snow and ice much farther equatorward than they occur today.

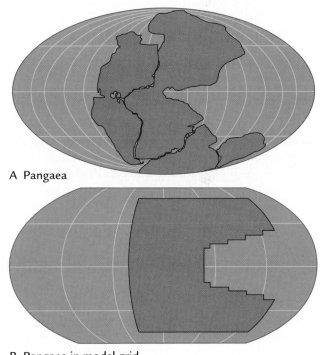

A Pangaea

B Pangaea in model grid

FIGURE 5-12 **The supercontinent Pangaea** Geographic reconstructions of the interval around 200 Myr ago show all the continents joined in a single landmass called Pangaea (A). Climate modelers have simplified this configuration into an idealized continent symmetric around the equator (B). (A: adapted from J. E. Kutzbach and R. G. Gallimore, "Megamonsoons of the Megacontinent," *Journal of Geophysical Research* 94 [1989]: 3341–57; B: from J. E. Kutzbach, "Idealized Pangean Climates: Sensitivity to Orbital Change," *Geological Society of America Special Paper* 288 [1994]: 51–55.)

FIGURE 5-13 **Pangaean trees** Modern gingko trees are descended from similar forms that first evolved some 200 Myr ago. (Courtesy of Mike Bowers, Blandy Farm, Boyce, Va.)

Yet evidence from Pangaea argues against a colder world. No ice sheets existed on Pangaea 200 Myr ago, even though its northern and southern limits lay within the Arctic and Antarctic circles (Figure 5-12A). Today landmasses at similar latitudes are either permanently ice-covered (Greenland) or alternately ice-covered and ice-free through time (North America, Europe, Asia). The absence of polar ice suggests that Pangaea's climate was somewhat warmer than Earth's climate is today.

Fossil evidence of vegetation on Pangaea leads to the same conclusion. Except for a few surviving types such as the ginkgo tree (Figure 5-13), evolution has greatly altered Earth's vegetation since Pangaean times, and comparisons between plant types now and then have to be based on types with similar appearances rather than on identical species. Several kinds of palmlike vegetation that would have been killed by hard freezes existed on Pangaea at latitudes as high as 40°. This suggests that the equatorward limit of hard freezes on Pangaea was 40°, poleward of the modern limit of 30° to 40°.

The most likely reason for a warmer (rather than a colder) Pangaea is that the CO_2 level 200 Myr ago was considerably higher than it is today, more than compensating for the weaker Sun. In the model experiment examined here, it was arbitrarily placed at a level of 1650 parts per million, almost six times the natural (preindustrial) value of 280 parts per million. As we will see, this choice not only produced temperature distributions consistent with the evidence from ice and from frost-sensitive vegetation but also simulated other cli-

matic features that match independent evidence from the Pangaean geologic record.

With the critical boundary conditions specified, the model simulation is ready to be run. After 15 years of simulated time to allow the model climate to come to a state of equilibrium (Chapter 3), the model uses the last 5 years for the actual simulation and runs through 5 full seasonal cycles, driven by strong solar radiation in summer and weak radiation in winter.

5-5 Output from the Model Simulation of Pangaean Climate

Because of its huge size, we might anticipate that the interior of Pangaea would have had an extremely dry continental climate, in the absence of the moderating influence of oceanic moisture. The climate model simulation confirms this expectation.

The model simulates widespread aridity at lower latitudes, especially in the Pangaean interior. Mean annual precipitation and soil moisture levels are low across large expanses of interior and western Pangaea between 40°S and 40°N (Figure 5-14). Precipitation values of 1–2 millimeters per day in these regions are

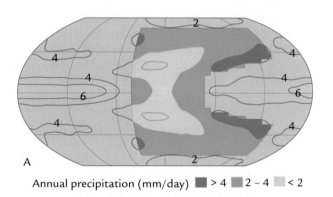

A

Annual precipitation (mm/day) ■ > 4 ■ 2 – 4 ▢ < 2

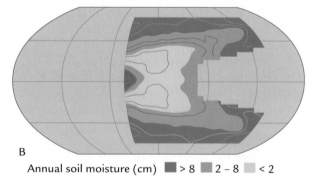

B

Annual soil moisture (cm) ■ > 8 ■ 2 – 8 ▢ < 2

FIGURE 5-14 **Precipitation on Pangaea** Climate models simulate patterns of annual mean precipitation (A) and annual soil moisture (B) on Pangaea. Broad areas of the tropics and subtropics are very dry. (Adapted from J. E. Kutzbach, "Idealized Pangean Climates: Sensitivity to Orbital Change," *Geological Society of America Special Paper 288* [1994]: 41–55.)

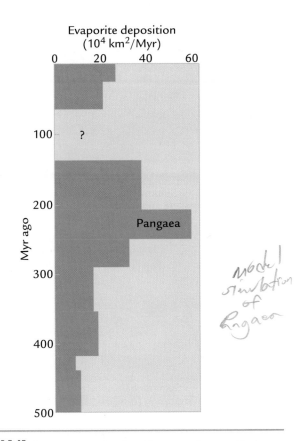

Evaporite deposition
(10⁴ km²/Myr)

FIGURE 5-15 **Pangaean evaporites** Greater volumes of salt deposits (evaporites) formed on Pangaea about 200 Myr ago than at any other time in the last 500 Myr. (Adapted from W. A. Gordon, "Distribution by Latitude of Phanerozoic Evaporites," *Journal of Geology* 83 [1975]: 671–84.)

equivalent to annual totals of 15–25 inches (35–70 cm) per year, comparable to those in semiarid grassland areas such as the western plains in the United States today (Figure 5-14A).

This pervasive aridity reflects two factors: (1) the great expanses of land at subtropical latitudes beneath the dry, downward-moving limb of the Hadley cell (Chapter 2) and (2) the large amount of land in the tropics, causing trade winds to lose most of their water vapor by the time they reached the continental interior. The ocean received far more rainfall than the land, even more than it does today.

Geologic evidence supports the model simulation of widespread Pangaean aridity. The clearest evidence comes from the distribution of **evaporite** deposits, salts that precipitated out of water either in lakes or in coastal margin basins with limited exchanges of water with the ocean. Evaporite salts form only in arid regions where evaporation far exceeds precipitation. More evaporite salt was deposited during the later phases of Pangaea, near 200 Myr ago, than at any time in the last several hundred million years (Figure 5-15), with many

deposits in the Pangaean interior. Evaporite deposits also occurred along the tropical east coast of Pangaea, a region the model simulates as less arid.

Because the moderating effects of ocean moisture did not reach much of Pangaea's interior, the continent was left vulnerable to seasonal extremes of heating by the Sun in summer and cooling during winter. As a result, the model simulates a huge seasonal temperature response (Figure 5-16). In some mid-latitude regions, summer daily mean temperatures of +25°C (77°F) alternated with winter daily mean temperatures of −15°C (+5°F).

The occurrence of extremely continental climates on Pangaea may help to explain the lack of ice sheets at high latitudes. The simulated winter temperatures were certainly cold enough to provide the snowfall needed for ice sheets to grow. But summers on Pangaea were hot, with simulated mean daily temperatures of 20°C even on the poleward margins of the landmass. Temperatures so far above freezing cause rapid melting of snow, and fast summer melting prohibits glaciation. Glaciation can develop more readily on smaller continents where summer temperatures are kept at more moderate levels by the greater influence of cool, moist summer winds off the ocean.

The model simulation also indicates that average daily temperatures in winter would have reached the freezing point as far equatorward as 40° latitude (Figure 5-16), closely matching the low-latitude limit of frost-

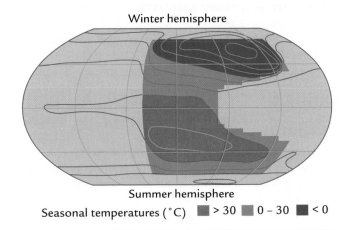

Winter hemisphere

Summer hemisphere

Seasonal temperatures (°C)   ▨ > 30   ▨ 0 – 30   ■ < 0

FIGURE 5-16 **Temperatures on Pangaea** Climate models simulate seasonal temperatures on Pangaea. Extreme seasonal contrasts are simulated for the summer hemisphere, warmed by solar radiation, and the winter hemisphere, which loses heat by longwave back radiation. The patterns shown switch back and forth between geographic hemispheres with changes in season. (Adapted from J. E. Kutzbach, "Idealized Pangean Climates: Sensitivity to Orbital Change," *Geological Society of America Special Paper* 288 [1994]: 41–55.)

sensitive vegetation on Pangaea. But with winter nights likely to have been colder than the daily mean, the model's results may actually disagree to some extent with the vegetation evidence. Despite the very high CO_2 values used as input to the simulation, freezing still occurs farther south in the model than the vegetation indicates. The same kind of mismatch between data and models recurs in subsequent intervals (Chapter 6).

Another fundamental feature of the climate of Pangaea was the strong reversal between summer and winter monsoon circulations. Monsoon circulations are driven by the different rates of response of the land and the oceans to solar heating in summer and radiative heat loss in winter (see Chapter 2). The large seasonal swings in land temperature and small seasonal changes in ocean temperature reflect these contrasting responses of land and ocean.

Strong solar heating over the part of Pangaea situated in the summer hemisphere causes heated air to rise over the land and a strong low-pressure cell to develop at the surface (Figure 5-17A). The rising of heated air causes a net inflow of moisture-bearing winds from the ocean, especially in the subtropics, bringing heavy rains to the subtropical east coast (Figure 5-17B).

The situation in the winter hemisphere is exactly the reverse. The weak radiative heat from the Sun and strong heat loss by back radiation cause cooling over the interior of Pangaea. The cooling causes air to sink toward the land surface, building up high pressures over the continent and causing cold, dry air to flow out to the ocean. As a result, precipitation over the land is reduced.

Note that the winds on the eastern margins of Pangaea from 0° to 45° latitude reverse direction between the seasons: warm summer monsoon winds blow from the sea onto the land, but cold winter monsoon winds blow from the land out to sea. As a result, the subtropical margins of Pangaea were places of enormous contrast in seasonal precipitation, alternating between extremely wet summers and dry winters.

Geologic evidence of seasonal moisture contrasts on Pangaea comes from the common occurrence of **red beds**, sandy or silty sedimentary rocks stained various shades of red by oxidation of their iron minerals. Red-colored sediments accumulate today in regions where the contrast in seasonal moisture is strong. The process of oxidation is analogous to rust that forms when metal tools are left out in the rain. In a geologic context, the wet season provides the necessary moisture, and the rust forms during the dry season or shorter dry intervals. Red beds were more widespread on Pangaea than during other geologic intervals, and this finding is consistent with the model simulation of highly seasonal changes in moisture between wet summer and dry winter monsoons.

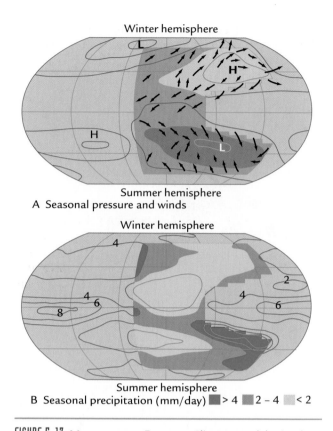

FIGURE 5-17 Monsoons on Pangaea Climate models simulate seasonal changes in surface pressure and winds (A) and monsoonal precipitation (B) on Pangaea. Summer heating creates a low-pressure region (L) and draws in moist oceanic winds, which drop heavy precipitation along the subtropical east coast. Winter cooling creates a high-pressure cell (H) that sends dry air out from land to sea and reduces precipitation. (Adapted from J. E. Kutzbach, "Idealized Pangean Climates: Sensitivity to Orbital Change," *Geological Society of America Special Paper* 288 [1994]: 41–55.)

Tectonic Control of CO_2 Input: The BLAG Spreading Rate Hypothesis

Our examination of both the polar position hypothesis and the climate of Pangaea suggests that changes in Earth's geography alone cannot explain the climatic trends of the last 550 Myr, and that changes in atmospheric CO_2 are also an important factor. In the remainder of this chapter we examine two competing hypotheses proposed to explain why CO_2 has changed through time, causing the observed variations between warm CO_2-rich greenhouse climates and cold CO_2-poor icehouse climates. The first hypothesis emphasizes changes in CO_2 input by volcanoes; the other focuses on changes in CO_2 removal by weathering.

A hypothesis published in 1983 proposed that climate changes during the last several hundred million years have been driven mainly by changes in the rate of

CO_2 input to the atmosphere (and ocean) by plate tectonic processes. This hypothesis is often referred to as the **BLAG hypothesis**, based on the initials of its authors (the geochemists Robert *B*erner, Antonio *L*asaga, and Robert *G*arrels). We will also refer to it as the **spreading rate hypothesis**.

5-6 Control of CO_2 Input by Seafloor Spreading

In our world of active plate tectonic processes, carbon cycles constantly between Earth's interior and its surface. It moves from the deep rock reservoirs to the surface mainly as CO_2 gas associated with volcanic activity along the margins of Earth's tectonic plates. CO_2 is expelled to the atmosphere today mainly at two kinds of sites (Figure 5-18): (1) at the margins of convergent plates, where portions of subducting plates melt and form molten magmas that rise to the surface in moun-

tain belt and island arc volcanoes, carrying CO_2 and other gases from Earth's interior to its atmosphere; and (2) at the margins of divergent plates (ocean ridges), where hot magma carrying CO_2 erupts directly into ocean water. These two plate margins where most CO_2 is expelled from Earth's interior are the most common types of plate boundaries found on our planet.

Some volcanoes emit CO_2 away from plate boundaries, at sites where thin plumes of molten material rise from deep within Earth's interior and reach the surface at volcanic **hot spots** (Figure 5-18, bottom). Most of these plumes penetrate from Earth's depths to the surface within the interior of the rigid plates. Finally, some additional CO_2 is released to the atmosphere by the slow oxidation of old organic carbon in sedimentary rocks eroded at Earth's surface (Chapter 4).

The centerpiece of the BLAG hypothesis is the concept that changes in the rate of seafloor spreading over

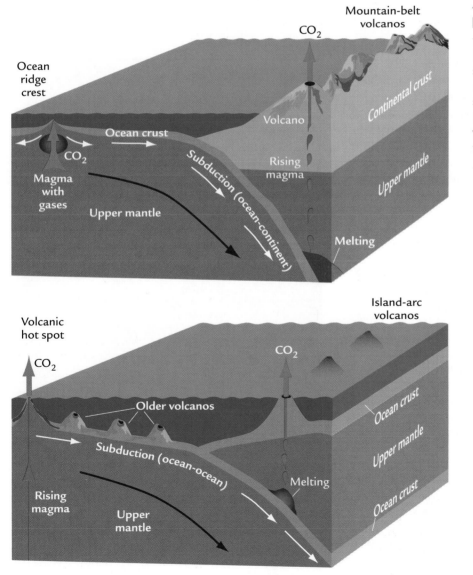

FIGURE 5-18 Sites of CO_2 input CO_2 is transferred from rocks in Earth's interior to the atmosphere-ocean system mainly at two kinds of plate margins: ocean ridges (top left) and subduction zones (top and bottom right). A smaller input of CO_2 occurs when volcanoes erupt at hot spots in the middle of plates (bottom left).

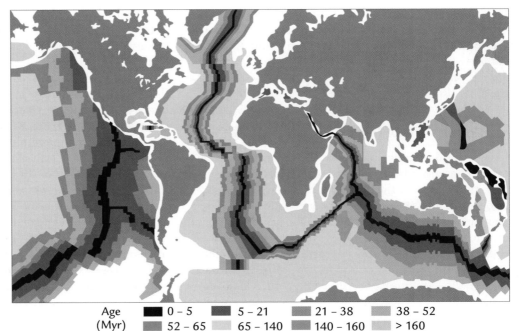

FIGURE 5-19 Age of the seafloor The age of the crust in all ocean basins is known; some crust dates as far back as 175 Myr ago. Spreading rates are as much as ten times faster in the Pacific than in the Atlantic. (Modified from S. Stanley, *Earth System History*, © 1999 by W. H. Freeman and Company, after W. C. Pitman et al., Map and Chart Series MC-6 [Boulder, Colo.: Geological Society of America, 1974].)

Age (Myr)			
■ 0 – 5	■ 5 – 21	■ 21 – 38	■ 38 – 52
■ 52 – 65	■ 65 – 140	■ 140 – 160	■ > 160

millions of years control the rate of delivery of CO_2 to the atmosphere from the large rock reservoir of carbon, with the resulting changes in atmospheric CO_2 concentrations controlling Earth's climate. Dated magnetic lineations show that ocean ridges have been spreading at widely varying rates for millions of years, and satellite measurements of recent plate movements confirm that these rates have continued during the last few decades. The ridge in the South Pacific Ocean spreads up to ten times faster than the Mid-Atlantic Ridge (Figure 5-19).

Similarly, through time intervals such as the present, when the globally averaged rate of seafloor spreading is slow, have alternated with times when the average spreading rate was faster, as it was 100 million years ago. The BLAG hypothesis focuses on these changes in the *globally averaged* rate of seafloor spreading.

Changes in the rate of spreading over tectonic intervals of tens of millions of years affect the transfer of CO_2 from Earth's rock reservoirs to its atmosphere at two locations: spreading ocean ridges and subduction zone volcanoes (Figure 5-18). These plate margins are intimately linked to the process of seafloor spreading: ridges are the sites where new ocean crust is created, and subduction zones are the sites where older crust is buried.

The BLAG hypothesis proposes that faster rates of spreading at ridge crests not only create larger amounts of new ocean crust but also cause more frequent releases of magma, which deliver greater amounts of CO_2 to the ocean (Figure 5-20). Faster spreading also causes more rapid subduction of crust and sediment in ocean

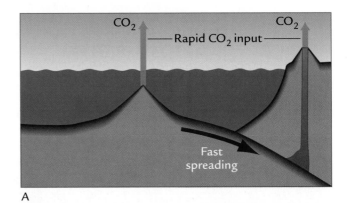

A

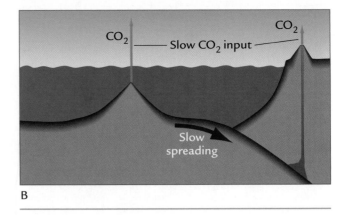

B

FIGURE 5-20 Spreading rates and CO_2 input The BLAG spreading rate hypothesis asserts that atmospheric CO_2 and global climate are driven by the global mean rate of seafloor spreading, which controls the global rate of CO_2 input at ocean ridge crests and subduction zones.

trenches and delivers larger volumes of carbon-bearing sediment and rock for subsequent melting and CO_2 release through volcanoes. Conversely, slower rates of spreading should reduce both kinds of CO_2 input to the atmosphere.

Although the BLAG hypothesis focuses on changes in spreading rates as the driver of climate change, it also calls on chemical weathering for negative feedback to moderate these changes, invoking the same process discussed in Chapter 4. Increased volcanic CO_2 input because of faster seafloor spreading leads to higher atmospheric CO_2 levels and a warm climate (Figure 5-21, top). This initial shift toward a greenhouse climate activates the combined effect of temperature, precipitation, and vegetation in speeding up the rate of chemical weathering and causes CO_2 to be drawn out of the atmosphere at a faster rate. The resulting CO_2 removal opposes and reduces some of the initial warming driven by faster spreading rates and higher CO_2 concentrations.

Chemical weathering feedback also works in reverse, offsetting some of the impact of cooling caused by slower volcanic input of CO_2 (Figure 5-21, bottom). In effect, the BLAG hypothesis calls on chemical weathering to moderate any fluctuation in climate driven by changes in volcanic CO_2 input.

The BLAG hypothesis further proposes that much of the cycling of carbon between the deeper Earth and the atmosphere occurs in a closed loop (Figure 5-22). Carbon taken from the atmosphere during chemical weathering is initially stored in dissolved HCO_3^- ions that flow in rivers to the sea. As we have seen, marine plankton use this dissolved carbon to form $CaCO_3$ shells, and the shells are deposited in ocean sediments when the organisms die. The movement of carbon through this part of the cycle is rapid, occurring in just a few years.

The $CaCO_3$-bearing sediments are then carried by seafloor spreading toward subduction zones at continental margins. There some sediment is scraped off at the ocean trenches, but most is carried downward in the subduction process (Figure 5-22). Much of the sediment initially scraped off at the trenches is later eroded, again carried down into the trench bottoms, and finally subducted. The slow movement of carbon-bearing sediments across the ocean floor and down into the trenches takes tens of millions of years.

Most of the $CaCO_3$ (and other carbon) carried down into Earth's interior by subduction melts at the hot temperatures found at great depths or is transformed in other ways by high pressure and temperature.

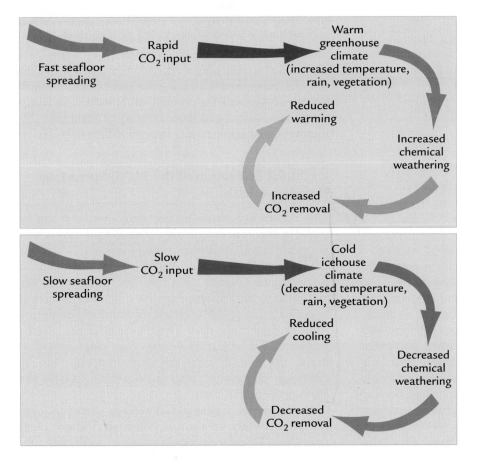

FIGURE 5-21 Negative feedback in BLAG The BLAG hypothesis invokes chemical weathering as a negative feedback that partially counters the changes in atmospheric CO_2 and global climate driven by changes in rates of seafloor spreading.

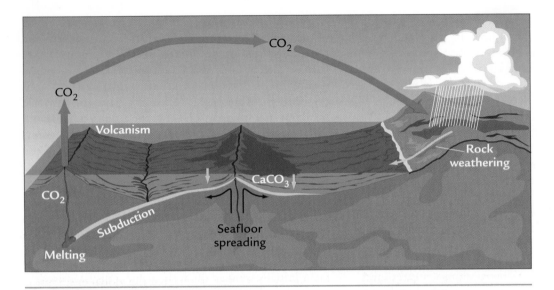

FIGURE 5-22 **Carbon cycling** In the BLAG hypothesis, carbon cycles continuously between rock reservoirs and the atmosphere: CO_2 is removed from the atmosphere by chemical weathering on land, deposited in the ocean, subducted, and returned to the atmosphere by volcanic activity. (Adapted from W. F. Ruddiman and J. E. Kutzbach, "Plateau Uplift and Climate Change," *Scientific American* 264 [1991]: 66–75.)

These processes eventually return CO_2 to the atmosphere through volcanoes and complete the cycle. Almost none of the carbon subducted is carried deep into the mantle. Movement of carbon through this deeper part of the cycle takes tens of millions of years.

The two reactions that summarize the basic chemical changes involved at the beginning and end of this tectonic-scale carbon cycle are mirror opposites:

Chemical weathering on land
$$CaSiO_3 \; + \; CO_2 \; \rightarrow \; CaCO_3 \; + \; SiO_2$$
Silicate rock   Atmosphere   Plankton   Plankton

Melting and transformation in subduction zones
$$CaCO_3 \; + \; SiO_2 \; \rightarrow \; CaSiO_3 \; + \; CO_2$$
Ocean sediments   Silicate rock   Atmosphere

Although the two reactions combined indicate a complete (closed) cycle, the cycle actually is completed only after a delay of many tens of millions of years. The delay is caused by the slow spreading and subduction of seafloor, the slow transformation of $CaCO_3$ at depth, and slow delivery of CO_2 to volcanoes. Despite this delay, changes in spreading rates can alter the rate of melting and CO_2 release to the atmosphere with little or no delay because carbon-bearing sediment is already in the pipeline. At any interval in time, carbon-bearing sediments are in the process of being subducted into Earth's interior, and changes in the global average rate of subduction soon result in faster melting of this material.

The BLAG hypothesis proposes that this cycling of carbon provides long-term stability to the climate system by moving a roughly constant amount of total carbon back and forth between the rocks and the atmosphere over long intervals of time. As a result, atmospheric CO_2 levels are constrained to vary only within moderate limits. But the long delays between weathering and burial do permit small imbalances between the rate of CO_2 burial and its return to the atmosphere, and these imbalances could drive climate changes over intervals of tens of millions of years.

5-7 Initial Evaluation of the BLAG Spreading Rate Hypothesis

The BLAG hypothesis makes predictions that can be tested over part of the geologic past. We saw earlier that magnetic lineations preserved on today's seafloor can be used to date the ocean crust. Each strip of dated crust lies a specific distance from the adjacent strip of crust, and dividing this distance by the difference in crustal ages yields the spreading rate at this one location during some interval in the past. These local rates can then be integrated across Earth's entire ocean ridge system to derive the globally averaged spreading rate, and by extension the relative rate of release of CO_2 to the atmosphere at that time.

Calculations of past global average spreading rates can be used only for the last 100 Myr. Seafloor older than about 175 Myr no longer exists; it has all been subducted in ocean trenches. The amount of seafloor

TABLE 5-2   Evaluation of the BLAG Spreading Rate (CO_2 Input) Hypothesis

Time (Myr ago)	Ice sheets present?	Spreading rates	Hypothesis supported?
100	No	Fast	Yes (high CO_2)
0	Yes	Slow	Yes (low CO_2)

remaining with ages between 175 and 100 Myr (see Figure 5-19) does not allow reliable calculations of global average spreading rates. Steadily increasing preservation of seafloor younger than 100 Myr does permit calculation of spreading rates, with increasing reliability toward the present day.

Calculations suggest that the global mean spreading rate was as much as 50% faster 100 Myr ago than it is at present, although the *exact* amount cannot be calculated with certainty. On the basis of this estimate, the BLAG theory predicts that the rate of input of CO_2 from the rocks to the atmosphere should have been higher 100 Myr ago than it is today, and that the resulting higher CO_2 levels in the atmosphere should have produced a warmer Earth.

This prediction of a warmer Earth 100 Myr ago is consistent with geologic data (Table 5-2). Ice sheets were totally absent at that time, even on the combined landmass of Antarctica, India, and Australia, located directly over the South Pole. The warmer climate and CO_2-rich atmosphere predicted by the BLAG theory provide one possible explanation for the lack of polar ice. We will examine the causes of this warm greenhouse world further in Chapter 6.

Tectonic Control of CO_2 Removal: The Uplift Weathering Hypothesis

A second hypothesis advanced to explain how plate tectonic processes might control atmospheric CO_2 levels emerged from work by the marine geologist Maureen Raymo and her colleagues in the late 1980s. Parts of this concept date back to work by the geologist T. C. Chamberlain a century ago. This **uplift weathering hypothesis** looks at chemical weathering as the active driver of climate change, rather than as a negative feedback that moderates climate changes.

5-8   Rock Exposure and Chemical Weathering

The BLAG hypothesis views chemical weathering as responding to three climate-related factors: tempera-

ture, precipitation, and vegetation. Although these factors do affect chemical weathering, they are not the only factors capable of doing so.

The uplift weathering hypothesis starts from a different perspective. It asserts that the global mean rate of chemical weathering is heavily affected by the availability of fresh rock and mineral surfaces that the weathering process can attack, and that this exposure effect can override the combined effects of the three climate-related factors both in some regions and globally.

One simple conceptual example of the importance of rock exposure is shown in Figure 5-23. It starts with a large cube of rock whose 6 surfaces consist of squares 1 meter across, each having a surface area of 1 m². This cube has a total surface area of 6 m², easily calculated from the total areas of the 6 sides.

Next we slice this cube into equal halves along all three of its major axes. This slicing creates 8 smaller cubes, each 0.5m on a side, so that each side has a surface area of 0.25m². The total surface area of these smaller cubes is 12m²:

$$(8 \text{ cubes}) \times (6 \text{ sides each}) \times (0.25\text{m}^2 \text{ of surface area per side})$$

Note that simply cutting the large cube into smaller cubes has doubled the surface area of the rock without changing its volume, at least for the imaginary laser-sharp cut in this example. The fragmentation has created more exposed surface area for the weathering process to attack.

The process can be carried on to finer grain sizes, with similar results. Ten sequential halvings of the rock's dimensions will produce over 1 billion cubes each 1 mm on a side, about the same size as grains of sand on a beach. Together these tiny cubes have a total surface area 1000 times larger than the original 1-m³ block, yet still retain the same volume. If fragmentation down to even smaller sizes occurred, it would expose still more surface area.

With over 1000 times more surface area to act on, chemical weathering of this fragmented material would increase by a factor of 1000 or more just because of this

Volume = l × w × h
Surface area = (l × w) × number of faces

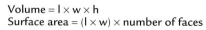

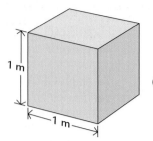
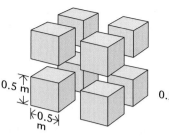
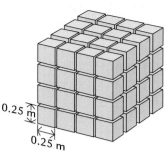
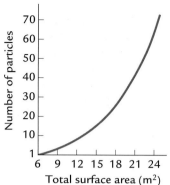

1 cube, 6 faces
Volume = 1 m³
Surface area = 6 m²

8 cubes, 48 faces
Total volume = 1 m³
Surface area of 1 cube = 1.5 m²
Total surface area = 12 m²

64 cubes, 384 faces
Total volume = 1 m³
Surface area of 1 cube = 0.375 m²
Total surface area = 24 m²

Dimensions of cube faces (m)	Number of cubes	Total surface area of cubes (m²)	Total volume of cubes (m³)
1.0	1	6	1
0.5	8	12	1
0.25	64	24	1
0.125	512	48	1
0.062	4096	96	1
0.031	32,768	192	1
0.016	262,144	384	1
0.008	2,100,000	768	1
0.004	16,800,000	1536	1
0.002	134,000,000	3072	1
0.001	1,100,000,000	6144	1

FIGURE 5-23 **Fragmentation of rock** Each time a cube-shaped rock is sliced into smaller cubes (with sides half as long as before), the total surface area of rock doubles, even though the volume remains the same. (D. Merritts et al., *Environmental Geology,* © 1997 by W. H. Freeman and Company.)

one effect. Such a proportional increase of weathering far exceeds the changes estimated to result from changes in temperature, precipitation, and vegetation. Clearly these climate-related factors are not the only processes to consider in evaluating chemical weathering.

Case Study: The Wind River Basin of Wyoming
Direct evidence of the importance of rock exposure in chemical weathering emerges from a recent study of a drainage basin in the Wind River Mountains of Wyoming. As all the bedrock in this basin consists of granite, the kind of silicate rock most typical of continental crust, this watershed is reasonably representative of the average response of continental rocks to weathering.

The Wind River Mountains have been glaciated repeatedly over the last several hundred thousand years, and each glaciation has left moraines, deposits of unsorted debris, in the foothills of the valleys below (Chapter 3). Because some of the older deposits have not been overridden by later glacial advances, several undeformed moraines of various ages can be found in the same valley. These moraines range in age from 200 years to 130,000 years.

The Wind River moraines provide an opportunity to quantify the amount of weathering of ground-up debris that is identical in composition but differs widely in age. The amount of weathering is determined by analyzing soils developed on the moraines. Chemical weathering turns the glacial debris into soils that gradually lose their major cations (Mg^{+2}, Na$^+$, K$^+$, Ca^{+2}, etc.) during the weathering process. The cumulative amount of chemical weathering that has occurred since each moraine was deposited can be determined by measuring the total loss of these cations. Dividing this total amount of weathering by the time elapsed since the moraine was deposited yields the *average* rate of chemical weathering over that entire interval.

The Wind River deposits show an exponential decrease in the mean rate of weathering versus their time of exposure (Figure 5-24). The younger moraines have average rates of weathering that are at least a factor of 100 faster than the older ones. The older moraines weathered much faster during a brief interval after deposition, but much more slowly later on. Why would the younger glacial deposits weather so much faster?

One explanation is that freshly ground rock has more weatherable material—that is, the kinds of fresh, unweathered silicate grains that are most vulnerable to the weathering process. These vulnerable minerals have been removed from older deposits through time. With only the more resistant minerals left, rates of weathering are slower.

Another part of the explanation relates to the effect of grain sizes on weathering (Figure 5-23). Finer grain sizes expose more surface area and cause faster weathering early in the process, but the finer sizes also disappear earlier as weathering quickly consumes them. The

coarser grain sizes that are left weather more slowly because they expose less surface area per unit of volume. Coarser fragments may also develop an outer coating or "rind" of weathering-resistant material that protects fresh material in their interiors and slows the weathering attack.

5-9 Uplift and Chemical Weathering

The uplift weathering hypothesis begins with the evidence that exposure of fragmented and unweathered rock is an important factor in the intensity of chemical weathering. It then links this evidence to the fact that exposure of freshly fragmented rock is enhanced in regions of tectonic uplift.

Several factors increase rates of exposure of fresh rock in uplifting areas. Mountains and plateaus are by definition high-elevation regions, with steep slopes both on their margins and in mountain valleys between high peaks. Erosional processes known as **mass wasting** are unusually active on steep slopes. Mass-wasting processes include rock slides and falls, flows of water-saturated debris, and a host of other processes that dislodge everything from huge slabs of rock to loose boulders, pebbles, and soil. Every event that removes overlying debris exposes fresh bedrock and unweathered material. Many high-mountain slopes consist almost entirely of fresh debris in frequent downslope movement (Figure 5-25).

Another important factor is the frequency of earthquakes. Mountains and high plateaus are built by tectonic forces that push together and stack huge slivers of faulted rock at the margins of converging plates. This stacking process is accompanied by earthquakes that generate large amounts of energy, shake the ground, and dislodge debris perched on unstable steep slopes. Even more fresh rock is exposed as a result.

A third important characteristic of steep slopes is that they are focal points for precipitation. Precipitation falls on these slopes when warm air is forced up and over high terrain, where it cools and condenses. Precipitation is heaviest on slopes facing the direction from which the wind is blowing, especially winds that carry moisture from the ocean (see Figure 2-20). Heavy precipitation favors chemical weathering.

Two kinds of high terrain generate heavy precipitation on their slopes. High but narrow mountain belts intercept much of the moisture carried by tropical easterlies and mid-latitude westerlies. In addition, large plateaus such as the Tibetan Plateau create their own monsoonal circulations by pulling moisture in from adjacent oceans (see Chapter 7).

Glacial ice also enhances chemical weathering in high terrain. Uplift can elevate rock surfaces to high

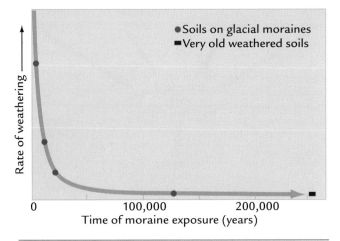

FIGURE 5-24 Weathering and exposure time Glacially eroded and fragmented granite debris weathers quickly soon after deposition, but much more slowly by 100,000 years later. (Adapted from J. D. Blum, "The Effect of Late Cenozoic Glaciation and Tectonic Uplift on Silicate Weathering Rates," in *Tectonic Uplift and Climate Change*, ed. W. F. Ruddiman [New York: Plenum Press, 1997].)

[handwritten margin notes:]
older deposits already lost their more vulnerable minerals

all based on tectonics, mt building, forcing mass wasting etc

FIGURE 5-25 **Debris on steep slopes** Steep slopes of actively uplifting mountains typically consist of highly fragmented debris periodically dislodged downslope. (Photosphere Images/Picture Quest.)

altitudes where temperatures are cold enough for mountain glaciers to form. Mountain glaciers pulverize blocks of underlying bedrock, carry the debris to lower elevations, and deposit it in moraines. As we saw in the study of the Wind River Range, glacial grinding can greatly enhance regional rates of chemical weathering.

All these factors (steep slopes, mass wasting, earthquakes, heavy precipitation, and mountain glaciation) are inherent in the existence of high mountains and plateaus. The uplift weathering hypothesis proposes that uplift in specific regions creates these conditions and accelerates chemical weathering (Figure 5-26). Faster weathering in turn draws CO_2 out of the atmosphere at a more rapid rate and cools global climate toward icehouse conditions. Conversely, during times when uplift is less prevalent, chemical weathering is slower, and CO_2 stays in the atmosphere and warms the climate, producing greenhouse conditions.

Two kinds of plate tectonic processes cause uplift, with different implications for the uplift weathering hypothesis. The first process, subduction of ocean crust underneath continental margins, is an integral part of plate movements and a process that is continually active in many regions on Earth. Because subduction occurs relatively steadily over time, the total amount of high mountain terrain on Earth is likely to be relatively constant through time, even though the locations and heights of individual mountain ranges vary.

The second process that creates high terrain is the collision of continents. These events are far more sporadic. Collision between India and Asia over the last 55 Myr created the Tibetan Plateau of southern Asia, but no plateau-like feature remotely close in size to Tibet had existed on Earth between 240 and 55 Myr ago, because no major continental collisions occurred.

Earlier, between 325 and 240 Myr ago, the collisions between Gondwana and other continents, which created the supercontinent Pangaea, produced moderate-sized plateaus and high mountain ranges in several regions, including the Appalachians of eastern North America and other regions in eastern Europe and northwestern Africa.

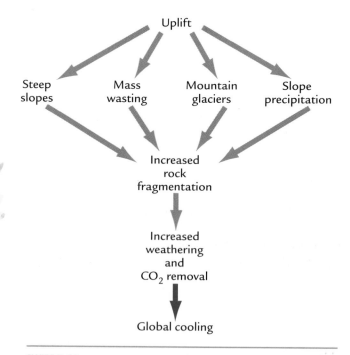

FIGURE 5-26 **Uplift weathering hypothesis** Active tectonic uplift produces several tectonic and climatic effects that cause strong weathering of freshly exposed and fragmented rock. This process removes CO_2 from the atmosphere and cools global climate.

of weathering processes, buried hundreds of meters beneath a protective cover of highly weathered clays. These clays are the end products of slow bedrock weathering over many millions of years, and they have little weatherable material left. As a result, the average rate of chemical weathering in this region is extremely low.

In contrast, the physical impacts of active uplift in the Andes (steep slopes, earthquakes, mass wasting, heavy precipitation, and glacial erosion) combine to generate a continual supply of fresh, finely ground rock debris for weathering. Some of this weathering occurs on the steep upper slopes of exposed high terrain even in the absence of much vegetation or soil cover. Much of it probably occurs in basins lower in the mountains, where soils and vegetation succeed in gaining a tenuous foothold, and yet the supply of fresh unweathered rock debris from higher-elevation streams and rivers is continuous.

The lack of obvious visible chemical weathering in the Andes has two explanations. First, chemical weathering products such as clays are continually overwhelmed by the much larger supply of physically fragmented debris cascading down the steep slopes. And second, the fine clays and other products of weathering are continually removed and carried down to the ocean by streams and rivers.

The Amazon Basin studies confirm that the rate of chemical weathering is rapid in the Andes, and presumably in many of Earth's other high-elevation regions as well, even though the visible effects of chemical weathering are not apparent. These studies also show that some warm, wet, vegetated regions may be places of surprisingly slow chemical weathering.

The evidence of slow weathering in the lower Amazon Basin might seem to indicate that climate has *no* significant control over chemical weathering, but such a conclusion would not be justified. Many other low-elevation regions with warm, wet climates and lush vegetation lack the protective mantle of highly weathered soil found in the lower Amazon Basin. As a result, weathering is faster in these regions, and the effect of climate on weathering is stronger. Climate remains a potentially significant control on chemical weathering across much of Earth's surface.

5-10 Weathering: Climate Forcing and Feedback

The original uplift weathering hypothesis left an important issue unresolved. It did not specify a negative feedback that would act as a thermostat and moderate the climatic effects produced by rapid uplift. Without such a thermostat, what would stop rapid uplift from accelerating chemical weathering to the point where Earth would freeze? And why wouldn't Earth overheat during times when uplift was minimal?

One possible mechanism that could moderate the degree of uplift-induced climate change is the amount of fresh rock exposed at Earth's surface. Plate tectonic processes can cause uplift across only a limited amount of Earth's surface at any one time because of the small length of plate margins involved in continental collisions, along with the limited regions actively involved in subduction processes on continental margins. These natural tectonic limits on the geographic extent of uplift should limit the amount of exposure of fresh rock at any one time and could thereby set a natural limit on the intensity of cooling that can be caused by uplift.

How this explanation would limit the amount of climate change that occurs when uplift slows is less clear. One possibility is that the ongoing subduction process in many regions always provides just enough uplift and fragmentation of fresh rock to keep weathering rates high enough so that climate does not become excessively hot. Climate then remains in a warm greenhouse condition until the next major continental collision and subsequent cooling.

A more intriguing source of negative feedback may be the chemical weathering process itself. Because chemical weathering may react to different factors in different regions, it may function both as the driver that initiates climate change and as the thermostat that moderates it. In actively uplifting regions, uplift could cause increased chemical weathering, pull more CO_2 out of the atmosphere, and cool global climate. But the regions of rapid uplift form only a small percentage (perhaps 1% to 2%) of the total area of Earth's land surface. Across the remaining 98% or more of Earth's continents, chemical weathering might well slow with the onset of colder, drier climates and reduction in vegetation cover. A slowing of the rate of CO_2 removal would leave more CO_2 in the atmosphere and moderate the overall cooling. In the end, the uplift-induced weathering increase would succeed in causing a net global cooling, but not nearly so large a cooling as would have occurred without the negative weathering feedback.

In summary, two hypotheses attempt to explain varying atmospheric CO_2 levels and resulting changes in climate during the last several hundred million years: the BLAG (spreading rate) hypothesis and the uplift weathering hypothesis. Both seem consistent with the major icehouse-greenhouse changes of climate over the last 400 million years (Tables 5-2, 5-3). In Chapter 7 we will revisit and further test both hypotheses by examining in greater detail the sequence of changes from the warm greenhouse climate of 100 Myr ago to the modern icehouse climate.

KEY TERMS

continental crust (p. 104)

ocean crust (p. 104)

mantle (p. 104)

lithosphere (p. 104)

asthenosphere (p. 105)

tectonic plates (p. 105)

divergent margins (p. 105)

convergent margins (p. 105)

subduction (p. 105)

continental collision (p. 106)

transform fault margins (p. 106)

magnetic field (p. 106)

paleomagnetism (p. 106)

magnetic stratigraphy (p. 108)

magnetic lineations (p. 108)

seafloor spreading (p. 108)

polar position hypothesis (p. 109)

Gondwana (p. 110)

Pangaea (p. 110)

evaporites (p. 115)

red beds (p. 116)

BLAG (spreading rate) hypothesis (p. 117)

hot spots (p. 117)

uplift weathering hypothesis (p. 121)

mass wasting (p. 123)

REVIEW QUESTIONS

1. How thick are the lithospheric plates that move across Earth's surface?

2. Does each plate correspond to an individual continent or ocean basin?

3. What kind of physical behavior in Earth's deeper layers allows its tectonic plates to move?

4. How does paleomagnetism tell us about rates of spreading at ocean ridges?

5. How does paleomagnetism tell us about past latitudes of continents?

6. Do glaciations generally occur when continents are located in polar positions?

7. Do glaciations *always* occur when continents are located in polar positions?

8. Describe three general characteristics of the climate of Pangaea.

9. What is the central concept behind the BLAG (spreading rate) hypothesis?

10. What role does chemical weathering play in the BLAG hypothesis?

11. Write a reaction showing how chemical weathering removes CO_2 from the atmosphere.

12. How soon after deposition does freshly fragmented debris undergo most chemical weathering?

13. What factors control rates of chemical weathering in the uplift weathering hypothesis?

14. Why is chemical weathering faster in the eastern Andes than in the Amazon lowlands?

15. How could chemical weathering be both the driver and the thermostat of Earth's climate?

ADDITIONAL RESOURCES

Advanced Reading

Berner, R. A. 1999. "A New Look at the Long-Term Carbon Cycle." *GSA Today* 9:1–6.

Blum, J. D. 1997. "The Effect of Late Cenozoic Glaciation and Tectonic Uplift on Silicate Weathering Rates and the Marine $^{87}Sr/^{86}Sr$ Record." In *Tectonic Uplift and Climate Change*, ed. W. F. Ruddiman. New York: Plenum Press.

Chamberlain, T. C. 1899. "An Attempt to Frame a Working Hypothesis of the Cause of Glacial Periods on an Atmospheric Basis." *Journal of Geology* 7:545–84, 667–85, 751–87.

Kutzbach, J. E. 1994. "Idealized Pangaean Climates: Sensitivity to Orbital Parameters." *Geological Society of America Special Paper* 288:41–55.

Parrish, J. T., A. M. Ziegler, and C. R. Scotese. 1982. "Rainfall Patterns and the Distribution of Coals and Evaporites in the Mesozoic and Cenozoic." *Palaeogeography, Palaeoclimate, Palaeoecology* 40:67–101.

Raymo, M. E., W. F. Ruddiman, and P. N. Froelich. 1986. "Influence of Late Cenozoic Mountain Building on Ocean Geochemical Cycles." *Geology* 16:649–53.

Robinson, P. L. 1971. "A Problem of Faunal Replacement on Permo-Triassic Continents." *Paleontology* 14:131–53.

Sclater, J. G., L. Meinke, A. Bennett, and C. Murphy. 1985. "The Depth of the Ocean Through the Neogene." *Geological Society of America Memoir* 163:1–19.

Stallard, R. F., and J. E. Edmond. 1983. "Geochemistry of the Amazon 2: The Influence of the Geology and Weathering Environments on the Dissolved Load." *Journal of Geophysical Research* 88:9671–88.

CHAPTER 6

Greenhouse Earth

Earth's geography is much easier to reconstruct in the interval that begins 100 Myr ago than it is in earlier intervals. The positions of all the continents are known, as are the location of the edge of the sea along continental margins, the shapes of ocean basins, and the regional and global average rates of seafloor spreading. Climate scientists also know much about Earth's climate 100 Myr ago, including the revealing fact that it was warm enough at the poles to keep ice sheets from forming. This was a greenhouse world. With this array of evidence before us, scientists can ask (and hope to answer) important questions about this interval: Do climate models simulate the warmth of this greenhouse climate? And if so, did a high level of atmospheric CO_2 cause it?

In this chapter, we first examine the greenhouse world and evaluate explanations for its warmth. One result of this exercise will be to find out what lessons this past greenhouse world holds for our future climate in a world warmed by rising levels of greenhouse gases. Next, we explore the reasons why sea level 100 Myr ago was some 200 meters higher than it is today, and we evaluate the effects of high sea level on climate. Finally, we investigate the climatic and environmental effects of the impact of a giant asteroid that collided with Earth 65 Myr ago, during the later stages of warm greenhouse conditions.

What Explains Greenhouse Warmth 100 Myr Ago?

Around 175 Myr ago, the giant single continent of Pangaea began to break apart. By 100 Myr ago, most of the present-day continents had separated from one another, producing a very different-looking Earth consisting of a half-dozen smaller continents. In addition, for reasons that we will explore later in this chapter, global sea level stood at least 200 meters higher than it does today (a little higher than the Washington Monument). A shallow layer of ocean water flooded far up onto low-lying continental margins and into low-lying continental interiors.

The net geographic effect of this flooding was to fragment the existing continents into even smaller areas of land (Figure 6-1), making the geography of this greenhouse world even more unlike the single Pangaean landmass. Geologists call this interval the middle **Cretaceous**, a word meaning "abundance of chalk," because limestone deposited by these high seas is common around the world.

This interval is important to climate scientists for a number of reasons, the most important of which is that it is the most recent example in the geologic record of a warm greenhouse world. Geologic records from this interval contain no evidence of permanent ice anywhere on Earth, even on the parts of the Antarctic continent situated right over the South Pole (Figure 6-1).

In the northern hemisphere, evidence for unusual warmth comes from several fossil indicators, including warm-adapted vegetation such as broad-leafed evergreen trees that hold their leaves throughout the year except for a brief interval of leaf fall and regrowth. Other evidence includes warm-adapted animals such as dinosaurs, turtles, and crocodiles, all existing north of the Arctic Circle (Figure 6-2). This evidence contrasts markedly with the cold climatic conditions at such latitudes today.

The rest of Earth's surface also was warmer than it is today. Dinosaurs lived even on those parts of Australia and Antarctica lying south of the Antarctic Circle. And at lower latitudes, coral reefs indicative of warm tropical ocean temperatures extended some 10° of latitude farther from the equator than they do today (40° vs. 30°).

6-1 Model Simulations of a Greenhouse World

In order to understand the reasons for this greenhouse warmth, we can turn to climate models. Because the geography of this greenhouse world is well known, we can run model simulations based on a world that actual-

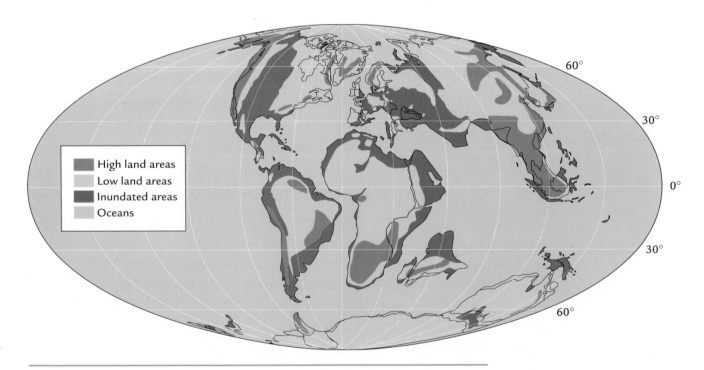

FIGURE 6-1 **The world 100 Myr ago** By 100 Myr ago, plate tectonic processes had broken the giant Pangaean continent into separate smaller continents, and flooding of these continents by shallow seas further reduced the extent of dry land. (D. Merritts et al., *Environmental Geology*, © 1997 by W. H. Freeman and Company.)

A

B

FIGURE 6-2 **Evidence of greenhouse warmth 100 Myr ago** Vegetation and animals that appear to have been warm-adapted lived in both polar regions 100 Myr ago: fossils of breadfruit trees like those that live today in the tropics (A) and dinosaurs (B), many species of which lived poleward of the Arctic and Antarctic circles. (A: Swedish Museum of Natural History, Yvonne Arremo, Stockholm; B: T. Steuberth/Institut für Geologie und Mineralogie der Universität Erlangen, Nuremberg.)

ly existed and expect to produce reasonably accurate simulations of its climate.

Climate scientists (particularly the climate modeler Eric Barron and his colleagues) pioneered the use of GCMs to explore the greenhouse climates of the Cretaceous. They first defined a "target signal" for the climate simulations. The target they used is an estimate of surface temperatures 100 Myr ago taken from compilations of the available climate information—past distributions of animals and vegetation, as well as geochemical evidence.

The temperature estimates at many locations were averaged around lines of latitude and plotted as a zonal mean trend (Figure 6-3). This kind of latitude-averaged plot is a useful shorthand way of capturing the general temperature trend across Earth's surface. Latitudes in Figure 6-3 (and subsequent figures) are plotted in such

a way as to correct for the much larger area of Earth's surface at low latitudes than in polar regions. For example, the temperature value at 0° latitude is an average based on the entire 40,000-kilometer circumference of Earth, while the 90° latitude average applies only to the single point centered on the pole. Earth's present-day temperature trend is also shown in Figure 6-3 for comparison with that of the Cretaceous.

The plot in Figure 6-3 shows that temperatures near the equator 100 Myr ago were a few degrees centigrade warmer than those today, but as much as 20° to 40°C warmer in the polar regions. Much of the huge temperature difference at the South Pole is due to the high elevation of the modern-day Antarctic ice sheet, with its surface at an altitude of 3 to 4 kilometers. As a result of the lapse-rate cooling of the atmosphere with elevation (Chapter 2), the modern surface temperature

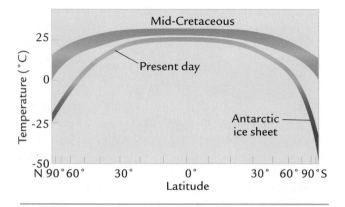

FIGURE 6-3 **The greenhouse target signal** Climate scientists have used geologic data (faunal, floral, and geochemical) to compile an estimate of temperatures 100 Myr ago. Temperatures were warmer than they are today at all latitudes, but especially in polar regions. (Adapted from E. J. Barron and W. M. Washington, "Warm Cretaceous Climates: High Atmospheric CO_2 as a Plausible Mechanism," in *The Carbon Cycle and Atmospheric CO_2: Natural Variations, Archaean to Present*, ed. E. T. Sundquist and W. S. Broecker, Geophysical Monograph 32 [Washington, D.C.: American Geophysical Union, 1985].)

at high southern latitudes is much cooler simply because the ice protrudes so high into the cold atmosphere. In addition, ice-albedo feedback from the ice sheet and the extensive cover of sea ice in winter further cool the south polar region today.

Barron and his colleagues chose to start the modeling process by running a very simple experiment, using the fewest and best-justified up-front assumptions, and then introducing additional assumptions one at a time. This sensitivity test approach (Chapter 3) is the best way to separate out the contributions of each input factor to the match (or mismatch) of the model simulations with the estimates based on geologic data.

Each kind of target signal estimate of past temperatures includes some range of uncertainty. The colored region lying between the minimum and maximum estimates in Figure 6-4 shows the range of possible target signal values.

The first experiment run in an attempt to simulate this target signal used the Cretaceous geography (land-sea distribution and mountain elevations) as the input to the model simulation. For this experiment, the tropical temperatures simulated by the model lie well within the range of estimated target temperatures, but the temperatures simulated in regions poleward of about 40° latitude are considerably colder than the geologic estimates (Figure 6-4).

Why does the model fail to simulate the warmth at the poles? Before trying to answer this question, we

need to take a closer look at the modeling process. Part of the reason that the temperatures simulated for the polar regions 100 Myr ago were warmer than those today is simply the absence of ice sheets in the input to the model run. We learned in Chapter 2 that GCMs can be run only long enough to simulate a few years of elapsed time, not nearly long enough to grow or melt ice sheets. Because of this limitation, the presence or absence of ice sheets must be specified *in advance* as input to the simulations. Since geologic evidence shows that no ice sheets existed 100 Myr ago, none were used as a boundary condition for the simulations.

But the absence of ice sheets in the input to the model does not answer the question we just asked. We need to figure out why the model still simulates polar climates colder than those observed.

As we learned in Chapters 4 and 5, a second obvious possibility is that CO_2 levels were higher 100 Myr ago. This kind of evidence led Barron and his colleagues to investigate the effect of higher CO_2 levels in simulations of climate conditions 100 Myr ago. Although atmospheric CO_2 concentrations from 100 Myr ago are generally thought to have been somewhere in the range of four to ten times those today, the precise value is not

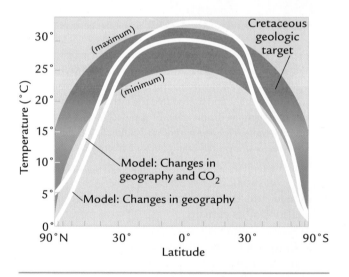

FIGURE 6-4 **Data-model comparisons** Climate model simulations attempt to match the temperature target signal for 100 Myr ago provided by geologic data. Simulations based on the altered geography of 100 Myr ago and other simulations incorporating higher CO_2 values reproduce some of the target signal, but simulate a steeper pole-to-equator gradient. (Adapted from E. J. Barron and W. M. Washington, "Warm Cretaceous Climates: High Atmospheric CO_2 as a Plausible Mechanism," in *The Carbon Cycle and Atmospheric CO_2: Natural Variations, Archaean to Present*, ed. E. T. Sundquist and W. S. Broecker, Geophysical Monograph 32 [Washington, D.C.: American Geophysical Union, 1985].)

known. Barron and his colleagues chose a CO_2 value slightly more than four times the modern (preindustrial) level, a conservative choice at the lower end of this possible range (Box 6-1).

The second simulation with altered geography *and* CO_2 levels four times higher warms the planet by an additional 4°C on average. In a global average sense, this greater warmth improves the match with the estimates from Earth's climate records, but the match is not significantly better in the zonal average plots (Figure 6-4). The temperatures simulated for mid-latitude and polar regions remain slightly too cold, and now the tropics have become slightly warmer than the geologic estimates.

Additional simulations could be run with even higher CO_2 levels (Box 6-1) in order to warm the poles even more and produce a better match with the evidence at higher latitudes. But this approach would further warm the tropics and worsen the emerging mismatch between the model and observations at low latitudes. At this point, we must conclude that higher CO_2 levels have not resolved the problem.

This mismatch between the geologic evidence and the model simulations of the Cretaceous has received an enormous amount of attention from climate scientists. One reason it is so important to resolve this discrepancy is that climate scientists would like to use this greenhouse world of 100 Myr ago as a source of insight into the climates we will face in the fossil-fuel greenhouse of the future. This cannot be done until this discrepancy is resolved.

6-2 What Explains the Data-Model Mismatch?

As is the case for any study of past climates, a persistent mismatch between model simulations and the geologic data used as the target signal forces climate scientists to reexamine all their basic methods and assumptions. The explanation for any mismatch could lie either in the geologic data and how they have been interpreted (or misinterpreted), in the climate model simulation and its possible shortcomings, or in a combination of the two.

Possible Problems with the Data One possibility is that gradual but cumulative evolutionary changes in the past 100 Myr have altered the basic responses of similar-looking organisms to climate through time. If so, attempts to reconstruct Cretaceous climate based on the similarity of fossil organisms to their modern counterparts will fail. More specifically, this explanation would require a pervasive and gradual shift on the part of many organisms toward a lower tolerance of cold temperatures since 100 Myr ago. Such a shift would make it appear that Cretaceous organisms lived in climates warmer than those in which their modern coun-

terparts live, when in fact they were actually successfully surviving at cool temperatures. The problem with this idea is that there is no obvious reason why such a thing would happen, especially for so many unrelated organisms. This evolutionary explanation is unlikely.

A second possible problem with the geologic evidence from 100 Myr ago is that most of it comes from the coastal margins of continents flooded by shallow seas, which tend to be regions of maritime climate, with mild winters moderated by the presence of ocean waters. Remains of organisms that lived in arid interior regions are much less likely to be preserved because of the drier conditions. This could mean that regions with harsh continental climates, and particularly cold winters, may be underrepresented in the geologic record. This underrepresentation could bias reconstructions of Cretaceous climates toward erroneously warm values. This possible explanation needs to be evaluated by additional research.

A third potential problem is postdepositional alteration of materials in the geologic record. These 100-Myr-old deposits have been lying buried under the seafloor or deep in sedimentary basins on the continents long enough for some of their original properties to have been altered. This problem particularly applies to the kinds of isotopic geochemical measurements discussed in Chapter 7. Postdepositional alteration could invalidate geochemical measurements, along with climatic interpretations based on these older deposits.

Several recent studies have shown that estimates of surface-ocean temperatures from the last 100 Myr have been altered in such a way that they now indicate tropical temperatures cooler than those that actually prevailed. Because the estimated tropical temperatures used as the latitude-averaged target signal are based mainly on ocean sediments, actual tropical temperatures 100 Myr ago may have been as much as 5°C warmer than the warmest temperatures shown in Figures 6-3 and 6-4. If additional research confirms this conclusion, it could lead to a possible solution to this problem. Using CO_2 levels well above four times the modern value in further model simulations would warm the poles without overheating the tropics, and thereby produce an acceptable overall match with the geologic evidence.

Possible Problems with the Models A second general group of explanations looks for the source of the data-model mismatch in the climate models. One major problem with the current generation of GCMs is that the treatment of ocean circulation is still very crude. The processes of upwelling of cool water along coastlines and near the equator are not included in global-scale ocean models, and deep-water circulation

BOX 6-1 CLIMATE INTERACTIONS AND FEEDBACKS

The Effect of CO_2 on Climate

One way to understand the effect of past CO_2 levels on Earth's climate is to run a series of sensitivity tests with GCMs using different levels of CO_2. Several climate scientists have run these kinds of experiments using Earth's present-day geography as common input to all simulations, and allowing the level of atmospheric CO_2 to vary from one experiment to another. The CO_2 levels in these experiments ranged from values as low as 100 ppm to as high as 1000 ppm. The preindustrial value (280 ppm) lies in the lower part of this range.

The full set of model simulations shows global average temperature rising with increasing CO_2 levels. Although a progressive warming is evident throughout the entire range of CO_2 changes, the relationship is not linear (directly proportional). Instead, Earth's temperature reacts strongly to small changes in CO_2 values at the lower end of the range (<200 ppm), but changes much less at the high end of the range (>800 ppm).

What explains the shape of this curve? One important explanation is positive feedback effects from snow and sea ice. At low (<200 ppm) CO_2 values, sea ice and snow advance well past their average limits today and cover a relatively large fraction of Earth's high-latitude surfaces in summer and winter. These bright surfaces reflect incoming solar radiation and send it back out into space, cooling the planet. With the planet covered by so much snow and ice, even small changes in CO_2 can have a relatively big impact in altering the area of Earth covered by snow and ice, especially in winter.

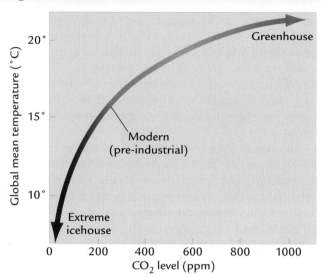

Effect of CO_2 on global temperature Climate model simulations of the effects of changing atmospheric CO_2 levels on global temperature show greater warmth for higher CO_2, but with far more temperature sensitivity at lower CO_2 values than at higher concentrations. (Adapted from R. J. Oglesby and B. Saltzman, "Sensitivity of the Equilibrium Surface Temperature of a GCM to Changes in Atmospheric Carbon Dioxide," *Geophysical Research Letters* 17 [1990]: 1089–92.)

Small CO_2 increases greatly reduce the extent of this bright reflective area, while small decreases in CO_2 greatly enlarge it. This albedo-temperature feedback makes the climate system in an icehouse world react more strongly to changes in CO_2.

---

is handled with less than total success in the few models that make the attempt. Even atmospheric models still have problems simulating the effects of clouds on climate. In view of these shortcomings, climate scientists may simply be asking more from models than they can deliver at this point in their development. As a result, the models could be the source of the mismatch.

One possibility that fascinates many climate scientists is that the Cretaceous ocean operated in a fundamentally different way from the present-day ocean. It makes sense that the ocean could be an important factor in the data-model mismatch: ocean water accounts for

70% of the Earth's surface used to calculate the global mean temperature, and today the surface ocean transports about half as much heat poleward as the atmosphere. If the ocean carried roughly twice as much heat poleward in the Cretaceous as it does today, two problems in the data-model mismatch could be resolved: (1) the poles would be warmed by greater heat influx, and (2) the tropics would be cooled (or would not heat up so much) because of the greater export of heat. The pole-to-equator temperature gradient might then come into agreement with the target signal in Figure 6-4. This concept, which has been proposed in many forms

BOX 6-1 CONT.   CLIMATE INTERACTIONS AND FEEDBACKS

In contrast, the much higher CO_2 values that warm our planet also reduce the average amount of snow and ice present at high latitudes. At CO_2 levels at and above 1000 ppm, the only permanent sea ice present in the Arctic is in a tiny region centered on the North Pole, although thin seasonal ice briefly develops around this permanent ice pack each winter. As a result, only very small surface areas of snow and ice are available to provide positive feedback to changes in CO_2. Further warming caused by even higher CO_2 levels has little snow or ice to melt, and large decreases in CO_2 can cause only small advances of snow or ice. The climate system in such a greenhouse world is much less sensitive to changing CO_2 levels.

A second factor is saturation of the atmosphere with CO_2. As CO_2 concentrations rise, the atmosphere gradually begins to approach a point where further CO_2 increases do not have much effect in trapping additional back radiation from Earth's surface. This effect, **CO$_2$ saturation,** also contributes to the slower rise of temperature at higher CO_2 levels.

A third factor is water vapor feedback. The warm atmosphere produced by CO_2 values of 1000 ppm can hold much more water vapor than the cold atmosphere produced by CO_2 values of 100 ppm. In this case, however, the feedback effect of water vapor on temperature grows stronger in warmer, higher-CO_2 conditions, in contrast to the diminished effect of albedo-temperature feedback.

These sensitivity experiments show that large ice sheets cannot exist anywhere on Earth when CO_2 concentrations exceed about 1000 ppm. At these higher CO_2 values, with a global mean temperature well above 20°C, mean annual temperatures in polar regions increase to above 0°C. Because ice sheets require mean annual temperatures near −10°C to persist through the summer melting season, they cannot survive in high-CO_2 greenhouse warmth.

These modeling results lead to an important conclusion. If no ice sheets existed even at the South Pole 100 Myr ago, and if model experiments indicate that CO_2 values above 1000 ppm are incompatible with the presence of ice sheets, then we can conclude that CO_2 levels in the Cretaceous must have been at least four times higher than the preindustrial level of 280 ppm to keep ice sheets from forming, and may well have been considerably higher.

This conclusion needs to be qualified somewhat. Different GCMs simulate different levels of temperature sensitivity to changes in CO_2. These variations reflect how well each model simulates the effects of such features as snow and sea ice, which provide strong positive feedback, and clouds, which have feedbacks that are still poorly known. For the control-case simulation of modern-day climate, some models simulate too much snow and ice and so are too sensitive in experiments involving past climates. The opposite is true of models that simulate too little snow and ice in the control case: they show too small a temperature response to CO_2 changes. The modeling example chosen here lies within the middle range of model sensitivities.

---

by different climate scientists, is part of the **ocean heat transport hypothesis.**

One way to alter ocean heat transport is by changing the way water moves through the deep ocean (Figure 6-5). Today the deep ocean is filled with cold dense water that forms and sinks in polar regions when it is strongly cooled in winter (Chapter 2). Most of the deep ocean receives water from just two polar source areas (Figure 6-5A). In addition, a small amount of water at intermediate depths in the Atlantic comes from the sinking of salty water that forms in the northern Mediterranean Sea and flows out into the Atlantic. This warmer water owes its density more to the higher salinity it gains by strong evaporation than it does to winter chilling.

Several climate scientists, beginning with the geologist T. C. Chamberlin in 1906, have cited the present-day deep Mediterranean circulation as a rough analog for a very different kind of ocean circulation in the past. They propose that the deep ocean 100 Myr ago was filled with **warm saline deep water** formed in the tropics or subtropics, rather than with cold water from high latitudes (Figure 6-5B). Like Mediterranean water today, this water was dense mainly because of its high

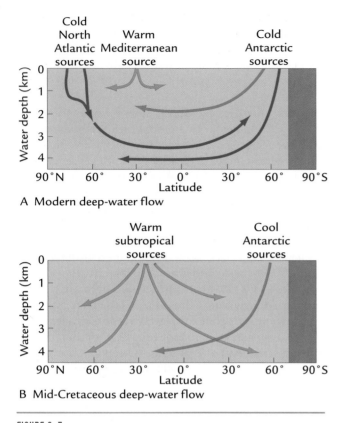

A Modern deep-water flow

B Mid-Cretaceous deep-water flow

FIGURE 6-5 **A different way of forming deep water?** All the deep water that fills the ocean basins today forms in two cold polar regions (A), but 100 Myr ago more of it may have formed in shallow, salty subtropical seas (B). The brown shading represents Antarctica.

salt content, and not because of winter chilling. The concept is that subtropical water might have become dense enough to sink deeper than any water formed near the poles, which were not very cold 100 Myr ago.

The configuration of the continents 100 Myr ago makes this kind of circulation plausible (Figure 6-1). A large seaway spanned much of the northern tropical and subtropical latitudes, including the region just underneath the sinking air of the northern branch of the Hadley cell. The dry air subsiding in this region could have provided the excess of evaporation over precipitation needed to increase the salinity of the ocean's surface and possibly form warm saline deep water.

Experiments with ocean models capable of crudely simulating deep-water flow indicate that deep water could have formed in the northern subtropics in regions where salinities exceeded 37‰ (Figure 6-6), but also indicate that freshwater runoff from the continents would have slowed or stopped the process. No deep water would have formed in marginal seas or coastal regions close to river mouths, because the freshwater from the rivers would have lowered the density of the

surface water enough to stop it from sinking. Away from the river inflows, warm saline deep water might have formed and sunk to the seafloor.

If warm salty bottom water did form, it could have had major implications for global climate. A strong flow of warm deep water from the tropics to the poles (Figure 6-5B) could have contributed to the poleward heat flux needed to warm the poles and to resolve the data-model mismatch. Also, by reducing the large temperature (and density) contrast that exists today between the surface and deep ocean, warm salty deep water could have caused faster overturning of the deep ocean and greater poleward heat transport.

Several model simulations have been run in which the amount of heat carried poleward by the oceans is fixed in advance as part of the input to the model. For obvious reasons, these simulations made the poles warmer (and the tropics less warm), thereby improving the match between the geologic data and the models, but the improvement is somewhat artificial. That is, it would have been preferable if the model had *solved for* any increase in poleward heat transport as part of its simulation of actual processes internal to the climate system, rather than having the heat transport *imposed* in advance as a fixed constraint.

A second general kind of data-model mismatch occurs in simulations of greenhouse climates 100 Myr ago, and it persists through model simulations of subsequent greenhouse climates over the next several tens of millions of years. Data from warm-adapted vegetation (early palmlike trees) and from fossil reptiles suggest that continents at high and middle latitudes had moderate climates and did not freeze in winter. In contrast, all the GCM experiments described above (those using altered geography, higher CO_2 levels, or increased

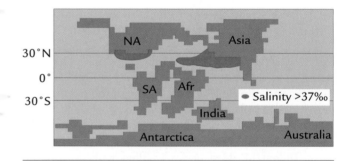

FIGURE 6-6 **Model simulations of past ocean salinities** Model simulations of ocean circulation near 100 Myr ago indicate that deep water may have formed in northern tropical and subtropical seas where estimated salinities were greater than 37‰. (Adapted from C. Johnson et al., "Middle Cretaceous Reef Collapse Linked to Ocean Heat Transport," *Geology* 24 [1997]: 376-88.)

ocean heat transport, alone or in combination) simulate hard freezes in winter across the interiors of the northern hemisphere continents.

We have already examined one possible explanation for this mismatch: the relative scarcity of representative fossil remains from the continental interiors due to poor preservation. More data coverage might well show us that many continental interiors froze. Another possible explanation is the fact that the lakes and small seas that existed in interior regions of the continents are too small to be represented in the large model grid boxes. These bodies of water might have provided moisture to produce a more maritime climate with milder winters in continental interiors, in closer agreement with the existing geologic data.

At this point, attempts to model the greenhouse climate of 100 Myr ago have been only a partial success. Climate scientists do not yet have a full explanation for the apparent existence of much warmer poles and only slightly warmer tropics 100 Myr ago. The simplest explanation may be that the tropical temperatures indicated in the target signal (Figure 6-3) were actually as much as 5°C warmer than those shown, but that subsequent alteration of fossils in the tropical oceans has produced incorrect estimates. CO_2 values up to ten times present-day levels could then produce model simulations matching the warmer tropics in the revised target signal and yet preventing interiors of mid-latitude continents from freezing in winter.

Much more evidence is needed to test whether a higher CO_2 level was the key cause of Earth's greater warmth 100 Myr ago. If it was, the Cretaceous greenhouse world may hold answers to our likely climatic future as greenhouse-gas levels rise. If not, other factors such as increased ocean heat transport will need to be considered.

Sea Level Changes and Climate

Over tectonic time scales, the average level of the world ocean has risen and fallen by several hundred meters against the margins of the continents. These rises and falls of the sea in relation to the land are called marine **transgressions** and **regressions**. Although these changes have been small in relation to the >4000-meter average depth of most ocean basins, they could potentially have significant effects on climate in those regions that were alternately flooded and exposed.

Many continental margins are flat, with changes of just 1 part in the vertical for every 1000 in the horizontal: a rise in elevation of just 1 meter can translate into a 1-kilometer shift inland. As a result, the vertical sea level changes of hundreds of meters that have occurred

in the past can translate into horizontal movements of the coastline measured in hundreds of kilometers. During times of high sea level, the ocean can flood shallow interiors of continents and form large seas. These long-term sea level changes also deposit and erode the thick coastal sedimentary sequences that hold much of the evidence of long-term changes in sea level and in climate.

When sea level is low (as it is today), the coastline tends to be situated near the break between the relatively flat continental shelf and the steeper continental slope that falls off abruptly into the deep ocean basins (Figure 6-7A). During these times, erosion prevails on continental margins, and most of the eroded sediment is carried out to the continental slope and down into the deep ocean. When sea level is high, the ocean floods the low-gradient continental margin to depths of 200 meters or more (Figure 6-7B). During these times, sediment is deposited on the submerged continental shelf, and some of this sediment will later survive erosion and remain as part of the geologic record.

Local tectonic factors that cause uplift or subsidence of the land can also affect the relative vertical position

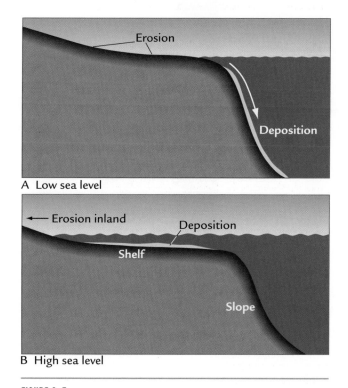

A Low sea level

B High sea level

FIGURE 6-7 **Sea level** When sea level is low (A), the coastline lies near the base of the continental shelf and sediment is carried out and deposited on the continental slope. When high sea levels flood the continental margin (B), more sediment is trapped on the submerged shelf and deposited there.

of the ocean margin against the land, even in the absence of changes in global sea level. These regional processes include mountain building, broad-scale warping of Earth's surface connected to deep-seated heating, and local depression and rebound of the land caused by the weight of ice sheets. In this chapter, we ignore these local effects on sea level and focus on **eustatic** changes—changes that are global in scale.

The most persuasive evidence for high global sea level in the past comes from the presence of marine sediments simultaneously deposited on the coastal margins and in the shallow interiors of many continents at levels well above present sea level. Deposition of marine sediments on several continents at the same time indicates that the changes in sea level are global in scale, not just local features.

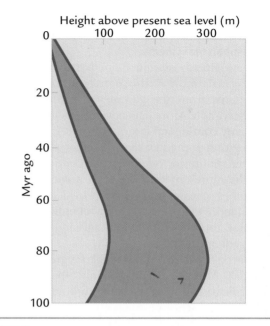

FIGURE 6-9 **Sea level in the last 100 Myr** Quantitative estimates of sea level change over the last 100 Myr vary widely, but all show a progressive drop to the low sea level of today. The blue area defines the range of estimates based on different methods. (Adapted from M. Steckler, "Changes in Sea Level," in *Patterns of Change in Earth Evolution*, ed. H. H. Holland and A. F. Trendall [Berlin: Springer-Verlag, 1984].)

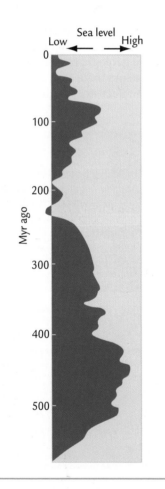

FIGURE 6-8 **550 Myr of sea level change** Estimates of sea level in the distant geologic past are based on evidence of simultaneous flooding or exposure of the margins and interiors of many continents. Sea level inferred from this evidence shows gradual changes over hundreds of millions of years. (Modified from A. Hallam, "Pre-Quaternary Sea Level Changes," *Annual Reviews of Earth Planetary Science* 12 [1984]: 205–43.)

Several scientists have attempted to estimate global sea level during the last several hundred million years. In broad outline, curves like the one shown in Figure 6-8 reveal high sea levels from 500 to 425 and about 120 to 80 million years ago, and low sea levels near 200 million years ago and again at the present. The best-constrained estimates of past sea level are those spanning the drop that has occurred during the last 100 million years (Figure 6-9), because the geologic record of this interval is relatively complete. But even through this interval, estimates of eustatic sea level changes are uncertain to within a factor of 2 or more because of the complex interactions of many factors: local and regional tectonic uplift and subsidence; sediment deposition, compaction, and erosion; and subsidence and rebound of continental margins due to sediment loading and removal.

6-3 Causes of Tectonic-Scale Changes in Sea Level

In the Cretaceous world of 100 to 80 million years ago, coastlines and interiors of most continents were flooded by an ocean that stood some 200 meters higher than it does today (Figure 6-1). The ocean flooded much of North Africa and Europe (Figure 6-10), and seaways penetrated northward into the interior of North America

FIGURE 6-10 **Marine limestone exposed on land** Marine limestone deposits that today form the coasts of southern England and northern France are evidence of higher sea levels 100 Myr ago. (Andrew Ward/Life File/PhotoDisc.)

from the Gulf of Mexico and southward from the Arctic Ocean. Since 80 million years ago, sea level has fallen from this long-term maximum to its modern position, close to the lowest level on record (Figure 6-9).

This large eustatic drop in the level of the ocean in the last 80 million years is mainly the result of two categories of factors: (1) tectonically driven changes in the volume of the ocean basins and their capacity to hold water and (2) changes in the volume of water in the ocean as a result of changes in climate.

Changes in the Volume of the Ocean Basins

1. *Changes in the volume of ocean ridges* Ocean ridges owe their high elevations to unusual heating from below, including hot molten material located just below the surface of the ocean crust. Heating of rocks in this region causes them to expand, and expansion of the rock causes the surface of the ocean crust to rise.

The degree of expansion-induced elevation of ocean ridges varies through time in response to changes in the rates of seafloor spreading, and the changes in the height of the ridges in turn alter the volume of ocean

basins and their capacity to hold water (Figure 6-11). The basic concept is that ocean water is displaced up onto the continents during times when ridges spread rapidly and produce wide, high-elevation ("fat") profiles, but the sea withdraws from the continents during times when the ocean ridges spread more slowly and produce narrower, low-elevation ("thin") profiles that displace less water.

The profiles (and volumes) of ocean ridges existing in the past can be reconstructed on the basis of a systematic relationship observed in modern ocean basins. All ocean ridges today have crests at an average depth of 2500 meters below the sea surface. Away from the crest, the subsurface depth profiles of these ridges follow a simple equation:

$$\text{Ridge depth} = 2500\text{ m} + 350\text{ (crustal age)}^{1/2}$$
(in meters)   (at 0 age)   (in Myr)

This equation describes a ridge crest that starts at an initial depth of 2500 meters below the sea surface and gradually deepens with age away from the ridge crest as the heated rock cools and contracts (Figure 6-12). The seafloor ages used to derive this relationship were obtained from the paleomagnetic age data examined in Chapter 5.

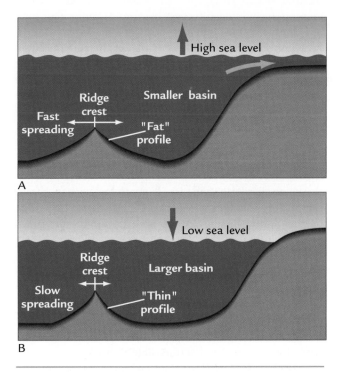

FIGURE 6-11 **Spreading rates and sea level** Rates of spreading at ocean ridges vary widely. Fast spreading creates wider ridge profiles that reduce the volume of the ocean basins and displace more water onto the continents (A). Slow spreading produces narrower profiles and creates larger ocean basins that can hold more seawater (B).

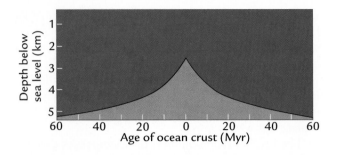

FIGURE 6-12 **Subsidence of ocean ridges with time** All ocean ridges show the same average profile of age (time since formation) vs. depth. Heat elevates the ridge crests high above the rest of the seafloor to a depth of 2500 meters below the ocean surface. As the crust spreads away from the ridge crest, it ages, cools, and contracts, rapidly at first and then more slowly. The crust eventually reaches a stable depth of more than 5000 meters below sea level. (Adapted from J. G. Sclater et al., "The Depth of the Ocean Through the Neogene," *Geological Society of America Memoir* 163 [1985]: 1–19.)

Ridge crests initially stand high above the rest of the seafloor because of anomalously strong heating associated with formation of new crust from molten magma. At first ocean ridge elevations subside rapidly while moving away from the crest because of rapid heat loss, but the later subsidence is more gradual as the rate of heat loss slows on the lower flanks of the ridges. By 60 million years after formation, almost all the excess heat is gone, and the ridge elevations have reached a nearly stable depth of almost 5500 meters (Figure 6-12). Local variations in depth of a few hundred meters occur at ridge crests and down their flanks as a result of small-scale tectonic irregularities, but the mean values of ocean ridge depths follow the equation remarkably well throughout the world's oceans.

Paleomagnetic evidence from today's ocean shows that different ridges spread at different rates. The ridge in the South Pacific Ocean spreads up to ten times faster than that in the North Atlantic (see Figure 5-19). Because all ridge depths are constant with age (as shown by the preceding equation), crust of a given age (and more critically a particular depth below sea level) will have been carried much farther from the ridge crest in a given amount of time in the fast-spreading South Pacific than in the slow-spreading North Atlantic. Fast spreading gives the Pacific ridge a "fatter" elevation profile than that of the Atlantic (Figure 6-11), and the wider Pacific ridge profile displaces more water for each kilometer of its length than does the narrow Atlantic ridge.

Through time, this contrast in thin versus fat ridge profiles may also vary not just from ridge to ridge but on a *globally averaged* basis. At times like the present,

globally averaged rates of seafloor spreading are slow, mean ridge profiles are relatively thin, and little water is displaced onto the continents. And at times like 100 million years ago, average spreading rates were faster, mean ridge profiles were relatively wide, and more water was pushed up onto the continents.

The net effect of these changes in global average spreading rates on global sea level can be investigated throughout the last 80 to 100 million years. Note that using the above equation to calculate spreading rates at any time in the past requires resetting the ages of the ridge crests to zero for the time being examined and recalculating the past ages of the ridge flanks accordingly.

Mean spreading rates 80 million years ago appear to have been as much as 50% higher than they are today, although this estimate is not altogether certain. We cannot be sure of the rate of spreading in those parts of the small seaway in the northern tropics. Ocean crust from that basin was destroyed during the convergence of Africa and Europe and the collision of India and Asia. There is also some uncertainty about the age of very old crust in the western North Pacific. Slowing of global mean spreading rates since 80 million years ago is estimated to have produced ocean basins large enough to hold an extra 200 to 300 meters of water.

2. *Collision of continents* Most plate tectonic movements do not change the net area of either the oceans or the continents: creation of new ocean crust at ocean ridge crests is balanced by destruction of ocean crust subducting into trenches, leaving the area of the ocean basins constant. However, collision of continents does alter the area of the ocean basins and also affects sea level.

Because continental crust is low in density, two colliding continents tend to float near Earth's surface rather than be pushed or pulled deep down into Earth's mantle (Figure 6-13). In the region where they collide, continental crust thickens from its normal value of 30 kilometers to about twice that value, building a high plateau that rises well above sea level and at the same time thickening the subsurface low-density root of this plateau down to 60 or 70 kilometers below Earth's surface. In the upper 15 kilometers of Earth's crust, the thickening that creates the plateau occurs by movements along faults that cause thin slivers of crust to shear off and stack up on top of each other. Below a depth of 15 kilometers, thickening occurs when slow flow causes rock layers to be squeezed and folded.

Because collision drives two continents into each other to form a plateau with a double-thick crust, this thickening must result in a net loss in the area of continental crust. To a first approximation, the area of plateau across which the crust doubles in thickness must equal the net loss of area of continental crust. This decrease in area of the continents requires an equivalent

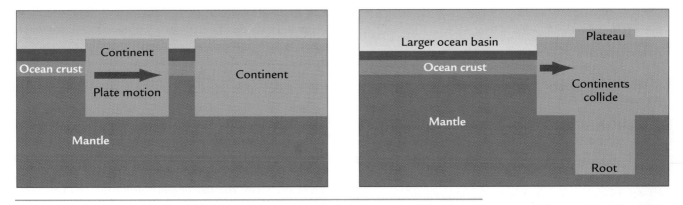

FIGURE 6-13 **Continental collisions and sea level** When continents collide, the continental crust doubles in thickness and creates a high plateau with a thick low-density crustal root. This thickening reduces the original area of the continents and increases the area of the ocean basins. The increased area and volume of the ocean basins causes sea level to fall.

increase in area of the ocean basins (Figure 6-13). With a larger area of ocean to fill, the water level in the ocean must drop.

As we saw in Chapter 5, continental collisions have occurred only sporadically through geologic time. The only major collision that has occurred since 100 Myr ago began when northern India first made contact with southern Asia, some 55 Myr ago. This collision, still in progress, has increased the area of the ocean by some 2 million km² over the last 55 million years. As seawater has flowed in to fill this new area of ocean basin, global sea level has fallen about 40 meters below the level of 100 Myr ago, when no collisions were occurring.

3. *Construction of volcanic plateaus in the ocean* The remains of a huge region of volcanic plateaus that formed between 110 and 80 million years ago lie buried beneath a thin blanket of sediment in the western tropical Pacific Ocean. When the plateaus first formed, heat associated with their volcanic origins caused them to

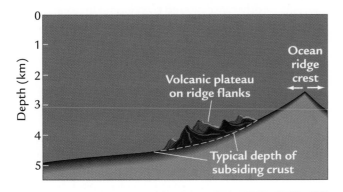

FIGURE 6-14 **Volcanic plateaus and sea level** Construction of volcanic plateaus elevated above the average level of the ocean ridges can displace ocean water onto the continents.

expand high above the nearby seafloor (Figure 6-14). Subsequent cooling and contraction have dropped them to much lower levels, but they still stand well above the surrounding ocean basin of the same age.

Because large volcanic plateaus were more common 80 million years ago than they are today, they displaced a larger volume of seawater than they do today. The volcanic plateaus that existed 80 Myr ago may have displaced a volume of water equivalent to between 10 and 40 more meters of sea level than the amount displaced today.

Climate Factors and Sea level

4. *Water stored in ice sheets* Continent-sized ice sheets several kilometers in thickness and thousands of kilometers in lateral extent can extract enormous volumes of water from the ocean and store it on land (Figure 6-15). Because no ice sheets existed 100 to 80 million years ago, no water was stored on the land as ice. Today the Antarctic ice sheet holds the equivalent in seawater of 66 meters of global sea level, and the ice sheet on Greenland has another 6 meters. The Antarctic ice sheet has come into existence and grown to its present size within the last 35 million years, the Greenland ice sheet within the last 7 to 3 million years. Together these ice sheets have extracted a volume of ocean water equivalent to 72 meters of global sea level.

5. *Thermal contraction of seawater* Ocean water has a limited capacity to expand and contract with temperature changes. The **thermal expansion coefficient** of water (the fractional change in its volume per degree of change in temperature) averages about 1 part in 7000 for each 1°C of temperature change. Because of this property, even a constant amount of seawater would have lost volume during the cooling of the last 80 million years. The temperature of low-latitude surface waters has cooled slightly over the last 80 to 100 Myr,

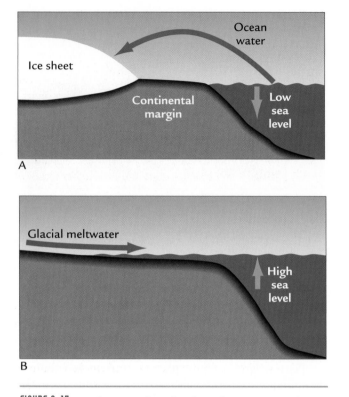

FIGURE 6-15 **Ice sheets and sea level** Ice sheets covering large parts of continents hold volumes of water equivalent to many tens of meters of global ocean level. Sea level falls when ice sheets are present on the land (A) and rises when they melt (B).

but the high-latitude surface ocean and the deep ocean have both cooled by 10°–15°C since 100 million years ago. The contraction of seawater caused by this cooling has reduced global sea level by roughly 7 meters.

Sea Level Adjustment Estimates of the quantitative effects of several of these factors on global sea level have to be adjusted for several complications (Box 6-2). One problem is the fact that water moving in (or out) of the ocean basins represents a large weight added to (or removed from) the underlying ocean crust, which sags (or rebounds) accordingly. This response of the ocean crust decreases the net magnitude of the change in sea level. The other complication has to do with translating a change in the volume of ocean water or in the volume of the ocean basins into actual movement of sea level against the complex shapes of the world's continental margins.

Taken together, and adjusted as discussed in Box 6-2, the five factors discussed here can account for a fall of some 300 to 440 meters in sea level since 80 million years ago (Table 6-1). This wide range of estimated fall exceeds the drop of 100 to 300 meters inferred from marine deposits on continental margins. In addition to the large uncertainties in the effects of spreading rates and volcanic plateaus on past sea levels, other factors not considered here may contribute to this apparent mismatch.

One example is an increase in the total volume of sediment in the global ocean since 80 million years ago.

BOX 6-2   LOOKING DEEPER INTO CLIMATE SCIENCE

Calculating Changes in Sea Level

Water transferred between the continents and oceans represents weight added to (or removed from) the bedrock underlying the ocean basins. When water is added to the ocean, the underlying bedrock sags under the load. Similarly, the bedrock rebounds if the load is removed.

This bedrock response reduces the change in sea level that would otherwise have occurred. For example, adding meltwater to the oceans raises sea level, but the depression of ocean bedrock under the load of the added water cancels about 30% of the sea level change. This 30% reduction is a direct result of the difference in density between water ($1 g/cm^3$) and bedrock ($3.3 g/cm^3$): $1 \div 3.3 = 0.3$.

The second complication is that the margins of the oceans have complex profiles of elevation versus distance that vary widely from region to region. Some have simple low-gradient profiles, but others are steepened by mountain building on tectonically active margins or otherwise altered by regional factors.

As a result, changes in the total amount of water in the ocean, or in the volume of water the ocean can hold, need to be translated into actual net changes in the level of the global ocean produced by all the complications contained in these varying local profiles. All regional profiles are summed into a single **hypsometric curve,** a graph that displays the proportions of Earth's surface that lie at various altitudes above and depths below sea level. Such a graph is easily constructed for today's Earth, but the exercise becomes more speculative for past intervals of plate tectonic configurations.

TABLE 6-1 Factors Contributing to Sea Level Fall in the Last 80 Million Years

Cause of sea level change	Estimated change (meters)
Decrease in ocean ridge volume	−200 to −300
Collision of India and Asia	−40
Decrease in volcanic plateau volume	−10 to −40
Water stored in ice sheets	−50
Thermal contraction of seawater	−7
All factors	−300 to −440

Sediments may be deposited at higher rates during times of lower sea level, such as today, because of greater exposure of land to subaerial erosion and to rapid erosion of high plateaus and mountains. The sediment delivered by these increased influxes lies on top of the ocean crust, displaces additional ocean water, and raises sea level, but quantitative estimates of this effect are uncertain, for reasons explained in Chapter 7.

6-4 Effect of Sea Level Changes on Climate

Climate scientists have occasionally cited sea level as a potential factor in long-term climate changes, although often in varying and even contradictory ways. The most likely effect of sea level changes on climate is linked to the very different thermal responses of land and water (Chapter 2).

The shallow (100–200 m) layer of ocean that overlaps the continental margins and invades the interior seaways has the large heat capacity typical of ocean water, in contrast to the small heat capacity typical of land surfaces. As a result, flooding of the land will moderate continental extremes of climate and produce milder, more maritime winters and cooler summers. Withdrawal of the sea should have the opposite effect. For large eustatic (global) sea level changes, the synchronous invasion and withdrawal of the sea on many continents should result in simultaneous fluctuations between harsh continental and mild maritime climates in many regions around the world.

On continental margins flooded by rising sea level, the maritime climates of the coastal regions may simply shift landward, displacing formerly continental climates with more maritime conditions. Given the low (1:1000) gradients typical of some continental margins, such changes in climate may affect large regions.

Continental interiors flooded by newly created inland seas, such as those shown for the mid-Cretaceous in Figure 6-1, may undergo more striking climatic changes. Unflooded continental interiors are likely to have arid, harsh climates because of their great distance from the ocean and because mountain chains cast rain shadows within continental interiors. The invasion of extensive seas into these low-lying interior regions of continents should produce far milder maritime climates, especially in regions lying just downwind of the seaways.

Decades ago some climate scientists thought that sea level might be a controlling factor in the long-term succession of glacial (icehouse) versus nonglacial (greenhouse) climates. In this view, high sea levels caused warm climates by moderating the harsh winters, and low sea levels caused cold climates by permitting the very cold winters typical of continental conditions. Although the timing of the high sea levels and nonglacial climate of 100 Myr ago and the low sea level and glacial climate of today match this explanation, this correlation in part simply reflects the fact that glaciation is one of the *causes* of low sea level (because of storage of ocean water in ice sheets) rather than a *result* of it.

The major criticism of the idea that sea level controls glaciation centers on the fact that summer-season ablation is a powerful factor determining the extent of snow and ice (Chapter 2; see also Chapter 10). The logic of this criticism runs as follows: low sea levels and withdrawal of the ocean from continental interiors will lead to more extreme continental climates, including very hot summers. Hot summers should easily melt any snow that accumulated during very cold continental winters and thereby oppose glaciation. Conversely, high sea levels should cause cooler, more maritime summers that favor the persistence of snow and ice through the summer ablation season at very high latitudes.

Yet the record of the last 100 Myr shows just the opposite trend. The high sea levels of 100 Myr ago should have aided glaciation, but none occurred. The low sea levels of today should oppose glaciation, but ice sheets are present. As a result of this fundamental mismatch, the hypothesis that sea level controls long-term glaciation finds little or no support today.

Asteroid Impacts

The greatest catastrophes known to have affected Earth in its long history are the very rare but massive impacts of large extraterrestrial space debris, notably asteroids and comets. An inverse relationship exists between the sizes of these extraterrestrial objects and the frequency with which they hit Earth, with the largest bodies (more than 10 km in diameter) arriving only every 50 to 100 Myr, but resulting in much greater environmental effects than the smaller, more frequent impacts.

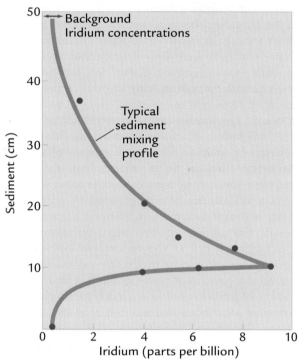

A Iridium spike

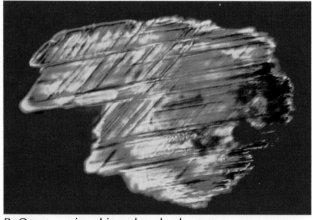

B Quartz grain subjected to shock waves

FIGURE 6-16 **Evidence of an asteroid impact** Ocean sediments with a layer enriched in the element iridium (A) are evidence of a large asteroid impact 65 Myr ago. Sediments deposited in Montana 65 Myr ago (B) contain grains of quartz crisscrossed by multiple lineations produced by high-pressure shock waves from an asteroid impact. (A: adapted from W. Alvarez et al., "Extraterrestrial Cause for the Cretaceous-Tertiary Extinction," *Science* 280 [1095–1108]; B: Glenn Izett, Williamsburg, VA.)

The last extremely large event was an impact 65 Myr ago that coincided with a global-scale extinction of some 70% of the species and 40% of the genera living at that time, including all the dinosaurs and all but one of twenty-five species of planktic foraminifera. The geologic evidence for this impact includes the world-wide distribution of a thin layer of sediment enriched in iridium (Ir), an element that is rare on Earth but 10,000 times more abundant in some kinds of meteorites (Figure 6-16A). The iridium was deposited in a thin layer that has since been mixed over several tens of centimeters by burrowing animals.

Other evidence for an impact event includes small grains of quartz with distinctive textures called "shock lamellae" that are formed by the shock wave of sudden pressures much larger than those found on Earth, even in highly explosive volcanoes (Figure 6-16B). The best candidate for the site of the impact 65 Myr ago is a crater in eastern Mexico on the Yucatán Peninsula, between the Caribbean Sea and the Gulf of Mexico (Figure 6-17).

Climate scientists infer that asteroid impacts would have affected Earth on time scales ranging in duration from instantaneous to as long as a few hundred or even a few thousand years (Figure 6-18). The instantaneous effects were those caused when the asteroid blasted a hole through Earth's atmosphere. The tremendous speed of the incoming asteroid, 20 kilometers per second, created a shock wave that moved outward at the same speed, flattening objects for hundreds of miles around the site of impact and heating Earth's atmosphere. Seismic waves sent through Earth's interior are thought to have been equivalent to those caused by an earthquake that would have measured 11 on the Richter scale, 100 to 1000 times stronger than the strongest earthquakes in recorded human history.

In addition, water and rock in the vicinity of the impact were instantly vaporized by an explosion equivalent to four times the energy of all currently existing nuclear weapons. Some of this material, along with part of the fireball created by the heat of the impact, was blasted back out into space through the hole in the atmosphere initially created by the incoming object, while some hot debris remained in Earth's atmosphere, heating it still further. The combined heating caused large-scale (possibly global) wildfires that ignited much of the aboveground vegetation and sent a thick layer of soot into the atmosphere.

Over the slightly longer term of days to years, the dust and soot that was raised by the impact but retained within the atmosphere spread around the planet, blocking much of the incoming solar radiation (Figure 6-18). The debris injected into the lower atmosphere (the troposphere) would probably have been removed over a period of days to at most weeks, because rainfall quickly

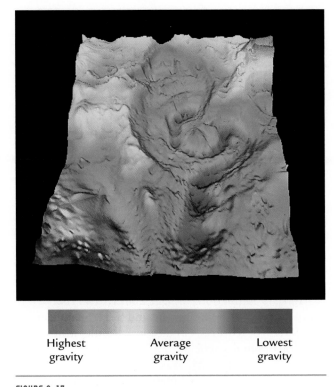

Highest   Average   Lowest
gravity   gravity   gravity

FIGURE 6-17 **A 65-Myr-old impact crater?** Mexico's Yucatán Peninsula has a circular area more than 200 km in diameter that is a good candidate for the site of the asteroid impact 65 Myr ago. The pattern shown is based on measurements of Earth's gravity that can detect low-density pulverized rock (in blue) and higher-density rock (in green and yellow). (B. Sharpton, Lunar and Planetary Institute.)

By any standard used, this impact event was an enormous short-term environmental catastrophe, but what were the long-range effects on Earth's climate? Surprisingly, there appears to have been little long-term climatic effect. Earth's climate was in a warm greenhouse state at the time of the impact, and there it remained afterward.

There are several problems in trying to determine the exact climatic effects of such an ancient impact event. One is that rapid changes in sedimentary records are blurred and smeared by burrowing animals and by bottom currents (Chapter 3). Distinctive impact-related features such as the iridium layer can still be detected despite this blurring (Figure 6-15A), because they leave geochemical fingerprints that contrast clearly with the material in which they are deposited. But for most kinds of sedimentary archives (Chapter 3), the signal from a single year (or even a decade or century) of climate that is warmer or cooler than normal will be mixed into and combined with the signals left by "normal" years and blurred beyond recognition.

A second problem with detecting the climatic effect of the impacts of even huge asteroids is that climate changes were already occurring for other reasons before the impact. Records from marine sediments show a large (3°–4°C) warming and then cooling during the 500,000 years before the impact, but little or no additional change after it. Some land vegetation records suggest a long-term warming for hundreds of thousands of years after the impact, but it has not been demon-

clears debris from that level of the atmosphere. But it would have taken several months or years for the dust and soot injected into the stratosphere to settle out. No rainfall occurs at these altitudes to remove debris, only the slow pull of gravity on small particles. As a result, stratospheric particles (particularly dark soot) would have blocked significant amounts of sunlight for a year or more and cooled Earth's climate. Another likely effect over the course of a few years would have been the partial acidification of the oceans due to the creation of nitric acid (a component of acid rain) from atmospheric nitrogen, oxygen, and water vapor by the heat of the impact.

On the longer term of decades to centuries, the abrupt injection of carbon biomass into the atmosphere by burning should have produced higher CO_2 levels (Figure 6-18). Those plants that recovered from the firestorm and began to grow again would have pulled some of the excess CO_2 out of the atmosphere, but the full recovery of many kinds of vegetation and the development of new forms to replace those that went extinct took longer. Thus any warming induced by higher CO_2 may have lasted for centuries or longer.

Time after asteroid impact	Minutes or less	Days to years	Decades to centuries
Effects on the environment	Shock waves Water & rock vaporized Tidal waves Firestorms	Soot & dust in stratosphere Acidification of lakes and ocean	Higher levels of CO_2 in the atmosphere
Climatic effects	Warming	Cooling	Warming

FIGURE 6-18 **Climatic and environmental effects of asteroid impacts** The asteroid impact 65 Myr ago is thought to have had major effects on Earth's environment, including the extinction of over two-thirds of the species then alive. The likely climatic effects vary with the time that elapsed after initial impact and appear to have been restricted to a few centuries at most.

strated that this warming had anything to do with the impact event. The increase in CO_2 levels caused by the immediate effects of the event seems unlikely to have persisted that long.

In summary, impact events such as the one at the Cretaceous-Tertiary boundary have clearly had apocalypse-like effects on the environment. Impact events are implicated in mass extinctions of organisms that suddenly transformed life on Earth. Despite this environmental apocalypse, the background state of the climate system 65 Myr ago seems to have been changed little or not at all.

KEY TERMS

Cretaceous (p. 130)

CO_2 saturation (p. 135)

ocean heat transport hypothesis (p. 135)

warm saline deep water (p. 135)

transgressions (p. 137)

regressions (p. 137)

eustatic (p. 138)

thermal expansion coefficient (p. 141)

hypsometric curve (p. 142)

REVIEW QUESTIONS

1. What evidence shows that the world was warmer 100 Myr ago than today?

2. Why does higher CO_2 make sense as one explanation for this greater warmth?

3. How well do model simulations capture the distribution of temperatures 100 Myr ago?

4. What are the possible causes of mismatches between the models and geologic observations?

5. Why do some climate scientists believe that increased ocean heat transport is required?

6. Why might deep water have formed in the subtropics 100 Myr ago rather than at the poles?

7. Which regions of the continents were flooded by high seas 100 Myr ago?

8. What major factors controlled the drop in sea level since 100 Myr ago?

9. How did higher sea levels affect global climate?

10. Do sea level changes explain past glaciations?

11. Did asteroid impacts have major effects on climate?

ADDITIONAL RESOURCES

Basic Reading

Alvarez, L. W., W. Alvarez, F. Asaro, and H. V. Michel. 1980. "Extraterrestrial Cause for the Cretaceous-Tertiary Extinction." *Science* 208: 1095–1108.

Barron, E. J., S. L. Thompson, and S. H. Schneider. 1981. "An Ice-Free Cretaceous? Results from a Model Simulation." *Science* 212:501–8.

Barron, E. J., and W. M. Washington. 1985. "Warm Cretaceous Climates: High Atmospheric CO_2 as a Plausible Explanation." In *The Carbon Cycle and Atmospheric CO2: Natural Variations, Archaean to Present*, ed. E. T. Sundquist and W. S. Broecker Geophysical Monograph 32, 546–53. Washington, D.C.: American Geophysical Union.

Hallam, A. 1984. "Pre-Quaternary Sea Level Changes." *Annual Review of Earth and Planetary Sciences* 12:205–43.

Advanced Reading:

Bice, K. I., E. J. Barron, and W. H. Peterson. 1997. "Continental Runoff and Early Cenozoic Bottom Waters." *Geology* 25:951–54.

Oglesby, R. J., and B. Saltzman. 1992. "Equilibrium Climate Statistics of a General Circulation Model as a Function of Atmospheric Carbon Dioxide." Pt. I. "Geographic Distribution of Primary Variables." *Journal of Climate* 5:66–92.

Steckler, M. 1984. "Changes in Sea Level." In *Patterns of Change in Earth Evolution*, ed. H. F. Holland and A. F. Trendall. Berlin: Springer-Verlag.

CHAPTER 7

Back into the Icehouse: The Last 55 Million Years

Today's cold icehouse climate, marked by the presence of ice sheets and extensive sea ice in both hemispheres, is the result of gradual cooling over many tens of millions of years. Earth's record of this transition is rich in information for climate scientists to examine, including the initial appearance of mountain glaciers and continental ice sheets and cold-adapted vegetation, as well as marine geochemical evidence of cooling ocean temperatures and growing ice volume. In this chapter we first examine evidence showing when this greenhouse-icehouse cooling occurred. Then we test whether or not this cooling can be explained by three hypotheses already examined in previous chapters: the BLAG spreading rate hypothesis, the uplift weathering hypothesis, and the ocean heat transport hypothesis. Finally, we investigate factors that will determine the slow changes in future climate over tectonic time scales.

Global Climate Change since 55 Myr Ago

Earth has undergone a profound cooling at both poles and across the lower latitudes of both hemispheres during the last 55 Myr. Both ice and vegetation have left abundant evidence of this cooling (Figure 7-1).

7-1 Evidence from Ice and Vegetation

A cooling climate causes two kinds of glacial ice to form on land. Small mountain glaciers and ice caps may appear on the tops and sides of high mountains, and large ice sheets may cover vast areas of the continents. Because average temperatures vary from region to region, the subfreezing conditions that permit ice to persist through warm summers will not appear at the same time everywhere across the globe. Geologic evidence from land can often date the first appearances of glacial ice (Chapter 3), but it can rarely document subsequent changes in the amount of ice in any detail.

In the southern hemisphere, no evidence exists for ice on Antarctica until 35 Myr ago, when ice-rafted debris was first dropped in ocean sediments on the nearby continental margin (Figure 7-1). Since that time, the size of the Antarctic ice sheet has increased irregularly toward the present, with a particularly large growth phase near 13 Myr ago. Increased amounts of ice-rafted debris in nearby ocean sediments suggest an additional increase in Antarctic ice near 7–5 Myr ago. Today more than 97% of Antarctica is buried under a permanent ice cover (Figure 7-2, left). In the lower middle latitudes of the southern hemisphere, the earliest evidence of mountain glaciers in the high Andes dates to between 7 and 4 Myr ago.

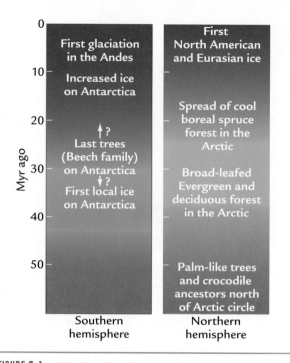

FIGURE 7-1 Global cooling for 55 Myr Gradual cooling during the last 55 Myr is demonstrated by the initiation of mountain glaciation and of continental-scale ice sheets, and by a progressive trend toward cold-adapted vegetation in both hemispheres.

In the northern hemisphere, glacial ice first developed on Greenland sometime between 7 and 3 Myr ago, although small mountain glaciers may have existed locally in mountains around the North Atlantic Ocean before that time. The first evidence of glaciers in the high coastal mountains of southern Alaska dates to about the same time, near 5 Myr ago. The first conti-

FIGURE 7-2 Cooling in Antarctica Today an ice sheet as much as 4 km thick covers most of Antarctica, although mountains locally protrude through the thinner cover around the margins (left), but 30 Myr ago *Nothofagus* trees (members of the beech family), like those living today at the southern tip of South America (right), still existed in some parts of Antarctica. (Left: Ward's Natural Science Establishment; right: courtesy of Calvin Heuser, Tuxedo, N.Y.)

FIGURE 7-3 **Cooling in the Arctic** Warm-adapted breadfruit trees lived above the Arctic Circle in Canada until 60 Myr ago (left), but the landmasses surrounding the Arctic Ocean today are covered by scrubby tundra vegetation grazed by caribou (right). (Left: Swedish Museum of Natural History, photo by Yvonne Arremo, Stockholm; right: Corbis.)

nental ice sheets of significant size appeared 2.7 Myr ago in North America and Eurasia. These ice sheets formed and melted in repeated cycles, and their maximum size increased after 0.9 Myr ago (Chapter 10). Although ice sheets developed more than 30 million years later in the northern hemisphere than in the southern, they are part of the same overall cooling trend into the present-day icehouse climate (Figure 7-1).

Fossil remains of vegetation also indicate a progressive cooling over the last 55 Myr (Figure 7-1). For example, a form of beech tree called *Nothofagus* (Figure 7-2, right) lived on Antarctica before 40 Myr ago, along with several types of ferns. This vegetation disappeared as ice gradually spread across Antarctica and as the polar climate became increasingly frigid. Today the only vegetation on Antarctica is lichen and algae found in summer meltwater ponds in ice-free regions of a few coastal valleys.

The same kind of long-term cooling trend is evident in the north polar regions (Figure 7-3). Palmlike and other broad-leafed evergreen vegetation lived in the Canadian Arctic at 80°N near 60 Myr ago (Figure 7-3, left), along with ancestors of modern alligators that appear to have been ill suited to withstand extreme cold. Even the coastal margins of the Arctic Ocean had temperate climates, with sea ice apparently absent.

Then gradually these warm conditions in the Arctic gave way to today's cold. The development of conifer forests of spruce and larch by 20 Myr ago indicates cooling, and the modern band of tundra that encircles

the Arctic Ocean developed only in the last few million years. The tundra environment consists of scrubby grasslike or shrublike vegetation that lives on thawed layers lying above deeper ground called **permafrost**, kept frozen by intense winter cold (Figure 7-3, right). Some scientists attribute the appearance of tundra and permafrost a few million years ago to the frigid winters brought on by expanding sea ice.

The shapes of tree leaves can also be used to reconstruct past climate (Figure 7-4). Leaves of trees living

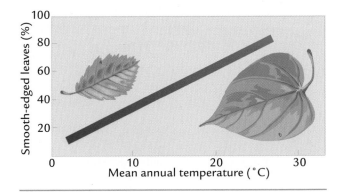

FIGURE 7-4 **Leaf outlines indicate temperature** Trees with smooth-edged leaves flourish today in the tropics, while trees with more jagged-edged leaves grow in colder climates. (Adapted from S. Stanley, *Earth System History*, © 1999 by W. H. Freeman and Company, after J. A. Wolfe, "A Paleobotanical Interpretation of Tertiary Climates in the Northern Hemisphere," *American Scientist* 66 [1978]: 994–1003.)

today in the warm tropics tend to have smoothly rounded margins, while leaves of trees in cooler climates generally have irregular edges, some of them jagged or serrated in outline. The reason for this relationship is not known, but the correlation with temperature in the modern vegetation is so strong that climate scientists have used this relationship to estimate past temperatures from assemblages of fossil leaves preserved in sedimentary rocks.

One record derived from leaf-margin evidence in western North America shows an ongoing cooling over 55 Myr (Figure 7-5). Although interrupted by small, short-lived warmings, the cooling trend resumes and reaches even lower temperatures over time.

7-2 Oxygen Isotope Data

Evidence of climate on continents over the last 55 Myr is incomplete. The first occurrences of ice on land are isolated events in time and space, and subsequent changes in size are difficult to detect. Lakes that accumulate remains of vegetation in their muddy sediments rarely persist for millions of years. In contrast, the deep ocean is a climate archive that has persisted for the

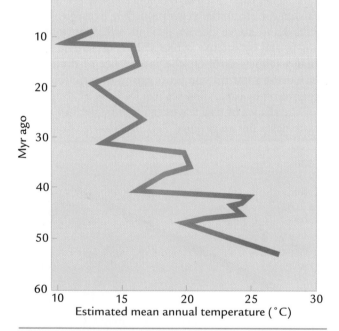

FIGURE 7-5 Cooling in western North America Temperature trends estimated from the outline shapes of fossil leaves indicate an erratic but progressive cooling of the middle latitudes of the northern hemisphere during the last 55 Myr. (Adapted from J. A. Wolfe, "Tertiary Climatic Changes in Western North America," *Palaeogeography, Palaeoclimatology, Palaeoecology* 108 [1994]: 195–205.)

entire 55-Myr interval and contains quantitative information about climate change.

Foraminifera that live in the ocean form shells made of calcite, or $CaCO_3$. The oxygen in these shells consists of two isotopes (^{18}O and ^{16}O) taken directly from the ocean water. When the foraminifera die and become part of the permanent geologic record on the seafloor, they form a history of changes in the relative abundance of these two isotopes of oxygen in the oceans (Box 7-1).

The isotopic composition of oxygen from ocean water recorded in foraminifera shells has changed over time, mainly in response to two important climate-related factors: (1) the total amount of ice existing on Earth in continental ice sheets and (2) the local temperature of the ocean water in which the shells form. This means that foraminifera shells in ocean sediments give us access to past changes in both ice volume and ocean temperature.

Two kinds of foraminifera live in climatically important parts of the world ocean. Planktic foraminifera live mainly in the upper 100 meters, and their shells contain oxygen taken from waters near the surface. Benthic foraminifera live on the seafloor and within the upper layers of ocean sediment, and their shells contain oxygen taken from deep water.

The temperature of ocean water is directly related to the $\delta^{18}O$ value recorded in the shells of foraminifera: for each 4.2°C increase in temperature, the $\delta^{18}O$ ratio decreases by 1‰ (that is, ^{18}O becomes less abundant in relation to ^{16}O). This relationship was initially determined by laboratory experiments in which organisms that form calcite shells were grown at different temperatures. It is also consistent with observations from the modern-day tropical ocean. For example, planktic foraminifera growing in relatively warm (21°C) subtropical surface waters today have $\delta^{18}O$ values of about −1‰, while benthic foraminifera growing in cold (2°C) deep waters below have $\delta^{18}O$ values of +3.5‰.

It would seem that this consistent temperature/$\delta^{18}O$ relationship provides scientists with an obvious opportunity: Can $\delta^{18}O$ be used as a thermometer to measure *past* temperatures by analyzing the $\delta^{18}O$ values of the shells of fossil planktic foraminifera? Yes, but only under certain conditions. The problem is that temperature is not the only factor that affects $\delta^{18}O$ values in foraminifera.

The second major factor that affects $\delta^{18}O$ values in foraminifera has to do with two other parts of the climate system: the atmosphere and the ice sheets. The tropical atmosphere is rich in water vapor (H_2O_v) evaporated from the warm tropical ocean (Chapter 2). Much of this water vapor is gradually transported to higher latitudes (and higher altitudes), where it condenses and falls back to Earth's surface as precipitation. Because the

BOX 7-1   TOOLS OF CLIMATE SCIENCE

Oxygen Isotope Ratios ($\delta^{18}O$)

Both ^{16}O and ^{18}O are stable (nonradioactive) isotopes of oxygen that occur naturally in Earth's water and air. The ^{16}O isotope forms 99.8% of all the oxygen present, and ^{18}O accounts for most of the rest. The ratio of ^{18}O to ^{16}O is about 1/400, equivalent to a value of 0.0025. Climate scientists who analyze the $CaCO_3$ shells of foraminifera in the oceans measure small *variations* around this average value.

Individual measurements of the $^{18}O/^{16}O$ ratio in calcite shells are reported as departures in parts per thousand (‰) from a laboratory standard:

$$\delta^{18}O_{(in\ ‰)} = \frac{(^{18}O/^{16}O)_{sample} - (^{18}O/^{16}O)_{standard}}{(^{18}O/^{16}O)_{standard}} \times 1000$$

All measurements are referenced to a powdered standard supplied by the U.S. National Bureau of Standards in order to establish a common reference point that links all measurements made in all laboratories. The measured ratio is multiplied by 1000 for convenience: it converts the extremely small variations in an already small ratio to a more workable numerical form. As a result of this multiplication, the $\delta^{18}O$ values actually measured in calcite shells from the oceans fall in the range between about +4 and −2.

Samples with relatively large amounts of ^{18}O (compared with ^{16}O) are said to have more positive $\delta^{18}O$ values and are referred to as ^{18}O-enriched (or ^{16}O-depleted). Samples with relatively small amounts of ^{18}O have more negative $\delta^{18}O$ values and are referred to as ^{18}O-depleted (or ^{16}O-enriched).

Sediment samples taken from the ocean are sieved to remove the finer mud and silt and to isolate the sand-sized fraction. Foraminifera are hand-picked from this sand fraction with a small brush with a moistened end. Typical $\delta^{18}O$ analyses require a few milligrams of sample, usually amounting to fewer than a dozen foraminifera. The $CaCO_3$ shells are dissolved in acid to produce CO_2 gas, which is then analyzed in a **mass spectrometer,** an instrument capable of detecting the small

difference in atomic mass between the ^{16}O and ^{18}O isotopes. Mass spectrometers can also analyze ice taken from glaciers. Samples are drilled out of the ice sheets and melted, and the water is vaporized to form the gas H_2O_v for analysis.

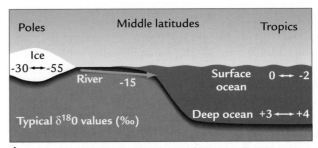

A

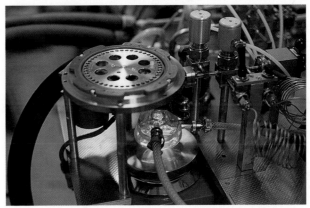

B

Measuring $\delta^{18}O$ values (A) In the modern ocean, typical $\delta^{18}O$ values vary from 0 to −2 ‰ in warm tropical surface waters to as much as +3 to +4‰ in cold deep ocean waters. In present-day ice sheets, in contrast, typical $\delta^{18}O$ values reach −30‰ in Greenland and −55‰ in Antarctica. (B) Mass spectrometers measure the relative abundances of two isotopes of oxygen (^{18}O and ^{16}O) in the calcite ($CaCO_3$) shells of marine plankton and in water melted from ice sheets. (B: courtesy of Steven Macko, University of Virginia.)

moving water vapor contains oxygen evaporated from the ocean, this process clearly has the potential to affect the isotope value of ocean waters (and the values recorded in the shells of foraminifera).

The process of gradual poleward transport of water vapor consists of many cycles of evaporation and pre-cipitation, and each of these cycles has an effect on the isotope composition of both the water vapor and the ocean. Because the lighter ^{16}O isotope evaporates more easily from the ocean, it tends to be preferentially extracted from the low-latitude ocean and sent toward higher latitudes, leaving the tropical ocean enriched in ^{18}O. Similarly, the heavier ^{18}O isotope is more easily removed from the atmosphere when condensation and precipitation occur, leaving the water vapor remaining in the atmosphere even more enriched in ^{16}O and the low-latitude ocean still more enriched in ^{18}O. Each step in this cycle decreases the $\delta^{18}O$ value of the water vapor by 10‰ in relation to that of ocean water left behind (Figure 7-6).

Because of these processes, water vapor in the atmosphere becomes progressively enriched in ^{16}O toward higher latitudes: that is, ^{16}O travels poleward more easily than ^{18}O. The same enrichment process, called **fractionation,** also occurs at higher altitudes, because high-altitude air has been through the same processes and has become similarly enriched in ^{16}O. This process of isotope fractionation is accompanied by the progressive removal of water vapor from the air, because cooler air holds much less water vapor than warmer air (Chapter 2). As a result, the air masses that are the most enriched in ^{16}O contain the least water vapor.

One result of this fractionation effect is that $\delta^{18}O$ values in today's surface ocean do not follow the trend that would be expected if temperature were their only

controlling factor. Modern tropical surface waters at 25°C have $\delta^{18}O$ values near 0‰. Using the temperature/$\delta^{18}O$ relationship just defined, high-latitude surface waters at temperatures of 0°C (just above the −1.8°C freezing point of seawater) should have $\delta^{18}O$ values as positive as +5‰. Instead, these waters have $\delta^{18}O$ values that are not very different from those of tropical surface waters.

The reason for these negative values is that high-latitude rivers carry water fed by precipitation with $\delta^{18}O$ values averaging near −15‰. The ‰ value of the rivers depends on the degree of fractionation that has occurred in the precipitation reaching its watershed. This annual delivery of ^{16}O-rich river water amounts to just a small fraction of the total volume of the high-latitude surface ocean, but it succeeds in driving the oceanic $\delta^{18}O$ composition toward more negative values. Coastal surface waters heavily affected by such rivers can be far more negative in $\delta^{18}O$ than the tropical ocean, and even high-latitude surface waters in regions well away from rivers have values comparable to those of the tropical ocean. This dilution effect by river water is also closely related to similar effects on ocean salinity: each 1.0‰ decrease in the $\delta^{18}O$ value of ocean water is accompanied by a 0.5‰ decrease in salinity due to delivery of fresh (nonsaline) water.

Another complication (but also an important opportunity for climate scientists) arises from the fact that some of the precipitation that falls on land is *not* returned quickly to the sea, but stays on land as ice. Precipitation at higher latitudes and elevations falls mainly as snow, and some of this snow falls on ice sheets and is stored there as ice. The result is that water vapor initially evaporated from warm tropical oceans ends up stored as high-latitude ice and is lost from the global ocean for many thousands of years until the ice melts.

Because of the many cycles of evaporation and condensation that occur during poleward transport of water vapor (Figure 7-6), the $\delta^{18}O$ composition of snow is highly enriched in ^{16}O, with values ranging between −15‰ for low-elevation continental interiors at middle latitudes to −55‰ for the very cold, high-elevation central domes of the Antarctic ice sheet. By contrast, the relatively warm snow that replenishes the Greenland ice sheet accumulates with a mean $\delta^{18}O$ value of about −35‰.

This long-term storage of freshwater taken from the ocean and stored as ^{16}O-rich ice leaves the ocean depleted in ^{16}O (enriched in ^{18}O). The $\delta^{18}O$ value of the entire world ocean becomes more positive than it would be if no ice were present.

A rough mass balance calculation reveals the size of this effect today. The volume of water stored in ice sheets today is roughly one-fiftieth of the volume of water in the oceans, and the two ice sheets combined

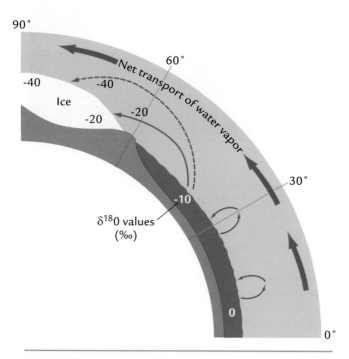

FIGURE 7-6 Isotope fractionation As water vapor moves from the tropics to high latitudes, it is enriched in the ^{16}O isotope during each step of evaporation and condensation. This fractionation process makes the $\delta^{18}O$ values of snow falling on (and stored in) ice sheets negative (^{16}O-rich).

have a mean $\delta^{18}O$ value of about $-50‰$. This should translate into a net effect on the mean $\delta^{18}O$ value of the ocean as follows: $1/50 \times 50‰ = 1‰$. This calculation indicates that the average $\delta^{18}O$ composition of today's ocean is about 1‰ more positive (more enriched in ^{18}O) than it would be if this large amount of ^{16}O-enriched water vapor had not been gradually extracted and stored in the present-day ice sheets.

We can now combine the effects of changes both in ocean temperature and in the amount of water stored in ice sheets on $\delta^{18}O$ values in the ocean to derive one equation that applies to the changes we may encounter over Earth's history:

$$T = 16.9 - 4.2\,(\delta^{18}O_c - \delta^{18}O_w)$$

where T = temperature in °C

$\delta^{18}O_c$ = $\delta^{18}O$ measured in the calcite shells of foraminifera

$\delta^{18}O_w$ = mean $\delta^{18}O$ value of ocean water when shells formed

In this equation, the value 16.9° is a constant, and the 4.2°C temperature effect per 1‰ change of ocean-water $\delta^{18}O$ appears directly as the term in front of the parentheses. The term $(\delta^{18}O_c - \delta^{18}O_w)$ indicates that the $\delta^{18}O$ values measured in the shells of foraminifera $(\delta^{18}O_c)$ have to be corrected to remove the effect of any changes in the mean ocean $\delta^{18}O$ value $(\delta^{18}O_w)$ caused by storage of ^{16}O-rich water in ice sheets. The modern ocean has a global mean value of 0‰, so that any change toward more or less ice than is present today needs to be corrected for in this equation. As a result, if ice sheets were growing or melting during an earlier time interval under study, this temperature relationship cannot be applied unless the changes in ice volume are known (which they rarely are).

A simpler and more functional form of this equation used by climate scientists is

$$\Delta\delta^{18}O_c = \Delta\delta^{18}O_w - 0.23\Delta T$$

where Δ means "change in." This equation tells us that measured changes in the $\delta^{18}O$ of foraminifera $(\Delta\delta^{18}O_c)$ are the result of changes in the mean $\delta^{18}O$ of the oceans $(\Delta\delta^{18}O_w)$ and of changes in the temperature of the water in which the shell formed (ΔT). The value 0.23 is the inverse of the value 4.2 in the previous equation: $\frac{1}{4.2}$.

One way to understand how this equation works is to imagine what would happen if we melted the modern ice sheets on Antarctica and Greenland over the next 10,000 years. The ice on both continents is rich in ^{16}O, leaving the modern ocean richer in ^{18}O (more positive $\delta^{18}O$ values). All this ice melted back into the ocean would deliver a large volume of ^{16}O-rich water. If we simply reverse the sense of the mass balance calculation made earlier, it tells us that melting all the modern-day

ice would gradually shift the ocean's average $\delta^{18}O$ value $(\delta^{18}O_w)$ from its present-day value of 0‰ to a value of about $-1‰$ over that 10,000-year interval.

This $-1‰$ change would be recorded in the shells of planktic and benthic foraminifera everywhere in the world ocean. If a climate scientist 10,000 years from now measured the shells of foraminifera deposited over the preceding 10,000 years but was unaware that the ice sheets had melted, the $-1‰$ shift recorded in the shells might be misinterpreted as a 4.2°C warming of the entire world ocean, rather than as a change in ice volume.

Trends in $\delta^{18}O$ over the Last 55 Myr Marine scientists have compiled a record of $\delta^{18}O_c$ from benthic foraminifera living on the floor of the deep ocean over the last 70 Myr (Figure 7-7). Although the trend is shown as a solid line, it is actually derived from hundreds of individual analyses that scatter around this line because of local temperature conditions specific to each ocean basin. This $\delta^{18}O_c$ curve shows an erratic trend toward more positive values beginning 55 Myr ago, with the fastest changes just before 36 Myr ago, near 13 Myr ago, and in the last 3 Myr.

As the preceding equations indicate, changes in $\delta^{18}O_c$ toward more positive values could be caused by

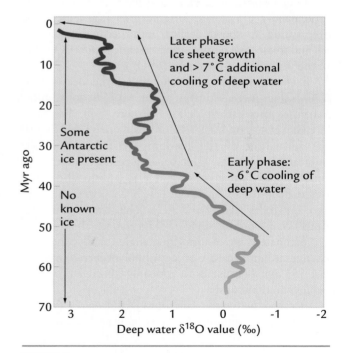

FIGURE 7-7 Long-term $\delta^{18}O$ trend Measurements of $\delta^{18}O$ in benthic foraminifera show an erratic long-term trend toward more positive values. From 55 to 40 Myr ago, the increase in $\delta^{18}O$ was caused by cooling of the deep ocean. After 40 to 35 Myr ago, it reflects both further cooling of the deep ocean and formation of ice sheets. (Adapted from K. G. Miller et al., "Tertiary Oxygen Isotope Synthesis: Sea Level History and Continental Margin Erosion," *Paleoceanography* 2 [1987]: 1–19.)

one or both of two factors: (1) a cooling of the deep ocean produced by cooler temperatures in the regions where deep water forms and (2) growth of ice sheets on land and increased storage of ^{16}O-rich water in the ice. Because both factors are critical aspects of the transition from a greenhouse to an icehouse climate, the $\delta^{18}O$ curve is an important index of global climate change, regardless of the relative importance of the two controlling factors.

We can disentangle the effects of temperature and ice volume in this curve to some extent. No evidence exists of ice on Antarctica or anywhere else on Earth between 70 and 40 Myr ago. The volume of any ice that did exist must have been negligible. During the interval between 55 and 40 Myr ago, $\delta^{18}O_c$ values increased from -0.75‰ to $+0.75$‰, a net change of $+1.5$‰. The temperature $-\delta^{18}O_c$ relationship tells us that deep waters cooled by over $6°C$ (1.5‰ x $4.2°C/$‰) between 55 and 40 Myr ago, before ice sheets appeared (Figure 7-7).

By 35 Myr ago, some ice had definitely appeared on Earth. Unfortunately, climate scientists have no way of knowing exactly what volume of ice existed at any time between 35 Myr ago and today. Another difficulty is deciding whether the ice that did exist in the past had a $\delta^{18}O$ composition similar to ice on modern-day Greenland (-35‰) or Antarctica (-55‰), or some other value. As shown earlier, both the total volume of ice and its isotope composition can affect the (^{18}O value of the water left in the ocean ($\delta^{18}O_w$).

Between 40 Myr ago and today, the deep-ocean $\delta^{18}O_c$ values increased from about $+0.75$‰ to $+3.5$‰, a further increase of 2.75‰ (Figure 7-7). We have already seen that about 1‰ of this increase was due to the formation of the modern ^{16}O-rich ice sheets. Subtracting this 1‰ value leaves a residual increase of roughly 1.75‰ that must be due to additional cooling of the deep ocean by more than $7°C$ between 40 Myr ago and today (1.75‰ $\times$ 4.2). This brings the total deep-ocean cooling since 55 Myr ago to about $14°C$. Unfortunately, we cannot disentangle exactly when the changes in temperature or ice volume occurred over the last 35 Myr, because both probably affected the $\delta^{18}O_c$ values simultaneously.

Because the temperature of today's deep ocean averages about $2°C$ and has cooled by at least $14°C$ over 55 Myr, the deep-ocean temperature of the greenhouse ocean 55 Myr ago must have been near $16°C$. If deep water formed mainly in high latitudes as it does today, the polar climates that sent such warm water into the deep ocean must have been much more temperate than they are today. This conclusion is consistent with the vegetation and glacier evidence discussed earlier.

Similarly, the evolving trend of the $\delta^{18}O$ signal in Figure 7-7 matches some of the evidence summarized in Figure 7-1: the first Antarctic glaciation 35 Myr ago

falls at the end of a long interval of increasing $\delta^{18}O$ values, and the most recent major $\delta^{18}O$ increase matches the spreading of tundra around the Arctic and the first northern hemisphere glaciation just after 3 Myr ago. In a general way, the erratic but progressive cooling indicated by the $\delta^{18}O$ trend also resembles the mid-latitude temperature curve derived from leaf-margin evidence (Figure 7-5).

Examined in more detail, however, the leaf-margin and $\delta^{18}O$ trends disagree in the timing of many shorter temperature fluctuations. This disagreement could reflect the lack of enough leaf-margin samples to resolve finer structure in that record, whereas the $\delta^{18}O$ record is rich in detail. Alternatively, it might simply result from the fact that no one signal can possibly be a complete representation of true *global* climate. Even today, changes in climate over decades vary from region to region rather than following a single pattern in lock-step, and past climate changes must have varied from region to region to some extent. One example of such regional variations (not shown here) is that $\delta^{18}O$ trends from tropical planktic foraminifera changed very little over the last 15 Myr, indicating little if any cooling of low-latitude surface waters during a time when high-latitude (and deep-ocean) temperatures were cooling markedly.

Still another cause of differences between the curves is that deep-ocean circulation may react to climatic forcing with threshold-like responses; that is, it may remain in one more or less stable mode for a long interval of time even though the climate system is being pushed in a new direction, but then abruptly switch to a different circulation mode once some critical threshold is reached. For example, several climatic indicators discussed earlier point to significant cooling between 7 and 4 Myr ago, yet no major change shows up in the deep-ocean $\delta^{18}O$ curve at that time, but a large steplike increase occurs after 3 Myr ago.

In summary, a wide array of evidence documents a progressive, if somewhat erratic, cooling of both poles and of broad areas of the middle latitudes over the last 55 Myr. Roughly comparable amounts of cooling appear to have occurred in the first and second halves of this long interval. Any hypothesis proposing to explain this greenhouse-to-icehouse transition must provide a climatic driving force that persists throughout the entire 55 million years.

Why Did Global Climate Cool over the Last 55 Myr?

How do we choose among the hypotheses that might explain the cause or causes of the progressive tectonic-scale cooling of the last 55 Myr? Fortunately, climate scientists have enough observations in hand to narrow the possibilities.

We have seen in previous chapters that the polar position hypothesis cannot explain this transition. Antarctica was located at the South Pole during the greenhouse climate of 100 Myr ago and is still there during the icehouse climate of today. In addition, the largest overall shift of the continents in relation to latitude during the last 55 Myr has been the northward movement of India and Australia into tropical latitudes and away from the South Pole (Figure 7-8).

Evidence presented in the last section indicates the need for a mechanism that cools climate at both poles over 55 Myr. At this point, many climate scientists look to falling CO_2 levels as an obvious means of cooling both poles at once. Some believe that a CO_2 decrease is the only mechanism, whether by slower input (BLAG) or by faster removal (uplift weathering). Many others infer that something else is needed: an increase in ocean heat transport toward the poles.

7-3 Evaluating the BLAG Spreading Rate Hypothesis

To be a credible explanation of the global cooling of the last 55 Myr, the spreading rate hypothesis must pass one critical test. It must be demonstrated that a slowing of global mean spreading and subduction rates occurred throughout that interval, leading to slower rates of CO_2 input to the atmosphere and causing global cooling.

It might seem initially that the spreading rate hypothesis has already passed this test. We saw in Chapter 6 that spreading rates 100 Myr ago appear to have been as much as 50% faster than they are today. Slower spreading today is consistent with a colder climate.

But when we look more closely at the last 55 Myr, this explanation is less satisfactory. Before 15 Myr ago, global mean spreading (and subduction) rates show a decrease consistent with the spreading rate hypothesis (Figure 7-9). But since 15 Myr ago, the mean rate of spreading and subduction have instead *increased* and actually returned to a present-day value equal to the one that existed almost 30 Myr ago.

The trend since 15 Myr ago is in direct disagreement with the spreading rate hypothesis. An increase in rates of spreading in the last 15 Myr should have put more CO_2 into the atmosphere and warmed global climate. Instead, the record shows that climate continued to cool, with all northern hemisphere ice (both mountain glaciers and ice sheets) first appearing during this interval, along with a substantial increase in size of the Antarctic ice sheet (Figure 7-1).

Could additional volcanic input of CO_2 at sites away from ocean ridges and subduction zones explain this discrepancy? During some intervals in the past, extra CO_2 has been released at oceanic and continental hot spots located well away from plate margins (Chapter 6). These episodes of volcanism have been radiometrically dated, and the estimated volume of volcanic rock they produced can be added to the amount produced by spreading and subduction (Figure 7-9). This adjustment does not change the basic picture; if anything, it makes the discrepancy worse. Inferred rates of volcanism and CO_2 input still increase during the last 15 Myr, and now the modern-day rate of CO_2 addition appears almost comparable to that 40 Myr ago, even though most of the greenhouse-to-icehouse cooling has occurred during exactly this time interval.

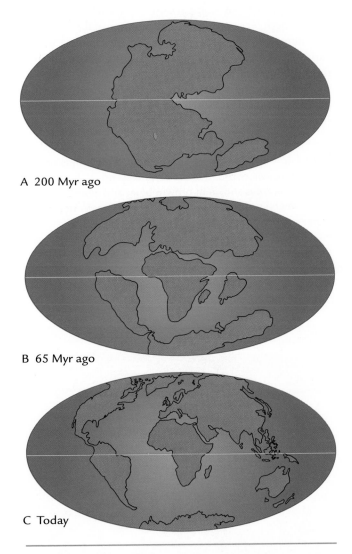

A 200 Myr ago

B 65 Myr ago

C Today

FIGURE 7-8 Continental movements since 200 Myr ago Since the time of Pangaea, 200 Myr ago (A), the Atlantic Ocean has widened, the Pacific Ocean has narrowed, and India and Australia have separated from Antarctica and moved northward to lower latitudes (B, C). (Modified from F. Press and R. Siever, *Understanding Earth*, 2nd ed., © 1998 by W. H. Freeman and Company.)

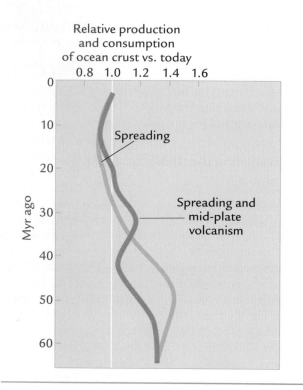

Relative production
and consumption
of ocean crust vs. today

FIGURE 7-9 **Changes in spreading rates** The average rate of seafloor spreading slowed until 15 Myr ago, and has since sped up again. Adding in the effects of generation of new crust by volcanism at hot spots away from plate margins does not change this basic trend. (Adapted from L. R. Kump and M. A. Arthur, "Global Chemical Erosion During the Cenozoic," in *Tectonic Uplift and Climate Change,* ed. W. F. Ruddiman [New York: Plenum Press, 1997].)

In summary, the evidence indicates that the spreading rate hypothesis may have been a cause of global cooling before 15 Myr ago, and particularly before 30 or 40 Myr ago. But it predicts a significant warming during the last 15 Myr, when a major cooling has actually occurred.

7-4 Evaluating the Uplift Weathering Hypothesis

To demonstrate that the uplift weathering hypothesis explains global cooling during the last 55 Myr, its three main predictions must be verified: (1) the amount of high-elevation terrain in existence today must be unusually large in comparison with earlier intervals; (2) this high terrain must be causing unusual amounts of fragmentation of rock; and (3) the fragmentation and exposure of fresh debris must be causing unusually high rates of chemical weathering.

To determine whether or not these geographic and environmental factors are unusual at the present time,

FIGURE 7-10 **Earth's high topography** Earth today has only a few regions where broad areas of land stand more than 1 km high (areas shown in brown, blue, and white). Except for the high ice domes on Antarctica and Greenland, the highest bedrock surfaces are the Tibetan Plateau and other high terrain in southern Asia, the Andes of South America, the Rocky Mountains and Colorado Plateau of North America, and the volcanic plateaus of eastern and southern Africa. (Courtesy of Peter Schloss, National Geophysical Data Center, Boulder, Colo.)

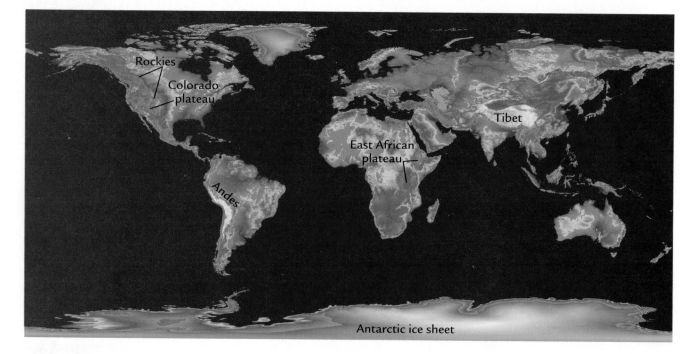

we have to compare them with some interval in the past. The last half of the Cretaceous interval, from 100 to 65 Myr ago, is a useful basis for comparison, for two reasons: (1) it is recent enough to have left abundant evidence in the geologic record and (2) it is an interval of full greenhouse climate (Chapter 6).

Prediction 1: Extensive High Terrain At first glance it appears obvious that uplift has been unusually active in most mountain ranges in the last few tens of millions of years. Marine sediments deposited at or below sea level 100 to 65 Myr ago are now found at high elevations in Tibet and the Himalayas of Asia, the South American Andes, the North American Rocky Mountains, and the European Alps (Figure 7-10). These sediments have all been uplifted from sea level to their present heights in 70 Myr or less. Other information confirms this conclusion, such as evidence that slivers of crust have been pushed together and stacked on top of each other during the same interval, thickening the crust and causing uplift.

Although much of today's high topography is geologically youthful, this evidence alone does not prove that modern elevations are *uniquely* high. Plate tectonic processes continually cause uplift in many regions throughout geologic time, while erosion continually attacks the highest topography and wears it down. As a result, the highest topography existing during any interval of geologic time is always recent in origin, just as it is today.

The strongest evidence that the amount of high terrain is indeed more massive today than in earlier geologic eras is the Tibetan Plateau, a feature more than 2 million square kilometers in area looming at an average elevation above 5 kilometers. Tibet formed after the initial collision of India and Asia 55 Myr ago, and the uplift of its surface has occurred since then (Figure 7-11A and

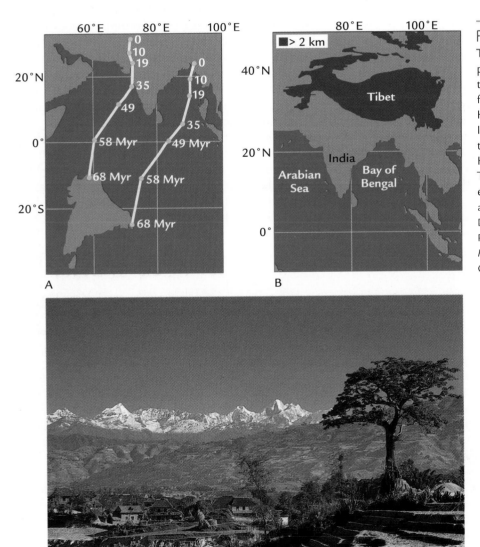

FIGURE 7-11 India-Asia collision and Tibet Collision of India and Asia (A) produced the Tibetan Plateau (B), the largest high-elevation rock feature on Earth today. (C) The Himalaya Mountains tower over the Indian subcontinent to the south (in the foreground). Behind the Himalayas to the north lies the vast Tibetan Plateau at an average elevation above 5000 m. (A and B: adapted from P. Molnar et al., "Mantle Dynamics, the Uplift of the Tibetan Plateau, and the Indian Monsoon," *Review of Geophysics* 31 [1993]: 357–96; C: Emil Muench/Photo Researchers.)

BOX 7-2 CLIMATE DEBATE

The Timing of Uplift in Western North America

Although it might seem surprising, climate scientists disagree about the age of uplift of most of the high terrain in western North America. All agree that mountains of some size have existed in Nevada and eastern California for 200 Myr or more because of ongoing subduction along the coast until about 30 Myr ago. They also agree that the Rocky Mountain West region (the modern-day High Plains, Colorado Plateau, and both the U.S. and Canadian Rocky Mountains) was flooded by an inland sea from 100 to 70 Myr ago and has since been uplifted to its present height. But everything else about the timing and amount of uplift in this region is in dispute.

One group emphasizes recent uplift. Their view is that the earlier mountain terrain near Nevada was a series of discontinuous low-elevation peaks, not a major topographic feature. This group infers that broad, large-scale upwarping of the *entire* West from the Sierra Nevada of California to the High Plains of Colorado and Wyoming began about 20 Myr ago because of some kind of deep-seated heating process. Travelers in the American West who read signs at national parks will notice that this view is widely promoted by the U.S. Park Service.

The other group interprets major uplift as occurring at an earlier time, followed by a more recent *loss* of elevation. In their view, the mountain belt in the Far West before 100 Myr ago was a continuous feature at 3–4 km elevation like the modern Andes, and that this region subsequently dropped to its modern height of 1–2 km. While this group acknowledges large-scale uplift of the Rocky Mountain West since 70 Myr ago, they infer that most uplift in that region actually occurred between 70 and 45 Myr ago during an interval of heightened tectonic activity and that most of the Rocky Mountain area has lost elevation since this activity ceased.

Temperature reconstructions based on leaf-margin types can be used to test these two opposing views. Temperatures are estimated from deposits of the same age in both higher-elevation mountain areas and nearby coastal regions. Because temperature decreases with elevation in a known way because of the prevailing lapse rate, the estimated temperature *differences* between coastal and high-elevation regions can be converted to estimates of past elevation. This method suggests a small decrease (or at least no increase) in elevation of most parts of the Rocky Mountain West during the last 40 or 50 Myr. The only area proved to have risen in the last 10 or 20 Myr is the region around the Yellowstone hot spot, domed upward by shallow subsurface heating.

B). In contrast, no continental collisions occurred from 100 to 65 Myr ago, and no such massive plateaus existed then, or for the preceding 150 million years. The presence of the Tibetan Plateau and Himalayan complex (Figure 7-11B and C) is a strong argument that modern topography is unusually high and massive.

Most other high-elevation regions on Earth (Figure 7-10) have been formed by subduction of ocean crust beneath continental margins. Because subduction is an ongoing process, this kind of mountain terrain must have existed more or less continuously through time, in contrast to plateaus produced by sporadic continental collisions.

The modern Andes and the central plateau called the Altiplano result from subduction along the west coast of South America. Because subduction has been under way there for more than 100 Myr, a mountain range has long existed in the western Andes. The central Altiplano and the eastern Andes were both created in the last 55 Myr, as part of a major eastward expansion of the Andean chain. This expansion apparently increased the total mass of the high Andes.

Subduction has also occurred along western North America for some 200 million years. Scientists are sharply divided about the history of uplift in this region over the last 100 Myr, but new evidence indicates that uplift of a large region in the Rocky Mountain area was offset by lowering of high terrain farther west, near Nevada (Box 7-2).

Another kind of high terrain is the extensive low plateau in eastern and southern Africa at an elevation of 1 kilometer (Figure 7-10). This plateau results from deep-seated heating that causes a broad upward doming and outpouring of volcanic lava. The East African

BOX 7-2 CONT. CLIMATE DEBATE

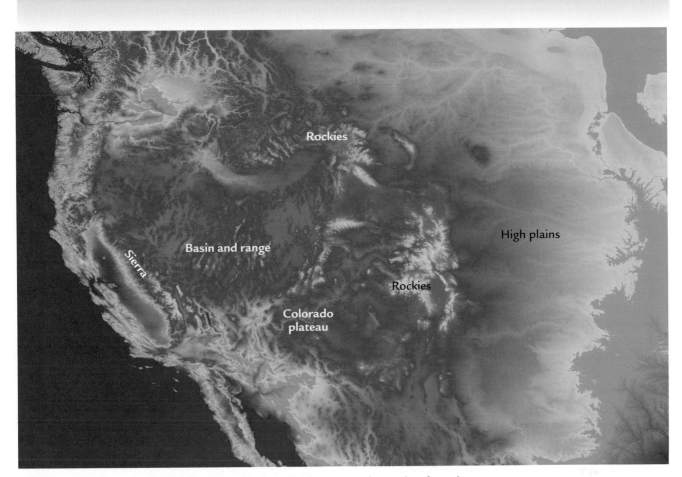

Topography in the American West A broad bulge of high topography reaches from the Sierra and the Basin and Range region in the far western United States to the Colorado Plateau, Rocky Mountains, and High Plains farther east. Low-elevation regions a few hundred meters above sea level are shown in green, with progressively higher elevations in yellow, brown, pale orange, and white. (Courtesy of Peter Schloss, National Geophysical Data Center, Boulder, Colo.)

plateau appears to have been built in the last 30 Myr, but similar or possibly greater amounts of plateau construction occurred on Africa near 100 Myr ago as the giant continent of Pangaea broke up. As a result, it is hard to argue that the present plateau is a unique feature on that continent, unlike anything that occurred in earlier intervals.

In summary, the existence of the massive Tibetan Plateau makes modern topography unusual in comparison with much of geologic time, consistent with the uplift weathering hypothesis. Regions of high youthful terrain also exist along subducting plate margins

and elsewhere, but it is difficult to prove that these are higher than similar features constructed during earlier intervals.

Prediction 2: Unusual Physical Weathering The second test of the uplift weathering hypothesis is whether or not today's high topography is causing higher rates of physical weathering and rock fragmentation than in the past. The best record of rates of erosion lies in sediments deposited in ocean basins by rivers. By far the largest mass of young sediment in the ocean today is found on the seafloor of the Indian Ocean south of the Himalayas. Deposition of this pile of sediment began

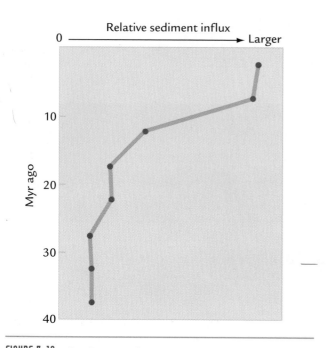

Relative sediment influx

FIGURE 7-12 Himalayan sediments in the Indian Ocean The rate of influx of sediments from the Himalayas and Tibet to the deep Indian Ocean has increased almost tenfold since 40 Myr ago. (Adapted from D. K. Rea, "Delivery of Himalayan Sediment to the Northern Indian Ocean and Its Relation to Global Climate, Sea Level, Uplift, and Seawater Strontium," in *Synthesis of Results from Scientific Drilling of the Indian Ocean,* ed. R. A. Duncan et al. [Washington, D.C.: American Geophysical Union, 1992].)

near 40 Myr ago, increased after 20 Myr ago, and accelerated rapidly near 10 Myr ago (Figure 7-12).

Climate model experiments indicate that this influx of sediment to the Indian Ocean is a result of two factors: (1) creation of steep terrain along the southern Himalayan margin of the Tibetan Plateau and (2) the fact that a plateau the size of Tibet in effect creates its own weather, including the powerful South Asian monsoon (Figure 7-13).

Monsoons result from different rates of heating of continents and oceans due to the different heat capacities of land and water (Chapter 2). In summer, more rapid heating of the continents than of the ocean causes rising motion over the land, which draws in moist ocean air and creates the wet summer monsoon. Plateau uplift intensifies this circulation by creating a high-elevation surface that releases more sensible heat through the thinner atmosphere. The increased heat loss further reduces air pressure over the plateau and draws in still more air from the ocean. In addition, the sides of the uplifted plateaus form obstacles to the incoming ocean air, forcing it to rise and its water vapor to condense. As a result, the steep slopes (in this case, the Himalayas on the southern margins of the plateau) become a natural focal point for strong monsoon rains, which release latent heat during the summer and fuel even more powerful monsoons.

Unfortunately, ocean sediments cannot give us a definitive estimate of global rates of physical weathering. Much of the sediment eroded from coastal mountain ranges and deposited in the nearby ocean is soon subducted into nearby trenches. The large amount of sediment lost in this way cannot be quantified. In addition, some sediments deposited on the seafloor are eroded and redeposited, and this reworking skews compilations of sediment deposition rates through time toward younger ages.

In summary, rapid deposition of huge amounts of Himalayan sediment supports the hypothesis that physical weathering is stronger today on a global basis than in earlier times, but this conclusion remains unproved because of sediment lost to subduction in ocean trenches. If erosion of sediments from mountains along subducting margins has occurred at a nearly constant global-average rate through time, then today's average physical weathering rate should be higher as a direct consequence of the extra sediment shed by the uplifting plateau.

Prediction 3: Unusual Chemical Weathering The final test of the uplift weathering hypothesis is

FIGURE 7-13 Tibet and the monsoon Heating of the Tibetan Plateau draws in moisture from the Indian Ocean and enhances the intensity of the warm, moist summer monsoon on its southern (Himalayan) margin.

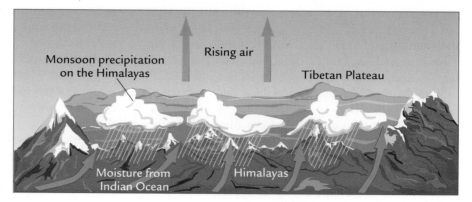

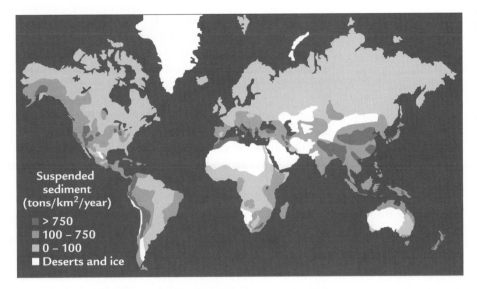

FIGURE 7-14 Sediments suspended in rivers The annual yield of suspended sediments is highest in two regions: the Himalayas of southeast Asia and the Andes of South America. (Adapted from D. E. Walling and B. W. Webb, "Patterns of Sediment Yield," in *Background to Paleohydrology*, ed. K. J. Gregory [New York: John Wiley, 1983].)

whether or not the global average rate of chemical weathering today is higher than it was in the past. Unfortunately, chemical weathering rates are difficult to determine even on regional scales, much less for the entire Earth.

Climate scientists quantify modern rates of chemical weathering on a regional basis by measuring the total amount of ions dissolved and transported in rivers. This chemical measure reflects the amount of weathering within the watershed drained by each river. But disturbances of natural weathering processes by humans complicate such studies, and it is also difficult to distinguish between the ions put into the rivers by slow weathering of silicate rocks (hydrolysis) and those resulting from rapid dissolution of carbonate rocks. Only hydrolysis affects the CO_2 balance in the atmosphere (Chapter 4). Another limitation is strategic: it is effectively impossible to study enough rivers to reach an accurate estimate of the global weathering rate, because too many rivers contribute significantly to the global total.

It is even more difficult to reconstruct rates of chemical weathering during earlier intervals. One potential solution examined later in this chapter would be to find chemical indices that integrate the entire global delivery of weathering products to the world ocean. But at this point, the strongest case for unusually rapid chemical weathering today is an argument based on inference. The height of the terrain in the Tibetan-Himalayan complex is unusual in comparison with elevations at earlier intervals, resulting in unusually large exposure of fresh rock on steep slopes. The steep slopes also receive some of the most intense rainfall on Earth today, including the monsoon rains generated by the Tibetan Plateau itself. These heavy rains produce the largest concentrations of suspended particles observed in rivers anywhere on Earth (Figure 7-14). By inference, this combination of several favorable factors should promote unusually rapid chemical weathering in southeast Asia, which should increase the global average rate of chemical weathering and cause CO_2 removal from the atmosphere. But inference is not proof.

7-5 Evaluating the Ocean Heat Transport Hypothesis

As we saw in Chapter 6, higher CO_2 is a likely cause of general greenhouse warmth 100 Myr ago, but the pole-to-equator temperature pattern simulated by models with higher levels of atmospheric CO_2 does not match the observed temperature pattern. As a result, some climate scientists invoke the ocean heat transport hypothesis, calling on stronger ocean heat transport to warm polar regions and cool the tropics during the greenhouse climates of 100 Myr ago.

This hypothesis has also been applied to the subsequent interval when Earth began to cool. As was the case for 100 Myr ago, some evidence indicates that the north polar regions remained anomalously warm and the tropics anomalously cool in relation to the patterns predicted by model simulations of the effects of atmospheric CO_2 alone. Many climate scientists infer that strong poleward heat transport continued during much of this long-term cooling but progressively decreased, and that the *loss* of this ocean heat contributed to the polar cooling of the last 55 Myr.

Several versions of the ocean heat transport hypothesis focus on the opening or closing of critical **oceanic gateways**. Gateways are narrow passages linking the major ocean basins, and changes in their configuration alter the amount of seawater exchanged between oceans, as well as the heat and salt carried by seawater. Over the last several decades, several climate scientists have proposed that gradual changes in key gateways caused glaciations by altering the poleward transport of heat and salt.

One reason for the interest in this explanation is that some major gateway changes correlate closely in time with large-scale changes in climate. For example, the emergence of the Isthmus of Panama, which closed an open seaway between North and South America near 4 Myr ago, preceded the appearance of large ice sheets in the northern hemisphere by 1 million years.

Although major changes in ocean gateways definitely can alter inter-ocean exchanges of heat and salt in a significant way, the question here is the extent to which these gateway changes affect the climate system on regional and global scales. One way to address this question is to run experiments using ocean general circulation models. O-GCMs have been used to study two major gateway changes during the last 55 million years.

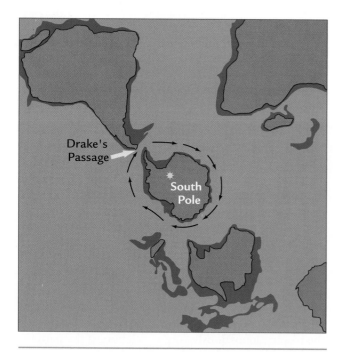

FIGURE 7-15 Opening of Drake's Passage Opening of an ocean gap between South America and Antarctica near 25 to 20 Myr ago allowed a strong Antarctic circumpolar current (arrows) to flow uninterrupted around the Antarctic continent. The passageway between Australia and Antarctica had opened 10 to 15 million years earlier. (Adapted from E. J. Barron et al., "Paleogeography: 180 Million Years Ago to the Present," *Ecologae Geologicae Helvatiae* 74 [1981]: 443–70.)

Gateway Case Study 1: Opening of Ocean Circulation around Antarctica During the last 55 million years, the last of the Gondwana continents connected to Antarctica split off and moved north, leaving Antarctica isolated and surrounded by a circumpolar ocean. In the late 1970s the marine geologist James Kennett proposed that this breakup caused glaciation to develop on Antarctica. Before the separation of the continents, flow of the ocean around the Antarctic was impeded by the land connection with South America and by the barrier of the northward-projecting Australian continent (Figure 7-8, middle). Kennett hypothesized that these barriers diverted warm ocean currents poleward from lower latitudes and brought Antarctica enough heat to prevent glaciation.

After separation of the continents, the land barriers opened and permitted a strong, unimpeded eastward flow around Antarctica. Without the warm poleward flow of heat, the continent cooled and glaciation began. The original hypothesis pointed to the separation of southern Australia from Antarctica as the critical event that initiated glaciation, while later versions have focused on the opening of Drake's Passage between South America and Antarctica near 25 to 20 Myr ago (Figure 7-15).

This hypothesis has been questioned by some climate scientists because of differences in timing between Antarctic glaciation and final opening of unrestricted circumpolar flow. The opening of Drake's Passage between South America and Antarctica near 25 to 20 Myr ago occurred 10 million years later than the first appearance of Antarctic ice 35 Myr ago and 10 million years earlier than the interval of intensified Antarctic glaciation and global cooling near 13 Myr ago (Figure 7-1).

Climate scientists have also used O-GCMs to examine this hypothesis with sensitivity tests. In a pair of experiments, Drake's Passage was alternately closed and left open, while all other features of Earth's geography were kept the same. The model results suggest that opening Drake's Passage did not significantly alter ocean (or atmospheric) temperatures near Antarctica. Instead, the model simulated a frigid climate over Antarctica both with and without unrestricted ocean flow around the continent. The results of this experiment suggest that gateway opening was not a significant factor in the onset and development of Antarctic glaciation.

Because ocean models are still at a relatively crude stage of development, further experiments with improved models are needed to test this conclusion. Smaller grid boxes would permit more realistic representation of an open Drake's Passage, which by necessity was represented in the model by an opening much larger than the size of the actual passage today.

Gateway Case Study 2: Closing the Central American Seaway During the last 10 million years, uplift in Central America closed a deep ocean passage that had separated North and South America in the region of Panama and created the Central American part of the Cordilleran mountain chain. Final closure of this gateway between the Atlantic and Pacific oceans occurred just before 4 Myr ago, and was followed by the first large-scale glaciation of North America near 2.7 Myr ago.

Many climate scientists have speculated that these two events are linked. They hypothesize that construction of the Isthmus of Panama blocked the strong westward flow of warm, salty tropical water that had previously been driven out of the tropical Atlantic Ocean and into the eastern Pacific by trade winds. They also suggest that the newly formed isthmus redirected this flow into the Gulf Stream and on toward high latitudes of the Atlantic. This strengthened northward flow of warm, salty water would suppress the formation of sea ice in north polar regions, because saltier waters resist freezing better than fresher water (Chapter 2). Researchers hypothesize that the reduced cover of sea ice made more moisture from the ocean available to nearby landmasses and triggered the growth of ice sheets there.

Scientists have also used ocean GCMs to test this hypothesis, again by running a pair of experiments with the Panama region configured first as an open gateway passage and then as a closed-off isthmus. These two configurations correspond to the end points of a gradual process of shallowing and final closure of the isthmus between about 10 and 4 Myr ago.

The results from the pair of simulations confirm the prediction of a major redirection of warm, salty water northward in the Atlantic (Figure 7-16). The simulations also indicate that closing the isthmus could have cut off a return flow of low-salinity Pacific water into the Atlantic, which had balanced the Atlantic-to-Pacific inflow of warmer, saltier water. Blockage of this low-salinity flow by the isthmus further increased the salinity of northward-flowing water in the Atlantic.

In other respects, however, the model simulations do not support, and actually contradict, the original gateway hypothesis. The strengthened northward flow of salty water in the Atlantic and resulting reduction of sea ice caused by the closing of the isthmus did not significantly alter precipitation patterns around the high-latitude North Atlantic. As a result, the hypothesized increase in moisture needed to grow ice sheets did not occur. Instead, the stronger northward flow transferred a large amount of heat to the atmosphere and warmed the regions where ice sheets eventually formed (Figure 7-16). This warming increased summer melting of snow and thereby opposed the conditions needed for glaciation to begin, contradicting the gateway hypothesis.

Assessment of Gateway Changes You may have noticed that the assumptions made by climate scientists about the effects of changing ocean circulation work in

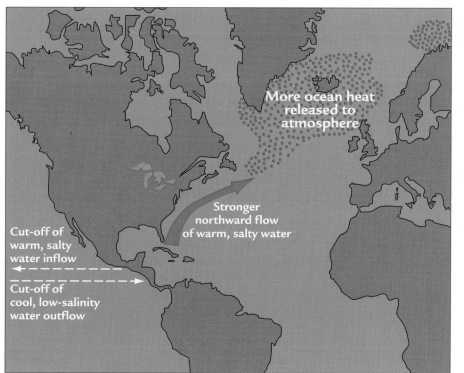

FIGURE 7-16 Closing of the Isthmus of Panama Simulations with ocean models indicate that gradual closing of the Central American isthmus between 10 and 4 Myr ago redirected warm salty water northward into the Atlantic Ocean, reduced the extent of high-latitude sea ice, and handed off additional heat to the atmosphere. (Adapted from E. Maier-Reimer et al., "Ocean General Circulation Model Sensitivity Experiment with an Open Central American Isthmus," *Paleoceanography* 5 [1990]: 349–66.)

exactly opposite directions in these two gateway case studies. For the Antarctic case study, scientists proposed that *reduced* poleward flow of warm ocean water caused cooling that promoted glaciation. For the Isthmus of Panama case study, they proposed that *increased* poleward flow of warm ocean water caused an increase in moisture flux that promoted glaciation.

These differing assumptions reflect disagreements among climate scientists about how the ocean actually affects the mass balance of ice sheets. Some climate scientists emphasize that a warmer ocean will release more *latent* heat (water vapor) to the atmosphere and thereby supply more moisture to aid the growth of ice sheets. Most climate scientists, and especially climate modelers, emphasize that a warmer ocean will release more *sensible* heat to the overlying atmosphere and thereby potentially melt more ice. The latter view is consistent with the strong increase in ablation on low-elevation margins of the Greenland ice sheet in comparison with the smaller range of variation in rates of accumulation (see Figure 2-31). Ablation is potentially a more powerful factor than accumulation.

In any case, neither set of model experiments supports the hypothesis that changes in the poleward flow of warm ocean water would have had an effect on climate large enough to initiate growth of ice sheets. This criticism needs to be tempered by the realization that ocean GCMs are still at an early stage of development.

Critics of the hypothesis that changes in gateways control ocean heat transport also point out that each opening or closing of a gateway is a discrete episode in time in a limited geographic region, rather than a continuous process. They question how these discontinuous regional events could combine to drive a progressive climatic cooling that has lasted 55 million years. For example, if the final closure of the Isthmus of Panama is called on to explain the onset of northern hemisphere glaciation just after 3 Myr ago, a completely different factor would then have to be invoked to explain the subsequent intensification of the size of glaciations near 1 Myr ago.

Whether or not gateway changes affect climate on a global scale, they definitely alter the production and flow of deep and bottom water through ocean basins. Major gateway changes redistribute heat and salt in ocean surface waters at sites where deep waters form, and the changes in surface-water properties affect formation of deep water.

For example, the model experiments in the second case study indicated that closing the Isthmus of Panama increased the formation of deep water in the high latitudes of the North Atlantic. Formation of deep water increased in this simulation because of the increased salinity of northward-flowing Gulf Stream waters (Figure 7-16). Higher-salinity surface water promotes stronger deep-water formation because the water is already dense when it encounters cold winter air masses at high latitudes. The results from this experiment agree with evidence that formation of North Atlantic deep water increased between 10 and 4 Myr ago, the interval over which the Central American seaway was gradually reduced and finally closed off by the emergence of the Central American isthmus.

7-6 Causes of Brief Tectonic-Scale Climate Change

The cooling that produced the present icehouse was not a one-way process but an erratic trend interrupted by shorter-term warming and cooling episodes, each lasting a few million years. The cause of these shorter oscillations is not certain, but several hypotheses have been proposed to explain climate coolings over these briefer intervals.

Volcanic Aerosols Explosive eruptions of volcanoes cool climate over intervals of a few years. Volcanoes erupt sulfur dioxide (SO_2) gas, which mixes with water vapor in the air and forms droplets and particles of sulfuric acid called **sulfate aerosols**. Highly explosive eruptions inject aerosols 20 to 30 kilometers into the stratosphere, where they block some of the incoming solar radiation from reaching the ground. When solar radiation is reduced, Earth's surface cools.

The locations of most volcanoes on Earth today are determined by plate tectonic processes. Most volcanoes occur in arcs surrounding the Pacific Ocean; others are in the northern Indian Ocean and a few in the Caribbean and Mediterranean and on ocean islands such as Iceland (Figure 7-17). In most of these areas, tectonic plates are converging, and ocean crust and lithosphere are subducting beneath continents or other oceanic plates.

Latitude determines the geographic extent of the effects of volcanic eruptions on climate. Volcanoes that erupt poleward of the tropics (between 23.5° and 90° latitude) produce particles that stay within the hemisphere where the eruption occurs, limiting their cooling impact to that hemisphere. In contrast, Earth's atmosphere can mix particles produced by volcanoes that erupt in the tropics into both hemispheres and produce a global impact on climate.

Another factor that determines the impact of eruptions is their explosivity. Ocean-island volcanoes with iron- and magnesium-rich compositions tend not to cause explosive eruptions but instead produce lava that flows easily across the land (as on Hawaii). Volcanic particles sent into the air by these small eruptions never reach the stratosphere but stay within the troposphere. Although these particles block sunlight and may produce disastrous environmental effects near the volcano,

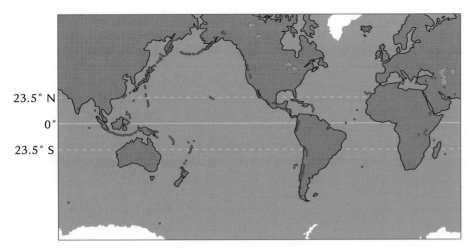

FIGURE 7-17 **Locations of Earth's volcanoes** Earth's active volcanoes mainly occur in regions of convergence of tectonic plates. Most lie along the margins of the Pacific Ocean. (Modified from F. Press and R. Siever, *Understanding Earth,* 2nd ed., © 1998 by W. H. Freeman and Company.)

they are brought back down to Earth within a few days by rain. With so brief a stay in the atmosphere, these particles cannot be widely enough distributed around the planet to produce hemispheric or global effects on climate.

In contrast, volcanoes along converging plate margins are fed by magmas richer in silica and other elements found in continental crust. Their eruptions are more explosive because the natural resistance of this kind of molten magma to flow causes internal pressures to build up to the point where volcanic particles can be injected into the stratosphere, above the level where precipitation can wash them out. Slow settling of these fine particles due to gravity takes years, long enough for them to be distributed within a hemisphere or across the entire planet.

Sulfate aerosol concentrations in the stratosphere reach maximum levels within months and then begin to decrease as gravity removes the particles. Their decrease follows an exponential trend: each year about half of the remaining particles settle out, and within a few years aerosol concentrations are much reduced (Figure 7-18, top). The cooling effect of these aerosols on climate follows the same trend, with a maximum initial cooling that soon fades away (Figure 7-18, bottom).

Climate scientists face difficulties in their efforts to reconstruct the effects of ancient volcanic explosions on climate. They may be able to get some idea of the magnitude of ancient eruptions from the sizes of the craters left by the explosion or the volume of volcanic ash deposited nearby. The geographic area over which the ash is distributed may also provide clues about the explosivity of the eruption and whether the volcanic particles might have reached the stratosphere. But sulfur forms only a small and variable fraction of the total volume of erupted material, and the volume of ash cannot directly be used to estimate the amount of sulfur

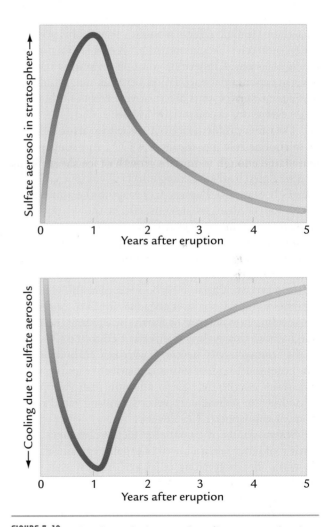

FIGURE 7-18 **Volcanic explosions and cooling** Large volcanic explosions launch sulfate aerosols into the stratosphere (top) and cool climate for a few years (bottom). (Adapted from R. S. Bradley, "The Explosive Volcanic Eruption Signal in Northern Hemisphere Continental Temperature Records," *Climate Change* 12 [1988]: 221–43.)

erupted. As a result, it is impossible to estimate the climatic effects of ancient eruptions reliably.

Some climate scientists argue that unusually large volcanic explosions might have played a significant role in longer-term climate cooling. They point out that the eruptions observed during the last century or two are much smaller than eruptions known from the geologic record and infer that much larger eruptions in the past would have caused stronger cooling.

Although large eruptions no doubt have caused significant cooling for a few years, it is more difficult to make a case for cooling over longer intervals. Even the large cooling from a massive eruption should last only the few years it takes for gravity to remove the sun-blocking sulfate aerosols from the stratosphere. To cool global climate significantly for even a century would take a dozen or more large eruptions every few years so that their climatic effects would overlap. This scenario seems unlikely: like many other geologic phenomena (such as asteroid impacts), volcanic eruptions have an inverse exponential correlation between magnitude and frequency: small events are relatively frequent, but very large events are exponentially less so.

The other problem faced by those who favor a longer-term cooling role for volcanoes is CO_2. Any longer-term cooling effect from volcanic sulfates will at some point be overwhelmed by the warming effects of volcanic CO_2 (the BLAG hypothesis). Because CO_2 stays in the atmosphere much longer than sulfates, its effects should dominate over intervals longer than a century or so.

Burial of Organic Carbon Another possible source of short-term climate changes at tectonic time scales is variation in the rate of burial and exposure of organic carbon. Organic carbon is a plausible driver of climate changes because it accounts for 20% of the carbon cycling into and out of Earth's sediments and rocks, and because this cycling is directly tied to the CO_2 level of the atmosphere (Chapter 4). Organic carbon also has the potential to affect climate relatively rapidly: under the right conditions, large amounts can be quickly buried in the sedimentary record, causing rapid reductions of atmospheric CO_2 levels.

Several kinds of climatic or tectonic changes could favor rapid increases in burial of organic carbon: changes in wind direction along a coastal region that cause increased upwelling and carbon production; an increase in the total amount of organic carbon and nutrients delivered to the ocean; or a change toward a wetter climate on continental margins, where flat topography naturally favors swamps and deposition of organic matter.

An increase in the rate of burial of organic carbon has been proposed as the cause of a cooling trend near 13 Myr ago. The large increase in deep-ocean $\delta^{18}O$ values at this time indicates some combination of deep-

water cooling and increase in size of the Antarctic ice sheet. These changes followed an interval when carbon-rich sediments were deposited in shallow waters around the margins of the Pacific Ocean, including the Monterey coast of California. The marine geologists Edith Vincent and Wolfgang Berger suggested that a major increase in coastal upwelling, perhaps driven by stronger winds caused by long-term climate cooling, buried enough organic carbon along the margins of the Pacific to reduce atmospheric CO_2, cool global climate, and allow ice to build up on Antarctica. They called this the **Monterey hypothesis**.

The Monterey hypothesis has been criticized because there was a lag of 2 to 3 million years between the onset of rapid carbon burial and the time of fastest cooling shown by the $\delta^{18}O$ record, although the fastest rates of carbon burial appear to have occurred closer to the cooling. Other climate scientists have suggested that the increased carbon burial in the Pacific could reflect a greater supply of carbon eroded from older sedimentary rocks on land (in the Himalayas). This explanation does not require a net loss of carbon from the ocean and atmosphere.

Burial of organic carbon on shallow continental margins also tends to produce its own negative feedback. Carbon-rich sediments deposited in shallow areas are later reexposed to the atmosphere if sea level falls because ice sheets grow. Exposure of this buried organic carbon allows it to be oxidized back to CO_2 and returned to the atmosphere. The return of CO_2 then causes the climate to warm.

Some climate scientists speculate that changing rates of weathering and of organic carbon burial in the deep ocean could be an important cause of longer-term cooling over tens of millions of years. If imbalances between rates of burial of organic carbon and rates of weathering and oxidation of old organic carbon persist for many millions of years, the result could be increases or decreases in the total amount of carbon in the ocean-atmosphere system and of the level of CO_2 in the atmosphere. Scientists are investigating whether the organic carbon subcycle has been adding or removing carbon (and CO_2) from the ocean-atmosphere system over long intervals.

Understanding and Predicting Tectonic Climate Change

The processes that have slowly driven Earth's climate into the icehouse continue today. Northern hemisphere ice sheets developed in just the last 3 million years, and they have reached their largest extent in the last million years during periodic glaciations caused by changes in Earth's orbit around the Sun, as we will see in Part III. This gradual cooling trend has reached the point where northeastern Canada has been free of ice sheets during only about 10% of the last half-million years. All this evidence indicates that we are slowly headed toward an even colder icehouse with more persistent glacial ice in the future. If climate scientists understood the cause of this cooling more clearly, they could better predict our longer-term climatic future.

What can scientists do to resolve the uncertainty about the cause of this long-term cooling? In the case of the ocean heat transport hypothesis, the greatest need is for improved ocean models with smaller grid boxes to replicate the configurations of the narrow gateway passages, as well as improved simulations of the processes that form deep water.

For hypotheses that invoke falling CO_2 levels as the main explanation of cooling, the most useful information would be some kind of index of the global mean rate of chemical weathering through time. Such an index is critically needed because the spreading rate and uplift weathering hypotheses make opposite predictions about the trend of chemical weathering over the last 55 Myr. The uplift weathering hypothesis predicts that weathering increased and drove the cooling, while the spreading rate (BLAG) hypothesis predicts that weathering slowed and moderated the cooling. A decade ago, some scientists thought they had found such an index in the geologic record, but later work showed this view to be too simplistic (Box 7-3).

Despite the wealth of evidence that climate has cooled over the last 55 million years, it is impossible to predict future tectonic-scale climatic changes. One reason is that scientists still disagree about whether the past cooling was driven by uplift, reduced ocean heat transport, slower seafloor spreading, or a combination of the three. More critically, future changes in all these plate tectonic processes (spreading, uplift, gateways) are at this point inherently unpredictable because the balance of driving forces behind plate tectonics is unclear. If we cannot accurately predict these tectonic processes, we cannot predict their climatic effects. The safest prediction—that tectonic-scale climate will continue to cool into the distant future—is not really a prediction at all, just a projection of past trends.

The climate of the distant future will also be determined by the nature of the feedback processes involved, both the positive feedbacks that amplify ongoing climatic change and the negative feedbacks that moderate it. The most comforting possibility is that chemical weathering feedback will work as a thermostat that moderates future tectonic-scale cooling.

We have already seen that negative feedback from chemical weathering is an integral part of the BLAG hypothesis (Chapter 5). A rough calculation shows that this negative feedback may have acted to offset any increase in chemical weathering and cooling of climate driven by uplift (Figure 7-19). In this calculation, we assume that uplift has affected an area equivalent to 1%

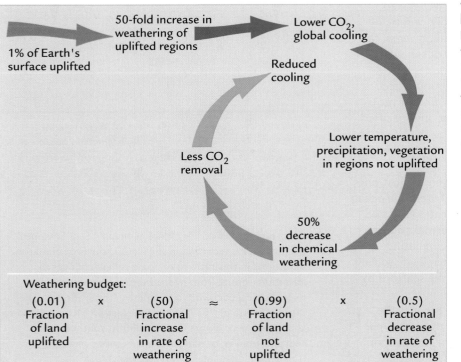

FIGURE 7-19 Chemical weathering balance? If increased chemical weathering in localized regions of uplift causes global climate cooling, the resulting reduction of chemical weathering in other regions may act as a negative feedback that moderates much of the cooling driven by uplift.

Weathering budget:				
(0.01) Fraction of land uplifted	x	(50) Fractional increase in rate of weathering	≈	(0.99) Fraction of land not uplifted	x	(0.5) Fractional decrease in rate of weathering

of the total area of continental crust, approximately the size of the Himalayas and the southern parts of the Tibetan Plateau. We also assume that uplift increased the rate of exposure of fresh bedrock in this small region by a factor of 50, broadly consistent with the relative differences in fluxes between mountains and lowlands in the Amazon Basin. A 50-fold increase in weathering over 1% of Earth's land surface would increase global chemical weathering by 50%.

What changes in weathering outside the uplifted region would it take to offset this increase within the uplifted region? As climate cools, the negative feedback role of weathering should begin to moderate climate across the globe because chemical weathering is slower

BOX 7-3   LOOKING DEEPER INTO CLIMATE SCIENCE

Is $^{87}Sr/^{86}Sr$ an Index of Chemical Weathering?

One way to determine global chemical weathering rates is to find a chemical index in the ocean that is sensitive to the global rate of delivery of chemical weathering products. One possible index of this kind is the ratio of two isotopes of the element strontium (Sr).

Strontium occurs in rocks at very small natural abundances. Some of the ^{87}Sr isotope is a product of radioactive decay; the stable ^{86}Sr isotope is not. Chemical weathering of continental rocks extracts both isotopes of strontium and sends them to the ocean as dissolved Sr^{+2} ions. In the ocean, trace amounts of the strontium isotopes are incorporated into the $CaCO_3$ shells of marine plankton as $SrCO_3$ in place of Ca^{+2} ions.

Strontium delivered by rivers is well mixed through the entire ocean. The residence time of strontium in the ocean is 3 to 5 million years, far longer than the 1500 years it takes ocean waters to mix. As a result, the relative amounts of the two strontium isotopes are well homogenized throughout the world ocean. The $^{87}Sr/^{86}Sr$ ratio characteristic of the entire ocean at any specified time can be determined by analyzing the $^{87}Sr/^{86}Sr$ ratio in any marine $CaCO_3$ shell formed in any area of the world ocean.

Chemical weathering on continents is just one of three sources of strontium that enters the ocean and determines its mean $^{87}Sr/^{86}Sr$ ratio. The two others are (1) input from **hydrothermal sources** (hot fluids) to deep ocean waters circulating through spreading ocean ridges and (2) dissolving of previously deposited calcite on the seafloor by corrosive bottom waters.

The mean ocean $^{87}Sr/^{86}Sr$ ratio records the balance among these three inputs through time. Each input is defined by a flux term (the rate of strontium input) and a $^{87}Sr/^{86}Sr$ ratio characteristic of that flux. The mean ocean $^{87}Sr/^{86}Sr$ ratio is the weighted mean value of these three fluxes:

$$\text{Mean ocean } ^{87}Sr/^{86}Sr \text{ ratio} = \frac{(\text{River flux})(Sr \text{ ratio}) + (\text{Hydrothermal flux})(Sr \text{ ratio}) + (\text{dissolution flux})(Sr \text{ ratio})}{\text{Total Sr flux}}$$

The mean ocean $^{87}Sr/^{86}Sr$ trend for the last 400 million years has varied as a result of changes in these input terms. Dissolution has had a negligible effect on long-term changes of the $^{87}Sr/^{86}Sr$ ratio because its size (flux) is small. The hydrothermal strontium flux is much larger and also has a ratio

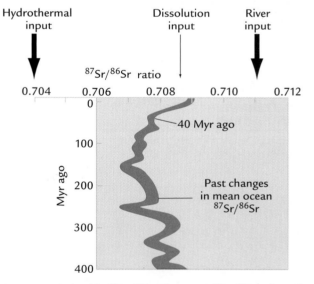

Long-term trend in $^{87}Sr/^{86}Sr$ The mean $^{87}Sr/^{86}Sr$ value of ocean water is recorded in the $CaCO_3$ shells of plankton. Variations in this ratio in the past reflect changes in strontium inputs from hydrothermal, dissolution, and river sources, each with a distinctive $^{87}Sr/^{86}Sr$ ratio. The rapid increase in $^{87}Sr/^{86}Sr$ over the last 40 Myr is related to changes in river inputs tied to uplift in southeast Asia. (Adapted from W. B. Harland et al., *A Geologic Time Scale* [Cambridge: Cambridge University Press, 1982].)

in cooler, drier, less vegetated conditions. A global temperature decrease of 3° to 4°C would be enough to drop weathering rates by 50% over the remaining 99% of the land surface and offset most of the localized increase in chemical weathering due to the effects of uplift. A global cooling of 3° to 4°C is also well within the range estimated for the last 55 Myr.

This calculation suggests that the moderating effects of the chemical weathering thermostat may have been in action throughout the last 55 Myr, *whatever* the ultimate cause of the cooling. But other considerations suggest we should not take too much comfort from this information. The opposite possibility is that the process of global cooling produces its own positive feedbacks

BOX 7-3 CONT.   LOOKING DEEPER INTO CLIMATE SCIENCE

lying well below the range of variation of the mean ocean curve. As a result, changes in this flux could be important in driving the mean ocean $^{87}Sr/^{86}Sr$ ratio. The hydrothermal flux of strontium is also thought to be a rough measure of the rate of seafloor spreading, because spreading rates control the rate of generation of new magma and the associated hydrothermal activity.

Strontium delivered by rivers from weathering on land is also an important cause of changes in the $^{87}Sr/^{86}Sr$ curve because the river flux of strontium is large and because its ratio lies well above the high end of the long-term trend. As a result, long-term changes in the $^{87}Sr/^{86}Sr$ ratio reflect a balance between changes in river and hydrothermal inputs of strontium.

The long-term $^{87}Sr/^{86}Sr$ trend shows a rapid increase in values in the last 40 Myr. If we rule out the small dissolution effect, one possible explanation for this trend is a decrease in hydrothermal input of strontium in the last 40 Myr. Slowing a flux with a low ratio has the same effect as increasing a flux with a high ratio: both cause the average $^{87}Sr/^{86}Sr$ value to increase. But spreading rates today are the same as they were 30 or even 40 Myr ago, and this fact argues against citing a slowing of hydrothermal input at spreading ridges as the major explanation for the rise in strontium isotope values during the last 40 Myr.

The only explanation left for the abrupt increase in $^{87}Sr/^{86}Sr$ ratio over the last 40 Myr is an increase in the river influx term. This conclusion is important because it could indicate that a major increase in chemical weathering over the last 40 Myr caused an increased supply of strontium flux by rivers. This interval also matches the time of uplift of Tibet.

But this conclusion is premature. An increase in the mean ocean $^{87}Sr/^{86}Sr$ ratio could have occurred either because of a greater river influx of strontium (with no change in the $^{87}Sr/^{86}Sr$ ratio), or because of an increase in the $^{87}Sr/^{86}Sr$ ratio carried by the rivers (but with no increase in total strontium influx). The first explanation requires increased chemical weathering, but the second does not.

Geochemists disagree about the relative contributions of these two factors to the $^{87}Sr/^{86}Sr$ increase during the last 40 Myr, although all agree that the increase is linked to the uplift of Tibet. Some argue that the $^{87}Sr/^{86}Sr$ ratio increased *only* because of production of unusually ^{87}Sr-rich rocks by tectonic processes along the zone of collision between India and Asia. If this is true, there is no need for an increase in either the total strontium influx or in the global rate of chemical weathering. Others acknowledge an increase in the average $^{87}Sr/^{86}Sr$ ratio of the weathered material but argue that an overall increase in total strontium input due to increased chemical weathering in that region explains as much as half of the observed $^{87}Sr/^{86}Sr$ increase.

Over at least the last 350 million years, the oceanic $^{87}Sr/^{86}Sr$ ratio appears to correlate to some extent with intervals of icehouse and greenhouse climate. The ratio was low during the long greenhouse interval between 240 and 35 Myr, but generally higher during the icehouse interval from 325 to 240 Myr ago, the time of the continental collisions that produced Pangaea. One explanation for this long-term correlation is that strong chemical weathering produces high river fluxes of ^{87}Sr-rich water to the ocean at the same time that it creates icehouse climates via the uplift weathering mechanism. An alternative explanation is that intervals of high $^{87}Sr/^{86}Sr$ values mark times of continental collision that created unusually ^{87}Sr-enriched rocks but did not significantly alter weathering rates or affect climate.

that keep driving climate toward ever colder icehouse conditions (Figure 7-20). These feedbacks result from increased amounts of freshly fragmented rock generated by the ice itself.

One such idea comes from work by the geophysicist Peter Molnar. He reasoned that global cooling increases the extent of glaciation in high-elevation mountains and plateaus, and the mountain glaciers grind large volumes of bedrock during oscillations toward colder climates. This grinding action could increase the rate of

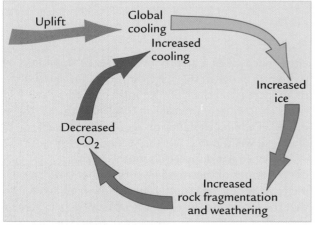

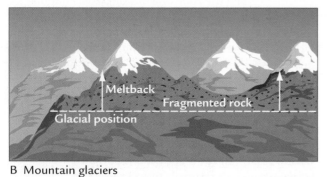

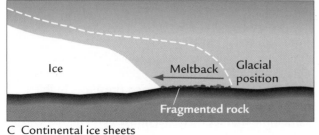

FIGURE 7-20 **Positive feedback from ice?** Global cooling produces more ice on Earth (A), and the ice increases rock fragmentation in high mountain terrain (B) and near ice sheets (C). Chemical weathering of this fragmented debris may cause further cooling by positive feedback.

exposure of freshly fragmented rock, speed up rates of chemical weathering, pull more CO_2 out of the atmosphere, and cause additional cooling (Figure 7-20A and B). Although this kind of feedback occurs in uplifting regions, as we saw in Chapter 5, it can also occur even in high terrain that is no longer being actively uplifted. In those regions, the atmosphere over the high terrain may cool enough to lower the threshold of freezing onto the highest peaks and initiate glaciation. If glaciation begins, it will fragment fresh rock in a way that had not happened during the warmer wetter climate regime that preceded glaciation.

An increase in weathering is particularly likely if the mountain glaciers move up and down the mountainsides frequently, alternately grinding fresh rock and then exposing it to the atmosphere for chemical weathering (Figure 7-20B). Molnar noted that these kinds of fluctuations result from the orbital-scale cycles of warming and cooling we will examine in Part III.

A similar positive feedback may occur in connection with large continental-scale ice sheets (Figure 7-20C). As global climate cools and vast ice sheets appear, they erode the preexisting cover of soil already weathered on the continents, expose fresh underlying bedrock, and grind the rock down to finer particles. The debris from these ice sheets is deposited in extensive moraine ridges at the ice margins.

If the ice sheets simply remain at their maximum extents for millions of years (as the Antarctic ice sheet has done during the last few million years), their net effect will probably be to slow overall rates of chemical weathering. Weathering may increase along the ice margins where piles of fragmented debris are exposed, but these localized increases would probably be overwhelmed by a regional decrease in weathering caused by ice covering much of the continent.

If, however, the ice sheets fluctuate in size over orbital time scales of tens of thousands of years, these variations should cause a net increase in chemical weathering. When shorter-term variations cause the ice sheets to melt back from their maximum extents or even disappear (as has happened in North America since 20,000 years ago), vast areas of finely ground glacial debris are exposed, along with freshly scraped bedrock surfaces (Figure 7-20C). This fresh debris can then be rapidly weathered during the warmer climates typical of intervals between glaciations. The large amount of weathering in these brief intervals between glaciations may far exceed the reduced weathering during intervals when ice sheets are large. The net result is an increase in chemical weathering.

In summary, feedback processes originating from the action of glacial ice may drive Earth's climate toward ever deeper icehouse conditions. In the view of some climate scientists, much of the cooling of the last few million years has been the result of these feedback processes, rather than any tectonic process (spreading, uplift, or gateways).

Obviously, further research is needed to assess whether the cooling of the last few million years is the result of tectonic processes or of internal feedbacks, and also to weigh the relative importance of all the possible positive and negative feedbacks. Whatever the answers, Earth's long-term forecast over tectonic time scales calls for colder temperatures and more ice.

Of course, this cooling is hardly imminent. All tectonic-scale processes and feedbacks operate at extremely slow rates, and the changes they produce become evident only over millions of years. Even though we are headed toward a colder future, Earth's climate won't be getting there soon enough for it to cause you or me any concern. And in the meantime, other factors will drive climate changes on shorter time scales more relevant to human concerns, including the orbital-scale changes explored in the next part.

KEY TERMS

permafrost (p. 149)
mass spectrometer (p. 151)
fractionation (p. 152)
oceanic gateways (p. 161)
sulfate aerosols (p. 164)
Monterey hypothesis (p. 166)
hydrothermal sources (p. 168)

REVIEW QUESTIONS

1. What kinds of changes in vegetation and ice show that Earth has cooled in the last 55 Myr?

2. What do changes in $\delta^{18}O$ values in the last 55 Myr tell us about climate changes?

3. How well does the spreading rate (BLAG) hypothesis explain the last 55 Myr of cooling?

4. How well does the uplift weathering hypothesis explain the last 55 Myr of cooling?

5. How well do changes in ocean gateways explain the last 55 Myr of cooling?

6. Explain how chemical weathering could drive the cooling and at the same time moderate it.

7. How could erosion of bedrock by ice have enhanced the cooling trend?

ADDITIONAL RESOURCES

Basic Reading/Viewing

Cracking the Ice Ages. 1996. NOVA Video. Boston: WGBH.

Ruddiman, W. F., and J. E. Kutzbach. 1991. "Plateau Uplift and Climatic Change." *Scientific American* (March), 66–75.

Advanced Reading

Berner, R. A. 1999. "A New Look at the Long-Term Carbon Cycle." *GSA Today* 9:1–6.

Crowley, T. J., and K. C. Burke. 1998. *Tectonic Boundary Conditions for Climatic Reconstructions.* New York: Oxford University Press.

Kennett, J. P. 1977. "Cenozoic Evolution of Antarctic Glaciation, the Circum-Antarctic Ocean, and Their Impact on Global Paleoceanography." *Journal of Geophysical Research* 82:3843–60.

Mikolajewicz, U., T. Maier-Reimer, T. J. Crowley, and K.-Y. Kim. 1993. "Effect of Drake and Panamanian Gateways on the Circulation of an Ocean Model." *Paleoceanography* 8:409–26.

Miller, K. G., R. A. Fairbanks, and G. S. Mountain. 1987. "Tertiary Oxygen Isotope Synthesis, Sea-Level History, and Continental Margin Erosion." *Paleoceanography* 2:1–19.

Molnar, P., and P. England. 1990. "Late Cenozoic Uplift of Mountain Ranges and Global Climate Change: Chicken or Egg?" *Nature* 346:2934.

Raymo, M. E., and W. F. Ruddiman. 1992. "Tectonic Forcing of Late Cenozoic Climate." *Nature* 359:117–22.

Ruddiman, W. F. (1997). *Tectonic Uplift and Climate Change.* New York: Plenum Press.

Vincent, E., and W. H. Berger. 1985. "Carbon Dioxide and Polar Cooling in the Miocene: The Monterey Hypothesis." In *The Carbon Cycle and Atmospheric CO_2: Natural Variations, Archaean to Present,* ed. E. T. Sundquist and W. S. Broecker, 455–68. Geophysical Monograph 32. Washington, D.C.: American Geophysical Union.

Wolfe, J. A. 1994. "Tectonic Climate Change at Middle Latitudes of Western North America." *Palaeogeography, Palaeoclimate, Palaeoecology* 108:195–205.

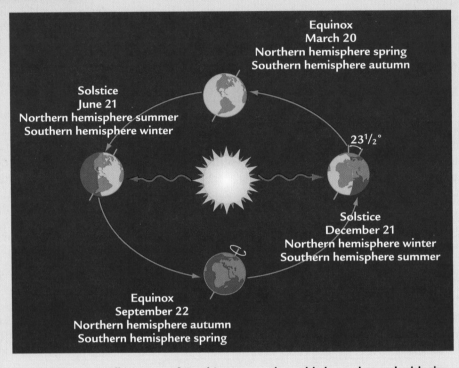

Orbital changes All aspects of Earth's present-day orbit have changed with time: the tilt of its axis, the shape of its path around the Sun, and the positions of the seasons on this path. These changes in orbit have driven climatic changes on Earth. (Adapted from F. K. Lutgens and E. J. Tarbuck, *The Atmosphere* [Englewood Cliffs, N.J.: Prentice-Hall, 1992].)

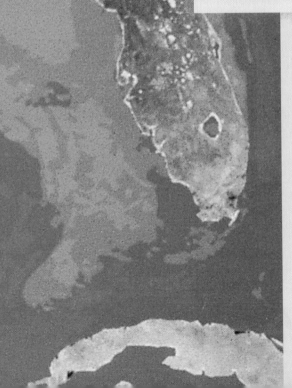

In Part III we move from tectonic-scale to *orbital-scale* climate changes, focusing on the last 3 million years, when continents and oceans were reaching their current positions. One reason for narrowing our focus is that well-dated climate records are available from more sites in this interval than in earlier times, and the increased regional coverage provides greater insights into possible connections among the various components of the climate system.

PART III

Orbital-Scale Climate Change

Most climate records covering orbital time scales come from ocean sediments. These records are dated by a combination of techniques: radiometric dating and correlation of magnetic reversals, uranium-series dating of corals, and orbital tuning. These techniques make it possible to resolve time to within a few thousand years in most records.

More important, because ocean sediments are multichannel recorders that carry several kinds of climate signals side by side, scientists can easily determine the relative timing of climatic responses of both the oceans (from shells of plankton) and the land (using windblown pollen and ice-rafted mineral grains). Over the most recent 400,000 years, researchers also gain access to climate signals in ice cores, including past fluctuations of greenhouse gases (CO_2 and methane) and of dust blown into the ice from arid continental source regions.

The focus of Part III is on climate changes caused by subtle shifts in Earth's orbit. Earth's present-day orbit around the Sun is not permanent. Three of its features have changed over time, varying at cycles ranging in length from 20,000 to 400,000 years: the tilt of Earth's axis, the shape of its yearly path of revolution around the Sun, and the changing positions of the seasons along that path. The timing of these cyclic changes in orbit is accurately known for the last several million years. Climate scientists have discovered these same orbital cycles in records of many important climatic responses on Earth: the strength of the low-latitude African and Asian monsoons, the size of north polar ice sheets, oscillations of high-latitude climate between cold/dry and warm/wet conditions, the circulation pattern of the deep ocean, and the concentrations of important greenhouse gases (CO_2 and CH_4) through time.

Although climate has changed on orbital scales throughout Earth's history, the fact that these changes are best resolved and dated in records from the last several million years gives climate scientists a unique opportunity to examine and unravel cause-and-effect relationships between the orbital changes that act as the climatic *forcing* and the changes within Earth's climate system that produce the climatic *responses* observed.

Changes in Earth's orbit affect climate by altering the amount of solar radiation received by season and by latitude (Chapter 8). The main task for climate scientists is to determine *how* these changes in incoming solar radiation have driven climate on orbital time scales. The major questions we examine in this part are:

- **How do orbital variations drive the strength of tropical monsoons?**
- **How do orbital variations control the size of northern hemisphere ice sheets?**
- **What controls orbital-scale fluctuations of atmospheric greenhouse gases?**
- **What is the origin of the 100,000-year climate cycle of the last 0.9 Myr?**

CHAPTER 8

Astronomical Control of Solar Radiation

Each year we feel the effects of Earth's orbit around the Sun through seasonal changes in the angle of the Sun's rays and their direct effects on temperature and other responses. We experience seasonal changes because Earth is tilted as it orbits the Sun, its tilt directed toward the Sun in summer and away from it in winter. The seasonal cycle is by far the largest climate-related signal humans experience in a lifetime.

In this chapter we examine much longer term changes in Earth's orbit that are equally important to the climate system: changes in the angle of tilt of Earth's axis of rotation, in the shape of its orbit as it revolves around the Sun, and in the timing of the seasons in relation to its noncircular orbit. These longer-term variations in Earth's orbit occur at cycles ranging from about 20,000 to 400,000 years in length, and they cause cyclic variations in the amount of solar radiation received at the top of the atmosphere by latitude and by season. Changes in solar heating driven by changes in Earth's orbit are the major cause of cyclic climate changes over time scales of tens to hundreds of thousands of years.

Earth's Orbit Today

The geometry of Earth's present-day solar orbit forms the basis for understanding past changes in Earth-Sun geometry. Much of our knowledge of Earth's orbit dates back to investigations in the seventeenth century by the astronomer Johann Kepler. The larger frame of reference for understanding Earth's present-day orbit is the plane in which it moves around the Sun, the **plane of the ecliptic** (Figure 8-1).

8-1 Earth's Tilted Axis of Rotation and the Seasons

Two fundamental motions describe the present-day orbit. First, Earth spins around on its axis once every day. We experience the result of this rotation as the daily rising and setting of the Sun, but of course that description is inaccurate. Days and nights result from Earth's rotational spin, which carries different regions into and out of the path of the Sun's radiation over a 24-hour cycle.

Earth rotates around an axis (or line) that passes through its poles (Figure 8-1). This axis is tilted at an angle of 23.5°, referred to as Earth's "obliquity," or **tilt**. This angle of tilt can be visualized equally well in either of two ways: (1) as the angle Earth's axis of rotation makes with a line perpendicular to the plane of the orbit or (2) as the angle that a plane passing through Earth's equator makes with the plane of the solar orbit.

The second fundamental motion that describes Earth's present-day orbit is its once-a-year revolution around the Sun. Again, we experience the results of this kind of motion as seasonal shifts between long summer days, when the Sun rises high in the sky and delivers intense radiation, and short winter days, when the Sun stays low in the sky and delivers weaker radiation. These seasonal differences culminate at the summer and winter **solstices**, which mark the longest and shortest days of the year (June 21 and December 21 in the northern hemisphere, the reverse in the southern).

But if we move outside our Earthbound perspective, we find that the main cause of the seasons, the solstices, and the changes in length of day and angle of incoming solar radiation is the changing *position* of the tilted Earth with respect to the Sun. During each yearly revolution around the Sun, Earth maintains a constant *angle* of tilt (23.5°) and a constant *direction* of tilt in space. When a hemisphere (northern or southern) is tilted directly toward the Sun, it receives the more direct radiation of summer. When it tilts directly away from the Sun, it receives the less direct radiation of winter. But at both times (and at all times of year) it keeps the same 23.5° tilt.

Now if we switch back to our Earthbound perspective, we see the Sun appearing to move back and forth

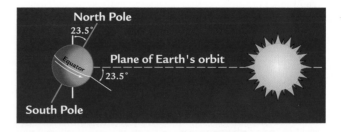

FIGURE 8-1 Earth's tilt Earth's rotational (spin) axis is currently tilted at an angle of 23.5° away from a line perpendicular to the plane of its orbit around the Sun.

through the year between the north tropic (Cancer) at 23.5°N and the south tropic (Capricorn) at 23.5°S. But as before, this apparent movement is actually just the result of Earth's revolution around the Sun with a constant 23.5° tilt.

Earth's 23.5° tilt also defines the 66.5° latitude of the Arctic and Antarctic circles by this relationship: 90° − 23.5° = 66.5°. Because of Earth's 23.5° tilt away from the Sun in winter, no sunlight reaches latitudes higher than 66.5° on the shortest winter day (winter solstice).

Between the winter and summer solstices, during intermediate positions in Earth's revolution around the Sun, the lengths of night and day become equal in each hemisphere on the **equinoxes** (which means "equal nights"—that is, equal to the days). Again, Earth's tilt angle remains at 23.5° during the equinoxes, and its direction of tilt in space stays the same. The only factor that changes is Earth's position in respect to the Sun.

The two equinoxes and two solstices are referred to as the four **cardinal points** in Earth's orbit. They are handy reference points for describing the most distinctive features of its orbit.

8-2 Earth's Eccentric Orbit: Changes in the Distance between Earth and Sun

Up to this point, everything we have learned would be true whether Earth's orbit was perfectly circular or not. But Earth's actual orbit (Figure 8-2) is not a perfect circle: it has a slightly eccentric or elliptical shape. The noncircular shape of Earth's orbit is the result of the gravitational pull of other planets on Earth as it moves in space.

Basic geometry tells us that ellipses have two focal points, in contrast to the single central focus of a circle. In Earth's case, the Sun lies at one of the two focal points in its elliptical orbit, as required by the physical laws of gravitation. The other focus is empty (Figure 8-2).

In this elliptically shaped orbit, the distance to the Sun changes with Earth's position in its orbit. As you might expect, this changing distance has a direct effect

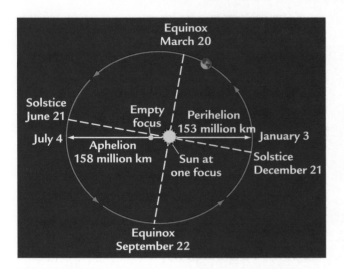

FIGURE 8-2 Earth's eccentric orbit Earth's orbit around the Sun is slightly elliptical. Earth is most distant from the sun at aphelion, on July 4, just after the June 21 solstice, and closest to the Sun at perihelion, on January 3, just after the December 21 solstice. (Modified from J. Imbrie and K. P. Imbrie, *Ice Ages: Solving the Mystery* [Short Hills, N.J.: Enslow, 1979].)

on the amount of solar radiation Earth receives. The simplest way to describe this elliptical orbit is to examine its two extremes. The position in which Earth is closest to the Sun is called **perihelion** (the "close pass" position, from the Greek meaning "near the Sun"), while the position farthest from the Sun is called **aphelion** (the "distant pass" position, from the Greek meaning "away from the Sun"). On average, Earth lies 155.5 million kilometers from the Sun, but the distance ranges between 153 million kilometers at perihelion and 158 million at aphelion. This difference is equivalent to a total range of variation of slightly more than 3% in relation to the mean value.

At present Earth is in the perihelion position (closest to the Sun) on January 3, near the time of the December 21 winter solstice in the northern hemisphere and the summer solstice in the southern hemisphere (Figure 8-2). The occurrence of this close-pass position in January causes winter radiation in the northern hemisphere and summer radiation in the southern hemisphere to be slightly higher than they would be in a perfectly circular orbit.

Conversely, Earth lies farthest from the Sun on July 4, near the time of the June 21 summer solstice in the northern hemisphere and the winter solstice in the southern hemisphere. The occurrence of this present-day distant-pass position in July makes summer radiation in the northern hemisphere and winter radiation in the southern hemisphere slightly lower than they would be in a circular orbit.

The effect of Earth's elliptical orbit on its seasons is small, enhancing or reducing the intensity of radiation

received by only a few percentage points. The main cause of the seasons is the direction of tilt of Earth's axis in its orbit around the Sun (Figure 8-1).

Another consequence of Earth's eccentric orbit is that the two time intervals between the equinoxes are not exactly equal: there are seven more days in the long part of the orbit, between the March 20 equinox and the September 22 equinox, than in the short part of the orbit, between September 22 and March 20. The greater length of the interval from March 20 to September 22 tends to compensate for the fact that Earth is farther from the Sun on this part of the orbit and thus receiving less solar radiation.

Long-Term Changes in Earth's Orbit

Astronomers have known for centuries that Earth's orbit around the Sun is not fixed over long intervals of time. Instead, it varies in a regular (cyclic) way because of the mass gravitational attractions among Earth, its moon, the Sun, and the other planets and their moons. These changing gravitational attractions cause cyclic variations in Earth's angle of tilt, eccentricity of orbit, and relative position of the solstices and equinoxes around the elliptical orbit (Box 8-1).

8-3 Changes in Earth's Axial Tilt through Time

We have already seen that the tilt of Earth's axis creates our seasons. Another way of understanding why this is so is to consider two hypothetical cases, using the sim-

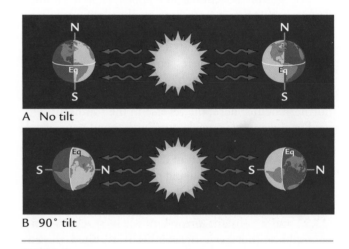

FIGURE 8-3 Extremes of tilt (A) If Earth's orbit were circular and its axis had no tilt, solar radiation would not change through the year and there would be no seasons. (B) For a 90° tilt, the poles would alternate seasonally between conditions of day-long darkness and day-long direct overhead Sun. (Adapted from J. Imbrie and K. P. Imbrie, *Ice Ages: Solving the Mystery* [Short Hills, N.J.: Enslow, 1979].)

BOX 8-1   TOOLS OF CLIMATE SCIENCE

Cycles and Modulation

Slow changes in Earth's orbit around the Sun occur in a cyclic or rhythmic way, as do the changes in strength of the incoming solar radiation they cause. The science of wave physics provides the terminology we need to describe these changes. The length of a cycle is referred to as its **wavelength.** Expressed in units of time, the wavelength of the cycle is called its **period,** a measure of the time span between successive pairs of peaks or valleys.

The opposite (or inverse) of the period of a cycle is its **frequency,** the number of cycles (or in this case fractions of one cycle) that occur in one year. If a cycle has a period of 10,000 years, its frequency is

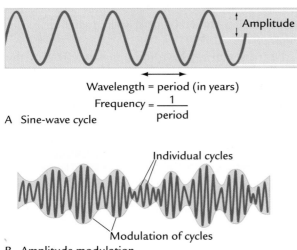

Wavelength = period (in years)

$$\text{Frequency} = \frac{1}{\text{period}}$$

A   Sine-wave cycle

Individual cycles

Modulation of cycles

B   Amplitude modulation

Description of wave behavior (A) Cyclic behavior can be represented by a sine wave with a particular period and amplitude. (B) Cycles may show variations in amplitude, called modulation.

0.0001 cycle per year (equivalent to one cycle every 10,000 years). In this book, we refer to cycles only in terms of their periods.

Another way of describing cycles focuses on the magnitude of their variations, rather than on their duration. The term for this characteristic is **amplitude,** a measure of the amount by which the cycles depart from their long-term average. Low-amplitude cycles barely depart from the long-term mean trend; high-amplitude cycles oscillate widely around the average long-term trend.

Not all cycles are perfectly regular. More commonly the sizes of peaks and valleys oscillate irregularly around the long-term mean trend over time. The behavior of amplitudes of peaks and valleys that change in a repetitive or cyclic way between high and low values is called **modulation,** a concept that lies behind the principle of AM (amplitude modulation) radio. Modulation creates a kind of envelope that describes the changing amplitudes of the cycle. Note that *modulation of a cycle is not in itself a cycle;* it simply adds amplitude variations to an existing cycle.

If variations in a particular signal are regular *both* in period and in amplitude, it is appropriate to use the term "cycle." For the case of perfect cyclicity, this behavior may be described as "sinusoidal" or **sine wave** changes. If the variations are irregular in period, it is not technically correct to use the term "cycle"; the term "quasi-cyclic" or "quasi-periodic" is preferable. We ignore this formal definition and informally use the term "cyclic" or "periodic" for climatic signals that are basically regular but may vary somewhat in wavelength or amplitude.

---

plifying assumption that Earth has a perfectly circular orbit around the Sun. For both cases, we examine the two seasonal extremes in Earth's orbit: the summer solstices, when solar radiation is most direct in each hemisphere, and the winter solstices, when solar radiation is least direct.

First we consider the case in which Earth's axis is not tilted (Figure 8-3A). With no tilt, incoming solar radiation is always directed straight at the equator throughout the year, and it always passes by the poles at a 90° angle. With no tilt, no seasonal changes occur

in solar radiation received at any latitude, including the polar regions. As a result, solstices and equinoxes do not even exist. A tilted axis is necessary for Earth to have seasons.

Next we consider the opposite extreme: a maximum tilt of 90° (Figure 8-3B). In this case, solar radiation is directed straight at the summer-season pole, while the winter-season pole lies in complete darkness. Six months later, the two poles have reversed position, moving between full-time darkness and sunlight. The difference between the two configurations shown in

Figure 8-3 suggests that seasonal differences in solar radiation due to tilt should be particularly striking at polar latitudes.

Over time, the angle of Earth's tilt actually varies in a narrow range, cycling back and forth between values as small as 22.2° and as large as almost 24.5° (Figure 8-4). These variations, caused mainly by the gravitational tug of large planets such as Jupiter, were first discovered by the French astronomer Urbain Leverrier in the 1840s. The present-day tilt (23.5°) is near the middle of the range, and the value is decreasing. Cyclic changes in tilt angle occur mainly at a period of 41,000 years, the time interval separating successive peaks or successive valleys. The cycles are fairly regular, both in period and in amplitude.

Changes in tilt cause long-term variations in seasonal solar insolation received on Earth, with the largest changes at high latitudes. The main effect of these changes is to amplify or suppress the seasons: increased tilt amplifies seasonal differences, decreased tilt reduces them.

Larger tilt angles turn the summer-hemisphere poles more directly toward the Sun and increase the amount of solar radiation received (Figure 8-5). The increase in tilt that turns the North pole more directly toward the Sun at its summer solstice on June 21 also turns the South Pole more directly toward the Sun at its summer solstice six months later (December 21). On

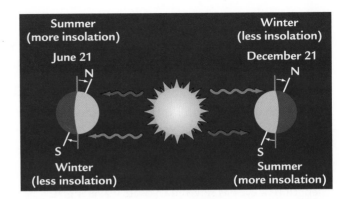

FIGURE 8-5 **Effects of increased tilt on polar regions** Increased tilt brings more solar radiation to the two summer-season poles and less radiation to the two winter-season poles.

the other hand, the increased angle of tilt that turns each polar region more directly toward the Sun in summer also turns each winter-season pole away from the Sun, reducing the amount of solar radiation received (Figure 8-5).

Decreases in tilt have the opposite effect of those shown in Figure 8-5: they diminish the amplitude of the seasonal differences. Smaller tilt angles put the Earth slightly closer to the configuration shown in Figure 8-3A, which has no seasonal differences at all.

8-4 Changes in Earth's Eccentric Orbit through Time

The shape of Earth's orbit around the Sun has also varied in the past, sometimes becoming nearly circular and at other times more elliptical (or "eccentric") than today. These variations, too, were discovered by Leverrier in the 1840s.

The shape of an ellipse can be described by reference to its two main axes: the "major" (or longer) axis and the "minor" (or shorter) axis (Figure 8-6). The degree of departure from a perfectly circular orbit can be described by

$$\epsilon = \frac{\sqrt{a^2 - b^2}}{a}$$

where ϵ is the **eccentricity** of the ellipse and a and b are half of the lengths of the major and minor axes (the "semimajor" and "semiminor" axes), respectively.

The eccentricity of the elliptical orbit increases as the lengths of these two axes become more unequal. At the opposite extreme, if the two axes become exactly equal ($a = b$), the eccentricity drops to zero because the orbit is circular ($a^2 - b^2 = 0$).

Eccentricity (ϵ) has varied over time between values of 0.005 and nearly 0.0607 (Figure 8-7), and today's

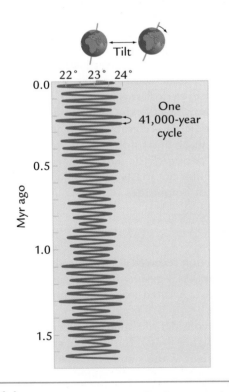

FIGURE 8-4 **Long-term changes in tilt** Changes in the tilt of Earth's axis have occurred on a regular 41,000-year cycle.

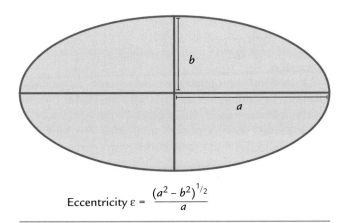

$$\text{Eccentricity } \varepsilon = \frac{(a^2 - b^2)^{1/2}}{a}$$

FIGURE 8-6 **Eccentricity of an ellipse** The eccentricity of an ellipse is related to half the lengths of its longer (major) and shorter (minor) axes.

value of 0.0167 lies well toward the lower end of this range (close to circular). These changes in orbital eccentricity are concentrated mainly at two periods, both of which are far more irregular than the 41,000-year tilt cycle we looked at earlier.

One eccentricity cycle shows up clearly as variations between maxima and minima at a period near 100,000 years (Figure 8-7). This cycle actually consists of four

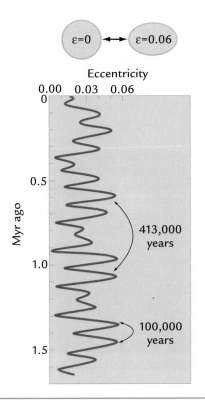

FIGURE 8-7 **Long-term changes in eccentricity** The eccentricity (ε) of Earth's orbit varies at periods of 100,000 and 413,000 years.

cycles of nearly equal strength and with periods ranging between 95,000 and 131,000 years, but these cycles blend and form a single cycle near 100,000 years.

The second major eccentricity cycle is a cycle at a wavelength of 413,000 years. This longer cycle is somewhat more difficult to see, but it shows up in the tendency of clusters of 100,000-year cycles to alternate between larger and smaller amplitudes (Figure 8-7). A third eccentricity cycle also exists at a period of 2.1 million years, but this cycle is much weaker in amplitude and not at all evident in Figure 8-7.

8-5 Precession of Solstices and Equinoxes around Earth's Orbit

The positions of the solstices and equinoxes in relation to the eccentric orbit have not always been fixed at the present-day locations shown in Figure 8-2. Instead, the solstices and equinoxes gradually shift position with respect to Earth's eccentric orbit, and also with respect to the perihelion (close-pass) and aphelion (distant-pass) positions. Although these changes were noticed by Hipparchus in ancient Greece, they came to be understood fully only in the eighteenth century, by the French mathematician, scientist, and philosopher Jean Le Rond d'Alembert. The importance of these motions is that *the distance from Earth to the Sun has varied with time for each of the seasons*, and these changes in distance have produced changes in solar radiation received on Earth.

The cause of these changes lies in a long-term wobbling motion of Earth, analogous to the motion of a child's top. Tops typically move with three superimposed kinds of motion (Figure 8-8). They spin very

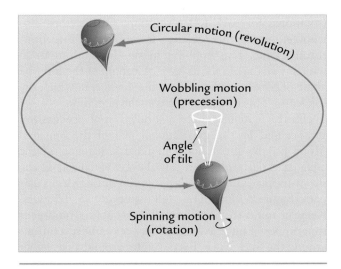

FIGURE 8-8 **Earth's wobble** In addition to its rapid (daily) rotational spin and its slower (yearly) revolution around the Sun, Earth wobbles slowly like a top, with one full wobble every 25,700 years.

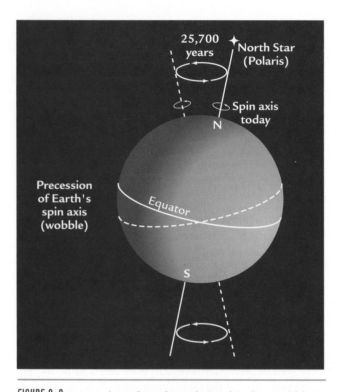

FIGURE 8-9 **Precession of Earth's axis** Earth's slow wobbling motion causes its rotational axis to point in different directions through time, sometimes (as today) toward the North Star, Polaris, but at other times toward other stars. (Adapted from J. Imbrie and K. P. Imbrie, *Ice Ages: Solving the Mystery* [Short Hills, N.J.: Enslow, 1979].)

rapidly around an axis tilted like that of Earth. They also tend to move with a slower near-circular (revolving) motion across the surface on which they spin, similar to Earth's orbit around the Sun, with many spins (rotations) for each full revolution. Finally, tops also wobble, gradually leaning in different directions through time. This wobbling motion has nothing to do with any change in the *amount* of leaning of the top (the angle of tilt), but is defined by the *direction* in which the top leans. This direction changes through time.

Earth's wobbling motion is called **axial precession**. It is caused by the gravitational pull of the Sun and Moon on the slight bulge in Earth's diameter at the equator. Axial precession can also be visualized as the slow turning of Earth's axis of rotation through a circular path (Figure 8-9), with one full turn every 25,700 years. Keep in mind that this very slow wobbling motion is superimposed on Earth's much faster yearly revolutions around the Sun. Today Earth rotates around an axis that points to the North Star (Polaris), but over time the wobbling motion causes the axis of rotation to point to other celestial reference points (Figure 8-9).

A second motion is known as **precession of the ellipse**. In this case, the elliptical shape of Earth's orbit

itself rotates, with the long and short axes of the ellipse turning slowly in space (Figure 8-10). This turning motion is even slower than the wobbling motion of axial precession.

The combined effects of these two precessional motions (wobbling of the axis and the slow turning of the ellipse) cause the solstices and equinoxes to move around Earth's orbit, completing one full 360° orbit around the Sun every 23,000 years (Figure 8-11). This combined movement, called the **precession of the equinoxes**, describes the absolute motion of the equinoxes and solstices in the larger reference frame of the universe. It consists of a strong cycle near 23,000 years and a weaker one near 19,000 years, the two together averaging one cycle every 21,700 years. For the rest of this book, we will refer mainly to the stronger 23,000-year cycle of precession.

Precession of the equinoxes obviously involves complicated angular motions in three-dimensional space, but we need a way to reduce these motions to a simple mathematical form that is easy to use, analogous to the plot of changes in tilt angle in Figure 8-7. To do this, we need to make use of the two main geometric characteristics of precessional motion.

The first of these characteristics is the angular form of the precessional motions around the Sun. We define

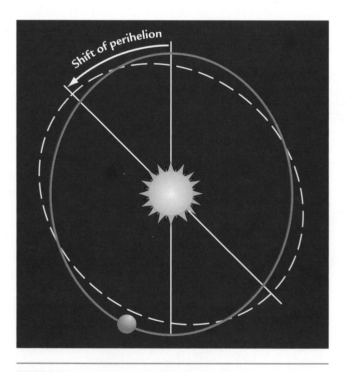

FIGURE 8-10 **Precession of the ellipse** The elliptical shape of Earth's orbit slowly precesses in space, so that the major and minor axes of the ellipse slowly shift through time. (Adapted from N. Pisias and J. Imbrie, "Orbital Geometry, CO₂, and Pleistocene Climate," *Oceanus* 29 [1986-87]: 43–49.)

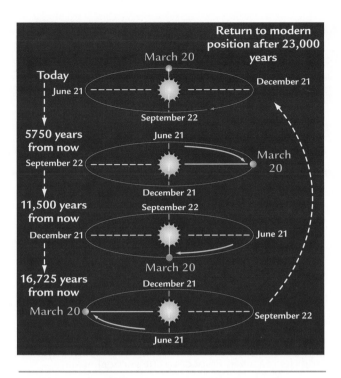

FIGURE 8-11 **Precession of the equinoxes** Earth's wobble and the slow turning of its elliptical orbit combine to produce the precession of the equinoxes. Both the solstices and equinoxes move slowly around the eccentric orbit in cycles that take 23,000 years. (Adapted from J. Imbrie and K. P. Imbrie, *Ice Ages: Solving the Mystery* [Short Hills, N.J.: Enslow, 1979].)

oscillating wave (a sine wave) in a *rectangular* coordinate system (Box 8-2).

Box 8-2 shows how sine-wave functions give us a method of projecting the sweeping motion of a radius vector around a circle onto a vertical coordinate system. This conversion allows a circular motion to be represented by a simple oscillating sine wave on an *x-y* plot. More important for our purposes here, it gives us a way of representing the angular movements involved in Earth's precession.

In a general sense, the function $\sin\omega$ represents the changes in angle between the rotating ellipse (Figure 8-10) and the seasonal solstices and equinoxes as they move along the elliptical orbit (Figure 8-11). More specifically, it is defined as the relative angular motion between the perihelion (close-pass) axis and the March 20 equinox. As Box 8-2 notes, $\sin\omega$ cycles back and forth between values of -1 and $+1$.

The second aspect of Earth's orbital motion that we need to consider is its eccentricity. The precessional movement of the solstices and equinoxes (Figure 8-11) results in long-term changes in the amount of solar radiation Earth receives because of the noncircularity

ω (omega) as the angle formed between two imaginary lines connecting Earth to the Sun (Figure 8-12A): (1) the line linking the Sun and the position of Earth at perihelion (its closest pass to the Sun) and (2) the line connecting the Sun to Earth's position at the March 20 equinox. One line is defined by the shape of Earth's orbit and the other by the position of the seasons.

We have already seen that the March 20 equinox moves around the Sun because of axial precession (Figure 8-9), and that Earth's perihelion and aphelion positions move around the Sun because of precession of the ellipse (Figure 8-10). The angle ω summarizes the combined *relative* effect of these two precessional motions: ω is the angle that opens up in the arc between these two moving lines (Figure 8-12A). Changes in the angle ω gradually sweep out a 360° arc around the Sun, starting at 0° and increasing to 90°, 180°, and 270°, before returning to 0° after completing a full 360° cycle (Figure 8-12B).

Next we need to reduce the complexities of these angular relationships between Earth and the Sun to a usable mathematical form. We can do this by using basic geometry and trigonometry to convert these *angular* motions of the orbit to the simpler form of an

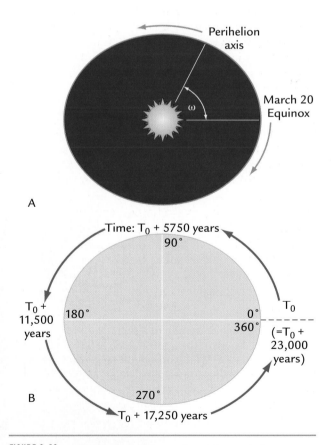

FIGURE 8-12 **Precession and the angle ω** The angle between lines marking Earth's perihelion axis and the vernal equinox (March 20) is called ω (A). The angle ω increases from 0° to 360° with each full 23,000-year cycle of precession (B).

BOX 8-2 LOOKING DEEPER INTO CLIMATE SCIENCE

Earth's Precession as a Sine Wave

We start with a basic triangle, for which the sine of an angle ω is defined as the length of the *opposite* side of the triangle over the length of the *hypotenuse*. Next we consider a circle whose radius is a vector r that sweeps around in 360° arcs in an angular motion measured by the angle ω. Note that the circular motion described by the angle ω is analogous to changes in Earth-Sun geometry.

Now we need to convert the angular motion around the circle into changes in the dimensions of a triangle tracking that motion. The radius vector r can also be thought of as the hypotenuse of a triangle whose sides are measured in the rectangular

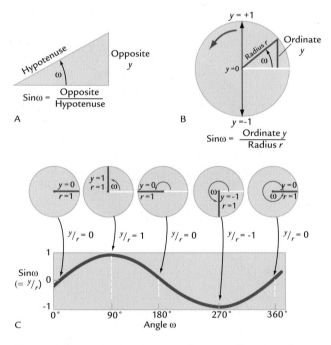

coordinate system laid out in the horizontal and vertical directions. We define the sine of the angle ω as the opposite side of the triangle divided by its hypotenuse, but in this case the hypotenuse is also the radius vector, r, for the circle.

The sweeping motion of the radius vector around the circle causes the shape of this triangle to change. The radius vector (r) will *always* have a value of +1, because its length stays the same and its sign is defined as a positive value in the angular coordinate system. But the length of the opposite side of the triangle is defined in terms of the rectangular coordinate system along the ordinate y. As the radius vector r sweeps around the circle, the length of the side of the triangle y increases and decreases in respect to the vertical scale, cycling back and forth between values of +1 and −1. When r lies in the top half of the circle, y has values above 0. When it lies in the lower half, y is negative.

Now we can convert the angular motion to a linear mathematical form. We do this by plotting the changes in sinω as the radius vector r sweeps out an entire 360° circle, with the angle ω increasing from 0° to 90°, 180°, 270°, and back to 360° (= 0°). As before, sinω is defined as the ratio of the opposite side y over the hypotenuse r.

The function sinω cycles smoothly between values of +1 and −1 for one complete revolution of the radius vector r. At the starting point (ω = 0°), the length of the opposite side y is 0 and the radius is +1, so the value of sinω is 0/1, or 0. As the angle ω increases, the length of the opposite side of the triangle (y) increases in relation to the (always constant) radius of the circle. When ω reaches 90°, sinω = 1 because the length of the opposite side and the radius (hypotenuse) are identical (1/1). At 180°, sinω has returned to 0 because the opposite side (y) is again 0. For angles greater than 180°, the sinω values become negative because the opposite side of the triangle falls in negative rectangular coordinates (below y = 0 on the vertical axis). Sinω values reach a minimum value of −1 at ω = 270° (−1/1). After that, sinω again begins to increase, returning to a value of 0 at ω = 360° (=0°).

Converting angular motion to a sine wave The sine of an angle is the length of the opposite side of a triangle over its hypotenuse (A). This concept can be applied to a circle where the hypotenuse is the radius (amplitude = 1) and the length of the opposite side of the triangle varies from −1 to +1 with reference to a vertical coordinate axis (B). As the radius vector sweeps out a full circle and ω increases from 0° to 360°, the sine of ω varies between −1 and +1, producing a sine wave representation of circular motion and of Earth's precessional motion (C) of Earth's orbit. With a perfectly circular orbit, the movement of the solstices and equinoxes around it would not alter the amount of sunlight received on Earth, because the distance to the Sun would remain constant throughout.

In an eccentric orbit, however, gradual movements bring the solstices and equinoxes into orbital positions at varying distances from the Sun, and these changes in distance alter the amount of solar radiation received on Earth. Consider the two extreme positions of the solstices in the eccentric orbit (Figure 8-13). As we noted earlier, the present-day position of the solstices has June 21 (the northern hemisphere's summer and the southern hemisphere's winter) occurring very near aphelion, the most distant pass from the Sun (Figure 8-13, top). This greater Earth-Sun distance slightly reduces the amount of solar radiation received at those seasons. Conversely, with December 21 (the northern hemisphere's winter and the southern hemisphere's summer) occurring today near perihelion, the closest pass to the Sun, solar radiation is higher at those seasons than it would be in a perfectly circular orbit.

Roughly 11,500 years ago, this configuration was exactly reversed (Figure 8-13, bottom). The June solstice occurred at perihelion, the closest pass to the Sun, causing greater insolation at that time of year. The December solstice was located at aphelion, the most distant pass, reducing solar radiation at that time of year.

Remember that the two positions shown in Figure 8-13 are only the extreme points in a continuous eccentric orbit. Precession also moves the solstices through intermediate orbital positions like those shown in Figure 8-11, and these positions lie at intermediate Earth-Sun distances. In the next 11,500 years, the solstices will move from their present position back to the position shown at the bottom of Figure 8-13.

Now we are ready to add the last complication to the full story of precession: the effect of changes in eccentricity. Recall that the eccentricity of Earth's orbit varies through time, ranging between 0.005 and 0.06. Changes in eccentricity will affect the extreme perihelion and aphelion positions shown in Figure 8-13 by further altering the distance between Earth and the Sun. With greater eccentricity, the differences in distance between a close pass and a distant pass will be magnified. With a nearly circular orbit, differences in distance will nearly vanish. In summary, changes in eccentricity magnify or suppress contrasts in Earth-Sun distance around the orbit. The changes in distance to the Sun in turn alter the amount of solar radiation received on Earth.

As a result, the complete expression for the effect of precession on solar radiation received on Earth must also incorporate a term that represents the effect of changing eccentricity. The complete term that results, $\epsilon\sin\omega$, is called the **precessional index**. The $\sin\omega$ part of this index is the sine-wave representation of the slow 360° movement of the equinoxes and solstices around the orbit (Figure 8-11). The eccentricity term, ϵ, introduces amplitude variations into the $\sin\omega$ index because of the direct multiplier effect of eccentricity ϵ on $\sin\omega$. The product of multiplying these two terms is the precessional index, $\epsilon\sin\omega$ (Figure 8-14).

Long-term variations in the precessional index show two major characteristics (Figure 8-15). First, they occur in cycles with periods near 23,000 years. This cycle is the direct result of the precessional motions represented by the $\sin\omega$ term. The second obvious feature of the $\epsilon\sin\omega$ curve is its large variations in amplitude (Figure 8-15). At times the cycle swings back and forth between extreme maxima and minima, while at other times the amplitude of changes is small. The

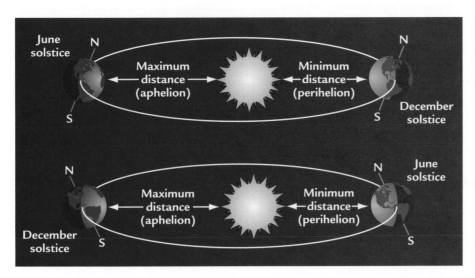

FIGURE 8-13 Extreme solstice positions Slow precessional changes in the attitude (direction) of Earth's spin axis produce changes in the distance between Earth and Sun as the summer and winter solstices move into the extreme (perihelion and aphelion) positions in Earth's eccentric orbit. (Modified from W. F. Ruddiman and A. McIntyre, "Oceanic Mechanisms for Amplification of the 23,000-Year Ice-Volume Cycle," *Science* 212 [1981]: 617–27.)

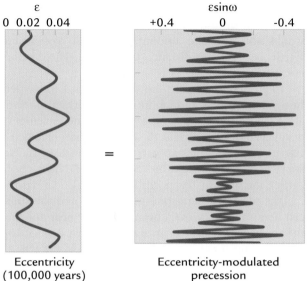

FIGURE 8-14 The precessional index
The precessional index, $\epsilon\sin\omega$, is the product of the sine wave function ($\sin\omega$) caused by precessional motion and the eccentricity (ϵ) of Earth's orbit.

Angular motion of precession (23,000 years)

Eccentricity (100,000 years)

Eccentricity-modulated precession (23,000 years)

source of this modulation (Box 8-1) is the eccentricity of Earth's orbit (the ϵ term in $\epsilon\sin\omega$). Eccentricity provides the longer-term amplitude modulation in the precessional index.

The modulation effect caused by eccentricity in the $\epsilon\sin\omega$ curve is not a real cycle. This statement may seem to go against your intuition, because eccentricity obviously varies at cycles of 100,000 and 413,000 years (Figures 8-7, 8-14), and because cycles of that length clearly shape the amplitude of the $\epsilon\sin\omega$ trend by creating the upper and lower envelopes of modulation in Figure 8-15. The key distinction is that envelopes of modulation are not the same thing as real cycles; in effect, offsetting effects between the upper and lower envelopes cancel each other out.

For example, you may argue that you can draw a curve showing a 100,000-year cycle by outlining the left-hand side of the envelope of *maximum* values in the $\epsilon\sin\omega$ curve in Figure 8-15. But this argument can be countered by the fact that it is possible to draw a curve with an exactly opposite trend outlining the right-hand envelope of *minimum* $\epsilon\sin\omega$ values. If the two opposing curves defined by these envelopes are added together, their values will exactly cancel and the *net* amplitude of changes at the 100,000-year and 413,000-year periods will be zero. As a result, no real 100,000-year or 413,000-year cycle exists in the $\epsilon\sin\omega$ trend, even though a modulation of the shorter (23,000-year) precession cycle obviously exists at those periods.

In summary, gradual changes in Earth's orbit around the Sun result in changes in solar radiation received by season and by hemisphere at two cycles: the tilt cycle with a wavelength of 41,000 years and the precession cycle at 23,000 (and 19,000) years. Eccentricity varia-

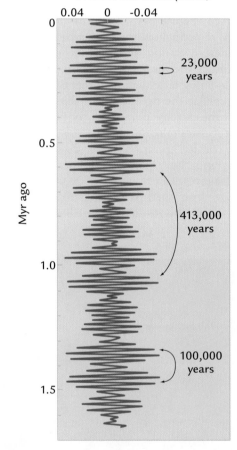

Precessional index ($\epsilon\sin\omega$)

23,000 years

413,000 years

100,000 years

FIGURE 8-15 Long-term changes in precession The precessional index ($\epsilon\sin\omega$) changes mainly at a cycle of 23,000 years. The amplitude of this cycle is modulated at the eccentricity periods of 100,000 and 413,000 years.

tions at 100,000 and 413,000 years modulate the amplitude of the precessional cycle.

Changes in Insolation Received on Earth

We now examine orbital variations from the perspective of changes in solar radiation received by latitude and by season on Earth. Climate scientists refer to the radiation arriving at the top of Earth's atmosphere as **insolation**. This is not necessarily the same as the changes received at Earth's surface, because clouds and other features within the climate system can alter the amount of radiation that actually penetrates the atmosphere. Still, these calculations of insolation are our best guide to the effects of orbital changes on Earth's climate.

8-6 Insolation Changes by Month and Season

The long-term trends of tilt (Figure 8-4) and of $\epsilon\sin\omega$ (Figure 8-15) contain all the information needed to calculate the amount of insolation arriving on Earth at any latitude and season. By convention, climate scientists usually show the amount of insolation (or departures of insolation from a long-term average) during the solstice months of June and December in W/m^2 (the equivalent form of calories per cm^2 per unit of time was commonly used earlier).

June and December insolation values over the last 300,000 years show a strong dominance of the 23,000-year cycle of precession at lower and middle latitudes, and even at higher latitudes during the summer season (Figure 8-16). Just like the $\epsilon\sin\omega$ precessional index, individual insolation cycles at lower latitudes occur at

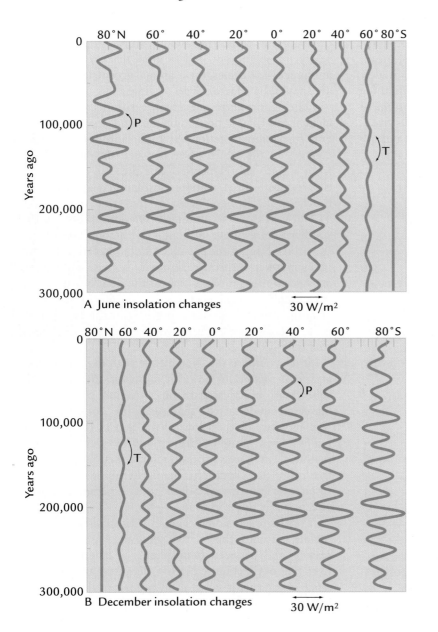

A June insolation changes

B December insolation changes

FIGURE 8-16 **June and December insolation variations** June and December monthly insolation values show the prevalence of precessional changes at low and middle latitudes and the presence of tilt changes at higher latitudes. Cycles of tilt and precession are indicated by T and P. The double arrows indicate variations of 30 W/m^2 for these signals.

wavelengths near 23,000 years, with modulation at periods of 100,000 and 413,000 years. The June and December monthly insolation curves in Figure 8-16 are opposite in sign at a given latitude (Figure 8-17). Both vary by as much as 12% (40 W/m²) around the long-term mean value (roughly 340 W/m²).

The 41,000-year cycle of tilt (obliquity) is not evident in the low-latitude signals. Instead, it shows up in variations of winter-season insolation at high latitudes (January in the northern hemisphere, June in the southern hemisphere), but at smaller amplitudes. The tilt cycle is also present, although much less obvious, in summer insolation signals at high latitudes. For example, two precession cycles evident near 50,000 years ago in the June insolation signal for latitude 20°N gradually blend and merge into a single tilt cycle at latitude 80°N (Figure 8-16).

The summer-season insolation changes at the tilt cycle actually exceed those in the winter season (although this excess is not evident in these precession-dominated plots). As a result, changes in annual mean insolation at the 41,000-year tilt signal at high latitudes have the same sign as (but a much lower amplitude than) the summer insolation anomalies. The insignificance of winter changes at polar latitudes results from the fact that these latitudes receive no insolation at all during long stretches of the polar winter, so winter contributes little or nothing to the annual mean.

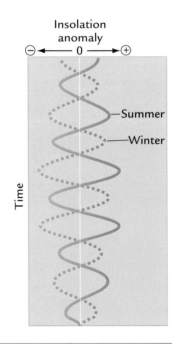

Insolation
anomaly
⊖ ◄─── 0 ───► ⊕

—Summer

—Winter

Time

FIGURE 8-17 **Opposing seasonal insolation trends** Seasonal insolation trends move in opposite directions in winter and in summer at any location. For precession, these opposing trends cancel each other out when they are added to calculate an annual average.

In summary, monthly seasonal insolation changes are dominated by precession at low and middle latitudes, with the effects of tilt more evident at higher mid-latitudes. Note that no cycle of insolation change at 100,000 (or 413,000) years is obvious in any of these latitudinal signals: eccentricity is *not* significant as a direct cycle of seasonal insolation change. Again, the major contribution of eccentricity to changes in seasonal insolation is in modulating the amplitude of the precession cycle.

Small variations in received insolation do occur in connection with Earth's eccentric orbit around the Sun, but these appear only as changes in the total energy received by the entire Earth, not as seasonal variations. These changes are governed by the term $(1 - \epsilon^2)^{\frac{1}{2}}$. We have already seen that ϵ varies between 0.005 and 0.0607. Using these values in $(1 - \epsilon^2)^{\frac{1}{2}}$, we can calculate that changes in insolation received vary by at most 0.002 (0.2%) around the long-term mean value because of changes in eccentricity. Compared to the changes in seasonal insolation of 10% or more that occur at the tilt and precession cycles, these annual eccentricity changes are smaller by a factor of 50 or more.

The hemispheric and seasonal patterns of insolation changes for tilt and precession are fundamentally different. Insolation variations at high latitudes caused by changes in tilt are *in phase* between the hemispheres from a seasonal perspective. For the case of increased tilt (Figure 8-18A), summer insolation maxima in the northern hemisphere occur at the same time in the 41,000-year cycle as summer insolation maxima in the southern hemisphere (although on the opposite side of the orbit). Higher tilt produces more insolation at both poles in their respective summers because both poles are turned more directly toward the Sun. For the same reason, more pronounced insolation minima also occur at both winter poles for a higher tilt: the two winter poles are tilted away from the Sun at the same time (Figure 8-18A).

If we compare the North Pole with the South Pole at a *fixed position* in the orbit, however, the two hemispheres are exactly out of phase (Figure 8-18A). The increased angle of tilt that turns the northern polar regions more directly toward the Sun in the northern hemisphere's summer also tilts the southern polar regions farther away from the Sun at that same place in the orbit (the southern hemisphere's winter). As a result, the North and South poles experience reverse effects at any point in time.

For the precessional changes that dominate low and middle latitudes, the relative sense of phasing between seasons and between hemispheres is exactly reversed from that of tilt (Figure 8-18B). Because Earth-Sun distance is the major control on precessional changes in insolation, Earth's position close to the Sun (at perihe-

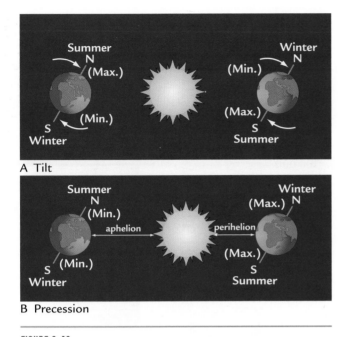

A Tilt

B Precession

FIGURE 8-18 **Phasing of insolation maxima and minima** Tilt causes in-phase changes for polar regions of both hemispheres in their respective summer and winter seasons (A). Precession causes out-of-phase changes between hemispheres for their summer and winter seasons (B).

lion) produces higher insolation than normal over *all* of Earth's surface. A precessional-cycle insolation maximum occurring at June 21 (or December 21) will be simultaneous everywhere on Earth, including the North and South poles. Distant-pass positions (at aphelion) will simultaneously diminish insolation everywhere on Earth.

Recall, however, that the seasons are reversed across the equator. As a result, an insolation maximum at June 21 is a *summer* insolation maximum in the northern hemisphere, but it is a *winter* insolation maximum in the southern hemisphere, where June 21 is the winter solstice. As a result of the seasonal reversal at the equator, insolation signals considered in terms of the season of the year are *out of phase* between the hemispheres for precession. This pattern is exactly opposite in sense to the in-phase pattern for tilt at high latitudes of both hemispheres.

Another way of looking at the relative phasing of precessional insolation is to track the changes between seasons within a single hemisphere. The orbital position shown on the left in Figure 8-18B, which produces minimum summer (June 21) insolation in the northern hemisphere because it occurs at a distant position from the Sun (aphelion), must also produce six months later a maximum in winter (December 21) insolation in the same hemisphere when Earth revolves around to the perihelion position. As a result, precessional variations

in insolation between the summer and winter seasons at any one location always move in opposite directions. In this one respect, precession and tilt are alike: winter and summer insolation trends (maxima vs. minima) are opposite at any latitude.

Precessional changes in insolation have an additional characteristic not found in changes caused by tilt: an entire family of insolation curves exists for each month and each season of the year. As a matter of convention, insolation changes are typically shown only for the extreme solstice months of June and December. However, precessional motion by its very definition involves the movement not just of the solstices but also of the equinoxes, as well as of all other times of the year. Each of these months and seasons precesses into parts of the eccentric orbit that are alternately farther from the Sun and closer to the Sun at the same 23,000-year cycle.

As a result, each of these months or seasons experiences the same cycle of increasing and decreasing insolation values relative to the long-term mean, but these anomalies (departures from the mean) are offset slightly in time from the preceding month or season (Figure 8-19). These offsets produce a family of monthly (and seasonal) insolation curves.

For example, we have already seen (Figure 8-11) that the June 21 summer solstice will move into the close-pass perihelion position once every 23,000 years because of Earth's slow precessional motion. But the same thing will happen for the September 22 equinox at a later time in the cycle of precession. Because the solstices and equinoxes divide Earth's orbit into four roughly equivalent parts, and because each season fin-

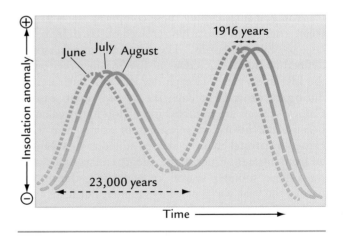

FIGURE 8-19 **Family of monthly precession curves** Because all seasons change position (precess) around Earth's orbit, each season (and month) has its own insolation trend through time. Monthly insolation curves are offset by slightly less than 2000 years (23,000 years divided by 12 months).

ishes a complete precessional orbit every 23,000 years, the September equinox position will pass through perihelion roughly 5750 years after the June position ($\frac{1}{4} \times 23,000 = 5750$).

As a result, a curve of precessional insolation anomalies for September will lag 5750 years behind a curve of insolation anomalies for June. The insolation anomaly curve for the December 21 solstice will lag another 5750 years, putting it 11,500 years behind the June curve and thus exactly out of phase. And the March 20 equinox will lag another 5750 years, putting it 16,250 years behind the June curve. A family of curves also exists for each month between the solstices and equinoxes (Figure 8-19), with each monthly anomaly curve offset from the preceding or following month by 1916 years ($\frac{1}{12} \times 23,000 = 1916$).

8-7 Insolation Changes According to Caloric Season

Calculations of monthly insolation are complicated by an additional factor related to the eccentricity of Earth's orbit. We have seen that Earth gradually moves through a 360° arc in its orbit around the Sun. In an eccentric orbit, however, this steady angular motion does not result in a constant rate of actual motion in space. Instead, Earth slows down as it nears the extreme perihelion and aphelion positions, and speeds up near orbital positions intermediate between perihelion and aphelion. As a result, as the solstices and equinoxes move slowly around the eccentric orbit, they gradually pass through regions of faster and slower absolute movement in space. These changes in speed cause changes in the lengths of the months and seasons in relation to a year determined by calendar time.

A way of sidestepping these complications is to calculate changes in insolation received on Earth by reference to "caloric seasons." The caloric summer season is defined as those 182.5 days of the 365-day year when the insolation received each day exceeds that received during the other 182.5 days (the caloric winter season). Caloric seasons are not fixed in relation to the calendar because the insolation variations of as much as 12% caused by orbital changes may be added to or subtracted from different times in the calendar year. As a result, the caloric summer season generally falls during the part of the year we think of as summer, but it may not be centered exactly on the June 21 summer solstice.

Changes in insolation viewed in reference to the half-year caloric seasons give a different portrayal of the relative importance of tilt and precession. Although the maxima and minima of low-latitude insolation anomalies are separated by the expected 23,000-year spacing caused by precession, the anomalies at higher latitudes through the winter caloric half-year follow a much

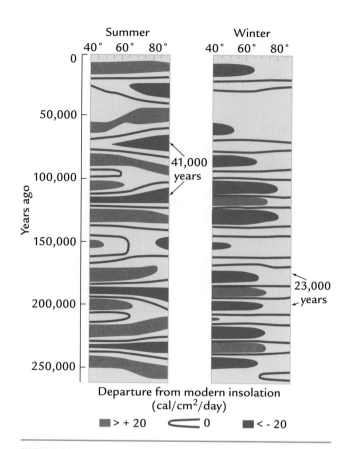

FIGURE 8-20 **Caloric season insolation anomalies** Plots of insolation anomalies for the summer and winter caloric half-year show a larger influence of tilt in relation to precession at higher latitudes than do the monthly anomalies. (Adapted from W. F. Ruddiman and A. McIntyre, "Oceanic Mechanisms for Amplification of the 23,000-Year Ice-Volume Cycle," *Science* 212 [1981]: 617–27.)

more obvious 41,000-year tilt rhythm (Figure 8-20). In this portrayal, tilt appears to be much more important than in the monthly insolation curves (Figure 8-16). Another aspect of caloric-season calculations is that the insolation values vary only by a maximum of about 5% around the mean, compared to variations as large as 10–15% for monthly insolation changes.

Looking for Orbital-Scale Changes in Climate Records

In succeeding chapters we will find abundant evidence that orbital-scale cycles are recorded in Earth's climate records. Many of these records contain two or even three orbital-scale cycles at the same time, and it can be difficult to disentangle them.

For example, consider the three cycles shown in Figure 8-21A, with periods of 100,000 years, 41,000 years, and 23,000 years. These three cycles are equivalent to the three most prominent cycles of orbital change, although for purposes of simplicity they are shown as perfect sine waves rather than the more complex forms of actual orbital variations (Figures 8-4, 8-7, 8-15).

We can combine these three cycles by adding them together in various ways, as in Figure 8-21B. When the 23,000-year and 100,000-year cycles are combined, it is obvious that the resulting signal is just a simple addition of the two separate cycles. The reason we can distinguish the two cycles is that they differ significantly in period (by a factor of more than 4: 100,000 divided by 23,000 = 4.3).

It becomes more difficult to detect the two original signals when we look at the combined 23,000-year and 41,000-year cycles. Because the periods of these two cycles are more similar, they reinforce and cancel each other in complicated ways. It takes an experienced eye

to see that the combined signal is really formed from just the two cycles. The task becomes even more complicated when all three cycles are combined, as shown in the bottom curve of Figure 8-21B. It is not really obvious at all that this signal is a combination of three perfect sine waves.

In the actual case of Earth's climate records, we face an even more complex situation in which all three cycles are not only superimposed but also have complex modulations of their amplitude through time (Figures 8-4, 8-7, 8-15). Obviously, it will be impossible to disentangle all this overlapping information just by eye.

8-8 Time Series Analysis

To simplify the complexities involved in exploring cyclic variations in insolation and in climate changes on Earth, climate scientists use the techniques of **time series analysis**. The term "time series" refers to records of climate change plotted against age (or time). A primary goal of time series analysis is to examine complicated climatic signals and extract any rhythmic cycles embedded within them.

The first step in time series analysis is to convert measurements of climatic indicators to a time framework. Individual measurements of some kind of climatic indicator are first made (for example, over a sequence of depths in a sediment core). Next, all available dating techniques are used to define the ages of different levels within this sediment sequence. A complete time scale can then be developed by interpolating the ages of all depths lying between the dated levels in the sequence. Finally, the time scale is used to plot measurements of climatic indicators (proxies) in the sequence against a time axis. The time scale transforms the climatic measurements from a depth scale to the time framework needed for time series analysis.

One such technique is **spectral analysis**. Modern techniques of spectral analysis are beyond the scope of this book, but we need at least a basic sense of how this technique detects cycles in records of past climate change. One way to visualize what happens in spectral analysis is to imagine taking a climate record plotted on a time axis and gradually sliding a series of sine waves of different periods across it. As this is done, the correlation between each sine wave used and the full climatic signal is measured. If the climate record being examined contains a strong cyclic signal, at some point in the sliding process the climate record will show a strong correlation with the sine wave representing that cycle length. This indicates that the climatic signal contains a strong cycle at that period. As this process is repeated for different sine waves having different periods, other cyclic signals embedded in the full climate record may also correlate strongly with other sine waves.

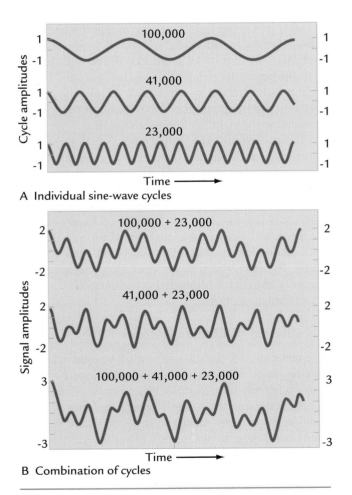

A Individual sine-wave cycles

B Combination of cycles

FIGURE 8-21 **Complications from overlapping cycles** If perfect sine-wave cycles with periods of 100,000, 41,000, and 23,000 years are added together so that they are superimposed on top of one another, the original cycles are almost impossible to detect by eye in the combined signal.

Now we return to the example of the three superimposed sine-wave cycles shown at the bottom of Figure 8-21B. A spectral analysis run on this signal will extract the three component cycles, and these can be displayed on a plot called a **power spectrum** (Figure 8-22). The horizontal axis shows a range of periods plotted on a log scale, with the shorter periods to the right. The vertical axis represents the amplitude of the cycles (see Box 8-1), also known as their "power." The height of the lines plotted on the power spectrum is related to the square of the amplitude of that cycle.

For the example shown in Figure 8-22, all three cycles that are detected by the spectral analysis technique plot as "line spectra," showing that their power is concentrated entirely at the three periods shown as solid vertical lines. In this idealized example, no power occurs anywhere else in the spectrum.

In actual studies of climate, power spectra are almost never this simple. One reason is that even the more regular-looking orbital cycles (such as the tilt changes in Figure 8-4) are not perfect sine waves but instead have variations spread out over a small range of periods. As a result, their power spectra look more like the dashed lines in Figure 8-22. The same kind of thing can happen if small errors occur in measuring or dating records of past climate change: such errors also spread the power over a broader range of periods than would be the case for perfectly measured and dated signals. In all such cases, the total amount of power associated with each cycle is the entire area under the dashed curves rather than the height of the line spectra.

Another reason real-world spectra are more complicated is that random noise exists in the climate system, consisting of irregular climatic responses not concentrated at orbital or other cycles. In most records, the effect of noise is spread over a range of periods in the spectrum.

In general, the amount of power also tends to be larger at longer periods because slower changes in climate tend to have more time in which to reach larger size. As a result, spectra from real-world climatic signals tend to look like the thick curved line in Figure 8-22. The spectral peaks that rise farthest above the baseline formed by the rest of the trend are the ones that are most significant (believable) in a statistical sense.

A second useful time series analysis technique is called **filtering**. This technique is used to extract the form of individual cycles at a specific period (or narrow range of periods) from the complexity of the total signal. This process is often referred to as "bandpass filtering" (filtering of a narrow band or range of the many periods actually present in a given signal). Filtering is analogous to using glasses with colored lenses to filter out all colors of the light spectrum except the one color (wavelength) we wish to see.

Filters are constructed directly from well-defined peaks in power spectra like those in Figure 8-22. The highest point on the spectral peak defines the central period of the filter, and the sloping sides of the spectral peak define the shape of the rest of the filter.

To understand the importance of filtering, we return once more to the example of the three hypothetical sine waves in Figure 8-21. We can create filters for these three cycles based on the peaks in the power spectrum shown in Figure 8-22. If we pass these filters across the combined signal at the bottom of Figure 8-21B, the filters will extract the exact original form of all three of the individual cycles (at 23,000, 41,000, and 100,000 years). In effect, the filtering operation lets us see the shapes of any cycles embedded in the complex shapes of actual climate records.

8-9 Aliasing of Climate Records

The technique of spectral analysis can be used only for a specific range of cycles within any one climate record. For a relatively long-period cycle to be well identified in time series analysis, it must be repeated four or more times in the original record (that is, the record must be four times longer than the cycle being analyzed). These repetitions confirm that the long-period cycle is indeed present.

At the other extreme, at least two samples for each relatively short-period cycle are needed to verify that a given shorter-term cycle is present in the original record. With fewer than two samples, time series analysis runs into the problem of **aliasing**, a term that refers

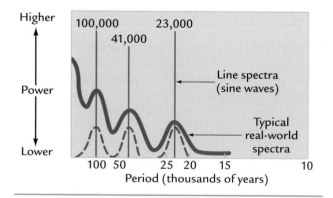

FIGURE 8-22 Spectral analysis Spectral analysis reveals the presence of cycles within complex climate signals. In this example, the original sine-wave cycles from Figure 8-21A form line spectra (vertical bars) whose heights indicate cycle amplitudes. Actual climate records have peaks that are spread over a broader range of periods (dashed lines and curving solid line).

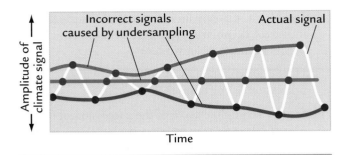

FIGURE 8-23 Aliasing (undersampling) of climate signals Undersampling of a climate signal (in this case one that is a direct response to changes in orbital precession) can produce aliased climate signals completely unlike the actual one.

to false trends generated by undersampling of the true complexity present in a climatic signal.

Consider the hypothetical case of a climatic signal that happens to take the form of the 23,000-year cycle of orbital precession, with its typical variations in amplitude (Figure 8-23). Assume that three scientists sample this signal but are unaware of its true form. Also assume that by chance all three choose to sample this record at an average spacing of 23,000 years, but that each begins the sampling process at a different place in the record.

If one scientist happened to start sampling exactly at a maximum in the signal, that scientist would by chance measure a record only of the successive maxima. If another scientist happened to start at a minimum, he or she would get a record only of successive minima. These two sampling attempts give completely different (in fact opposite) results because they are persistently biased toward different sides of the highly modulated cycle. A third scientist who happened to start sampling exactly at a crossover point between minima and maxima might extract a record suggesting that no signal exists at all.

These differences show the danger of aliasing. Although the example in Figure 8-23 is obviously selected to show this problem at its worst, aliasing is a problem to at least some degree in most climate records, and an especially big problem in sparsely sampled ones. It is easy to imagine from the example in Figure 8-23 that we could define the true climatic signal fairly accurately if we were to take five or six samples in each cycle. But we need at least two samples per cycle to sense that the cycle even exists.

8-10 Tectonic-Scale Changes in Earth's Orbit

Over time scales of hundreds of millions of years, some of Earth's orbital characteristics have slowly evolved. One place to find evidence of these long-term changes is in ancient corals. Corals are made of banded $CaCO_3$ lay-ers caused by changes in environmental conditions. The primary banding is annual and reflects seasonal changes in sunlight and water temperature. A secondary kind of banding follows the tidal cycles created by the Moon and Sun, which affect water depth and other factors in the reef environment that influence coral growth.

Corals from 440 million years ago show a larger number of tidal cycles per year than modern corals do, implying that Earth spun on its rotational axis 11% more times per year than at present. As a result, each year had 11% more days. Gradually over the last 440 million years, the spin rate and number of days decreased to their present levels. The cause of this gradual slowing in spin rate arises from the fact that the Moon and Sun produce tides on Earth, and the tides cause a frictional drag on Earth's rotation. The drag gradually slows Earth's rate of rotation.

Other changes in Earth's orbit that can also be inferred from this kind of information, such as changes in Earth-Moon distance, are thought to have affected the wavelengths of tilt and precession over tectonic-scale intervals. One estimate of the slow, long-term increases in the periods of tilt and precession toward their present-day values is shown in Figure 8-24. This slow evolution of orbital periods would occur only over extremely long intervals of time but should not be evident within the length of individual records, even over many millions of years.

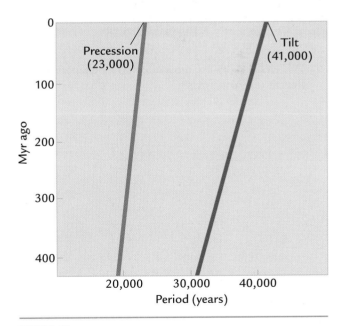

FIGURE 8-24 Tectonic-scale orbital changes Gradual changes in Earth's orbit over long tectonic time scales have caused a slow increase in the periods of the tilt and precession cycles. (Adapted from A. Berger et al., "Pre-Quaternary Milankovitch Frequencies," *Nature* 342 [1989]: 133–34.)

KEY TERMS

plane of the ecliptic
  (p. 175)
tilt (p. 175)
solstices (p. 175)
equinoxes (p. 175)
cardinal points (p. 175)
perihelion (p. 176)
aphelion (p. 176)
wavelength (p. 177)
period (p. 177)
frequency (p. 177)
amplitude (p. 177)
modulation (p. 177)
sine wave (p. 177)

eccentricity (p. 178)
axial precession (p. 180)
precession of the ellipse
  (p. 180)
precession of the
  equinoxes (p. 180)
precessional index
  (p. 183)
insolation (p. 185)
time series analysis (p. 189)
spectral analysis (p. 189)
power spectrum (p. 190)
filtering (p. 190)
aliasing (p. 190)

REVIEW QUESTIONS

1. Why does Earth have seasons?

2. When is Earth closest to the Sun in its present-day orbit? How does this position affect the amount of radiation received by Earth?

3. What is modulation of a cycle?

4. What aspect of Earth's orbit changes in cycles of 41,000 years? Of 100,000 years? Of 23,000 years?

5. Which latitudes are most affected by changes in the tilt of Earth's axis?

6. Earth's tilt is slowly increasing today. As the tilt increases, do polar regions receive more radiation or less in summer? in winter?

7. How does axial precession differ from the precession of the ellipse?

8. Sketch a complete cycle of the precession of the March 20 equinox in an eccentric orbit.

9. How does eccentricity combine with precession to affect long-term insolation on Earth?

10. Do changes in insolation during summer and winter have the same or opposite timing at any location?

11. Do the following occur at the same time in orbital cycles or on the opposite side of the cycle?
  (a) Summer insolation maxima at the poles caused by changes in tilt.
  (b) Summer insolation maxima in the tropics of both hemispheres caused by precession.

ADDITIONAL RESOURCES

Basic Reading

Imbrie, J., and K. P. Imbrie. 1979. *Ice Ages: Solving the Mystery*. Short Hills, N.J.: Enslow.

Advanced Reading

Berger, A. L. 1978. "Long-Term Variations of Caloric Insolation Resulting from the Earth's Orbital Elements." *Quaternary Research* 9:139–67.

CHAPTER 9

Insolation Control of Monsoons

Monsoon circulations exist on Earth today because the land responds to seasonal changes in solar radiation much more quickly than does the ocean. The reason for these differing rates of response is that the land has a far smaller capacity than the ocean for storing heat (Chapter 2). In this chapter we will examine evidence that changes in insolation over orbital time scales have driven major changes in the strength of the summer monsoons. Changes of 12% in the amount of insolation received at low latitudes have caused large changes in heating of tropical landmasses and in the strength of summer monsoons at a cycle near 23,000 years in length. Just 10,000 years ago, when high summer insolation drove a strong monsoon circulation in North Africa, the southern part of the present-day Sahara Desert was dotted with lakes, seasonally flowing rivers, and grassland vegetation. We will see that similarly large changes in monsoon strength have occurred throughout the last several hundred thousand years, and even back 200 Myr ago across the northern tropics of Pangaea.

Monsoon Circulations

In summer, strong solar radiation causes rapid warming of the land, but slower and much less intense warming of the ocean. Rapid heating over the continents causes air to warm, expand, and rise, and the upward movement of air creates an area of low pressure at the surface. Flow of air toward this low-pressure region causes in-and-up motion of warm, moist air from the oceans (Figure 9-1A). This inflow carries water vapor evaporated from the nearby ocean and contributes to monsoonal rainfall. Regions of the continents far from the ocean or protected by intervening mountain ranges may be beyond the reach of oceanic moisture and may bake under the strong summer radiation.

During winter, when solar radiation is weaker, air over the land cools off rapidly, increases in density in relation to air over the still-warm ocean, and sinks from higher levels in the atmosphere. This downward move-ment creates a region of high pressure over the land, in contrast to the relatively low pressure over the still-warm oceans. The overall atmospheric flow in winter is a down-and-out movement of cold, dry air from the land to the sea (Figure 9-1B).

Most of the strong summer monsoons today occur in the northern hemisphere, where landmasses are large (Asia and North Africa) and elevations are high (the Tibetan region of southeast Asia). Monsoons are weaker in the southern hemisphere, in part because the land-masses at tropical and subtropical latitudes are smaller.

Here we focus on past variations of the North African monsoon, for two reasons. First, North Africa lies far from the high-latitude ice sheets that might complicate the direct response of land surfaces to solar heating (as we will see in upcoming chapters). In addition, the oceans around North Africa yield a rich variety of climate records showing monsoon-related signals.

FIGURE 9-1 Seasonal monsoon circulations Seasonal changes in the strength of solar radiation affect the surface of the land more than the ocean. In summer, intense solar heating of the land causes an in-and-up circulation of moist air from the ocean (A). In winter, weak solar radiation allows the land to cool off and creates a down-and-out circulation of cold dry air (B).

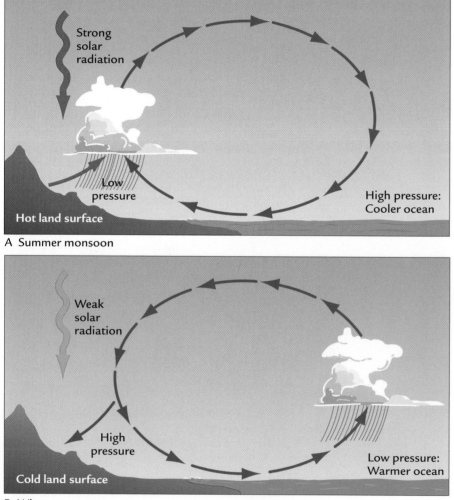

A Summer monsoon

B Winter monsoon

Africa is a deceptively large landmass, although this fact is obscured on Mercator maps, which project a spherical Earth onto a horizontal-vertical grid. It stretches from 37°N to 35°S, but far more land lies north of the equator, in fact an area almost twice the size of the U.S. mainland. This huge North African land surface is situated at tropical and lower subtropical latitudes that are strongly influenced by the overhead Sun, which reaches as far north as the Tropic of Cancer, at 23.5°N, on June 21.

As a result of strong solar heating in the northern hemisphere's summer, a low-pressure region develops over west-central North Africa, drawing moisture-bearing winds in from the tropical Atlantic to the south (Figure 9-2A). During an average summer, this monsoonal rainfall penetrates northward to about 17°N latitude (the southern edge of the Sahara Desert) before retreating southward later in the year.

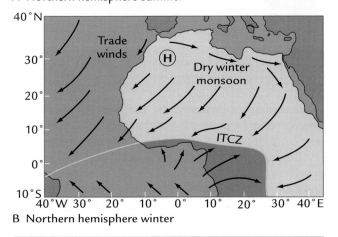

B Northern hemisphere winter

FIGURE 9-2 **Monsoon circulations over North Africa**
Seasonal changes cause a moist inflow of monsoonal air toward the low-pressure center over North Africa in summer (A), and a dry monsoonal outflow from the high-pressure center over the land in winter (B). (Adapted from J. F. Griffiths, *Climates of Africa* [Amsterdam: Elsevier, 1972].)

In the northern hemisphere's winter the overhead Sun is in the southern hemisphere, and solar radiation over North Africa is weaker. Cooling of the North African land surface by back radiation causes sinking of air from above, and a high-pressure cell develops at the surface over the northwestern Sahara Desert (Figure 9-2B). Strong and persistent winds associated with this high-pressure cell and with similar circulation over the adjacent North Atlantic blow southwestward from North Africa and across the tropical Atlantic. These are the northern trade winds.

The trade winds of the winter monsoon are dry, and winter precipitation is rare in North Africa. Only two areas receive much winter rain: the northernmost Mediterranean margin, where storms occasionally blow in from the Atlantic Ocean, and the tropical southwest coast (the Ivory Coast), where the moist intertropical convergence zone (ITCZ) remains over the land.

Because most of the rainfall in North Africa occurs in association with the summer monsoon, the distribution of major vegetation types reflects the monsoonal delivery of precipitation from the south (Figure 9-3). Rain forest in the year-round wet climate near the equator gives way northward to a sequence of progressively drier vegetation, first the tree-and-grass savannas of the Sahel region and then the desert scrub of the arid Sahara.

9-1 Orbital-Scale Control of Summer Monsoons

The idea that changing insolation could control the strength of monsoons over orbital time scales was proposed by the meteorologist John Kutzbach in the early 1980s, although it had been anticipated to some extent by Rudolf Spitaler in the late nineteenth century. And while this hypothesis has been widely accepted, it has not yet been given a name. We will refer to it as the **orbital monsoon hypothesis**.

The concept behind the orbital monsoon hypothesis is based on a simple logical extension of factors at work in the present-day seasonal monsoon circulations. Because modern monsoon circulations are linked to changes in the strength of solar radiation during summer and winter, longer-term, orbital-scale changes in the strength of summer and winter insolation should have affected the strength of the monsoons in the same manner in the past.

The basic concept behind the orbital monsoon hypothesis is shown in Figure 9-4. Departures from modern-day summer and winter insolation values are known to have occurred in the past (Chapter 8). If summer insolation was higher in the past than it is today, the summer monsoon circulation should have been stronger, with greater heating of the land, stronger

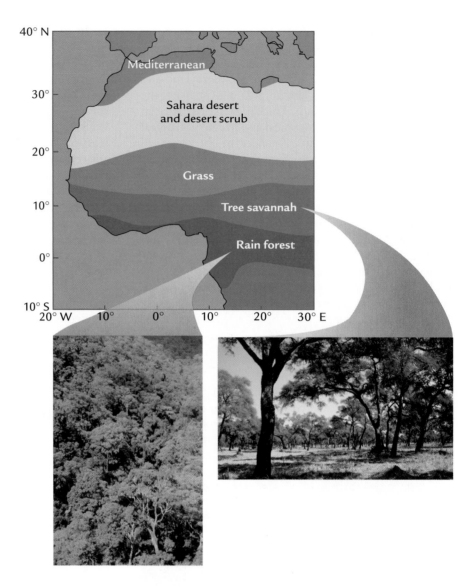

rising motion, more inflow of moist ocean air, and more rainfall (Figure 9-4A, B). Similarly, summer insolation values lower than those today should have driven a weaker summer monsoon in the past.

The same kind of reasoning applies to the strength of the winter monsoon. Winter insolation minima weaker than today's should have enhanced the cooling of the land surface, which should have driven a stronger down-and-out winter monsoon flow from land to sea (Figure 9-4C). Deeper winter insolation minima should have driven stronger (dry) monsoon flows.

Recall from Chapter 8 that more intense summer insolation maxima and deeper winter insolation minima *always* occur together at any location in their respective seasons. As a result, stronger in-and-up monsoon flows in summer should occur at the same times in the past as stronger down-and-out monsoon flows in winter.

At first it might seem that the climatic effects of these opposed insolation trends in the two seasons might cancel each other, but if we examine the amount of precipitation produced by the two seasonal monsoons, we find that this is not the case. In fact, one season has a far more important response than the other.

Because monsoonal winters are always dry, regardless of insolation levels, orbital-scale changes in winter insolation have no effect on annual rainfall. In contrast, because summer is the wet season, orbital-scale changes in summer insolation do affect rainfall during the wet summer monsoon. And because summer is the *only* wet season of the year, the amount of summer rainfall in turn determines the entire annual average.

This is an example of a **nonlinear response** of the climate system to insolation changes: the response measured by the amount of rainfall is highly sensitive to one season and not sensitive at all to the other. As a result,

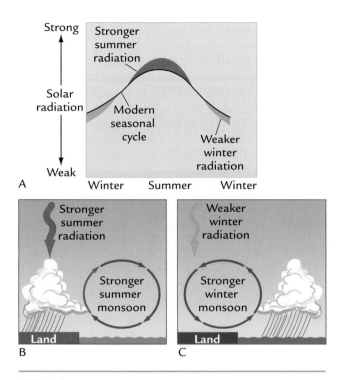

(Figure 9-5). Note that today's June insolation level is well below the long-term average of the last 140,000 years. The orbital monsoon hypothesis predicts that stronger summer monsoons should have occurred at those times in the past when summer insolation values were significantly larger than the value today.

The most recent instance of summer insolation values substantially higher than today's (by 8%) occurred near 10,000 years ago. Evidence we will examine in Part IV shows that lake levels across tropical and subtropical North Africa were at much higher levels during this most recent summer insolation maximum, and for some time afterward. In fact, many lakes that were filled to high levels at that time are completely dry today. This evidence indicates that a strengthened summer monsoon circulation reached much farther northward into North Africa nearly 10,000 years ago than it

FIGURE 9-4 **The orbital monsoon hypothesis** (A) Departures from the modern seasonal cycle of solar radiation have driven stronger monsoon circulations in the past. Greater summer radiation intensified the wet summer monsoon (B), and decreased winter insolation intensified the dry winter monsoon (C). (Adapted from J. E. Kutzbach and T. Webb III, "Late Quaternary Climatic and Vegetational Change in Eastern North America: Concepts, Models, and Data," in *Quaternary Landscapes,* ed. L. C. K. Shane and E. J. Cushing [Minneapolis: University of Minnesota Press, 1991].)

the system has a strong net response even though insolation trends in the two seasons might have been expected to cancel each other.

Evidence of Orbital-Scale Changes in Summer Monsoons

What kind of evidence would climate scientists look for to test whether or not Kutzbach's orbital monsoon hypothesis is valid? The answer to this question lies in a closer examination of the way insolation should drive changes in the strength of the wet summer monsoon.

At low and middle latitudes, where monsoon circulations are strongest, changes in the amount of incoming solar insolation follow the 23,000-year rhythm of orbital precession (Chapter 8). A June insolation curve from latitude 30°N covering the last 140,000 years clearly shows this 23,000-year precessional tempo

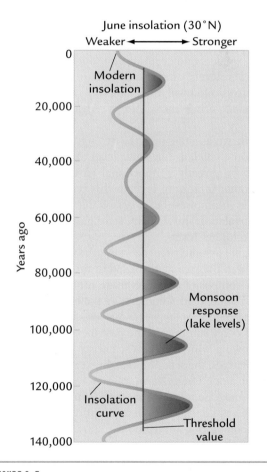

FIGURE 9-5 **Conceptual model of monsoon response to summer insolation** Increases in summer insolation heating to values above a critical threshold should drive a strong monsoon response at the 23,000-year tempo of orbital precession. The amplitude of monsoon response should be related to the amount of increase in summer insolation forcing.

does today. This evidence supports the orbital monsoon hypothesis.

But this is only one interval in the long history of changes in North Africa. What about the much longer term behavior of the summer monsoon? We can adapt the June insolation curve in Figure 9-5 to construct a simple conceptual prediction of how the summer monsoon may have varied with time. This predicted monsoon response is based on three assumptions.

First, we assume a **threshold insolation level** below which the monsoon response will be so weak that it will leave little or no evidence in the geologic record. An obvious example of the effect of such a threshold is the level of lakes. If insolation is weak and the summer monsoon produces little rain, the lakes may dry up. Once a lake has dried, it can no longer record changes toward even drier climates. It has reached the threshold limit of its ability to record climate.

This assumption has a good basis in fact. Many lakes in North Africa that existed 10,000 years ago during the strong monsoon interval are totally dry today, even though a weak summer monsoon does occur under today's lower levels of summer radiation. Apparently it takes a threshold insolation value well above modern-day levels to bring most North African lakes into existence.

Second, we assume that the strength of the monsoon response (such as the water level of North African lakes) will be directly proportional to the amount by which summer insolation exceeds this critical threshold value. This assumption has a reasonable physical basis: stronger insolation should drive stronger monsoons and fill lakes to higher levels.

Third, we assume that the strength of the monsoon at any time in the past as recorded in the record of lake levels is a composite of the average monsoon strength over many individual summers. Actually, this is more fact than assumption: the wet monsoon circulations that developed every summer in the past inevitably vanished every following winter. When scientists sample records of these changes, the separate year-by-year responses are blended into one longer-term average response characteristic of many hundreds or even thousands of years.

On the basis of these three assumptions, we arrive at the predicted monsoon response shown in green in Figure 9-5. In effect, this response exactly mimics the shape of the summer insolation curve, except that it is truncated (cut off) at the threshold value. Insolation values below the threshold leave no monsoon evidence at all in the geologic record (as in the example of dry lakes). Insolation maxima above the threshold produce a series of pulselike monsoon maxima at regular 23,000-year intervals, and these pulses vary in strength according to the amount by which summer insolation exceeds the threshold value.

Also note that the strongest predicted monsoon peaks in Figure 9-5 occur in a cluster of two or three during the interval between 85,000 and 130,000 years ago, when the summer insolation curve reached strong maxima because of modulation of precession by orbital eccentricity (Chapter 8). In contrast, the weaker insolation maxima near 35,000 and 60,000 years ago are predicted to create much weaker peak monsoons. Summer insolation minima (such as the one we are in today), as well as most of the intermediate insolation values, fall below the critical threshold and should produce no monsoon response that can be detected in climate records.

Now we examine actual climate records to see if we can find evidence of this predicted monsoon response. Because most of North Africa is arid and erosion of sediment is much more prevalent than deposition, its climate history is sparse (as well as difficult to date). Fortunately, nearby seas and oceans contain the kinds of continuous and well-dated records we seek.

9-2 "Stinky Muds" in the Mediterranean

The water that fills the Mediterranean Sea today has a relatively high oxygen content from top to bottom. Near-surface waters are well oxygenated because they continuously exchange oxygen-rich air with the atmosphere and because photosynthesis produces O_2. The deeper waters owe their oxygen content to sinking of oxygen-rich water from the surface in winter (Figure 9-6A). This sinking motion results from two factors discussed in Chapter 2: (1) the high salt content of the Mediterranean Sea, caused by the excess of summer evaporation over precipitation, and (2) winter chilling of salty water along the northern margins of the Mediterranean during incursions of cold air from the north. The dense, oxygen-rich surface waters that sink into the deep basin eventually exit westward into the Atlantic. As a result, the floor of the present-day Mediterranean Sea is covered by the kind of sediment typical of well-oxygenated ocean basins: mostly beige-colored silty mud containing shells of plankton that once lived at the sea surface and benthic foraminifera that lived on the seafloor.

Mediterranean sediments also contain occasional distinct layers of black organic-rich muds, called **sapropels**. One example of these deposits comes from a sequence of layers initially deposited in the Mediterranean Sea but since uplifted above the sea surface by tectonic activity along the Mediterranean coast (Figure 9-7). These sapropel muds show no sign of the shelled remains of bottom-dwelling organisms typical of sediments from well-oxygenated ocean basins. Their high organic carbon content indicates that they formed at times when the waters at the seafloor lacked the oxy-

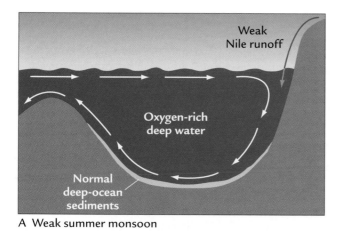

Weak
Nile runoff

Oxygen-rich
deep water

Normal
deep-ocean
sediments

A Weak summer monsoon

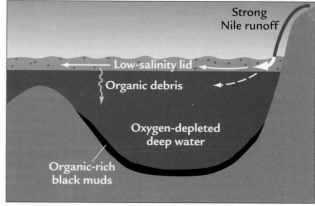

Strong
Nile runoff

Low-salinity lid

Organic debris

Oxygen-depleted
deep water

Organic-rich
black muds

B Strong summer monsoon

FIGURE 9-6 **Mediterranean circulation and monsoons** In today's typical Mediterranean circulation, salty surface water chilled by cold air in winter sinks and carries dissolved oxygen to deeper layers (A). At intervals in the past, strong summer monsoons in tropical Africa caused an increased discharge of fresh Nile River water into the eastern Mediterranean, creating a low-density surface-water lid that inhibited sinking of surface water and caused the deep ocean to lose its oxygen and deposit organic-rich black muds (B).

gen needed to convert (oxidize) organic carbon to inorganic form. This lack of oxygen led to stagnation of deep water and deposition of iron sulfides, giving the muds a "stinky" (rotten-egg) odor. The lack of oxygen also kept animals from living on the seafloor.

How was the deep Mediterranean basin deprived of oxygen during these intervals? One explanation proposed by the paleoecologist Martine Rossignol-Strick is that the sinking of oxygen-rich water was cut off when the sea surface was capped by a layer of low-density

freshwater brought in by rivers (Figure 9-6B). Although the atmosphere probably still sent cold air masses across the northern margins of the Mediterranean Sea in winter, the surface waters chilled by these cold air masses did not become dense enough to sink deep into the basin because of the low-salinity, low-density freshwater lid. As a result, the deep Mediterranean basin lost its supply of oxygen.

At the same time, production of planktic organisms continued at the surface, or may even have increased if

FIGURE 9-7 **Mediterranean "stinky muds"** Black organic-rich mud layers (sapropels) produced by strong monsoons occur interbedded with lighter well-oxygenated $CaCO_3$-rich sediments typical of deep-ocean environments. These sediments deposited in the northern Mediterranean have been exposed along the coast of Italy by subsequent uplift. (Courtesy of Tim Patterson, Carleton College, Ottawa.)

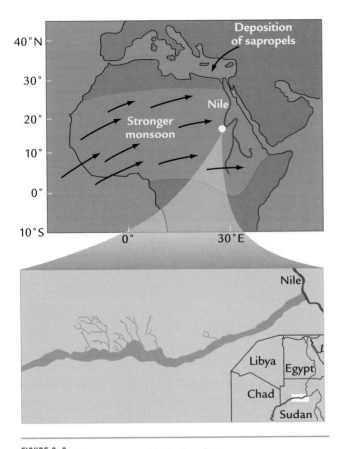

FIGURE 9-8 **Monsoons and Nile floods** Strong summer monsoons in tropical North Africa periodically caused large discharges of freshwater from the Nile northward to the arid Mediterranean region. Satellite sensors have detected riverbed deposits buried beneath younger sheets of sand now blowing across the hyperarid eastern Sahara Desert in Sudan (inset). (Inset: adapted from H.-J. Pachur and S. Kroplein, "Wadi Howar: Paleoclimatic Evidence from an Extinct River System in the Southeastern Sahara," *Science* 237 [1987]: 298–300.

the stronger river flow brought extra nutrients into the Mediterranean. The high productivity at the surface continually sent organic-rich remains of dead plankton toward the seafloor. Sinking and oxidation of this organic carbon continually depleted oxygen levels in the deep Mediterranean and produced stagnation and stinky muds on the seafloor.

The most recent sapropel in the eastern Mediterranean dates to 10,000 to 8000 years ago, the same interval when African lakes were high and the African summer monsoon was strong. Earlier layers of organic-rich mud deeper in Mediterranean sediment cores occur at regular 23,000-year intervals that coincide with times when summer insolation was higher than it is today. The sapropels were best developed (thickest and most carbon-rich) near the time of the strongest summer insolation maxima, were poorly developed during weaker insolation maxima, and were absent the rest of the time. This pattern matches very well the pattern predicted by the orbital monsoon hypothesis, shown in Figure 9-5, and indicates some kind of connection to the low-latitude monsoon over North Africa.

But some climate scientists initially questioned this explanation. The Mediterranean Sea is situated at high subtropical latitudes (30°–40°N), somewhat beyond the northernmost expansions of the tropical summer monsoon indicated by lake-level evidence from North Africa. If climate within the confines of the Mediterranean region never became truly monsoonal, how can the stinky muds in the Mediterranean Sea be a response to the North African monsoon?

The critical link is the Nile River (Figure 9-8), which gathers most of its water from the highlands of eastern North Africa at tropical latitudes. Even today these highlands receive summer rains from the relatively weak tropical monsoon. When summer insolation was much higher than it is today, the strengthened monsoon expanded northward and eastward, bringing much heavier rainfall to these high-elevation regions that were and are the headwater sources for the Nile. In effect, rainfall in the North African tropics exerts a kind of remote control on the salinity of the subtropical Mediterranean Sea during times of strong monsoons.

In support of this idea, satellite sensors have detected the buried remnants of streams and rivers that once flowed across Sudan (Figure 9-8) but are now covered by sheets of sand blowing across the hyperarid southeastern Sahara Desert. The fact that these streams flowed eastward and joined the Nile River indicates that lower-elevation regions also contributed to the Nile's stronger flow during major monsoons.

9-3 Freshwater Diatoms in the Tropical Atlantic

Direct evidence that the size of North African lakes fluctuates at the 23,000-year tempo of orbital precession can be found in sediment cores from the north tropical Atlantic Ocean (Figure 9-9). These Atlantic sediments have distinct layers containing concentrations of the opaline ($SiO_2 \cdot H_2O$) shells of a species of freshwater diatom (*Aulacoseira granulata*) that must have lived in freshwater lakes, not in the ocean. Because some of these ocean cores are thousands of kilometers away from land, the only way for lake diatoms to get there is by being blown in by winds. In arid and semiarid regions, winds scoop out (deflate) sediment from the beds of dried-out lakes and blow the fine debris far away, and some of it reaches the oceans.

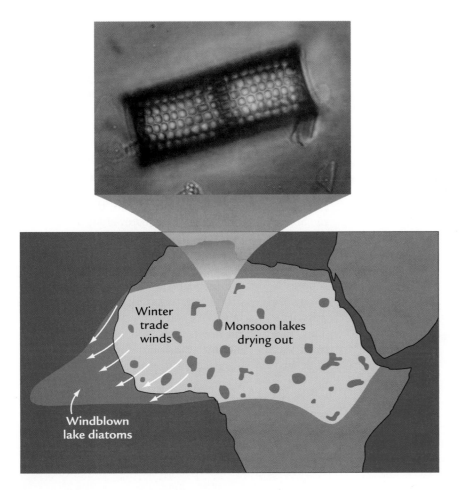

The lake diatoms in these tropical Atlantic cores must have come from North Africa, which lies directly upwind in the prevailing flow of the northern winter trade winds. The layers with freshwater diatoms in the Atlantic cores must mark times in the past when lakes were drying out across North Africa, exposing their muddy lake beds to the winter monsoon trade winds.

The records in the Atlantic cores show that lake diatoms were carried out in distinct pulses separated by 23,000 years. This 23,000-year tempo in the diatom influx, as in the Mediterranean sapropels, is a direct indication of a connection to the tropical monsoon on North Africa. In this case, however, each diatom pulse occurs considerably later than the insolation maximum, lagging behind it by 5000 to 6000 years (Figure 9-10).

This long delay makes good physical sense if we consider the sequence of events that should occur during a typical monsoon cycle. Lakes in North Africa fill to maximum size during the summer insolation maxima that drive the strong monsoons. These high lake levels produce lake-bed deposits rich in diatoms, but the high water levels in the lakes keep the lake beds from being exposed to the winds, eroded, and blown to the oceans.

Then, as summer insolation begins to decrease toward the next insolation minimum, the monsoon weakens, summers become drier, and lake levels start to drop. Eventually many lakes dry up completely and expose their lake muds to the winter winds, which pick up the diatom-bearing silts and clays and blow them out to the ocean. Once the lakes have dried out completely and their diatom-bearing sediments have all been blown away, transport of diatoms to the ocean stops, even though the monsoon continues to weaken as summer insolation continues to fall toward the next minimum.

If the sequence described in Figure 9-10 is correct, the diatom pulses sent to the ocean will lag behind the monsoon maximum because of the time needed to dry the lakes. But the diatom pulses should also precede the subsequent summer insolation minimum, because the lake beds become fully exposed partway into the drying trend. This sequence is consistent with the timing of the diatom pulses observed.

Another indication of a link to the North African summer monsoon comes from the amplitude of the diatom peaks. Each 23,000-year diatom pulse has the

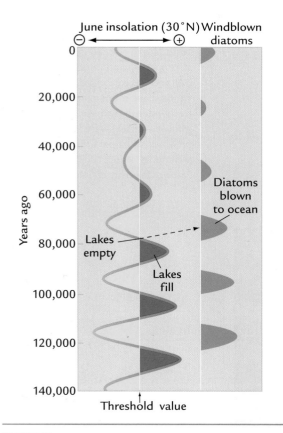

FIGURE 9-10 **Delayed diatom deposition in the Atlantic**
Monsoons created large North African lakes every 23,000
years (left), but diatom-bearing lake sediments were not
deposited in the tropical Atlantic Ocean until some 5000
years later (right). (Adapted from W. F. Ruddiman, "Tropical
Atlantic Terrigenous Fluxes Since 25,000 Years B.P.," *Marine Geology*
136 [1997]: 189–207; based on E. M. Pokras and A. C. Mix,
"Earth's Precession Cycle and Quaternary Climatic Changes in
Tropical Africa," *Nature* 326 [1987]: 486–7.)

same strength as the immediately preceding summer
insolation maximum (Figure 9-10). This pattern is consistent with a scenario in which stronger insolation drives stronger summer monsoon maxima, which create
larger lakes, and the larger lakes provide larger sources
of diatom-bearing sediments for transport to the ocean
when the monsoon subsequently weakens.

9-4　Upwelling in the Equatorial Atlantic

Atlantic sediments contain a second kind of evidence
consistent with the hypothesis that the North African
monsoon fluctuates at the 23,000-year tempo of orbital
precession. Atlantic cores lying along and just south of
the equator show that the structure of the upper layers
of water in this part of the ocean has changed at a
prominent 23,000-year rhythm.

On first consideration, this might seem unexpected.
Why would the ocean circulation in this part of the
Atlantic be sensitive to the strength of the summer
monsoon over North Africa? The answer is that strong
North African summer monsoons set up an atmospheric circulation pattern that overrides the normal (nonmonsoonal) ocean circulation.

When the North African summer monsoon is relatively weak (as it is today), trade winds along the equator
have a strong east-to-west flow (Figure 9-11A). The
strongest trade winds occur in the southern hemisphere's winter (July and August), and they blow from
the South Atlantic toward the equator along a southeast-
to-northwest trajectory. Part of this flow crosses the
equator and enters North Africa as the summer monsoon flow, but the weak African monsoon at present
suppresses this cross-equatorial wind flow.

Instead, a strong flow of trade winds toward the
west drives warm surface waters away from the equator,
This upper-ocean flow causes a shallowing of the seasonal thermocline, the transition zone between cooler,
nutrient-rich subsurface water and warmer, nutrient-

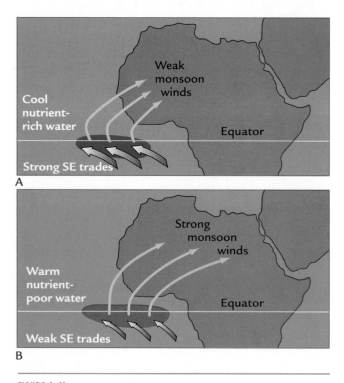

FIGURE 9-11 **Effects of monsoons on the southeast trade
winds** When monsoonal circulation over North Africa is
weak, strong southeasterly trade winds in the eastern
tropical Atlantic cause cool nutrient-rich waters to rise
close to the surface (A). A strong monsoon circulation over
North Africa weakens the trade winds and tropical waters
become warm and depleted of nutrients (B).

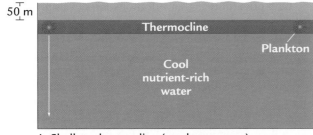

A Shallow thermocline (weak monsoon)

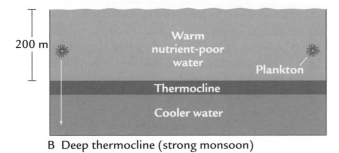

B Deep thermocline (strong monsoon)

FIGURE 9-12 **Effects of monsoons on the tropical Atlantic Ocean** During weak North African monsoon circulations like those of today, strong southern trade winds cause vigorous upwelling and the development of a shallow thermocline that favors cool-adapted plankton (A). During times of strong monsoons, which divert trade winds into North Africa, upwelling is weaker, the thermocline is deeper, and the surface ocean in which the plankton live is warmer (B).

poor near-surface water. As a result, cooler waters rich in nutrients rise nearer the sea surface just south of the equator (Figure 9-12A).

In contrast, when summer insolation was higher than it is today, the much stronger monsoonal flow into North Africa overrode this circulation pattern (Figure 9-11B). A much larger portion of the southeast trade-wind flow crossed the equator, turned to the northeast, and was drawn into North Africa in the summer monsoon circulation. This strengthening of the monsoon flow into Africa weakened the westward trade-wind flow along the equator, and the weaker trade winds reduced the intensity of upwelling. As a result, the thermocline deepened, and the thickened layer of water overlying the thermocline was warmed by the sun (Figure 9-12B).

Changes back and forth between these two patterns of surface-ocean circulation over time can be measured by examining variations in the relative amounts of planktic organisms that inhabit near-surface waters and leave shells in the sediments below. In equatorial Atlantic sediments, planktic foraminifera and coccoliths are the most common shelled organisms. Different

species of these two kinds of plankton prefer different environmental conditions near the sea surface.

When the monsoon is weak and the thermocline is shallow (Figures 9-11A, 9-12A), cooler nutrient-rich waters rise to shallow depths that receive plenty of sunlight for photosynthesis. These conditions allow species with a preference for cooler temperatures to proliferate. When the monsoon is strong and the thermocline is deep (Figures 9-11B, 9-12B), sunlight cannot penetrate down to the nutrient-rich subsurface waters, and the warmer near-surface waters are low in nutrients. Nevertheless, several species of plankton survive in these waters and their shells record these past environmental conditions in sediment layers on the seafloor.

Sediment cores from the Atlantic Ocean just south of the equator contain obvious 23,000-year cycles of alternating abundances in these two types of plankton (Figure 9-13). The presence of the familiar 23,000-year precessional cycle once again suggests a connection to the North African summer monsoon. Intervals containing plankton shells typical of warm, nutrient-poor water

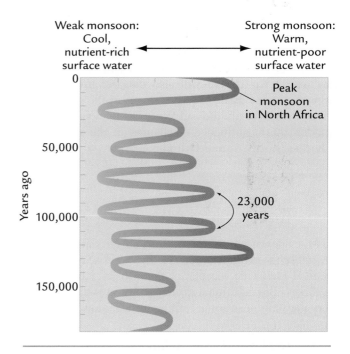

FIGURE 9-13 **Plankton and tropical Atlantic circulation** Changes in the relative abundance of environmentally sensitive plankton indicate that monsoonal vs. nonmonsoonal ocean circulation patterns in the tropical Atlantic occur mainly at the 23,000-year precessional cycle. (Adapted from A. McIntyre et al., "Surface Water Response of the Equatorial Atlantic Ocean to Orbital Forcing," *Paleoceanography* 4 [1989]: 19–56, and B. Molfino and A. McIntyre, "Precessional Forcing of Nutricline Dynamics in the Equatorial Atlantic," *Science* 249 [1990]: 766–69.)

and weak upwelling coincide with summer insolation maxima, indicating times when a strong monsoon circulation overrides the normal trade-wind pattern (Figure 9-12B). Intervals with shells typical of cool, nutrient-rich waters and strong upwelling occur during insolation minima, when there is little or no monsoonal overprint on equatorial Atlantic circulation (Figure 9-12A).

With all this evidence of orbital-scale changes in the North African summer monsoon, what do we know about the winter season? We learned in Chapter 8 that orbital-scale changes in winter-season insolation at low latitudes are also dominated by the 23,000-year tempo of orbital precession. This suggests that geological indices sensitive to the winter monsoon circulation might also respond at the 23,000-year rhythm. Do we see any signs of a winter-season signal?

Because the low-level monsoon flow in winter has a down-and-out pattern (Figure 9-1B), we might expect to find changes in the amount of fine debris transported by cool, dry winds as a record of fluctuations in low-latitude winter monsoons at the precessional rhythm. In fact, we have just seen that freshwater diatoms are transported by winter trade winds at this tempo (Figure 9-10). But we have also seen that the freshwater diatom signal is at least as much a result of lake filling by summer monsoon rains as it is of later drying and transport by winter winds. We will also discover in upcoming chapters that signals of past changes in the strength of *winter* winds at middle and lower latitudes can be overridden by the influence of changes in the sizes of ice sheets at northern mid-latitudes.

Refinements of the Orbital Monsoon Hypothesis

The evidence we have examined so far supports John Kutzbach's orbital monsoon hypothesis, which has gained such wide acceptance that it merits the status of a theory. But additional evidence indicates that the theory may still need some refining or elaboration.

9-5 Lag of Monsoons Behind Summer Insolation

On close inspection, climate records of the monsoon indicate that the peak development of the summer monsoon response in North Africa lags slightly behind the June 21 summer insolation signal cited as its driving force. The predicted monsoon response pattern shown in Figure 9-5 assumes an instantaneous response of monsoon intensity to insolation forcing, rather than a delay in the response. The lag appears to be 1000 to 2000 years. What could explain such a delayed response?

One possibility is that full monsoon development in the tropics and subtropics can be retarded for a while by changes occurring elsewhere in the climate system. One obvious source of such a retarding effect is ice sheets. Strong monsoons developed during intervals when northern hemisphere ice sheets were still extensive, although melting rapidly. The large-scale cooling caused by these lingering ice sheets could have retarded the full summer monsoon heating in North Africa for several thousand years, until the ice sheets finally became too small to have any impact on the climate of the tropics. If this explanation of an ice delay were proved correct, it would not disprove the orbital monsoon hypothesis, but would simply show that interactions with other parts of the climate system can complicate the operation of the monsoon.

A related explanation is that the tropical ocean may have retarded full development of the monsoon. The tropical ocean surface cooled during glaciations, and a cooler ocean surface is a weaker source of latent heat to fuel monsoons than a warm ocean. Evidence from the Indian Ocean indicates that the timing of the monsoon maximum over eastern Africa and across India could have been delayed by a cooler surface of the Indian Ocean.

Another possible explanation for the delayed response is that the true driving force behind the summer monsoon in North Africa is not the June 21 summer solstice insolation curve shown in Figure 9-5 but an insolation curve from a time somewhat later in the summer. As shown in Figure 8-19, Earth's gradual precessional motion around the sun produces an entire family of monthly insolation curves, each offset from the preceding month by one-twelfth of the 23,000-year cycle length, or just under 2000 years. This group of monthly insolation curves means that we could choose any one of them as the hypothetical orbital forcing of the summer monsoon. In effect, our choice of monthly driver determines in advance the size of the lag in years.

Peak development of the North African monsoon in climate records is in phase with the summer insolation curve from middle to late July rather than with the June 21 signal used in the original hypothesis. As a result, if we choose the July insolation signal as our hypothetical monsoon driver, the apparent lag between the insolation forcing and the monsoon response will disappear. This choice may also have a reasonable physical basis: today the summer monsoon reaches its peak expression in late July or early August, well after the June 21 summer solstice. It may be that July is a more sensitive part of the summer for long-term insolation changes to enhance monsoon strength.

At this point, climate scientists cannot prove which of these possible explanations (or combination of them) is correct. June 21 insolation may indeed be the critical driver of the summer monsoon, with ice sheets delaying the monsoon's response. Alternatively, the initial hypothesis may need to be modified slightly by picking a later part of the summer precession signal as the critical driver of the monsoon.

9-6 Clipped Monsoon Responses and Monsoon Harmonics

The truncation, or cutting off, of the summer monsoon response pattern at some critical threshold value (Figure 9-5) is referred to as a **clipped response**. As a result of clipping, monsoon responses recorded in actual climate records register only a portion of each 23,000-year cycle caused by orbital precession. For example, higher insolation values drive stronger monsoons and produce progressively higher lake levels, but insolation values that fall below the threshold level produce no monsoon-signal response at all because lakes have already dried out completely and cannot get drier when insolation drops even further. The lake level response to insolation is clipped at the threshold insolation value.

Clipping distorts the way the impacts of monsoon changes are recorded in the climate record. In effect, the clipping process produces false signals (artifacts) created by the biased recording mechanism, signals that are not actually present in the original orbital insolation forcing.

One artifact produced by clipping is that the record of lake levels (or similar monsoon-driven responses) contains significant cyclic responses at 100,000 and 413,000 years. The 30°N orbital insolation curve that drives monsoons contains no power at either of these two eccentricity cycles; these eccentricity changes act only to modulate the 23,000-year precession signal (Figure 9-14). But the lake levels respond only to one side of the 23,000-year cycle (the high-insolation part), rather than to the full range of insolation highs and lows. As a result, lake-level records are biased to high orbital insolation values, a good example of a nonlinear response in the climate system.

As a result of this nonlinear response, lake records contain not just the 23,000-year pulses but also a more subtle longer-term trend from higher to lower lake maxima at cycles of 100,000 and 413,000 years. The 23,000-year cycle that dominates the climate records reflects the original orbital forcing and the direct summer monsoon response, but the insolation forcing had no real cycles at periods of 100,000 or 413,000 years.

A second feature of monsoon response records is the presence of **harmonics** of the 23,000-year precession cycle. Harmonics are shorter cycles generated by clipping off one side of the 23,000-year response. They also result in part from the fact that some records of the

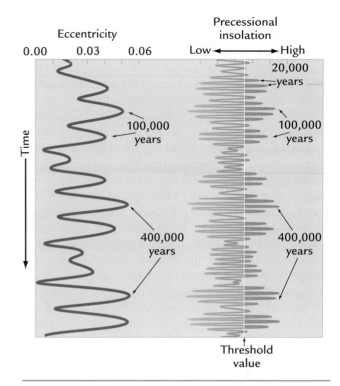

FIGURE 9-14 Monsoon signals recorded in sediments Monsoonal influences can be detected in older sediment sequences. High orbital eccentricity values (left) should amplify individual 23,000-year precession cycles every 100,000 and 400,000 years (right). The expected signal of a strong monsoon is shown by the red-shaded area to the right of the threshold insolation value.

23,000-year monsoon signal register abrupt input events, such as the brief pulses of freshwater diatoms blown into the tropical Atlantic Ocean.

For an orbital cycle with a period of N years, the harmonics have periods of $N/2$, $N/3$, $N/4$, and so on. The 23,000-year monsoon signal produces harmonic cycles of 11,500 years ($N/2$), 7600 years ($N/3$), 5750 years ($N/4$), and so on. Again, these harmonic cycles are not present in the original orbital signal or in the changes of strength of the summer monsoon, but are artifacts of the biased way the climate system records its response to orbital insolation changes.

Monsoon Forcing Earlier in Earth's History

Insolation forcing of summer monsoon circulations is a powerful concept that can be applied to environments that existed in the distant geologic past. This insight is based on two facts: (1) orbital-scale changes in summer insolation drive changes in monsoonal precipitation at the precession cycle, and (2) precipitation is an important driving force behind many processes that scientists can measure in ancient climate records.

Precipitation is obviously central to the process of erosion: heavy rainfall causes strong runoff, and running water carries sediment across the land and deposits it in quieter water. As a result, many sedimentary rocks hold information about the amount of precipitation that controlled their erosion, transport, and deposition.

For cases in which excellent time control is available, unmistakable imprints of ancient monsoons can be detected even at tectonic time scales. As in our examination of the more recent monsoon cycles in North Africa, we will be looking for evidence of a monsoon signature like the one shown in Figure 9-5. In this case, however, we are looking for such a pattern in sediments deposited over longer intervals of time. The expected pattern of the longer-term monsoon response is shown in Figure 9-14, in the signal on the right. Because we are looking further back in time, the periods of the orbital cycles will be slightly shorter than they are today because of long-term evolution of the cycle lengths (Chapter 8).

The expected monsoon response has cycles near 20,000 years in length, longer-term variations in cycle amplitudes driven by insolation, and a clipped response at the threshold value (Figure 9-14, right). Again, we expect the pattern of monsoon response to have clusters of two or three strong maxima, separated by clusters of two or three weak maxima, with each cluster repeating roughly every 100,000 years. These clusters reflect orbital eccentricity control of the amplitude of the precession curve (Chapter 8). And because in this case we are looking at much longer records, we also expect to see additional larger-scale clusters of cycles at a longer eccentricity period near 400,000 years. Any of these features may show up in a wide range of sediment indicators linked to precipitation, erosion, runoff, transport, and deposition.

9-7 Monsoons on Pangaea 200 Myr ago

Just before 200 million years ago, a chain of basins formed in a region that is now the eastern United States, but was then deep in the interior of the giant supercontinent Pangaea (Figure 9-15). These features were deep depressions in a region of generally high terrain and were formed by early precursors of the forces that would eventually rip the Pangaean continent apart and create the Atlantic Ocean. But this part of Pangaea would not break up until tens of millions of years later, and it would actually do so along an axis several hundred kilometers to the east of these basins (Chapter 6).

Sediments deposited in one of these depressions, the Newark Basin in present-day New Jersey, have been extensively investigated in recent decades. The fossil compasses provided by magnetic evidence indicate that the Newark Basin was located in the tropics 200 million years ago, about 10° of latitude north of the equator (Figure 9-15). Because of its tropical location, the Newark Basin would have been situated in a regime dominated by precessional insolation changes, similar to those on modern-day North Africa. Because the basin was far from the ocean, its climate was moderately arid, yet enough moisture arrived to create a lake that varied greatly in size over time.

Geologic evidence preserved in a thick (> 7000 m) sequence of lake sediments clearly shows that the size of this lake fluctuated at a tempo near 20,000 years. Several layers of molten magma that intruded into the lake-bed sequence and quickly cooled have been dated by radiometric methods. These rocks suggest that the lake-bed sequence was deposited over an interval of at least 20 million years centered near 200 Myr ago.

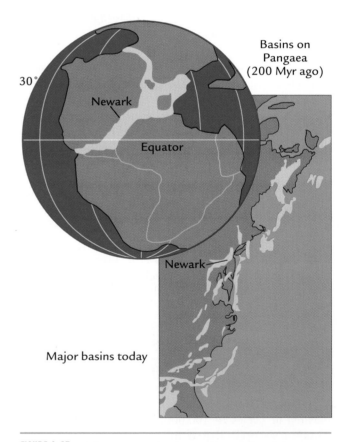

FIGURE 9-15 **Mid-Pangaean basins** In the middle of the Pangaean supercontinent 200 million years ago (top left), the Newark Basin developed in what is now present-day New Jersey, as one of a chain of basins of equivalent age (bottom right). (Adapted from P. E. Olsen and D. V. Kent, "Milankovitch Climate Forcing in the Tropics of Pangaea During the Late Triassic," *Palaeogeography, Palaeoclimatology, Palaeoecology* 122 [1996]: 1–26.)

FIGURE 9-16 **Evidence of changing lake levels** Dinosaur footprints in lake muds that have since hardened into rock show that the lakes on Pangaea occasionally dried out completely. These footprints are from a basin in Connecticut formed at the same time as the Newark Basin in New Jersey. (Dinosaur State Park, Rocky Hill, Conn.)

This estimate is confirmed by the presence of fine laminations (varves) in parts of the sequence. These varves are tiny (0.2–0.3 mm) couplets of alternating light and dark layers, with one light/dark pair deposited each year. Darker organic-rich layers were deposited in the summer and lighter mineral-rich layers in the winter. Dissolved oxygen concentrations must have been low or zero in the deeper levels of the lake when the organic-rich layers accumulated, preventing destruction of the delicate varves by animals moving across and within the sediments. Use of these varves as an internal chronometer to count elapsed time confirms that the total time of lake-sediment deposition was about 20 million years.

The type of sediment deposited in the Newark Basin varied widely in response to changes in lake depth. When the lake was deep (100 m or more), the sediments tended to be gray or black muds containing large amounts of organic carbon. These sediments often show finely laminated varves because they have not been disturbed by burrowing animals, and they are often rich in well-preserved remains of fish. Sediments deposited when the lake was shallower or entirely dried out tend to be red or purple because they have been oxidized (rusted) by contact with air, and they also contain

mud cracks due to occasional exposure and drying of the muds. Dinosaur footprints and the remains of plant roots are also common in sediments from the dried-out, vegetated parts of the lake beds (Figure 9-16).

The thick sequences preserved in the Newark Basin repeatedly cycle between sediments typical of deep lakes and those indicating shallowing or brief drying out of the lakes. Individual layers in these cyclic sequences are continuous over large areas, indicating that the wet-dry climate cycles affected the entire basin.

Extensive investigations show cyclic variations in the depths of these lakes over millions of years, and at several depth and time scales (Figure 9-17). The shortest cycles occur over rock thicknesses averaging 4 to 5 meters, equivalent to about 20,000 years in time based on the 0.2- to 0.3-millimeter average thickness of each

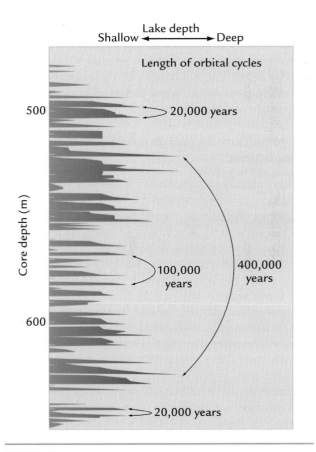

FIGURE 9-17 **Fluctuations of Pangaean lakes** Newark Basin lake sediments varied in depth from very shallow to over 100 m deep. Three scales of past cycles in lake depth are evident: individual cycles near 20,000 years occurring every 4–5 m, clusters of deep-lake maxima at intervals near 100,000 years every 20–25 m, and larger deep-lake clusters at intervals near 400,000 years every 90–100 m. (Adapted from P. E. Olsen and D. V. Kent, "Milankovitch Climate Forcing in the Tropics of Pangaea During the Late Triassic," *Palaeogeography, Palaeoclimatology, Palaeoecology* 122 [1996]: 1–26.)

annual varve. These shortest cycles correspond to individual monsoon cycles driven by precession. The evidence shows that monsoons filled and emptied these Pangaean lakes in response to cycles of orbital precession in just the same way that North African lakes have filled and emptied during much more recent times.

Two larger-scale groupings of cycle peaks are also evident. The amplitude of the 20,000-year peaks rises and falls roughly every five or six cycles, so that the intervals of deepest lakes occur in clusters of two or three peaks separated from similar clusters by an average of 20 to 25 meters of sediment, or a little less than 100,000 years of time. An even larger-scale change in the intensity of individual monsoon-cycle peaks is present (although harder to detect) roughly every 90 meters, or at a time interval of about 400,000 years.

These two larger-scale, longer-term patterns match the expected long-term monsoon signature shown in Figure 9-14 remarkably well. They reflect a modulation of the strength of the 20,000-year precession cycles by eccentricity changes at intervals of about 100,000 and 400,000 years. The full imprint of ancient monsoons is amazingly clear in the sediments of this basin despite the passage of 200 million years.

9-8 | Joint Tectonic and Orbital Control of Monsoons

We have already seen that tectonic changes can alter the intensity of monsoon circulations. Large landmasses such as Pangaea intensify monsoons by offering a larger area for the Sun to heat. Positioning of landmasses at lower latitudes is also important because solar radiation is more direct and albedos are much lower than at higher latitudes, where the land is generally covered with snow or ice. Topography is also a vital control over monsoon strength at tectonic time scales, because high-elevation regions focus strong monsoonal rains on their margins.

The processes that control monsoon intensity over tectonic time scales interact with those at orbital scales. Tectonic-scale processes alter the average strength of the monsoon over millions of years, while the orbital-scale changes in insolation dictate the shorter-term strength of the monsoon mainly at cycles near 20,000 years. But how do these two factors actually work in combination?

One way they might interact is suggested in Figure 9-18. On the left is a schematic version of the low-latitude summer insolation curve, with individual maxima and minima at the 20,000-year precessional cycle and modulation of this cycle every 100,000 and 413,000 years, but no long-term trend toward higher or lower values.

The smooth curve in the center of Figure 9-18 represents some kind of gradually changing tectonic-scale process, in this case a slow uplift that gradually intensi-

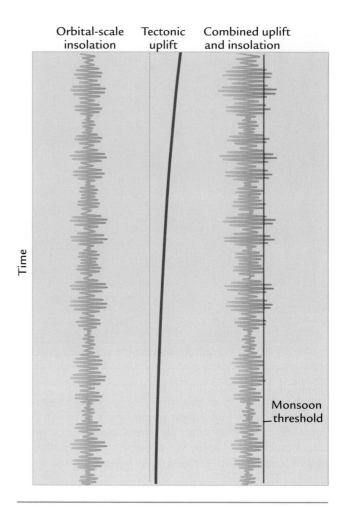

FIGURE 9-18 Combined tectonic and orbital forcing of monsoons Monsoons are driven by orbital-scale variations in insolation (left) and by slower-acting tectonic factors such as uplift (center). The tectonic factors cause long-term changes in the amplitude of orbital-scale monsoons (right) and produce monsoon responses that gradually exceed critical thresholds during insolation maxima. (Adapted from W. F. Ruddiman et al., "Late Miocene to Pleistocene Evolution of Climate in Africa and the Low-Latitude Atlantic: Overview of Leg 108 Results," Ocean Drilling Program Initial Reports 108B [1989]: 463–84.)

fies the average strength of the monsoon over millions of years. This slow tectonic-scale increase in monsoon strength combines with the orbital-scale monsoon cycles to produce the response shown on the right. The combined result is a *slow increase in the amplitude of the orbital-scale cycles* caused by the amplification provided by tectonic processes.

Once again we need to invoke the concept of a threshold value above which key climatic indices record monsoon responses but below which no such response is registered. If we apply this concept to the combined tectonic and orbital changes shown on the right in

Figure 9-18, we find that the tectonic influence was so weak during earlier intervals that the orbital-scale monsoon cycles never exceeded the critical strong-monsoon threshold. Later, as tectonic processes created conditions more favorable to monsoons, individual peaks in summer insolation would have driven monsoons that began to exceed the threshold by small amounts, and still later by steadily increasing amounts.

Something close to this kind of climatic response is thought to have occurred in Southeast Asia over the last 30 or 40 million years. A long-term tectonic increase in monsoon intensity due to uplift progressively intensified the amplitude of orbitally driven monsoon cycles in this region. Many scientists infer that uplift of the Tibetan Plateau and the Himalayas was concentrated in rapid spurts near 20 and 10 Myr ago, rather than occurring as a slow and progressive process. If this view is correct, faster tectonic uplift during these two intervals would have caused particularly rapid intensifications of monsoon strength at these times. This may have caused orbital-scale monsoon responses suddenly to exceed the threshold level, rather than to follow the more gradual trend shown in Figure 9-18.

In addition, the actual process of monsoon intensification may represent still another example of a nonlinear response. Simulations run on general circulation models indicate that the effects of orbital-scale insolation and uplift are not additive in a simple linear way. Instead, it appears that plateau uplift sensitizes the monsoon system to insolation forcing in such a way that the combined monsoon response to uplift and insolation is far stronger than a simple linear combination of the two effects.

Key Terms

orbital monsoon hypothesis (p. 195)
nonlinear response (p. 196)
threshold insolation level (p. 198)
sapropels (p. 198)
clipped responses (p. 205)
harmonics (p. 205)

Review Questions

1. Why does rain fall only in summer across most of North Africa today?

2. Why does the tropical monsoon vary at a cycle of 23,000 years?

3. Why does the intensity of 23,000-year monsoon peaks change at longer intervals of 100,000 and 413,000 years?

4. In what sense is the orbital monsoon hypothesis a simple extension of processes associated with present-day changes in seasonal radiation?

5. Describe two kinds of evidence showing variations in the strength of the tropical North African monsoon over the last several hundred thousand years.

6. How did the Mediterranean Sea acquire a freshwater lid during times when little precipitation was falling in that region?

7. What sequence of processes deposited lake diatoms in tropical Atlantic sediments?

8. Does peak monsoon strength lag behind summer insolation forcing, and if so, why?

9. What is a "clipped response" to insolation?

10. What similarities exist between the monsoon changes in Pangaea 200 Myr ago and those in North Africa during the last several hundred thousand years?

11. How do tectonic uplift and orbital variations combine to affect the long-term intensity of monsoons?

Additional Resources

Kutzbach, J. E. 1981. "Monsoon Climate of the Early Holocene: Climate Experiment with Earth's Orbital Parameters for 9000 Years Ago." *Science* 214:59–61.

Mcintyre, A., W. F. Ruddiman, K. Karlin, and A. C. Mix. 1989. "Surface Water Response of the Equatorial Atlantic Ocean to Orbital Forcing." *Paleoceanography* 30:349–62.

Olsen, P. E. 1986. "A 40-Million-Year Lake Record of Early Mesozoic Orbital Climatic Forcing." *Science* 234:842–48.

Pokras, E. M., and A. C. Mix. 1987. "Earth's Precession Cycle and Quaternary Climatic Changes in Tropical Africa." *Nature* 326:486–87.

Rossignol-Strick, M., W. Nesteroff, P. Olive, and C. Vergnaud-Grazzini. 1982. "After the Deluge: Mediterranean Stagnation and Sapropel Formation." *Nature* 295:105–10.

CHAPTER 10

Insolation Control of Ice Sheets

Ice covered northern North America, Europe, and Asia 20,000 years ago, burying the present-day locations of Toronto, New York, Chicago, Seattle, and London under hundreds of meters or more of ice. Later, the ice melted, and the last remnants disappeared 6000 years ago as human civilizations began. The tectonic-scale cooling described in Part II caused ice sheets to appear, but why did they grow and melt over shorter intervals? The answer is that they reacted strongly to orbital insolation changes.

In this chapter, we investigate how changes in summer insolation control the size of ice sheets by fixing the rate of ice melting. We also explore two critical lags in the ice response: the size of ice sheets lags thousands of years behind changes in insolation, and the depression of underlying bedrock lags thousands of years behind the size of ice sheets. Then we review a conceptual model of ice sheet responses over the last 3 Myr and compare it to oxygen-isotopic ($\delta^{18}O$) evidence of actual ice sheet cycles. Finally, we examine evidence of the past size of ice sheets from coral reefs, which record changes in sea level at orbital time scales.

What Controls the Size of Ice Sheets?

Continental ice sheets exist because the overall rate at which snow and ice accumulates across the entire pile of ice equals or exceeds the overall rate of ice loss or ablation. Snow accumulates and turns into ice in high-elevation and high-latitude regions where temperatures are cold enough not just to permit frozen precipitation but also to prevent melting in the warm summer season. For continent-sized ice sheets, which often exist at sea level, temperatures sufficiently cold to sustain ice sheets occur at higher latitudes.

Rates of ice accumulation and ablation vary with temperature, but the two relationships differ in a critical way. Ice (snow) accumulates at mean annual temperatures below 10°C, but the rates of accumulation stay below 0.5 meter per year at all temperatures (Figure 10-1A). At higher temperatures, more of the precipitation falls as rain and thereby limits net ice accumulation. At extremely low temperatures, all the precipitation falls as snow, but the frigid air carries little water vapor. As a result, rates of ice accumulation remain relatively low even at colder temperatures.

In contrast, ablation of glacial ice accelerates rapidly when temperatures are warm. Melting begins at mean annual temperatures above −10°C, equivalent to summer temperatures well above 0°C. Ablation can reach rates equivalent to 3 meters of ice or more per year, much larger than the maximum rates of accumulation.

Ablation can occur in several ways: by absorption of solar radiation, by uptake of sensible or latent heat delivered by warm air masses (and rain) moving across the ice, and by calving or shedding of icebergs to the ocean or to lakes. Calving differs from the other two ablation processes in that icebergs leave the main ice mass and move elsewhere to melt, often in a warmer environment than that near the ice sheet.

The total mass balance over an ice sheet is the net balance between accumulation and ablation (Figure 10-1B). The mass balance at very cold temperatures (below −20°C) is positive but small because while there is no ablation, there is also little snowfall. The mass balance at mean annual temperatures near −10° to −15°C is slightly more positive because snow accumulation rates are more rapid but ablation is not strong. The mass balance turns sharply negative at temperatures above −10°C because ablation accelerates and overwhelms accumulation. The boundary between the upper area of positive ice mass balance and the lower area of net loss of ice mass is called the **equilibrium line**.

If net accumulation and ablation balance out over an entire ice sheet, the ice sheet is said to be in a condition of stable equilibrium. The net accumulation high on the ice sheet is exactly balanced by ablation low on the ice sheet, and no net change in ice sheet size results. Ice

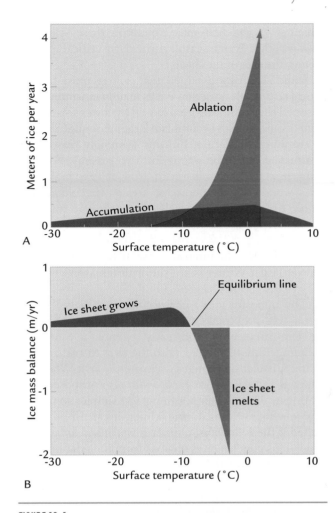

A

B

FIGURE 10-1 **Temperature and ice mass balance** Temperature is the main factor that determines whether ice sheets are in a regime of net ablation (negative mass balance) or net accumulation (positive mass balance). Ablation increases sharply at higher temperatures. (Modified from J. Oerlemans, "The Role of Ice Sheets in the Pleistocene Climate," *Norsk Geologisk Tidsskrift* 71 [1991]: 155–61.)

must flow within the ice sheet from accumulation areas to ablation areas to maintain overall mass balance, but the total mass of the ice sheet remains stable.

But our focus here is not on ice sheets in equilibrium. We want to understand what makes ice sheets grow and shrink.

10-1 Orbital-Scale Control of Ice Sheets: The Milankovitch Theory

Beginning with the Belgian mathematician Joseph Adhemar in the 1840s, scientists speculated that orbitally driven changes in solar insolation might in some way be linked to the growth and melting of continent-sized ice sheets. Because orbital changes alter the amount of insolation received on Earth in both winter and summer, as well as during intermediate seasons

(Chapter 8), the first major question these scientists faced was this: Is one season particularly critical in controlling the size of ice sheets?

Some scientists assumed that winter must be the critical season. Because snow falls mainly in winter, they reasoned that colder winters caused by lower solar insolation would help to accumulate larger amounts of snow and promote glaciation. But this seemingly reasonable idea turned out to be wrong, for two reasons. For one thing, new ice sheets begin to grow at very high latitudes where the temperatures are *always* cold in the winter, even during relatively warm climatic intervals like today. In addition, the Sun at these latitudes is always low in the sky, regardless of ongoing orbital changes, so that the intensity of solar radiation arriving is never large. Winter is not the critical season of sensitivity to insolation changes.

The opposite idea—*summer insolation control of ice sheets*—was proposed by several scientists working in the late nineteenth and early twentieth centuries, including Rudolf Spitaler (also the first to realize that summer insolation changes drive monsoons); Wladimir Köppen; and Alfred Wegener (who also proposed that continents drift). Their reasoning was simple: no matter how much snow falls during winter, it can all be easily melted if the following summer is warm and ablation is rapid. This conclusion is obvious from the relationships plotted in Figure 10-1.

As a result, these scientists concluded that low summer insolation is the critical factor that cools the climate enough to allow snow and ice to persist from one winter to the next. This idea gained popularity during the early and mid-twentieth century from work by the Serbian astronomer Milutin Milankovitch, who also first quantified in a systematic way the impact of astronomical changes on insolation received on Earth. The theory is now known as the **Milankovitch theory.**

Because the Milankovitch theory attempted to explain ice sheet changes in the northern hemisphere, it proposed that ice growth occurs during times when summer insolation is low in high northern latitudes (Figure 10-2A). Low summer insolation occurs during times when Earth's orbital tilt is small and its poles are pointed less directly at the sun. Low insolation also results from the fact that the northern hemisphere's summer solstice occurs when Earth is farthest from the Sun (in the aphelion or distant-pass position) and when the orbit is highly eccentric (increasing the Earth-Sun distance at this aphelion position). Milankovitch reasoned that the most sensitive latitude for low insolation values is 65°N, the latitude at which ice sheets first accumulated and last melted. Ice melts during the opposite orbital configuration (Figure 10-2B).

The amount of summer insolation arriving at the top of Earth's atmosphere at 65°N and nearby latitudes can vary by as much as ±12% around the long-term mean value. Changes in winter insolation also occur with the exact opposite timing (Chapter 8), but they are not considered important to ice sheet survival for the reasons just noted.

Although it is possible to calculate how much solar radiation arrived at the *top* of Earth's atmosphere at any latitude, we have no way of knowing how much actually makes it through the atmosphere to Earth's surface and the ice sheets. Complex regional-scale changes in atmospheric circulation, clouds, and water vapor also affect the amount of solar radiation that reaches Earth's surface. Milankovitch noted these complications but simply assumed that the amount of radiation penetrating to Earth's surface is determined by the amount arriving at the top of the atmosphere.

In summary, the Milankovitch theory proposes that when summer insolation is strong, more radiation is absorbed at Earth's surface at high latitudes, making the climate warmer. The warming accelerates ablation, melts more snow and ice, and either prevents glaciation or

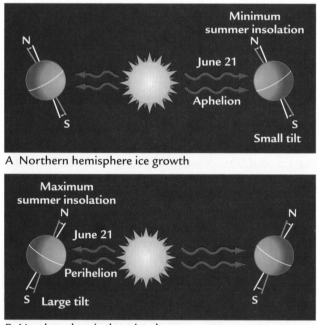

A Northern hemisphere ice growth

B Northern hemisphere ice decay

FIGURE 10-2 **Orbital changes and ice sheets** According to the Milankovitch theory, ice sheets grow in the northern hemisphere's summer at times when insolation is low because the tilt is low, and the northern hemisphere is in the aphelion (distant-pass) position (A). Ice melts in summer when insolation is high because the tilt is high, and the northern hemisphere is in the perihelion (close-pass) position (B). (Adapted from W. F. Ruddiman and A. McIntyre, "Oceanic Mechanisms for Amplification of the 23,000-Year Ice-Volume Cycle," *Science* 212 [1981]: 617–27.)

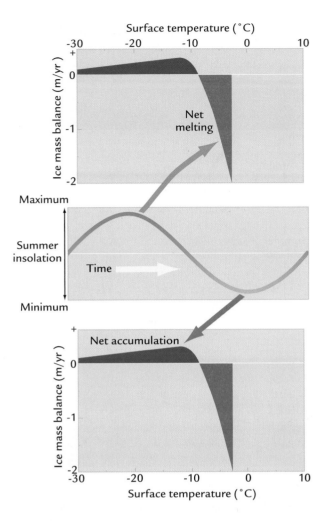

FIGURE 10-3 **The Milankovitch theory** The Milankovitch theory proposes that high summer insolation heats the land and results in greater ablation (top), while low summer insolation allows the land to cool enough so that snow accumulates and glaciers form (bottom). (Modified from J. Oerlemans, "The Role of Ice Sheets in the Pleistocene Climate," *Norsk Geologisk Tidsskrift* 71 [1991]: 155–61.)

reduces the rate of summer ablation and allows snow to accumulate and ice sheets to grow (Figure 10-3, bottom).

The critical link in the Milankovitch theory—summer insolation control of ice sheets—has a firm physical basis. Because the rate of ice melting accelerates rapidly as temperatures warm, summer should be the most sensitive season for controlling ice sheet size.

Modeling the Behavior of Ice Sheets

To gain more insight into how summer insolation changes control ice sheets in the northern hemisphere, climate scientists have developed two-dimensional numerical models based on an idealized ice sheet surrounding the Arctic Ocean (see Chapter 3). These models provide a simplified representation of the changes in ice sheets that occur across a typical high-latitude continent near the Arctic by ignoring possible changes from one sector of the Arctic to another (such as those in Europe vs. North America). This simplification is reasonably well justified because ice sheets have grown on all the continents around the Arctic. At the last glacial maximum 20,000 years ago, ice sheets existed in northern North America, northern Europe, Greenland, and parts of northern Asia (Figure 10-4).

shrinks existing ice sheets (Figure 10-3, top). Conversely, when summer insolation is weak, less radiation is delivered to Earth's surface at high latitudes, and the reduction in radiation cools the regional climate. This cooling

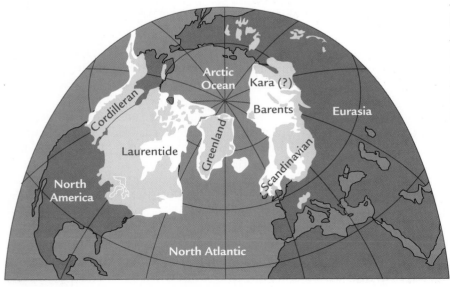

FIGURE 10-4 **Ice sheets around the Arctic Ocean** At the last glacial maximum, 20,000 years ago, ice sheets surrounded much of the Arctic Ocean. (Modified from G. Denton and T. Hughes, *The Last Great Ice Sheets* [New York: John Wiley, 1981].)

10-2 Insolation Control of Ice Sheet Size

How do changes in summer insolation control the size of ice sheets on the northern landmasses in these models? The basic idea follows directly from the Milankovitch theory: changes in summer insolation values cause changes in regional temperature that alter melting rates and change the overall mass balance of the ice sheet.

More specifically, the models represent this process in terms of changes in mass balance along a north-south line or transect (Figure 10-5). These transects have just two dimensions, one in a vertical direction (altitude) and the other in a north-south direction (latitude). Changes in the third dimension (longitude) are ignored.

The equilibrium line, the boundary between areas of net ice ablation and accumulation, slopes upward into the atmosphere toward the south at a low angle. This slope is consistent with conditions today: temperatures become colder toward higher latitudes and altitudes, and warmer toward lower latitudes and altitudes. As a result, subfreezing temperatures occur only at high latitudes and altitudes today.

Linked to this moving equilibrium line are parallel lines of ice sheet mass balance. These lines represent the total thickness in meters of ice that accumulates or melts each year (as before, snowfall is converted to an equivalent thickness of solid ice). Ice accumulates above the equilibrium line and toward the north because of the colder temperatures, and it melts below the equilib-

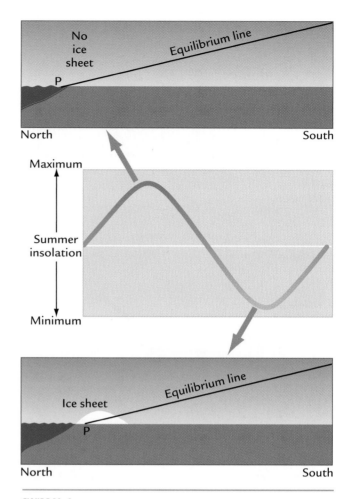

FIGURE 10-6 **Insolation changes displace the equilibrium line** When the equilibrium line is driven north by maximum values of summer insolation, the continents lie in a regime of net ablation and no ice can accumulate (top). When it is driven south by summer insolation minima, the northern landmasses lie in a regime of net accumulation and ice sheets grow (bottom). (P = climate point.) (Modified from J. Oerlemans, "The Role of Ice Sheets in the Pleistocene Climate," *Norsk Geologisk Tidsskrift* 71 [1991]: 155–61.)

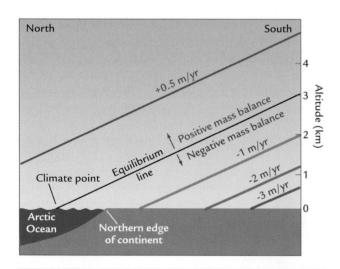

FIGURE 10-5 **Ice sheet models** Two-dimensional models of the development of ice sheets represent the northern hemisphere along a single north-south line. The equilibrium line separates northern (and higher) regions of net accumulation from southern (and lower) regions of net ablation, and it intersects Earth's surface at the climate point. (Adapted from J. Oerlemans, "Model Experiments of the 100,000-Year Glacial Cycle," *Nature* 287 [1987]: 430–32.)

rium line and toward the south. The rates of ice melting accelerate in the relative warmth at lower elevations, consistent with the relationship shown in Figure 10-1.

The equilibrium line intercepts Earth's surface somewhere in the higher latitudes at a location called the **climate point** (Figure 10-5). Ice sheet models use known orbital-scale changes in summer insolation to move this climate point (and the attached equilibrium line) north and south across the landmasses (Figure 10-6). The amount of north-south shift of the equilibrium line in the models is specified to be directly proportional to the amount of change in summer insolation. These shifts can amount to more than 10° of latitude over tens of thousands of years.

Changes in the amount of summer radiation received at high latitudes through time slowly shift the areas of net snow accumulation and net ice ablation back and forth across the land. Strong summer insolation warms the high-latitude landmasses in summer, shifts the climate point northward over the Arctic Ocean, and puts the northern landmasses entirely in an ablation regime where all of the winter snowfall melts each summer (Figure 10-6, top). As a result, no ice sheets form. Lower summer insolation cools the landmasses, shifts the climate point southward out of the Arctic Ocean, and sets up a positive mass balance over the northern edge of the continents. Permanent ice begins to accumulate on the land (Figure 10-6, bottom).

Once ice sheets begin to form, their vertical dimension (altitude) also comes into play in a powerful way (Figure 10-7). As the ice sheets grow upward into the atmosphere, their surfaces reach altitudes where the prevailing temperatures are much colder. Ice sheets can reach elevations of 2 to 3 kilometers, which translate into temperatures 12° to 19°C cooler than those at sea level (using an average lapse-rate cooling of 6.5°C per km of altitude, as in Chapter 2). This cooling increases the accumulation of snow and ice because the ice mass balance becomes more positive.

In effect, once ice sheets get started, they contribute to their own positive mass balance by growing upward into a net accumulation regime. This positive feedback (called **ice elevation feedback**) does not depend in any direct way on north-south shifts of the equilibrium line (Figure 10-7). Eventually the tops of ice sheets reach elevations at which snowfall is much lower, and the pos-

itive feedback effect of ice elevation is reduced in those regions.

Because the equilibrium line is sloped, its north-south movements raise and lower its position at specific locations. As a result, the moving equilibrium line interacts with the changing height of the ice sheet surface. Consider, for example, an ice sheet that has grown large, but summer insolation has begun its next cyclic increase. At this point the increasing insolation will cause the equilibrium line to shift back to the north, but this shift will put a larger area of the ice sheet's surface under a regime of net ablation and help to melt the ice.

10-3 Ice Sheet Lags behind Summer Insolation Forcing

Although ice sheets are driven by summer insolation changes, their response is far from immediate. We can get some feeling for this delay by imagining what would happen if climate suddenly cooled enough to permit snow to fall throughout the year and accumulate rapidly over all of Canada. Using values plotted in Figure 10-1, and assuming the mass balance of the region suddenly became highly positive, we can estimate that about 0.3 meters of ice might accumulate each year. But even under this most favorable condition, a full-sized ice sheet 3000 meters thick would take 10,000 years to form. In reality, ice would take much longer to accumulate, because orbital-scale cooling and growth of ice sheets develop gradually in response to slowly changing insolation.

The geologist John Imbrie and his colleagues have led the exploration of the link between insolation and ice volume. They use as an analogy the conceptual model examined in Chapter 1 and shown in Figure 10-8 (top): variations in the intensity of the flame in a Bunsen burner through time. Because water has a high heat capacity and reacts slowly to changes in heating, changes in water temperature lag behind changes in intensity of the heat source.

The sizes of ice sheets exhibit the same lagging response to summer insolation (Figure 10-8, bottom), but on a much longer time scale. As summer insolation declines from a maximum value, ice begins to accumulate at some point. The *rate* at which ice volume increases reaches a maximum when summer insolation falls to its lowest value, because the rate of ablation is slowest at that time. As summer insolation begins to increase, the ice sheet continues to grow because insolation values are still low (although increasing). The ice sheet does not reach its maximum *size* until insolation is just reaching the values that will begin to cause net ice ablation. As a result, the maximum volume of the ice lags thousands of years behind the curve of the summer insolation that drives its growth.

As summer insolation continues to increase, the rate of ablation speeds up (Figure 10-8, bottom). When

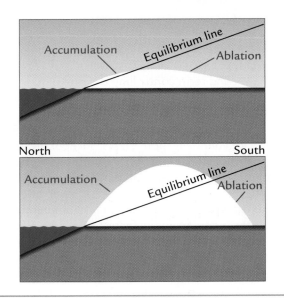

FIGURE 10-7 **Ice elevation feedback** As ice sheets grow higher, more of their surface lies above the equilibrium line in a regime of net accumulation.

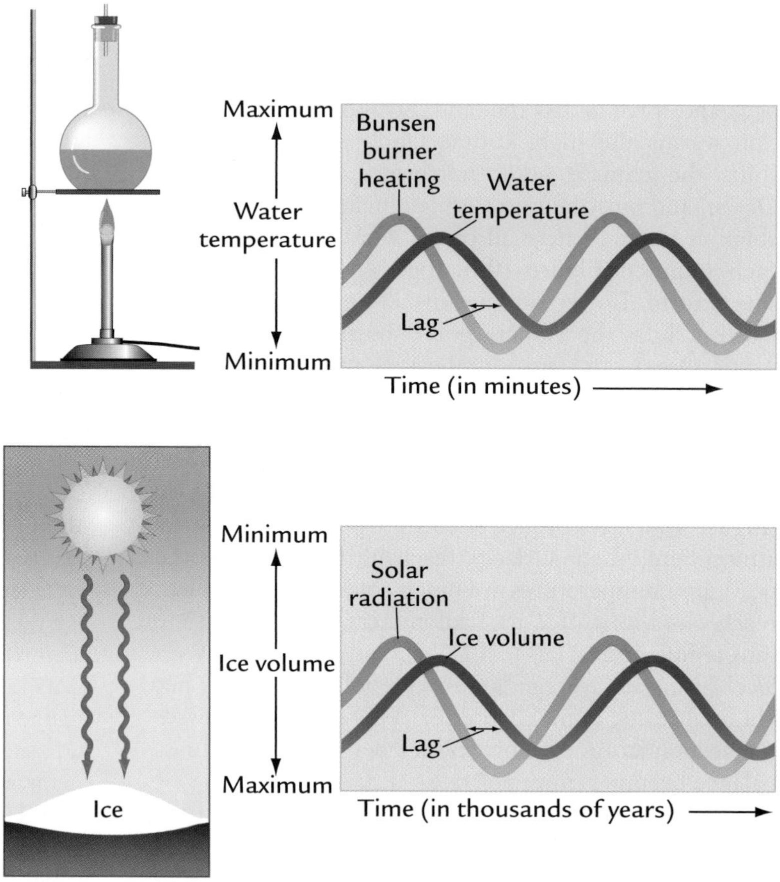

FIGURE 10-8 **Ice volume lags insolation** As the flame of a Bunsen burner is alternately turned higher and lower, the water heats and cools, but with a short time lag behind the changes in heating (top). Similarly, long-term increases and decreases in summer insolation cause ice sheets to melt and grow, but with lags measured in thousands of years (bottom). (Adapted from J. Imbrie, "A Theoretical Framework for the Pleistocene Ice Ages," *Journal of the Geological Society* (London) 142 [1985]: 417–32.)

BOX 10-1　LOOKING DEEPER INTO CLIMATE SCIENCE

Ice Volume Response to Insolation

The dependence of ice volume on summer insolation can be expressed by this equation:

$$\frac{d(I)}{d(t)} = \frac{1}{T}(S - I)$$

where I is ice volume

$\frac{d(I)}{d(t)}$ is the rate of change of ice volume per unit of time (t)

T is the response time of the ice sheet

S is the curve of changing summer insolation

This equation specifies that the rate of change of ice volume with time is a function of just two factors. One factor is T, the response time of ice sheets. Ice sheets are inherently sluggish and have long response times of at least 15,000 years. The equation specifies that the larger the time constant T of the ice sheet response, the slower the resulting rate of ice volume change, because it depends on the inverse value of T, or $1/T$.

The second term controlling the rate of ice volume change is $(S - I)$. This term in the equation is a measure of the degree of disequilibrium (or offset) between the summer insolation forcing (S) and the ice volume response (I). Conceptually, this disequilibrium can be thought of in this way: the ice volume response (I) is constantly chasing after the insolation forcing (S), but it never catches up. For example, when insolation is at a minimum, ice volume is growing at its fastest rate in an effort to catch up and reach its maximum value. But when ice volume finally does reach maximum size, the insolation curve has already turned and risen halfway toward its next maximum, causing the ice volume curve to reverse direction and decrease.

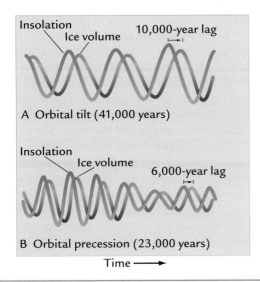

FIGURE 10-9 **Ice volume lags tilt and precession** At the 41,000-year cycle of orbital tilt, ice sheets lag roughly one-quarter wavelength (almost 10,000 years) behind changes in summer insolation (A). At the 23,000-year cycle of orbital precession, ice sheets lag roughly one-quarter wavelength (6,000 years) behind changes in summer insolation, and they also show the same modulation of amplitude (B).

insolation reaches its maximum value, the *rate* of ablation also reaches its highest value. Again, however, the minimum ice volume does not occur at the insolation maximum, but thousands of years later, about halfway through the drop in insolation. This relationship can be described by a simple equation (Box 10-1).

The persistent delay in ice volume response relative to the summer insolation curve shown in Figure 10-8 (bottom) is formally referred to as the **phase lag** between the two cycles. Remember that the period of a cycle is the interval of time separating successive peaks or successive valleys. The phase lag of ice volume behind summer insolation shown in Figure 10-8 (bottom) represents one-quarter of the full cycle length, or period.

This relationship also applies to the separate ice volume responses driven by the individual insolation cycles of orbital tilt (at 41,000 years) and precession (at 23,000 years). For the part of the insolation signal related to changes in orbital tilt at a cycle of 41,000 years, the ice volume response has the same regular sine-wave shape as the summer insolation signal, but it lags behind by about 8000 years, a little less than one-quarter of the 41,000-year wavelength (Figure 10-9A). For the part of the insolation signal tied to the 23,000-year orbital precession cycle, the ice-volume response lags behind by 5000 to 6000 years (again almost one-quarter of the wavelength), and it has the same amplitude modulation found in the precession signal (Figure 10-9B).

10-4 Delayed Bedrock Response beneath Ice Sheets

As ice sheets grow, the pressure of their weight on the underlying bedrock becomes significant. Although the density of solid ice (just less than $1 g/cm^3$) is much lower than that of the underlying rock (averaging 3.3 g/cm^3), ice sheets can reach thicknesses of 3000 meters or more, roughly equivalent to the weight of 1000 meters of solid rock. This ice load is enough to depress the underlying bedrock far beneath its original level.

One way to visualize this process is a thought experiment in which we instantaneously load an ice sheet 3.3 kilometers thick onto bedrock (Figure 10-10A). Although this could not actually happen instantaneously in nature,

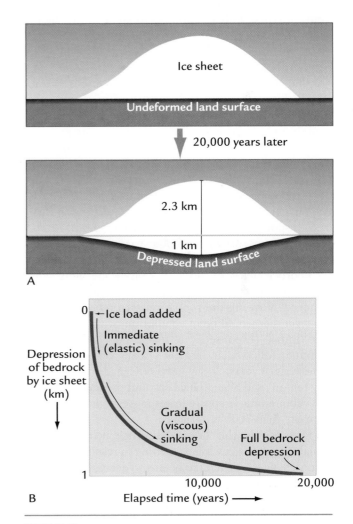

FIGURE 10-10 **Bedrock sinking** If an ice sheet 3.3 km thick were suddenly placed on the land, in time the bedrock would sink almost 1 km under the load (A). The initial sinking would be elastic and immediate, but the bedrock would later respond more slowly and in a viscous way, sinking half of the remaining distance toward eventual equilibrium every 3000 years (B).

the example helps to clarify the processes at work. Given enough time for the process to reach equilibrium, a 3.3-kilometer ice sheet would eventually depress the underlying bedrock by 1 kilometer. To place the importance of this vertical change in a climatic context, 1 kilometer of elevation change is equivalent to a 6.5°C change in temperature at Earth's prevailing lapse rate. For this reason, changes in bedrock elevation can have large effects on the temperature and mass balance at the surface of the overlying ice sheet.

Bedrock responds to a heavy ice load in two phases (Figure 10-10B). Its immediate reaction, to sag beneath the weight of the ice, is called the **elastic response** and represents about 30% of the total response. This rapid elastic depression of the bedrock is then followed by a phase of slower bedrock response spread over thousands of years. This **viscous response** is caused by the extremely slow flow of rock in the softer layer of the upper mantle (the asthenosphere; see Chapter 5) between 100 and 350 kilometers deep.

The viscous part of the bedrock response is characterized by a relatively fast initial rate of change, followed by a progressively slower approach toward final equilibrium. This viscous behavior has a response time (see Chapter 1) of about 3000 years: about half of the remaining response needed to reach a final equilibrium value occurs every 3000 years. The rate of change of the curve gradually slows because each successive 3000-year response time eliminates half of the remaining (unrealized) response (1 > 1/2 > 1/4 > 1/8, and so on). After 5 response times, or about 15,000 years, only a tiny fraction of the total bedrock depression remains unrealized.

Bedrock behavior works in exactly the same sense (but in the opposite direction) if the ice load is abruptly removed, allowing the rock surface to rebound toward the level that would be in equilibrium with no ice load. The initial response is a quick but partial elastic rebound, followed by a slow viscous rebound lasting thousands of years. Today parts of Canada (the Hudson Bay region) and Scandinavia (around the Baltic Sea) are still rebounding viscously in response to the melting of ice more than 7000 years ago.

How would this delayed bedrock behavior affect actual ice sheets growing or melting in response to insolation changes? When ice begins to accumulate on the land, the elastic part of the bedrock behavior immediately depresses the land and the ice sheet along with it. As the ice surface is shifted to lower elevations, it remains in a warmer climatic regime with faster ablation that tends to oppose the net accumulation of snow.

But the slow viscous response forms a much larger component of the bedrock's total behavior than the elastic response and it produces exactly the opposite feedback (Figure 10-11, top). The delay in bedrock sagging keeps the growing ice sheet at higher elevations,

where temperatures are colder, ablation is slower, and the ice mass balance is more positive. As a result, the effect of delayed bedrock sinking provides positive feedback to the growing ice sheet. Because the viscous behavior of bedrock exceeds the elastic response, the overall bedrock feedback to ice sheets is positive.

Bedrock also plays the same positive feedback role during times when the ice is melting (Figure 10-11, bottom). The weight of a large ice sheet that has existed for thousands of years creates a deep depression in the bedrock beneath it. As the ice begins to melt, the elastic part of the rebound quickly lifts the bedrock and eliminates a small part of this depression. But the much larger viscous part of the rebound leaves much of the depression intact for thousands of years. The ice sheet

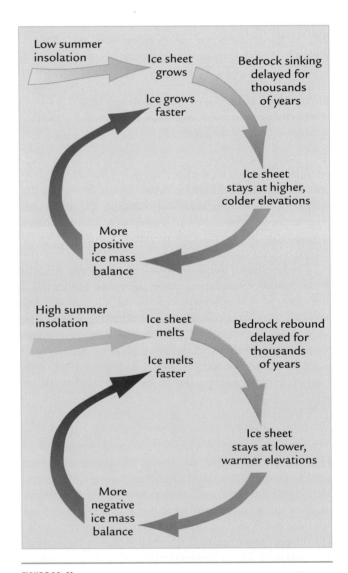

FIGURE 10-11 **Bedrock feedback to ice growth and melting**
Delayed bedrock sinking during ice accumulation (A) and delayed rebound during ice melting (B) provide positive feedback to the growth and decay of ice sheets.

is left sitting at lower elevations in the depression it created and in relatively warm air that helps it to melt.

10-5 Full Cycle of Ice Growth and Decay

We have examined three factors that control ice sheets: insolation control of ice sheet size, the initial lag of ice volume behind insolation, and the subsequent lag of bedrock depression and rebound behind ice loading and unloading. Now we explore the interactions of all three factors during a typical cycle of ice sheet growth and decay. Because of the long lags inherent in these responses, these factors interact in an intricate way to create and destroy ice sheets.

We start with an interglacial maximum such as today's, with the climate point *P* located in the Arctic Ocean and no ice sheet present on the northern continent (Figure 10-12A). As summer insolation begins to decrease from a previous maximum, the equilibrium line shifts south and the climate point gradually moves onto the land. Some snow survives summer ablation in the far north, and an ice sheet slowly begins to form (Figure 10-12B).

As the ice sheet thickens, it gradually grows to higher, colder elevations where accumulation dominates over ablation (Figure 10-12C). The ice also advances toward the south, partly because the equilibrium line is moving south and partly because of internal flow from the area of ice accumulation in the north to the area of ablation in the south.

The thickening ice sheet also slowly begins to weigh down the bedrock, but most of this bedrock depression lags several thousand years behind the accumulation of ice. This ongoing delay in the sagging of bedrock helps to keep the surface of the ice sheet at higher, colder elevations above the equilibrium line at altitudes where accumulation prevails over ablation.

The highest rate of ice accumulation occurs when summer insolation reaches a minimum value, so that the equilibrium line is displaced farthest south (Figure 10-12C). At this point the ice sheet has still not reached maximum size because of the lag of ice volume behind insolation. In addition, the rapid growth of new ice continues to weigh down the bedrock even more, again with a lag of thousands of years for each new increment of ice.

Summer insolation then begins to increase and shift the equilibrium line slowly back to the north, but for several thousand years afterward the ice sheet continues to grow to its maximum size (Figure 10-12D). Ice

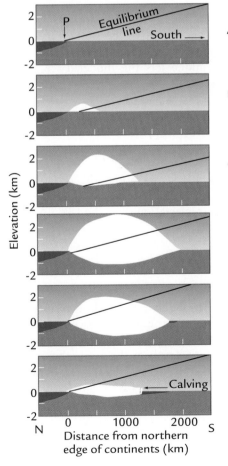

A No ice sheet (interglacial)

B Insolation drops, equilibrium line shifts south, ice sheet starts to grow

C Insolation at a minimum, ice sheet grows rapidly, ice depresses bedrock

D Insolation rises, equilibrium line moves north, ice sheet at maximum size, bedrock depression increases

E Insolation at a maximum, equilibrium line far to north, ice melts rapidly, bedrock starts to rise

F Insolation starts to drop, last ice remnants melt, bedrock rises rapidly

FIGURE 10-12 Cycle of ice sheet growth and decay A full cycle of ice sheet growth and decay incorporates both the delayed response of the ice sheets to summer insolation forcing as well as the delayed response of bedrock to the ice sheet load. (Adapted from J. Oerlemans, "The Role of Ice Sheets in the Pleistocene Climate," *Norsk Geologisk Tidsskrift* 71 [1991]: 155–61.)

growth continues because insolation levels are still relatively low and because most of the ice sheet's surface has by this point risen well above the equilibrium line, where it is initially protected from the slowly increasing levels of ablation. The last stages of ice growth are also helped by the slow response of the bedrock, which still remains at elevations slightly higher than those to which it will finally sink, keeping much of the surface of the ice sheet high and cold.

At some point the combined effects of the continued northward shift of the equilibrium line, along with the eventually large amount of bedrock depression, bring a larger area of the southern end of the ice sheet below the equilibrium line (Figure 10-12E). With enough of the ice sheet now undergoing ablation, the overall mass balance of the ice sheet turns negative, and ice volume begins to decrease. As the surface of the southern edge of the ice sheet sinks, more ice is exposed to warmer air and ablation accelerates. Melting is also aided by the delayed rebound of bedrock from the weight of the earlier (larger) ice sheet.

In effect, the southern edge of the ice sheet is now beginning to sag back entirely into the bedrock hole it has created, and the air temperatures in this hole are relatively warm. Melting continues because more of the ice sheet lies below the equilibrium line. Melting is also aided by the flow of ice from regions of higher elevation in the north to regions of lower elevation along the southern margin where ablation is rapid.

Eventually rising summer insolation drives the equilibrium line far enough to the north to move the climate point off the land and back over the Arctic Ocean (Figure 10-12F). Most of the remaining ice now lies in an ablation regime, and the last remnants may disappear several thousand years later. But if a small amount of ice survives intact through a summer insolation maximum, it will serve as a natural nucleus from which a larger ice sheet can quickly grow.

10-6 Ice Slipping and Calving

Ice can be transferred within the body of ice sheets by slow flow from colder, higher regions of net accumulation to lower, warmer regions of net ablation (see Chapter 2). This flow is usually modeled in a simplified way in the 2-D ice models just discussed by allowing a slow diffusion of ice from higher to lower elevations. Several other types of ice behavior are often omitted from these models because they are inherently less predictable.

Another important process of ice transfer is **basal slip** of ice sheets. Slipping occurs because meltwater present at the base of the ice sheet saturates soft sediments and creates a lubricated layer across which ice can slide. This process is usually not included in models

because of the difficulty of predicting when and where it will occur. We will see in Chapter 12 that basal sliding may help to determine the maximum size ice sheets can reach.

Iceberg calving along the ocean margins of ice sheets is another unpredictable process. Real-world ice sheets such as those that existed 20,000 years ago in North America and Scandinavia had margins that bordered on the ocean (see Figure 10-4), and they lost a substantial fraction of their total mass by calving icebergs to the ocean. This loss is also difficult to quantify accurately in models. We will see in Part IV that this process occurs in brief spurts separated by several thousand years, and for reasons that are extremely difficult to predict.

More complex models of ice sheet evolution can be constructed. One method is to couple the kind of two-dimensional (altitude/latitude) ice sheet model shown in Figure 10-5 to a simplified 2-D physical model of the atmosphere (Chapter 3). The goal is to simulate at least some of the actual changes in the atmosphere/ocean system caused by the presence of ice sheets.

Ice sheet models can also be coupled to 3-D GCM models by using a two-step approach. First, ice sheets of a specified size are used to drive GCMs for simulations of 20 years to produce climate output data (temperature and precipitation) near the ice sheets. Then these climatic data are combined with the slowly changing orbital insolation values to drive the simpler ice sheet model for simulations of a few hundreds or thousands of years. This technique lets the modeled ice sheets change in response to the new climatic configuration.

Northern Hemisphere Ice Sheet History

What we have learned so far about the orbital-scale behavior of ice sheets leads us to expect the history of northern hemisphere glaciation over the last 3 Myr to show a strong orbital influence. Glaciation cycles lasting tens of thousands of years should begin, intensify, wane, and end in response to orbitally driven changes in summer insolation.

We might also anticipate that a second factor would also be evident in the longer-term ice sheet history. We learned in Chapter 7 that Earth has been gradually cooling for at least 55 million years, probably because of a slow decrease in CO_2 levels in the atmosphere. This slow cooling appears to have caused the onset of northern hemisphere glaciation after 3 Myr ago, but it could also have gradually changed the character of the glacial cycles over that interval.

To explore these ideas, we first examine a conceptual model of how glacial cycles may have evolved during the last 3 Myr as global cooling intensified. Then we investigate evidence showing how the glacial cycles actually developed.

10-7 Conceptual Model: Evolution of Ice Sheet Cycles

The conceptual model of glaciation history is based on two major features (Figure 10-13). One is a curve of summer insolation values at high northern latitudes. This curve incorporates the combined influence of cycles of 41,000 years (tilt) and 23,000 years (precession). Changes in orbital eccentricity at periods of 100,000 and 413,000 years do not appear as cycles of insolation, but they do modulate the amplitude of the 23,000-year cycle (Chapter 8).

Insolation has varied around a constant long-term mean and has followed the same tilt and precession cycles for millions of years. Consequently, its fundamental character has stayed the same. As a result, we can use the same insolation curve throughout the full 3-Myr history of northern hemisphere glaciation illustrated by this conceptual model.

The second factor included in the conceptual model is a threshold temperature value at high latitudes where ice sheets develop (Figure 10-13). This critical threshold is represented by a line that separates temperatures cold enough to permit ice accumulation (glaciation) from those warm enough to cause snow to melt and prevent ice from forming. In effect, this line functions as an equilibrium line through time. This threshold lies

at summer temperatures near 0°C and mean annual temperatures near −10°C (see Figure 10-1).

The ice sheet response over any interval of time results from the interaction between the slowly changing equilibrium-line threshold and the more rapidly changing curve of summer insolation. When insolation values fall below the equilibrium-line threshold, ice sheets can grow. When they rise above it, ice sheets will melt. Both the growth and melting lag thousands of years behind the insolation forcing.

We examine four intervals in the development of northern hemisphere glaciation (Figure 10-14). In the first interval, the preglaciation phase, the equilibrium-line threshold lies near the conditions necessary for glaciation to develop, but even the deepest summer insolation minima failed to reach the critical threshold (Figure 10-14A). As a result, high latitudes remain too warm for ice sheets to form.

By the time of the second, small glaciation phase, additional global cooling has altered the position of the equilibrium-line threshold so that it now begins to interact with the summer insolation curve. At major summer insolation minima separated by intervals of 41,000 (or 23,000) years, small ice sheets begin to grow when summer insolation drops below the equilibrium line. With insolation minima at the 41,000-year cycle lasting about twice as long as those at the 23,000-year cycle, ice sheets have more time to grow at the tilt cycle. The growth of ice sheets at both cycles lags behind the insolation signals by several thousand years. The ice sheets persist until the insolation curve again rises above the equilibrium line, and then all the ice melts over an interval of several thousand years.

As slow global cooling continues and the equilibrium line again shifts, more of the insolation curve lies in a regime of ice accumulation rather than ice melting (Figure 10-14C). At some point, even some of the smaller insolation maxima remain in the regime of ice accumulation. As a result, ice sheets do not disappear during these minor insolation maxima but persist until a stronger insolation maximum is reached. Because the ice sheets now have a longer interval in which to grow, they will also become larger. This is the large glaciation phase. During this phase, ice sheets should persist longer than the 23,000-year and 41,000-year cycles of insolation, but it is not at this point obvious how long they would last.

The last conceptual phase is one in which further cooling causes the equilibrium line to lie completely above the range of the summer insolation curve (Figure 10-14D). For this phase, all points on the insolation curve fall within the regime of positive ice mass balance, and even the strongest insolation maxima fail to reach the ablation regime. As a result, permanent ice sheets remain on the continents. This is the permanent glaciation phase: ice sheets never disappear.

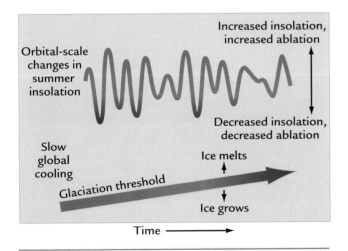

FIGURE 10-13 **Factors in long-term evolution of ice sheets** The long-term evolution of ice sheets reflects the interaction of two factors: changes in summer insolation that drive shorter-term changes in ice sheet mass balance (top) and a much more gradual global cooling represented by a slowly changing glaciation threshold (bottom). Ice sheets accumulate when summer insolation falls below a critical glaciation threshold and melt when it rises above it.

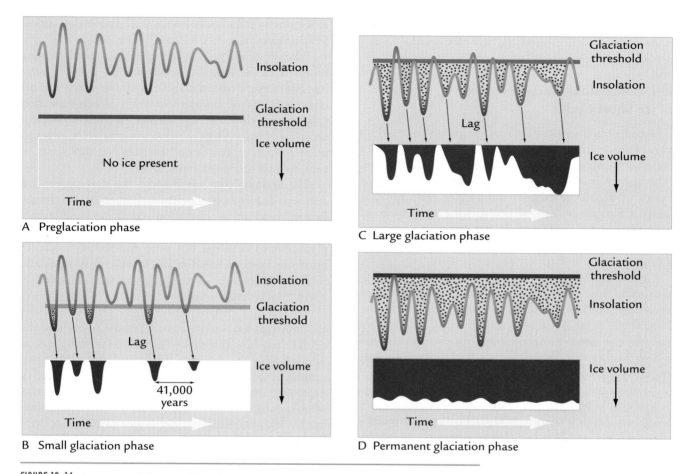

FIGURE 10-14 Conceptual phases of ice sheet evolution Four phases should occur in the long-term evolution of ice sheets: (A) a preglaciation phase, when no ice can accumulate; (B) a small glaciation phase, when ice accumulates during individual summer insolation minima but then melts entirely during the next insolation maximum; (C) a large glaciation phase, when ice persists through weaker summer insolation maxima; and (D) a permanent glaciation phase, when ice persists through all insolation maxima.

10-8 Evidence from δ¹⁸O: How Ice Sheets Actually Evolved

What does the geologic record tell us about the actual history of ice sheets? At first thought, it might seem that the best record of glaciation would come from those parts of the continents where sheets existed. Ice sheets erode underlying sediments and bedrock and then deposit long, winding moraine ridges that contain unsorted sediment called till (Chapter 3). Unfortunately, these deposits are of little use in compiling the longer-term history of glaciation. The main problem is that each successive glaciation erodes and destroys most of the sediment from the previous glaciations, erasing much of the evidence needed to compile a complete history. The few undisturbed deposits that remain from older glaciations are usually difficult to date because they represent brief fragments of time beyond the reach

of radiocarbon dating. The ages of some older deposits can be estimated by dating ash layers or determining magnetic polarity in the intervening layers of sediment, but these opportunities are rare.

In fact, the best records of glacial history come from the ocean, where the deposition of sediment is generally uninterrupted. Ocean sediments contain two key indicators of past glaciations: (1) ice-rafted debris, a mixture of coarse and fine sediments delivered to the ocean by melting icebergs that calve off from ice sheet margins, and (2) δ¹⁸O records from the shells of foraminifera, which provide a quantitative measure of the combined effects of changes in ice volume and in the temperature of the water in which the foraminifera grew (Box 7-1). Both kinds of records accumulate layer by layer on the seafloor and record long-term glacial history. Marine geologists retrieve sediment cores containing these records from the ocean and date them by defining the pattern of the magnetic reversals contained in the sediments.

The marine scientists Cesare Emiliani and Nick Shackleton pioneered the use of oxygen-isotope ratios in marine foraminifera to study past climates. In the 1950s and 1960s Emiliani analyzed the first $\delta^{18}O$ records extending back as far as a few hundred thousand years and interpreted them as mainly a record of past temperature changes. In the late 1960s Shackleton found that $\delta^{18}O$ signals over that interval are instead an indicator mostly of changing global ice volume, with a smaller overprint caused by temperature changes. Using micromass spectrometers, Shackleton was able to analyze small numbers of foraminifera in longer core records and publish detailed $\delta^{18}O$ histories extending far into the past. In 1976 James Hays and John Imbrie joined Shackleton in writing the first paper that conclusively linked changes in $\delta^{18}O$ (ice volume) to changes in orbital insolation.

The first continuous $\delta^{18}O$ record of the entire 2.75-Myr glacial history of the northern hemisphere was compiled in the late 1980s by analysis of the shells of benthic foraminifera that once lived on the floor of the North Atlantic (Figure 10-15). The core from which this $\delta^{18}O$ record was taken also contains ice-rafted debris that directly marks those times when ice sheets were present on adjacent continents.

The full 2.75-Myr $\delta^{18}O$ record shows two kinds of variations: (1) numerous cyclic-looking oscillations from positive to negative $\delta^{18}O$ values and (2) a gradual drift toward more positive values. Both of these features in the $\delta^{18}O$ record incorporate the combined effects of changes in temperature and ice volume. Changes toward more positive $\delta^{18}O$ values represent some combination of more ice on the land and a colder deep ocean. More negative $\delta^{18}O$ values indicate less ice and warmer temperatures. Because a cooling of the deep ocean is likely to be tied to the high-latitude cooling that causes ice sheets to grow, we can regard this $\delta^{18}O$ curve as a record of a net glacial cooling effect.

Before 2.75 Myr ago, the $\delta^{18}O$ values were relatively negative (< 3.5‰) and no ice-rafted debris was present. During this interval, either northern hemisphere ice sheets did not exist or they never reached the size needed to send icebergs into the central North Atlantic. The short-term variations in $\delta^{18}O$ values were small and probably reflected minor temperature changes in the deep waters. This was the preglaciation phase for the northern hemisphere.

Beginning at 2.75 Myr ago, significant amounts of ice-rafted debris appear for the first time in the record, indicating that ice sheets were now at least sporadically present. This debris accumulates during intervals of positive $\delta^{18}O$ values, which come and go mainly at a cycle near 41,000 years (Figure 10-15). This part of the record suggests that ice sheets were now forming as

some snow and ice survived during intervals of low summer insolation, but then probably disappeared each time ablation increased during the subsequent summer insolation maximum. This pattern is very much like the response expected during a small glaciation phase. This phase persisted from 2.75 to 0.9 Myr ago, or the first two-thirds of the full interval of northern hemisphere glaciation. Depending on how finely you choose to subdivide this curve, almost fifty discrete $\delta^{18}O$ cycles can be detected during this interval. In addition, the $\delta^{18}O$ signal drifts gradually and irregularly toward more positive values, indicating an underlying shift toward a colder, more glacial world.

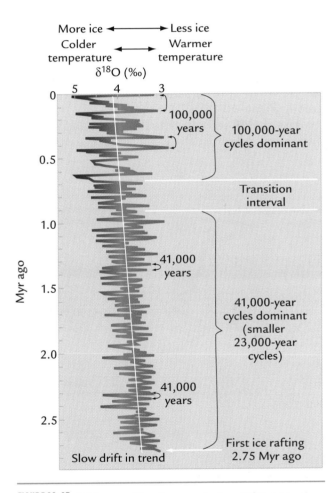

FIGURE 10-15 Evidence of ice sheet evolution: $\delta^{18}O$ A North Atlantic Ocean sediment core holds a 3-Myr $\delta^{18}O$ record of ice volume and deep-water temperature change. After a preglaciation phase with no major ice sheets before 2.75 Myr ago, small ice sheets grew and melted at cycles of 41,000 and 23,000 years until 0.9 Myr ago, and then large ice sheets grew and melted at a cycle of 100,000 years. The diagonal white line shows a gradual long-term $\delta^{18}O$ trend toward more ice and colder temperature. (Adapted from M. E. Raymo, "The Initiation of Northern Hemisphere Glaciation," *Annual Reviews of Earth and Planetary Sciences* 22 [1994]: 353–83.)

Beginning near 0.9 Myr ago and becoming more obvious by 0.6 Myr ago, the fundamental character of the $\delta^{18}O$ record changes (Figure 10-15). Maximum $\delta^{18}O$ values not only increase but also appear to be spaced farther apart, indicating that ice sheets persisted for longer intervals of time and grew larger. These characteristics of the more recent $\delta^{18}O$ record are similar to those anticipated in the large glaciation phase. Rapid $\delta^{18}O$ decreases indicate abrupt episodes of melting of these large ice sheets. Over the last 0.6 Myr, there have been six large $\delta^{18}O$ maxima followed by abrupt deglaciations, spaced at average intervals of 100,000 years.

Although it is hard to see in the highly compressed record plotted in Figure 10-15, the smaller 41,000-year and 23,000-year oscillations do not disappear during the last 0.9 Myr, but persist as secondary oscillations largely hidden underneath the new 100,000-year cycle. To get a better idea of the character of all cycles present, we can zoom in on the most recent 150,000 years of record (Figure 10-16).

This shorter record begins during the late stages of a large-scale glaciation. The end of that glaciation near 130,000 years ago is marked by an abrupt $\delta^{18}O$ decrease. The interval from roughly 130,000 to 120,000 years ago is similar to today, with negative $\delta^{18}O$ values and no ice-rafted debris present. This 10,000-year interval represents peak interglacial conditions, with no ice sheets present in the northern hemisphere except perhaps on Greenland.

Between 125,000 and 80,000 years ago, the $\delta^{18}O$ signal oscillates between heavier values indicative of small accumulations of ice (and confirmed by the presence of ice-rafted debris) and lighter values indicating smaller ice sheets. These three oscillations occur at roughly a 23,000-year spacing, confirming that the orbital precession signal persists into this interval.

Two additional $\delta^{18}O$ maxima occur near 20,000 and 60,000 years ago. The spacing of about 40,000 years between these peaks indicates that the orbital tilt signal also persists into the recent part of the $\delta^{18}O$ record. The $\delta^{18}O$ maximum 20,000 years ago is known as the last glacial maximum, which we will examine closely in Part IV.

The $\delta^{18}O$ values decrease rapidly between 15,000 and 10,000 years ago, indicating another abrupt deglaciation like the earlier one 130,000 years ago. These deglaciations that suddenly end the long (100,000-year) cycles of glacial buildup are called **terminations**. This most recent termination led to the warm interglacial climate of today.

Milankovitch, who died in 1958, did not live to see these $\delta^{18}O$ records published. During his lifetime, the only records available were those from glacial moraines on land. Most of the moraines that exist today in the high latitudes of the northern hemisphere were deposited either during the most recent glaciation, 20,000 years ago, or in the subsequent phase of ice retreat from that maximum. Almost all older moraines from previous glaciations had been eroded by repeated ice advances earlier.

As a result, glacial geologists of Milankovitch's time thought there had been at most 4 or 5 glaciations, and this view persisted well into the 1960s. Yet the first complete 3-Myr marine $\delta^{18}O$ records revealed more than fifty glacial maxima, depending on how many of the separate $\delta^{18}O$ peaks are considered worth counting.

What would Milankovitch have thought of the record shown in Figure 10-15? He probably would not have been surprised and indeed may well have been delighted with the first two-thirds of the record, the interval between 2.75 and 0.9 Myr. The existence of discrete glacial intervals, each lagging just behind individual summer insolation minima and ending during the subsequent summer insolation maximum, is essentially what the Milankovitch theory predicted (equivalent to the small glaciation phase discussed earlier).

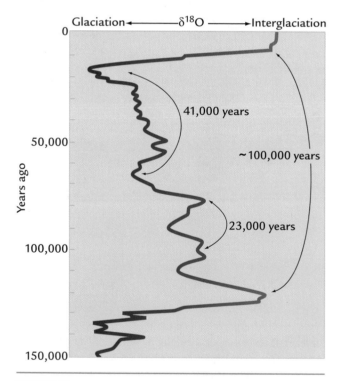

FIGURE 10-16 Ice sheet $\delta^{18}O$ changes over the last 150,000 years A close look at the last 150,000 years of the $\delta^{18}O$ record shows 23,000-year and 41,000-year oscillations almost hidden by the dominant 100,000-year cycle. (Adapted from D. Martinson et al., "Age Dating and the Orbital Theory of the Ice Ages: Development of a High-Resolution 0 to 300,000-Year Chronostratigraphy," *Quaternary Research* 27 [1987]: 1–29.)

But the most recent third of this record, extending from 0.9 Myr ago through the present day, would almost certainly have been a surprise. Although he did anticipate the 23,000-year and 41,000-year glacial cycles, Milankovitch did not anticipate the dominant 100,000-year cycle. A second surprise would have been the sawtoothed character of these 100,000-year cycles, with a slow oscillatory buildup of large ice sheets over most of the cycle followed by rapid deglaciations in about 10,000 years (Figure 10-16).

We can use the method of power-spectra analysis introduced in Chapter 8 to demonstrate why this result is so surprising (Figure 10-17). A power-spectrum plot shows that summer insolation at 65°N varies entirely at the periods of tilt (41,000 years) and precession (mainly at 23,000 but also at 19,000 years). Changes in orbital eccentricity do not show up because they act only as a multiplier effect on the size of the precession changes: they are not a true cycle.

If we analyze the record of $\delta^{18}O$ changes from 2.75 to 0.9 Myr ago to quantify the cyclic changes in ice volume (and temperature), we find similarities to and differences from the insolation signal. The same orbital cycles of tilt and precession are present, but the 41,000-year cycle is larger in the $\delta^{18}O$ signal than in the insolation signal, and the precession peaks are smaller. This difference may be explained by the fact that the greater length of the 41,000-year insolation cycles gives ice sheets more time to grow than does the shorter length of cycles of precession.

Over the last 0.9 Myr, the $\delta^{18}O$ signal continues to show comparably strong cycles of tilt and precession. But a new cycle centered at 100,000 years now emerges as the strongest ice volume response, even though summer insolation changes have no 100,000-year cycle. Something more complicated must be happening to make ice sheets grow and melt at a 100,000-year rhythm during this entire interval. We will return to this problem in Chapter 12.

Note that the record of northern hemisphere ice volume does not reach a permanent glaciation phase like that shown in Figure 10-14D. Global climate has not yet grown cold enough to create ice sheets that can survive intact over North America, Europe, or Asia through the strongest peaks of summer insolation. But the $\delta^{18}O$ signal in Figure 10-15 does show a slow, long-term trend toward more positive values, and also toward successively larger glacial maxima. An interval of permanent glaciation may still lie in the distant future for North America and Eurasia. For Greenland and Antarctica (Box 10-2), such an interval has long since arrived.

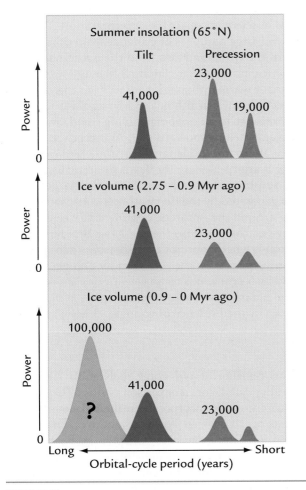

FIGURE 10-17 Spectral analysis: insolation and ice volume
Summer insolation changes at 65°N occur only at the orbital rhythms of tilt (41,000 years) and precession (23,000 and 19,000 years) (top). Between 2.75 and 0.9 Myr ago, the $\delta^{18}O$ signal (northern hemisphere ice volume and deep-ocean temperature) contains the same rhythms as orbital insolation (middle). The large 100,000-year cycle evident in the $\delta^{18}O$ (ice volume) signal since 0.9 Myr ago is not present in the insolation signal (bottom).

10-9 Confirming Ice Volume Changes: Coral Reefs and Sea Level

The combination of $\delta^{18}O$ signals and the presence of ice-rafted debris gives scientists a clear sense of when ice sheets were present through the last 3 Myr, as well as a good approximation of changes in their relative size, but neither signal quantifies the amount of ice in a precise way. We still need some kind of independent confirmation that the $\delta^{18}O$ signal is a valid representation of ice volume.

Once again, the oceans provide such a signal, this time in the form of the fossil remains of coral reefs. Today most coral reefs grow in warm tropical ocean water, and they prefer the clear water near small islands

BOX 10-2 CLIMATE DEBATE

Antarctic Deglaciation 3 Myr Ago?

Ice appeared on Antarctica by 35 Myr ago, and the Antarctic ice sheet has apparently become a larger, more permanent feature since that time. Today, even at a point in the orbital-scale climate cycles when most high-latitude regions in both hemispheres are in a brief interlude of interglacial warmth, the Antarctic continent is still stuck in a condition of permanent glaciation. Apparently Antarctica is simply too cold to be much affected by orbital-scale changes in solar insolation.

Some scientists believe that Antarctica was not in this deep-frozen state just 3 Myr ago. They base their view on the fact that glacial moraines perched high on the mountains of the Transantarctic Range at an elevation of 2.5 km contain small amounts of fossil shells of marine diatoms mixed in with a jumble of unsorted glacial debris. The ages of these marine diatoms range as young as 3 Myr, implying that the glacial deposits in which they lie are also no older than 3 Myr. This evidence raises an intriguing question: How did marine diatoms formed at sea level 3 Myr ago come to rest in glacial deposits 2.5 km up a mountainside?

One view is that the diatoms were initially deposited in shallow marine bays on low-lying parts of the Antarctic continent during a time when global climate was much warmer, the Antarctic ice sheet was much smaller, and the melted ice raised global sea level high enough to flood low-lying parts of

Antarctica. Later, when climate cooled and Antarctica entered its current state of permanent glaciation, ice sheets formed on the locations of the previous embayments, scooped up the marine sediments containing diatom shells, and dumped them in the rubble found high on the Transantarctic Range. Because ice sheets cannot carry sediments far uphill, this explanation also requires extremely rapid and large-scale uplift of the Transantarctic Range in the last 3 Myr.

Other scientists reject this interpretation. They note that large amounts of evidence from other parts of the world and also from the ocean near Antarctica indicate that climate 3 Myr ago was only marginally warmer than it is today. From their perspective, any warmth near 3 Myr ago was just a minor reversal in an ongoing cooling trend, and not capable of melting much (if any) Antarctic ice.

The alternative explanation for the presence of marine diatoms high on the Transantarctic Range is wind. Shipboard observations reveal that large waves generated by the fierce storms constantly raging around the Antarctic continent can easily toss live diatoms up into the overlying air. The main question is whether or not these diatoms can remain suspended in the turbulent air and be carried by storm winds to an elevation of 2.5 km.

The plausibility of airborne transport of diatoms is shown by the presence of young ($<$3 Myr) marine

---

(Figure 10-18) rather than water muddied by river runoff on the margins of large continents. Coral reefs grow near sea level, and the species of coral most useful to climate scientists (such as *Acropora palmata*) grow only at sea level or just a few meters below. Reefs have strong structural frameworks that remain intact long after individual coral organisms have died, and so they preserve records of past sea level positions.

As sea level rises and falls through time, coral reefs slowly follow these changes by migrating upslope and downslope. Thus ancient reefs function as dipsticks that measure the past level of water in the world ocean. Over orbital cycles of tens to hundreds of thousands of years, fluctuations in sea level result mainly from changes in the amount of water extracted from the ocean and stored in ice sheets on land. As a result, the sea level history record-ed by the coral reefs is a direct record of ice volume.

The remains of old coral reefs must be radio-metrically dated for comparison with $\delta^{18}O$ signals. Fortunately, corals incorporate in their skeletons small amounts of ^{234}U, which slowly decays to ^{230}Th and can be used for dating purposes (Chapter 3). Because this dating technique is best suited for use during the last several hundred thousand years, we return to the last 150,000 years of $\delta^{18}O$ record (Figure 10-16) to see whether dated coral reefs confirm the observed $\delta^{18}O$ changes.

Many ocean islands situated in stable regions such as Bermuda, far from the disturbances of plate tectonic activity, have prominent fossil coral reefs that date to about 125,000 years ago and currently lie about 6 meters above present-day sea level. Reefs of this age are unique on tectonically stable islands for the interval of the last 150,000 years: no other reefs mark past sea lev-

BOX 10-2 CONT.   CLIMATE DEBATE

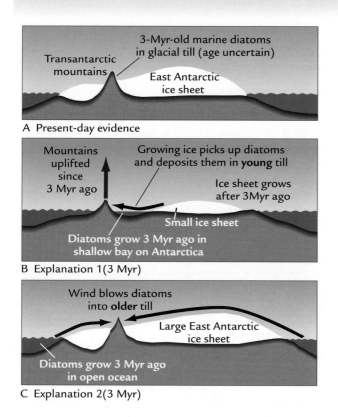

A Present-day evidence

B Explanation 1(3 Myr)

C Explanation 2(3 Myr)

Marine diatoms in Antarctic mountains Marine diatoms as young as 3 Myr are found today in glacial moraines high on the Transantarctic Mountain Range (A). One explanation is that the marine diatoms were deposited at sea level 3 Myr ago, when Antarctica was much warmer and the ice sheet was smaller, and then were picked up by the growing ice sheet and deposited on the rapidly uplifting mountains (B). The other explanation is that the diatoms were lifted from the ocean during storms and blown onto the mountains by strong winds (C).

diatoms on the surfaces of rocks known to be at least 200 Myr old. Because these diatoms clearly lived long after the rocks formed, they must have

been plastered onto the rock surfaces in the last 3 Myr by wind. If so, the glacial deposits high on the Transantarctic Range cannot be dated to 3 Myr simply because diatoms of that age are present: the diatoms could be later additions, and the glacial deposits could be much older. By this interpretation, scientists have no need to call on smaller ice sheets and marine embayments on Antarctica at 3 Myr, or on large-scale uplift of the Transantarctic Range since that time.

This debate continues. The main issue appears to be whether the marine diatoms are simply plastered onto the surfaces of the glacial tills by strong winds or are embedded deep in the tills by ice. The evidence is not yet good enough to resolve this issue definitively.

---

els above the modern sea level position. This first line of evidence exactly confirms the $\delta^{18}O$ record in Figure 10-16: the only $\delta^{18}O$ minimum comparable to the modern value within the last 150,000 years is the one dated in ocean cores to 125,000 years ago. Both types of evidence agree that only at this time was there as little ice as there is today (in fact, even less). Sea level would rise 6 meters if all present-day ice on Greenland melted, or 10% of the ice now on Antarctica, or some combination of the two.

This initial evidence supports the idea that the $\delta^{18}O$ signal can be used as an indicator of ice volume, but unfortunately no other coral reefs younger than 150,000 years are exposed on tectonically stable islands for comparison with other $\delta^{18}O$ oscillations. The problem is that more ice existed at *all other times* during the last 150,000 years than exists today, and any other coral

reefs that grew during this interval must have formed below modern-day sea level. If so, they are still submerged on the underwater slopes of these tectonically stable islands.

Instead, we can turn to ocean islands in a different tectonic setting: areas where tectonic uplift has raised older reefs that formed below modern-day sea level and exposed them above sea level (Figure 10-19). Two well-studied islands of this kind are Barbados, in the eastern Caribbean, and New Guinea, in the western Pacific. Both islands are rimmed by prominent coral reef terraces lifted out of the ocean and made easily available for scientific study.

One of the major reefs in New Guinea (Figure 10-20) dates from the last interglacial period, 125,000 years ago. Two other reefs date from about 104,000 and 82,000 years ago. The ages of the two younger reefs

FIGURE 10-18 **Coral reefs** Coral reefs form in clear, shallow waters in warm tropical seas. (Ian Cartwright/PhotoDisc.)

−110 meters near 20,000 years ago (Chapters 13 and 14). This low sea level correlates with the most prominent recent $\delta^{18}O$ maximum in Figure 10-16 and provides further confirmation that $\delta^{18}O$ is a reliable indicator of ice volume.

We can also use the −110-meter level at 20,000 years ago and the +6-meter level at 125,000 years ago as extreme values to anchor the full range of sea level change over the last 150,000 years. If we do this, we find that the −17-meter reefs at 82,000 and 104,000 years ago (Box 10-3) fall about 15% of the way between full interglacial and full glacial values. This percentage change is close to the relative change between the 125,000-year $\delta^{18}O$ minimum and the 20,000-year $\delta^{18}O$ maximum in

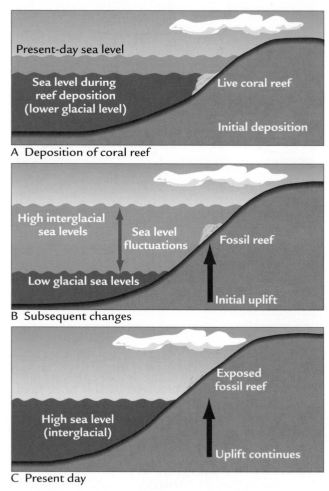

FIGURE 10-19 **Gradual uplift of coral reefs** Coral reefs may initially form on the edge of an uplifting island at times when sea level lies below its modern position (A). As time passes, uplift steadily raises the island and the fossil reef toward higher elevations, while sea level moves up and down against the island in response to changes in ice volume (B). Today old fossil reefs may lie well above sea level as a result of uplift (C).

match two prominent minima in the $\delta^{18}O$ signal in Figure 10-16, suggesting that they record high sea levels caused by ice melting that was less complete than at full interglacial periods, such as today or 125,000 years ago.

If these islands have been slowly rising out of the ocean because of tectonic activity, we need to devise a way of eliminating the overprint of uplift in order to reconstruct the level of the sea at the times the two reefs originally grew. The effects of uplift can be eliminated if we assume that the rate of uplift of each island has been constant over the interval of time we are examining. Using that assumption, scientists calculate that the 82,000-year reef (and the 104,000-year reef as well) formed when sea level was 15 to 20 meters below its modern position (Box 10-3).

Recent technological advances that permit drilling of deeply submerged reefs have shown that the lowest global sea level reached in the last 125,000 years was

FIGURE 10-20 **Uplifted coral reef terraces** Terraces formed by erosion-resistant coral reefs lie well above sea level on the island of New Guinea, in the Western Pacific. (Courtesy of Arthur Bloom, Cornell University.)

Figure 10-16. This agreement lends still more support to the idea that $\delta^{18}O$ is a good index of ice volume, despite the temperature effects known to be present in the $\delta^{18}O$ signal. Each 10-meter change in global sea level results in a roughly 0.1‰ change in $\delta^{18}O$.

In summary, coral reefs formed in the last 150,000 years confirm that the $\delta^{18}O$ signal is a reasonably faithful proxy of ice sheet size. The ages of the most prominent $\delta^{18}O$ minima correspond to the ages of prominent coral reefs that record high stands of sea level caused by reduced ice volume. And the absolute sea level values estimated from these reefs also correspond to the sizes projected from relative changes in the $\delta^{18}O$ signal.

In addition, the coral reefs confirm two important responses of ice sheets to orbital insolation noted earlier in this chapter. The approximate 22,000-year separation between the successive high sea levels dating to 125,000, 104,000, and 82,000 years ago is clearly a response to orbital forcing at the precession rhythm. In addition, each of these three sea level highs (ice volume lows) follows several thousand years behind the preceding peak in summer insolation, just as we would expect for a slow ice volume response. This persistent and dependable time relationship between insolation and ice volume is now the basis for a new method of dating any ocean-sediment core that contains a $\delta^{18}O$ signal.

10-10 Using Astronomical and $\delta^{18}O$ Signals as a Chronometer

The link between the summer insolation driving force and the ice-volume response has in recent years become the basis for a new way of dating marine sedimentary records known as **orbital tuning**. The method is based on the constant relationship in time between the astronomically determined insolation signal and the ice-volume response. We know the timing of the orbital insolation changes, and we can measure the ice-volume responses in ocean sediments. If the ice-volume response always lags behind the orbital forcing by a constant amount, we can date the climate records in ocean sediments *in relation to* the known timing of the orbital changes.

A clear demonstration of the validity of orbital tuning came from studies of ocean cores containing high-quality $\delta^{18}O$ signals along with records of reversals of Earth's magnetic field. Magnetic reversals occur irregularly over intervals separated by tens to hundreds of thousands of years, they can be dated by radiometric methods, and they provide initial constraints on the ages of sediment sequences (Chapter 3). When scientists first used the reversals to set an initial time scale for these cores, the observed oscillations in the $\delta^{18}O$ signals were found to fall at periods close to 20,000, 40,000, and (in the last 0.9 Myr only) 100,000 years.

This finding led climate scientists to make the bold assumption that orbital forcing at cycles of 41,000 and 23,000 years has produced such regular and dependable climatic responses that these two climate cycles can be used as chronometers independent of further radiometric dating. The way these chronometers work is already implicit in the relationships found in Figure 10-9: the portions of the $\delta^{18}O$ signals that change at cycles of 41,000 and 23,000 years have the same shape as the summer insolation signals at these cycles and lag behind them by a consistent length of time. Recall that this phase-lag relationship was also confirmed by the dating of coral reefs.

BOX 10-3 LOOKING DEEPER INTO CLIMATE SCIENCE

Sea Level on Uplifting Islands

Coral reefs on ocean islands that are slowly being lifted by tectonic processes can be used as sea level dipsticks after removal of the overprint caused by uplift. Ancient coral reefs sit today on the flanks of modern islands at elevations that are defined with reference to modern-day sea level. The present-day elevation of old coral reef terraces in relation to *modern* sea level is a result of two factors: (1) the amount of uplift of the island since the reef formed and (2) the position of sea level at the time the reef formed in relation to the modern position.

The amount of uplift of each reef will depend on its age (more uplift for older reefs) and the rate of uplift of the island on which it forms (more uplift for a reef of a given age on a fast-rising island than on a slow-rising island). Keep in mind that even though the present-day elevations of these older reefs are cited in relation to present-day sea level, modern sea level is just one position in a pattern of continual rises and falls through time.

First we calculate the average rate of uplift (U) for each island under study. To do this, we use the last interglacial reef on each island as a reference point because both its age (125,000 years) and the true level of the sea at that time (6 m higher than it is today) are well determined from tectonically stable islands.

$$U = \frac{(H_t - 6)}{125}$$

where U = the mean uplift rate in meters per 1000 years

H_t = the present-day elevation (in meters) of the 125,000-year reef on the island under study

This calculation subtracts the 6-meter difference in sea level between today and 125,000 years ago in order to isolate the effect of uplift. With the mean uplift rate determined, and assuming it has remained constant through the last 125,000 years, we can now correct for the amount of uplift that has occurred since a reef of any intermediate age formed in order to derive an estimate of sea level (S) at the time it formed.

$$S = h - (U)(t)$$

where S = relative sea level at the time the older reef formed (in meters)

h = the present-day elevation of the older reef (in meters)

t = the time elapsed since the reef formed (in thousands of years)

U = the mean uplift rate in meters per 1000 year

Using these equations, we can remove the different effects of uplift that have occurred on each island. For example, New Guinea is uplifting at a rate close to 2 meters per 1000 years, while Barbados is uplifting at a rate of just 0.3 meters per 1000 years. As a result, a coral reef dated to 82,000 years ago has been uplifted by 150 meters on New Guinea since 82,000 years ago, but by only 25 meters at Barbados. After correcting for the different local uplift effects, we find that the 82,000-year reef formed when global sea level was about 17 meters lower than it is today. The same kind of calculation shows that the 104,000-year reef formed at roughly the same level.

Because insolation forcing at the 100,000-year orbital eccentricity cycle is negligible, it cannot be assumed to drive the prominent 100,000-year cycle measured in $\delta^{18}O$ signals over the last 0.9 Myr. As a result, eccentricity changes cannot be used as a chronometer, but the 41,000-year tilt cycle and 23,000-year precession cycle are used in the tuning process.

Astronomical calculations provide the absolute time scale of orbital variations in tilt and precession reaching millions of years back into the past (Chapter 8). These variations form a tuning target, as shown on the left in Figure 10-21 (for the interval of the last 400,000 years).

The goal of the tuning exercise is to find the one time scale for the $\delta^{18}O$ signal on the right that best matches its component cycles to the orbital signals of tilt and precession. This comparison requires separating the orbital insolation and $\delta^{18}O$ signals into their 41,000-year (tilt) and 23,000-year (precession) components using the filtering technique discussed in Box 8-2.

With the magnetic reversals used as the initial starting point for age control, the $\delta^{18}O$ signal is shifted back and forth in the time domain by a few percentage points in relation to the initial time scale provided by the radiometric ages of the reversals. At some point, one

BOX 10-3 CONT.   LOOKING DEEPER INTO CLIMATE SCIENCE

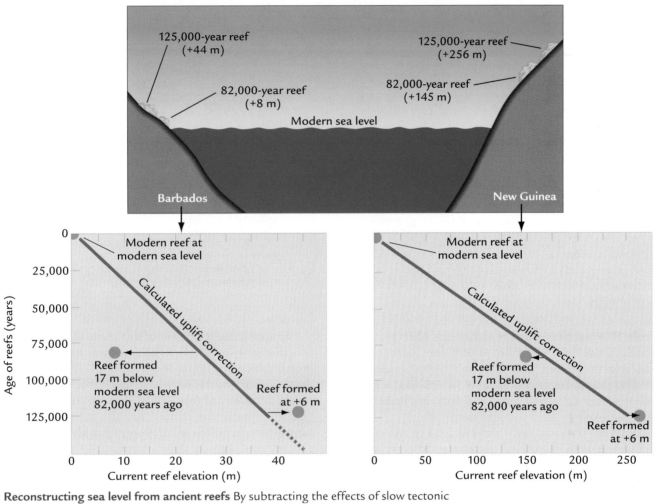

Reconstructing sea level from ancient reefs By subtracting the effects of slow tectonic uplift of the islands of Barbados and New Guinea, scientists can reconstruct sea level at the time coral reefs were formed, tens of thousands of years ago.

time scale optimizes the match between the orbital insolation and $\delta^{18}O$ signals at the cycles of tilt and precession. This match produces similar-looking signals with a consistent time-lagged relationship (Figure 10-21, middle).

Some scientists question the uniqueness of a time scale derived in this way. Their skepticism is especially appropriate for those parts of the $\delta^{18}O$ record earlier than 0.9 Myr ago that consist mainly of 41,000-year $\delta^{18}O$ oscillations (see Figure 10-15). Because the 41,000-year cycles of orbital insolation due to changes in tilt have so little variation in amplitude, one cycle

tends to look very much like another, and it would be easy to miscorrelate by one or more tilt cycles because of gaps in records or incomplete sampling.

Miscorrelations can be avoided if a clear 23,000-year precession signal also appears in the $\delta^{18}O$ record. The precession signal has such a strong amplitude modulation that the pattern of successive minima and maxima provides a distinctive tuning target. Clusters of two or three large-amplitude cycles alternate with clusters of two or three smaller cycles every 100,000 years, with other distinctive amplitude changes occurring every 413,000 years. Thus scientists have to find the one time

FIGURE 10-21 **Orbital tuning** The 41,000-year and 23,000-year cycles from astronomically dated insolation curves (left) provide tuning targets against which to match (and date) similar cycles embedded in δ¹⁸O (ice volume) curves (right). (Adapted from J. Imbrie et al., "On the Structure and Origin of Major Glaciation Cycles. 2. The 100,000-Year Cycle." *Paleoceanography* 8 [1993]: 699–735.)

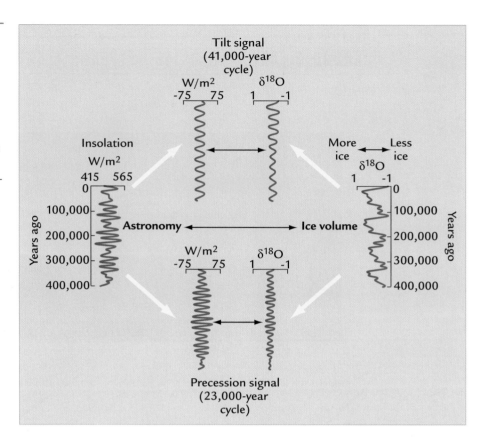

scale that simultaneously matches the observed δ¹⁸O changes not just with the 41,000-year astronomical changes in tilt but also with the highly modulated 23,000-year changes in precession. This added requirement makes the tuning process a far more demanding exercise likely to yield a unique time scale.

In fact, the time scale derived for the last several million years by the orbital tuning chronometer is now widely regarded as *more* accurate than the one based on radiometric dating. Until a decade ago, the most recent magnetic reversal had been dated by radiometric methods to 730,000 years ago. The orbital tuning method then placed the age of this reversal at 785,000 years ago. When geochemists re-examined their K/Ar radiometric ages from basalts formed at the time of the reversal, they found systematic evidence that reversal ages younger than 785,000 years ago were an artifact of loss of argon gas, either by natural escape of argon from the rocks or because of incomplete laboratory extraction of argon from the rocks. The age of the reversal was soon found to match that derived from orbital tuning.

With a time scale for the δ¹⁸O variations confirmed, climate scientists can apply it to any ocean sediment core containing a δ¹⁸O record, because all such records should contain a common ice volume signal, along with smaller, locally variable overprints due to temperature changes. Orbital tuning is now the main method of dating ocean cores.

To appreciate the power of this method, look back at the monsoon-response signals shown in cores near North Africa in Figures 9-10 and 9-13. The age scale for these records was derived from orbital tuning based on the δ¹⁸O record (the lagged response of ice volume behind summer insolation). Yet the monsoon-driven signals revealed in these cores occur mainly at a 23,000-year cycle.

This is an impressive demonstration that the tuning process provides a unique and accurate time scale. In effect, scientists first devise a time scale based on tuning the response of high-latitude ice sheets (δ¹⁸O) to orbital variations. But when they transfer this time scale to low-latitude ocean cores containing proxy signals of past changes in monsoon strength, they detect a different but physically reasonable orbital-scale monsoon response at a dominant 23,000-year cycle, and without the large lag associated with the ice sheet 23,000-year cycle. The fact that these physically reasonable monsoon responses fall out from an ice sheet tuning exercise makes the orbitally tuned δ¹⁸O time scale look believable indeed.

For intervals before the onset of glaciation in the northern hemisphere 2.75 Myr ago, records of the tropical monsoon are widely used as the tuning target. The distinctive changes through time in the intensity of individual monsoon maxima are matched to similar maxima in low-latitude summer insolation every 23,000 years.

Key Terms

equilibrium line (p. 211)

Milankovitch theory (p. 212)

climate point (p. 214)

ice elevation feedback (p. 215)

phase lag (p. 217)

elastic response (p. 218)

viscous response (p. 218)

basal slip (p. 220)

termination (p. 224)

orbital tuning (p. 229)

Review Questions

1. Which season determines whether or not ice sheets grow or melt? Why?

2. What is the equilibrium line? Why is it important?

3. How did Milankovitch's ideas differ from earlier ideas about orbital control of ice sheets?

4. Why does the size of a growing or melting ice sheet lag well behind changes in insolation?

5. How does the delay in bedrock response to ice loading act as a positive feedback to ice volume?

6. What does the 2.75-Myr $\delta^{18}O$ history tell us about ice sheets in the northern hemisphere?

7. Why are northern ice sheets more responsive to insolation changes than Antarctic ice?

8. Which parts of the Milankovitch theory have proved to be accurate? Which parts are insufficient?

9. How do corals indicate the volume of water tied up in ice on land?

Additional Resources

Basic Reading

Hays, J. D., J. Imbrie, and N. J. Shackleton. 1976. "Variations in the Earth's Orbit: Pacemaker of the Ice Ages." *Science* 194:1121–32.

Imbrie, J., And K. P. Imbrie. 1979. *Ice Ages: Solving the Mystery.* Short Hills, N.J.: Enslow.

Mesolella, K. J., R. K. Matthews, W. S. Broecker, and D. L. Thurber. 1969. "The Astronomical Theory of Climate Change: Barbados Data." *Journal of Geology* 77:250–74.

Advanced Reading

Imbrie, J. 1982. "Astronomical Theory of the Ice Ages: A Brief Historical Review." *Icarus* 50:408–22.

———. 1985. "A Theoretical Framework for the Ice Ages." *Journal of the Geological Society* (London) 142:417–32.

Milankovitch, M. 1941. *Kanon der Erdbestrahlung und seine Andwendung auf das Eiszeitenproblem.* Belgrade: Königlich Serbische Akademie. Published in English as *Canon of Insolation and the Ice-Age Problem.* 1969. Translation by Israel Program for Scientific Translations. Washington, D.C.: U.S. Department of Commerce and National Science Foundation.

Oerlemans, J. 1991. "The Role of Ice Sheets in the Pleistocene Climate." *Norsk Geologisk Tidsskrift* 71:155–61.

Shackleton, N. J., A. Berger, and W. R. Peltier. 1990. "An Alternative Astronomical Calibration of the Lower Pleistocene Timescale Based on ODP Site 677." Transactions of the Royal Society of Edinburgh. *Earth Science* 81:251–61.

Weertman, J. (1964). "Rate of Growth or Shrinkage of Non-equilibrium Ice Sheets." *Journal of Glaciology* 5:145–58.

CHAPTER 11

Orbital-Scale Changes in Carbon Dioxide and Methane

Earth provides scientists with another important archive of climate change over the last 400,000 years: long cores drilled from ice sheets. One of the most remarkable discoveries in recent years is that two important greenhouse gases, methane (CH_4) and carbon dioxide (CO_2), have varied over orbital time scales. Methane levels have fluctuated mainly at the 23,000-year orbital rhythm of precession, and we will evaluate the hypothesis that these changes are linked to fluctuations in the strength of monsoons in Southeast Asia. During glaciations, atmospheric CO_2 values have repeatedly dropped by 30% from the levels typical of warm interglaciations, such as those of today, but the complex reasons for these CO_2 changes have not yet been fully sorted out. Although glacial cooling of ocean surface waters can explain part of this observed difference, most of the explanation for lower glacial CO_2 appears to be tied to a transfer of carbon into the deep ocean.

Ice Cores

Cores from present-day ice sheets date back tens to hundreds of thousands of years. These records are invaluable archives of climatic signals unavailable from other sources. Two of the most important signals are the greenhouse gases carbon dioxide (CO_2) and methane (CH_4).

11-1 Drilling and Dating Ice Cores

Scientists searching for the oldest ice in an ice sheet drill down from the top of its highest domes (Figure 11-1). They avoid the higher edges because the ice there flows relatively quickly toward the ice margins, where it melts. In contrast, ice that accumulates on the highest domes flows slowly down into the interior of the ice sheet, where it is stretched and thinned but not melted. As a result, the oldest ice in an ice sheet sits under the middle of it.

Because winter weather on top of ice sheets is totally inhospitable, drilling is done in the summer season (Figure 11-2). On Greenland a warm summer day may reach temperatures a few degrees below freezing (0°C), but on top of the Antarctic ice sheet temperatures rarely warm even to −20°C in summer. Drilling takes place in structures that provide some protection from the elements. Hundreds of ice cores, each a few meters in length, are retrieved as drilling proceeds down through

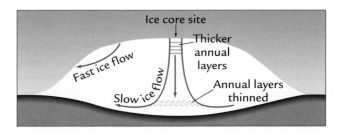

FIGURE 11-1 **Ice coring location** The best place on an ice sheet to take ice cores is at the top, because ice flows slowly down into the ice sheet and old ice is preserved at the bottom.

several kilometers of ice. It may take more than one summer to drill all the way through an ice sheet and retrieve the entire record.

As we saw in Chapter 3, many ice cores can be dated by counting annually deposited layers. Annual layering is recorded in several properties of ice cores, the most obvious of which are layers of dust easily visible to the eye. Dust is usually deposited at the end of cold, dry windy winters. The count starts with the top layer, representing the year the coring operation began, and it proceeds downward through the ice sheet as far as annual layers remain detectable. Eventually the natural stretching of ice layers by flow deep in the ice sheet blurs the layers, often beyond recognition.

A   B   C

FIGURE 11-2 **Ice coring operations** Ice drilling during cold "summer" conditions retrieves sequences of ice cores thousands of meters thick (A, B). Scientists may also examine upper ice layers in pits dug into the ice (C). (Courtesy of Paul Mayewski, University of Maine.)

This layer-count method works best for ice sheets on which snow accumulates rapidly. For the Greenland ice sheet, where ice accumulates at a rate of 0.5 meters or more per year, piling up more than the height of a man every 4 years, annual layers may be detected thousands or even tens of thousands of years into the past. In contrast, on the moisture-starved central domes of the Antarctic ice sheet, where ice accumulates at only a tenth that rate (less than 0.05 m, the length of a finger, each year), annual layering is barely detectable even in the snow at the surface, and any detectable initial layering quickly disappears in the stretched and thinned ice below.

For cores without annual layering, the most common technique used to date ice is to construct an **ice flow model** based on the physical properties of the ice sheet and assuming a smooth steady flow of ice below the surface, like that shown in Figure 11-1. These models produce fairly good estimates of the age of the ice but are not as accurate as annual layer counts.

11-2 Trapping Gases in the Ice

Snow that falls on the high central domes of ice sheets is light and fluffy, and air circulates easily among its upper layers (Figure 11-3). As additional snow accumulates, the underlying layers are gradually buried and compressed by the weight of overlying snow and ice and the pressure eventually turns the snow to crystals of ice. Little by little, this process shuts off the flow of air within the subsurface ice, creating a zone where air can only slowly diffuse within the ice. At depths of 50 meters or more below the surface of the ice sheet, air no longer circulates. Air that had been slowly diffusing down to these depths is sealed off as small bubbles and trapped in the ice, a process called **sintering**. Air sealed in ice forms a permanent record of the atmosphere at the time the sintering occurred.

At the time the air bubbles are permanently sealed off, their average age is greater than that of the overlying atmosphere and the snow falling on top of the ice sheet (because of the slow diffusion of air) but younger than the ice in which the bubbles are trapped. The difference in age between the air bubbles and the surrounding ice varies with the rate at which ice accumulates. If deposition of ice is fast (0.5–1 m/year), the age difference between the bubbles and the ice enclosing them will be only a few hundred years. If deposition is slow (0.05–0.1 m/per year), the age offset can be as large as 1000 to 2000 years.

Before interpreting records of greenhouse gases trapped in ice cores, scientists first need to verify that the techniques they use to extract and measure the gas concentrations are reliable. To do so, they measure gas bubbles deposited in the upper layers of ice in cores taken from sheltered pockets on ice sheets where snow

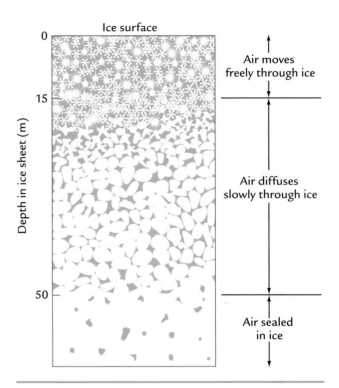

FIGURE 11-3 Sintering: Sealing air bubbles in ice Air moves freely through snow and ice in the upper 15 m of an ice sheet, but flow is increasingly restricted below this level. Bubbles of old air are eventually sealed off completely in ice 50 to 100 m below the surface. (Adapted from D. Raynaud, "The Ice Core Record of the Atmospheric Composition: A Summary, Chiefly of CO_2, CH_4, and O_2," *in Trace Gases in the Biosphere,* ed. B. Moore and D. Schimel [Boulder, Colo.: UCAR Office for Interdisciplinary Studies, 1992].)

accumulates more rapidly than in other regions. Short ice cores taken from these sites provide measurements of CO_2 and methane values from recent centuries, with each analysis representing an average of 5 to 10 years.

Records obtained from ice cores indicate that CO_2 values were about 280 ppm (parts per million by volume) for thousands of years until the middle of the eighteenth century. By late in the century, CO_2 concentrations began to rise at rates that increased toward (and past) 300 ppm in the early and middle twentieth century (Figure 11-4A). This accelerating trend in the ice core CO_2 measurements merges smoothly with a record of atmospheric CO_2 based on instrumental analyses of actual samples of air taken at the Mauna Loa Observatory in Hawaii by the atmospheric chemist David Keeling since 1958. The instrumental measurements show CO_2 levels accelerating from 315 ppm in 1958 to the present-day value above 365 ppm. The smooth merge of the ice core and instrumental trends shows climate scientists that the CO_2 measurements in ice cores are reliable.

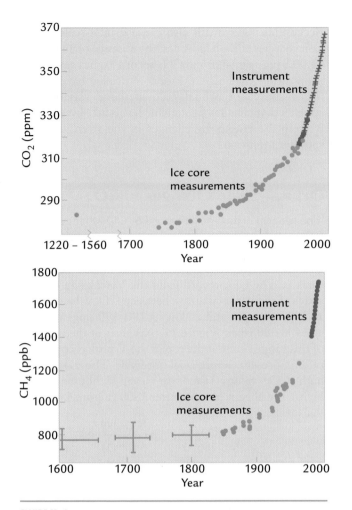

FIGURE 11-4 **Ice core and instrumental CO$_2$ and CH$_4$ measurements** Measurements of CO$_2$ (top) and methane (bottom) taken on bubbles in ice cores merge perfectly with measurements of the atmosphere in recent decades. (Top: adapted from H. H. Friedli et al., "Ice Core Record of the ^{13}C/^{12}C Ratio of Atmospheric CO$_2$ in the Past Two Centuries," *Nature* 324 [1986]: 237–38; bottom: adapted from M. A. K. Khalil and R. A. Rasmussen, "Atmospheric Methane: Trends over the Last 10,000 Years," *Atmospheric Environment* 21 [1987]: 2445–52.)

Ice core measurements of methane over the last few centuries blend in a similar way with instrumental measurements made in the late twentieth century (Figure 11-4B). Ice core values varied within the range of 700 to 750 ppb (parts per billion by volume) for thousands of years before the Industrial Revolution; they have accelerated sharply since the late nineteenth and early twentieth centuries as human influences have become detectable, reaching modern levels in excess of 1700 ppb.

With the CO$_2$ and CH$_4$ records in ice cores each proved valid by their close match to recent instrumental measurements, the longer-term records of these gases extracted from ice cores can also be accepted as reliable. We first look at long-term methane changes, which

have a simple and plausible explanation, and then later explore the CO$_2$ signal, which is more complicated.

Orbital-Scale Changes in Methane

One of the most impressive accomplishments in ice core drilling in recent years has been the recovery of a long, continuous sequence of ice by Russian engineers and scientists at a site high on the Antarctic ice sheet. This sequence is called the Vostok ice record.

The record of methane from this site shows a series of cyclic variations between maxima of 550 to 700 ppb (parts per billion) and minima of 350 to 450 ppb (Figure 11-5, left). The ice flow model constructed for this part of the ice sheet indicates that the high methane values at the top of the ice core date from the modern interglacial maximum of the last 10,000 years, and that the low values of 350 ppb in the immediately preceding interval date to the last glaciation maximum, 20,000 years ago. The ice flow model also indicates that the only previous interval with methane values comparable to, or even higher than, those of the present interglaciation dates very near the time of the last interglacial period, 125,000 years ago (Chapter 10), and that the low values just before it date from the preceding glaciation.

If we accept that this time scale is approximately correct, the six CH$_4$ cycles shown at the left in Figure 11-5 are separated by an average interval of 23,000 years (125,000 − 10,000 = 115,000 years divided by 5 cycles = 23,000 years/cycle). This cycle length points to a likely connection with changes in orbital precession that dominate insolation changes at lower latitudes (Chapter 8).

The methane signal also closely resembles the monsoon response signal discussed in Chapter 9 (Figure 11-5, right). The peak values of methane match the expected peaks in monsoon intensity not only in timing but also in amplitude. Stronger methane peaks line up near strong insolation maxima at 30°N, and methane minima match insolation minima. This match greatly strengthens the case for a connection between CH$_4$ concentrations and the monsoon.

But how would changes in the strength of low-latitude monsoons produce changes in atmospheric methane concentrations? One possible link is the impact monsoon fluctuations have on the amount of precipitation that falls in Southeast Asia. Heavy rainfall in such regions saturates the ground, reduces its ability to absorb water, and thereby increases the amount of standing water in bogs. Decaying vegetation uses up any oxygen in the water and creates the oxygen-free conditions needed to generate methane. The extent of these boggy areas must have expanded during wet monsoon maxima and shrunk during dry monsoon minima.

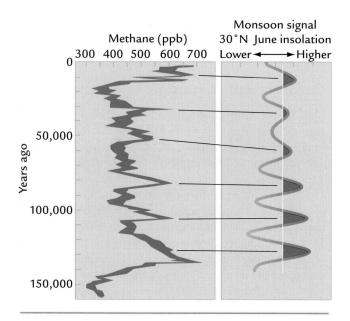

FIGURE 11-5 **Methane and the monsoon** The methane record from Vostok ice in Antarctica shows regular cycles at intervals near 23,000 years (left). This signal closely resembles the monsoon-response signal driven by low-latitude insolation (right). (Left: adapted from J. Chapellaz et al., "Ice Core Record of Atmospheric Methane over the Last 160,000 Years," *Nature* 345 [1990]: 127–31; right: adapted from D. Martinson et al., "Age Dating and the Orbital Theory of the Ice Ages: Development of a High-Resolution 0 to 300,000-Year Chronostratigraphy," *Quaternary Research* 27 [1987]: 1–29.)

An alternative explanation linking orbital precession and atmospheric CH_4 concentrations calls on changes occurring in higher latitudes and Arctic regions. CH_4 can also be generated in high-latitude bogs during the brief summer warming. In addition, large amounts of CH_4 are stored at high latitudes in frozen ground beneath the bogs and in shallow marine sediments on continental shelves in the Arctic. During colder intervals, all this subsurface methane is frozen and unavailable to the atmosphere, but during exceptionally warm intervals it may thaw and be released to the atmosphere.

Because insolation changes at the rhythm of orbital precession are felt even on the far northern portions of circum-Arctic landmasses, this explanation seems plausible. In these locations, the response of the land surface to orbital insolation changes occurs as cycles of warming and cooling by the sun, rather than as the cyclic changes in monsoon strength seen in the tropics. These cycles of warming and cooling every 23,000 years have the potential to enhance and suppress the amount of methane released to the atmosphere.

One argument against a northern origin for the methane changes is the absence of any obvious CH_4 fluctuations at the 41,000-year period of orbital tilt.

Because changes in tilt also cause orbital-scale insolation changes at high latitudes, the absence of an obvious 41,000-year signal in the CH_4 record argues against the proposition that high northern latitudes have been a major source of orbital-scale methane variations. It appears that changes in methane levels are driven mainly from the tropics by changes in the strength of the summer monsoon.

Orbital-Scale Changes in CO_2

The longest orbital-scale record of CO_2 changes comes from the Vostok site. After five summers of drilling at this location, Russian engineers recently reached the base of the Antarctic ice sheet 4000 meters below the surface. The CO_2 record from the Vostok site shows a series of regular oscillations between CO_2 values as high as 280–300 ppm and as low as 180–190 ppm over the last 400,000 years (Figure 11-6). Values at the high end of this range occur at the top of the Vostok record within the present interglacial interval. These relatively high CO_2 values lasted for several thousand years before the abrupt increase after 1800 (Figure 11-4A).

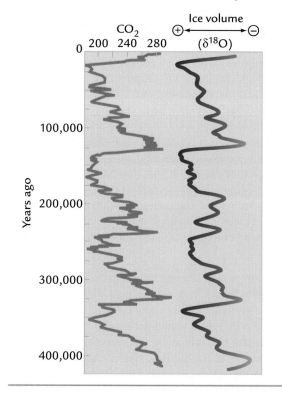

FIGURE 11-6 **Long-term CO_2 changes** A 400,000-year record of CO_2 from Vostok ice in Antarctica shows four large-scale cycles at a period of 100,000 years similar to those in the marine $\delta^{18}O$ (ice volume) record. (Adapted from J. R. Petit et al., "Climate and Atmospheric History of the Past 420,000 Years from the Vostok Ice Core, Antarctica," *Nature* 399 [1999]: 429–36.)

Unlike the methane record in Figure 11-5, the 23,000-year cycle of orbital precession is not especially prominent in these CO_2 variations. Instead, four major cycles are apparent over the 400,000 years of record, a clear indication that the major cycle of CO_2 variation over that interval has been at or near 100,000 years.

We learned in Chapter 10 that the dominant cycle of ice sheet variations over the last several hundred thousand years has also been 100,000 years. Do the 100,000-year variations in atmospheric CO_2 match those of the ice sheets? Yes, they do.

Over the 400,000-year length of the ice core record, the major cycles of CO_2 change line up well with the best indicator of ice volume changes, the marine $\delta^{18}O$ record (Figure 11-6). The two signals share not only the 100,000-year signal but also its asymmetric shape. Abrupt increases in CO_2 occur during times of rapid ice melting ($\delta^{18}O$ decreases), while slower CO_2 decreases match the irregular and slower phases of ice volume buildup ($\delta^{18}O$ increases).

This basic match also holds if we examine the changes over the last 150,000 years in more detail (Figure 11-7). Even smaller, shorter-term fluctuations (including some at the much smaller 41,000-year and 23,000-year orbital rhythms) appear to match up fairly well between the CO_2 and $\delta^{18}O$ (ice volume) signals, although not in all cases.

The obvious conclusion—that a *general* match exists between CO_2 and ice volume—tells us that some kind of cause-and-effect relationship exists between these two components of the climate system. But what is the nature of this relationship? The complete answer to this question requires that we know much more precisely the correlation between changes in CO_2 and ice sheets. Are the two signals truly correlative in detail, or do

minor leads and lags exist between them? If leads and lags exist, do CO_2 changes occur first and drive ice sheet size? Or do changes in ice volume occur first and in some way control CO_2 levels?

Firm answers to these questions require accurate dating of both records, and especially of the CO_2 records in the ice cores. Unfortunately, the extremely slow rates of ice accumulation high on the Antarctic ice sheet limit the dating and resolution of its records of greenhouse gases, as we noted earlier. To add to the complications, the atmospheric CO_2 signal is at least 2000 years younger than the ice layers in which it is measured, and the amount of this offset may vary through time.

The Greenland ice sheet might seem to be a more promising place to obtain a well-dated CO_2 record. Ice accumulates fast enough on Greenland to develop annual layering that can be detected back tens of thousands of years and can be counted to provide accurate dating. Unfortunately, the windblown dust that helps define these annual ice layers is rich in fine $CaCO_3$ particles eroded from northern hemisphere continents and blown to the ice sheet by high-level winds. If even a small amount of the old carbon in this $CaCO_3$ dust reacts chemically with the CO_2 bubbles in the cores, it will contaminate the air bubbles and produce artificially high CO_2 values.

As a result of these problems, the precise timing between changes in CO_2 levels and in ice volume is still uncertain. But the dating is certainly good enough to show that the two signals correlate in an overall sense and that some link between them must exist. Before we return to the crucial issue of the cause-and-effect relationship between these two signals in the next chapter, we first face a more basic problem: How could CO_2

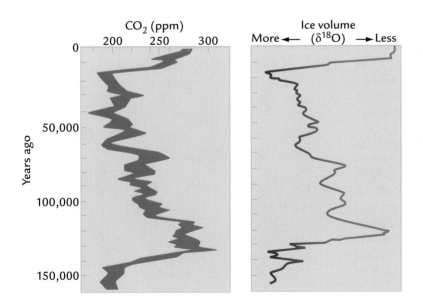

FIGURE 11-7 **The most recent CO_2 cycle** A record of the last 160,000 years of CO_2 variations from Vostok ice in Antarctica (left) resembles the marine $\delta^{18}O$ (ice volume) record (right), both in the asymmetric shape of the 100,000-year cycle and in several of the shorter oscillations. (Left: adapted from J. M. Barnola et al., "Vostok Ice Core Provides 160,000-Year Record of Atmospheric CO_2," *Nature* 329 [1987]: 408–14; right, adapted from D. Martinson et al., "Age Dating and the Orbital Theory of the Ice Ages: Development of a High-Resolution 0 to 300,000-Year Chronostratigraphy," *Quaternary Research* 27 [1987]: 1–29.)

vary by 30% or more over orbital time scales? What factors could explain the observed 90-ppm drop in CO_2 levels during glacial intervals from the levels observed in interglacial intervals?

11-3 Physical Oceanographic Explanations of CO_2 Changes

One possibility is that changes in the physical oceanographic characteristics of the surface ocean—its temperature and salinity—during glaciations might alter the chemical solubility of CO_2 in seawater and affect the amount of CO_2 left in the atmosphere. Because CO_2 dissolves more readily in colder seawater, atmospheric CO_2 levels will drop by 9 ppm for each 1°C of ocean cooling. As we will see in Chapter 13, surface-ocean temperatures during the well-studied last glacial maximum 20,000 years ago cooled by an amount estimated at between 1.5° and about 4°C, with a central estimate of 2.5°C. If these estimates are correct, the glacial cooling of the ocean surface would have caused CO_2 to drop by 22 ppm (Table 11-1).

The altered salinity of the glacial ocean would also have affected atmospheric CO_2, but in the opposite direction. CO_2 dissolves more easily in seawater with a lower salinity, but the average glacial ocean was saltier than it is today because of the amount of freshwater taken from the ocean and stored in ice sheets. Although some high-latitude ocean surfaces (such as the North Atlantic) became less salty during glaciations, the average salinity of the entire ocean increased by about 1.1‰, enough to cause an estimated glacial CO_2 increase of 11 ppm (Table 11-1).

This 11-ppm CO_2 increase caused by higher salinity would have canceled half the 22-ppm decrease caused by ocean cooling, reducing the net effect of these two factors to an 11-ppm drop in atmospheric CO_2, equivalent to only a small part of the 90-ppm change actually observed. As a result, we are still left with almost 90% (79 ppm) of the CO_2 decrease to account for by other mechanisms. If physical mechanisms explain only a small fraction of the glacial drop in atmospheric CO_2 concentrations, what other explanations remain?

One way to explore this question is to come at the problem from another direction: track the movement of carbon through the climate system during glaciations. We know that carbon is removed from the glacial atmosphere, but where does it go? If we can answer this question, we should gain new insights into the cause of lower glacial CO_2. Tracking glacial carbon requires that we revisit the major carbon reservoirs and examine the kinds of exchanges that could occur among them over orbital time scales.

11-4 Orbital-Scale Carbon Reservoirs

We learned in Section II that gradual changes in atmospheric CO_2 levels over tens of millions of years are caused by extremely slow exchanges between the large reservoir of carbon stored in sediments and rocks and the much smaller surface carbon reservoirs in the atmosphere, ocean, and vegetation. Over extremely long intervals of time, the effects of these slow exchanges can accumulate and cause large changes in CO_2.

But now we face a situation in which CO_2 concentrations in the atmosphere changed by 30% or more over time spans of just a few thousand years (Figure 11-7). CO_2 changes this rapid cannot be caused by carbon entering or leaving the rock reservoirs, because the exchange rates are far too slow. We need to look elsewhere in the carbon system to explain these much faster orbital-scale CO_2 changes. Carbon can move rapidly among the air, water, and vegetation reservoirs because they all have much faster rates of carbon exchange than do the slow interactions with the rocks (Figure 11-8).

The most obvious place to look for the CO_2 carbon removed from the atmosphere during times of glaciation is in the vegetation-soil reservoir. Scientists have abundant information on the amount of carbon stored on land in living vegetation and in soils during the last glaciation, 20,000 years ago. But, as we will see in more detail in Part IV, the continents had less net vegetation cover and as a result held less carbon during maximum glaciations than they do during interglacial intervals such as today.

The main reason for the decrease in glacial carbon was the expansion of ice sheets across large areas of North America and Eurasia that today are mostly covered by forests of coniferous and deciduous trees. South of the ice sheets, many regions that are forested during warmer interglacial climates such as ours were instead covered by steppes and grasslands with low levels of carbon biomass during glaciations. The forested regions that did exist during full glaciations were much smaller than those today.

TABLE 11-1 Physical Causes of Lower CO_2 Levels in the Glacial Atmosphere

Properties of ocean surface waters	CO_2 change (parts per million)
Cooling by ~2.5°C	−22
Increase in salinity by 1.1‰	+11
All physical properties	−11

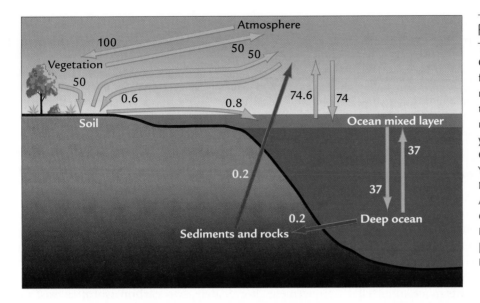

FIGURE 11-8 Exchanges of carbon The large changes in atmospheric CO_2 in ice cores over intervals of a few thousand years must involve rapid exchanges of carbon among the near-surface reservoirs. (Modern rates are shown in gigatons per year.) (Adapted from J. Horel and J. Geisler, *Global Environmental Change* [New York: John Wiley, 1997], and from National Research Council Board on Atmospheric Sciences and Climate, *Changing Climate*, Report of the Carbon Dioxide Assessment Committee [Washington, D.C.: National Academy Press, 1993].)

Many regions of the continents that were not covered by ice sheets had lakes with sediments containing pollen that can tell us about past changes of the nearby vegetation. The picture that emerges from these lake-core records is that most regions on Earth were drier and less vegetated during maximum glaciations than they are today. Even in the tropics, the rain forests were less extensive. The only region where a large-scale increase in glacial vegetation may have occurred was in the area north of Australia. There the fall of sea level caused by storage of water in glacial ice sheets exposed huge expanses of now-submerged continental shelf, and these regions are thought to have been covered by tropical rain forests.

Assembling all this information, climate scientists estimate that the total amount of vegetation on land was reduced by roughly 25% (from 2160 to 1630 gigatons)

during glacial maximum times like the one that occurred 20,000 years ago (Figure 11-9). Estimates of the exact decrease of biomass are still uncertain, ranging between 15% and 30%, but in any case the decrease was substantial. We can say for certain that the carbon removed from the atmosphere did not go into land vegetation, and now we have even more carbon whose fate we have to account for during glaciations.

The only remaining site where the missing carbon could be stored is the ocean. But which ocean reservoir took up the carbon, the small surface reservoir or the much larger deep reservoir?

We can eliminate the surface ocean as an option. Ocean surface waters exchange all their carbon with the atmosphere within just a few years. Because of this rapid exchange of CO_2 gas, most areas of the surface ocean today have CO_2 values within 30 ppm of the value in the

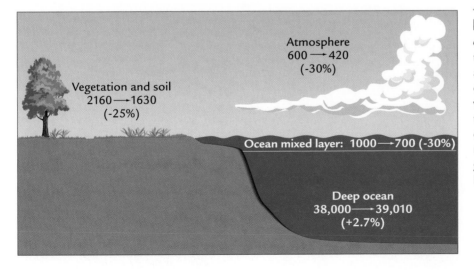

FIGURE 11-9 Interglacial-glacial changes in carbon reservoirs During the glacial maximum 20,000 years ago, large reductions of carbon occurred in the atmosphere, in vegetation and soils on land, and in the surface ocean. The total amount of carbon removed from these reservoirs (more than 1000 gigatons) was added to the much larger reservoir in the deep ocean.

overlying atmosphere. Such rapid exchanges mean that if CO_2 values were 30% lower in the glacial atmosphere, they must have been lower by nearly the same average amount in the glacial surface ocean as well (ignoring the changes caused by temperature and salinity already discussed). We can conclude that the missing glacial carbon could not have been stored in the surface ocean. And now we have still more unaccounted-for carbon.

Where does this leave us? The only remaining reservoir available to take up the unaccounted-for carbon is the deep ocean. The glacial carbon removed from the air, from the vegetation, and from the surface waters must have been stored or sequestered in the deep ocean, in effect hidden from the atmosphere.

We can roughly calculate a mass balance to assess the total changes in all four carbon reservoirs during glaciations (Figure 11-9). CO_2 measurements in ice cores tell us that the atmospheric reservoir decreased to 190 ppm, equivalent to a loss of 180 gigatons of carbon. The estimated 25% reduction of vegetation on land amounts to 530 gigatons, if we assume that soil carbon was reduced by the same fraction as the vegetation biomass. And the 30% decrease in size of the surface ocean reservoir adds another 300 gigatons to the total carbon reduction.

Adding up the reductions in these three carbon reservoirs, we arrive at a total loss of some 1010 gigatons of carbon (Figure 11-9). Roughly this amount of carbon must have been added to the deep ocean during glacial-maximum climates like the one 20,000 years ago. Because the deep ocean is so large a carbon reservoir (38,000 gigatons), this additional carbon would result in only a 2.7% increase in its size in comparison with nonglacial climates.

Earlier we found that temperature and salinity changes of surface waters can explain only a little over 10% of the total glacial reduction in atmospheric CO_2 levels, leaving almost 90% yet to be explained. And now we know that the atmospheric (and other) carbon was stored in the deep ocean. The next obvious question is this: How did the carbon get into the deep ocean? Or, put differently, why was more carbon stored in the deep ocean during glacial times than today? Something about the pattern of ocean circulation must have changed in such a way as to store less carbon in the surface waters and to hide more of it at depth.

11-5 Tracking Carbon through the Climate System

To explore further the question of how the different kinds of carbon have moved among its reservoirs in the climate system, we need a quantitative way to track its movement. Fortunately, two **carbon isotopes** exist in nature, and different types of carbon have distinctive carbon isotope ($\delta^{13}C$) ratios (Box 11-1). These carbon isotope values give scientists a way of tracking carbon transfers within the climate system, particularly the transfers into and within the oceanic reservoirs that hold the answer to glacial atmospheric CO_2 changes.

Carbon moves through Earth's reservoirs in one of two forms (Chapter 2): organic carbon, which consists of both living and dead organic matter, and inorganic carbon, which consists mainly of ions dissolved in water (HCO_3^-, CO_3^{-2}) but also includes atmospheric CO_2. Distributions of these two types of carbon in the major reservoirs are shown in Figure 11-10, along with their abundances and typical $\delta^{13}C$ values.

Both organic and inorganic carbon are preserved in the geologic record. The $CaCO_3$ shells of marine foraminifera are formed from inorganic carbon dissolved in seawater, typically with values near 0‰. In

FIGURE 11-10 **Carbon reservoir $\delta^{13}C$ values** The major reservoirs of carbon on Earth have varying amounts of organic and inorganic carbon (shown in gigatons of carbon), and each type of carbon has characteristic carbon isotope ($\delta^{13}C$) values.

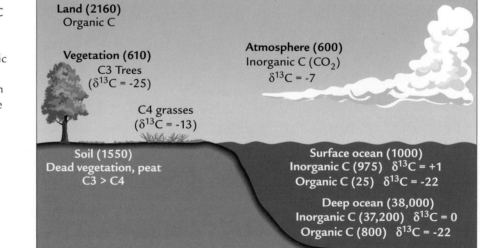

Land (2160)
Organic C

Vegetation (610)
C3 Trees
($\delta^{13}C = -25$)

C4 grasses
($\delta^{13}C = -13$)

Soil (1550)
Dead vegetation, peat
C3 > C4

Atmosphere (600)
Inorganic C (CO_2)
$\delta^{13}C = -7$

Surface ocean (1000)
Inorganic C (975) $\delta^{13}C = +1$
Organic C (25) $\delta^{13}C = -22$

Deep ocean (38,000)
Inorganic C (37,200) $\delta^{13}C = 0$
Organic C (800) $\delta^{13}C = -22$

BOX 11-1   TOOLS OF CLIMATE SCIENCE

BOX 11-1   TOOLS OF CLIMATE SCIENCE

Carbon Isotope Ratios ($\delta^{13}C$)

Both ^{13}C and ^{12}C are stable (nonradioactive) isotopes of carbon that occur naturally in Earth's vegetation, water, and air. The ^{12}C isotope accounts for more than 99% of all the carbon present on Earth, and ^{13}C accounts for most of the rest. We will ignore the much scarcer (and radioactive) ^{14}C isotope here. Geochemists who analyze material for its carbon isotope composition measure small variations around the average $^{13}C/^{12}C$ ratio of less than 0.01.

Similar to the convention used for oxygen isotopes, measurements of $^{13}C/^{12}C$ ratios are reported as departures in parts per thousand (‰) from a laboratory standard:

$$\delta^{13}C \atop (‰) = \frac{\left(^{13}C/^{12}C\right)_{sample} - \left(^{13}C/^{12}C\right)_{standard}}{\left(^{13}C/^{12}C\right)_{standard}} \times 1000$$

All measurements are referenced to standards supplied by the National Bureau of Standards for use as a common reference point. Like the oxygen isotope ratios, all carbon isotope ratios are multiplied by 1000 to convert the very small measured variations in an already small ratio to a more workable numerical form. As a result, $\delta^{13}C$

values fall in a range between $-25‰$ or more for some of the vegetation on land to $+2‰$ for ocean surface waters in some regions.

Carbon samples with relatively large amounts of ^{13}C compared with ^{12}C have more positive $\delta^{13}C$ values and are referred to as ^{13}C-enriched or ^{12}C-depleted. Samples with relatively small amounts of ^{13}C compared with ^{12}C have more negative $\delta^{13}C$ values and are referred to as ^{13}C-depleted or ^{12}C-enriched. Terrestrial vegetation with $\delta^{13}C$ values ranging between $-11‰$ and $-28‰$ is said to be ^{13}C-depleted in relation to ocean waters with values between $-1‰$ and $+2‰$.

The $\delta^{13}C$ composition of a wide range of materials can be analyzed on mass spectrometers. The same handpicked $CaCO_3$ shells of marine foraminifera that yield oxygen for $\delta^{18}O$ analyses also provide carbon for $\delta^{13}C$ analyses. Acid can be used to dissolve the $CaCO_3$ shells and produce the CO_2 for analysis. Living vegetation or its organic residue in soils as well as organic matter in plankton and ocean sediments can also be analyzed on mass spectrometers after the organic carbon is burned to produce CO_2 gas.

---

addition, many kinds of geologic deposits on land and in the ocean contain small amounts of organic carbon, usually with much more negative $\delta^{13}C$ values.

The $\delta^{13}C$ values of inorganic and organic carbon differ because of changes that occur during photosynthesis. During this process, plants take inorganic carbon and turn it into organic carbon (Chapter 2). In doing so, they incorporate the ^{12}C isotope into their living tissue more easily than the ^{13}C isotope. This discrimination (or fractionation) in favor of the ^{12}C isotope shifts the $\delta^{13}C$ of the organic matter produced toward more negative ^{12}C-enriched values than the inorganic carbon that formed the initial source. Fractionation of carbon within living systems can be compared with the *physical* fractionation of oxygen isotopes by evaporation and condensation (Chapter 7).

For example, plant plankton in the ocean take inorganic carbon with initial $\delta^{13}C$ values near $0‰$ from seawater and convert it to organic carbon with $\delta^{13}C$ values near $-22‰$, a net fractionation effect of $-22‰$ (Figure 11-11). Some of the organic carbon is sent to the deep ocean, but much of it is oxidized back to inor-

ganic form and recycled within or back up into surface waters. The net export of a small fraction of ^{12}C-rich organic carbon to the deep ocean leaves the remaining surface water enriched in ^{13}C, with $\delta^{13}C$ values of $1‰$ or higher. Overall, organic carbon forms a small fraction of the total carbon reservoir in the ocean, where inorganic carbon predominates. The combined effect of the two reservoirs (a small amount of organic carbon with $\delta^{13}C$ values near $-22‰$ and a large amount of inorganic carbon with a $\delta^{13}C$ value near $+1‰$ yields a mean $\delta^{13}C$ value of $0‰$ for all of the carbon in the ocean.

On land, atmospheric inorganic CO_2 carbon with a $\delta^{13}C$ value of $-7‰$ is the carbon source used in photosynthesis of plant matter with a wide range of negative $\delta^{13}C$ values (Figure 11-11). All trees as well as most shrubs and cool-climate grasses use a type of photosynthesis called the C_3 pathway, which produces organic tissue with $\delta^{13}C$ values in the range of $-21‰$ to $-28‰$ (averaging $-25‰$). Some shrubs and most grasses that grow in hot climates during the summer season use a different kind of photosynthesis called the C_4 pathway,

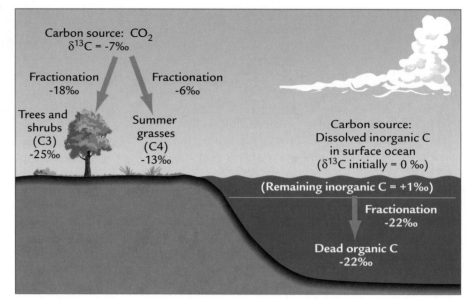

FIGURE 11-11 Photosynthesis and carbon isotope fractionation Photosynthesis on land and in the surface ocean converts inorganic carbon to organic form and causes large negative shifts in $\delta^{13}C$ values of the organic carbon produced.

which produces vegetation with less negative $\delta^{13}C$ values, ranging between $-11‰$ and $-15‰$ (averaging $-13‰$). By far the largest amount of organic biomass on Earth's land surfaces resides in C_3 trees, and as a result the average $\delta^{13}C$ of vegetation on land (and of litter in soils) is close to $-25‰$.

11-6 Can $\delta^{13}C$ Evidence Detect Glacial Changes in Carbon Reservoirs?

How can climate scientists be certain that $\delta^{13}C$ values are useful for tracking the movement of carbon among reservoirs during the past? One way is to see if changes in $\delta^{13}C$ values can be used to detect the glacial transfer of carbon into the deep ocean that we calculated earlier.

Organic carbon in land vegetation is tagged with negative $\delta^{13}C$ values (averaging $-25‰$), whereas the large amount of inorganic carbon in the ocean has an average $\delta^{13}C$ value of $0‰$. Because of this contrast, we should be able to track the transfer of organic carbon from the land into the ocean during glaciations. Although a small fraction of the transferred organic carbon remains as organic matter in the ocean, most of it is converted to inorganic carbon still tagged with the very negative $-25‰$ $\delta^{13}C$ composition.

We can use a mass balance calculation to estimate the effect of adding very negative (^{12}C-enriched) carbon to the inorganic carbon already present in the deep sea:

$$\underset{\substack{\text{Inorganic C} \\ \text{in ocean}}}{(38,000)} \; \underset{\substack{\text{Mean} \\ \delta^{13}C}}{(0‰)} + \underset{\substack{\text{C added} \\ \text{from land}}}{(530)} \; \underset{\substack{\text{Mean} \\ \delta^{13}C}}{(-25‰)} = \underset{\substack{\text{Glacial ocean} \\ \text{carbon total}}}{(38,530)} \; \underset{\substack{\text{Mean} \\ \delta^{13}C}}{(x‰)}$$

where x is the $\delta^{13}C$ value of glacial inorganic carbon in the ocean, and the size of the carbon reservoirs are in

gigatons (10^{15} grams) of carbon. Solving for x, we find that the mean $\delta^{13}C$ value of inorganic carbon in the glacial ocean should have shifted from $0‰$ to $-0.34‰$ because of the addition of ^{12}C-rich carbon from the land.

This estimated $-0.34‰$ shift can be tested by comparing it with actual data from the ocean. Marine geochemists have analyzed $\delta^{13}C$ values in the shells of benthic foraminifera taken from the last glacial level in cores distributed across large areas of the world ocean and over a wide range of water depths. These bottom-dwelling organisms should record regional deep-water $\delta^{13}C$ values in their $CaCO_3$ shells.

When all analyses are combined to calculate a global average, the estimated $\delta^{13}C$ change in the entire ocean turns out to be $-0.35‰$ to $-0.4‰$. In view of the uncertainties in estimates of the amount of carbon transferred from the land to the ocean and the incomplete coverage of the ocean provided by the cores available, the good agreement between the two methods indicates that carbon isotopes are useful tracers of past carbon shifts among Earth's reservoirs.

This evidence from the last glacial maximum also indicates that we should be able to trace carbon transfers during earlier glacial cycles using $\delta^{13}C$ changes measured in the shells of benthic foraminifera. Because the Pacific is by far the largest of Earth's oceans, scientists have used changes in its deep-water $\delta^{13}C$ values through time as the best single index of average $\delta^{13}C$ changes in the global ocean.

The longer-term $\delta^{13}C$ changes in Figure 11-12 show a series of oscillations between more positive values like those in the region today ($0‰$ or greater) and more negative values like those at the last glacial maximum ($-0.5‰$ or lower). The most negative $\delta^{13}C$ values occur in cycles of 100,000 years during the last 0.9 Myr, and in cycles of 41,000 years before 0.9 Myr ago. Although the correlation between the two signals is far from perfect, time series analysis confirms that the $\delta^{13}C$ (carbon transfer) and $\delta^{18}O$ (ice volume) signals share the same major cycles, and these cycles vary with the same approximate timing.

The results in Figure 11-12 confirm just what we would expect: large amounts of organic carbon from the land with negative (−25‰) isotope values were transferred into the ocean every time ice sheets grew large, and then were returned to the land when the ice melted. This glacial transfer of ^{12}C-enriched carbon from the land to the deep ocean was opposite in sense to the transfer of ^{16}O-enriched water vapor from the ocean to the ice sheets (Figure 11-13). As a result, the $\delta^{13}C$ signal became more negative each time the $\delta^{18}O$ signal became more positive.

On closer inspection, many of the $\delta^{13}C$ changes shown in Figure 11-12 exceed the −0.35 to −0.4‰ shift that can plausibly be attributed to transfers of carbon from land to sea. This unexpectedly large $\delta^{13}C$ response tells us that additional factors not yet examined

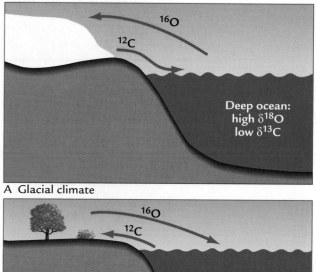

A Glacial climate

B Interglacial climate

FIGURE 11-13 **Glacial transfers of ^{12}C and ^{16}O** During glaciations (A), ^{12}C-enriched organic matter is transferred from the land to the ocean at the same time that ^{16}O-enriched water vapor is extracted from the ocean and stored in ice sheets. During interglaciations (B), ^{12}C-rich carbon returns to the land as ^{16}O-rich water flows back into the ocean.

must have been affecting oceanic $\delta^{13}C$ values through time. These additional factors provide other ways of tracking the movement of carbon through the ocean.

11-7 Pumping of Carbon into the Deep Ocean during Glaciations

Allowing for the physical effects of surface ocean temperature and salinity on glacial CO_2 levels noted earlier, we have found that we still have almost 90% of the glacial CO_2 decrease left to explain. We also know that the glacial surface ocean lost carbon (lower CO_2) while the deep ocean gained carbon. But how could such a transfer of carbon from surface to deep water occur?

One hypothesis, proposed by the geochemist Wally Broecker, is that carbon was exported from surface waters to the deep ocean by higher rates of photosynthesis and biologic productivity. This concept is called the **ocean carbon pump hypothesis**.

Photosynthesis and organic productivity occur in the surface ocean because sunlight and nutrients, such

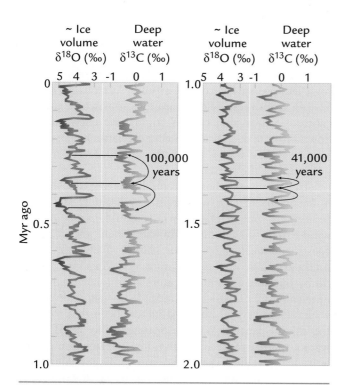

FIGURE 11-12 **Carbon transfer during glaciations** A sediment core from the deep Pacific Ocean shows more negative $\delta^{13}C$ values during glaciations ($\delta^{18}O$ maxima), in part because ^{12}C-enriched carbon is transferred from the land into the ocean. (Adapted from D. W. Oppo et al., "A $\delta^{13}C$ Record of Upper North Atlantic Deep Water During the Past 2.6 Million Years," *Paleoceanography* 10 [1995]: 373–94.)

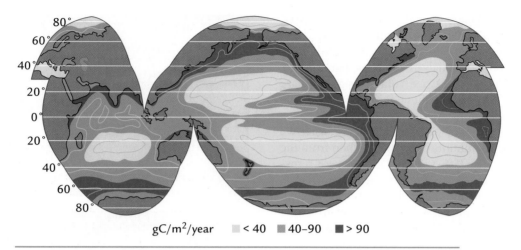

gC/m²/year ⬜ < 40 ⬛ 40–90 ⬛ > 90

FIGURE 11-14 **Annual carbon production in the modern surface ocean** Primary production of carbon (g/m²/yr) is highest in shallow coastal regions, in high-latitude oceans (especially the Southern Ocean), and across equatorial upwelling belts, but lower in central ocean gyres. (Adapted from W. H. Berger et al., "Ocean Carbon Flux: Global Maps of Primary Production and Export Production," in *Biogeochemical Cycling and Fluxes Between the Deep Euphotic Zone and Other Oceanic Realms*, National Undersea Research Program Report 88-1 [Asheville, N.C.: NOAA, 1987].)

as phosphorus and nitrogen, are available for energy (Chapter 2). Photosynthesis extracts CO_2 from surface waters and incorporates it in organic tissue (CH_2O), some of which sinks to the deep ocean:

$$CO_2 + H_2O \rightarrow CH_2O + O_2$$

If the rate of photosynthesis increases, CO_2 concentrations in surface waters will decrease because of greater downward export of carbon in dead organic matter. The pumping hypothesis asserts that greater downward removal of carbon occurs because more nutrients become available to stimulate greater removal of organic carbon from surface waters by photosynthesis.

What provides the source of added nutrients to stimulate greater photosynthesis? Most of the surface ocean is starved for nutrients because organisms readily use any that are available (Chapter 2). In contrast, the deep ocean is loaded with high nutrient concentrations, because the nitrogen and phosphorus contained in organic matter that falls from the surface waters are oxidized back to mineral form. But in the darkness of the deep sea, these vital nutrients remain unused, unless they are delivered back to the surface waters by upwelling or other processes.

Large-scale changes in biological pumping of carbon to the deep sea can occur only in relatively productive regions. The central gyres of today's warm, low-latitude oceans are starved for nutrients, and carbon export in these regions is low (Figure 11-14). Because these regions appear to have remained nutrient-starved during glaciations, they could not have been places of

increased photosynthetic pumping of carbon to the deep sea.

The most productive regions in the world ocean are in areas of upwelling along coastal margins and near the equator (especially in the Pacific Ocean) and in the high-latitude surface ocean around Antarctica (Figure 11-14). Both these regions are locations where significantly more organic carbon could have been pumped down into the deep ocean during glaciations than today, leaving surface waters with reduced CO_2 levels in comparison with the present.

Two kinds of changes could increase delivery of nutrients to surface waters in low-latitude upwelling regions. Stronger glacial winds could cause greater wind-driven upwelling of nutrient-rich waters along coasts and near the equator. Alternatively, upwelling might continue at the same rate as during interglaciations, but the upwelling waters might have a higher nutrient content. Most low-latitude upwelling draws on waters from several hundred meters deep that form in the middle latitudes of the Southern Ocean. If these waters carried more nutrients during glaciations, low-latitude upwelling could tap into a larger source of nutrients. Another possibility for increasing nutrient levels in the glacial surface ocean is by greater delivery of critical elements from the continents by stronger winds (Box 11-2).

The second region in which increased carbon biopumping might occur is near Antarctica, but under conditions different from those at lower latitudes. Solar radiation is weak at higher latitudes, and active photo-

BOX 11-2 CLIMATE DEBATE

Do Winds Fertilize the Glacial Ocean?

Much of the productivity in the surface ocean is a response to upwelling of cool subsurface waters rich in nutrients into nutrient-starved surface waters. In several regions of the ocean, both along coasts and in regions subject to trade winds, scientists have found evidence that winds blew stronger during the last glacial maximum, causing greater upwelling and bringing up more nutrients from the deep ocean to the surface.

The marine scientist John Martin has proposed a different way to fertilize the surface ocean and stimulate greater glacial photosynthesis and productivity. He has suggested that in addition to nitrogen and phosphorus, the element iron is critical to marine life (as it is to humans). Because the main source of iron for the oceans is erosion of the land, he suggested that ocean regions well away from the coasts receive their iron boost from iron-rich dust blown in by winds, a concept called the **iron fertilization hypothesis.**

Because of the iron fertilization effect, stronger glacial winds blowing off semiarid and arid continental areas should have carried greater amounts of iron to coastal and perhaps remote ocean areas than they do today, stimulating even greater productivity than changes caused by increases in wind-driven upwelling alone. Martin further hypothesized that extra iron blown from the southern tip of South America during glaciations stimulated greater productivity across a broad region

of the Southern Ocean and drew down atmospheric CO_2 levels.

Martin's hypothesis is still being debated. Oceanic field tests over areas of 10 to 20 square kilometers have shown that adding iron to surface waters can stimulate greater short-term productivity of ocean plankton. But it remains to be shown whether or not this iron fertilization effect stimulated significantly greater glacial productivity in the Southern Ocean and altered atmospheric CO_2.

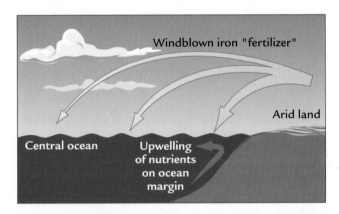

Iron fertilization of ocean surface waters The basic premise of the iron fertilization hypothesis is that dust rich in such elements as iron is blown to the ocean during glaciations by strong winds from arid continental interiors. The addition of this critical nutritional supplement stimulates productivity in mid-ocean regions far from coastal upwelling.

---

synthesis is limited to a brief summer season. In addition, the large-scale deep-ocean circulation in the Antarctic brings abundant nutrients up into the surface waters. Because of the short growing season and high nutrient levels today, photosynthesis cannot utilize most of the available nutrients, and high concentrations persist even during productive seasons. During glacial intervals in the Antarctic region, increased carbon pumping out of surface waters could occur through increased efficiency in the photosynthetic use of carbon in surface waters, either because nutrient-rich waters remained at the surface longer so that more of their carbon was used or because the even higher glacial nutrient levels stimulated more productivity.

How can we use the $\delta^{13}C$ method to test the hypothesis that the oceanic carbon pump sent more carbon to

the seafloor during glaciations? One promising avenue to follow is tied to the fact that the photosynthesis process that extracts nutrients from surface waters and sends them to the deep ocean also preferentially removes ^{12}C-enriched carbon and sends it to the deep ocean in dead organic matter (see Figure 11-11). Because photosynthesis removes both nutrients and ^{12}C from the surface ocean and exports them to the deep ocean, $\delta^{13}C$ and nutrient values are closely correlated throughout today's ocean (Figure 11-15). We cannot measure past nutrient values directly in the ocean (although proxy geochemical techniques to estimate them have been developed), but we can measure past $\delta^{13}C$ values in the ocean by analyzing the shells of foraminifera.

As we noted earlier, fractionation and removal of ^{12}C-rich carbon leaves inorganic carbon in surface

FIGURE 11-15 **Link between nutrients and δ¹³C values** Photosynthesis in ocean surface waters sends ¹²C-rich organic matter to the deep sea, leaving surface waters enriched in ¹³C (left). At the same time, photosynthesis extracts nutrients like phosphate (PO_4^{-2}) from surface waters and sends them to the deep sea. As a result, seawater δ¹³C values and phosphate concentrations are closely correlated (right). (Left: adapted from J. Kroopnick, "The Distribution of ¹³C in the World Oceans," *Deep-Sea Research* 32 [1985]: 57–64; right: adapted from E. A. Boyle, "Paired Carbon Isotope and Cadmium Data from Benthic Foraminifera: Implications for Changes in Oceanic Phosphorus, Oceanic Circulation, and Atmospheric Carbon Dioxide," *Geochimica Cosmochimica Acta* 50 [1986]: 265–76.)

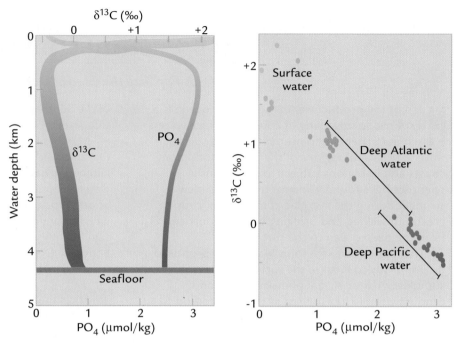

waters enriched in ¹³C (its δ¹³C values become more positive, ranging as high as +1 to +2 ‰), while the carbon that is oxidized back to inorganic form in deep waters causes them to become enriched in ¹²C (its δ¹³C values become more negative). These opposing effects of photosynthesis on δ¹³C values in surface waters and deep waters give us a way to measure the strength of the ocean carbon pump. Because planktic foraminif-

era record surface-water δ¹³C values and benthic foraminifera record deep-water δ¹³C values, we can use this pair of δ¹³C measurements to monitor the changing strength of the carbon pump (Figure 11-16).

For example, if greater nutrient delivery to surface waters increases the rate of photosynthesis and strengthens the ocean carbon pump, the result should be an increase in the *difference* in δ¹³C ratios between surface and deep waters (more positive values in surface waters and more negative values below). In contrast, when the carbon pump is weaker, less ¹²C-rich carbon is sent to the deep sea, and the δ¹³C difference between surface and deep waters is smaller. We should be able to detect these changes by analyzing δ¹³C in the shells of planktic and benthic foraminifera.

Ocean sediment records of the difference between surface- and deep-ocean δ¹³C values show suggestive (but not perfect) correlations in the predicted sense with ice core CO_2 trends (Figure 11-17). Such trends have led some investigators to claim that a stronger glacial carbon pump in the tropical oceans can explain as much as 25 ppm of the CO_2 lowering (Table 11-2), but other scientists infer a smaller impact. Evidence from the Southern Ocean does not indicate increased carbon pumping during glaciations, although more work needs to be done in this critical region to verify this conclusion.

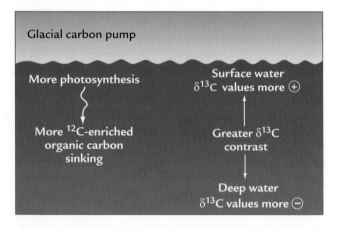

FIGURE 11-16 **Measuring changes in the ocean carbon pump** Greater photosynthesis in surface waters during glaciations would pump more organic carbon to the deep sea and reduce atmospheric CO_2 levels. Past changes in this process can be measured by the increased difference between ¹³C-enriched carbon (causing higher δ¹³C values) in the shells of planktic foraminifera living in surface waters and ¹²C-enriched carbon (causing lower δ¹³C values) in the shells of benthic foraminifera living on the deep ocean floor.

11-8 Changes in the Circulation of Deep Water during Glaciations

A second possible cause of observed variations in oceanic δ¹³C values during orbital cycles is changes in the pat-

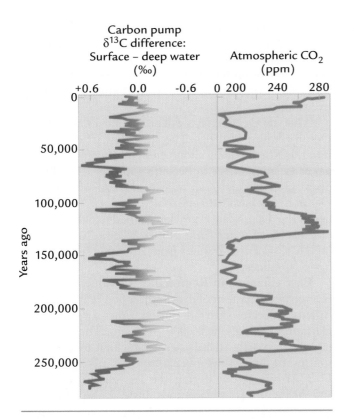

FIGURE 11-17 **Past changes in the ocean carbon pump** If the ocean carbon pump affects atmospheric CO_2 levels, the net difference between surface and deep-water $\delta^{13}C$ values should increase when CO_2 levels are low. Measured $\delta^{13}C$ differences show some correlation with past changes in atmospheric CO_2. (Adapted from W. B. Curry and T. J. Crowley, "The $\delta^{13}C$ of Equatorial Surface Waters: Implications for Ice Age pCO$_2$ Levels," *Paleoceanography* 2 [1987]: 489–517.)

TABLE 11-2 **All Causes of Lowering of Glacial CO_2 Levels by 90 Parts per Million**

Cause	CO$_2$ Change (parts per million)
Physical properties of surface waters	
Cooling by ~2.5°C	−22
Increase in salinity by 1.1‰	+11
All physical properties	−11
Biochemical-chemical redistribution of ocean carbon	
Carbon pumping	
Antarctic Ocean	?
Tropical upwelling regions	−25?
Increased CO$_3^{-2}$	
Global deep water	−10?
Antarctic surface water	−40
All biochemical-chemical redistribution	>−75
Total estimated CO$_2$ change	>−86

terns of deep-water flow. Varying degrees of photosynthesis in different areas of the ocean give surface waters distinctively different nutrient contents and $\delta^{13}C$ values. As a result, deep waters that form from these surface waters in different areas of the ocean start their trip into the deep ocean with their own distinctive nutrient and $\delta^{13}C$ values. Because nutrient concentrations and $\delta^{13}C$ values have opposite trends in surface waters, deep waters of the world ocean show the same negative correlation (Figure 11-15). Low nutrient concentrations correlate with high $\delta^{13}C$ values, and vice versa.

Present-Day Controls on Regional $\delta^{13}C$ Values
Regional variations of $\delta^{13}C$ values measured in deep-water inorganic carbon are shown in Figure 11-18. At one extreme are the relatively high $\delta^{13}C$ values (> 0.8‰) of waters in the North Atlantic at depths near 3000–4000 meters. These values result from the fact that surface waters greatly enriched in ^{13}C by photosynthesis at subtropical latitudes move north to high latitudes and sink as North Atlantic deep water tagged with

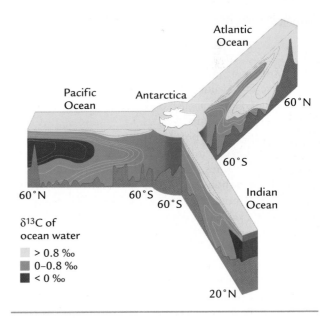

FIGURE 11-18 **Modern deepocean $\delta^{13}C$ patterns** In today's ocean, photosynthesis and carbon isotope fractionation leave $\delta^{13}C$ values higher in surface waters and lower in deep waters. (Adapted from C. D. Charles and R. G. Fairbanks, "Evidence from Southern Ocean Sediments for the Effect of North Atlantic Deepwater Flux on Climate," *Nature* 355 [1992]: 416–19.)

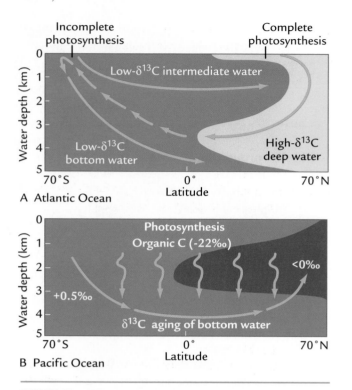

A Atlantic Ocean

B Pacific Ocean

FIGURE 11-19 Regional δ¹³C differences (A) Deep water formed in the North Atlantic is high in δ¹³C and low in nutrients, while intermediate and bottom waters formed in the Southern Ocean near Antarctica are low in δ¹³C but rich in nutrients. (B) Highly negative δ¹³C values in the deep North Pacific result from gradual sinking of ¹²C-rich carbon from low-latitude surface waters into deep waters slowly moving northward from the Antarctic.

more positive δ¹³C values (Figure 11-19A). The plume of positive δ¹³C values flowing south at depths of 2000 to 4000 meters defines the core of this flow. This water is also poor in nutrients.

In contrast, the bottom water and intermediate-depth water that form in the Antarctic region have less positive δ¹³C values, < 0.5‰. In the Antarctic, fractionation of carbon isotopes is incomplete because of the limited amount of photosynthesis, and extremely positive δ¹³C values do not develop in surface waters. As a result, Antarctic bottom water sinks with low δ³C values and high concentrations of nutrients. The contrast between high-δ¹³C North Atlantic water and low-δ¹³C Antarctic water makes them easy to trace.

A second process also affects the regional pattern of modern δ¹³C values. As deep waters flow from their source regions to other parts of the world ocean, their initial δ¹³C values gradually become more negative, an effect sometimes called **δ¹³C aging**. The cause of this gradual shift toward negative δ¹³C values is the downward rain of ¹²C-rich organic carbon (δ¹³C = −22‰) produced by photosynthesis in surface waters. Most of

the sinking carbon gradually oxidizes to inorganic form but retains its very negative δ¹³C values. As it mixes with the large amount of inorganic carbon already present in deep waters (and having δ¹³C values closer to 0‰), it shifts the deep-water δ¹³C toward slightly more negative values.

This aging effect is evident mainly in the Pacific Ocean (Figure 11-19B). The slow flow of deep water into the North Pacific allows enough time for the gradual δ¹³C aging effect to become apparent. In contrast, deep water circulates so quickly through the Atlantic Ocean that it has little time to age. As a result, δ¹³C values in the deep Atlantic are determined mainly by physical mixing of water from North Atlantic sources with high δ¹³C values and water from Antarctic sources with low δ¹³C values.

Past Changes in Regional δ¹³C Values A major change in the pattern of distribution of δ¹³C in the deep Atlantic Ocean occurred during times of maximum glaciation like the one 20,000 years ago. The tongue of high-δ¹³C North Atlantic deep water that today moves southward at depths of 2000–4000 meters (Figure 11-20A) was absent during the last glacial maximum and during previous glaciations. Instead, the water with highest δ¹³C values in the Atlantic was limited to depths above 1500 meters (Figure 11-20B).

This change in δ¹³C distribution indicates a fundamentally different pattern of deep circulation in the Atlantic during glaciations. We can interpret this flow pattern by assuming that waters coming from the north and south polar regions were both tagged with δ¹³C values similar to those they have today (although the values need not have been identical to modern values). The glacial pattern shown in Figure 11-20B suggests that water chilled in northern latitudes did not become dense enough to sink as deep as it does today, but instead reached only intermediate depths (< 1500 meters). The lower δ¹³C values at depths below 2000 meters indicate the presence of a water mass that originated near Antarctica and flowed north. This change represents a relative glacial increase in water from Antarctic sources compared to conditions at present.

We can track these changing deep-water circulation patterns in the Atlantic Ocean back through time by looking at records of δ¹³C changes in benthic foraminifera from sediment cores. The tropical Atlantic seafloor at 3000 to 4000 meters is a particularly good place for this kind of study, because it lies in the geographic region and at the water depths most influenced by δ¹³C changes between the glacial and interglacial pattern (Figure 11-20).

Measurements of δ¹³C values in the shells of benthic foraminifera from the tropical Atlantic can be used to calculate the relative contributions of deep water arriving from two major source regions: water with high

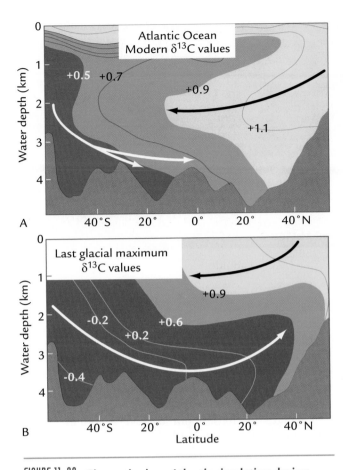

A

B

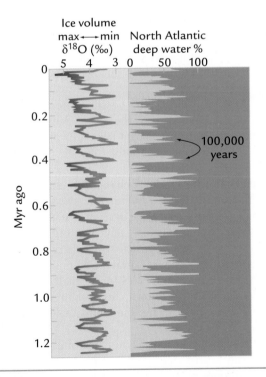

dominate during glaciations ($\delta^{18}O$ maxima). The measured fluctuations in deep-water sources have a prominent 100,000-year cyclicity after 0.9 Myr ago.

The presence of larger amounts of low-$\delta^{13}C$ Antarctic source water during glaciations shifts the $\delta^{13}C$ values in the deep Atlantic Ocean toward more negative values. This shift pushes $\delta^{13}C$ values in the same direction as the additions of ^{12}C-rich carbon from the continents during glaciations. This added effect explains a result noted earlier—the fact that some deep-ocean cores show glacial $\delta^{13}C$ changes larger than those predicted by the land-sea transfers alone.

The carbon isotope trend from the tropical Atlantic provides clear evidence of an important link between the size of northern hemisphere ice sheets and the formation of deep water in the North Atlantic nearby (Figure 11-21). Less deep water formed in the North Atlantic every time ice sheets grew larger at the 100,000-year cycle after 0.9 Myr ago. Conversely, more northern-source deep water formed when ice sheets were small or absent.

Changes in Ocean Chemistry But how would changes in the type of water masses reaching the remote ocean depths have contributed to lower CO_2 concentra-

FIGURE 11-20 Change in deep Atlantic circulation during glaciation In contrast to the modern $\delta^{13}C$ pattern in the Atlantic Ocean (A), the last glacial maximum (B) shows no plume of high-$\delta^{13}C$ water sinking to depths of 2000–4000 m in the North Atlantic and flowing south. (Modified from J.-C. Duplessy and E. Maier-Reimer, "Global Ocean Circulation Changes," in *Global Changes in the Perspective of the Past*, ed. J. A. Eddy and H. Oeschger [New York: John Wiley, 1993].)

$\delta^{13}C$ values formed in the North Atlantic, and water with low $\delta^{13}C$ values formed near Antarctica. Changes in initial $\delta^{13}C$ values through time at each source area can also be determined by analyzing benthic foraminifera in cores from the source regions. Then $\delta^{13}C$ values measured in foraminifera at the tropical Atlantic site can be expressed in terms of the percentage contributions from the two sources. This kind of analysis removes the $\delta^{13}C$ changes common to all oceans caused by carbon transfers from land to sea and lets scientists quantify the relative contributions from northern and southern sources through time at any site.

A long record from one tropical Atlantic core shows cyclic changes in the percentage of northern-source and southern-source water (Figure 11-21). In general, northern-source waters dominate during interglaciations ($\delta^{18}O$ minima), while Antarctic-source waters

FIGURE 11-21 Changing sources of Atlantic deep water The percentage of deep water originating in the North Atlantic and flowing to the equator during the last 1.25 Myr has been consistently lower during glaciations than during interglaciations. (Adapted from M. E. Raymo et al., "The Mid-Pleistocene Climate Transition: A Deep-Sea Carbon Isotopic Perspective," *Paleoceanography* 12 [1997]: 546–59.)

FIGURE 11-22 Carbon system controls on CO₂ in the glacial atmosphere Among the changes in the distribution of ocean carbon that could reduce glacial levels of atmospheric CO_2 are faster downward organic carbon pumping in areas of coastal or tropical upwelling (1) and in the Antarctic (2), changes in the chemistry of Antarctic surface waters toward higher CO_3^{-2} content (3), and changes in the chemistry of shallow subsurface waters originating from southern latitudes (4).

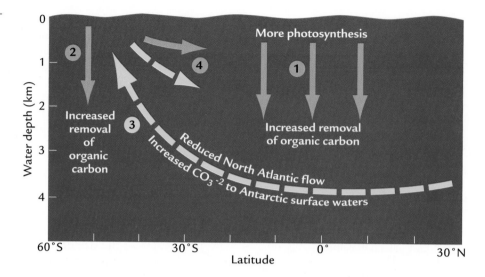

tions in the glacial atmosphere? One likely link is through the carbonate chemistry of deep-ocean water. One of the major factors that determines CO_2 levels in surface waters of the world ocean is the amount of the carbonate ion CO_3^{-2}. These ions are produced when corrosive bottom water dissolves $CaCO_3$ on the seafloor. When these ions are later returned to surface waters by ocean circulation, they can combine chemically with CO_2 in the surface waters and produce the bicarbonate ion HCO_3^-. As a result, increasing the amount of CO_3^{-2} ions dissolved by and carried in deep water can have the effect of reducing CO_2 concentrations in surface water.

Initial explorations of the possible effect of glacial changes in the chemistry of the entire deep ocean on atmospheric CO_2 pointed to a large (40 ppm) reduction in CO_2. But scientists later came to the conclusion that deep-ocean regions experienced different and often opposing kinds of changes in corrosiveness during glaciations (increases in some areas, decreases in others). Because these changes in $CaCO_3$ dissolution to some extent canceled each other out at a global scale, at most a small global-scale increase in the CO_3^{-2} content of deep or surface waters seems likely to result from this factor. With only small changes in surface-water CO_3^{-2} content, little change in atmospheric CO_2 (−10 ppm by one estimate) would have occurred (Table 11-2).

The conclusion that *average* carbonate dissolution rates in the world ocean changed by only a small amount still does not rule out the possibility of important regional-scale CO_3^{-2} ion changes. A recent hypothesis by Wally Broecker and his colleague Tsung-Hung Peng suggests that changes in the CO_3^{-2} content of surface waters in the Southern Ocean may have had a disproportionately large impact on atmospheric CO_2. Antarctic surface waters are particularly vulnerable to

changes in deep-ocean CO_3^{-2} content because they are in direct communication with deep water arriving from depths of 3 to 4 kilometers in the Atlantic. North Atlantic deep water rises to the surface in the Southern Ocean during interglaciations like today, but this flow was suppressed or even eliminated entirely during glaciations (Figure 11-20).

Changes in the amount of CO_3^{-2} ion in glacial Antarctic surface waters could have occurred as a result of different chemical properties of the deep waters rising to the surface. Because modern-day North Atlantic deep water is a relatively noncorrosive water mass, it dissolves relatively little $CaCO_3$ on the seafloor and delivers relatively small concentrations of CO_3^{-2} ions to the Antarctic sea surface, allowing surface-water CO_2 to remain at relatively high levels.

But we have seen that the deep circulation in the Atlantic Ocean was entirely different during glaciations. Noncorrosive North Atlantic deep water was replaced through most of the deep Atlantic basin by water from southern sources, which today are much more corrosive and dissolve $CaCO_3$ much more rapidly. If deep water richer in CO_3^{-2} surfaced in the Southern Ocean during glaciations, the resulting increase in CO_3^{-2} content of Antarctic surface waters would have reduced their CO_2 content. Broecker and Peng estimate that this mechanism may explain 40 ppm of the observed decrease of 90 ppm in atmospheric CO_2 (Table 11-2).

Changes in the chemistry of Antarctic surface waters may also be transmitted to other parts of the ocean during glaciations. One possibility is that waters following the northward path of modern-day Antarctic intermediate water at depths at or above 1500 meters (Figure 11-19A) were richer in CO_3^{-2} and lower in CO_2 than they are today. If this water welled up in the tropics, it should have reduced CO_2 levels in surface waters and may have resulted in lower atmospheric CO_2.

In summary, at least six factors may contribute to atmospheric CO_2 changes during glacial intervals (Table 11-2). Four of the factors are specifically tied to the carbon cycle through changes in nutrient upwelling or through changes in Antarctic surface-water chemistry (Figure 11-22). The factors that at this point appear to explain the largest portion of the 90-ppm drop in glacial CO_2 levels are lower surface-ocean temperatures, stronger photosynthetic pumping of carbon, and increased CO_3^{-2} ion concentrations linked to changes in deep-ocean circulation. Climate scientists continue to explore the many mechanisms that could be responsible for the large drop in glacial CO_2 levels.

Key Terms

ice flow model (p. 236)

sintering (p. 236)

carbon isotopes (p. 242)

ocean carbon pump hypothesis (p. 245)

iron fertilization hypothesis (p. 247)

$\delta^{13}C$ aging (p. 250)

Review Questions

1. How are ice cores dated?

2. Why are air bubbles in ice cores younger than the ice in which they are sealed?

3. What features of the ice core CH_4 signal suggest a link to tropical monsoons?

4. To what extent does a cooler glacial ocean surface explain lower CO_2 in the glacial atmosphere?

5. Which reservoir has the most carbon: the atmosphere, vegetation, or the ocean?

6. Where did the carbon (CO_2) removed from the atmosphere go during glaciations?

7. What effect does fractionation have on carbon-isotope values?

8. Explain how carbon isotopes ($\delta^{13}C$) trace shifts of carbon from the land to the ocean.

9. How does the ocean carbon pump reduce atmospheric CO_2?

10. What evidence indicates a different glacial deep-water flow pattern in the North Atlantic?

11. How could a change in deep circulation in the Atlantic alter atmospheric CO_2?

Additional Resources

Basic Reading

Barnola, J. M., D. Raynaud, Y. S. Korotkevich, and C. Lorius. 1987. "Vostok Ice Core Provides a 160,000 Year Record of Atmospheric CO_2." *Nature* 329:408–14.

Chappellaz, J., J. M. Barnola, D. Raynaud, Y. S. Korotkevich, and C. Lorius. 1990. "Ice Core Record of Atmospheric Methane over the Last 160,000 Years." *Nature* 345:127–31.

Advanced Reading

Broecker, W. S., and T.-H. Peng. 1989. "The Cause of the Glacial to Interglacial Atmospheric CO_2 Change: A Polar Alkalinity Hypothesis." *Global Biogeochemical Cycles* 3:215–39.

Martin, J. H. 1990. "Glacial-Interglacial CO_2 Change: The Iron Hypothesis." *Paleoceanography* 5:1–13.

Petit, J. R., et al. 1999. "Climate and Atmospheric History of the Past 420,000 Years from the Vostok Ice Core, Antarctica." *Nature* 399:429–37.

Pisias, N. D., and N. J. Shackleton. 1984. "Modeling the Global Climate Response to Orbital Forcing and Atmospheric Carbon Dioxide Changes." *Nature* 310:757–59.

Prell, W. L., and J. E. Kutzbach. 1987. "Monsoon Variability over the Past 150,000 Years." *Journal of Geophysical Research* 92:8411–25.

Raynaud, D., J. Jouzel, J. M. Barnola, J. Chappellaz, R. J. Delmas, and C. Lorius. 1993. "The Ice Record of Greenhouse Gases." *Science* 259:926–34.

CHAPTER 12

Orbital-Scale Interactions in the Climate System

Which came first, the chicken or the egg? Scientists studying climate often face questions of this kind: Which factors drive changes in climate and which are just internal responses? Earlier, you learned that cyclic changes in orbital insolation are important as the *ultimate* driver of changes in monsoons, ice sheets, and greenhouse gases over cycles of tens of thousands of years. But important questions about the orbital-scale operation of the climate system remain unanswered.

In this chapter, we examine many orbital-scale responses in Earth's climatic archives, including changes in continental vegetation, ocean temperatures, and windblown dust. These responses vary at orbital time scales, but why? Some of these signals are responses to the changing size of ice sheets rather than direct responses to changes in insolation. Then we return to the central problem of interactions between atmospheric CO_2 and the ice sheets and attempt to figure out which drives which (the most crucial chicken-and-egg question). Finally, we look at a major mystery at orbital time scales: What explains the 100,000-year ice sheet cycles during the last 0.9 Myr?

Orbital-Scale Forcing and Response Revisited

The basic concepts of forcing, response, and response times in the climate system presented in Chapter 1 have reappeared in connection with specific issues in Chapters 8–11. These concepts are central to the orbital-scale changes we review in this chapter.

A broader application of forcing-and-response concepts to orbital-scale climate changes is summarized in Figure 12-1. Consider two parts of the climate system, each responding to the insolation driving force (Figure 12-1A). One part of the climate system has a fast response time, measured in months or years, and the other has a slow response time, measured in many thousands of years. The fast-responding component should closely track the insolation changes that drive it, while the slow-responding part should lag many thousands of years behind. In fact, we have already looked at two parts of the climate system with these kinds of responses to insolation: the fast-responding monsoon system over North Africa (Chapter 9) and the slow-responding ice sheets at high northern latitudes (Chapter 10).

Now we add a new ingredient to this analysis: *interactions* among components within the climate system. When different components of Earth's climate system are in close geographic proximity, they are likely to interact. And when they do, the slow-responding parts of the climate system often set the tempo for the fast-responding parts. In effect, the fast-responding parts have no thermal inertia and so are easily influenced (forced) by other parts of the climate system. In contrast, the slow-responding components have a large thermal inertia. As a result, their response gains some degree of independence from the rest of the climate system.

Consider the margin of a glacial-maximum ice sheet 20,000 years ago, just before deglacial melting began. Air temperatures along this ice margin are likely to have been determined more by the presence of the huge mass of nearby ice than by local insolation changes. Even if insolation had risen to values almost comparable to those of today, the refrigeration provided by the nearby ice sheet would still have been the main factor controlling local air temperatures.

If we had a sequence of long-term measurements of the temperature of the air in this region, we would probably find that it tracks the size of the ice sheet fairly closely (Figure 12-1B). Every time the ice sheet advances, the air chills. Every time the ice retreats, the air warms. In this case, a slow-responding internal part of the climate system (the ice) is, in effect, acting as the regional climatic driver or forcing, while a second internal component (the air) is quickly responding to that internal source of forcing.

We could try out an alternative interpretation of the trends shown in Figure 12-1B. Because the fast-response (air temperature) record shows cyclic behavior at orbital time scales, we could argue that insolation *directly* drives air temperature. But to make this argument, we would have to explain why the temperature of the air, with its characteristic response time of days to weeks, would lag thousands of years behind the insolation changes proposed as the driving force. In the end this argument makes no sense and can be rejected.

The explanation that does make sense is that changes in insolation drive the slow-responding ice sheet with a large lag, and the ice sheet in turn drives the air temperature with no perceptible lag. In this first chicken-or-egg example, the most obvious interpretation is that the ice sheet's response came first, followed by the response of the nearby air. This example is typical of the orbital-scale evidence that scientists must interpret. By comparing the relative responses of different components of the climate system, and by doing so with an awareness of their different characteristic response times, we can learn much about the climate system.

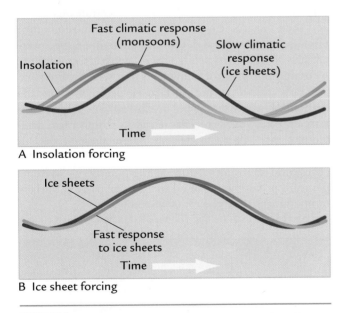

A Insolation forcing

B Ice sheet forcing

FIGURE 12-1 **Slow and fast responses to insolation and ice** (A) When insolation drives climate change on Earth, fast-responding parts of the climate system, such as air temperature, may track along with insolation, while slow-responding parts will lag behind. (B) Slow-responding ice sheets may also set the tempo of response for fast-responding nearby parts of the climate system.

Ice-Driven Climate Responses

Ice sheets act as important drivers of climate change from *within* the climate system because of several physical characteristics. One is their height: ice sheets protrude thousands of meters into the air and form massive

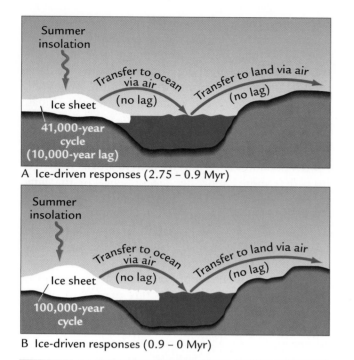

A Ice-driven responses (2.75 – 0.9 Myr)

B Ice-driven responses (0.9 – 0 Myr)

FIGURE 12-2 **Ice-driven responses** Orbital-scale ice sheet rhythms may be quickly transferred to other parts of the climate system via the atmosphere and ocean. (Adapted from W. F. Ruddiman, "Northern Oceans," in *North America and Adjacent Oceans During the Last Deglaciation*, ed. W. F. Ruddiman and H. E. Wright, Geological Society of America DNAG vol. K-3 [Boulder, Colo.: Geological Society of America, 1987].)

obstacles to the free circulation of winds in the lower atmosphere, thereby rearranging the flow of air. Another property is their bright surfaces, which reflect most incoming sunlight, creating a major albedo contrast with ice-free ground, which absorbs most incoming radiation. Ice sheets also calve icebergs and deliver meltwater to nearby oceans.

Any fast-response part of the climate system driven directly by changes in ice sheet size should react with the same pattern of climate changes as the ice (Figure 12-1B). We will call these **ice-driven responses**. For the last 3 Myr, the major changes in Earth's ice sheets have occurred in the northern hemisphere. The simple conceptual model we are using suggests that the ice-driven responses should show the same major orbital rhythms as the ice sheets—mainly 41,000-year cycles before 0.9 Myr ago and 100,000-year cycles since that time—and they should track the ice sheet cycles without any obvious lag (Figure 12-2).

12-1 Ice-Driven Responses in High Northern Latitudes

The kinds of ice-driven responses noted above are most apparent across regions in relatively close proximity to the northern hemisphere ice sheets. Scientists have traced evidence of ice-driven responses across the high northern latitudes from the North Atlantic Ocean to Europe to Asia and into the western North Pacific (Figure 12-3).

Ocean Surface Temperatures The North Atlantic Ocean is an obvious place to expect an ice-driven response. The high-latitude North Atlantic is almost surrounded by the great ice sheets of North America and Eurasia, and it might be expected to track the changing size of ice sheets with no perceptible lag, because the surface ocean is a fast-responding part of the climate system.

As discussed in Chapter 3, sea surface temperatures from the last million or more years of Earth's history can be reconstructed by analysis of assemblages of plankton (in this case, foraminifera) that live in surface waters and respond to environmental conditions there, and then die and leave fossil shells in ocean sediments. Climate scientists first determine the temperature preferences of modern assemblages of planktic foraminifera by examining shells of individuals that recently died and are now found at the surface of sediments on the seafloor. These assemblages are linked to average ocean temperatures over recent decades from atlas data. Counts are then made of older populations of foraminifera at regular intervals through sediment cores, and the information on their modern-day temperature preferences is used to estimate past sea surface temperatures from the older assemblages.

FIGURE 12-3 **Regions of ice-driven responses** Mid- and high-latitude regions of the northern hemisphere show evidence of climate responses controlled by changes in the sizes of ice sheets.

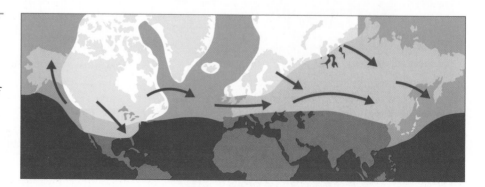

Past sea surface temperatures estimated in this way correlate almost perfectly with $\delta^{18}O$ (~ ice volume) in a core from the North Atlantic Ocean (Figure 12-4). The records shown span an interval from 1.5 to 1.2 Myr ago, a time when the 41,000-year cycle of orbital tilt dominated the ice sheet ($\delta^{18}O$) response. The $\delta^{18}O$ record in this sediment core cycles at an obvious 41,000-year rhythm. The North Atlantic sea surface temperature signal tracks the $\delta^{18}O$ (ice volume) signal peak for peak during this entire 300,000-year interval, with warm ocean temperatures at times when ice sheets were small and cool temperatures when ice sheets were large.

This match is exactly the kind of signal we expect to see if orbital-scale changes in sea surface temperature are driven by the ice sheets rather than by insolation (Figure 12-1). In this part of the North Atlantic at this time, the temperature of the sea surface was largely

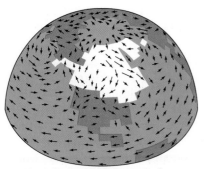

A Glacial winter surface winds

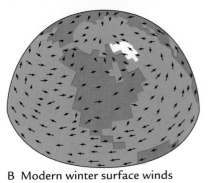

B Modern winter surface winds

FIGURE 12-5 **Ice sheet sensitivity test** A sensitivity test with a GCM shows that inserting an ice sheet over North America (A) causes a clockwise flow of cold winds out over the nearby North Atlantic Ocean, whereas the flow of air there is warm today (B). (Courtesy of P. Behling, University of Wisconsin, Madison.)

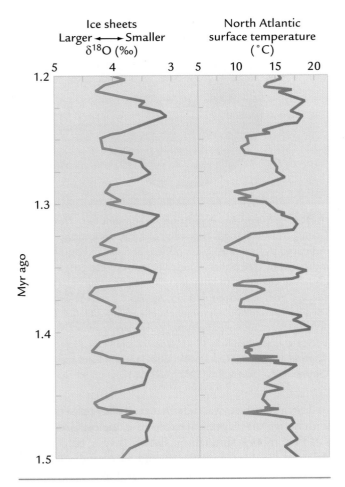

FIGURE 12-4 **North Atlantic surface response to ice** Climate signals in a North Atlantic sediment core show that sea surface temperatures between 1.5 and 1.2 Myr ago closely tracked $\delta^{18}O$ (ice volume) fluctuations. (Adapted from W. F. Ruddiman et al., "Pleistocene Evolution: Northern Hemisphere Ice Sheets and North Atlantic Ocean Climate," *Paleoceanography* 4 [1989]: 353–412.)

under the control of the nearby ice sheets at orbital time scales. The same kind of relationship holds for the interval after 0.9 Myr, when the dominant 100,000-year oscillations in ice sheets were matched by simultaneous 100,000-year cycles of sea surface temperature.

How do the ice sheets transfer their climatic signal to the ocean? One mechanism might be icebergs that calve off from the margins of the ice sheets bordering the ocean and float out into the ocean to melt. The melting icebergs cool the ocean surface.

Another way for ice sheets to influence nearby regions is through the atmosphere. Climate scientists have run sensitivity-test experiments in which large ice sheets were inserted into general circulation models (GCMs) to isolate their effects on atmospheric circulation. The huge ice sheet over North America had the largest influence on nearby ocean temperatures. A clockwise flow of winds occurred around the central dome of the North American ice sheet (Figure 12-5A). This flow sent very cold winds from the northern part of the ice sheet toward the southeast and out over the western North Atlantic. This cold, dry air replaced the warmer air that blows mainly from the southwest during interglacial times like today, when ice sheets are absent (Figure 12-5B). A second important effect (in summer) was the presence of a high-albedo ice surface at latitudes where snow does not fall today.

As a result of these ice-related changes, the model simulated a 5° to 10°C cooling in the high latitudes of the North Atlantic, similar to the cooling estimated from planktic foraminifera (Figure 12-4). The evidence from the climate record and the model simulation indicates that ice sheets can transfer a large orbital-scale signal to the ocean via the atmosphere, with no lag.

Climate in Northern Europe and Asia We can also track this signal eastward into Europe, where sediment cores from lakes and bogs show dramatic changes in pollen over the last 130,000 years (Figure 12-6). Pollen grains deposited during warm, moist interglacial climates like today's are mostly from trees like those in present-day European forests. In contrast, levels deeper in the same cores consist mainly of grass and herb pollen, indicating much drier and colder conditions in the past. When climate scientists dated the organic carbon in these deposits, the cold, dry levels turned out to be glacial in age. The record of European pollen fluctuations over the last 130,000 years correlates well with the $\delta^{18}O$ (ice volume) changes recorded in ocean cores. Both records clearly show one longer-term 100,000-year cycle and several shorter orbital-scale oscillations, and the dating suggests no large leads or lags between the two signals.

This evidence tells us that central Europe was covered by cold, dry, tundralike vegetation during glaciations and that a barren polar desert with almost no vegetation at all existed along the margins of the ice sheet in northern Europe. How did this remarkable change from the moist, temperate conditions of today occur?

Part of the explanation is that cold, dry winds blowing southward from the Scandinavian ice sheet at higher latitudes altered climate over Europe, but the changes over the Atlantic Ocean just west of Europe were also important. Today the inflow of warm, moist air from the southwest off the Atlantic Ocean keeps Europe's climate temperate at the same latitudes where winters are bitterly cold in North America and Asia. During glaciations, with the North Atlantic chilled by cold winds and icebergs from the North American ice sheet, and with some of the ocean surface probably frozen over completely during especially cold winters, this source of maritime moisture and warmth for Europe in winter must have been choked off.

Another GCM sensitivity test has been run to evaluate how large an effect ocean temperature has on Europe's climate. For this model experiment, the only change in the boundary condition input was the temperature of the North Atlantic Ocean. Ocean temperatures north of 20°N were reduced to their full glacial values in the experiment (Figure 12-7A), while the rest

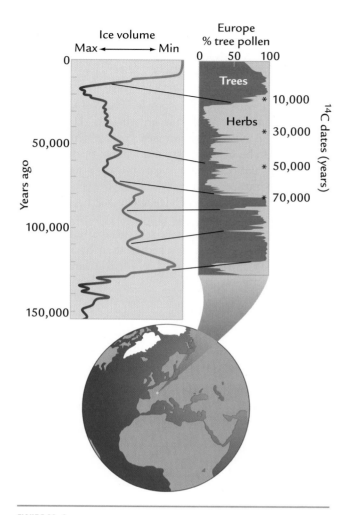

FIGURE 12-6 European vegetation A sediment core from France shows changes between cold, dry conditions, indicated by herb and grass pollen, and warm conditions, indicated by tree pollen. These vegetation responses correlate closely with changes in $\delta^{18}O$ (ice volume). (Left: adapted from D. Martinson et al., "Age Dating and the Orbital Theory of the Ice Ages: Development of a High-Resolution 0 to 300,000-Year Chronostratigraphy," *Quaternary Research* 27 [1987]: 1–29; right: adapted from G. M. Woillard and W. G. Mook, "Carbon-14 Dates at Grande Pile: Correlation of Land and Sea Chronologies," *Science* 215 [1982]: 159–61.)

of Earth's surface was left in exactly its present state, with no ice sheets present except the one on Greenland.

The model simulation shows that a cold North Atlantic sea surface produces cold air temperatures not just over the North Atlantic but also downwind into the western maritime parts of Europe, and even well into the Eurasian continent (Figure 12-7B). Precipitation in these chilled regions of Europe and Asia also fell significantly in the simulation.

The combination of evidence from the climatic record and the model simulation indicates that an ice sheet signal initially transferred to the North Atlantic can also be transferred into Europe and even Asia. Both of these signal transfers occurred without any significant lags.

We can trace this ice-driven response farther eastward into east-central Asia. Remarkable features called loess plateaus are found in western China (see Figure 3-3). These plateaus consist of two kinds of material lying in layered sequences hundreds of meters thick. One is the type of weathered clay-rich soil produced by warm, moist climates like that found in southern China today. The other kind of deposit consists of silt-sized loess grains deposited by winds under dry, cold conditions, which allowed little or no chemical weathering. The loess grains may have been produced in central Asia by mountain glaciers and other physical weathering processes and then lifted by winds and transported south by strong winds to the loess plateau.

The loess plateau deposits also contain a history of reversals of Earth's magnetic field that can be used to date the entire sequence. Until 2.75 Myr ago, only weathered clays were deposited in the loess plateau region. Afterward, beginning at the same time as the first cycles of northern hemisphere glaciation, layers of silty loess began to be sandwiched between layers of clay-rich soil. Over the last several hundred thousand years, alternations between soils and loess have occurred at a 100,000-year cycle, with each loess layer deposited during a time of major glaciation (Figure 12-8, left). Once again we can see the familiar imprint of the tempo of ice sheet fluctuations, but now in a region far from the North American and Eurasian ice sheets.

How the ice sheet signal penetrates so far into eastern Asia is not entirely clear. Part of the reason may be the loss of Atlantic moisture, some of which penetrates far into central Eurasia during warm climates like today's. With a chilled and partly ice-covered North Atlantic (Figure 12-7), this source of moisture would have been choked off. Another explanation indicated by GCM experiments is that the Siberian high-pressure center, even today the winter-season source of strong cooling and powerful winds in northern Asia (Chapter 2), may have strengthened during glacial climates and lasted for a much longer part of the year. These winds would have blown greater amounts of loess particles created in central Asia southward to the loess plateaus.

Additional evidence for dry, windy conditions during glaciations comes from the western North Pacific Ocean. There sediment cores contain records of fine-grained, wind-transported silt from east-central Asia (Figure 12-8, right). Maximum values in windblown silt occur during maxima in $\delta^{18}O$ (ice volume) every 100,000 years over the last 500,000 years, indicating some combination of drier source regions and stronger winds during glaciations.

By this point, we have tracked the characteristic signal of an ice-driven response more than halfway around the middle latitudes of the northern hemisphere. All the evidence we have seen points to immediate transfer of the orbital-scale ice sheet signal to the air, to the water, and to the land surfaces of this region. We can even track this ice-driven imprint all the way around to the Greenland ice sheet. Dust particles in glacial-age layers of Greenland ice come mainly from Asia, lifted by strong winds and blown eastward at jet-stream elevations in the atmosphere. This brings us full circle around the northern hemisphere middle latitudes.

In summary, the ice volume signal can be transferred far from the immediate proximity of the ice sheets by altered wind patterns and by resulting changes in air and surface-ocean temperatures and in precipitation over land. As a result, northern hemisphere ice sheets have clearly been the main control on orbital-scale signals of climate change across the northern mid-latitudes over the last 2.75 million years, and especially during the

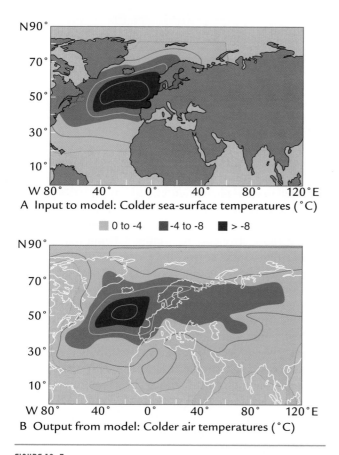

A Input to model: Colder sea-surface temperatures (°C)

0 to -4   -4 to -8   > -8

B Output from model: Colder air temperatures (°C)

FIGURE 12-7 **Surface-ocean sensitivity test** A sensitivity test with a GCM shows that inserting cold glacial ocean temperatures into an interglacial world (A) produces colder air temperatures downwind, over Europe and Asia (B). (Adapted from D. Rind et al., "The Impact of Cold North Atlantic Sea Surface Temperatures on Climate: Implications for the Younger Dryas," *Climate Dynamics* 1 [1986]: 3–33.)

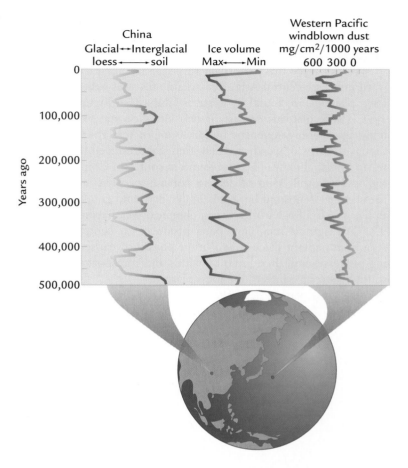

FIGURE 12-8 Responses of windblown debris in East Asia to ice volume Alternating layers of windblown loess and soils in Southeast Asia (left) and the amount of fine dust blown out to the western Pacific (right) both show changes that closely match δ¹⁸O (ice volume) changes (middle). (Left: adapted from G. Kukla et al., "Pleistocene Climates in China Dated by Magnetic Susceptibility," *Geology* 16 [1988]: 811–14; middle and right: adapted from S. A. Hovan et al., "Late Pleistocene Continental Climate and Oceanic Variability Recorded in Northwest Pacific Sediments," *Paleoceanograpy* 6 [1991]: 349–70.)

large 100,000-year cycles of the last 0.9 Myr. So far, the chicken-and-egg issue does not seem to be a difficult problem: ice sheets respond to orbital forcing and drive numerous other responses in high northern latitudes.

But how far across the planet would we expect the physical influences of the ice sheets (those transmitted by changes in winds, albedo, icebergs, and meltwater) to reach? What happens when we look farther from the regions we expect to be controlled in this way by the ice sheets?

12-2 Orbital Cycles in Regions Remote from Northern Hemisphere Ice

We have already seen that ice sheets do not drive orbital-scale climatic responses everywhere on Earth. In Chapter 9 we found that the North African monsoon fluctuates mainly at a cycle of 23,000 years, a response pattern quite different from the ice sheet signal with its dominant 100,000-year rhythm and prominent 41,000-year cycle. Obviously the ice sheets are not the major control on the North African monsoon.

One reason for the small effect of changes in ice volume on the African summer monsoon is that North Africa is situated far from the ice sheets of the northern hemisphere. A more important explanation may lie in

seasonal sensitivity (Chapter 9). The wet monsoon that fills North African lakes every 23,000 years is mainly a summer-season phenomenon, driven by strong local insolation changes. In contrast, the winds that transfer the ice sheet signal elsewhere in the northern hemisphere blow strongest in winter. Changes that occur during the winter season have only a small effect on African lake levels because winters in Africa are always dry, regardless of orbital configuration. As a result, strong summer insolation forcing of the monsoon overwhelms any signal sent from the ice sheets.

Still, this diminished influence of northern ice sheets over North Africa seems to predict even less influence at high southern latitudes still farther away. Model simulations in which ice sheets are inserted into an otherwise interglacial world (as in Figure 12-5) indicate no significant transfer of cold air from the ice sheets across the equator.

Yet, contrary to this expectation, the northern hemisphere ice sheet signal shows up in many faraway regions. In the western Indian Ocean, dust is blown out to sea from the Arabian Desert at a tempo that closely matches the ice sheet cycles (Figure 12-9) . More dust arrives during the glacial intervals defined by δ¹⁸O maxima, and less dust during interglacial intervals like today.

This pattern continues into other areas. A long sediment core from a lake in the eastern Colombian Andes of South America shows cyclic alternations between pollen produced by trees and by mountain grasslands (Figure 12-10). Dating of this record by radiocarbon, volcanic ash layers, and magnetic reversals indicates that the major cycle of pollen changes occurs at 100,000 years. The pollen cycles in this mountainous region again match the familiar ice volume pattern of slow cooling (increased grass pollen) followed by rapid warming (increased tree pollen).

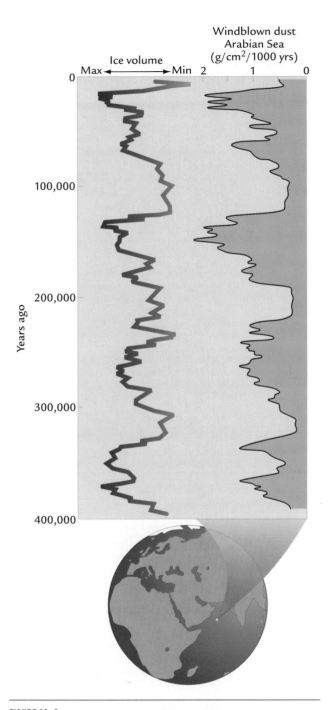

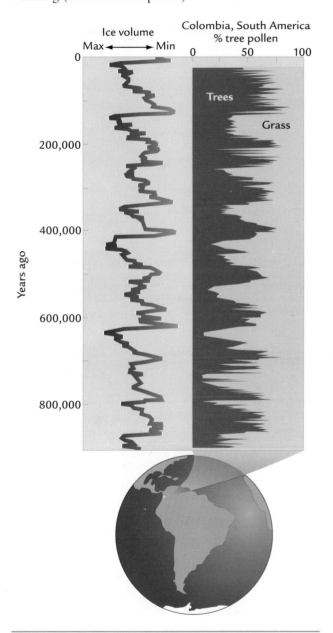

FIGURE 12-9 **Responses of windblown debris in the Arabian Desert to ice volume** The amount of dust blown from the Arabian Desert eastward into the Indian Ocean follows the timing of ice growth and melting. (Adapted from S. C. Clemens and W. L. Prell, "Late Pleistocene Variability of Arabian Sea Summer Monsoon Winds and Continental Aridity: Eolian Records from the Lithogenic Components of Deep-Sea Sediments," *Paleoceanography* 5 [1990]: 109–45.)

FIGURE 12-10 **Pollen responses in South America to ice volume** A long lake core from the eastern Andes in Colombia shows major shifts between forest and grassland pollen that match 100,000-year glacial-interglacial ice volume changes in the northern hemisphere. (Adapted from H. Hooghiemstra et al., "Frequency Spectra and Paleoclimatic Variability of the High-Resolution 30–1450 Kyr Funza I Pollen Record," *Quaternary Science Reviews* 12 [1993]: 141–56.)

Still farther south, a marine sediment core from east of New Zealand holds a similar history of cyclic variations between tree pollen and mountain grassland types (Figure 12-11). The dominant cycle of pollen changes on this distant oceanic island matches the 100,000-year ice volume ($\delta^{18}O$) cycles recorded in the same core.

As a final example, marine sediment cores from the Southern Ocean contain assemblages of radiolarians (marine plankton made of opaline silica; see Chapter 3) that are sensitive to extreme chilling and sea-ice cover. Estimates of past sea surface temperature from a western Indian Ocean core at 43°S match in a general way the changes in ice volume ($\delta^{18}O$) recorded in the same core (Figure 12-12).

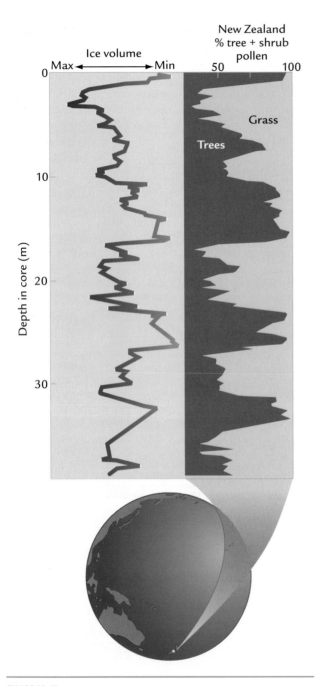

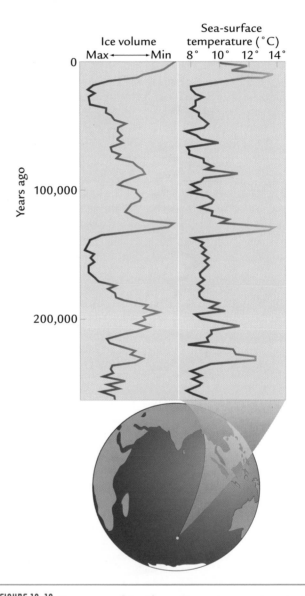

FIGURE 12-11 **Pollen responses in New Zealand to ice volume** A marine sediment core from the east coast of New Zealand shows major 100,000-year shifts between forest and grassland pollen that match glacial-interglacial ice volume ($\delta^{18}O$) signals. (Adapted from L. E. Heusser and G. van der Geer, "Direct Correlation of Terrestrial and Marine Paleoclimatic Records from Four Glacial-Interglacial Cycles-DSDP Site 594, Southwest Pacific," *Quaternary Science Reviews* 13 [1994]: 275–82.)

FIGURE 12-12 **Response of Southern Ocean temperature to ice volume** Sediment cores from the Southern Ocean show changes in estimated sea surface temperature (based on radiolarian plankton) that resemble changes in ice volume ($\delta^{18}O$). (Adapted from J. D. Hays, "A Review of the Late Quaternary History of Antarctic Seas," *Antarctic Glacial History and World Paleoenvironments*, ed. E. M. Van Zinderen Bakker [Rotterdam: A. A. Balkema, 1978].)

Except for the region of tropical summer monsoons, we find to our surprise that the northern hemisphere ice volume signal is the major response pattern all over the world! This observation brings us face to face with one of the biggest problems in climate science: Why do both polar regions (as well as most other regions on Earth) have such similar orbital-scale responses during the last several hundred thousand years?

We already know enough to infer part of the answer to this question. Although these regional responses occur partly at orbital-scale cycles of 41,000 and 23,000 years, they do not have the timing of the local (overhead) insolation signals. Instead, the regional responses even in the remote southern hemisphere all resemble the δ¹⁸O pattern of the ice sheets, lagging thousands of years behind the insolation forcing. Apparently insolation has affected climate in faraway regions of the southern hemisphere by driving changes in slow-responding ice sheets that are then transmitted elsewhere as ice-driven responses. The immediate driver of these regional climatic responses must then be ice sheets, but *which* ice sheets?

Northern or Southern Ice Sheet Forcing? Could the huge ice sheet on Antarctica have been the source of some of these regional responses, rather than the northern hemisphere ice sheets? This initially seems to be a logical explanation for climatic responses situated in the higher latitudes of the southern hemisphere near Antarctica (Figures 12-11 and 12-12).

But the answer to this question is no. The Antarctic ice sheet already covers 97% of the continent during the present warm interglacial interval, and it cannot have gotten much larger than this during glaciations. The available evidence indicates expansion of the ice sheet to the new continental margins uncovered by falling sea level, but no major expansion in volume.

In contrast, several kinds of evidence (primarily glacial moraines on land and ice-rafted debris dumped into ocean sediments) convincingly show that the northern hemisphere ice sheets fluctuated enormously in size between cold glacial intervals such as 20,000 years ago and warm interglacial times like today, when ice has largely disappeared from North America and Eurasia. These large changes in volume make the northern ice sheets a far more likely candidate as the critical center of action in the global climate system.

A second argument favoring northern hemisphere ice sheets as the key control on climatic responses across much of the planet draws on the phase relationships between summer insolation and δ¹⁸O (ice volume) responses in marine sediment cores. Those insolation changes driven by tilt are in phase between the hemispheres, because both poles are affected in the same way when tilt increases or decreases (Chapter 8). In contrast, seasonal changes in precession are opposite in phase

between the two hemispheres: if one hemisphere is closer to the Sun for a given season, the other hemisphere must be farther away for that same season, because it occurs on the opposite side of the orbit (see Figure 8-13).

Because of the differences caused by precession, insolation changes at high latitudes of the northern and southern hemisphere differ considerably (Figure 12-13). For example, 11,000 years ago, insolation values reached a maximum at high northern latitudes (65°N) but a minimum at the same latitude in the southern hemisphere. The extra radiation provided by the northern hemisphere summer insolation maximum provides a good explanation for the rapid decrease in global ice volume shown by the δ¹⁸O curve. In contrast, the summer insolation minimum in the southern hemisphere should have promoted ice sheet growth, the opposite of the trend observed in the δ¹⁸O signal.

This same phase mismatch occurs much farther back in time. Summer insolation in the northern hemisphere has the right timing to explain the observed δ¹⁸O response: it consistently leads the ice volume (δ¹⁸O) signal by almost 10,000 years at the 41,000-year tilt cycle and by almost 6000 years at the 23,000-year precession cycle (Figure 12-14A). Both of these relationships are consistent with the slow response to insolation forcing expected from ice sheets.

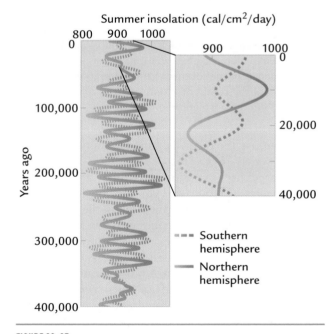

FIGURE 12-13 Out-of-phase summer insolation between the hemispheres Seasonal insolation changes at 65° in the northern and southern hemispheres look different because the phase of the precession cycle is reversed between hemispheres. (Adapted from W. S. Broecker, "Terminations," in *Milankovitch and Climate,* ed. A. L. Berger et al. [Dordrecht: D. Reidel, 1984].)

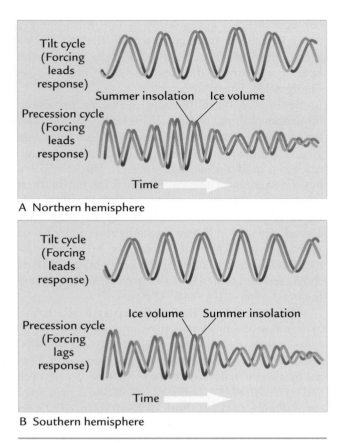

A Northern hemisphere

B Southern hemisphere

FIGURE 12-14 Phasing of insolation vs. ice volume (A) Both the 41,000-year and 23,000-year components of δ¹⁸O (ice volume) curves lag behind northern hemisphere summer insolation by a physically reasonable amount. (B) The same ice volume lag behind 41,000-year insolation changes occurs in the southern hemisphere, but the 23,000-year δ¹⁸O component occurs ahead of the supposed summer insolation forcing.

In contrast, the relative timing of summer insolation forcing versus δ¹⁸O response argues against the Antarctic ice sheet as a critical control on global ice volume changes (Figure 12-14B). Although insolation changes at the 41,000-year tilt cycle in the south polar region do have the right timing in relation to ice volume (δ¹⁸O), this match does not hold for the 23,000-year insolation changes caused by precession. Summer insolation changes in the south polar region at the precessional cycle lag thousands of years *behind* the δ¹⁸O (ice volume) response. Like putting the cart before the horse, this sequence is not reasonable.

This mismatch confirms that the south polar ice sheet was not the means of transferring the ice-driven climatic signal to the rest of the world, or even into the nearby middle and upper latitudes of the southern hemisphere. A transfer from the northern ice sheets to the south polar ocean makes better sense.

The ice volume signal transferred from the northern ice sheets includes not just the smaller 41,000-year and 23,000-year signals, initially driven in an understandable way by summer insolation, but also (during the last 0.9 Myr) the dominant 100,000-year signal, whose origin we have not yet examined. We know from ice rafting and other regional evidence that northern hemisphere ice sheets have varied at this rhythm since 0.9 Myr ago, even though at this point we haven't found out why or how. Somehow this mysterious signal is also transferred all around the world.

Global Transfer of Signals from Northern Hemisphere Ice Sheets How could the global transfer of all three ice sheet cycles happen? One way is by changes in sea level tied to storage and release of ocean water in ice sheets. Changes in sea level affect climate in coastal regions alternately flooded and exposed by vertical movements of shallow seas. Climates in and near such areas may vary between relatively temperate maritime and harsh continental conditions as the ocean rises and falls. As we noted above, the Antarctic ice sheet may also expand as falling sea level opens up new ground for ice to grow on and then retreats as sea level rises.

A second way of connecting the polar regions of the two hemispheres is by changes in deep-water circulation. The reduction in deep-water formation in the North Atlantic that occurs during glaciations (see Figure 11-20) should either reduce the amount of this relatively warm and salty water sent south to the Southern Ocean or reroute it to different regions. These changes could affect the temperature and salt content of Antarctic surface waters and the extent of sea ice, although it is difficult to predict exactly what climatic changes would occur near Antarctica.

These hemispheric connections (via sea level and deep water) take place within the oceans, so they might be expected to affect parts of the land lying close to the sea. But neither connection explains evidence from high-elevation interior regions of the continents located far from the sea, such as the pollen record from the high eastern Andes of Colombia (Figure 12-10). Such responses suggest the need for another kind of link with a more global reach.

Atmospheric CO_2, which varies with more or less the same pattern as ice volume, provides a mechanism to transmit climatic signals from northern hemisphere ice sheets to the rest of the globe. Lower glacial CO_2 concentrations in the atmosphere have the potential to cool every part of the planet and to reduce the amount of water vapor held in the atmosphere everywhere around the globe. CO_2 changes would also affect interior continental regions such as the Colombian Andes. Sensitivity test experiments with general circulation models suggest that the 30% CO_2 reductions that occurred during maximum glaciations would even cool the south polar ocean and increase its sea-ice cover. Changes in CO_2 appear to be a promising way to link climatic responses across the globe. But now we have to confront one of the most important and difficult chicken-and-egg questions in climate science.

CO_2 Level and Ice Volume: Which Drives Which?

Records of atmospheric CO_2 and $\delta^{18}O$ (ice volume) during the last several hundred thousand years both contain obvious orbital-scale cycles (Figure 12-15; see also Figure 11-6). We can conclude that both ice volume and CO_2 levels are in an ultimate sense driven by changes in Earth's orbit. In addition, the strong correlation between CO_2 levels and ice volume strongly implies that the two signals are linked in some way. But what is the link? Is CO_2 driving ice volume or is ice volume driving CO_2?

We might make a reasonable first guess that atmospheric CO_2 is driving the ice sheet cycles: after all, we already know that CO_2 levels affect temperature, and temperature determines ice sheet mass balance. But if this is the critical link between the two signals, we would expect ice volume to lag thousands of years behind the CO_2 signal because of the inherently slow response time of ice sheets (Chapter 10). Yet there is no persistent lag of this kind in the two signals shown in Figure 12-15, arguing against CO_2 as a powerful driver of ice volume. Instead, as we have seen, ice volume lags behind the driving force of summer insolation in the north (Figure 12-14A).

What about the opposite possibility? Could ice volume drive changes in CO_2 level? This choice is more consistent with the relationships shown in Figure 12-15. The CO_2 signal generally tracks the changes in ice volume quite closely, much like the ice-driven signals considered earlier in this chapter. This similar timing suggests that orbital-scale changes in ice sheets somehow drive cycles of atmospheric CO_2 with little or no lag.

Evidence examined in Chapter 11 suggested that orbital-scale CO_2 changes are connected to changes in circulation and carbon storage in the deep ocean. Because ocean circulation is capable of responding relatively quickly (within hundreds of years) to changes in climatic forcing, relatively little time lag should occur between the ice sheet forcing and the deep-ocean CO_2 response. The generally similar timing of the $\delta^{18}O$ and CO_2 signals matches this expectation.

Even though we conclude that CO_2 changes are driven by ice sheets, CO_2 may still have a positive feedback effect on ice volume. If growth of the ice sheets causes CO_2 concentrations to decrease, the lower CO_2 levels should further cool climate and encourage further growth of ice sheets. And if initial melting of ice sheets causes CO_2 levels to increase, climate should warm and the remaining ice should melt faster. In effect, insolation drives the ice sheets, the ice sheets largely determine CO_2 levels, and CO_2 then amplifies the size of ice sheet changes.

In addition to providing positive feedback, the glacial-interglacial changes in CO_2 levels also carry the ice volume signal to other parts of the climate system far from the ice sheets. This transfer should cause changes in temperature and associated moisture levels in all regions on Earth, with more or less the same

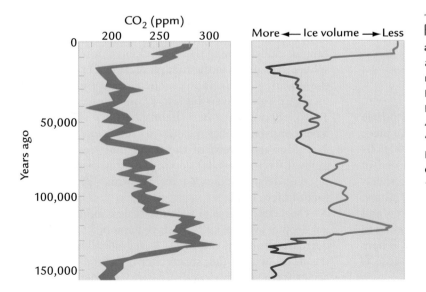

FIGURE 12-15 Relative timing of ice volume ($\delta^{18}O$) and changes in CO_2 The ice core record of atmospheric CO_2 closely resembles the $\delta^{18}O$ record of ice volume. (Left: adapted from J. M. Barnola et al., "Vostok Ice Core Provides 160,000-Year Record of Atmospheric CO_2," *Nature* 329 [1987]: 408-14; right: adapted from D. Martinson et al., "Age Dating and the Orbital Theory of the Ice Ages: Development of a High-Resolution 0 to 300,000-Year Chronostratigraphy," *Quaternary Research* 27 [1987]: 1–29.)

signal as the ice sheets far to the north. In fact, the ice-driven responses examined at the start of this chapter are no doubt partly a response to lower glacial CO_2 concentrations, in addition to the physical factors discussed earlier.

The full story of the link between CO_2 and ice volume is almost certainly more complex than just simply one of positive CO_2 feedback. At this point, scientists do not know the relative timing of changes in CO_2 versus changes in ice volume with a high degree of accuracy (Chapter 11). The timing is well enough known so that we can be certain of the general correlation between the two signals, but not of the details. A close look at the timing indicated in Figure 12-15 suggests that CO_2 changes may either lead or lag changes in ice volume on specific transitions. If real, these leads and lags suggest that CO_2 does not simply track ice volume changes, but is to some degree independent.

For example, atmospheric CO_2 apparently remained at high levels during two major ice-growth transitions, near 115,000 and 75,000 years ago. If this was indeed the case, then CO_2 initially would not have provided positive feedback to aid ice growth during those transitions. Reduced summer insolation, perhaps aided by other feedbacks, must have done the job.

In contrast, the CO_2 level appears to have risen before the rapid deglaciation 130,000 years ago, and again slightly before the more recent deglaciation 15,000 years ago. If this relative timing is correct, high CO_2 values may have been important in actively driving early melting of the ice sheets, rather than simply acting as a positive feedback in aiding deglaciation.

The full story of the role of atmospheric CO_2 in ice volume changes will not be understood until climate scientists recover CO_2 records in ice cores containing annual laminations for more accurate dating. A program to drill such cores in the ice sheet on west Antarctica has begun.

The Mystery of the 100,000-Year Cycle

As we gradually assemble the complicated jigsaw puzzle of climate change at orbital time scales, one crucial piece of the puzzle is still missing: the origin of the 100,000-year climate cycles of the last 0.9 Myr. We learned in Chapter 10 that large northern hemisphere ice sheets first came into existence near 2.75 Myr ago. For almost the next 2 million years, these smaller ice sheets grew and melted at cycles of 41,000 and 23,000 years in response to summer insolation changes at high latitudes. These changes in size of the early ice sheets make good sense as a direct response to overhead summer insolation.

Since 0.9 Myr ago, the 41,000-year and 23,000-year cycles of ice volume change have continued, driven by summer insolation variations with the same characteristic lags of several thousand years. But after 0.9 Myr ago these cycles were overridden by larger and more dominant fluctuations of ice sheets at a period of 100,000 years.

This transition in ice sheet behavior near 900,000 years ago is a mystery for several reasons. For one thing, insolation forcing at the 100,000-year cycle is negligible; the only significant cycles of insolation change at high northern latitudes are at 41,000 and 23,000 years (Figure 10-17). As a result, the main tempo of ice volume cycling for almost a million years and extending right to the present has had at most a trivial direct driving force from Earth's orbital changes.

In addition, no obvious change occurred in the tempo of external insolation cycles during this major shift in ice volume response. Both before and after 0.9 Myr ago, insolation held to the same long-term mean, and variations around this mean through time were similar in average size. If anything, the minuscule amount of insolation change at the 100,000-year rhythm grew slightly *smaller* after 0.9 Myr ago, just as the ice sheets adopted this rhythm as their major cycle. Changes in the basic character of insolation clearly were not the cause of this shift in ice sheet response.

What *is* the explanation? Most ideas proposed by climate scientists to explain this shift concentrate on the two major trends evident in long-term $\delta^{18}O$ records. First, why did more ice accumulate after 0.9 Myr ago? And second, why did these larger ice sheets melt rapidly every 100,000 years?

12-3 Why Have Ice Sheets Grown Larger since 0.9 Myr Ago?

The most likely reason ice sheets grew larger beginning 0.9 Myr ago was that they were responding to the continuing global cooling that had initiated northern hemisphere glaciations in the first place. A $\delta^{18}O$ record from the Pacific Ocean shows a persistent trend toward increased values over the last 4.5 million years (Figure 12-16). Because northern hemisphere ice sheets could have contributed significantly to this trend only after 2.75 Myr ago, much of the trend must result from a gradual (tectonic-scale) cooling, in this case a cooling of the deep Pacific waters in which the benthic foraminifera lived.

One likely source of tectonic-scale cooling is a gradual drop in average CO_2 levels (Chapters 5 and 7). This decrease probably occurred as a gradual background trend on which the shorter-term orbital-scale CO_2 cycles were superimposed. By 2.75 Myr ago, this slow

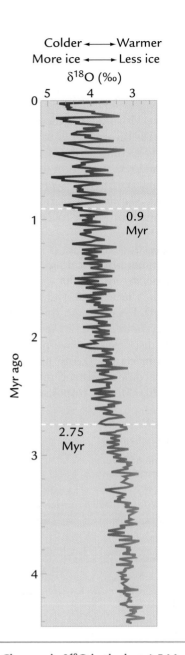

FIGURE 12-16 Changes in δ¹⁸O in the last 4.5 Myr A 4.5-Myr δ¹⁸O record from a sediment core in the eastern Pacific Ocean shows a slow increase in δ¹⁸O values. The δ¹⁸O increases before the onset of northern hemisphere glaciation at 2.75 Myr require cooling of deep water. (Adapted from A. C. Mix et al., "Benthic Foraminifer Stable Isotope Record from Site 849 [0–5Ma]: Local and Global Climate Changes," *Ocean Drilling Program, Scientific Results* 138 [1995]: 371–412.)

peared completely during insolation maxima. As a result, ice sheets grew only at cycles of 41,000 or 23,000 years during this phase.

But if falling CO_2 levels continued to cool climate gradually, at some point the ice sheets would have reached another important threshold. They would begin to survive weak insolation maxima without completely melting, and the head start provided by the surviving ice remnants would enable them to grow larger during the subsequent decreases in summer insolation—the *large glaciation phase* (see Figure 10-14C). This carry-over of some ice into the next interval of ice growth would then provide two key requirements for explaining the 100,000-year signal of the last 0.9 Myr: ice sheets began to grow larger, and they persisted longer. But this explanation still does not yet tell us why ice sheet cycles occurred specifically at a rhythm of 100,000 years.

A second possible explanation contributing to larger ice volume cycles after 0.9 Myr ago has nothing at all to do with changes in climate. Instead, it centers on the character of the materials over which ice sheets move and the way the movement of the ice sheets affects their thickness.

Glacial geologists have found several ice-deposited moraines in Iowa and Nebraska lying beyond the geographic limits of the large ice sheet that existed at the most recent glacial maximum, 20,000 years ago. Layers of volcanic ash date these glacial moraines to about 2 Myr ago. during the very early part of the interval of 41,000-year ice volume cycles (Figure 12-16). These deposits surprised many climate scientists by showing that the smaller-volume 41,000-year ice sheets in east-central North America were already reaching maximum extents comparable to those of the later large-volume 100,000-year ice sheets. This evidence (small volume, large area) suggests that these earlier ice sheets on North America may have been quite thin.

Why would the thickness of the earlier ice sheets have been so different? One hypothesis proposed by the glacial geologist Peter Clark is that the earlier ice sheets accumulated on top of ancient soils tens of meters thick that had been developing for tens of millions of years before glaciation in the northern hemisphere. The weight of overlying ice melted the bottom ice layers, and the meltwater trickled down into the underlying soils. Soils saturated with water are more likely to be deformed by movement of the overlying ice and can cause the ice to slip. This kind of motion is much faster than the slow flow that occurs within ice sheets, and it can send large amounts of ice sliding toward the ice sheet margins and also southward to warm latitudes, where ice ablation rates are high. Between 2.75 and 0.9 Myr ago, sliding may have kept the ice sheet on North

drop in CO_2 first reached the threshold at which significant ice sheets could exist in the northern hemisphere. For this *small glaciation phase* (see Figure 10-14B), ice sheets probably were able to grow only to moderate size during major summer insolation minima and disap-

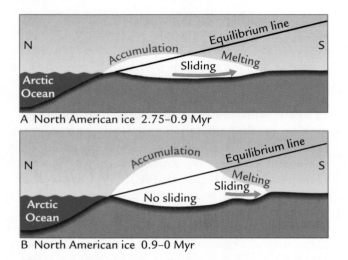

A North American ice 2.75–0.9 Myr

B North American ice 0.9–0 Myr

FIGURE 12-17 **Ice slipping may affect ice sheet volume** (A) During earlier glaciations of North America, ice sheets may have been thin because they slid on water-saturated soils toward lower elevations and warmer temperatures. (B) Later, after ice sheets stripped off most of the underlying soil, their central regions could grow higher because they no longer slid.

America low and thin, subject to greater ablation, and smaller in volume (Figure 12-17A). Thin ice sheets would be easier to destroy during insolation maxima.

Ice sheets slowly erode landscapes through time, and in the process gradually strip soil cover. At present none of the original soil cover is left beneath the central part of Canada. The surface is mostly bare bedrock, with scattered areas of coarse ice-eroded debris left by melting ice. With so little soft sediment left, ice sheets cannot slide easily in this region, and the absence of sliding could allow ice sheets to grow much thicker (Figure 12-17B). Reconstructions of ice sheet thickness at the last glacial maximum 20,000 years ago indicate a broad interior region where ice was thick (and sliding was unlikely) surrounded by a thinner ice margin where sliding occurred because of loose glacial sediments deposited by previous ice advances. A thick ice sheet should survive through weak insolation maxima into the next ice-growth interval.

12-4 What Causes Abrupt Deglaciations (Terminations)?

Slow global cooling, perhaps aided by less sliding beneath ice sheets, helped to create thicker and more extensive ice sheets beginning 900,000 years ago. But two critical questions are still unanswered. What causes the abrupt melting of large amounts of ice that brings these large glaciation cycles to an end? And

why do these terminations occur at intervals of 100,000 years?

Several scientists have proposed that changes in summer insolation are the ultimate cause of rapid melting roughly every 100,000 years. This explanation is consistent with the Milankovitch theory, which calls on summer insolation as the major control of ice sheet size. But at first it would seem to contradict the fact that only a trivial amount of direct insolation forcing exists at the 100,000-year orbital eccentricity cycle. With such small changes, how can summer insolation be invoked to explain the terminations that occur every 100,000 years?

The proposed connection lies in a different way of looking at the summer insolation signal: the emphasis is now placed on the strong 100,000-year modulation of the amplitude of the 23,000-year precession cycle (Figure 12-18, left). Even though this modulation does not form a true cycle in a physical sense, it is nevertheless present as a distinct repetitive pattern in the precessional insolation signal. If any part of the climate system were to become sensitized in some way to this modulation, a 100,000-year response could obviously be created.

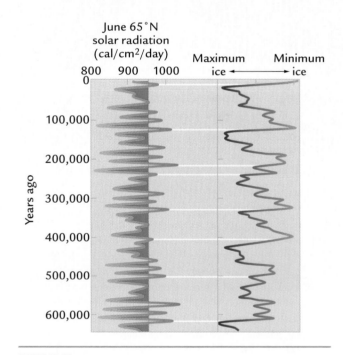

FIGURE 12-18 **Strong summer insolation peaks pace rapid deglaciations** Strong summer insolation peaks at the precession cycle resulting from eccentricity modulation (left) match rapid deglacial terminations indicated by δ¹⁸O (ice volume) signals (right). (Adapted from W. S. Broecker, "Terminations," in *Milankovitch and Climate*, ed. A. L. Berger et al. [Dordrecht: D. Reidel, 1984].)

A clear hint that such a sensitization exists lies in the fact that the most rapid deglaciations of the last 0.9 Myr (Figure 12-18, right) have generally occurred at times of unusually large amplitude swings (strong modulation) of the summer insolation signal. This correlation in time strongly suggests that near 0.9 Myr ago something in Earth's climate system began to become locked on to (sensitive to) these large-amplitude swings in summer insolation.

Because ice ablation occurs at high insolation values, it obviously makes sense to look at the large-amplitude swings of summer insolation toward unusually high values as the cause of rapid glacial terminations. The largest insolation peaks occur in groups of two or three maxima separated by intervals of roughly 100,000 years. This timing is a result of the modulating effect of eccentricity on the 23,000-year precession cycle. The proposed link between ice volume and insolation at a cycle of 100,000 years is this: rapid deglaciations should occur during times when eccentricity modulation creates unusually large summer insolation peaks, as happens every 100,000 years (indicated by the white lines in Figure 12-18).

In effect, rapid ice melting is created at a cycle of 100,000 years from an insolation curve that cycles only at 23,000 years; this is possible because ice melting is sensitive only to one side of the envelope of modulation. The prominent insolation minima on the opposite side of the envelope of precessional insolation changes are irrelevant to ice melting.

Armed with this information, we can return to the conceptual model of the *large glaciation phase* of ice sheet development. As the climate system cools and enters this phase, ice sheets begin to survive through some of the weaker peaks in summer insolation. This permits ice sheets to grow large because they remain on Earth over longer intervals. But how long?

The timing of the terminations provides the answer. Rapid melting events that occur every 100,000 years set the limit to the slow buildup of large ice sheets. Large ice sheets gradually accumulate over tens of thousands of years, but then melt rapidly every 100,000 years.

We can clarify this concept by tracking more closely the relationship between insolation and ice volume in Figure 12-18. Within this sequence lie several long intervals when a sequence of weak summer insolation peaks occurred in succession as a result of modulation of low amplitude by eccentricity. The most recent of these intervals occurred between 75,000 and 15,000 years ago. During this interval, strong summer insolation maxima were absent, ablation must have remained low, and ice sheets were able to grow larger over a long interval. Similar intervals occurred earlier, such as the one between 400,000 and 325,000 years ago.

Each longer-term interval of ice growth was eventually ended by a larger insolation peak, like the one that occurred 10,000 years ago. The sequences shown in Figure 12-18 indicate that it is the *first* high-insolation peak in each cluster of two or three large peaks that melts the large ice sheets. These peaks at the precession rhythm are also closely aligned with insolation peaks caused by maximum values of orbital tilt. Their combined effects initiate rapid melting of a large volume of ice in a short time, but the crucial 100,000-year timing comes from the modulation of precession by eccentricity.

How does so much ice melt at these times? One obvious answer is that a large volume of ice has built up and is available for melting. If only a small volume of ice were present, the total amount of melting would be small even if all the ice disappeared. Another factor is the large warming likely to be produced by the simultaneous summer insolation maxima occurring at the 41,000-year tilt and 23,000-year precession cycles. Still another factor is the delay in bedrock rebound under the melting ice sheet; the ice melts back into the depression created by its own weight. Temperatures are warmer at the lower elevations in this depression, and rates of ice ablation increase exponentially with rising temperature (Figure 10-1). For all these reasons, rapid melting of large ice sheets occurs during major deglaciations every 100,000 years.

Drawing on these basic observations of ice sheet responses to insolation, climate scientists have constructed simple numerical models of ice volume change through time in response to summer insolation in the northern hemisphere. Their basic assumption is that rising insolation causes ice sheets to melt at rates several times faster than falling insolation causes them to grow. These models succeed in capturing much of the observed ice sheet response over the last 900,000 years, including the 100,000-year cycles of gradual ice buildup and rapid ice melting (Figure 12-19). The generally close match between the models and the observations supports the idea that ice sheets will vary at cycles of 100,000 years if melting rates are sensitive to high summer insolation.

But these models fail in one respect. Eccentricity modulates the precessional insolation signal not just at the 100,000-year period but also at the longer cycle of 413,000 years (Chapter 8). This longer modulation causes unusually large summer insolation peaks to occur in clusters spaced at intervals averaging 413,000 years. The last two such clusters occurred near 200,000 and 600,000 years ago. Groups of relatively weak summer insolation peaks occurred near 800,000 and 400,000 years ago, and again near the present (Figure 12-19).

If high summer insolation peaks are indeed responsible for intervals of unusually rapid ice melting, we

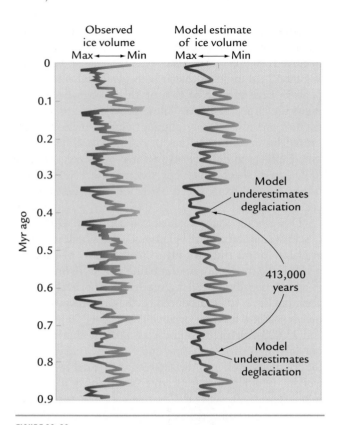

Myr ago

Model underestimates deglaciation

413,000 years

Model underestimates deglaciation

FIGURE 12-19 Numerical model: insolation control of ice volume Numerical models assume that ice sheets melt four times faster when 65°N summer insolation is rising than they grow when insolation is falling. This model shows that insolation changes (occurring only at cycles of 41,000 and 23,000 years) can simulate the observed 100,000-year fluctuations in ice sheets. (Adapted from J. C. Bassinot et al., "The Astronomical Theory of Climate and the Age of the Brunhes-Matuyama Magnetic Reversal," *Earth and Planetary Science Letters* 126 [1994]: 91–108.)

would expect to see unusually strong and complete deglaciations occurring roughly every 400,000 years in the ice volume ($\delta^{18}O$) record. But no such signal is apparent (Figures 12-18 and 12-19). Interglacial ice volume minima tend to reach similar values, regardless of the size of the preceding summer insolation maxima.

To some extent, the similar size of ice volume minima in $\delta^{18}O$ signals may simply reflect the fact that when all the ice has been melted from North America and Asia, no easily melted ice is left on Earth. The response of the vulnerable ice sheets has reached its natural limit.

But a serious problem still remains. One of the largest of the 100,000-year deglaciations recorded in the $\delta^{18}O$ record occurred just before 400,000 years ago, yet the summer insolation peak at that time was unusually small because eccentricity was at a minimum in its

413,000-year cycle (Figure 12-18). Similarly, the most recent deglaciation, near 10,000 years ago, was large and rapid, yet it occurred during a relatively weak peak in summer insolation. How could such small insolation peaks produce such large deglaciations?

The model based on increased ice melting during times of unusually high summer insolation does such a good job in reproducing the overall appearance of the observed ice volume ($\delta^{18}O$) response in Figure 12-19 that rejecting it outright makes no sense. But how do we explain its shortcomings at these critical times?

One possible explanation is that the pace of the 100,000-year cycles (and particularly the rapid deglaciations) is set *externally* by insolation changes, but that the strength of the 100,000-year cycles depends in part on critical *internal* interactions within the climate system. In effect, the timing of the 100,000-year ice volume cycles is locked on to the timing provided by the modulation of precession by eccentricity (Figure 12-18), but the actual amplitude of the fluctuations in ice volume is only partly related to insolation and also reflects internal interactions.

Large Ice Sheets Are More Vulnerable to Fast Melting One possibility is that the internal interactions that produce 100,000-year cycles came into being 0.9 Myr ago as a direct result of the initial appearance of larger ice sheets. In effect, the larger ice sheets themselves produced internal interactions within the climate system that hastened their own destruction every 100,000 years, during intervals whose basic timing was paced by changes in summer insolation. These internal interactions that destroyed the ice sheets must have acted as positive feedbacks that accelerated ice melting processes already under way because of increasing summer insolation.

One proposed explanation calls on the greater depression of bedrock by the weight of the larger ice sheets (see Figure 10-10). When the southern margins of large ice sheets began to melt as summer insolation rose, they retreated into the deep bedrock holes they had created and melted rapidly in the warm temperatures at lower elevations. The delay in bedrock rebound kept the ice sheets in a warmer environment and accelerated rates of melting (see Figure 10-11).

A major problem with this idea is that models of Earth's internal behavior now suggest that bedrock rebounds *faster* after large ice sheets melt than after smaller ones do. Faster rebound would carry the remaining ice to higher, colder altitudes and reduce the positive feedback available to explain rapid deglaciations at the 100,000-year cycle.

Another proposed positive feedback to hasten ice melting is the presence of marine ice sheets on continental shelves. These ice sheets can respond more quickly to climate changes (such as rising insolation) than the more sluggish ice sheets on land. Evidence from sediments in the North Atlantic Ocean suggests that large marine ice sheets first appeared in the Arctic margins north of Norway near 0.9 Myr ago. Other evidence indicates that one of these ice sheets was the first to melt during the most recent deglaciation. The early disappearance of this marine ice might have helped hasten the melting of slower-responding ice sheets on the northern continents.

A third possibility is that rising CO_2 levels during deglaciations provide positive feedback to help melt ice sheets. This is a particularly promising choice because atmospheric CO_2 appears to have risen very near the time of the last two deglaciations (Figure 12-15).

Conceptual Model of the 100,000-Year Climate Cycle   The marine geologist John Imbrie, working in the 1990s with colleagues on a project called SPECMAP, synthesized all available orbital-scale climatic evidence into a conceptual model to explain the 100,000-year ice sheet cycle. This model relies on orbital insolation changes as the basic pacemaker of climate changes over the last several hundred thousand years but also calls on internal feedbacks within the climate system to create the 100,000-year cycle (Figure 12-20).

The SPECMAP model uses Milankovitch's insight that changes in summer insolation in the northern hemisphere drive ice sheets at cycles of 41,000 and 23,000 years. Changes in ice volume from 2.75 to 0.9 Myr ago can be accounted for by this relatively simple model of direct insolation forcing (Figure 12-20A). The ice volume signals produced at cycles of 41,000 and 23,000 years during this interval were quickly transferred to other locations in the climate system by several mechanisms, including changes in winds, sea level, and deep-water circulation.

This earlier regime came to an end near 0.9 Myr ago with the first appearance of ice sheets that exceeded some critical size threshold (Figure 12-20B). Changes in summer insolation at 41,000 and 23,000 years continued to cause moderate variations in the sizes of ice sheets at those cycles, but now the larger size of the ice sheets themselves also became a factor. The larger ice sheets produced far more extensive ice-driven responses within the rest of the climate system through changes in winds, ocean circulation, sea level, and other factors.

For the first time these larger ice-driven responses also triggered strong feedbacks within the climate system that produced the 100,000-year signal. Potentially the strongest of these internal feedbacks is the linked system of deep-water circulation and surface-water nutrient levels that controls atmospheric CO_2 (Chapter 11). Imbrie and his colleagues proposed that large

northern hemisphere ice sheets began to exert the main control on this system primarily by altering low-level winds in regions where deep water forms in the North Atlantic. In this way, the ice sheets altered the size and timing of the CO_2 cycle in such a way as to provide positive feedbacks to their own growth and melting. Other feedbacks (marine ice sheets, for example) also occurred.

The greatest remaining uncertainty in ice-age hypotheses centers on the timing of responses observed in the southern hemisphere, especially the Antarctic Ocean. As we noted earlier, climate responses in these regions appear to be generally similar to those of northern hemisphere ice sheets. But when we examine them in more detail, we find that air and ocean responses in the southern hemisphere tend to occur slightly before those of the northern hemisphere ice sheets (for example, see Figure 12-12). If climate responses in the south

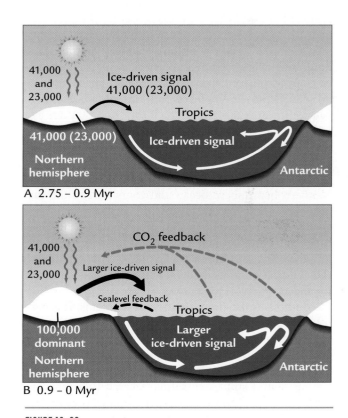

FIGURE 12-20 Conceptual (SPECMAP) model of ice-driven climate changes (A) Before 0.9 Myr ago, insolation changes drove slow-responding ice sheets at the 41,000-year and 23,000-year cycles, and these signals were quickly transferred to other parts of the climate system. (B) When ice sheets began to exceed a critical size threshold, the signals they sent to the rest of the climate system created powerful feedbacks that produced a 100,000-year response not present in the direct insolation forcing.

polar region are claimed to be driven from the north, why is the south polar cart moving slightly before the north polar horse?

Imbrie and his colleagues note two possible answers to this question. One is that the critical first changes in the climate system do in fact occur in the northern hemisphere, but initially in the form of changes in air and ocean temperatures. These early local changes in the northern hemisphere then lead to two other kinds of changes: (1) fast responses of ocean and air temperatures in the south polar regions and (2) slow changes in the sizes of northern ice sheets over thousands of years. As a result, while the early south polar air and ocean responses do lead the slower northern ice responses, both may in fact be responses to the initial changes in air and ocean temperature in the northern hemisphere. In this case, the slight lead of the south polar responses over ice volume does not indicate a direct cause-and-effect relationship.

Signs of an early-response cooling in north polar regions before glaciation remain scarce, although evidence favoring such a response is gradually accumulating. Critics of this part of the SPECMAP hypothesis also note convincingly that atmospheric CO_2 cannot provide positive feedback during some phases of ice growth, because CO_2 levels fall long after ice growth begins (for example, near 115,000 and 75,000 years ago in Figure 12-15).

The other explanation proposed for the slightly early responses in the southern hemisphere is that something within the climate system centered on the southern hemisphere ocean and connected to the carbon cycle reacts to orbital forcing earlier than the ice sheets and largely independently of them. In this view, the carbon cycle (including atmospheric CO_2 levels) functions as a separate player in the climate system with the ability to influence global climate independently of ice sheets.

But this explanation has its own problems. It is hard to formulate a convincing hypothesis to explain why such a separate oceanic control on the carbon system and on atmospheric CO_2 would respond to orbital insolation forcing independently of ice sheets and yet with a timing that is still quite similar overall (Chapter 11). The central problem for such a hypothesis is to explain why an ocean-centered response would lag many thousands of years behind insolation forcing at cycles of 41,000 and 23,000 years, even though the slowest turnover of deep-ocean circulation occurs within 1500 years.

Finally, some scientists believe that Earth's climate system has a natural resonance at long orbital-scale wavelengths. They use the analogy of a bell that rings with a characteristic period or frequency, regardless of the exact time when it is struck. This behavior is characteristic of a **resonant response**. They hypothesize that any force that disturbs Earth's climate (changes in insolation, or perhaps something else) will produce a characteristic internal resonant response. Critics of this hypothesis ask why resonant behavior at a rhythm of 100,000 years appeared only 0.9 Myr ago.

In summary, although complex interactions within the climate system raise a number of confusing chicken-and-egg (and cart-and-horse) questions, climate scientists are close to a full theory of ice sheet variations over the last 2.75 Myr. For the first two-thirds of this interval, growth and melting of northern hemisphere ice sheets were controlled by changes in summer insolation at rhythms of 41,000 and 23,000 years, as Milankovitch predicted. By 900,000 years ago, global cooling permitted ice sheets to begin growing larger. The dominant 100,000-year rhythm of change in these ice sheets was paced by changes in summer insolation but ultimately governed by internal feedbacks produced by the ice sheets, including changes in atmospheric CO_2.

KEY TERMS

ice-driven responses (p. 256)
resonant response (p. 272)

REVIEW QUESTIONS

1. In what sense are ice sheets both a climatic response and a source of climatic forcing?

2. Name an ice-driven response and cite evidence that it originates from changes in ice sheet size.

3. How could responses driven by northern hemisphere ice sheets be found in the southern hemisphere?

4. Cite two kinds of evidence indicating that changes in northern hemisphere rather than Antarctic ice sheets are a major player in orbital-scale climate changes.

5. What evidence suggests that orbital-scale changes in northern hemisphere ice volume drive changes in atmospheric CO_2, rather than vice versa?

6. How could 100,000-year cycles in the size of northern hemisphere ice sheets be paced by summer insolation changes occurring only at cycles of 41,000 and 23,000/19,000 years?

ADDITIONAL RESOURCES

Basic Reading

Broecker, W. S. 1984. "Terminations." In *Milankovitch and Climate*, ed. A. L. Berger et al., 687–98. Dordrecht: Reidel.

Broecker, W. S., and G. H. Denton. 1990. "What Drives Glacial Cycles?" *Scientific American* (January), 48–56.

Manabe, S., and A. J. Broccoli. 1985. "The Influence of Continental Ice Sheets on the Climate of an Ice Age." *Journal of Geophysical Research* 90:2167–90.

Rind, D., D. Peteet, W. S. Broecker, A. McIntyre, and W. F. Ruddiman. 1986. "The Impact of Cold North Atlantic Sea Surface Temperatures on Climate: Implications for the Younger Dryas Cooling (11–10K)." *Climate Dynamics* 1: 3–33.

Advanced Reading

Imbrie, J., et al. 1992–93. "On the Structure and Origin of Major Glaciation Cycles." Pts. 1 and 2. *Paleoceanography* 7:701–38, 8:699–735.

Weyl, P. K. 1968. "The Role of the Oceans in Climate Change: A Theory of the Ice Ages." *Meteorological Monographs* 8:37–62.

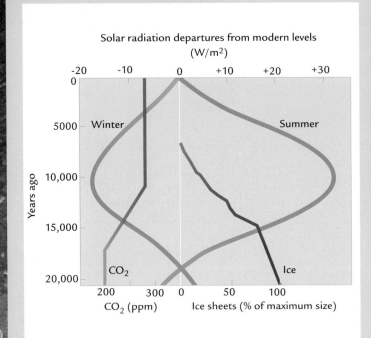

Solar radiation departures from modern levels (W/m²)

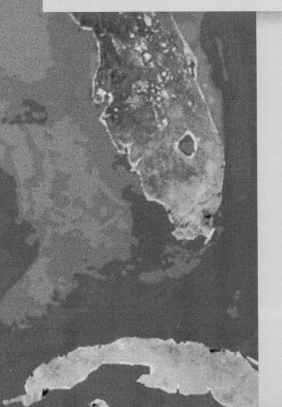

Deglacial climatic forcing   For the past 21,000 years, model experiments based on known climatic forcing (insolation, ice sheets, and CO_2) can be compared with observations from the geologic record. (Adapted from J. E. Kutzbach et al., "Climate and Biome Simulations for the Past 21,000 Years," *Quaternary Science Reviews* 17 [1998]: 473–506.)

The last several tens of thousands of years have seen enormous climate changes, many of them occurring within the time span of recorded human civilization. These changes all lie within the time span of ^{14}C dating, and the thousands of ^{14}C-dated records from the land complete the regional coverage of Earth's surface available from the oceans. A long list of organic remains can be ^{14}C-dated, including wood, leaves, twigs, acorns, charcoal, bones, and dispersed organic carbon. Most ^{14}C dates come from lakes or bogs, where watery environments poor in oxygen preserve organic carbon. Lake sediments hold two widely used kinds of climate signals: records of changing lake levels and fluctuations in the abundance of pollen grains.

Climate changes decades to hundreds of years in length can be resolved in lake records, and this level of resolution enables scientists to detect the relative rates of response of different parts of the climate system during rapid climatic changes. As we will find out in Chapter 14, ^{14}C dates cannot be used directly as true ages, but must first be converted to calendar years, which for some

PART IV

Deglacial and Millennial Climate Changes

intervals are a few thousand years older. Throughout this part, all ages are in calendar (actual) years unless ^{14}C ages are specified.

The ability to reconstruct regional climate responses across Earth's entire surface gives climate scientists the opportunity to look at climate change from a geographic perspective. Mapping climate changes at specific times ("time slices") gives researchers insights into specific regional processes that alter climate and permits them to compare the output of models with data. They can use boundary conditions based on observations from the geologic record to run model simulations of climate at times in the past, and then compare the results (output) from these climate simulations with independent geologic data assembled into maps. The major boundary conditions that have driven climate changes during the last 21,000 years have been the changes in the size of ice sheets, in seasonal insolation, and in the levels of greenhouse gases in the atmosphere.

In climate archives with sufficiently high resolution, climatic oscillations lasting thousands of years have recently been detected, especially near the North Atlantic Ocean. These fluctuations can rarely be dated accurately enough to permit firm correlations among archives. These changes require an explanation other than orbital forcing, but their cause is as yet unknown.

In this part we address the following important questions:

- **To what extent does the timing of ice melting in the last 21,000 years support the Milankovitch theory that ice sheets are forced by orbital insolation?**

- **To what extent do changes in tropical moisture during the last 21,000 years support the Kutzbach theory that orbital insolation controls monsoons?**

- **What does the cooling of the tropics at the glacial maximum tell us about Earth's sensitivity to the concentration of atmospheric CO_2?**

- **How has Earth's climate system responded to insolation changes during the several thousand years since the ice sheets melted?**

- **Why do modern-day tide gauges still have a lingering memory of the melted ice sheets?**

- **What are the origin and significance of brief oscillations of climate at time scales much shorter than changes in Earth's orbital configurations?**

CHAPTER 13

The Last Glacial Maximum

When the last glacial cycle culminated 21,000 years ago, the world was very different from the one we know today. Ice 2 or more kilometers high covered Canada, the northern United States, northern Europe, and parts of Eurasia. Global sea level was 110 meters (365 feet!) lower, joining modern islands between Asia and Australia and connecting Britain to mainland Europe.

South of the ice sheets, conditions were cold, windy, and dirty. Forests of modern North America, Europe, and Asia were replaced by tundra or grasslands, and lower levels of atmospheric CO_2 and CH_4 caused cooling and drying across the tropics and the southern hemisphere.

In this chapter we first focus on physical aspects of this glacial world, such as the extent and thickness of glacial maximum ice sheets and the debris they produce. Then we explore how changes in the distribution of life forms, especially land vegetation and ocean plankton, allow us to test climate simulations from models. Finally, we examine a controversy with implications for future climate change: How cold were the tropics at the last glacial maximum?

FIGURE 13-1 **Mammals of the glacial maximum** Many mammals that are now extinct once roamed the glacial maximum world of 21,000 years ago, including woolly mammoths, saber-toothed tigers, and giant ground sloths. (Courtesy of Smithsonian Institution, Washington, D.C.; painted by Jay Matternes.)

The Glacial World: More Ice and Less Gas

The glacial maximum world of 21,000 years ago has long fascinated climate scientists. Part of the reason is the wild animals that lived on Earth so recently. Large mammals now extinct, such as mammoths, saber-toothed tigers, and giant ground sloths, roamed these cold, dry, windy landscapes (Figure 13-1). Another reason is the human connection: in Spain and southern France, our ancestors sheltered in caves and covered the walls with hauntingly beautiful paintings and charcoal sketches of the game they hunted (Figure 13-2). In a

FIGURE 13-2 **A cave painting of the glacial era** Despite the harsh glacial climate, our ancestors left beautiful, almost modern-looking paintings on the walls of caves in southern Europe. (Ferrero/Labat/Auscape International.)

sense, this icy, cold, dry, windy, sparsely vegetated world is an alternative version of our own present-day Earth, a giant experiment run by the climate system in response to forcing factors that are reasonably well known. As we saw in Part III, the three factors with the greatest potential to account for departures from modern climate conditions were the larger ice sheets, the lower CO_2 levels, and the changes in seasonal insolation.

But surprisingly, seasonal insolation levels 21,000 years ago were nearly identical to those today. This seeming contradiction can be explained if we recall that intervals of low summer insolation had helped to build the ice sheets many thousands of years earlier, and the ice sheets had responded by growing sluggishly to maximum size by 21,000 years ago. But by the time the ice reached maximum size, summer insolation had risen close to today's levels and was headed toward the higher levels that would soon begin to melt the ice (Figure 13-3).

Because insolation levels 21,000 years ago were essentially the same as those today, insolation cannot have been a major factor in the differences between the climate that prevailed during this glacial snapshot in time and our own. Only two factors are left to explain the colder and drier glacial maximum climates: the larger ice sheets themselves and the lower values of greenhouse gases. This chapter focuses on a simple question: How thoroughly do large ice sheets and lower greenhouse gases explain global climate at the last glacial maximum?

13-1 Project CLIMAP: Reconstructing the Last Glacial Maximum

Cooperative efforts to reconstruct climate began in the 1970s, when a large interdisciplinary effort called the **CLIMAP** (*Cli*mate *Map*ping and *P*rediction) **Project**

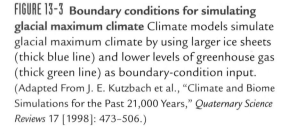

FIGURE 13-3 **Boundary conditions for simulating glacial maximum climate** Climate models simulate glacial maximum climate by using larger ice sheets (thick blue line) and lower levels of greenhouse gas (thick green line) as boundary-condition input. (Adapted From J. E. Kutzbach et al., "Climate and Biome Simulations for the Past 21,000 Years," *Quaternary Science Reviews* 17 [1998]: 473–506.)

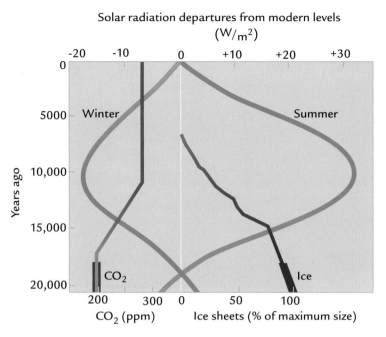

reconstructed the surface of Earth as it was during the last glacial maximum. This group was led by the marine geologists John Imbrie and Jim Hays and the geochemist Nick Shackleton. Although its primary focus was on records in ocean sediments, CLIMAP also drew on the expertise of scientists with specialized knowledge of ice sheets, windblown deposits, vegetation, and the still-new field of climate modeling.

In the years before CLIMAP, climate scientists were few, and they usually worked on individual research projects using widely differing techniques. CLIMAP was remarkably successful in bringing these scientists together and combining their skills into a single interdisciplinary effort. CLIMAP helped to teach a generation of climate scientists in many disciplines to see Earth's climate as a whole and to work together to achieve a larger goal. Today interdisciplinary alliances are common in the study of climate science.

The CLIMAP group published its first map of the ice-age Earth in 1976 and a revised version in 1981 (Figure 13-4). These maps show conditions at Earth's surface during typical seasons (in this case summer) at the last glacial maximum. Because individual years and even centuries could not be resolved in glacial-age records, the maps show an *average* northern summer during the millennium or so centered on the glacial maximum.

The most striking features of the CLIMAP reconstruction are the great continent-sized ice sheets covering North America as far south as 37°N latitude and Scandinavia to 48°N, as well as large parts of the Arctic coast of Eurasia. Today ice sheets on Antarctica and Greenland cover a combined area of about 14.2 million square kilometers, equivalent to just under 3% of Earth's surface area and about 10% of its land surface.

The CLIMAP reconstruction of Earth at the last glacial maximum shows ice sheets covering an area of 35 million square kilometers, equal to 7% of Earth's total surface and 25% of the area of the continents.

The North American ice sheet was by far the largest of the northern hemisphere ice sheets, accounting for over 55% of the amount of ice in excess of that existing on Earth today (Table 13-1). This great ice sheet was roughly equivalent in volume to the ice sheet now covering all of Antarctica. Most of the ice in North America was in the **Laurentide ice sheet**, centered on east-central Canada, with the rest in the much smaller **Cordilleran ice sheet**, over the Rockies in the American West. The **Scandinavian ice sheet,** in northern Europe, and the **Barents ice sheet**, on the northern Eurasian continental shelf, accounted for about 22% of the excess ice. The remaining 23% is accounted for by expansion of ice sheets on Antarctica and Greenland across new land exposed by the fall in sea level. Expansion of mountain glaciers dramatically transformed the appearance of high terrain around the world but contributed little to the global change in ice volume. In some regions, such as Patagonia, in southernmost South America, mountain glaciers merged into small ice caps.

The glacial world in the CLIMAP reconstruction was almost 4°C colder than Earth today. The North Atlantic cooled by 8°C or more, and sea ice was more extensive than it is today (Figure 13-4B). Farther from the ice sheets, the North Pacific cooled by a lesser amount, with expanded winter sea ice in the northwest. Sea ice also advanced well beyond its present limits in the Southern Ocean, with a band of cooler ocean surface temperatures north of the expanded limit of sea ice

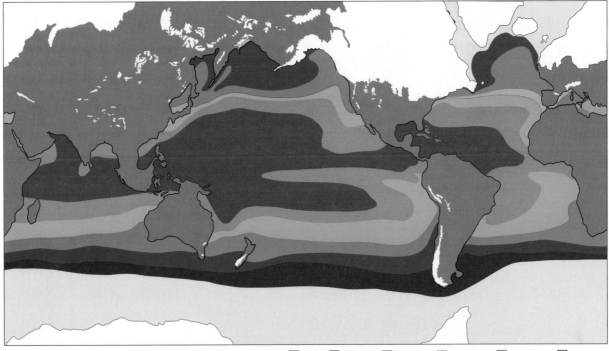

A August ocean temperature 21,000 years ago (°C)   ■ < 5   ■ 5 – 10   ■ 10 –15   ■ 15 – 20   ■ 20 – 25   ■ > 25

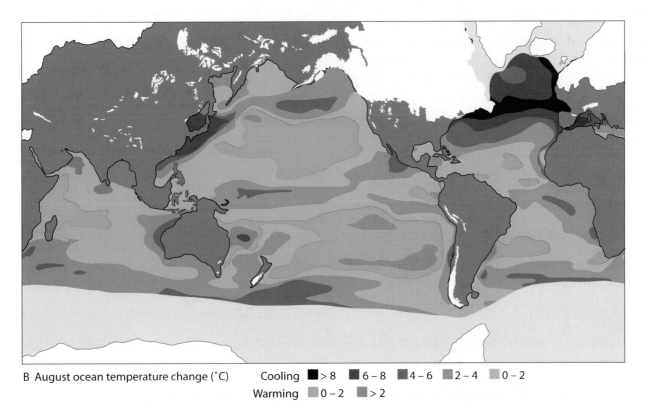

B August ocean temperature change (°C)   Cooling ■ > 8   ■ 6 – 8   ■ 4 – 6   ■ 2 – 4   ■ 0 – 2
                                            Warming ■ 0 – 2   ■ > 2

FIGURE 13-4 The CLIMAP reconstruction of glacial maximum ocean temperatures CLIMAP produced the first global-scale reconstruction of Earth's surface at the most recent glacial maximum, including ice sheet size and ocean surface temperature. These maps show conditions during an average summer during the glacial maximum (A) and the changes in temperature between then and today (B). (Adapted from CLIMAP Project Members, *Seasonal Reconstruction of the Earth's Surface at the Last Glacial Maximum,* Map and Chart Series MC-36 [Boulder, Colo.: Geological Society of America, 1981].)

TABLE 13-1 Approximate Volumes of Ice and Amounts of Water Stored in Glacial Ice Sheets Shown by Lowering of Sea Level beneath Today's Level

Ice sheet	Location	Excess ice volume (million km³)	Sea level Amount (m)	Sea level Change (m)*
Laurentide	East-central Canada	25–34†	72–100	50–70
Cordilleran	Western North America	1.8	5	3.5
Greenland	Greenland	2.6‡	7	5
Britain	England, Scotland, Ireland	0.8	2	1.5
Scandinavian	Northern Europe	7.3	21	15
Barents/Kara	Shelf north of Eurasia	6.9	20	14
East Antarctic	Eastern Antarctica	+3.3§	9	6
West Antarctic	Western Antarctica	+6.5§	18	13
Others	Various	1.2	3	2
All ice sheets		55–64	155–183	109–129

*Net sea level changes are 30% smaller than the volumes of seawater removed from the ocean because ocean bedrock rises when the weight of water is removed.

†The higher estimate shown is for a thick ice sheet like that in the CLIMAP maximum reconstruction; the lower estimate is for a thin ice sheet.

‡Present-day volume of ice on Greenland is 3 million km³.

§Present-day volume of ice on Antarctica is 29 million km³.

Source: Adapted from G. H. Denton and T. J. Hughes, The Last Great Ice Sheets (New York: John Wiley, 1981).

around Antarctica. The CLIMAP reconstruction indicates low-latitude ocean temperatures only slightly cooler than today's, and in some regions even a bit warmer.

In the decades since this reconstruction was published, several of its features have been challenged, and a few of them are almost certainly wrong. Yet, as a continuing measure of its significance to climate science, the CLIMAP reconstruction of the ice-age Earth is still the standard against which all challenges are directed.

13-2 How Large Were the Ice Sheets?

What have scientists learned about the ice sheets since the CLIMAP reconstruction? The glacial geologists George Denton and Mikhail Grosswald and the glaciologist Terry Hughes headed the CLIMAP ice sheet reconstructions. Of the two ice sheet reconstructions actually published, one broke with conventional thinking by portraying the ice sheets at their maximum plausible size (the "maximum reconstruction"). Three aspects of this reconstruction proved highly controversial.

One debate arose over the lateral extent of the ice sheets, which in most places were shown extending to the maximum limits that ice had reached at any time in the several million years of northern hemisphere ice oscillations (Figure 13-4). Although the southern limits of most continental ice sheets at the last glacial maximum had not been in much doubt for many years before the CLIMAP reconstruction, larger uncertainties remained about whether or not higher-latitude ice margins actually reached the ocean. Many glacial geologists argued that the glacial maximum ice sheets in the north were less extensive than they had been during earlier glaciations. The main reason for this ongoing uncertainty was that the scarcity of organic carbon in cold, dry Arctic environments made it difficult to find ^{14}C to date the glacial moraines.

Estimates of the northern ice margins generally disagreed by up to a few hundred kilometers, amounts that were important to arguments about the physical state of the ice sheets in specific regions but small in comparison with the full lateral dimensions of ice sheets (thousands of kilometers) and with the size of grid boxes used in climate modeling (hundreds of kilometers on a side). As it has turned out, careful ^{14}C dating along the northern margins of the ice sheets has since confirmed that most were indeed at or near their maximum limits 21,000 years ago.

A second major disagreement about the extent of glacial maximum ice sheets centered on the marine ice sheets, those built on shallow continental shelves with their bases below sea level (Chapter 2). The CLIMAP reconstruction placed two marine ice sheets along the northern margin of Eurasia, a large one over the Barents Sea, north of Scandinavia, and a smaller one over the Kara Sea, north of western Russia. This aspect of the reconstruction caused years of heated controversy, some of which persists today.

Later, marine geologists surveying these shallow seas collected evidence (glacial debris in sediment cores and depth soundings showing the extent of submerged moraines) that proved that a large marine ice sheet did exist in the Barents Sea and perhaps in the westernmost Kara Sea at the glacial maximum. This aspect of the CLIMAP reconstruction was also largely confirmed.

Another major source of controversy in the CLIMAP reconstruction centered on the thickness (and height) of the glacial maximum ice sheets. Height and thickness are usually related in a simple way: as an ice sheet grows, it weighs down the underlying bedrock by an amount equal to 30% of its thickness (Chapter 10). As a result, the height of the ice above the surrounding landscape usually represents about 70% of its total thickness. Unfortunately, no simple method exists to measure or reconstruct changes in the thickness or height of ice through time.

The CLIMAP maximum reconstruction showed relatively tall (thick) ice sheets, especially over North America. The assumptions used in making this reconstruction were that the ice sheets had existed at or near their maximum extent long enough to build up slowly to full equilibrium thickness and that they consisted of stiff ice frozen to the underlying bedrock. These two assumptions produced ice sheets with profiles that rose steeply from the outer margins of the ice and reached high elevations and great thicknesses in their central regions (Figure 13-5).

This aspect of the CLIMAP reconstruction has not held up to two decades of further scrutiny. One line of evidence against thick ice sheets at the glacial maximum came from measurements of the amount global sea level fell as a result of the ice sheets. As we saw in Chapter 10, coral reefs can be used as dipsticks to measure past changes in sea level. Drilling of deeply submerged reefs that date to the glacial maximum detected a drop of 110 meters in sea level, significantly less than the 129 m or so inherent in the thick ice sheets of the CLIMAP reconstruction (Table 13-1).

If the ice sheets were as extensive as those CLIMAP proposed but if they also held a smaller volume of water from the ocean, one or more of the ice sheets must have been thinner than the CLIMAP reconstruction indicated. Scientists focused their attention on the North American ice sheet because of its huge size: if this ice sheet were thinner by about 30%, it would reconcile the disagreement between the volume of ice and the observed drop in sea level.

Independent evidence supports the possibility of thinner North American ice. One line of evidence

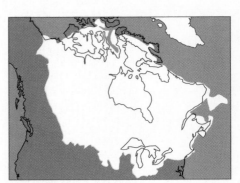

A Ice sheet extent

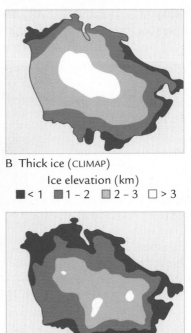

B Thick ice (CLIMAP)

Ice elevation (km)

■ < 1   ■ 1 – 2   ■ 2 – 3   □ > 3

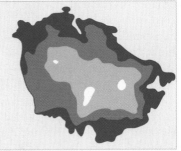

C Thin ice

FIGURE 13-5 **How thick were the ice sheets?** The limits of the North American ice sheet are well known (A), but the CLIMAP reconstruction showed a thick high-elevation ice sheet (white) during the glacial maximum (B) while recent reconstructions favor a thinner ice sheet (C). (Adapted from P. Clark et al., "Numerical Reconstruction of a Soft-Bedded Laurentide Ice Sheet During the Last Glacial Maximum," *Geology* 24 [1996]: 679–82.)

focuses on the conditions at the base of the ice sheet. As we saw in Chapter 10, ice sheets can remain frozen to hard bedrock bases, but they can also slip and slide across soft water-laden sediments and unconsolidated sedimentary rock. Only the central part of the North American ice sheet rested on hard bedrock, while the outer southern and southwestern margins lay on easily deformed sediments at lower, warmer elevations and latitudes. As a result, the southern margins of the ice sheet were probably thin because of continual sliding (Figure 13-5C). This sliding should also have thinned the inner portion of the ice sheet somewhat by drawing ice out toward the margins.

The second line of evidence comes from the amount that bedrock has rebounded since the ice sheets melted. Bedrock weighed down by ice sheets has a slow viscous response, and some of its rebound occurs long after the ice melts (Chapter 10). As a result, bedrock has a kind of memory of how much ice was once on the land and when the ice melted. Scientists can exploit this memory effect by examining regions in which the land is slowly rising out of the sea, leaving a trail of fossil beach shorelines up the side of the rising land. By ^{14}C dating shells that formed when these beaches were at sea level and measuring the present elevations of the old beaches above sea level, scientists can reconstruct the rate at which bedrock has risen out of the sea (see Chapter 14).

These rebound trends can then be used to project back into the past to estimate how thick the ice sheet was in each area. This technique has produced estimates of a thinner North American ice sheet than CLIMAP predicted. As we will see later in this chapter, a thinner glacial maximum ice sheet has consequences for nearby climate changes simulated by models.

13-3 Glacial Dirt and Winds

The ice sheets were prolific producers of debris in all sizes from large boulders to fine clay. Ice sheets grind across the landscape, scraping and dislodging soils and relatively unconsolidated sedimentary rocks. The weight of the ice sheets provides a pressure force that uses debris carried in the bottom layer of ice to grind and gouge out small pieces of even the hardest bedrock. In areas where basal layers of ice alternately freeze and thaw through time, water trickles down into cracks in bedrock when the ice melts and then expands when it freezes again, breaking off large chunks of bedrock. This freeze-thaw process quarries large slabs of bedrock and incorporates them in the ice for further grinding and fragmentation.

These and other processes at work in ice sheets erode large volumes of bedrock debris of all sizes. The ice sheets carry this material out toward their margins

and deposit it along their edges, where the ice melts. Much of the unsorted debris is piled into moraines (Chapter 2). Running water from melting ice or local precipitation reworks the rest of the debris, producing finer and better-sorted sediments called glacial **outwash**. Ice margins have little vegetation, both because the constant supply of new debris buries new vegetation and because meltwater inundates the region in summer. The lack of vegetation exposes the debris to further erosion.

Winds then rework these deposits, creating a gradation of grains away from the ice margins. The coarsest debris (boulders, cobbles, and pebbles) remains in place, but strong winds can transport medium to fine sand over short distances. Winds also lift and carry finer silt-sized sediment farther from source regions, leaving loess deposits that become thinner and finer away from the glacial outwash (Figure 13-6). The loess patterns suggest that winds carried this silty debris mainly from the west-northwest to the east-southeast in both North America and Europe.

Winds can carry even finer (clay-sized) dust completely around the world. Glacial-age layers in the Greenland ice sheet contain ten times as much fine dust as interglacial layers. Chemical analysis of this dust indicates that the main source region was Asia rather than nearby North America. Glacial ice also contains more Na^+ and Cl^- ions, an indication that far more salt was lifted from stormy glacial sea surfaces and deposited in the ice than is the case today.

Dust transport was also greater at lower latitudes during the last glacial maximum. Today the arid North African and Arabian deserts and their semiarid margins produce some of the largest dust storms on Earth

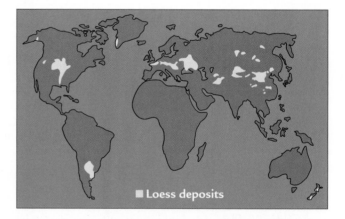

■ Loess deposits

FIGURE 13-6 Glacial maximum loess Ice sheets and mountain glaciers eroded large amounts of debris of all sizes and carried it to their margins. Winds picked up silt-sized loess and deposited it downwind of these sources. (Adapted from K. Pye, "Loess," *Progress in Physical Geography* 8 [1984]: 176–217.)

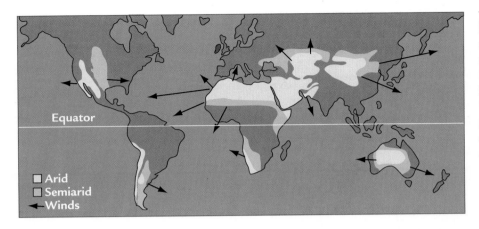

FIGURE 13-7 **Desert dust** Low-latitude arid and semiarid regions are major sources of windblown dust today. Desert margins in Arabia, the southern Sahara, and Australia produced even larger amounts of dust during the last glacial maximum. (Adapted from K. Pye, *Aeolian Dust and Dust Deposits* [New York: Academic Press, 1987], after G. Coude-Gaussen, "Le cycle des poussières éoliennes désertiques actuelles et la sédimentation des loess péridésertiques quaternaires," *Bull. Centre Rech. Explor. Product. Elf-Aquitaines* 8 [1984]: 167–82.)

(Figure 13-7). By comparison, sediment cores from the Indian Ocean east of the Arabian Peninsula show that dust accumulated five times faster during the last glacial maximum and several previous glaciations than it does today (Chapter 12). Cores from the equatorial Atlantic Ocean reveal that glacial dust was deposited at higher rates there as well.

The extremely arid cores of the deserts were also affected. Regions of Arabia and North Africa identified as deserts on modern maps actually vary in moisture level through time: moving sand dunes and loose soil are prevalent during extremely arid intervals, but the dunes are stabilized by sparse desert vegetation during slightly more moist periods. In both these areas and in Australia as well, moving sand dunes were much more extensive during the last glaciation than they are today because the winds were stronger and the climate was drier (Figure 13-8).

Even the South Pole was dustier. Glacial layers of ice cores from near the top of the Antarctic ice sheet contain more than ten times as much dust as modern interglacial layers. Geochemical fingerprinting of likely source areas suggests that this dust came from the southern tip of South America. The increased flux of dust probably resulted both from greater production of debris at the source and from a more turbulent atmosphere that carried dust farther and higher than it travels today.

Almost everywhere we look, the glacial world shows evidence that more debris was blowing across Earth's surface. One way of testing climate model simulations of the last glacial maximum is to examine the distribution of various kinds of debris carried by winds, ranging from desert sands to windblown silts (loess) to fine (clay-sized) dust. Because climate models can simulate the strength and direction of winds from the surface up to jet stream altitudes, the potential exists to compare model simulations with the observed patterns of windblown glacial debris.

Unfortunately, even though the current generation of climate models does a fairly good job of simulating the large-scale circulation of the atmosphere, this success does not yet extend to the smaller scales needed to simulate dust transport. The models are less successful at simulating the processes that actually lift and transport silt and dust from Earth's surface, such as local

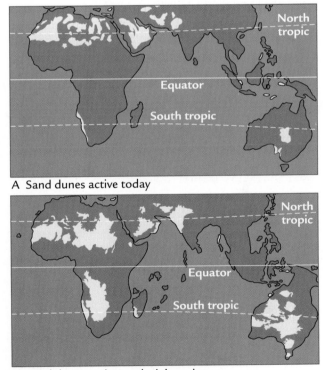

A Sand dunes active today

B Sand dunes active at glacial maximum

FIGURE 13-8 **Glacial maximum sand dunes** Moving sand dunes occur today in Africa, Arabia, and Australia (A). At the last glacial maximum, drier climates and stronger winds created more extensive sand dunes (B). (Adapted from M. Sarnthein, "Sand Deserts During Glacial Maximum and Climatic Optimum," *Nature* 272 [1978]: 43–46.)

wind gusts along frontal systems or small-scale eddies of wind. Models will have to be developed before they accurately simulate wind transport of glacial debris.

Testing Model Simulations Against Biotic Data

So far, we have examined only the physical aspects (ice and dirt) of the glacial maximum world. But living organisms were a part of this world as well, and they also have a story to tell. More important, they also let us test the performance of climate models on a world very different from ours. Such tests are critical because the same kinds of climate models are used to simulate future climate during times when greenhouse gases will increase.

13-4 Project COHMAP: Data-Model Comparisons

During the 1980's, an interdisciplinary project called **COHMAP** (*Co*operative *H*olocene *Ma*pping *P*roject) used a combined data-model approach to examine the last glacial maximum and the subsequent changes leading toward conditions today. Led by the meteorologists John Kutzbach and Tom Webb, the paleoecologist Herb Wright, and the geoscientist Alayne Street-Perrott, COHMAP brought together scientists from countries around the world to pool information from hundreds of individual ^{14}C-dated records of lake levels and pollen in

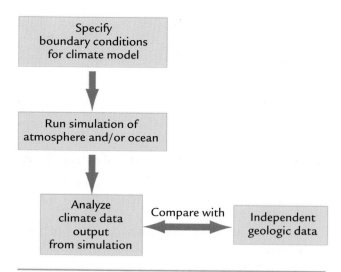

FIGURE 13-9 **Data-model comparisons** Past climates can be estimated by running climate model simulations with boundary conditions different from those of today and comparing the model output against estimates derived from pollen in lake sediments or other climatic data. (Adapted from J. Kutzbach et al., "Climate and Biome Simulations for the Past 21,000 Years," *Quaternary Science Reviews* 17 [1998]: 473–506.)

lake sediments. The individual records were compiled into maps to explore regional-scale patterns.

The first step in the COHMAP approach was to assemble records of the changing boundary conditions that have driven climate over the last 21,000 years (Figure 13-9). As we noted earlier, the boundary conditions most different from those of today at the last glacial maximum were the larger size of ice sheets and the lower levels of CO_2 (and CH_4) (see Figure 13-3).

The COHMAP researchers then ran model simulations of climate at intervals of several thousand years between the glacial maximum and the present to determine how changes in the major boundary conditions drove regional patterns of climate change. The COHMAP team focused on smooth orbital-scale changes in climate over thousands of years, rather than on the shorter-term fluctuations that will be the focus of Chapter 15.

The climate data produced as output from these model simulations were then tested against climate reconstructions based on ^{14}C-dated records of pollen from lake cores and plankton shells from ocean sediment cores. All organisms respond to their environment, especially to climatic variables such as temperature and precipitation. Modern relationships between the abundances of species and climatic variables can be measured, quantified, and used to reconstruct past climates from fossil organisms. By comparing these fossil-based estimates of climate with the changes simulated by the models, scientists can test the reliability of both approaches (Figure 13-9).

13-5 Pollen: An Indicator of Climate on the Continents

Vegetation responds strongly to climate. Precipitation and temperature determine the larger-scale vegetation units, such as forests, grasslands, and deserts, and also the distribution of species within those units. The pollen produced by vegetation is preserved in lake sediments that can be ^{14}C-dated. Changes in pollen through time, then, tell us about changes in climate.

Pollen is carried mainly by winds, and to a lesser extent by water and insects. Some pollen comes to rest in lakes and settles into the mud, where its resistant outer layer aids preservation. The pollen preserved in a lake reflects the average composition of vegetation over a region extending tens of kilometers beyond the lake. The pollen percentages are generally similar to those of the actual vegetation, although overproducers such as pine trees leave disproportionately large amounts of pollen in the lake muds in comparison with such underproducers as maples. Climate scientists can adjust for this kind of disproportionate representation.

The northern midwestern states are a useful region for showing how climate controls vegetation (Figure 13-10). The percentage of pollen from prairie grasses

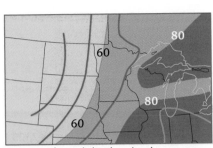

A Annual precipitation (cm)

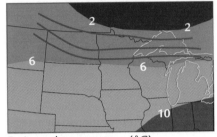

B Annual temperature (°C)

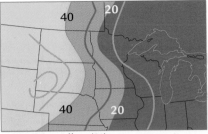

C Prairie pollen (%)

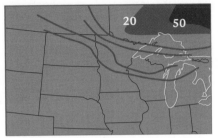

D Spruce pollen (%)

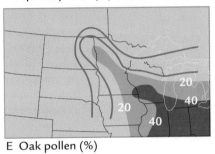

E Oak pollen (%)

FIGURE 13-10 **Pollen distributions and climate** Today the distribution of pollen in the northern midwestern United States reflects control by precipitation (A) and temperature (B). Prairie grasses and herbs are most abundant where rainfall is low (C), and tree pollen is more common in wetter eastern regions. Spruce trees proliferate in the colder north (D), oak in the warmer south (E). (Adapted from T. Webb III, "Holocene Palynology and Climate," in *Paleoclimate Analysis and Modeling*, ed. A. Hecht [New York: John Wiley, 1985].)

and herbs is higher in modern lake sediments of the semiarid West than in the wetter, tree-dominated region east of the Mississippi River. Within the wetter eastern forest, cold-tolerant spruce pollen is more abundant in the north, while oak pollen is more abundant in the warmer southern latitudes. These climatic controls can also be demonstrated by plotting pollen percentages against different combinations of seasonal and annual temperature and precipitation (Figure 13-11).

These modern relationships are a useful basis for understanding the past. The bottom layers of sediment in a ^{14}C-dated core from the upper Midwest (Minnesota) are of late glacial age, dating from just after the time the North American ice sheet melted back from this region, while the upper layers of mud record the postglacial climate of the present interglaciation (Figure 13-12). Most of the pollen in the older layers is from spruce trees, indicating conditions colder than today's. An abrupt switch from spruce pollen to warm-tolerant oak pollen near 10,000 years ago indicates a rapid warming in this region. During the subsequent interglaciation, changes in the relative amounts of herb and grass pollen, with maximum values in dry-adapted vegetation near 6000 years ago, indicate a climate somewhat drier than today's.

This core is just one of many hundreds examined in North America, along with additional hundreds in Europe and elsewhere in the world. Viewed together, these records provide a larger geographic perspective on the pattern of pollen (and vegetation) distribution at the last glacial maximum and during the deglaciation. This larger map perspective can be compared with map patterns produced by model simulations.

13-6 Using Pollen for Data-Model Comparisons

Data-model comparisons usually focus on the distribution of pollen at specific intervals in the past across specific geographic regions. Counts of pollen percentages in lake sediments within these regions produce mapped patterns of "observed" pollen abundance. These observed patterns are then compared with pollen distributions predicted by climate models for the same interval in the past.

These predicted pollen distributions result from a series of steps. First, boundary conditions are chosen and used in climate-model simulations to produce estimates of past temperature and precipitation. Then the model estimates of temperature and precipitation are used to generate estimates of the percentage of abundance of each type of pollen based on modern-day relationships between climate and pollen such as those shown in Figure 13-11. Each estimate of annual precipitation and mean July temperature simulated for a specific grid box in the model yields a specific estimate of the percentage of oak, spruce, and prairie pollen for that particular location. The map patterns of pollen abundance estimated in this way can then be compared directly with the map patterns based on pollen counts from lake cores.

One example of such a comparison is spruce pollen. Observations today show maximum amounts of spruce

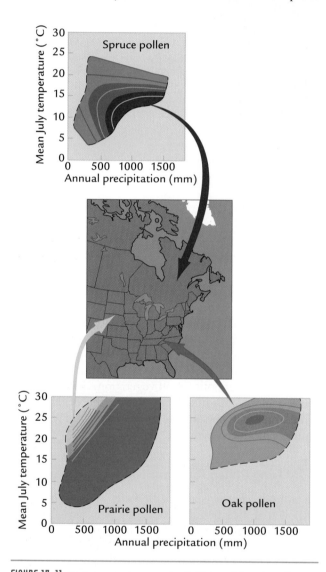

FIGURE 13-11 **Pollen percentages and climate** The abundances of spruce, oak, and prairie pollen follow distinct temperature and precipitation patterns. Colors indicate the same pollen abundances as in Figure 13-10. (Adapted from T. Webb III et al., "Climatic Change in Eastern North America During the Past 18,000 Years: Comparisons of Pollen Data with Model Results," in *North America and Adjacent Oceans During the Last Deglaciation,* ed. W. F. Ruddiman and H. E. Wright [Boulder, Colo.: Geological Society of America, 1987].)

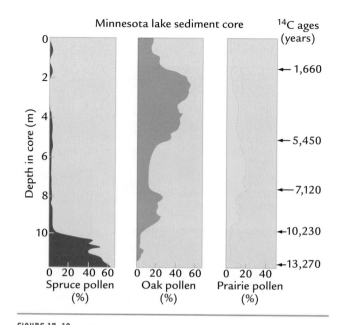

FIGURE 13-12 **Pollen in a lake core** A ^{14}C-dated sediment core from a Minnesota lake shows a transition in climate near 10,000 years ago, from colder conditions (abundant spruce) to a warmer climate (abundant oak). High percentages of prairie grasses near 6000 years ago indicate a drier climate. (Adapted from H. E. Wright et al., "Two Pollen Diagrams from Southeastern Minnesota: Problems in the Late- and Postglacial Vegetation History," *Geological Society of America Bulletin* 74 [1963]: 1371–96.)

pollen in northeastern Canada (Figure 13-13A). Counts of pollen in lake sediments of the last glaciation show spruce concentrated in the east-central United States just south of the ice sheet (Figure 13-13B). These observations agree closely with climate-model predictions of spruce at the glacial maximum (Figure 13-13C). But observations do not always match the model simulations this well, as we shall see.

Comparisons with pollen data can also be made with **biome models** (Figure 13-14). In order to estimate the vegetation that would have been present in different regions, the biome method again makes use of climatic variables simulated by GCMs for times in the past when boundary conditions differed from today's. The method is carried out in two steps. The first step uses broad temperature and precipitation constraints from the GCM to narrow the possible range of major vegetation types (for example, no trees can occur in model grid boxes with hyperarid climates, but grass and desert scrub vegetation can).

In the second step, the surviving vegetation units within each grid box compete for the resources necessary for growth and reproduction, such as water, nutrients, and light. Both steps are based on present-day relationships between vegetation and the environment.

Distribution of spruce pollen

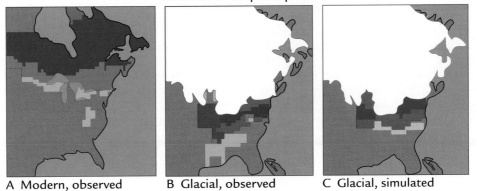

A Modern, observed   B Glacial, observed   C Glacial, simulated

■ < 1%   ■ 1 – 5%   ■ 5 – 20%   ■ > 20%

FIGURE 13-13 **Modern and glacial maximum spruce** Today spruce pollen is most abundant in the cold climate of northeastern Canada (A). At the glacial maximum, spruce pollen is found mainly in lake sediments from the northern United States (B). Model simulations confirm that the large North American ice sheet produced temperatures cold enough for spruce to flourish in the northeastern United States (C). (Adapted from T. Webb III et al., "Late Quaternary Climate Change in Eastern North America: A Comparison of Pollen-Derived Estimates with Climate Model Results," *Quaternary Science Reviews* 17 [1998]: 587–606.)

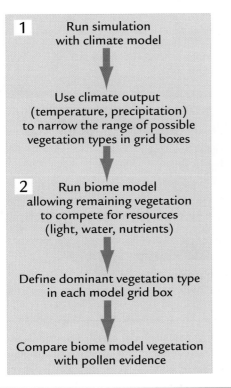

1 Run simulation with climate model

↓

Use climate output (temperature, precipitation) to narrow the range of possible vegetation types in grid boxes

↓

2 Run biome model allowing remaining vegetation to compete for resources (light, water, nutrients)

↓

Define dominant vegetation type in each model grid box

↓

Compare biome model vegetation with pollen evidence

FIGURE 13-14 **Biome models** Biome models work in two steps. First, large-scale temperature and precipitation restrictions from a climate model are used to narrow the range of vegetation in each region. Then the model allows the remaining vegetation to compete for light, water, nutrients, CO_2, and other resources.

Because the first step in the biome method includes all the major vegetation groupings on Earth, this approach can simulate changing vegetation patterns anywhere on Earth.

Data-Model Comparisons of Glacial Maximum Climates

In this section we examine the match—or mismatch—between model simulations of the climate of the last glacial maximum and several kinds of observations from the climate record.

13-7 Model Simulations of Glacial Maximum Climates

Glacial ice sheets are a critical boundary condition for simulations of glacial climate (Chapter 12). The central domes of these ice sheets protruded upward as massive icy plateaus, blocking and redirecting the flow of air. Climate model simulations suggest that the large ice sheet over North America could have split the winter jet stream into two branches at the glacial maximum.

In modern winters, a single jet stream enters North America near the border between Canada and the United States. Storms associated with this jet bring wet winters to Oregon, Washington State, and British Columbia (Figure 13-15A). During glacial times, the jet stream split into a northern branch located along the

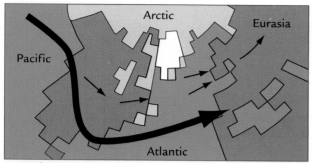

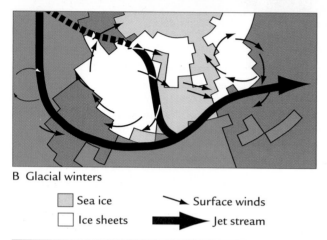

A Modern winters

B Glacial winters

▨ Sea ice		➚ Surface winds	
☐ Ice sheets		➤ Jet stream	

FIGURE 13-15 GCM simulation of climate near the northern ice sheets Climate model simulations reproduce the modern-day path of the winter jet stream over North America (A). For the last glacial maximum (B), model simulations using a high-elevation ice sheet over North America split the jet stream into two branches, one south and one north of the ice. At the surface, cold winds flow down off the North American and Scandinavian ice sheets and spiral in a clockwise pattern. (Adapted from COHMAP Project Members, "Climatic Changes of the Last 18,000 Years: Observations and Model Simulations," *Science* 241 [1988]: 1043–52.)

northern flank of the ice sheet and a southern branch over the American Southwest (Figure 13-15B).

Ice sheets did not literally poke high enough into the atmosphere to block the flow of the jet and cause it to split in two. Ice sheets reach elevations of only 2 to 3 kilometers, whereas winter jet streams flow at altitudes of 10 to 15 kilometers. But the ice did block lower-level atmospheric flow, and the effects of this disruption extended higher into the atmosphere. These effects, along with the tendency of jets to flow above regions of strong temperature change at Earth's surface, caused the split jet. Model simulations using a high ice sheet (the CLIMAP maximum reconstruction) split the jet to a much greater degree than do simulations based on lower-elevation ice.

Climate model simulations also indicate major changes in atmospheric circulation at Earth's surface caused by the ice sheets (Figure 13-15B). The models

simulate a clockwise spiral of cold air moving down, off, and around the ice sheets in winter. Cold air flowing eastward along the northern flank of the North American ice sheet as part of this circulation blew southeastward over the western North Atlantic, chilling the ocean surface. A narrow layer of cold winds blew westward across the northern United States, reversing the west-to-east wind flow that dominates that region today. In Alaska, the clockwise pattern produced a south-to-north wind flow during the glacial maximum that may have prevented climate in the ice-free Alaskan interior from becoming even harsher than it is today.

A similar clockwise spiral of winds over the Scandinavian ice sheet brought cold, dry air southward into Europe (Figure 13-15). In addition, a strong

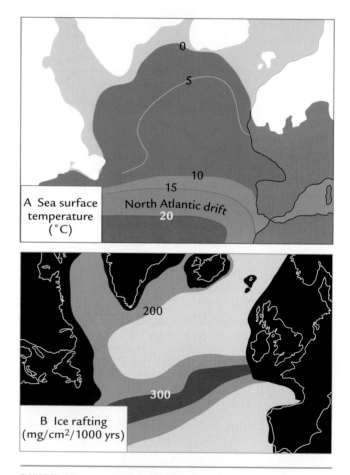

A Sea surface temperature (°C)

North Atlantic drift

B Ice rafting (mg/cm²/1000 yrs)

FIGURE 13-16 A cooler glacial North Atlantic Ocean The largest ocean cooling in the CLIMAP reconstruction is in the northern North Atlantic, surrounded by ice sheets (A). Highest rates of deposition of ice-rafted debris occurred near 50°N, where southward-floating icebergs first encountered warm waters and melted (B). (A: adapted from A. McIntyre et al., "Glacial North Atlantic 18,000 Years Ago: A CLIMAP Reconstruction," *Geological Society of America Memoir* 145 [1976]: 43–76; B: adapted from W. F. Ruddiman, "North Atlantic Ice Rafting: A Major Change at 75,000 Years B.P.," *Science* 196 [1977]: 1208–11.)

A

FIGURE 13-17 **The glacial Southwest was wetter** Today most basins in the southwestern United States, such as Death Valley, are dry or occupied only occasionally by temporary lakes (A). At the last glacial maximum, lakes filled hundreds of basins because the southward displacement of the jet stream from Canada brought increased rain and cloud cover (B). (A: Peter Kresan; B: adapted from G. I. Smith and F. A. Street-Perrott, "Pluvial Lakes of the Western United States," in *Late Quaternary Environments of the United States*, ed. S. C. Porter [Minneapolis: University of Minnesota Press, 1983].)

basically confirmed by independent geologic evidence. Those that disagree do so only in degree and have pointed the way to additional research.

The CLIMAP reconstruction based on the shells of planktic organisms shows the largest changes in estimated surface-ocean temperatures occurring in the North Atlantic Ocean (see Figure 13-4). Frigid water filled the higher latitudes, and sea ice reached farther south than it does today. The warm waters of the Gulf Stream and North Atlantic Drift flowed eastward toward Portugal instead of penetrating northeastward toward Scandinavia, as they do today (Figure 13-16A).

The flow of cold winds off the North American ice sheet was one important cause of this glacial cooling of the North Atlantic (Chapter 10). Climate models that allow the ocean surface to react to these cold winds simulated changes in sea surface temperature similar to those estimated by CLIMAP, although the models used for this comparison were crude.

Other factors may have contributed to the North Atlantic cooling. Coarse ice-rafted debris deposited in deep-ocean sediments across a broad band near 50°N latitude shows that icebergs that broke off from the continental ice sheets drifted southward through the cold North Atlantic until they encountered warm waters and melted (Figure 13-16B). In addition, the slower glacial rates of formation of deep water at high latitudes of the North Atlantic (Box 13-1) reduced the northward flow of warm water and cooled the ocean surface.

Changes in North America An impressive example of agreement between observations and model simulations for the last glacial maximum comes from the southwestern United States. Today this area is arid semidesert, except for deep winter snow and trees high on mountainsides, and small lakes maintained by meltwater runoff from the mountain snows. Most runoff is trapped in low-lying interior basins and never reaches the ocean (Figure 13-17A).

B

upper-level jet stream crossed the Atlantic Ocean along latitudes between 45° and 50°N and entered Europe well south of the ice sheet.

13-8 Climate Changes Near the Northern Ice Sheets

The most dramatic changes in climate at the glacial maximum were those in regions closest to and most directly influenced by the ice sheets. As we will see, most of the climate changes simulated by models are

---

BOX 13-1 LOOKING DEEPER INTO CLIMATE SCIENCE

Ventilating the Glacial Ocean

Formation of deep water in the Atlantic slowed during times when ice sheets were large. The availability of ^{14}C for dating gives researchers a way to evaluate the rate of ventilation (or replacement of water) in the deep ocean to see whether rates of deep-water replacement during the last glacial maximum were affected by this major change in circulation.

Measuring the ^{14}C age of dissolved inorganic carbon in modern seawater shows that deep water gradually "ages" as it moves from the North Atlantic into the Pacific Ocean. The ^{14}C ages shown are not the literal ages of these deep waters, but rather the ^{14}C age *difference* between surface waters, arbitrarily set at zero age, and the ages of local deep water, measured by radioactive decay of ^{14}C atoms as the water flows. Using this surface-to-deep ^{14}C age difference corrects for the fact that deep waters sink with a nonzero ^{14}C age.

A similar kind of analysis can be applied to the last glacial maximum. Scientists can measure the ^{14}C age of glacial surface waters from the $CaCO_3$ shells of planktic foraminifera and the ^{14}C age of deep waters from the shells of benthic (bottom-dwelling) foraminifera. The difference between these ages should be a measure of the age of the deep water at any location.

Measurements of this kind in the South China Sea of the western tropical Pacific show an average age difference between surface and deep waters for the last glacial maximum of about 1670 years, only slightly larger than the present value of 1600 years. Apparently glacial deep water in the Pacific was being replaced about as rapidly as it is today.

A larger change occurred in the tropical Atlantic at depths between 1.5 and 3.5 km. Instead of the present-day age difference of 350 years, the glacial difference was 675 years. This near-doubling in age suggests that deep Atlantic water was replaced more slowly at the glacial maximum than it is today. This result agrees with the evidence for reduced glacial rates of formation and southward flow of deep water in the North Atlantic.

At the last glacial maximum, this region was strikingly transformed, with hundreds of new lakes where none exist today and immense expansions of lakes that today are small (Figure 13-17B). The most prominent was glacial Lake Bonneville, near Salt Lake City in northwest Utah. Lake Bonneville was ten times larger than today's Great Salt Lake, and dissolved salt precipitated out into the lake muds to create the Bonneville salt flats.

Climate-model simulations of the last glacial maximum provide an explanation for this regionally wetter climate. The southern branch of the split jet stream entered North America over south-central California (Figure 13-17B) and produced two responses favorable to a moister climate and expanded lakes: more precipitation caused by winter storms following the path of the jet stream, and reduced evaporation caused by greater cloud cover and cooler temperatures.

In contrast to the wetter southwest, the climate of the Pacific Northwest was colder and drier during the glacial maximum. In this region today, frequent winter storms from the Pacific Ocean bring moisture that sustains lush forests, including rain forests on the western slopes of the Olympic Mountains in coastal Washington.

At the last glacial maximum, this region was covered by grass and herb vegetation indicative of much drier conditions and a loss of moisture from the Pacific.

The climate model simulations suggest at least two reasons for this change in the Pacific Northwest (Figure 13-15B). First, the shift of the winter jet stream to the southwestern United States displaced the main storm track and associated precipitation away from this region. In addition, the clockwise flow of cold, dry winds around the North American ice sheet produced more frequent low-level winds blowing westward from the dry mid-continent and replacing the flow of moist westerly winds from the Pacific.

The region with the most extensive coverage of lake cores and pollen data for testing climate models is eastern North America, today an area of temperate deciduous forests. In this region scientists can test the performance of climate models in a more quantitative way by checking for data-model agreement or disagreement over the *magnitude* of climate changes, not just their direction.

East-central North America south of the ice sheet had a mixture of spruce trees, scattered deciduous trees, and grasses and herbs during the last glacial maximum.

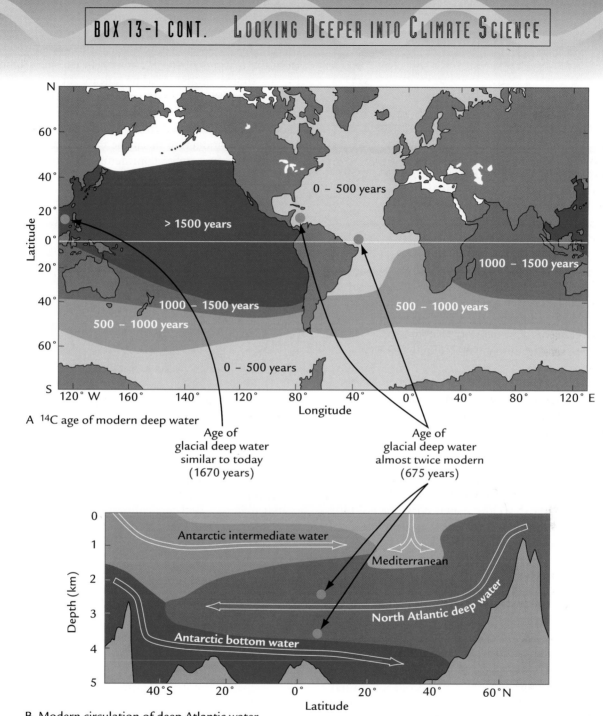

BOX 13-1 CONT. LOOKING DEEPER INTO CLIMATE SCIENCE

A ^{14}C age of modern deep water

Age of glacial deep water similar to today (1670 years)

Age of glacial deep water almost twice modern (675 years)

B Modern circulation of deep Atlantic water

Older Atlantic deep water during glaciations The ^{14}C age of deep water increases from the Atlantic to the Pacific today (A). At the last glacial maximum, deep western Pacific water had the same age it does today, but deep Atlantic water was almost twice as old (B). (A: adapted from M. Andrée et al., "Accelerator Radiocarbon Ages on Foraminifera Separated from Deep-Sea Sediments," in *The Carbon Cycle and Atmospheric CO$_2$: Natural Variations, Archean to Present,* ed. E. T. Sundquist and W. S Broecker, Geophysical Monograph 32 [Washington, D.C.: American Geophysical Union, 1985]; B: adapted from E. Berner and R. Berner, *Global Environment* [Englewood Cliffs, N.J.: Prentice-Hall, 1996].)

Distribution of elm pollen

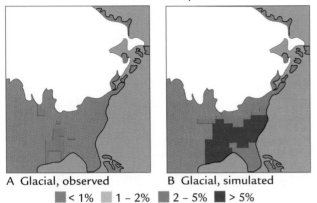

A Glacial, observed   B Glacial, simulated

■ < 1%   ■ 1 – 2%   ■ 2 – 5%   ■ > 5%

FIGURE 13-18 **Data-model mismatch in the southeastern United States** Observed abundances of warm-adapted deciduous pollen such as elm in the southeastern United States during the glacial maximum (A) are smaller than the amounts simulated by climate models (B). (Adapted from T. Webb III et al., "Late Quaternary Climate Change in Eastern North America: A Comparison of Pollen-Derived Estimates with Climate Model Results," *Quaternary Science Reviews* 17 [1998]: 587–606.)

This mixture indicates a region of discontinuous tree cover interrupted by grassy openings. The more continuous forest cover south of 35°N was a mixture of pine and various deciduous trees.

We saw earlier that the model-simulated and observed patterns of spruce in the northern United States at the glacial maximum match extremely well (see Figure 13-13), but this close match does not hold for several pollen types farther to the south. Pollen produced by deciduous trees such as oak and elm is much less abundant (or even nearly absent) in lake sediments from this region (Figure 13-18A) than the levels predicted by the climate models (Figure 13-18B).

This mismatch suggests that the cooling simulated by the climate model for the southeastern United States underrepresents the cooling that actually occurred, permitting too many warm-adapted trees. The cause of this mismatch may have been an unusual geographic configuration in which cold meltwater from the southern margin of the great Laurentide ice sheet flowed down the Mississippi River and emptied directly into the Gulf of Mexico at subtropical and tropical latitudes. If the boundary conditions entered into the model had incorporated this cold inflow, the model would have simulated cooler temperatures across a broad region of the southeastern United States influenced by air masses from the nearby Gulf. These early data-model disagreements point the way toward improvements needed for more realistic model simulations.

Changes in Eurasia Europe was completely transformed during the glacial maximum. The conifer and deciduous forests typical of today's interglacial climate (Figure 13-19A) were eliminated all the way southward from the margins of the Scandinavian ice sheet almost to the shores of the Mediterranean. In their place, grass-covered steppes and herb-covered tundra vegetation covered much of the continent, with a few remnants of forest scattered in the south (Figure 13-19B). In effect, the moderate maritime climate of today was replaced by a far harsher continental climate, more like

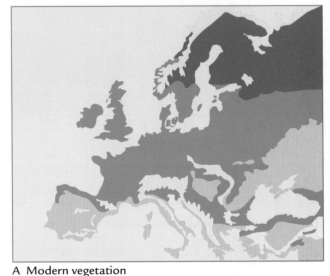

A Modern vegetation

☐ Ice   ■ Boreal forest   ▨ Mediterranean scrub
☐ Tundra and   ■ Deciduous   ▨ Prairie-steppe
   mountain   and conifer forest

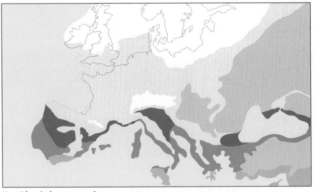

B Glacial vegetation

FIGURE 13-19 **Glacial north-central Europe was treeless** Vegetation in present-day Europe (A) is dominated by forest, with conifers in the north and deciduous trees to the south. At the glacial maximum (B), Arctic tundra covered a large area south of the ice sheet, with grassy steppe farther south and east, and forests reduced to patches near the southern coasts. (Adapted from R. F. Flint, *Glacial and Quaternary Geology* [New York: John Wiley, 1971].)

that of present-day northern Asia. These changes in vegetation agree with the dry, windy conditions indicated by the greater prevalence of windblown loess.

Biome models simulate glacial vegetation in Europe similar to that observed, with tundra replacing forest in the north near the ice sheets and grassy steppe vegetation prevailing farther to the south and east. One reason for this harsh glacial climate was the clockwise outflow of cold winds from the Scandinavian ice sheet (Figure 13-15B). A second reason was the large chilling of the North Atlantic Ocean. A climate-model sensitivity test examined in Chapter 12 showed that the moderating influence of the ocean on modern-day winters in Europe is greatly reduced when Atlantic temperatures are cooler, as they were at the glacial maximum.

Winter temperatures closer to the Mediterranean Sea were probably moderated by frequent storms arriving along the track of a southward-shifted jet stream from the Atlantic Ocean (Figure 13-15B). The precipitation delivered by these storms produced high lake levels, but steppe vegetation persisted in much of this area because summers were extremely dry.

One of the most striking features of the last glacial maximum was the vast extent of the steppe and tundra that covered much of northern Asia (Figure 13-20). A region covered today by forests of larch, birch, and alder trees was then a treeless expanse of grasses and herbs. Forests were completely absent from the entire northern part of Asia, an indication that this region had an even harsher continental climate than it does today.

The harsh winter cold and sparse snow cover froze the ground tens of meters deep forming **permafrost**, but in most regions the frozen surface layer thawed in summer, allowing grasses and herbs to proliferate. South of the regions of year-round and seasonal permafrost were extensive grassy steppes (Figure 13-20A). The greater area covered by tundra and steppe (rather than modern forest) made the ground surface much more reflective in the snowy winter season and further cooled this region.

Climate-model simulations suggest that the main reason for this colder, drier glacial climate in Asia was a stronger high-pressure cell in Siberia during winter that produced an increased outflow of cold, dry air southward (Figure 13-20B). In addition, the icy conditions in the North Atlantic cut off much of the source of moisture for the Asian interior.

The effects of these harsh winters in northern Asia extended well into southeastern Asia. Today the influence of the warm, moist summer monsoon allows forests to extend northward along the Pacific coasts of China and Japan. At the glacial maximum, the stronger and more persistent winter monsoon sent cold air southward from Siberia and pushed the northern forest

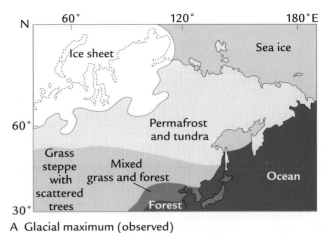

A Glacial maximum (observed)

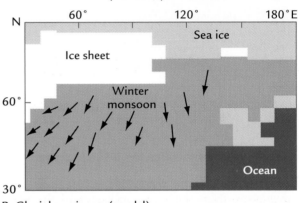

B Glacial maximum (model)

FIGURE 13-20 Glacial northern Asia was treeless At the last glacial maximum, Asia was covered by permafrost and tundra in the north and steppe in the south, with little forest left (A). Climate models indicate that this distribution of vegetation resulted from a much stronger and colder winter high-pressure cell in northern Asia, with stronger cold winds blowing to the south (B). (A: adapted from V. P. Grichuk, "Late Pleistocene Vegetation History," in *Late Quaternary Environments of the Soviet Union*, ed. A. A. Velichko et al. [Minneapolis: University of Minnesota Press, 1984]; B: adapted from J. E. Kutzbach et al., "Simulated Climatic Changes: Results of the COHMAP Climate Model Experiments," in *Global Climates Since the Last Glacial Maximum*, ed. H. E. Wright et al. [Minneapolis: University of Minnesota Press, 1993].)

limit far to the south (Figure 13-20A). The North Pacific was also colder than it is today, with much more extensive winter sea ice along the coast of Asia and in the Bering Sea because of cold Siberian air blowing out over the ocean.

Despite the harshness of this environment, the steppe and tundra supported a diverse population of large mammals, including woolly mammoths and rhinoceroses. It is not clear how these creatures found enough food to survive during the cold winters. Our ancestors in northern Asia used mammoth bones to

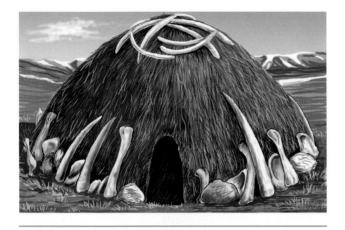

FIGURE 13-21 **Early human buildings** Humans living in northern Asia during the last glacial maximum constructed domed dwellings of hides draped over intricately linked mammoth bones. Other bones served as anchors.

build domed dwellings capped with animal hides (Figure 13-21).

13-9 Climate Changes Far from the Northern Ice Sheets

Farther away from the direct influence of the great northern hemisphere ice sheets, climate changes were less dramatic. In these regions the lower glacial atmospheric CO_2 value was probably the major cause of climate changes.

Large changes occurred in the Antarctic, where CLIMAP estimated that the winter limit of sea ice expanded northward by several degrees of latitude in the far-southern Atlantic and Indian oceans (Figure 13-22). Associated with this shift in sea ice was a northward displacement of the region of strongest upwelling and greatest surface-water productivity, but this change was offset by decreased productivity in regions nearer Antarctica where a cover of sea ice persisted over the summer.

Simulations using climate models with relatively simplistic representations of ocean circulation suggest that the CO_2 levels in the lower atmosphere during the glacial maximum may have cooled the high southern latitudes enough to explain this advance of Antarctic sea ice. Another possible factor in the cooling of the Southern Ocean may have been a reduced inflow of warm deep water from the North Atlantic (Chapter 12).

The record of climate changes on arid southern hemisphere continents remains sparse. Expanded desert dunes in Australia (see Figure 13-8) suggest an even more arid climate and an intensification of the modern counterclockwise wind flow on that continent (Figure 13-23A). Lake levels and pollen data also provide sup-

porting evidence for a drier interval near the glacial maximum. One factor that contributed to greater glacial aridity in northern Australia was the withdrawal of the ocean from a vast area just to its north (see Figure 13-4).

Climate scientists disagree about the cause of greater aridity in southern Australia. Although scientists initially proposed that the ocean cooling and advance of sea ice around Antarctica would have displaced climate belts to the north and made southern Australia cooler and wetter, climate model simulations suggest that the main belt of storms actually shifted to the south (Figure 13-23A), in accord with the observed drying. Biome models also indicate that lower atmospheric CO_2 levels cooled the atmosphere and contributed to drying over Australia.

Climate in much of South America is heavily influenced by winds from nearby oceans (Figure 13-23B). Most ocean moisture is dropped in the Amazon rain forest and along the eastern flanks of the Andes. Scattered records from the Amazon Basin hint at the possibility that the rain forest may have been fragment-

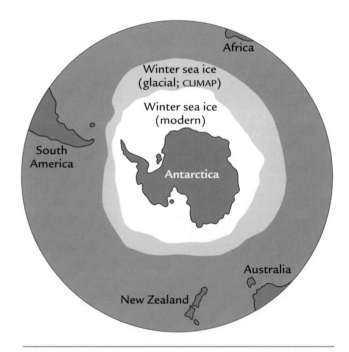

FIGURE 13-22 **Glacial Antarctica was surrounded by more sea ice** The CLIMAP glacial maximum reconstruction indicated that the seasonal maximum limit of sea ice in late winter and early spring expanded northward around Antarctica. (Adapted from J. D. Hays, "A Review of the Late Quaternary History of Antarctic Seas," in *Antarctic Glacial History and World Paleoenvironments,* ed. E. M. Van Zinderen Bakker [Rotterdam: A. A. Balkema, 1978]; and from L. H. Burckle et al., "Diatoms in Antarctic Ice Cores: Some Implications for the Glacial History of Antarctica," *Geology* 16 [1988]: 326–29.)

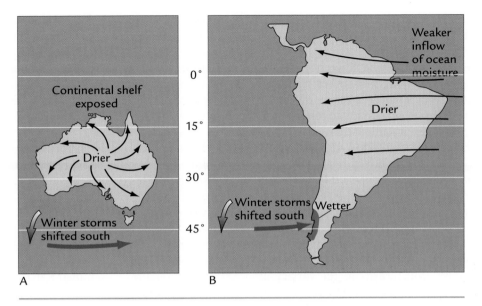

FIGURE 13-23 **Glacial Australia and South America were drier** At the last glacial maximum, lower atmospheric CO_2, cooler tropical oceans, and broader regions of exposed continental shelf made both Australia (A) and South America (B) drier. Climate model simulations indicate that the axis of the winter storm track shifted southward to latitude $40°–45°$ S, matching pollen evidence of a small zone of wetter climates in South America at this latitude. (A: adapted from J. M. Bowler, "Aridity in Australia: Age Origins and Expression in Aeolian Landforms and Sediments," *Earth Science Reviews* 12 [1976]: 279–310; B: adapted from V. Markgraf, "Climatic History of Central and South America Since 18,000 Years B.P.: Comparison of Pollen Records and Model Simulations," in *Global Climates Since the Last Glacial Maximum*, ed. H. E. Wright et al. [Minneapolis: University of Minnesota Press, 1993].)

ed into smaller pieces than the massive area forested today.

Along the Andes, where most lake-sediment records have been found, pollen data generally indicate drier conditions at the glacial maximum. This drying is probably the combined result of reduced extraction of water vapor from the cooler oceans, the lowering of sea level by 110 meters, and the cooler land temperatures resulting from lower CO_2 levels in the atmosphere.

Pollen data from latitudes $40°$ to $45°$ S indicate glacial climates wetter than today's west of the Andes, but drier conditions to the east (Figure 13-23B). This pattern suggests a narrow band of strengthened flow of moist westerly winds, along with drying in the rain shadow east of the Andes. Climate model simulations show a southward shift of the axis of westerly winds and moisture-bearing storms, in agreement with the pollen evidence and similar to that in Australia.

Because most of the tropics were more arid at the last glacial maximum, rain forest vegetation in both South America and Africa was probably less extensive than it is today. Yet despite this drier climate, tropical biomass may actually have increased. The 110-meter drop in global sea level exposed vast expanses of new land across continental shelves of southeast Asia (see Figure 13-4). Because this region lay within the moist intertropical convergence zone, biome models indicate that it would have supported tropical rain forest vegetation. The increase in rain forest in this one region may have offset the loss of biomass elsewhere in the tropics, but it could not offset the enormous decrease in forest biomass at high northern latitudes. As a result, the total glacial biomass on Earth's continents was reduced by about 25% (Chapter 11).

How Cold Were the Glacial Tropics?

For almost two decades climate scientists have argued about the amount of temperature change in the tropics and subtropics indicated by the CLIMAP glacial maximum reconstruction. Tropical sea surface temperatures reconstructed by CLIMAP on the basis of fossil shells of ocean plankton averaged just $1°$ to $2°$ C cooler than they are today, and in some regions such as the southern subtropical Pacific were more than $1°$ C warmer. Estimated coolings of $3°$ C or more occurred only in localized areas of wind-driven upwelling.

In contrast, other evidence suggests that temperatures over the land and at least parts of the ocean may have been 4° to 6°C cooler than at present, far larger than the CLIMAP reconstruction. The difference in these estimates is in itself large enough to bother scientists, but this discrepancy also has much larger ramifications in respect to Earth's basic sensitivity to changes in atmospheric CO_2 and other greenhouse gases.

The issue of tropical sensitivity to CO_2 levels arises from an analysis of cause-and-effect relationships at the last glacial maximum. The tropics lie far from the immediate thermal impact of the ice sheets, such as the direct effects transmitted by the circulation of the atmosphere and the surface ocean in high northern latitudes. The tropics are also far from any climatic influence from deep water that forms near the northern ice sheets, flows south, and returns to the surface at high southern latitudes. The vast area of the tropics is isolated from both of these high-latitude cooling effects. GCM experiments over the last two decades confirm that these direct effects of ice sheets do not extend to the tropics.

If the cooling of the tropics cannot be caused by direct thermal effects from the ice sheets, then what is its cause? Another possible explanation could be changes in solar radiation. But we have already seen that solar insolation values at the last glacial maximum were so close to those today that insolation could not have been a major factor in the cooling of glacial climates in the tropics.

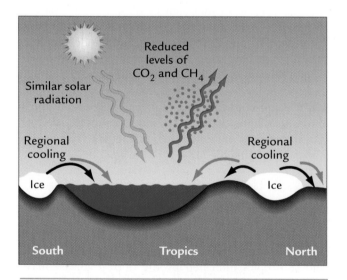

FIGURE 13-24 **Lower CO_2 and CH_4 levels cooled the glacial tropics** The tropics were too distant from the glacial ice sheets to feel their direct influence, and insolation values in summer and winter were close to those today. Lower levels of atmospheric CO_2 and CH_4 were the main cause of tropical cooling.

What is left to explain the glacial cooling of the tropics? By a process of elimination, the main cause must have been the 30% lower (190 ppm) levels of CO_2, along with the 50% drop in methane, a far less abundant greenhouse gas (Figure 13-24). When levels of greenhouse gases are low, less outgoing back radiation from Earth is trapped in the atmosphere and the temperature falls.

If this reasoning is correct, the amount of glacial cooling in the tropics should be a direct measure of the sensitivity of this part of the climate system to changes in atmospheric CO_2 and methane. Because half of Earth's surface area lies between 30°N and 30°S, the amount of glacial cooling in the tropics and lower subtropics is a measure of the fundamental sensitivity of the total climate system.

In Chapter 6 we examined Earth's response to higher levels of CO_2 in a greenhouse world. The last glacial maximum now provides a complementary perspective on the same relationship: Earth's response to lower levels of CO_2 in an icehouse world. Together the greenhouse and icehouse cases help climate scientists define Earth's basic sensitivity to CO_2 in terms of the change in global temperature per unit of change in CO_2 level. This analysis is directly relevant to future climate change, because we face a warming in the future caused by human-induced increases in atmospheric CO_2 and methane, and we need to know how large this warming will be.

13-10 Evidence for a Small Tropical Cooling

The evidence for a small tropical cooling in the CLIMAP reconstruction is based on the changes in planktic fauna and flora in the low-latitude oceans. CLIMAP's technique for reconstructing sea surface temperatures was based on the assumption that the distribution of species and assemblages of plankton is in large part determined by the temperature of the water in which they live. At higher northern latitudes during the glacial maximum, cold-adapted species moved into areas where warm-adapted species prevail today, indicating a large cooling in these regions. Across most low-latitude regions, in contrast, the species that existed at the glacial maximum were not much different from the warm-adapted forms found there today (Figure 13-25). This lack of change in tropical plankton led directly to the CLIMAP assessment that ocean temperatures did not cool much in the tropics (by an average of only 1.5°C).

Evidence obtained from the biochemical composition of plankton shells supports the CLIMAP estimates. One technique is based on the relative abundance of complex organic molecules called **alkenones**. Alkenones constitute small fractions of tiny plant plankton called

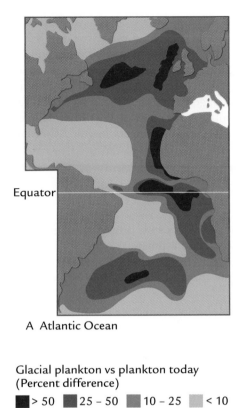

Equator

A Atlantic Ocean

Glacial plankton vs plankton today
(Percent difference)

■ > 50　■ 25 – 50　■ 10 – 25　■ < 10

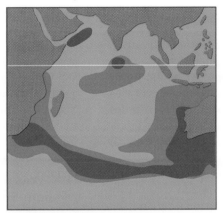

B Indian Ocean

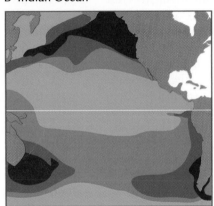

C Pacific Ocean

FIGURE 13-25 **Planktic fauna of the glacial maximum vs. that of today** The CLIMAP method of reconstructing glacial maximum ocean temperatures was based on temperature-sensitive plankton assemblages. Planktic assemblages in most low-latitude regions of the Atlantic (A), Indian (B), and Pacific (C) oceans differed only slightly from those of today, indicating little glacial cooling. (Adapted from T. C. Moore et al., "The Biological Record of the Ice-Age Ocean," *Palaeogeography, Palaeoclimatology, Palaeoecology* 35 [1981]: 357–70.)

coccolithophores, and the past abundances of these molecules can be measured in small $CaCO_3$ plates (coccoliths) deposited in ocean sediments (see Figure 3-17). The relative amounts of two types of alkenone molecules are sensitive to temperature in the modern ocean and can be used to reconstruct past temperatures.

In most of the regions where the CLIMAP temperature estimates have been compared with alkenone measurements, the two techniques show a roughly comparable glacial cooling. In a north–south transect of cores across the western Indian Ocean, the cooling indicated by both methods is generally less than 2°C (Figure 13-26). Basic (although not perfect) agreement between these methods persists at a dozen other tropical and mid-latitude sites in other oceans.

Oxygen isotope measurements on the $CaCO_3$ shells of plankton provide climate scientists with another way of testing the CLIMAP results. By measuring the difference between modern $\delta^{18}O$ values and those of the last glacial period, and by subtracting out the 1.1‰ signal caused by storage of ^{16}O-rich water in glacial ice sheets, scientists can use the remaining $\delta^{18}O$ difference as an estimate of temperature changes at the glacial maximum. This method suggests that most of the low-latitude Pacific Ocean cooled very little during the last

glaciation, while the Atlantic surface cooled by an average of 2° to 3°C. Both estimates are generally in agreement with CLIMAP in the regions examined.

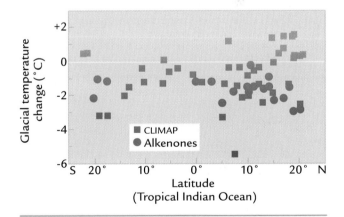

FIGURE 13-26 **Independent confirmation of small ocean cooling** A biochemical (alkenone) method of estimating past sea surface temperatures indicates a small cooling of the tropical Indian Ocean, similar to the values found by CLIMAP. (Adapted from E. Bard et al., "Interhemispheric Synchrony of the Last Deglaciation Inferred from Alkenone Paleothermometry," *Nature* 385 [1997]: 707–10.)

13-11 Evidence for a Large Tropical Cooling

A different view emerges from several other indicators, most of which come from continental records. The most compelling evidence is the descent of the lower limit of mountain glaciers by 600 to 1000 meters of elevation throughout the tropics and middle latitudes (Figure 13-27). This drop in the ice line has been interpreted as requiring a cooling of 4° to 6°C over tropical mountains.

The lower limit of mountain glaciers today is determined mainly by summer temperature and secondarily by factors such as the amount of precipitation and the degree to which local mountain topography shelters the glaciers from direct sunlight. Glaciers exist today on tropical mountains higher than 5 kilometers because the atmosphere cools by 6.5°C or more per kilometer of elevation, resulting in subfreezing temperatures at higher elevations. This relationship (the lapse-rate cooling; see Chapter 2) can be used to calculate a cooling of 4° to 6°C for the observed lowering of tropical mountain glaciers by 600 to 1000 meters during the glacial maximum.

Additional evidence for larger glacial cooling comes from the descent of the upper tree limit and other kinds of vegetation high on tropical mountains. In the harsh conditions on the upper flanks of mountains, temperature limits the growth of many kinds of vegetation, and the vertical drop in high-altitude vegetation limits during the last glaciation equals or even exceeds that of the mountain glaciers (Figure 13-27).

Other methods based on new temperature-sensitive chemical techniques also suggest a large tropical cooling. Analyses of trace amounts of the element strontium, which substitutes for calcium in the $CaCO_3$ shells of tropical corals, suggests a 5°C ocean cooling near Barbados, yet the CLIMAP method estimated only a 2°C drop in temperature. A technique based on the temperature-dependent concentrations of the noble gases xenon, krypton, argon, and neon dissolved in glacial-age groundwater suggests that the southwestern United States and the southeastern coast of Brazil both cooled by 5°C, whereas CLIMAP estimated cooling of 2°C or less for nearby ocean surfaces.

13-12 Was the Actual Tropical Cooling Medium-Small?

What are we to make of this wide divergence among the kinds of evidence used to estimate the size of the tropical cooling at the glacial maximum? Some climate scientists try to argue that one or the other view is completely wrong.

Critics of the CLIMAP evidence for small cooling note that although plankton are indeed sensitive to temperature changes at higher latitudes, they are relatively insensitive at lower latitudes, where most of the common species of plankton show relatively little systematic variation in abundance when ocean temperatures are warmer than 20°C. The critics also note that food may be a more important limitation than temperature to the survival of plankton at low latitudes. Most of the low-latitude surface ocean is depleted in nutrients, and plankton are forced to adopt strategies for surviving where food is scarce. In such a world, temperature is likely to be less important as a controlling factor than it is at high latitudes.

Another criticism of the CLIMAP reconstruction focuses on the Pacific Ocean, where the glacial cooling is smallest and broad regions even show a warming. The Pacific is the most difficult ocean in which to apply the CLIMAP techniques: extensive dissolving of $CaCO_3$ on the seafloor removes or alters most foraminifera and coccoliths, and the remnants of siliceous organisms (radiolaria and diatoms) left in the sediments are very different from those that actually lived in the surface waters. Estimates of temperatures at the surface of the Pacific derived from these different types of plankton often disagree, an indication that at least some are unreliable.

The large-cooling view also has its critics. Some suggest that the drier climate in most of the tropics at the glacial maximum might steepen the lapse rate from its present-day 6.5°C/km toward the 9.8°C/km rate typical of very dry air. A lapse rate that steepened by less than 1°C/km could reconcile the large (4° to 6°C) cooling on tropical mountaintops with the small (1° to 2°C) cooling at sea level. Independently, climate model simulations suggest that a steeper lapse rate in the drier

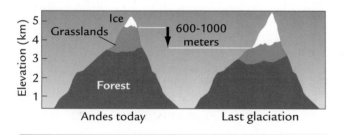

FIGURE 13-27 **Descent of tropical mountain glaciers and forests** The limits of mountain glaciers in the Andes were 600 to 1000 meters lower during the last glaciation than they are today, and the upper limits of forests were similarly lower. These major shifts indicate a tropical cooling of at least 5°C, much larger than the 1°–2°C suggested by CLIMAP. (Adapted from T. van der Hammen, "The Pleistocene Changes of Vegetation and Climate in Tropical South America," *Journal of Biogeography* 1 [1974]: 3–26.)

glacial tropics could account for up to half of the discrepancy between ocean and land evidence.

Another criticism of the large-cooling view focuses on the fact that the evidence from mountain glaciers is poorly dated. Despite pervasive evidence that glacier limits were once 600 to 1000 meters lower than they are today, only a handful of regions have ^{14}C dates that actually constrain this lowering specifically to the time of the glacial maximum, 21,000 years ago. Some glacial moraines initially thought to date from the glacial maximum later turned out to have formed before 30,000 years ago, and apparently during a time of cooler but also wetter climates. Lower glacier limits at such times could be caused at least in part by greater snowfall.

Critics have also suggested that the descent of vegetation down tropical mountains may result in part from lower atmospheric CO_2 levels. Greenhouse experiments show that vegetation cannot exist at CO_2 values below 100 ppm, a level not far from the 190 ppm glacial maximum value. Because trees, shrubs, and cool-season grasses are all adapted to high levels of CO_2, some part of the descent of the trees and other vegetation down the sides of mountains could result from lower CO_2 levels rather than cooler temperatures. Biome models suggest that lower glacial CO_2 may explain a large part of the descent of tropical vegetation.

The most likely resolution to the controversy over the glacial tropical cooling is not that one or the other view is completely wrong but that the truth lies somewhere in the middle. On the one hand, the good agreement between the CLIMAP estimates, the alkenone data, and the $\delta^{18}O$ data is confirmation that the small-cooling view is probably still valid in many ocean regions. On the other hand, it seems likely that the lower ice lines on most tropical mountains will indeed turn out to date at or near the glacial maximum and that the large-cooling view is valid over these regions of the continents.

How do we resolve these differences among the various kinds of evidence? Part of the CLIMAP reconstruction is likely to be in error, particularly that of the low-latitude Pacific Ocean, where fossil plankton are so poorly preserved. The glacial cooling of the lower subtropics and the eastern tropical Pacific upwelling region could have easily been larger than CLIMAP indicated.

The CLIMAP estimates are also likely to be in error along ocean margins and nearly enclosed seas. In many of these regions, such as the western margins of the Pacific Ocean, CLIMAP had little or no core coverage and simply extrapolated the small temperature changes found in the open ocean into the coastal regions. For others, such as the Gulf of Mexico and Mediterranean

Sea, the glacial maximum planktic assemblages had such unusual combinations of species that it is risky to apply techniques based largely on modern-day plankton from the open ocean.

If the marginal basins were generally cooler than CLIMAP estimated, then many apparent discrepancies between the land and sea would disappear: the open ocean cooled only a little, the land cooled much more, and the regions along the coast cooled by an intermediate amount. This explanation also makes physical sense: small ocean basins are far more easily influenced by air masses flowing out from the (colder) continents than are the more remote areas of open ocean. The stronger outflow of cold, dry winter winds in the strengthened winter monsoons typical of glacial times would be far more likely to affect coastal waters than mid-ocean regions.

13-13 Relevance of Glacial Tropical Temperatures to Future Climate

What do these differences in tropical temperatures imply about Earth's sensitivity to CO_2 and other greenhouse gases? We know from ice core measurements that the glacial atmospheric CO_2 value of 190 ppm was 90 ppm lower than the 280 ppm preindustrial level (and that CH_4 values were 50% lower as well). This information tells us the changes in greenhouse gases that caused the cooler tropical temperatures. We also know from observational data that the tropical cooling lay somewhere between 1.5°C (CLIMAP) and 5°C (land evidence), with a plausible medium-small cooling of perhaps 3°C. By comparing the changes in both greenhouse gases and glacial temperature, we can assess Earth's sensitivity.

As we will see in Part V, this range of possible tropical cooling in response to lower CO_2 matches almost exactly the range of uncertainty about Earth's CO_2 sensitivity indicated by simulations with GCMs. The small tropical cooling proposed by CLIMAP lies at the lower end of the range of CO_2 sensitivity derived from GCM simulations; the large tropical cooling lies near the upper end of the range; and the medium-small cooling falls almost exactly in the middle.

For obvious reasons, future efforts to narrow the range of uncertainty about the amount of tropical cooling at the last glacial maximum could narrow the range of uncertainty in our knowledge of Earth's basic sensitivity to CO_2. The basic success of climate models in reproducing most of the features of the alternative world at the last glacial maximum also gives climate scientists confidence that these models are useful in forecasting future climate changes resulting from increases in atmospheric CO_2 and other greenhouse gases.

KEY TERMS

CLIMAP (Climatic Mapping and Prediction) Project (p. 277)

Laurentide ice sheet (p. 278)

Cordilleran ice sheet (p. 278)

Scandinavian ice sheet (p. 278)

Barents ice sheet (p. 278)

outwash (p. 282)

COHMAP (Cooperative Holocene Mapping Project) (p. 284)

biome models (p. 286)

permafrost (p. 293)

alkenones (p. 296)

REVIEW QUESTIONS

1. How did Earth's surface at the last glacial maximum differ from its surface today?

2. What is the major uncertainty about the size of ice sheets at the glacial maximum?

3. In what ways did ice sheets make the glacial world a "dirty" place?

4. How does the composition of pollen in lake sediments tell us about climate?

5. How did the glacial climate of the southwestern United States differ from the climate there today? Why?

6. How did the glacial climate of Europe differ from the climate there today? Why?

7. How did the glacial climate of northern Asia differ from the climate there today? Why?

8. What caused the cooling of the tropics in the last glacial period?

9. Describe the evidence for a small versus a large tropical cooling during the glaciation, and explain why it is important for us to know the actual amount of cooling.

ADDITIONAL RESOURCES

Basic Reading

Clark, P. U., J. M. Licciardi, D. R. MacAyeal, and J. W. Jenson. 1996. "Numerical Reconstruction of a Soft-Bedded Laurentide Ice Sheet During the Last Glacial Maximum." *Geology* 24:679–82.

CLIMAP Members. 1981. *Seasonal Reconstruction of the Earth's Surface at the Last Glacial Maximum.* Map and Chart Series MC-36. Boulder, Colo.: Geological Society of America.

COHMAP Members. 1988. "Climatic Changes of the Last 18,000 Years: Observations and Model Simulations." *Science* 241:1043–62.

Moore, T. C., Jr., W. H. Hutson, N. Kipp, J. D. Hays, W. L. Prell, P. Thompson, and G. Boden. 1981. "The Biological Record of the Ice-Age Ocean." *Palaeogeography, Palaeoclimatology, Palaeoecology* 35: 357–70.

Rind, D., and D. Peteet. 1985. "Terrestrial Conditions at the Last Glacial Maximum and CLIMAP Sea-Surface Temperature Estimates: Are They Consistent?" *Quaternary Research* 24:1–22.

Advanced Reading

Broecker, W. S. 1985. "Oxygen Isotope Constraints on Surface Ocean Temperatures." *Quaternary Research* 26:121–34.

Fairbanks, R. G., and P. H. Wiebe. 1980. "Foraminifera and Chlorophyll Maximum: Vertical Distribution, Seasonal Succession, and Paleoceanographic Significance." *Science* 209:1524–26.

Peltier, W. R. 1994. "Ice Age Paleotopography." *Science* 265:195–201.

Stute, M., M. Forster, H. Frischkorn, A. Serejo, J. F. Clark, P. Schlosser, W. S. Broecker, and G. Bonani. 1995. "Cooling of Tropical Brazil (5°C) During the Last Glacial Maximum." *Science* 269:379–83.

CHAPTER 14

Climate During and Since the Last Deglaciation

E arth was transformed after the last glacial maximum. The melting ice sheets sent enough water to the ocean to raise global sea level by 110 meters. The rising ocean submerged links between continents and islands and flooded basins earlier cut off from the sea. Meltwater lakes formed in bedrock holes left by the retreating ice. Lakes dammed by lobes of ice sent catastrophic floods across the land when the dams were breached. Forests and tundra moved north to occupy broad regions abandoned by the ice, in some regions penetrating beyond their present-day limits before retreating. Tropical monsoons strengthened until 10,000 years ago and later weakened.

Abundant, well-dated records permit testing of theories proposed to explain these changes: the Milankovitch theory that insolation controls ice sheets and the Kutzbach theory that insolation controls monsoons. In general, the data confirm both theories: rising summer insolation in the northern hemisphere initiated melting of high-latitude ice sheets and strengthened tropical monsoons. Subsequent weakening of monsoons and cooling of high northern latitudes during the last 7000 years also match these theories. Something new is also evident in detailed climate records: shorter oscillations indicative of a different kind of behavior within the climate system.

Fire and Ice: A Shift in the Balance of Power

For the world of 21,000 years ago, the main factors that explain why climates were different from those today were the larger ice sheets and the lower atmospheric CO_2 levels. For the deglacial interval that followed, an important shift occurred in the balance of power among the factors that control global climate. Insolation values that had been near modern levels at the last glacial maximum began to change. By 10,000 years ago, the angle of tilt of Earth's axis had reached a maximum at the same time that Earth's precessional motion moved it closest to the Sun on June 21. These orbital changes combined to produce a summer insolation maximum at all latitudes of the northern hemisphere.

The rise in summer insolation at higher northern latitudes triggered melting of the huge northern ice sheets. As the ice sheets melted, their ability to influence climate diminished, and the insolation anomalies (their departures from modern insolation levels) became more important (Figure 14-1). The last deglaciation is mainly a story of this shift in the balance of power from ice (sheets) to fire (insolation). A second important change during the last deglaciation was the increase in atmospheric CO_2 from 190 to near 280 ppm (Figure 14-1), along with a doubling of methane levels. The increases in greenhouse gases more or less coincided with the melting of the ice.

14-1 When Did the Ice Sheets Melt?

The abundance of evidence available from the last deglaciation gives us the opportunity to gain detailed insight into how deglaciations actually occur. The Milankovitch theory (Chapter 10) predicts that the orbitally produced maximum in summer insolation near 10,000 years ago should have caused significantly higher rates of ice melting. Did it?

At first it seems that the most obvious way to try to quantify the rate of ice melting during the deglaciation is to measure the gradual retreat of the ice sheet margins. Organic carbon found in, under, or atop hundreds of moraines deposited by the ice can be used to date by radiocarbon the gradual shrinking of the ice limits. Radiocarbon dates show the retreat of the large ice sheet in North America beginning near 15,000 ^{14}C years ago, reaching a midpoint near 10,000 ^{14}C years ago, and ending by 6000 ^{14}C years ago (Figure 14-2). In Scandinavia, ^{14}C dates show that the smaller ice sheet retreated and disappeared a few thousand years earlier than the one in North America. These patterns support the predictions of the Milankovitch theory.

Knowing the area covered by the retreating ice is a good start, but in order to complete the analysis we need to convert these measurements into ice *volume*. To make this conversion, we need to know the thickness of the ice as it retreated (area × thickness = volume). But as we saw in Chapter 13, the thickness of an ice sheet can vary according to the conditions in its basal layer. Portions of ice sheets that repeatedly slide on their bases are thin and therefore relatively low in volume for a given area; portions that are frozen to their beds are thicker and larger in volume. Because of this uncertainty about thickness, having a record of changes in ice area through time does not guarantee a valid record of changes in ice volume.

FIGURE 14-1 **Causes of climate changes during deglaciation** During the deglacial interval between 17,000 and 6000 years ago, climate changes were driven by rising summer insolation and by increased concentrations of CO_2 in the atmosphere; as the ice sheets shrank, their ability to influence climate diminished. (Adapted from J. E. Kutzbach et al., "Climate and Biome Simulations for the Past 21,000 Years," *Quaternary Science Reviews* 17 [1998]: 473–506.)

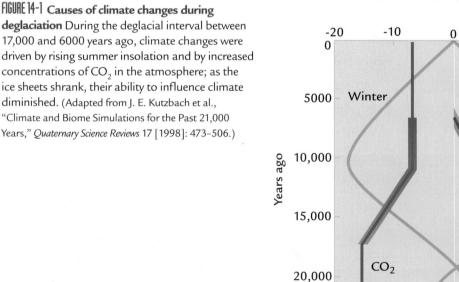

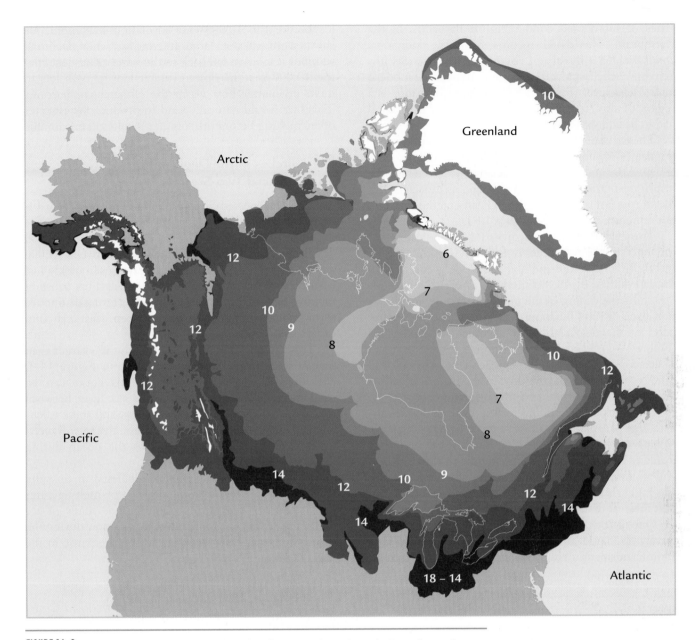

FIGURE 14-2 Retreat of the North American ice sheets Radiocarbon dating of organic remains shows that the margins of ice sheets in North America began to retreat near 14,000 ^{14}C years ago, and the ice disappeared completely shortly after 6000 years ago. The numbers indicate ^{14}C-dated ice limits in thousands of years. (Courtesy of Arthur Dyke, Geological Survey of Canada, Ottawa.)

14-2 Coral Reefs: A Measure of Rising Sea Level

The best record of ice sheet melting comes from tropical coral reefs far from the polar ice sheets. Because several coral species grow just below sea level, we can use the current elevation of older reefs built by living corals as a measure of changes in past sea level, assuming that we can correct for any local tectonic movements of bedrock after the death of the corals.

Changes in sea level are directly related to changes in global ice volume because continental ice sheets are made of water taken from the sea. Measurements made on coral reefs of lower sea level during the last glacial maximum and the subsequent deglaciation can be converted directly to a record of global ice volume, with a 1-meter rise of sea level equivalent to 0.4 million cubic kilometers of ice. This method provides the total global volume of ice, although not of individual ice sheets.

As the ice sheets melted and returned water to the ocean, they submerged the reefs of the last glacial and deglacial periods. In the late 1980s the marine geochemist Richard Fairbanks drilled and ^{14}C-dated a series of submerged coral reefs off Barbados, in the Caribbean. The dated records from the individual corals yielded a history of rising sea level from its low extreme at the last glacial maximum to near the modern interglacial level (Figure 14-3). Because Barbados is a region of slow tectonic uplift, the depth of each coral had to be adjusted by a few meters to remove this effect.

The curve of ^{14}C-dated deglacial sea level at Barbados (Figure 14-3) supports the Milankovitch theory in a general way: the middle of the deglaciation occurred near the insolation maximum 10,000 years ago, as expected. But the story is not that simple: the ^{14}C dates on the corals do not represent their true ages (Box 14-1).

The evidence surveyed in Box 14-1 indicates that the dates of the rise in deglacial sea level produced by the thorium/uranium method are more accurate than those produced by the ^{14}C method. The Th/U chronology shifts the timing of the middle part of the deglaciation back in time by about 2000 years, and the earlier parts of the deglaciation by as much as 3500 years (Figure 14-3). In contrast, the summer insolation signal is accurately fixed by the independent astronomical time scale. As a result, the time of the major rise in sea level shifts back in time from the insolation curve.

Does this earlier timing for the deglaciation invalidate the Milankovitch theory? In a larger sense, no. Milankovitch chose summer as the critical season of insolation control of ice sheets, and the last deglaciation still occurred during a time when summer insolation was higher than it is now, although somewhat earlier in that interval than the Milankovitch theory predicts.

One factor that contributes to this early response is simply the large amount of ice available to melt. As summer insolation rose to values high enough to melt ice before the 10,000-year insolation maximum, ice sheets were still very large. In contrast, when declining summer insolation fell back to the same values just *after* the 10,000-year maximum, much less ice was left to melt, no matter how warm the climate had become. This bias would naturally tend to produce fastest rates of ice melting before the summer insolation maximum.

14-3 Glitches in the Deglaciation: The Deglacial Two-Step

Another unexpected feature of the coral reef record is that sea level did not rise smoothly through the deglaciation (Figure 14-3). It rose quickly before 14,000 (Th/U or calendar) years ago, slowed between 14,000 and 12,000 years ago, and then again rose quickly just before the remaining ice slowly melted. This pause in melting rates gives the deglaciation pattern a distinctive form known as the **deglacial two-step**, with a rhythm of fast-slow-fast.

The pause in the rate of ice melting was larger than it may seem from the sea level curve in Figure 14-3. Measured sea level values along this curve can be used to calculate the *differences* in sea level between successive 1000-year intervals of time, and these calculated changes provide a measure of the net rate of meltwater flow from the ice sheets to the ocean during deglaciation.

The meltwater influx signal calculated in this way (Figure 14-4) shows a large slowing of melting rates during the middle of the deglaciation. Rates of ice melting were at least four to five times faster during the earlier and later intervals than during the pause in the middle.

FIGURE 14-3 Deglacial rise in sea level
Submerged corals off Barbados, in the Caribbean, show the deglacial history of the rise in sea level caused by the return of meltwater from the ice sheets to the ocean. (Adapted from R. G. Fairbanks, "A 17,000-Year Glacio-eustatic Sea Level Record: Influence of Glacial Meltwater on the Younger Dryas Event and Deep-Ocean Circulation," *Nature* 349 [1989]: 637–42; and from E. Bard et al., "Calibration of the [14]C Time Scale over the Past 30,000 Years Using Mass-Spectrometric U-Th Ages from Barbados Corals," *Nature* 345 [1990]: 405–10.)

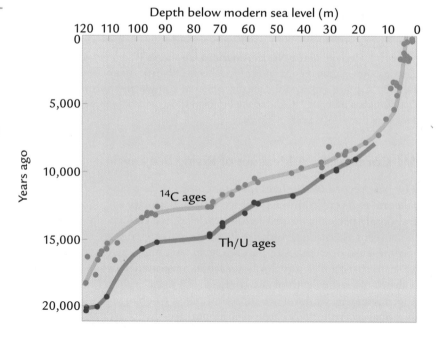

BOX 14-1   TOOLS OF CLIMATE SCIENCE

Deglacial ^{14}C Dates Are Too Young

When the same Barbados corals that had been dated by the ^{14}C method were dated by the thorium/uranium technique, those ages turned out to be older than the ^{14}C ages by an amount that increased back in time. For samples with a ^{14}C age of 8000 years, the Th/U age was older by 1000 years; for samples with a ^{14}C age of 18,000 years, the age difference increased to 3500 years.

With both sets of ages giving consistent-looking trends, scientists faced the problem of deciding which (if either) was correct. Fortunately, independent evidence was available from earlier work on the rings in long-lived trees. Individual rings in these trees had been dated both by the ^{14}C method and by counting backward year by year from the modern rings to the older rings.

The ages derived from counting individual rings turned out to be older than those from ^{14}C analyses. Near 8000 to 9000 years ago, the tree ring counts yielded ages older than the ^{14}C analyses by 1000 years, the same amount by which the Th/U coral ages were offset from the ^{14}C coral ages. This agreement between the Th/U and tree ring ages suggests that the ^{14}C ages are in error: they are younger than the actual ages.

The main reason the ^{14}C ages are too young is that the rate of production of ^{14}C atoms in Earth's atmosphere varied in the past. The ^{14}C dating method is based on the assumption that ^{14}C has been produced in the atmosphere at a constant rate through time. Atoms of ^{14}C are produced when cosmic particles from galaxies in outer space transform ^{14}N atoms in our atmosphere (nitrogen is the most abundant gas in Earth's atmosphere) into radioactive ^{14}C atoms, which then slowly decay away.

But the rates of cosmic bombardment were higher during glacial times than they are now, so more ^{14}C atoms were produced. Although many of those ^{14}C atoms have undergone radioactive decay, some of them still remain in the material used for ^{14}C dating. This surviving excess of ^{14}C atoms makes it look as if less ^{14}C has decayed (and so less time has elapsed) than is actually the case.

Why would cosmic bombardment have been higher in the past? Earth is partly shielded from cosmic rays by its magnetic field. If this shield had been weaker than it is today, it would have provided less protection against cosmic bombardment. Two observations confirm this explanation of the anomalously young ^{14}C ages. First, other elements

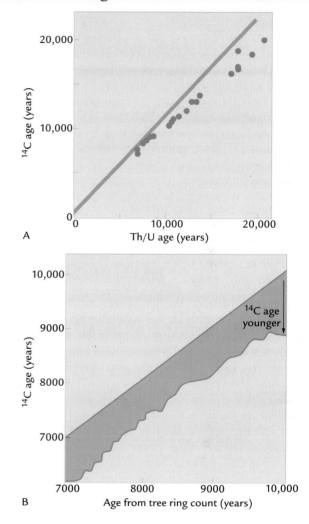

A

B

Offset of ^{14}C ages (A) Th/U dates from Barbados corals indicate that the ^{14}C ages are too young by as much as 3500 years. (B) Ages derived by counting tree rings backward in time are offset from ^{14}C dates on the same layers by a similar amount. (A: adapted from E. Bard et al., "Calibration of the ^{14}C Time Scale over the Last 30,000 Years Using Mass Spectrometric U-Th Ages from Barbados Corals," *Nature* 345 [1990]; 405–10; B: adapted from M. Stuiver et al., "Climatic, Solar, Oceanic, and Geomagnetic Influences on Late-Glacial and Holocene Atmospheric ^{14}C/^{12}C Change," *Quaternary Research* 35 [1991]: 1–24.)

known to be produced by cosmic bombardment are also more abundant before 7000 years ago, confirming weaker magnetic shielding. Second, direct measurements of the past magnetic field in Earth's rocks and sediments show that it was weaker before 7000 years ago.

FIGURE 14-4 Influx of deglacial meltwater to the oceans The rate of the deglacial rise in sea level determined from submerged coral reefs can be used to calculate the rate at which water flowed into the oceans from melting ice sheets. The flow of meltwater slowed significantly during a pause in the deglaciation between 14,000 and 12,000 years ago. (Adapted from R. G. Fairbanks, "A 17,000-Year Glacio-eustatic Sea Level Record: Influence of Glacial Meltwater on the Younger Dryas Event and Deep-Ocean Circulation," *Nature* 349 [1989]: 637–42; and from E. Bard et al., "Calibration of the ¹⁴C Time Scale over the Past 30,000 Years Using Mass-Spectrometric U-Th Ages from Barbados Corals," *Nature* 345 [1990]: 405–10.)

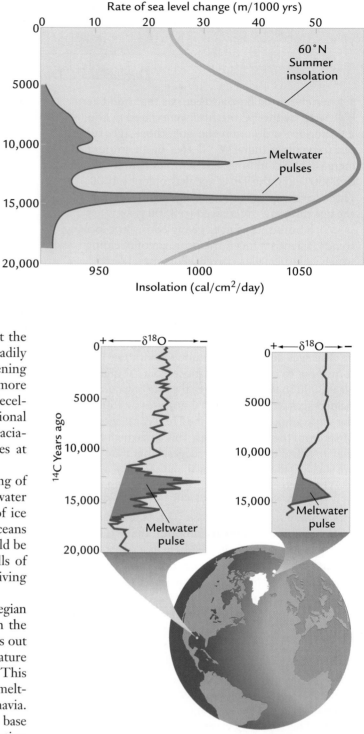

This two-step deglaciation pattern tells us that the glacial ice sheets were not just giant ice cubes steadily melting in warmer air masses under a strengthening summer sun. Instead, the ice sheets exhibited more complex behavior, with abrupt accelerations and decelerations in rates of melting. If we look at regional records of what was actually happening during deglaciation, we can gain some insight into the processes at work on a local scale.

Meltwater Pulses One way to monitor melting of individual ice sheets is to look for pulses of meltwater delivery to the oceans. Because the $\delta^{18}O$ values of ice sheets are at most $-30‰$, whereas those of the oceans are closer to $0‰$, major influxes of meltwater should be recorded as pulses of low $\delta^{18}O$ values in the shells of plankton living in ocean regions next to the arriving meltwater.

Planktic foraminifera in the northeastern Norwegian Sea record unusually negative $\delta^{18}O$ values early in the deglaciation (Figure 14-5), and other evidence rules out the possibility that a major fluctuation in temperature could have caused the fluctuation in $\delta^{18}O$ values. This $\delta^{18}O$ oscillation is the result of an episode of early melting of the nearby Barents ice sheet, north of Scandinavia. Apparently this marine-based ice sheet, with its base lying below sea level, was vulnerable to early destruction when summer insolation began to rise.

A similar low-$\delta^{18}O$ pulse is found in cores from the Gulf of Mexico (Figure 14-5). This pulse indicates an increase in the amount of meltwater from the North American ice sheet flowing down the Mississippi River and into the Gulf.

Iceberg Influxes Cores of ocean sediment southwest of Ireland contain a distinctive layer of sediment deposited 17,000 to 14,500 years ago, rich in ice-rafted

FIGURE 14-5 Local meltwater pulses CaCO₃ shells of ocean plankton from the Norwegian Sea and the Gulf of Mexico record pulses of low-$\delta^{18}O$ meltwater delivered from nearby ice sheets. (Top left: adapted from A. Leventer et al., "Dynamics of the Laurentide Ice Sheet During the Last Deglaciation: Evidence from the Gulf of Mexico," *Earth Planetary Science Letters* 59 [1982]: 11–17; top right: adapted from G. Jones and L. D. Keigwin, "Evidence from Fram Strait (78°N) for Early Deglaciation," *Nature* 336 [1988]: 56–59.)

sand grains but nearly barren of the planktic foraminifera and coccoliths normally found in that region. This "barren zone" is evidence of a large influx of icebergs to the North Atlantic early in the deglaciation. This influx arrived during the first pulse of rapid sea level rise in the deglacial two-step (Figure 14-4). This evidence could mean that the major continental ice sheets were losing a substantial amount of their mass by calving icebergs to the ocean. The calving process would have helped to accelerate the rapid loss of ice volume early in the deglaciation.

Mid-Deglacial Cooling: The Younger Dryas The mid-deglacial pause in ice melting was accompanied by a brief climatic oscillation especially evident in records near the subpolar North Atlantic Ocean. Temperatures in this region had warmed part of the way toward interglacial levels, but this reversal brought back almost full glacial cold (Figure 14-6).

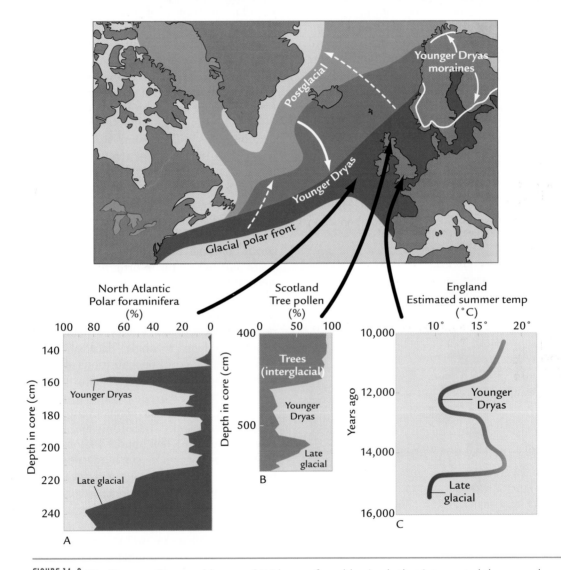

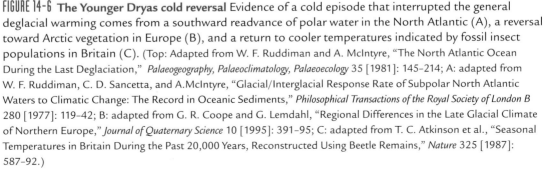

FIGURE 14-6 The Younger Dryas cold reversal Evidence of a cold episode that interrupted the general deglacial warming comes from a southward readvance of polar water in the North Atlantic (A), a reversal toward Arctic vegetation in Europe (B), and a return to cooler temperatures indicated by fossil insect populations in Britain (C). (Top: Adapted from W. F. Ruddiman and A. McIntyre, "The North Atlantic Ocean During the Last Deglaciation," *Palaeogeography, Palaeoclimatology, Palaeoecology* 35 [1981]: 145–214; A: adapted from W. F. Ruddiman, C. D. Sancetta, and A. McIntyre, "Glacial/Interglacial Response Rate of Subpolar North Atlantic Waters to Climatic Change: The Record in Oceanic Sediments," *Philosophical Transactions of the Royal Society of London B* 280 [1977]: 119–42; B: adapted from G. R. Coope and G. Lemdahl, "Regional Differences in the Late Glacial Climate of Northern Europe," *Journal of Quaternary Science* 10 [1995]: 391–95; C: adapted from T. C. Atkinson et al., "Seasonal Temperatures in Britain During the Past 20,000 Years, Reconstructed Using Beetle Remains," *Nature* 325 [1987]: 587–92.)

The first evidence for this event came from pollen records in Europe. As the ice sheet melted back to the north, scattered trees had begun to reoccupy northern Europe. Near 15,000 years ago, cold-tolerant herb tundra reinvaded the landscape, displacing trees back toward southern Europe. Because an Arctic plant called *Dryas* arrived during this episode, European scientists called it the **Younger Dryas** event.

Later work on sediments in the North Atlantic Ocean also detected a clear Younger Dryas imprint: a rapid oscillation in the regional extent of icy polar water (Figure 14-6A). During the glacial maximum, polar water had spread across the North Atlantic southward to 45°N, with its southern margin defined by the **polar front**, a zone of rapid transition with the more temperate waters to the south. Early in the deglaciation, near 15,000 years ago, the polar front had abruptly swung back toward the northwest as if hinged at a point near Newfoundland. This retreat allowed warm water to flow northward along the European coast, warming climate in Europe enough to permit trees to begin a northward advance.

Near 13,000 years ago the polar front advanced southward again, almost to its glacial position, and remained there for several hundred years. At the same time, cold-adapted vegetation (including *Dryas*) returned to northern Europe and trees retreated south (Figure 14-6B). Later, near 11,700 years ago, the polar front abruptly retreated to the north, and forests again moved northward in Europe.

The readvance of the polar front during the Younger Dryas event represents a major reversal in circulation patterns from a near-interglacial to an almost fully glacial pattern, and then back again to a fully interglacial pattern. The estimated change in sea surface temperature in the Atlantic Ocean west of Ireland during this event was at least 7°C, close to the difference between fully glacial and interglacial extremes. A similar cooling has been estimated from changes in the fossils of temperature-sensitive insect populations in England just downwind of the Atlantic (Figure 14-6C).

Ice cores from Greenland contain a remarkably detailed record of the Younger Dryas event (Figure 14-7). During fully glacial climates, snow accumulated slowly, but the rates increased abruptly near 15,000 years ago, when the North Atlantic Ocean warmed. Accumulation rates then slowed during the transition into the Younger Dryas event and abruptly increased as it ended, 11,700 years ago. Some of these transitions occurred in less than a century, with much of the change concentrated in a single decade. Similar changes occurred in ice-core concentrations of windblown dust, which peaked during the cold, dry, windy climate of the Younger Dryas.

During the Younger Dryas event, the ice sheets in Scandinavia stopped retreating and in some regions even readvanced a few hundred kilometers (see Figure 14-6). These pauses or small advances are thought to have been a response to the large regional cooling rather than a cause of it. The deglacial sea level curve derived from coral reefs (Figure 14-4) shows that global ice volume continued to shrink during the Younger Dryas, although at a slower rate.

What caused the Younger Dryas cold oscillation? A hypothesis advanced by the geochemist Wally Broecker calls on changes in the path of meltwater flow from the North American ice sheet (Figure 14-8). He proposed that an abrupt diversion of the meltwater route from the Gulf of Mexico to the North Atlantic Ocean during the Younger Dryas delivered a pulse of low-salinity water that rearranged the basic circulation of the North Atlantic by briefly cutting off the formation of deep water. Because ocean surface waters give off heat when deep water forms, cutting off the deep-water flow

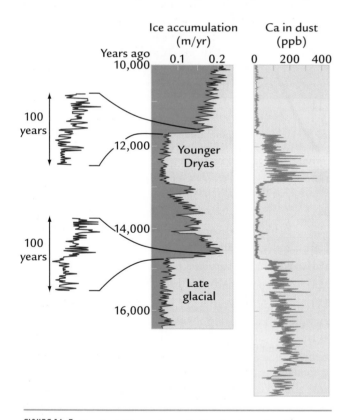

FIGURE 14-7 **Deglacial ice accumulation in Greenland** Rates of accumulation of ice in the Greenland ice sheet abruptly decreased during the Younger Dryas cold event and then increased when it ended, with the major changes occurring within 100 years. Concentrations of windblown dust increased during the Younger Dryas and decreased afterward. (Modified from R. B. Alley et al., "Abrupt Increase in Greenland Snow Accumulation at the End of the Younger Dryas Event," *Nature* 362 [1993]: 527–29).

FIGURE 14-8 **Routes of meltwater flow** During deglaciation, the direction of drainage of the North American ice sheet changed: southward to the Gulf of Mexico early in the deglaciation, then east to the Atlantic Ocean (briefly) during mid-deglaciation, and finally north into Hudson Bay and the Arctic Ocean late in the deglaciation. (Adapted from J. Teller, "Meltwater and Precipitation Runoff to the North Atlantic, Arctic, and Gulf of Mexico from the Laurentide Ice Sheet and Adjacent Regions During the Younger Dryas," *Paleoceanography* 5 [1990]: 897–905.)

would cool climate in the North Atlantic and surrounding regions.

At first this explanation looked promising. Before the Younger Dryas event, the major pathway of meltwater had been down the Mississippi, into the Gulf of Mexico, and out into the Atlantic through the Gulf Stream. This early meltwater influx is recorded in sediments in the Gulf of Mexico by a pulse of negative $\delta^{18}O$ values (see Figure 14-5). During the Younger Dryas event, the meltwater flowed directly eastward into the Atlantic through the St. Lawrence region of eastern Canada, and the low-$\delta^{18}O$ pulse in the Gulf of Mexico disappeared. Immediately after the Younger Dryas, most of the direct flow eastward to the North Atlantic ceased.

One criticism of this hypothesis is that the Younger Dryas episode occurred at the same time that the rate of global melting slowed by a factor of 5 or more (see Figure 14-4). With the overall meltwater flow to the oceans sharply diminished, it is harder to argue that the diversion of meltwater flow would have had a major impact on circulation in the North Atlantic. At the same time, simulations with ocean models suggest that the formation of deep water is much more sensitive to melt-water delivered to the high-latitude ocean than to the Gulf of Mexico.

Other debates about the Younger Dryas are under way. Some scientists note that the clearest evidence for a Younger Dryas cooling comes mainly from around the North Atlantic and infer that its ultimate cause must lie in that region. Others point out that brief oscillations in climate that seem to coincide with the Younger Dryas event can be found far from the North Atlantic: an interval when herblike vegetation replaced forests in the Pacific Northwest, a pulse of drying of North African lakes, an advance of mountain glaciers in New Zealand, and an interval of colder surface waters in the northwestern Pacific Ocean. These other records suggest either that changes originating in the North Atlantic are quickly transmitted to distant regions or that the cause of the Younger Dryas event lies in a mechanism that originates at a larger (possibly global) scale, rather than one centered near the North Atlantic.

This debate is not likely to be settled soon. Many records are difficult to date accurately, and small dating errors can invalidate interpretations. For example, Antarctic ice cores that were once interpreted as showing a small cooling during the Younger Dryas event have since been dated more accurately and now indicate a slow warming throughout the Younger Dryas interval.

All of these short-term oscillations (the early-deglacial influxes of meltwater and icebergs and the mid-deglacial melting pause and regional cooling) are evidence that the climate system is capable of behavior different from anything we have yet seen. These abrupt fluctuations, some of which last less than 1000 years and can switch between extreme states in less than 100 years or even a decade, cannot be the direct result of changes in solar insolation at cycles of 20,000 years or longer. In Chapter 15 we will explore the origin of these short-term oscillations.

14-4 Positive Feedbacks to Insolation Melting?

Climate scientists basically agree that rising summer insolation values caused by changes in Earth's orbital tilt and precession set in motion the melting of the great northern hemisphere ice sheets near 17,000 years ago, with the last of the ice disappearing near 6000 years ago. But this consensus leaves unanswered an important question we first explored in Chapter 12: How did so small an insolation maximum melt so much ice so quickly?

The answer to this question must be that positive feedbacks accelerated the loss of ice. These feedbacks must have been working most effectively when sea level was rising (ice was melting) most rapidly (Figure 14-4). The first of the two fast-melting phases may be the time of most active feedbacks, because the rapid rise in sea level (loss of ice volume) at that time occurred well before summer insolation had reached its peak.

What might explain such rapid melting? Several possible processes are suggested by the records we have already examined. The negative $\delta^{18}O$ pulse recorded in Norwegian Sea cores is evidence that a substantial part of the marine ice sheet over the Barents Sea, north of Scandinavia, melted early in the deglaciation (see Figure 14-5). Some scientists infer that the resulting rise in sea level destabilized the marine margins of other ice sheets.

A second important observation is the major influx of icebergs to the North Atlantic early in the deglacial sequence, from 17,000 to 14,500 years ago. The icebergs arrived at a time when the limits of the major ice sheets had not melted back far from their glacial positions, but when global ice volume was rapidly decreasing in the first of the two deglacial steps. Taken together, these observations indicate that the major landbased ice sheets were shrinking significantly in thickness but less so in area covered.

Such behavior is possible if lobes of ice were repeatedly surging (flowing rapidly) toward the margins of the ice sheets. Rapid flow of ice through ice streams can draw ice out of the central domes of an ice sheet and transfer it to the ice margins. Ice that arrives at the ocean's margins calves into the ocean as icebergs and melts elsewhere. Ice that flows to the southern margins of the ice sheets over land melts relatively rapidly because it lies at lower elevations in the bedrock depression left behind by the once-thicker ice. Insolation warming of the nearby land also helps to melt ice rapidly. Yet the ice limits remain extensive because ice streams keep replenishing the melting (or calving) ice. So the interior of the ice sheet can thin without any large-scale retreat of the margins.

The mid-deglacial melting pause and the coincident Younger Dryas cooling may represent an interval when positive feedback processes slowed their impacts on the climate system, or when negative feedbacks took over. Because summer insolation was still rising, melting continued, but more slowly. The second step of rapid ice melting and sea level rise presumably reflects the return of the various positive feedbacks to a more active role.

Other climate scientists contest parts of this north-centered view of the deglaciation. Some note that evidence exists for even earlier deglacial warming responses in the southern hemisphere near Antarctica. Such evidence can be interpreted to indicate that the south polar regions act as sensitive early triggers of melting in the north polar regions, and that the deglaciation cannot be understood by looking only at the northern hemisphere.

Eventually climate scientists will learn enough about the full sequence of events during deglaciation to fit the details of the story together and detect all the critical feedback processes. When the full story is written, it will probably combine two basic themes: the gradual changes in climate driven or paced by changes in summer insolation (Chapter 12) and the abrupt changes that signal the action of feedbacks and interactions within the climate system (Chapter 15).

14-5 Deglacial Lakes, Floods, and Sea Level Rise

As the ice sheets melted back, the land directly in front of them remained depressed for some time, rather than immediately rebounding to its former level (Chapter 10). Into these depressions poured meltwater from the retreating ice sheets, forming **proglacial lakes**. Because of the large volumes of meltwater arriving each summer, the lakes frequently cut new channels and overflowed into other lakes and then into rivers that carried water away from the ice sheets.

The proglacial lakes existed in a highly dynamic landscape (Figure 14-9). As deglaciation proceeded, the ice margins fluctuated, with lobes of ice occasionally sliding forward but then gradually shrinking farther and farther back over time. Each time the ice lobes retreated, they

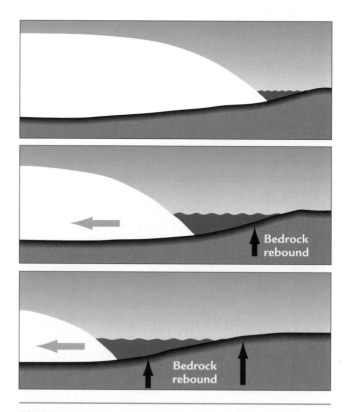

FIGURE 14-9 Proglacial lakes moving north Proglacial lakes develop in bedrock depressions left by melting ice sheets. Over time the lakes move north behind the ice sheets, while the land farther south rebounds toward its undepressed elevation.

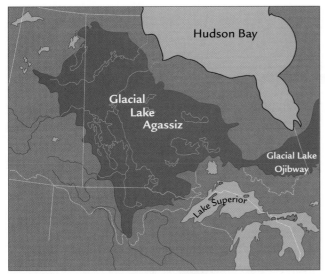

A Total area covered by deglacial lakes

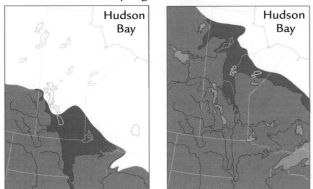

B Lakes during deglaciation

FIGURE 14-10 **Glacial Lake Aggasiz** During its several thousand years of existence, glacial Lake Agassiz flooded a total of more than 500,000 km² in western Canada (A), but with smaller areas flooded at specific times (B). (A: adapted from J. Teller, "Lake Agassiz and Its Contribution to Flow Through the Ottawa–St. Lawrence System," Geological Association of Canada Special Paper 35 [1987]: 281–89; B: adapted from J. Teller, "Glacial Lake Agassiz and Its Influence on the Great Lakes," Geological Association of Canada Special Paper 30 [1985]: 1–16.)

ice sheets to grow shallow and eventually disappear. Through time, the locations of the proglacial lakes moved north across the landscape, following the wave of depressed bedrock left just south of the retreating ice.

The largest of the proglacial lakes in North America was Lake Agassiz, in western Canada. Over its entire existence, it flooded an area greater than 500,000 square kilometers (Figure 14-10A), but smaller areas at specific intervals during the ice-retreat sequence (Figure 14-10B). At its maximum instantaneous size, it covered an area of more than 200,000 square kilometers to a depth of 100 meters or more, forming a reservoir of 20,000 cubic kilometers. Even this amount represented only a tiny fraction of the tens of millions of cubic kilometers of water stored in the glacial maximum ice sheets. Still, the water impounded in proglacial lakes occasionally did enormous damage to the landscape (Box 14-2).

The deglacial rise in sea level also changed Earth's surface on a large scale. Many regions of the world's continental shelves were exposed when sea level dropped by more than 110 meters at the glacial maximum, with especially large expanses where the slope of the land was low (Figure 14-11). The return of some 44

left behind new bedrock holes that formed the deepest parts of a proglacial lake. All the while, the slow rebound of bedrock caused the parts of the lakes well south of the

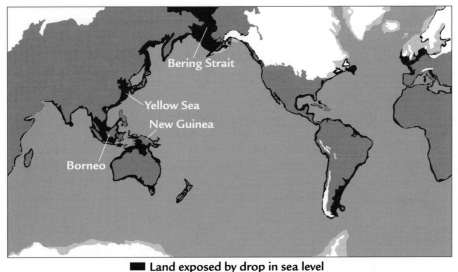

■ Land exposed by drop in sea level
■ Land created under ice by drop in sea level

FIGURE 14-11 **Deglacial flooding of coastlines** The regions shown in dark brown were exposed at the last glacial maximum but flooded by the 110-m rise in sea level when the ice melted. (Adapted from CLIMAP Project Members, *Seasonal Reconstruction of the Earth's Surface at the Last Glacial Maximum*, Map and Chart Series MC-36 [Boulder, Colo.: Geological Society of America, 1981].)

BOX 14-2   CLIMATE INTERACTIONS AND FEEDBACKS

Giant Deglacial Floods

In an unusual landscape called the **channeled scablands** in east-central Washington State and Idaho, bedrock consists of thick sequences of basalt deposited by lava flows emplaced during heightened volcanic activity near 15 Myr ago. As North American ice sheets melted during the much more recent deglaciation, the surface of these ancient lava flows was eroded into shapes suggesting the violent action of water on an immense scale. In the scablands, deep canyons with nearly vertical walls were gouged into bedrock. At some locations, these now-dry channels abruptly plunge over steep cliffs into larger channels with depressions like those found at the base of modern waterfalls (but much larger). Huge boulders and displaced gravel and sand lie in the channels, but upland areas nearby have a thin cover of windblown loess typical of much of the Pacific Northwest outside the scabland region.

The geologist Harlen J. Bretz, who worked there in the 1920s and 1930s, came to the conclusion that these erosional features resulted from a flood of immense proportions, one that within a few days must have carried a volume of water equivalent to all of Earth's rivers today. He suggested that the water in this flood ran wildly across the landscape, gouging and eroding the lower terrain but leaving the higher areas untouched before eventually flowing down the Columbia River and out into the Pacific Ocean. Bretz inferred that the source of all this water was a rapid melting event on the southern margin of the Cordilleran ice sheet, covering western Canada southward into the northwesternmost United States. He speculated that a volcano erupting beneath the ice margin had caused rapid

melting and the sudden release of an enormous torrent of water.

For decades Bretz's ideas were rejected by geologists. At that time, most geologists took too literally the principle of uniformitarianism, the concept that slow geologic processes working today have, over immense spans of time, shaped and molded all of the features we see on Earth. Overenthusiastic application of this otherwise useful concept left little room for infrequent catastrophic phenomena; these events were rejected because they were not seen at work today. The problem with this view is that a human life span is extremely short in relation to the age of the Earth, and our perspective on "normal" processes is accordingly narrow. With much longer life spans, we would naturally take a much broader view of what is normal.

In the 1950s, aerial photography showed that the scablands were covered by giant ripple marks and gravel bars more than 20 feet high and spaced at intervals of 400 feet. Working on foot, Bretz had not recognized these enormous features because they were masked by scrubby vegetation. This new evidence convinced most geologists that the scablands had indeed been flooded, and further studies suggested that a discharge of some 25,000 to 30,000 cubic meters (750,000 cubic feet) per second flowing at 80 kilometers (50 miles) per hour was required to carve such a landscape. All the features Bretz had noted were indeed the result of rushing water on an unimaginably large scale.

A likely source of the water was glacial Lake Missoula, a proglacial lake ponded against the side of the ice sheet in Idaho. Although Bretz proposed only a single flood, the multiple layers of sediment

---

million cubic kilometers of meltwater to the oceans raised sea level and transformed coastlines around the world. In flat coastal plain regions, sea level rises as large as 25 meters per 1000 years during the fastest deglacial melting pulses advanced inland by 25 meters per year along coastal gradients as low as 1:1000 (1m rise in sea level = 1000 m inland incursion).

Lowered sea level at the glacial maximum had exposed land connections between many continents or ocean islands (Figure 14-11). One such area was the huge expanse of dry land that joined Australia with New

Guinea, to the north. Land connections also linked the present-day southeast Asian mainland with islands as far south as Borneo and joined northeastern Asia (Siberia) and westernmost Alaska across the present-day Bering Strait. England and Scotland were linked to the European mainland during the glacial maximum just south of the ice sheet. The rise in sea level caused by deglacial meltwater submerged these land corridors.

The lower level of the glacial ocean also transformed smaller seas around the margins of the oceans, especially in the western Pacific (Figure 14-11). The present-day

BOX 14-2 CONT.   CLIMATE INTERACTIONS AND FEEDBACKS

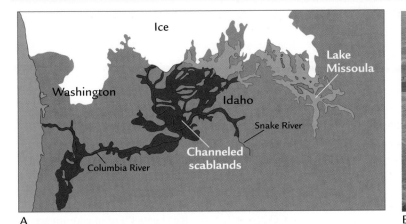

A

B

C

The channeled scablands Massive erosion by giant deglacial floods affected a large region in eastern Washington State and Idaho. One source of meltwater for the floods may have been glacial Lake Missoula (A), ponded against a retreating lobe of the ice sheet in western Canada. Evidence of erosion in the channeled scablands by floods of meltwater includes huge boulders, gravel, and sandbars on an immense scale (B) and channels that lead to the edges of cliffs (C). (A: adapted from G. A. Smith, "Missoula Flood Dynamics and Magnitudes Inferred from Sedimentology of Slackwater Deposits on the Columbia Plateau, Washington," *Geological Society of America Bulletin* 105 [1993]: 77–100; B and C: Victor Baker, University of Arizona.)

left by the waters implied dozens of floods. One possibility is that each time a lobe of Cordilleran ice advanced far enough to the south to act as a dam, Lake Missoula filled up and released water in catastrophic bursts when the blocking ice lobe melted back. Geologists have also cited evidence that large lakes hundreds of kilometers wide and tens of meters deep existed underneath the thin western lobes of the ice sheet in the Canadian provinces of Alberta and Saskatchewan, and that these lakes could have periodically released large volumes of water from under the ice.

Yellow Sea was dry land, and other seas in the western Pacific were more isolated from the open ocean because the sea level was lower. Rising sea level reflooded these seas and rejoined them to the open ocean.

Climate Changes During and Since Deglaciation

Two of the most important climate changes during and since the last deglaciation were the strength of tropical monsoons and the warmth of summers in north polar latitudes. Scientists have investigated these changes by comparing model simulations with observations from climate archives. Monsoons grew stronger and summers warmer as summer insolation rose toward the maximum values 10,000 years ago and as the effects of the ice sheets and the greenhouse gases diminished. By 6000 years ago, with the ice sheets melted and the greenhouse gases at or close to fully interglacial levels, the major factor left to influence climate was the drop in summer insolation and rise in winter insolation toward modern values (Figure 14-12).

FIGURE 14-12 Causes of climate changes since deglaciation During the last 6000 years, with the ice sheets melted and CO_2 levels stabilized at or near interglacial levels, the main orbital-scale factor affecting climate was the gradual change in solar insolation toward present-day values. (Adapted from J. E. Kutzbach et al., "Climate and Biome Simulations for the Past 21,000 Years," *Quaternary Science Reviews* 17 [1998]: 473–506.)

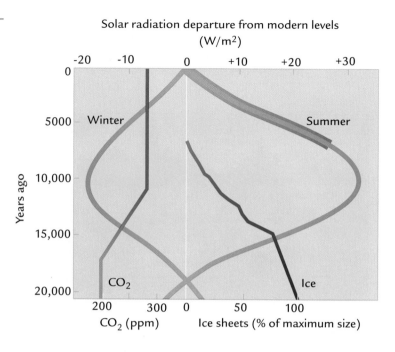

14-6 Stronger, Then Weaker Monsoons

Monsoons were strong near 10,000 years ago because of Earth's orbital configuration (Chapter 9). Summer insolation values 10,000 years ago over tropical and subtropical landmasses of the northern hemisphere were 8% higher than those today (see Figure 14-1). According to the Kutzbach theory, increased insolation should have driven a stronger summer monsoon circulation and produced higher tropical lake levels near 10,000 years ago than those today.

The COHMAP project compared climate-model simulations with observed data from this interval. COHMAP model simulations used summer insolation values from 9000 years ago, by which time northern hemisphere ice sheets had been greatly reduced. The model simulated stronger monsoons across the entire north-tropical region of North Africa, southern Arabia, and southern Asia (Figure 14-13A).

Geologic evidence supports this simulation (Figure 14-13B and C). Lake levels determined by ^{14}C dating of ancient lake muds were substantially higher than they are today across most of North Africa between 15° and 30°N, in the southern half of Arabia, and over southeastern Asia between 10,000 and 7500 ^{14}C years ago. Lake Chad, in northern Africa, expanded to 300,000 cubic kilometers, an area comparable in size to the modern Caspian Sea.

Evidence from sediment cores retrieved from the Arabian Sea indicates that much stronger monsoon winds blew along the coast of Somalia, in eastern Africa, and along the southeast coast of Arabia 9000 years ago (Figure 14-14). These intensified winds drove more

water offshore and caused greater upwelling, recorded by the percentage of upwelling species of planktic foraminifera in ocean cores.

In addition, scientists have discovered evidence that ancient rivers flowed across regions that today are hyperarid desert. In central Arabia, a river flowed more than 500 kilometers north-eastward through the present-day Arabian desert (Figure 14-15). Many parts of North Africa and Arabia that are now extremely dry were once grassy river valleys dotted with freshwater lakes and occupied by hippopotamuses, crocodiles, turtles, rhinoceroses, giraffes, and buffaloes.

Although climate models and evidence from the geologic record confirm the Kutzbach theory of stronger north-tropical monsoons 10,000 years ago, a closer comparison raises some questions. For example, why did lakes in North Africa reach their maximum sizes 9000 to 7000 ^{14}C years ago (Figure 14-13C), some 1000 to 3000 years after the maximum in summer insolation at 10,000 years ago?

Obviously, at least some part of the answer lies in the offset between ^{14}C and Th/U dates. Because ^{14}C years underestimate true ages, the evidence for high lake levels from 9000 to 7000 ^{14}C years ago actually dates to between 10,500 and 8000 calendar years ago. This recalibration of the ^{14}C dates puts the geologic evidence closer in line with the model simulations based on astronomical calculations of insolation changes in calendar years (Figure 14-13C).

A second question is why the North African lakes averaged about 25% larger in volume than they do today, when insolation values were only about 8% higher. These numbers imply too large a climatic response

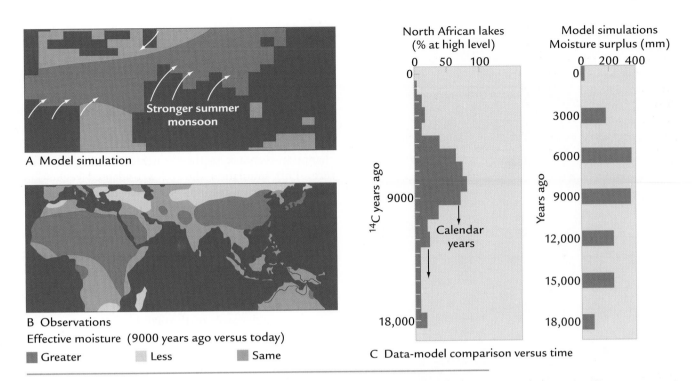

A Model simulation

B Observations

Effective moisture (9000 years ago versus today)

■ Greater   ■ Less   ■ Same

North African lakes
(% at high level)

Model simulations
Moisture surplus (mm)

C Data-model comparison versus time

FIGURE 14-13 **Tropical monsoon maximum** Climate model simulations of stronger summer monsoons in the north tropics near 9,000 years ago (A) agree with evidence in the climate record, such as higher lake levels (B, C). (A and B: adapted from COHMAP Project Members, "Climatic Changes of the Last 18,000 Years: Observations and Model Simulation," *Science* 241 [1988]: 1043–52; C: adapted from J. E. Kutzbach and F. A. Street-Perrott, "Milankovitch Forcing of Fluctuations in the Level of Tropical Lakes," *Nature* 317 [1985]: 130–34.)

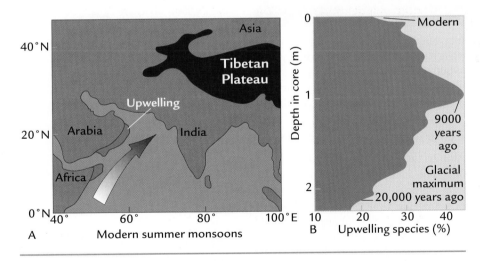

A Modern summer monsoons

B Upwelling species (%)

FIGURE 14-14 **Upwelling in the Arabian Sea** Climate model simulations of stronger summer monsoons over India 9000 years ago are supported by evidence from ocean cores indicating greater upwelling along the coasts of East Africa and Arabia in response to strong winds pushing surface waters offshore. (A: adapted from W. L. Prell, "Monsoonal Climate of the Arabian Sea During the Late Quaternary: A Response to Changing Solar Radiation," in *Milankovitch and Climate*, ed. A. Berger et al. [Dordrecht: D. Reidel, 1984]; B: W. L. Prell, "Variation of Monsoonal Upwelling: A Response to Changing Solar Radiation," in *Climate Processes and Climate Sensitivity*, ed. J. E. Hansen and T. Takahashi [Washington, D.C.: American Geophysical Union, 1984].)

FIGURE 14-15 **Rivers in the Arabian Desert** During the strong monsoons between 9000 and 5000 years ago, rivers ran across regions of central Arabia that are now hyperarid. (Courtesy of Farouk El-Baz, Center for Remote Sensing, Boston University.)

(lake levels) in relation to the amount of climatic forcing (summer insolation). But lake levels do not have to respond to insolation changes precisely in tandem: an increase in summer insolation by 1% might well result in an increase in lake levels by 2 or 3% because of feedbacks and thresholds of accelerated change.

Initial climate-model simulations by the COHMAP group indicated that summer insolation intensified the monsoon by only about half as much as climate records indicate. More recent modeling results point to positive feedback processes that can recycle water vapor in regions where moisture is sparse and increase the overall intensity of the monsoon. The feedbacks result from changes in vegetation and soil types (Figure 14-16).

First, a climate model was run with 8% higher summer insolation to determine how much rainfall would

increase in summer monsoons in response to insolation changes alone, without feedbacks. This simulation produced wetter soils and higher lake levels than we know today, but the simulated lakes were not as high as those actually observed in the geologic record.

Next, the simulated increases in rainfall and soil moisture were used in a second experiment to find out how the distributions of vegetation and soil types would change in response to the greater availability of moisture. This simulation showed wet-adapted vegetation expanding northward into North Africa: trees invaded grassland regions, and grasslands invaded scrub deserts.

Finally, the changes in vegetation and soil were entered into a third climate-model experiment to determine how the vegetation changes affected the cycling of water vapor. In regions where trees replaced grass or grass replaced desert scrub, the plant roots drew more moisture out of the soil and transferred it to the atmosphere through leaves or through blades of grass. The additional water vapor transferred to the atmosphere was then recycled farther into the interior of North Africa in an expanded monsoonal system (Figure 14-16). As a result, this positive feedback from recycling of water by vegetation (Chapter 2) produced a landscape wetter and greener than it would otherwise have been. The simulated lake levels also rose to levels that were in better (but still not complete) agreement with the geologic evidence.

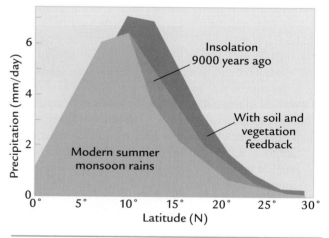

FIGURE 14-16 **Vegetation-moisture feedback** Climate model simulations indicate that higher summer insolation 9000 years ago caused stronger summer monsoons and greater northward penetration of moisture into Africa. Additional model experiments show that positive moisture feedback from wetter soils and increased vegetation caused even greater penetration of moisture into the continental interior. (Modified from J. E. Kutzbach et al., "Vegetation and Soil Feedbacks on the Response of the African Monsoon to Orbital Forcing in the Early to Middle Holocene," *Nature* 384 [1986]: 623–26.)

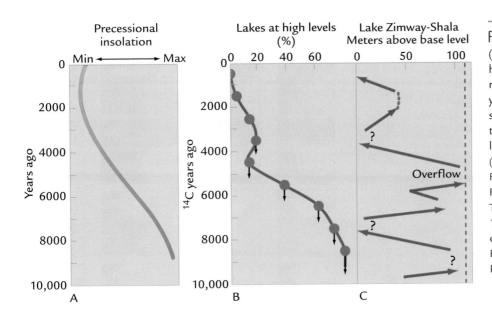

FIGURE 14-17 Weakening monsoons
(A) Low-latitude summer insolation has slowly decreased since the maximum value reached 10,000 years ago. (B, C) The decrease in summer insolation has weakened the summer monsoons and caused lake levels in North Africa to fall. (B: adapted from J. E. Kutzbach and F. A. Street-Perrott, "Milankovitch Forcing of Fluctuations in the Level of Tropical Lakes," *Nature* 317 [1985]: 130–34; C: adapted from R. Gillespie et al., "Postglacial Arid Episodes in Ethiopia Have Implications for Climate Prediction," *Nature* 306 [1983]: 680–83.)

Other positive feedback factors could further enhance the recycling of water in semiarid regions. One feedback not yet incorporated in models could come from small lakes and low swampy regions near rivers. These features are too small to represent in climate-model grid boxes but have the potential to recycle still more monsoonal water vapor into the African interior.

After reaching a peak near 10,000 years ago, summer insolation values at lower latitudes of the northern hemisphere have since fallen continuously (Figure 14-17A). This decrease has occurred because Earth's precessional motion has carried it from a June 21 position close to the Sun 10,000 years ago to a June 21 position far from the Sun today (Chapter 8).

By 6000 years ago, summer insolation values in the northern tropics were still about 5% higher than the modern levels but were slowly declining toward modern values. This slow decrease should have produced a corresponding decline in the strength of the tropical monsoons.

Direct observations and ^{14}C dates (converted to calendar ages) of lakes across North Africa confirm a major drop in water levels in the last 9000 years (Figure 14-17B). Lakes are lower now than they were between 9000 and 6000 years ago, and many have completely dried out.

Examined one by one, however, most lake-level histories in North Africa and India show many large and abrupt changes during the overall transition from higher to lower levels (Figure 14-17C). As was the case for the irregular rates at which the northern ice sheets melted, these repeated short-term changes in lake levels represent a type of climate response that cannot be directly attributed to the smooth, gradual forcing provided by changes in summer insolation. Other factors operating over shorter time intervals must cause these rapid changes.

14-7 Warmer, Then Cooler North Polar Summers

At the glacial maximum, the main controls on climate at high northern latitudes were the regional cooling effects of the ice sheets and the global cooling caused by low CO_2 values. As deglaciation began, rising summer insolation values increasingly warmed land areas located far from the ice sheets. During deglaciation, increasing warmth from rising summer insolation gradually overcame the cooling effects of the shrinking ice sheets. Summer insolation values reached a peak 10,000 years ago, with ice sheets smaller but still present.

Northward Shifts in Vegetation The last deglaciation dramatically transformed the landscapes of continents in the northern hemisphere. In North America, cold-tolerant spruce trees retreated from their glacial position in the central United States south of the ice sheet to their modern position in northeastern Canada (Figure 14-18A). Warm-tolerant trees such as oak moved a smaller distance from their glacial location in the far southeastern United States to their modern concentrations in the mid-Atlantic states (Figure 14-18B).

The climate changes that occurred midway through the deglaciation produced unusual mixtures of plants called **no-analog vegetation**, because no similar combination exists today. For example, spruce trees grew with hardwood deciduous trees (such as ash) in the northern Midwest of the United States early in the deglaciation, yet today ash and other deciduous trees are rare in regions where spruce trees grow.

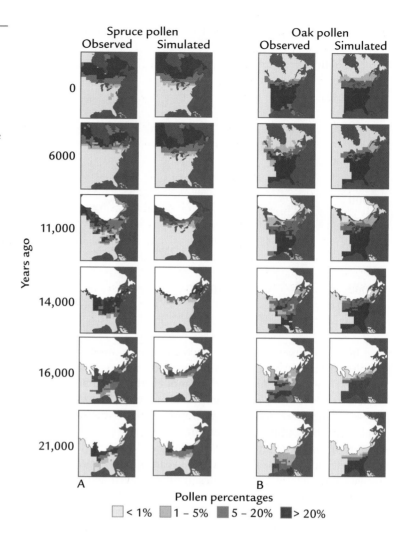

FIGURE 14-18 Data-model vegetation comparisons
Pollen in lake sediments indicates large-scale changes in the distribution of (A) spruce and (B) oak pollen during the last deglaciation. Model simulations of climate and vegetation reproduce most aspects of these observed patterns. (Adapted from T. Webb III et al., "Late Quaternary Climate Change in Eastern North America: A Comparison of Pollen-Derived Estimates with Climate Model Results," *Quaternary Science Reviews* 17 [1998]: 587–606.)

No-analog mixtures developed because each vegetation type responded to a combination of environmental variables different from those that controlled the other types. These individualistic responses make it impossible to analyze past vegetation changes by lumping pollen together into larger communities or assemblages; each vegetation type has to be analyzed on its own.

The distributions of spruce and oak pollen simulated by climate models are also shown in Figure 14-18 for comparison with the observed distributions. Both sets of maps show the same large-scale northward relocation of spruce, and they even agree on the existence of a mid-deglacial interval when spruce became rare throughout eastern North America. Both sets of maps show a similar northward expansion of oak, but the model simulates more oak in the southeast during deglaciation than is observed from pollen in lakes, a mismatch similar to those noted for the glacial maximum (Chapter 13).

Peak Warmth Once CO₂ values had risen to full interglacial levels and only remnants of the ice sheets were left, summer insolation values became the main control on climate responses, particularly for vegetation. The high summer insolation values that had triggered ice melting remained greater than they are today but had already begun to decrease toward modern values (see Figure 14-12). As a result, the warmest temperatures of the last several thousand years were registered immediately after the regional chilling effect of the ice sheets was removed, and renewed cooling followed immediately. Some climate scientists refer to this warmer-than-modern interval as the *hypsithermal*, but the time of greatest warmth actually varies widely from region to region, depending on when the nearby ice melted and its cooling effects were removed.

In response to summer insolation values that were still 5% higher than those today, the northern limit of boreal (spruce and larch) forest in Siberia and west-central North America moved as much as 300 kilometers north of its modern position by 6000 years ago, narrowing the fringe of tundra bordering the Arctic Ocean. This expansion of forest beyond its modern limits confirms that summer temperatures were warmer than they are today on the northern continents. Winter insolation values lower than today's may have produced cooler winter temperatures, but the northern limit of boreal forest is mainly sensitive to temperature in the summer growing season.

Climate models have been used to assess this high-latitude summer warming 6000 years ago. For summer insolation values 5% higher than today's, the models simulate a summer warming in northern Canada by as much as 2° to 3°C, but less warming in central Asia (Figure 14-19A). Additional model simulations have incorporated the positive feedback effect of vegetation (Figure 14-19B). In regions where boreal forest advanced northward beyond its modern limits, the low albedo (25%) typical of these dark-green trees replaced the high albedo (60%) typical of scrubby tundra vegetation under snow cover. The lower albedo of the trees absorbed more sunlight and further warmed the climate. Because far northern regions are snow-covered for much of the year, this additional warming extended through most of the year and affected broad areas of northern Canada and Asia (Figure 14-19B). The net effect was nearly to double the initial insolation warming of high northern latitudes to a combined total of 3°C (Figure 14-19C).

Sea ice contributed to the climate changes just described. High summer insolation caused the sea ice margin in the model to thin and retreat northward, and this change propagated into the rest of the yearly cycle, with delayed refreezing of seasonal sea ice in autumn, thinner and less extensive sea ice in winter, and earlier melting of sea ice in spring. As a result, even though winter insolation values were considerably lower than those today, the reduced sea ice cover in winter and larger areas of open water near the coasts moderated the winter cooling of the landmasses. As a result, the model simulates an annual average warming.

Renewed Cooling in the Last Several Thousand Years Summer insolation has been decreasing at high northern latitudes during the last 6000 years, falling from the elevated levels that drove deglacial melting of ice sheets. Diminished summer insolation at high latitudes has been caused in part by the movement of the summer solstice toward the aphelion (distant-pass) position and in part by the slow decrease of Earth's tilt from a maximum 10,000 years ago to an intermediate value of 23.5° today. These orbital changes have produced a 5% decrease in summer insolation and a 5% increase in winter insolation at high latitudes since 6000 years ago (see Figure 14-12).

The orbital glaciation hypothesis of John Imbrie and his colleagues (Chapter 12) proposes that one of the earliest steps toward the next glaciation will be a cooling

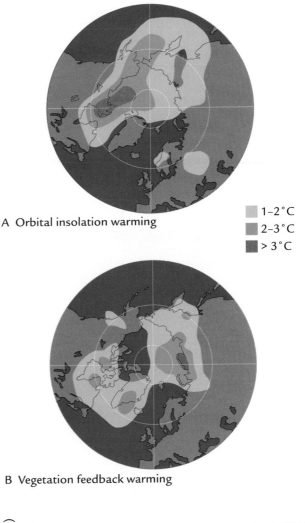

A Orbital insolation warming

    ■ 1–2°C
    ■ 2–3°C
    ■ > 3°C

B Vegetation feedback warming

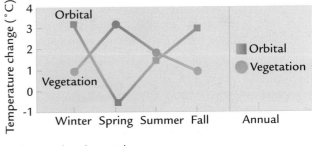

C Seasonal and annual averages

FIGURE 14-19 **Peak deglacial warmth** Climate model simulations indicate that higher insolation 6000 years ago warmed high latitudes, especially in Canada in the summer (A). Additional simulations show that positive feedback caused by northward expansion of low-albedo spruce forest into high-albedo tundra may have almost doubled this regional warming, with the largest changes in Asia (B) and during the spring (C). (Adapted from TEMPO (J. E. Kutzbach et al.), "Potential Role of Vegetation in the Climatic Sensitivity of High-Latitude Regions: A Case Study at 6000 Years B.P.," *Global Biogeochemical Cycles* 6 [1996]: 727–36.)

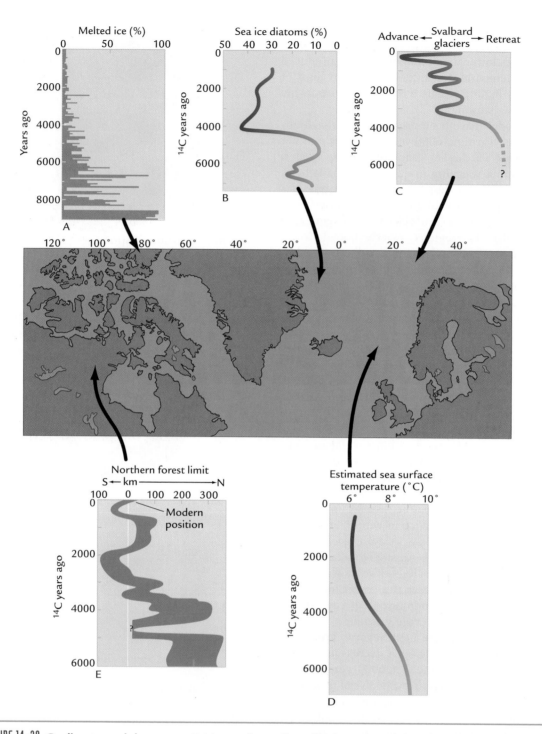

FIGURE 14-20 Cooling toward the present Evidence of a cooling of high northern latitudes during the last several thousand years includes (A) less frequent summer melting episodes in ice caps on Arctic islands; (B) more frequent sea ice off Greenland; (C) advances of ice caps on Arctic islands north of Europe; (D) lower temperatures in the Atlantic Ocean west of southern Norway; and (E) a southward shift of the boundary between tundra and spruce forest in northern Canada. (A: adapted from R. M. Koerner and D. A. Fisher, "A Record of Holocene Summer Climate from a Canadian High-Arctic Ice Core," *Nature* 343 [1990]: 630–31; B and D: adapted from N. Koc et al., "Paleoceanographic Reconstructions of Surface Ocean Conditions in the Greenland, Iceland, and Norwegian Seas Through the Last 14 Ka Based on Diatoms," *Quaternary Science Reviews* 12 [1992]: 115–40; C: adapted from J. Lubinski, S. L. Forman, and G. H. Miller, "Holocene Glacier and Climate Fluctuations on Franz Joseph Land, Arctic Russia," *Quaternary Science Reviews* 18 [1999]: 87–109; E: adapted from H. Nichols, "Palynological and Paleoclimatic Study of the Late Quaternary Displacement of the Boreal Forest-Tundra Ecotone in Keewatin and MacKenzie, N.W.T.," Institute of Arctic and Alpine Research Occasional Paper no. 15 [Boulder, Colo., 1975].)

of the ocean and nearby land at high latitudes of the northern hemisphere because of falling summer insolation values. In confirmation of this hypothesis, summer temperatures have fallen significantly during the last several thousand years in several regions at high northern latitudes (Figure 14-20).

One indication of cooler summers comes from ice cores taken from small ice caps near sea level in several parts of the Arctic. Ice from the tiny Agassiz ice cap on Ellesmere Island, in far northern Canada, shows that summer melting episodes were far more frequent before 5000 years ago than they have been since then (Figure 14-20A). This evidence supports a trend toward cooler summers, less favorable to ice melting.

Another region where cooling is evident in the last several thousand years is the high-latitude Atlantic Ocean off the coast of Greenland (Figure 14-20B). Today this region has a winter cover of sea ice. Sediment cores from this area contain shells of diatoms (small plant plankton made of opal or SiO_2) that once lived in these surface waters. Before about 5000 years ago, the diatom species present indicate that sea ice was absent or scarce in this region.

Small glaciers on islands in the Arctic Ocean have also increased in size in recent millennia. Glacier margins on Arctic islands such as Franz Josef Land were located well back from their modern positions between 8000 and 3500 years ago (Figure 14-20C), and some ice caps had melted entirely. These glaciers have reappeared or grown since 3500 years ago, oscillating toward a progressively larger size, presumably because of cooler summer temperatures and less melting.

Farther south, off the southwest coast of Norway, changes in the abundance of temperature-sensitive diatom species have been used to reconstruct estimated changes in sea surface temperature (Figure 14-20D). These trends show a gradual cooling after 6000 years ago.

The boundary between tundra to the north and boreal forest to the south is another sensitive climatic indicator in Asia and North America. This boundary in northern Canada was well north of its present limit 6000 years ago and has since advanced southward by as much as 300 kilometers (Figure 14-20E). This shift has been attributed to progressively cooler summers and a shorter growing season. In accord with several other records in Figure 14-20, the tree line shows shorter-term oscillations superimposed on the longer-term (orbital-scale) trend.

Still another indication of summer cooling in the last several thousand years comes from mountain glaciers. Like ice sheets, mountain glaciers melt if summer insolation increases (Chapter 10). As a result, the drop in summer insolation levels over the last several thousand years in the northern hemisphere should have caused mountain glaciers to advance to lower elevations. Mountain glaciers can respond to climate changes within just a few decades.

Because hundreds of mountain glaciers exist on Earth today, climate scientists have condensed a huge amount of information into curves representative of general glacier advances and retreats. Shown in Figure 14-21 are two reconstructions of glacier changes published in the last two decades.

The common trend evident in both signals is an overall long-term tendency toward larger and more persistent glaciers. Before 5000 years ago, only a few advances occurred, an indication that mountain glaciers were generally small. In contrast, mountain glaciers have on average remained at or near the lower limit of their ranges for much of the last 5000 years. The size of the advances has also increased in an irregular way, with the largest advances in most regions occurring in the last 1000 years. This gradual increase in the size and persistence of mountain-glacier advances is consistent with a long-term decrease in summer insolation, and with the evidence of progressive summer cooling already examined. Short-term oscillations at intervals of a few thousand years are also prominent in mountain-glacier signals.

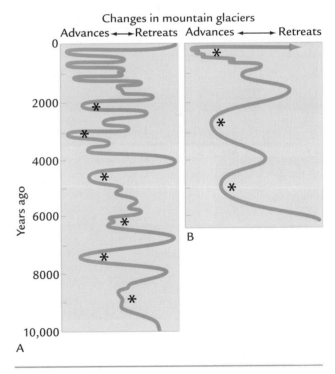

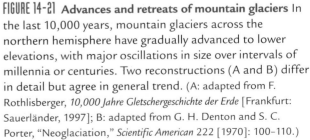

FIGURE 14-21 **Advances and retreats of mountain glaciers** In the last 10,000 years, mountain glaciers across the northern hemisphere have gradually advanced to lower elevations, with major oscillations in size over intervals of millennia or centuries. Two reconstructions (A and B) differ in detail but agree in general trend. (A: adapted from F. Rothlisberger, *10,000 Jahre Gletschergeschichte der Erde* [Frankfurt: Sauerländer, 1997]; B: adapted from G. H. Denton and S. C. Porter, "Neoglaciation," *Scientific American* 222 [1970]: 100–110.)

Fading Memories of Melting Ice

Even though the last remnants of the glacial maximum ice sheets melted near 6000 years ago, Earth's deeper bedrock retains a memory of their existence. This legacy of the ice sheets shows up in present-day patterns of the rising and falling of Earth's surface.

Beginning as early as the late eighteenth century and occurring more commonly by the late nineteenth, seaport towns and cities installed tide gauges to measure sea level changes caused by tides and large storms. The immediate goal of these efforts was to understand sea level changes well enough to build structures to protect the communities against flooding. Tide gauges were most common in seaports in Europe and along the East

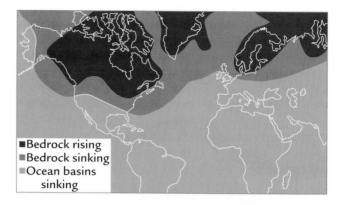

■ Bedrock rising
■ Bedrock sinking
□ Ocean basins sinking

FIGURE 14-23 **Patterns of sea level change** Relative sea level today is changing in several ways regionally because of bedrock movement. Bedrock is rapidly rising in areas formerly covered by thick ice and sinking in regions surrounding the former ice sheets. Farther from the ice sheets, ocean basins are sinking under the added weight of meltwater. (Adapted from A. M. Tushingham and W. R. Peltier, "Ice-3G: A New Global Model of Late Pleistocene Deglaciation Based upon Geophysical Predictions of Post-Glacial Relative Sea-Level Changes," *Journal of Geophysical Research* 96 [1991]: 4497–4523.)

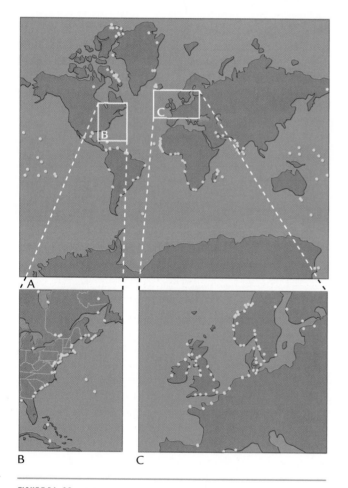

B   C

FIGURE 14-22 **Tide gauge stations** Tide gauge records spanning several decades to as much as two centuries are available from hundreds of coastal locations on Earth's surface (A). The largest concentrations are in eastern North America (B) and northwestern Europe (C). (Adapted from A. M. Tushingham and W. R. Peltier, "Ice-3G: A New Global Model of Late Pleistocene Deglaciation Based upon Geophysical Predictions of Post-Glacial Relative Sea-Level Changes," *Journal of Geophysical Research* 96 [1991]: 4497–4523.)

Coast of the United States, areas that had begun to industrialize early (Figure 14-22).

Some 100 to 200 years later, these tide-gauge records show not only short-term changes caused by tides and storms but also longer-term histories of sea level change in the decades and centuries since their installation. These records show a confusing hodgepodge of trends from region to region: some show sea level dropping quickly, others show it rising very slowly, and still others indicate somewhat larger rises.

Why would all these different trends occur? The world ocean is one interconnected body of water, and we would expect true global (eustatic) sea level to rise or fall by the same amount everywhere. How can Earth's one ocean rise in one place and fall in another over intervals of decades or centuries?

The answer is that tide gauges do not record just the global change in sea level; they also record relative sea level changes that vary from region to region because the land is behaving differently in different areas. If the land is rising rapidly in one area for some reason, this motion can overwhelm any smaller change in global sea level and cause a drop in the *relative* sea level sensed by a tide gauge in that particular location. The rapid rise in the land makes it look as if the sea is falling. The opposite is true in regions where the land is sinking rapidly.

The patterns of long-term change in relative sea level recorded by this global array of tide gauges fall into well-defined geographic groups (Figure 14-23).

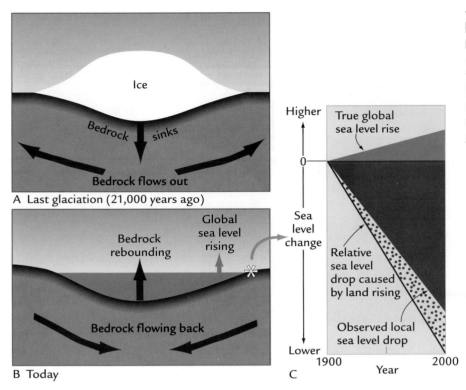

FIGURE 14-24 Bedrock rebound and sea level fall In regions where ice sheets existed, relative sea level is rapidly falling today. Bedrock in these areas is still rebounding in response to the earlier melting of ice, and the rebound of the land overwhelms the true global rise of sea level.

Within three major regions, the tide gauges show similar responses, and each of these responses reflects a different memory of the glacial ice sheets still held in the deep bedrock.

One geographic group of tide gauges shows rapid drops in relative sea level in recent decades (Figure 14-24). All tide gauges in this group are located in regions that were once directly beneath ice sheets, such as the Hudson Bay region of Canada and most of Scandinavia. The explanation for a relative fall of sea level in these regions is that the bedrock is rapidly rising, still rebounding from the effect of the removal of the ice sheet load thousands of years ago.

Because bedrock at depths of 100 to 200 kilometers has a slow viscous component of response, it takes many thousands of years to recover fully from loads imposed upon it (Chapter 10). The glacial maximum ice sheets were an enormous load that depressed bedrock surfaces by as much as 1 kilometer, causing deep rock to flow slowly out from under the center of the ice sheet at great depths. When the ice melted, rock began to flow back into this region, causing the depressed bedrock to rise gradually toward its former elevation. Even now, thousands of years after the last ice melted, the rates of relative sea level fall caused by ongoing bedrock rebound are as high as 10 millimeters a year, much larger than the true rate of global sea level rise of about 1.5 to 2 millimeters a year.

As a result of this slow bedrock rebound, ancient beach ridges surround the lower-lying parts of Hudson

Bay (Figure 14-25A). The modern-day beach is at sea level, and the beach ridges grow progressively older at successively higher elevations away from Hudson Bay (Figure 14-25B). The highest-elevation ridge dates to 7000 ^{14}C years ago, shortly after the last ice had melted from this area and ocean water had flooded back into the depression left in the region of Hudson Bay. The stair-step series of beach ridges shows that the land beneath the former ice sheet has been rising for thousands of years, although at rates that have slowed as the memory of the ice sheet load has weakened.

The second major geographic group of tide-gauge responses shows a relatively fast rate of sea level rise, typically near 2.5 millimeters a year. This group clusters in a halo pattern surrounding the former ice sheets but extending well beyond the ice margins (see Figure 14-23). In North America, this region includes the east coast from southern New England south to Georgia, and in Europe it includes a narrow band across England, France, and Germany.

The relatively rapid present-day rise of sea level in these regions is also caused by a memory of the glacial maximum ice sheets, even though these regions were not located directly beneath the ice loads. During glacial times, the deep rock displaced by the heavy load of the ice sheets had to go somewhere. Flowing out from under the ice and beyond the margins of the ice sheets, it caused a slight increase in the elevation of the land, called a **peripheral forebulge** (Figure 14-26). A rough analogy is a person sitting on a partly inflated air

FIGURE 14-25 Old beach ridges (A) A series of old beach ridges surrounds Hudson Bay. (B) The beaches increase in age with elevation because the land has been slowly rising for 7000 years. (A: courtesy of Claude Hillaire-Marcell, University of Quebec, Montreal; B: adapted from W. R. Peltier and J. T. Andrews, "Glacial-Isostatic Adjustment. I. The Forward Problem," *Geophysical Journal of the Royal Astronomical Society* 46 [1976]: 605–46.)

mattress in water: the person's weight depresses the center of the air mattress into the water, but the excess air pushed to the edges of the mattress makes the edges bulge slightly up out of the water.

After the ice sheets melted, the rock displaced beyond the ice margins gradually flowed back into the region where the ice had been. This return flow caused the peripheral bulge to collapse, very gradually dropping the land surface (Figure 14-26). Because the land is still dropping today at rates of as much as 1 millimeter a year, this effect adds to the true rise of global sea level of 1.5 to 2 millimeters a year and produces a relatively fast rise of relative sea level in these regions (2.5 mm/year or more). Dating of older (now submerged) beaches in these regions indicates that this pattern of rapid rise in relative sea level has persisted for thousands of years in these areas.

The third group of tide-gauge responses comes from coastlines located far from the northern hemisphere ice sheets (see Figure 14-23). Relative sea level in these regions is rising at rates averaging just under 1.5 millimeters a year, a value slightly less than the true global rate of sea level rise. It might seem that regions so far from the ice sheets should be free of memory effects from the ice, but they are not. The return of

glacial meltwater to the oceans has added an extra load on the bedrock beneath the ocean floor in these regions.

At maximum size, the ice sheets extracted a layer of water almost 110 meters thick from the world ocean. With this water load removed, the average level of the crust in the ocean basins rose over 30 meters more than on continents not directly affected by the weight of the ice sheets. When this layer of water returned to the oceans during deglaciation, it once again loaded down the ocean crust and caused it to sink (Figure 14-27). This slow sinking of ocean crust is still going on today, and it counteracts part of the true rise of global sea level, producing a slightly smaller rise in relative sea level.

Even though the great ice sheets of the last glaciation are gone, they are not forgotten, at least not by the bedrock and the shorelines. In fact, bedrock memories of these shifting loads of ice and water are the major obstacles to determining the true rate of the global rise in sea level in the past century. It is particularly unfortunate that many of the longest and most reliable tide-gauge records happen to have been located in just those regions where these overprints from the ice sheets were largest.

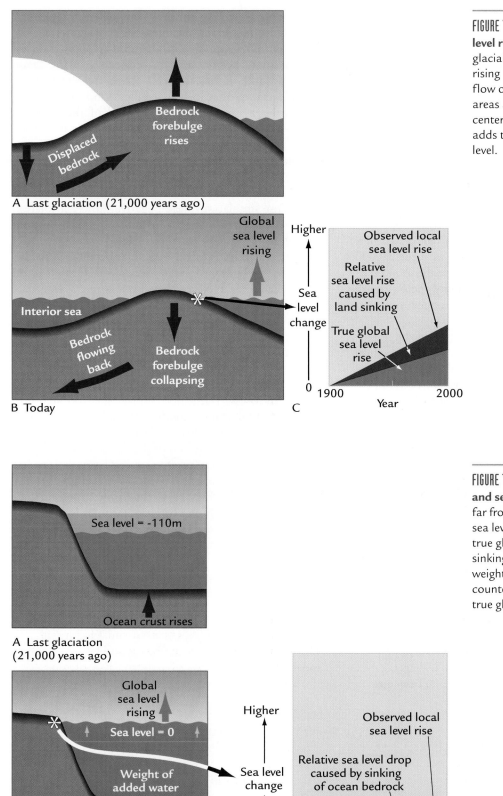

A Last glaciation (21,000 years ago)

B Today

C

FIGURE 14-26 Bedrock sinking and sea level rise In regions surrounding glacial ice sheets, relative sea level is rising rapidly today. The continued flow of bedrock away from these areas and into the former ice sheet centers causes the land to sink and adds to the true global rise of sea level.

FIGURE 14-27 Ocean bedrock sinking and sea level rise In coastal regions far from glacial ice sheets, relative sea level is not rising as fast as the true global average. The continued sinking of ocean bedrock under the weight of 110 m of meltwater counteracts a small fraction of the true global rise in sea level.

Predicting Future Orbital-Scale Changes

Astronomy tells us the changes in Earth's orbit that will occur over the next several thousand years. Having learned how insolation has driven climate change in the past, we can use future orbital changes to predict the general course of future climate.

One firm prediction is that summer monsoons in the northern hemisphere will once again intensify in response to rising summer insolation. Today June 21 occurs near the aphelion (most distant) position in Earth's eccentric orbit around the Sun. In another 10,000 years, Earth will have returned to the opposite configuration, and June 21 will occur at perihelion, when Earth is closest to the Sun, just as it did 10,000 years ago. This orbital shift will increase the amount of summer insolation in the northern hemisphere (Figure 14-28, left) and once again drive a stronger monsoon over North Africa and Asia.

Predicting the next cycle of glaciation in the northern hemisphere is far more difficult. Summer insolation is the main control on the size of northern hemisphere ice sheets (Chapter 10), but the insolation trends during the next 10,000 years move in opposite directions at high and low latitudes of the northern hemisphere.

Changes in precession will cause summer insolation at low and middle latitudes of the northern hemisphere to increase for the next 10,000 years (Figure 14-28, left), and this change should tend to oppose glaciation. But at the high northern latitudes that are more strongly influenced by the tilt of Earth's axis, summer insolation will continue to fall from the maximum value reached 10,000 years ago to the next minimum, 10,000 years in the future (Figure 14-28, right). The insolation drop at high latitudes should help to promote the next glaciation.

Climate scientists have attempted to determine when these insolation changes might trigger the next glaciation. One approach is to run GCM simulations of the conditions needed to start ice sheets growing again. The last time northern hemisphere ice sheets were absent everywhere (except Greenland) and then began a new cycle of growth was at the end of the previous interglaciation, 115,000 years ago. Modelers have used the low summer insolation values from that time as a boundary condition to see if northern hemisphere continents were cold enough to allow snow to persist through the summer ablation season. Initial results from these simulations were negative. Lower insolation values did cool the high northern latitudes, but not enough to prevent snow from melting during the brief summer warmth (Figure 14-29A). Snow did not persist through the summer, and no ice sheets formed (even though the record tells us they did).

Recent experiments have shown that part of the reason these initial experiments failed to form ice sheets was that the earlier models lacked critical feedbacks that could amplify the initial cooling caused by insolation. One important feedback results from the increase in albedo caused by the advance of tundra southward into regions of forest. With this feedback incorporated into climate and vegetation models, the additional cooling was sufficient to initiate glaciation over northern Canada and around the rim of the Arctic Ocean (Figure 14-29B). Despite this progress, GCM simulations are not yet sensitive enough to predict the next glaciation.

A second way of trying to determine when new ice sheets will grow at high northern latitudes is to project the repetitive long-term climate cycles recorded during the past several hundred thousand years forward into the future. Records of $\delta^{18}O$ changes covering the last 900,000 years are known to have been dominated by changes in ice volume, but contain a temperature over-

---

FIGURE 14-28 Future summer insolation trends During the next 10,000 years, precession-dominated insolation at low latitudes of the northern hemisphere will return to another maximum (left), while tilt-dominated insolation at very high northern latitudes will continue to fall toward the next minimum (right). (Adapted from A. Berger and M.-F. Loutre, "Modeling the Climatic Response to Astronomical and CO₂ Forcings," *C. R. Acad. Sci. Paris* 323 [1996]: 1–16.)

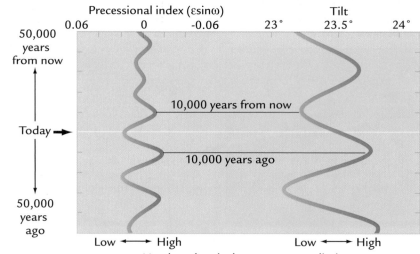

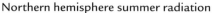
Northern hemisphere summer radiation

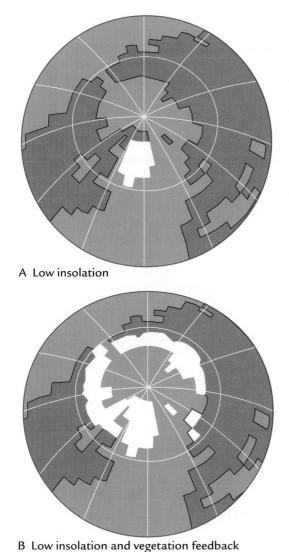

A Low insolation

B Low insolation and vegetation feedback

FIGURE 14-29 **Vegetation–temperature feedback** Climate models simulate some high-latitude cooling because of low summer insolation at the end of the last interglaciation, 115,000 years ago, but not enough to grow ice sheets (A). Model sensitivity tests that include albedo feedback caused by the southward shift of the tundra-spruce boundary simulate additional cooling that initiates circumarctic glaciation in high northern latitudes (B). (Adapted from R. G. Gallimore and J. E. Kutzbach, "Role of Orbitally Induced Changes in Tundra Area in the Onset of Glaciation," *Nature* 381 [1996]: 503–5.)

print as well. These signals show that the current interval of low $\delta^{18}O$ values (minimal ice volume and warm temperatures) has lasted just about as long as any previous interglaciation, implying that ice sheets should start growing soon (relatively speaking).

To project orbital-scale $\delta^{18}O$ cycles into the future, climate scientists first establish physically reasonable

quantitative links between past changes in summer insolation and past changes in $\delta^{18}O$, and then assume that these same relationships will continue into the future. One such attempt (Figure 14-30) was based on several assumptions consistent with what we learned about ice sheets in Chapters 10 and 12. In the past, summer insolation changes at the 23,000-year precession cycle and 41,000-year tilt cycle drove ice volume ($\delta^{18}O$) responses that lagged several thousand years behind because of the slow response times typical of ice sheets.

Climate scientists can use these lagged responses in $\delta^{18}O$ to project forward into the future, using astronomical calculations of future summer insolation as the driving force. This kind of model also stipulates that ice sheets melted four times faster when summer insolation was increasing than they grew when insolation was decreasing. This assumed behavior is based on the fact that ablation of ice can be much more rapid than accumulation (Chapter 10). Adding this rapid melting behavior helps to simulate the abrupt deglaciations that have occurred at intervals of 100,000 years during the last 0.9 Myr. This behavior was also incorporated into projections of future $\delta^{18}O$ (ice volume) behavior.

The predicted ice volume ($\delta^{18}O$) signal in Figure 14-30 shows the present-day climate system poised at the end of the interglacial conditions and even beginning the plunge into the next glaciation. It looks as if the next glaciation must be close or even at hand.

But two problems complicate attempts to use these $\delta^{18}O$ signals to predict the future. One problem is that for most of the $\delta^{18}O$ record we don't know exactly what fraction of the $\delta^{18}O$ changes was caused by changes in ice volume and what fraction by changes in temperature. We know that about two-thirds of the difference between $\delta^{18}O$ levels at the last glacial maximum and those today was caused by changes in ice volume and the other third by changes in temperature, but the relative contributions of these two factors may not have stayed exactly the same through the entire length of the curve.

For example, the evidence just examined showing a cooling of high northern latitudes during the last several thousand years or so tells us that the earliest phases of increasing $\delta^{18}O$ values at the end of interglacial intervals might be caused by cooler temperatures in the deep ocean at high northern latitudes rather than by increases in ice volume. This could explain why no glaciation has yet begun, even though the model simulation in Figure 14-30 shows it just getting under way.

The second factor that impedes predictions of the future is that the dating (age) of most of the $\delta^{18}O$ signal is uncertain to within a few thousand years. Although this is a remarkably small uncertainty compared with the total length of the $\delta^{18}O$ record, it is still a serious complication. If past changes in $\delta^{18}O$ are imprecisely

FIGURE 14-30 Projections of future δ¹⁸O
Simple numerical models can transform summer insolation in the northern hemisphere (left) into estimates of ice volume (center) that closely match δ¹⁸O signals (right) that mainly record global ice volume. Projection of the model predictions into the future (top) shows growth of northern hemisphere ice sheets. (Modified from J. Imbrie and J. Z. Imbrie, "Modeling the Climatic Response to Orbital Variation," *Science* 207 [1980]: 943–53.)

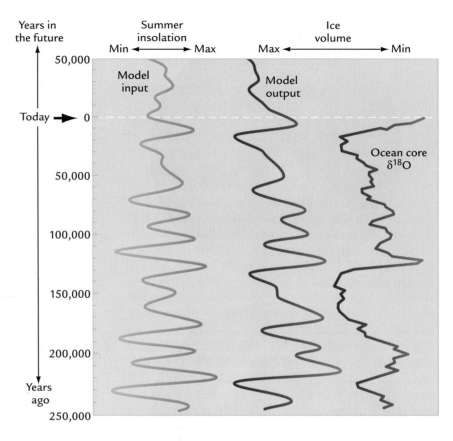

fixed within a time frame, projections of future δ¹⁸O changes are uncertain to exactly that same extent.

Most climate scientists conclude from these kinds of studies that orbital-scale changes will lead to another cycle of glaciation in the geologically near future (probably within the next few thousand years), but even the experts do not really know precisely when it will occur or how large the ice sheets will grow.

Predictions are also complicated by other problems that we have not yet examined. Abrupt millennial-scale climatic oscillations will also occur (Chapter 15), and they will combine with the slower orbital-scale changes to determine the total pattern of natural climate change in the future. In addition, burning of fossil fuels will cause a significant unnatural warming that will last for many centuries (Part V).

Key Terms

deglacial two-step (p. 304)

Younger Dryas (p. 308)

polar front (p. 308)

proglacial lakes (p. 310)

channeled scablands (p. 312)

no-analog vegetation (p. 317)

peripheral forebulge (p. 323)

Review Questions

1. What is the best method of measuring the melting of ice sheets over the last 17,000 years?

2. Does the timing of ice sheet melting support the Milankovitch theory that orbital insolation controls the sizes of ice sheets?

3. Do changes in the intensity of summer monsoons in the last 17,000 years support the Kutzbach theory that orbital insolation controls the intensity of monsoons?

4. What evidence suggests that variations in orbital insolation were not the only cause of climate changes during the last 17,000 years?

5. Describe how proglacial lakes travel slowly across the landscape behind melting ice sheets.

6. Why were summer temperatures at high northern latitudes warmer 6000 years ago than they are today?

7. List evidence from ice, land, and water for a cooling in north polar regions since 6000 years ago.

8. How do long-melted ice sheets complicate attempts to measure modern changes in sea level?

9. What do orbital trends imply about future changes in monsoons and northern ice sheets?

ADDITIONAL RESOURCES

Basic Reading

Bard, E., B. Hamelin, R. G. Fairbanks, and A. Zindler. 1990. "Calibration of the ^{14}C Time Scale over the Last 30,000 Years Using Mass Spectrometric U-Th Ages from Barbados Corals." *Nature* 345:405–10.

Broecker, W. S., J. P. Kennett, B. P. Flower, J. S. Teller, S. Trumbore, G. Bonani, and W. Wolfli. 1989. "Routing of Meltwater from the Laurentide Ice Sheet During the Younger Dryas Cold Episode." *Nature* 341:318–20.

COHMAP Members. 1988. "Climatic Changes of the Last 18,000 Years: Observations and Model Simulations." *Science* 241:1043–62.

Dyke, A. S., and V. K. Prest. 1987. "Late Wisconsinan and Holocene History of the Laurentide Ice Sheet." *Géographie Physique et Quaternaire* 41:237–63.

Fairbanks, R. G. 1989. "A 17,000-Year Glacio-eustatic Sea Level Record: Influence of Glacial Melting on the Younger Dryas Event and Deep-Ocean Circulation. "*Nature* 342:637–42.

Flint, R. F. 1971. *Glacial and Quaternary Geology.* New York: John Wiley.

Kutzbach, J. E., and F. A. Street-Perrott. 1985. "Milankovitch Forcing of Fluctuations in the Level of Tropical Lakes from 18 to 0 Kyr B.P." *Nature* 317:130–34.

Roberts, N. 1998. *The Holocene.* Oxford: Blackwell.

Teller, J. T. 1987. "Proglacial Lakes and the Southern Margin of the Laurentide Ice Sheet." In *North America and Adjacent Oceans During the Last Deglaciation*, ed. W. F. Ruddiman and H. E. Wright. Geology of North America, K-3. Boulder, Colo.: Geological Society of America.

Wright, H. E., Jr., J. E. Kutzbach, T. Webb III, W. F. Ruddiman, F. A. Street-Perrott, and P. J. Bartlein. 1993. *Global Climates Since the Last Glacial Maximum.* Minneapolis: University of Minnesota Press.

Advanced Reading

Rind, D., D. Peteet, and G. Kukla. 1989. "Can Milankovitch Orbital Variations Initiate the Growth of Ice Sheets in a General Circulation Model?" *Journal of Geophysical Research* 94D:12851–71.

Webb, T., III. 1998. "Late Quaternary Climates: Data Synthesis and Model Experiments." *Quaternary Science Reviews* 17:587–606.

Webb, T., III, S. Howe, R. H. W. Bradshaw, and K. M. Heide. 1981. "Estimating Plant Abundances from Pollen Percentages: The Use of Regression Analysis." *Review of Paleobotany and Palynology* 34:269–300.

CHAPTER 15

Millennial Oscillations in Climate

In this chapter we examine climate oscillations over intervals as short as 1000 years. Some of these fluctuations are rapid enough to be relevant to human concern. Because they occur over millennial intervals, they are called **millennial oscillations** (rather than cycles). These oscillations have been detected in Greenland ice cores, in North Atlantic sediment cores, and in records from many other regions. These changes were larger when glacial ice sheets existed, but smaller during interglacial climates like today.

In earlier chapters, evidence of climate change was presented as a hypothesis to be tested against observations. In this chapter we invert this approach. We assemble the evidence of these oscillations until a picture of their possible origin emerges, in effect repeating the journey climate scientists have taken during the last decade. We also draw on insights gained from orbital time scales (Part III) to address possible explanations of these oscillations: Are they global in extent or tied to a specific region? Are they driven by fluctuations in ice-sheet margins, by internal interactions between ice sheets and the ocean, or by external changes in the Sun's strength? By evaluating the available evidence, we become participants in the process of discovery, still under way today.

Millennial Oscillations During Glaciations

The first critical clue that the climate system is capable of large changes over intervals relevant to human life spans came from studies of the deglacial Younger Dryas event, which lasted about 1500 years and began and ended very abruptly (Chapter 14). More recently, evidence has emerged that an ongoing series of similar short-term oscillations is superimposed on the entire sequence of longer orbital-scale cycles. Because these short-term fluctuations are largest and best defined during glacial intervals, we examine this evidence first.

15-1 Oscillations Recorded in Greenland Ice Cores

Long ice cores first taken on Greenland in the 1970s recovered records spanning much of the last 100,000-year glacial cycle (Figure 15-1). The upper (interglacial) parts of these records were dated by counting annual layers, while the age of the lower parts of the sequence was estimated by using theoretical models of the flow of ice deep within ice sheets.

Early ice core studies in the 1970s concentrated on two climate signals: ice core oxygen isotope ($\delta^{18}O$) ratios and dust. Signals of $\delta^{18}O$ recorded in ice cores are entirely different from those measured in foraminifera from ocean cores: $\delta^{18}O$ signals in ice reflect changes in the oxygen isotope ratio of the water vapor that falls as snow and consolidates into ice (Box 15-1).

The $\delta^{18}O$ and dust signals in Greenland ice reveal two distinct trends in past climate changes (Figure 15-1). One is a gradual underlying trend from low dust concentrations and more positive $\delta^{18}O$ values at the top and bottom of this record to a region of more dust and more negative $\delta^{18}O$ values in the middle of the section. Although dating of this ice core record was imprecise, the upper section was found to span approximately the last 10,000 years, with the last glacial interval lying below and the previous interglacial interval near the bottom. The gradual underlying trends defining these larger-scale divisions in the ice core sequence were interpreted as the result of orbital-scale changes in Earth's climate between colder, drier, dustier glaciations and warmer, wetter, less dusty interglaciations, including the modern interglacial climate.

But the ice core record also revealed a second, distinctly different kind of signal: rapid oscillations between high and low dust concentrations and between negative and positive $\delta^{18}O$ values over intervals much shorter than the orbital-scale changes (Figure 15-1). The $\delta^{18}O$ fluctuations of 5‰ to 6‰ are equivalent to more than half the difference between fully glacial and fully interglacial values. Each fluctuation toward more negative

(glacial) $\delta^{18}O$ values is matched by an abrupt increase in dust concentrations. These are the millennial-scale oscillations.

For a long time, scientists paid relatively little attention to these oscillations in Greenland ice cores, viewing them as climatic responses limited to the immediate region of the ice sheet. In the 1970s and 1980s, the main focus was on orbital-scale changes.

This situation changed in the early 1990s, when two new long ice cores were drilled on the summit of the

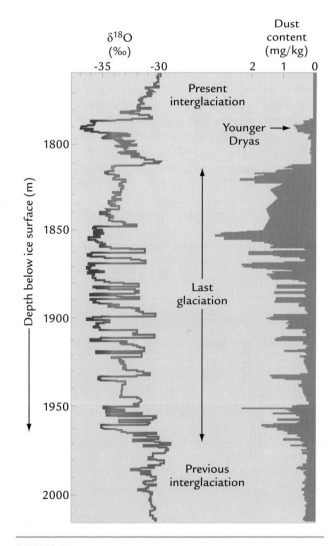

FIGURE 15-1 **Millennial oscillations in ice cores** An ice core drilled in the 1970s through the Greenland ice sheet contains records of $\delta^{18}O$ and dust concentrations. Large oscillations occur in the glacial portion of both signals (from depths of 1780 to 1960 m in the ice core), but not in the present interglaciation, above 1780 m, or in the previous one, below 1960 m. (Adapted from W. Dansgaard et al., "North Atlantic Climatic Oscillations Recorded by Deep Greenland Ice Cores," in *Climate Processes and Climate Sensitivity*, ed. J. E. Hansen and T. Takahashi [Washington, D.C.: American Geophysical Union, 1984].)

BOX 15-1 TOOLS OF CLIMATE SCIENCE

δ¹⁸O Fluctuations in Ice Cores

The mean $\delta^{18}O$ value of the ocean changes in response to the amount of ^{16}O-rich water vapor extracted from seawater and stored in ice sheets on the land. The ocean is left with relatively ^{18}O-rich water from which planktic foraminifera extract oxygen for their shells.

A second kind of $\delta^{18}O$ signal important to studies of climate is changes in the $\delta^{18}O$ value of ice layers within ice sheets. These values can vary by 5‰–10‰ over time: more negative values are typical of colder climates and less negative values of warmer climates. In Greenland, interglacial values are typically between −30‰ and −35‰, while glacial values may fall to between −35‰ and −40‰. As marine $\delta^{18}O$ values become more positive because of increased extraction of ^{16}O-rich water vapor, $\delta^{18}O$ values in ice cores become more negative.

These $\delta^{18}O$ changes reflect several influences. One is the temperature of the snow precipitating on the ice sheets. "Warm" snow may form at temperatures near freezing, whereas cold, powdery snow forms at much lower temperatures. The $\delta^{18}O$ value of snow falling on Greenland today becomes 0.7‰ more negative with each 1°C drop in the temperature of the air in which it forms. This temperature/$\delta^{18}O$ relationship holds constant today

TABLE 15-B1 Causes of δ¹⁸O Changes Recorded in Ice Cores

Negative◄──	Change in δ¹⁸O ──►values	Positive
Colder	Air temperature over ice	Warmer
Distant	Proximity of source region	Close
Low δ¹⁸O	δ¹⁸O composition of source	High δ¹⁸O
High	Elevation of ice	Low
Winter	Primary season of precipitation	Summer

for seasonal changes in temperature of the air masses delivering the snow, as well as for cooling of air at higher elevations on the ice sheet. But it did not always apply in the past: $\delta^{18}O$ values of glacial-age ice suggest temperatures about 10°C colder than those today, but measurements of the temperature of the ice itself indicate a cooling of 15°C or more.

Several other factors can also affect the $\delta^{18}O$ values of ice cores, including the source of the moisture, its path of transport to the ice sheets, and the season of precipitation (Table 15-B1). The water vapor that supplies snow to ice sheets comes from several sources, each with a different initial $\delta^{18}O$ value, and it follows different paths of transport. The longer the distance the water vapor travels, the more negative its $\delta^{18}O$ value, because it has evaporated and condensed repeatedly along the way. Changes in the relative amounts of water vapor coming from different sources can alter the mean $\delta^{18}O$ value of the snow that falls on the ice through the span of a year. Changes in the seasonal balance of water vapor delivery can also affect mean annual $\delta^{18}O$ values in the ice: more snow in the colder winter season results in lower $\delta^{18}O$ values.

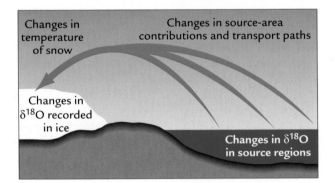

Controls of variations in ice core δ¹⁸O The $\delta^{18}O$ values recorded in ice cores vary with changes in the local temperature of snow precipitated over the ice, changes in the relative contributions among source areas of the water vapor, and $\delta^{18}O$ changes in the source regions.

Greenland ice sheet at the GISP/GRIP sites (Figure 15-2). These sites were carefully chosen in areas of smooth underlying bedrock to minimize the impact of changes in ice flow that can disturb the deeper ice layers, and

they were drilled within 30 kilometers of each other to see whether they would reveal similar climate histories.

Both of these new ice core sequences yielded nearly identical long-term climate records to within 200 meters of bedrock, a level in the ice dated to 110,000 years ago. Because the two cores recorded the same signal over the last 110,000 years, scientists had no doubt that both were reliable records of climate, not the results of local complications in ice flow.

Equally important, these two cores contain annual layering extending far back in time and permitting better dating than the theoretical ice flow models used previously. Even in these sequences, stretching and thinning of annual layers deep in the ice introduces a small uncertainty in identifying annual climate signals. Still, the combined use of several annually deposited signals (dust, $\delta^{18}O$, and others) lessens the chance of miscounting each annual layer. Estimated errors in annual-layer counts are as small as a few decades for ice that is 10,000 years old but grow to several thousand years at levels 50,000 years and older.

These two new records immediately confirmed two findings from the earlier records. They showed the same underlying trend of gradual glacial-interglacial changes in ice core $\delta^{18}O$ as the marine $\delta^{18}O$ signal (Figure 15-2). But more critically, they confirmed that shorter climatic oscillations occurred at intervals of a few thousand years throughout the last glaciation.

These millennial-scale oscillations are often informally referred to as **Dansgaard-Oeschger cycles** in honor of the geochemists Willi Dansgaard and Hans Oeschger, who found and studied them in the earlier ice core records. Because these oscillations vary widely in spacing and amplitude through the record, they are not true cycles, although obviously they are repetitive.

Even from the beginning, it was apparent that these millennial-scale oscillations come in all shapes and sizes. The original work by Dansgaard and Oeschger suggested that many of the oscillations were 2000 to 3000 years long, but some were spaced at intervals longer than 5000 years. Their average spacing was close to 4000 years, but with a wide range of variation around that value. Recent ice core studies at higher resolution and with greater analytical precision have also detected fairly regular-looking oscillations close to 1500 years in length.

Chemical analysis of the dust deposited during millennial oscillations indicates that the main source region is not in nearby North America but in more distant northern Asia. The transport path is unclear, but it could have been along the northern branch of the split jet stream that moved across the Canadian margin of the North American ice sheet (see Figure 13-15B).

The size of the dust particles is larger in the cold intervals than in the milder intervals, indicating that strong winds lifted and transported the dust from continental source regions when climate was colder. The colder intervals also contain larger amounts of sea salt (Na^+ and Cl^- ions) because strong winds over the oceans plucked more of these ions out of the salty sea spray tossed up above the rough sea surface and carried them to the ice.

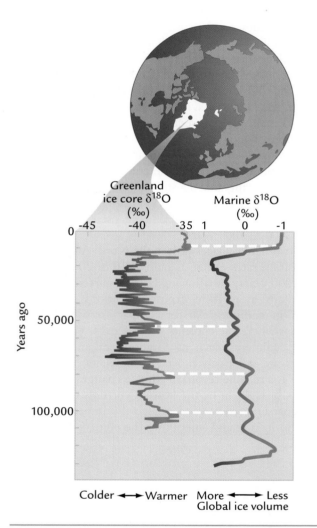

FIGURE 15-2 **New ice core evidence of millennial oscillations** Ice cores recently drilled in Greenland yielded identical $\delta^{18}O$ records reaching back 110,000 years. The longer-term changes in ice core $\delta^{18}O$ (left) are similar to orbital-scale changes in marine $\delta^{18}O$ (right), but the abrupt $\delta^{18}O$ changes in the ice reveal a different kind of behavior in the climate system. (Left: adapted from P. Grootes et al., "Comparison of Oxygen Isotope Records from the GISP and GRIP Greenland Ice Cores," *Nature* 366 [1993]: 552–54; right: adapted from D. Martinson et al., "Age Dating and the Orbital Theory of the Ice Ages: Development of a High-Resolution 0 to 300,000-Year Chronostratigraphy," *Quaternary Research* 27 [1987]: 1–29.)

15-2 Detecting and Dating Millennial Oscillations in Other Regions

Much of this chapter follows the search for millennial oscillations in regions other than Greenland and in archives other than ice, a search still going on today. Scientists doing this research face several pitfalls in their efforts to prove a connection to the oscillations documented in Greenland ice. Two questions recur: (1) Is the climatic archive being examined capable of recording such brief oscillations? (2) How accurately can the oscillations that are detected be dated?

Resolution of climate signals varies from archive to archive and from region to region for a given type of archive (see Chapter 3). The ideal archive for exploring millennial oscillations would be one that allowed resolution of annual changes and provided a record stretching back into the last glaciation or beyond. Unfortunately, few archives combine annual resolution with long records (see Figure 3-14). Fortunately, some archives that record climate changes lasting tens or hundreds of years do exist, allowing resolution of millennial-scale changes.

The second problem is dating any oscillations that are detected accurately enough to determine their correlation (if any) with those found in Greenland ice. The ^{14}C method that is used to date most continental records has inherent analytical uncertainties of several thousand years for glacial-age material, even after ^{14}C years are corrected to calendar years. Because these dating errors are comparable in size to the length of the oscillations being dated, it is often impossible to prove the true nature of the correlation.

Take an example in which an oscillation lasting for 2000 years is perfectly dated in one region but has a dating uncertainty of ± 1000 years in another (Figure 15-3). Within this uncertainty, these two signals could be (1) changing synchronously, (2) leading or lagging each other, or (3) responding in a totally opposite sense. We have already seen that leads and lags between climate signals provide critical clues to cause-and-effect relationships at orbital time scales (Chapter 12). Scientists who examine millennial-scale changes face the same problems and need the same information, but dating uncertainties are inherent in *all* millennial-scale signals.

Fortunately, it is much easier to show that millennial-scale oscillations are simply present or absent in a given climate record, whatever their exact ages. In a few cases we will examine, the millennial oscillations detected in various regions so obviously match the pattern of the changes in Greenland that no doubt can exist that these are the same millennial oscillations. But even in such cases, their precise relative timing usually remains uncertain.

15-3 Oscillations Recorded in North Atlantic Sediments

In the 1980s and 1990s, scientists began to find millennial oscillations in North Atlantic sediments. Hints of these oscillations had been detected in earlier studies as intervals called "barren zones" because they were devoid of the coccoliths and planktic foraminifera typically found in this region during warmer intervals. The barren zones consisted mainly of ice-rafted debris.

Normally, ocean sediments are not a promising archive in which to find short-term climatic fluctuations. Most deep-sea sediments are deposited at rates no greater than 1 or 2 centimeters per 1000 years, and small burrowing animals stir and mix the sediments to depths of 5 to 10 centimeters (Chapter 3). As a result, mixing usually obliterates climate oscillations shorter than 2500 to 5000 years.

Fortunately, places exist in the North Atlantic where deposition rates can be as high as 10 or 20 centimeters per 1000 years. Fast-flowing bottom currents carry fine sediments away from exposed locations and deposit them in **sediment drifts** (similar to snowdrifts) in regions where the currents slow. Because coarser sand-sized sediments such as foraminifera and ice-rafted debris are not easily moved by currents, they stay in place as a reliable record of climate changes. Finer sediments brought in by the bottom currents quickly bury the coarser particles and reduce the impact of organisms in mixing and smoothing the climate signals they carry.

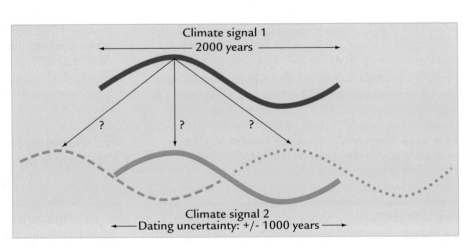

FIGURE 15-3 Uncertainties in dating millennial oscillations Typical dating uncertainties of 2000 years make it difficult to determine whether millennial oscillations in two climate records are synchronous, are exactly opposite in timing, or are leading or lagging each other.

In the 1980s, studies of rapidly deposited sediments in the North Atlantic gradually began to detect shorter climate oscillations. The marine geologist Hartmut Heinrich first discovered major episodes of ice rafting separated by as little as 5000 years but as much as 15,000 years or more. These are sometimes referred to as **Heinrich events**.

Soon afterward, marine geologists found even shorter-term variations in two kinds of climatic proxies: (1) the relative amounts of polar foraminifera shells versus grains of ice-rafted sand and (2) the percentage of polar species of foraminifera out of the total population. As in the case of orbital-scale changes, more ice-rafted debris and higher percentages of polar foraminifera indicated colder waters with more icebergs present.

Changes in the percentage of polar foraminifera in North Atlantic cores closely match the pattern of $\delta^{18}O$ changes in Greenland ice (Figure 15-4). The sense of the match is that times of cold air (negative $\delta^{18}O$) over Greenland correlate with times of cold ocean temperatures (more polar plankton) in the North Atlantic. Dating of the younger ice core cycles by annual layer counts and of the ocean cores by ^{14}C dating and then adjusting these ages to calendar years has confirmed the correlation for the part of the record younger than 30,000 years. But even in the younger part of the record, small leads or lags could still exist between the two signals within the dating uncertainties.

Both records show repeated slow drifting toward colder, more glacial conditions followed by abrupt shifts back to warmer conditions. It is difficult to count the exact number of oscillations in each record, because they vary in both length and amplitude. The ice-rafting events detected by Heinrich occurred at times when climate had been cooling for several millennia, and each ice-rafting episode was followed by a rapid return to warmer temperatures (Figure 15-4). Not all cooling sequences culminated in major ice-rafting episodes.

This new evidence from the North Atlantic shows that millennial oscillations during the last glaciation were not restricted just to changes in air temperature and circulation over the Greenland ice sheet. They also involved other important parts of the climate system, including the surface waters of the North Atlantic, where planktic foraminifera lived, and the Atlantic margins of the ice sheets, which supplied the icebergs that carried the coarse debris.

What were the sources of the debris deposited during the major ice-rafting events? Although most of the ice sheets surrounding the North Atlantic contributed debris to the largest ice-rafting events, a large fraction of the grains deposited in the major ice-rafting zone centered at 45° to 50°N latitude came from the northeastern margin of the North American Laurentide ice sheet (Figure 15-5). Abundant limestone fragments

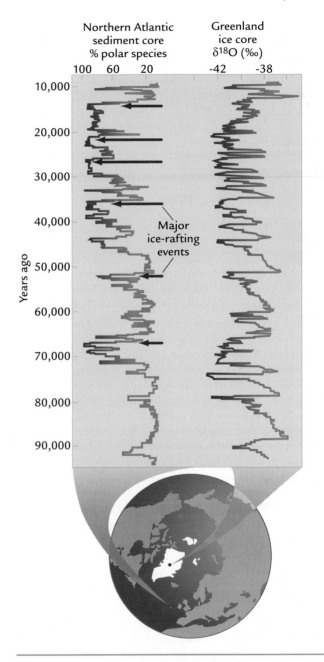

FIGURE 15-4 Millennial changes in the North Atlantic Ocean Millennial-scale fluctuations in composition of North Atlantic polar foraminifera and ice-rafting influxes (left) closely match $\delta^{18}O$ changes in Greenland ice cores (right). (Modified from S. Stanley, *Earth System History*, © 1999 by W. H. Freeman and Company, after G. Bond et al., "Correlations Between Climatic Records from North Atlantic Sediments and Greenland Ice," *Nature* 365 [1993]: 143–47.)

from source rocks in and north of Hudson Bay first suggested that North America was a major source of this debris. Evidence confirming this conclusion came from analysis of individual ice-rafted mineral grains: distinctive geochemical (isotopic) tracers point to bedrock sources north and east of Hudson Bay.

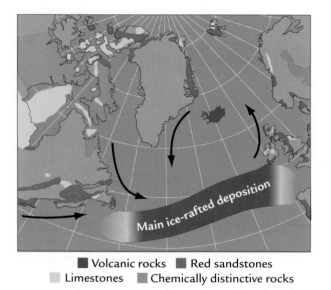

■ Volcanic rocks ■ Red sandstones
Limestones ■ Chemically distinctive rocks

FIGURE 15-5 **Sources and deposition of ice-rafted debris**
Highest rates of deposition of ice-rafted debris occur in the
North Atlantic Ocean between 45° and 50° N. During smaller
ice-rafting episodes, sources of debris include volcanic rocks
on Iceland and red sandstone rocks on several coastal
margins. During large ice-rafting events, massive amounts of
material are delivered from eastern North America, including
fragments of limestone from Hudson Bay and rock fragments
from other regions with distinctive chemical signatures.
(Adapted from G. Bond et al., "Evidence of Massive Discharges of
Icebergs into the North Atlantic During the Last Glacial Period,"
Nature 360 [1992]: 245–29, and from W. F. Ruddiman, "Late
Quaternary Deposition of Ice-Rafted Sand in the Subpolar North
Atlantic (Lat. 40° N to 65° N)," *Geological Society of America Bulletin*
88 [1977]: 1813–27.)

But were the *relative* increases in the amount of ice-
rafted debris in comparison with the abundance of
planktic shells caused by faster delivery of ice-rafted
debris, slower deposition of foraminifera, or both?
Radiocarbon dating of the $CaCO_3$ shells of foraminifera
(adjusted to calendar years) in the younger ice-rafting
events showed up to tenfold increases in the rate of
deposition of ice-rafted debris, along with smaller
decreases (generally by less than half) in the rate of
deposition of foraminifera. Greater iceberg delivery
must have been the main factor.

Much less ice-rafted debris was deposited during
smaller millennial-scale fluctuations. This debris came
from several source regions, at least two of which left
distinctive tracers as evidence. Fragments of clear and
dark volcanic glass originated mainly from eruptions on
Iceland (Figure 15-6A). Iron-stained quartz grains may
have come from several red sandstone source regions
(sandstones stained by oxidation of iron) around the
Atlantic margins (Figure 15-6B and C). Other regions
around the Atlantic margins also delivered debris during smaller climatic oscillations.

Detailed sampling of major ice-rafting events shows
that the first ice-rafted debris deposited came not from
the ice sheet on North America but from the smaller ice
sheets that contributed volcanic fragments and iron-
stained quartz grains to each small ice-rafting episode
(Figures 15-5, 15-6). Only later did the distinctive limestone debris from North America arrive.

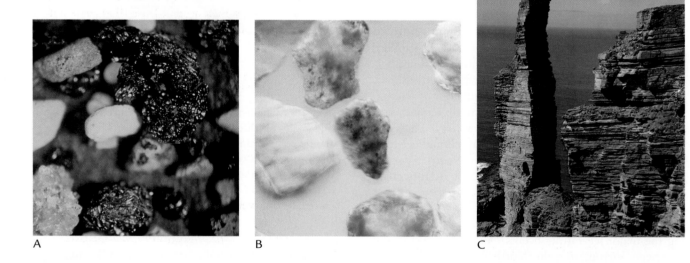

A   B   C

FIGURE 15-6 **Ice-rafted grains** Sand-sized grains ice-rafted into the North Atlantic include
(A) volcanic debris from Iceland and (B) red-stained quartz grains from red beds around
the Atlantic margins. (C) Sources of red-stained quartz grains include red sandstones
from the Orkney Islands, off northern Scotland. (A and B: courtesy of G. Bond, Lamont-
Doherty Earth Observatory of Columbia University; C: John Forbes/PEP.)

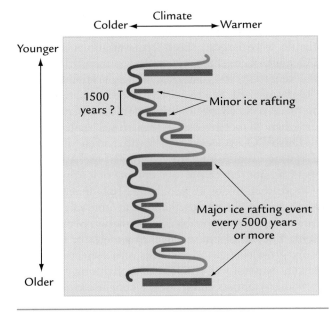

FIGURE 15-7 Millennial-scale North Atlantic cycles? One view of millennial-scale changes in the North Atlantic is that short cooling cycles 1500 years in length gradually drift toward colder conditions and occasionally culminate in major ice-rafting episodes, followed by an abrupt return to warmer conditions.

This evidence from the North Atlantic has been interpreted in widely varying ways. Some scientists infer that the record shows *true cycles* of 1500 years that gradually build toward cooler conditions and end in major ice-rafting episodes (Figure 15-7). In this view, the small ice sheets over Iceland and on other coastal margins sent icebergs into the Atlantic at intervals of 1500 years even when the climate was not extremely cold. Then, at some point in the gradual cooling, one of the 1500-year cycles suddenly reached a threshold that triggered a large influx of icebergs from the margins of the North American ice sheet. These major ice-rafting events occurred at intervals that were sometimes as short as 4500 years (three 1500-year cycles) but at other times occurred only after ten or more 1500-year cycles. At the end of each major ice-rafting episode, air and ocean temperatures quickly warmed, giving the signals from ice cores and ocean sediments an asymmetric appearance.

Other scientists interpret the same data differently. They do not believe that the records provided by ice cores and ocean sediments can be dated accurately enough to prove that the oscillations are truly cyclic, or even to determine their exact correlation in time.

Some even question whether the basic characteristics of the millennial oscillations across the North Atlantic have yet been adequately determined. They cite evidence from sediments in regions outside the main area of ice-rafted deposition near 45°–50°N. For example, the pattern (timing) of ice-rafting episodes north of Iceland at 65°N differs significantly from that shown in Figure 15-4, and major ice-rafting episodes in that region contain debris mainly from Scandinavia rather than North America. Recent studies of sediments from the western North Atlantic off the coast of Newfoundland at 45°N have found surprising evidence that water temperatures in that region were actually *warmer* during major ice-rafting events, rather than colder. Apparently we still have much to learn about these oscillations in the North Atlantic.

Long sequences of ocean sediments recovered by the Deep-Sea Drilling Project show that millennial fluctuations also occurred earlier. Earlier orbital-scale cycles of glaciation at the 100,000-year rhythm show similar oscillations, as do the still earlier 41,000-year glaciation cycles before 0.9 Myr ago. In each case, the millennial oscillations occurred during glacial climates but were weaker or absent during fully interglacial climates.

The North Atlantic sediment cores that have revealed fluctuations in ice rafting and plankton species also contain shells of bottom-dwelling (benthic) foraminifera that can be used to monitor past rates of deep-water formation. The method exploited here is the same one used to measure similar changes at orbital scales: more negative $\delta^{13}C$ values in the $CaCO_3$ shells of foraminifera mark times of slower deep-water formation in the North Atlantic (see Box 11-1). Because these deep-water $\delta^{13}C$ signals come from the same cores used to monitor conditions at the sea surface, the *relative* timing of all these changes is well determined, even without accurate dating.

Results from these studies suggest that deep-water formation became even weaker during major ice-rafting events than the already reduced glacial rates. Rates of deep-water formation slowed at least 1000 years before the maximum pulses of ice rafting began and then returned to high levels more than 1000 years after the ice-rafting influxes ended. Scientists have not been able to determine whether deep-water formation slowed during smaller millennial oscillations.

In summary, millennial oscillations occur in the Greenland–North Atlantic region, and they involve changes in air and surface-ocean temperature, in ice sheet margins and ice rafting, and in deep-water formation. This package of linked responses in the North Atlantic is similar to the linked orbital-scale changes we termed "ice-driven responses" in Chapter 12.

15-4 Where Else Did Millennial Oscillations Occur?

Because millennial oscillations are similar in several respects to changes found at orbital time scales, some scientists infer that they could have had a similar origin. This line of reasoning raises three questions: Could the millennial oscillations be the result of abrupt changes in the size of northern hemisphere ice sheets? Or are they the result of climate changes centered on the North Atlantic Ocean and then propagated elsewhere in the climate system? Alternatively, are the millennial oscillations broader (perhaps global) in extent and produced by larger-scale factors operating outside the North Atlantic region?

Changes in Ice Volume The first obvious issue to pursue is the total volume of ice involved in millennial oscillations. The deposition of ice-rafted debris tells us that ice sheets on the fringes of the North Atlantic sent huge numbers of icebergs to melt in the North Atlantic. But how much did these episodes reduce the total volume of ice present on the northern continents?

No method is available to determine exactly how much ice was involved in these episodes, but scientists can put an upper limit on the amount by examining marine $\delta^{18}O$ records from the deep tropical Pacific Ocean, far from the meltwater effects of the northern ice sheets. In this area, the $\delta^{18}O$ signals recorded in the shells of benthic foraminifera are controlled mainly by

global ice volume. During the last glaciation, when millennial fluctuations in Greenland and the North Atlantic were largest, short-term variations in Pacific $\delta^{18}O$ values were small, generally less than 0.1‰. This value represents less than 10% of the glacial–interglacial $\delta^{18}O$ change of 1.1‰ to 1.2‰ caused by changes in ice volume. A $\delta^{18}O$ change of 0.1‰ would translate into no more than 10 meters of change in global sea level.

This $\delta^{18}O$ evidence suggests that even the largest ice-rafting events along the margins of the Atlantic ice sheets did not reduce the total volume of ice on the land by more than about 10%. Changes in the sizes of ice sheets during shorter oscillations at intervals of 2000 years or less were probably much smaller, perhaps producing fluctuations in sea level of a meter or two. As a result, gross changes in the sizes of ice sheets seem an unlikely cause of most millennial oscillations.

Some scientists think such a conclusion is premature. In the midwestern United States, far from the Atlantic margins of the Laurentide ice sheet, several ice lobes appear to have fluctuated in rough synchrony with

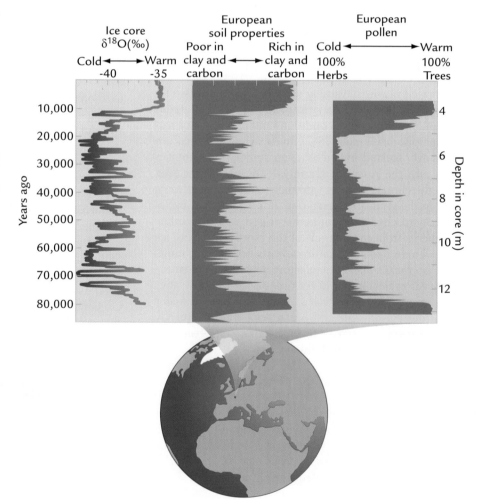

FIGURE 15-8 Millennial-scale climate changes in Europe Changes of $\delta^{18}O$ in ice cores from Greenland (left) show some resemblance to millennial-scale fluctuations in European soil type (middle) and to ^{14}C-dated changes in European pollen (right). (Left: adapted from P. Grootes et al., "Comparison of Oxygen Isotope Records from the GISP and GRIP Greenland Ice Cores," *Nature* 366 [1993]: 552–54; middle: adapted from N. Thouveny et al., "Climate Variations in Europe over the Past 140 Kyr Deduced from Rock Magnetism," *Nature* 371 [1994]: 503–6; right: adapted from G. M. Woillard and W. G. Mook, "Carbon-14 Dates at Grande Pile: Correlation of Land and Sea Chronologies," *Science* 215 [1982]: 159–61.)

the larger ice-rafting episodes, hinting at a possible link between the land and ocean margins. But how could these regions be linked? Did changes along the Atlantic margins of the ice sheet allow ice streams to pull enough ice out of the center of the ice sheet to somehow cause other ice margins over the land to adjust to a new configuration? This would seem hard to do if the overall changes in ice volume were only 1–10%. Or were both kinds of ice margins responding to larger-scale changes in climate?

Other Evidence from the Northern Hemisphere Because the size of northern hemisphere ice sheets varied only a little during most millennial oscillations, some scientists have turned to a second explanation that draws on knowledge gained from study of orbital changes. Could these millennial oscillations be produced by some kind of interaction between the North Atlantic and the ice sheet margins, with the changes in the North Atlantic then propagated through the atmosphere to other regions? At orbital time scales, changes in the temperature of the North Atlantic's surface can influence temperature and precipitation in Europe and Asia (see Figure 12-7).

Millennial-scale fluctuations have been detected in several climate records from Europe near the North Atlantic. The same kind of short-term oscillations that appear in $\delta^{18}O$ changes in Greenland ice (Figure 15-8, left) also appear in changes in the character of European soils (Figure 15-8, middle). These soils were richer in organic material during warmer episodes and almost free of organic carbon during colder oscillations.

In addition, the European pollen records that showed orbital-scale changes from interglacial forests to glacial tundra also show short-term fluctuations within glacial intervals from full tundra to mixed grass steppe and forest vegetation (Figure 15-8, right). The younger pollen fluctuations dated by ^{14}C appear to match the age of the ice core changes, with cold-adapted vegetation in Europe during times of colder air over Greenland, although the exact correlations are uncertain.

Short-term fluctuations have also been discovered in the glacial sections of windblown loess deposits in China. These oscillations register as changes between coarse loess-rich layers that indicate physical weathering during colder intervals and finer, clay-rich soils that indicate greater chemical weathering during warm episodes. Although the Asian soil/loess sequences are not well dated, changes in that region may also match oscillations in and around the North Atlantic.

The evidence from Europe and Asia provides some support for the idea that changes centered on the North Atlantic can explain other regional millennial oscillations. But other climate scientists have been searching for, and finding, millennial oscillations elsewhere in the northern hemisphere, some of them at locations that lie outside the region where the influence of the North Atlantic is quite so obvious (Figure 15-9).

A pollen record from Florida shows fluctuations during the last glaciation between pine pollen, indicative of wetter conditions, and grass and oak pollen, indicative of dry conditions (Figure 15-10, left). Several ^{14}C-dated fluctuations toward wetter conditions in Florida appear to correlate with major ice-rafting episodes in the North Atlantic (Figure 15-10, right), although the dating is not accurate enough to specify the relative timing.

What would link millennial oscillations in Florida to those in the northern ice sheets and high-latitude North Atlantic? One possibility is temperature changes in the Gulf of Mexico, today a major source of warm, humid air for nearby regions and for vegetation in nearby Florida. Short-term changes in rates of melting of the southern margin of the North American ice sheet would have sent varying amounts of cold meltwater down the Mississippi River, alternately cooling and

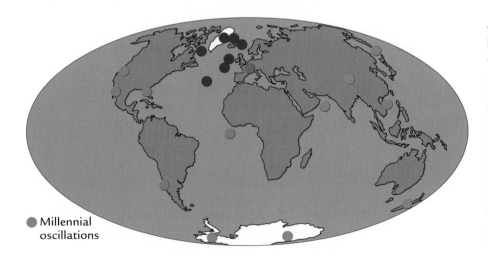

Millennial oscillations

FIGURE 15-9 Other millennial-scale oscillations In addition to the millennial oscillations in Greenland ice cores and North Atlantic sediments (dark red dots), scientists have detected similar oscillations elsewhere in the northern and southern hemispheres (lighter red dots). (Adapted from D. C. Leuschner and F. Sirocko, "The Low-Latitude Monsoon Climate During Dansgaard-Oeschger Cycles and Heinrich Events," *Quaternary Science Reviews* 19 [2000]: 243-54.)

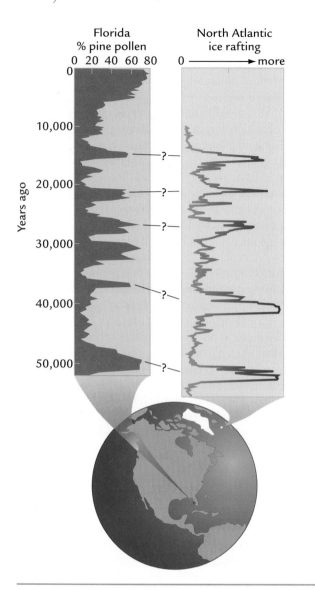

FIGURE 15-10 Millennial fluctuations in Florida Changes in the percentage of pine pollen in Florida (left) appear similar in timing to large-scale ice-rafting events in the North Atlantic (right). (Adapted from E. C. Grimm et al., "A 50,000-Year Record of Climate Oscillations from Florida and Its Temporal Correlation with the Heinrich Events," *Science* 261 [1993]: 198–200.)

warming the Gulf. The explanation for the short-term climate changes in Florida could still lie in the changes occurring at high latitudes.

But many scientists interpret this evidence differently. They infer that climate changes in Greenland, the North Atlantic, and farther south may be separate regional responses to a more pervasive cause of climate change acting at hemispheric and possibly even global scales.

Evidence that may support this interpretation comes from the Santa Barbara Basin, along the Pacific coast of North America at 35°N. Several kinds of cli-

mate signals recorded in this basin match fluctuations in the Greenland ice sheet's $\delta^{18}O$ signal in remarkably convincing detail. Oscillations in $\delta^{18}O$ values measured in the shells of planktic foraminifera indicate large (4°C or more) short-term temperature changes in near-surface waters (Figure 15-11, left). These $\delta^{18}O$ oscillations closely match the $\delta^{18}O$ pattern in the Greenland ice core records (Figure 15-11, middle).

In addition, the type of sediment deposited in the Santa Barbara Basin fluctuates between layers mixed by burrowing animals and layers with varvelike layering still intact (Figure 15-11, right). For such delicate layering to be preserved, oxygen must have been absent from the deep basin during warmer climates, eliminating small creatures that would otherwise have burrowed through the sediment and mixed it. In contrast, sediment mixing occurred during the colder oscillations, indicating that burrowing activity was vigorous. The obvious pattern match between the Santa Barbara Basin records and the Greenland ice core leaves little room for doubt that very similar kinds of millennial-scale oscillations have affected both regions.

Other indications of millennial-scale oscillations in western North America come from fluctuations of glacial Lake Bonneville in Utah (see Figure 13-17B): the younger ^{14}C-dated lake-level maxima appear to correlate with major ice-rafting events in the North Atlantic. Also, millennial-scale advances of mountain glaciers have been found in the Sierra Nevada of California, the Cascades of Oregon and Washington, and the Colorado Rockies.

Evidence of millennial fluctuations has also been found in the northern tropics, although in areas still within the possible influence of changes originating in the North Atlantic: fluctuations in the amount of dust delivered to the Arabian Sea, in the level of North African lakes, and in the abundance of plankton in the equatorial Atlantic.

The most persuasive demonstration that millennial-scale oscillations were worldwide would be evidence that they were present well into the southern hemisphere. Geologists have found evidence of mountain glacier oscillations in the South American Andes at 40°–45°S latitude and in the Alps of New Zealand, and some have interpreted these changes as proof that these oscillations are a global phenomenon. Others reject this conclusion as premature: most of the ^{14}C-dated ice advances that have been detected are not present at both locations.

Ice cores from Antarctica are another obvious place to look for millennial oscillations. Antarctic ice cores contain short-term $\delta^{18}O$ oscillations with some resemblance to the Greenland ice record, but the amplitude of the Antarctic $\delta^{18}O$ changes (1‰–2‰) is smaller (Figure 15-12). Several attempts to correlate ice core signals from Greenland and Antarctica suggest (but do

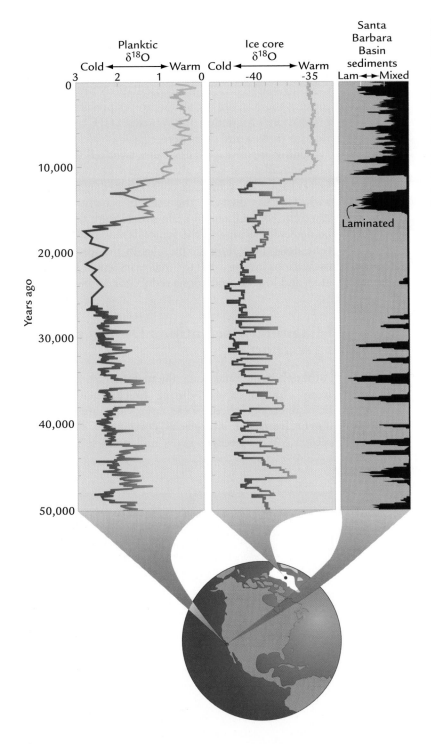

FIGURE 15-11 **Millennial fluctuations in the Santa Barbara Basin** The pattern of δ^{18}O changes in planktic foraminifera that lived in the surface waters of California's Santa Barbara Basin (left) closely matches δ^{18}O changes in Greenland ice (middle). Changes between varved sediments in the Santa Barbara Basin and intervals disturbed by burrowing animals show the same pattern (right). (Adapted from I. L. Hendy and J. P. Kennett, "Latest Quaternary North Pacific Surface-Water Responses Imply Atmosphere-Driven Climatic Instability," *Geology* 27 [1999]: 291–94.)

not yet conclusively prove) that temperature oscillations over Antarctica have at times been *opposite* in timing from those over Greenland. This interpretation is best justified for the more recent Younger Dryas fluctuation, because ice core dating of this event is more precise than for older fluctuations (see Chapter 14). This event, a prominent cold reversal in the North Atlantic, was a time of relative warmth at the South Pole.

What process in the climate system would make temperatures cold in the north at the same time that the south warmed? One hypothesis, proposed by the marine geologist Tom Crowley and independently confirmed by the climate modeler Tom Stocker, focuses on the northward conveyor-belt flow of warm water through the North Atlantic as the critical factor.

When deep-water formation in the North Atlantic is vigorous (Figure 15-13A), the northward flow of warm water removes heat from the Southern Ocean and cools it, but adds heat to the North Atlantic and warms it. Then, when deep-water formation slows or

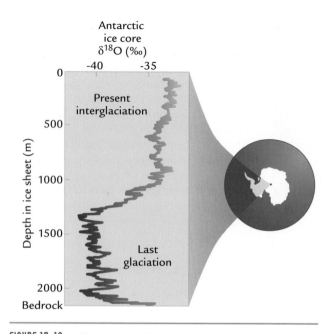

FIGURE 15-12 Millennial oscillations in Antarctic ice? Records from Antarctic ice cores show millennial-scale $\delta^{18}O$ fluctuations less than half as large as those in Greenland ice cores. (Adapted from C. Lorius et al., "A 30,000-Year Isotope Climatic Record from Antarctic Ice," *Nature* 280 [1979]: 644–48.)

stops and the conveyor belt weakens, the heat left in the southern hemisphere makes the ocean warmer, while the reduction of heat carried north makes the North Atlantic cooler (Figure 15-13B). Simulations with several oceanic GCMs suggest that such a hemispheric seesaw effect is likely.

With this surprising finding, we have come full circle back to the possibility that even the most distant millennial oscillations could have been driven from the North Atlantic region after all, as a response to changes in deep-water formation. But now we also have to keep in mind the possibility that the very nature of the climatic footprint of these oscillations can vary from region to region, even reversing sign in the south.

One final line of evidence not yet examined is greenhouse gases. Did these gases vary over millennial time scales, and could they have caused or contributed to these oscillations?

Greenland ice cores show definite millennial-scale changes in methane (Figure 15-14). CH_4 concentrations varied by 100 to 200 parts per billion, or as much as half the difference between fully glacial (350 ppb) and fully interglacial (700 ppb) values. Because air bubbles in Greenland ice were sealed off from the atmosphere some 70 meters below the surface, they are about 200 years younger than the ice in which they are trapped (Chapter 11). After adjusting for this offset, scientists concluded that the CH_4 fluctuations lag about three decades behind the $\delta^{18}O$ changes. Low

CH_4 levels occurred 30 years after the air grew colder (shown by more negative $\delta^{18}O$ values), and higher CH_4 levels were reached three decades after the air warmed. This relationship implies that methane values were driven by millennial oscillations, rather than causing them. Cold air temperatures presumably suppressed the production and release of CH_4 from boggy regions in Asia and elsewhere.

Unfortunately, ice cores cannot tell us whether or not millennial-scale changes in CO_2 occurred. In Greenland, CO_2 values are contaminated by interacting with $CaCO_3$ dust in the ice. In slowly deposited Antarctic ice, millennial-scale events were too brief to be detected because of the long (3000 years or greater) time delay in sealing air bubbles in the ice. Any CO_2 oscillations lasting 1000 to 3000 years would be obliterated, and longer oscillations reduced in amplitude.

Millennial Oscillations During the Last 8000 Years

Although millennial-scale fluctuations are evident in many climate records during times when northern hemisphere ice sheets were large, they are muted or absent from the interglacial portions of these same records. A few records show millennial-scale fluctua-

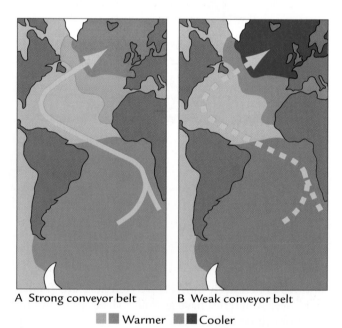

A Strong conveyor belt   B Weak conveyor belt
███ Warmer   ███ Cooler

FIGURE 15-13 Opposite hemispheric responses caused by the conveyor belt? (A) When the conveyor belt's flow in the Atlantic is strong, it warms the North Atlantic region intensely but cools a large area of the southern hemisphere by a small amount. (B) When it weakens, the temperature responses are reversed.

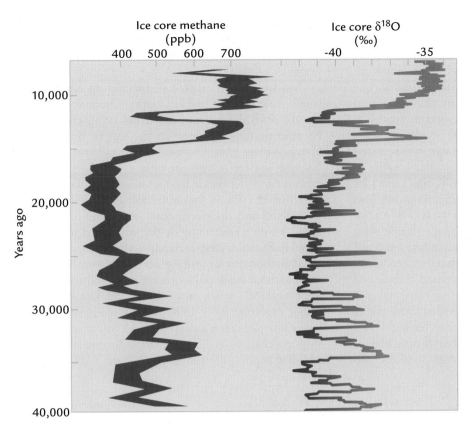

Ice core methane (ppb)

400 500 600 700

Ice core δ¹⁸O (‰)

-40 -35

Years ago

10,000

20,000

30,000

40,000

FIGURE 15-14 **Millennial-scale CH$_4$ changes** Methane concentrations in Greenland ice (left) fluctuate by 100–200 ppb at levels nearly equivalent to rapid changes in ice core δ¹⁸O (right). (Adapted from J. Chapellaz et al., "Ice Core Record of Atmospheric Methane over the Last 160,000 Years," *Nature* 345 [1990]: 127–31.)

tions during the last 8000 years of interglacial climate. Those fluctuations that have been detected vary in character from place to place, sometimes showing an apparent 1500-year cyclicity, but in other records either no cyclicity at all or a weak tendency toward a cycle of about 2500 years.

15-5 Greenland and North Atlantic Records

Millennial-scale δ¹⁸O oscillations are not obvious in Greenland ice cores during the last 8000 years (see Figure 15-4), but small fluctuations do occur in the amount of dust from the continents and in sea salt (Na$^+$ and Cl$^-$) from the ocean (Figure 15-15). These trends are interpreted as indicating changes in wind strength. Today salty spray from the sea surface is lifted by strong winter and spring winds and carried high onto the ice sheet, where it is deposited along with dust from continental sources. The peaks in the concentration of dust and sea salt have been interpreted as showing a 2600-year cycle, but any such cyclicity is weak at best.

Similarly, sediments from the North Atlantic indicate no major episodes of ice rafting or plankton reduction during the last 8000 years, at least not of the size of those during the glacial interval. This finding is not a total surprise: the only large northern hemisphere ice sheet left in existence during the last 6000 years is the one on Greenland.

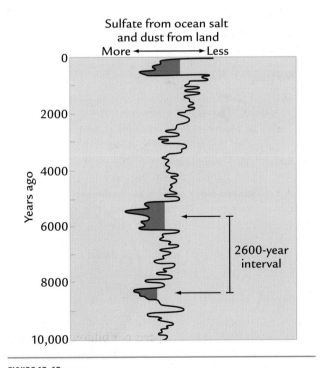

Sulfate from ocean salt and dust from land

More ⟵⟶ Less

0

Years ago

2000

4000

6000

8000

10,000

2600-year interval

FIGURE 15-15 **Changes in sea salt** Over the last 10,000 years, ice cores from Greenland show nearly identical short-term fluctuations in the amount of sea salt (Na$^+$ and Cl$^-$ ions) and in dust from the continents, with a faint suggestion of a cycle near 2600 years. (Adapted from P. A. O'Brien et al., "Complexity of Holocene Climate as Reconstructed from a Greenland Ice Core," *Science* 270 [1995]: 1962–64.)

Close examination of North Atlantic sediments has recently detected intervals with slight increases in concentrations of sand-sized mineral grains carried in by ice rafting (Figure 15-16). Even though these grains are 1000 to 100,000 times less abundant than the concentrations reached during glacial intervals, they show that small pulses of ice rafting did exist, apparently spaced at an interval of about 1500 years (± 500 years). For sediments deposited in the last 8000 years, errors associated with ^{14}C dating are small in relation to the lengths of the oscillations. These younger ice-rafting episodes contain two distinctive types of mineral grains also found in the larger ice-rafting pulses typical of glacial intervals (see Figure 15-6): fragments of volcanic glass from Iceland and grains of iron-stained quartz from red sandstone rocks around the Atlantic margins, including the east coast of Greenland and the coastal margins of the Arctic.

The kind of ice transport that delivered these grains is still unclear. Although icebergs have been far less abundant during the last 8000 years than during glaciations, a small number of the icebergs shed by the Greenland ice sheet can float as far south as 40°N (as the people aboard the *Titanic* found out). Icebergs from Greenland could have carried the red-stained quartz grains.

Sea ice is also capable of picking up and carrying small amounts of debris, either because sediment freezes onto the bottom ice layers along coastlines or because material is dumped on top of the ice from above, such as glass fragments from volcanic eruptions. Because sea ice is at most only a few meters thick, it cannot carry debris far into a warm ocean before melting. But sea ice is common along the east coast of Greenland and the north coast of Iceland today, and it could have carried many of the sand-sized grains measured in the core near Iceland.

Whether this ice rafting occurred in icebergs or on sea ice, these small pulses reveal increased discharges of ice every 1500 years or so during the last 8000 years. Other evidence for a 1500-year oscillation during this interval comes from a deep-sea core just west of Iceland showing changes in sediment size indicative of a decrease in the strength of bottom currents every 1500 years.

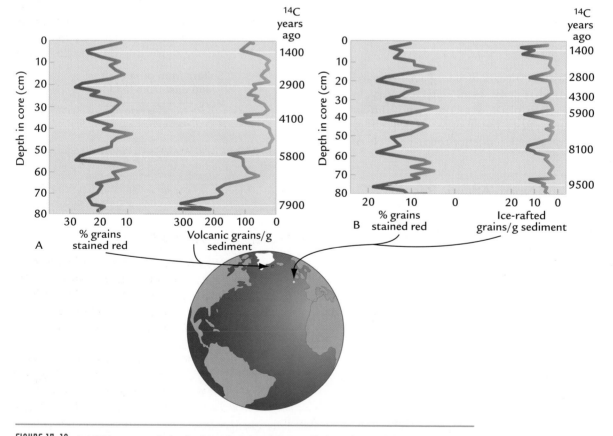

FIGURE 15-16 A 1500-year cycle in the North Atlantic? Detailed analysis of the last 10,000 years from two widely separated sediment cores shows small maxima in ice-rafted debris at intervals near 1500 years. Each ice-rafting peak contains volcanic glass fragments from Iceland, as well as iron-stained quartz and other grains from regions farther north. (Adapted from G. Bond et al., "A Pervasive Millennial-Scale Cycle in North Atlantic Holocene and Glacial Climates," *Science* 278 [1997]: 1257–66.)

15-6 Other Millennial-Scale Changes During Interglaciations

Millennial-scale advances and retreats of mountain glaciers have occurred over the last 8000 years, superimposed on a slow, erratic drift toward colder conditions (Figure 15-17). One compilation of glacial advances can be interpreted as showing a 1500-year cyclicity, although the spacing of the younger advances looks much less regular (Figure 15-17A). In contrast, an earlier study of this time interval found a 2500-year interval between advances (Figure 15-17B).

The differences between these two reconstructions in part reflect the difficulty in obtaining reliable ^{14}C (or other) dates that closely constrain both the upper and lower age limits of each ice advance in most mountain regions. At this point, researchers can say only that mountain glaciers do oscillate over millennial time scales, but not whether the fluctuations are cyclic (either at 1500 years, 2500 years, or some other cycle length).

Some lower-latitude regions also show what look like millennial-scale fluctuations during the last 8000 years. North African lake levels have fluctuated markedly during that time (see Figure 14-17C), although these changes are difficult to date accurately.

Climate fluctuations found in high-resolution records of the last 8000 years seem to disagree as much as they agree. This could be a result of the difficulty in detecting small climatic changes during this interval, or it could indicate that many of the components of Earth's climate system acted independently in the absence of large fluctuations in ice sheets to keep them closely coupled.

Causes of Millennial-Scale Oscillations

Investigations such as those summarized above have given climate scientists some idea (no doubt still incomplete) of the types and distributions of millennial oscillations, and of the possible linkages among parts of the climate system. But what about the ultimate cause of these fluctuations? Why would the climate system oscillate in such a way?

The hypothesis that successfully explains these oscillations must address several key questions. What initiates these oscillations? How are they transmitted to those parts of the climate system where they have been observed? Why are they stronger during glaciations than during interglaciations?

Three hypotheses view millennial-scale changes as:

1. The natural oscillations inherent in the internal behavior of northern hemisphere ice sheets.

2. The result of internal interactions among several parts of the climate system.

3. A response to solar variations external to the climate system.

15-7 Processes Within Ice Sheets

One possible explanation is that northern hemisphere ice sheets are the source of millennial climate oscillations as a result of their own natural internal variations. The margins of those ice sheets that send pulses of icebergs into the North Atlantic obviously merit close attention.

Considered as a whole, ice sheets have very slow response times of many thousands of years (Chapter 10), but only in the overall behavior of the bulk of the ice. The margins of ice sheets are capable of faster changes, both on land in areas where soft sediments lie below (Chapter 13) and along the oceans. In these regions ice sheets are thinner and capable of much more rapid interactions with their surroundings.

Along the marine margins of ice sheets, ice flows over bedrock with irregular bumps and depressions (Figure 15-18A). The bottom layers of ice scrape

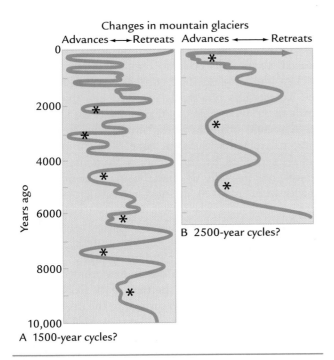

FIGURE 15-17 Millennial-scale oscillations of mountain glaciers Two attempts to synthesize advances and retreats of mountain glaciers over the last several millennia have produced different interpretations. The one on the left hints that advances may have occurred at intervals of 1500 years, but the one on the right indicates advances separated by about 2500 years. (A: adapted from F. Rothlisberger, *10,000 Jahre Gletschergeschichte der Erde* [Frankfurt: Sauerländer, 1997]; B: adapted from G. H. Denton and S. C. Porter, "Neoglaciation," *Scientific American* 222 [1970]: 100–110.)

against bedrock bumps called **bedrock pinning points**, and the friction slows the flow of ice. The bottom layers of ice can freeze to the bedrock and slow the flow even more. Ocean water can produce the opposite effect: ice floats in water, and the ocean can lift the ice off these pinning points.

One idea tied to the margins of marine ice sheets is that the natural ongoing release of small amounts of heat from Earth's interior can melt the lower ice layers (Figure 15-18B). The resulting meltwater trickles into the soft underlying sediments and makes them unstable, causing the ice margins to surge forward into the ocean. These surges release icebergs, which float away out into the ocean, leaving thinner ice sheet margins well inland from their initial positions. These new margins then slowly thicken and advance until the buildup of heat again destabilizes them.

A second idea focuses on a different kind of interaction between ice margins and bedrock (Figure 15-18C).

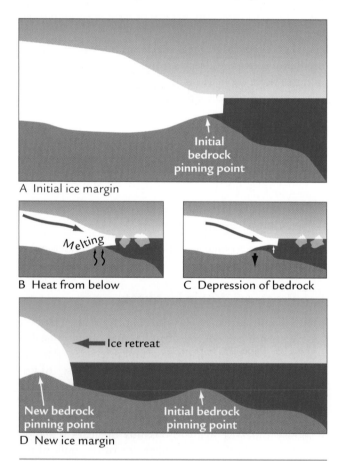

A Initial ice margin

B Heat from below   C Depression of bedrock

D New ice margin

FIGURE 15-18 **Natural oscillations of ice margins** (A) Marine margins of ice sheets end in thin ice shelves flowing across upward-protruding bedrock knobs. This ice can be dislodged from these pinning points either by Earth's heat escaping from below and melting ice (B) or by the gradual weighing down of bedrock under the heavy load of growing ice (C). Either way, the ice margin retreats inland and stabilizes over another bedrock pinning point (D).

Over time, as the ice margins thicken, their weight depresses the underlying bedrock, which slowly sinks over several thousand years. At some point, depression of the bedrock causes the ice to sink far enough into the coastal ocean so that it can be lifted and floated by sea-water. Because the ice margin is no longer anchored to the bedrock, it flows faster, and the ice streams release icebergs into the ocean. Once this outward flow of ice is exhausted, the ice stream retreats to another bedrock pinning point farther upstream (Figure 15-18D).

Both hypotheses are consistent with ice-rafting evidence from the North Atlantic. Millennial-scale ice-rafting pulses occurred mainly during times when large ice sheets existed. In addition, the largest ice-rafting episodes occurred when the air and ocean were already cold, rather than during warmer intervals that might otherwise be expected to cause ice margins to collapse.

But these ice-related hypotheses have limitations. The measured intervals of 5000 to 10,000 years or more between major ice-rafting events appear to be long enough to allow marine ice margins to recover from previous collapses and grow large again. But it is harder to argue that ice margins can grow significantly during the shorter oscillations of 1000 to 2000 years.

And if further work were to confirm that a 1500-year cycle exists, other questions would arise. Why would coastal ice margins cycle so regularly? Why, with the background climatic state constantly changing, would the ice margins repeatedly grow to just the size that made them susceptible to collapse in exactly 1500 years?

A second problem is explaining why so many ice margins reacted at the same time. The composition of debris deposited by the icebergs from many distinct source regions tells us that the margins of most ice sheets were involved in most ice-rafting events, but why?

One possible link is sea level (Figure 15-18C). If one ice margin surges and sends icebergs into the ocean, the icebergs will displace water and melt, raising sea level. A rise in sea level may then destabilize other coastal ice margins by floating ice shelves off bedrock pinning points and causing them to surge into the ocean as well. The problem with this idea is that the actual rises in sea level may have been too small to produce this result, particularly the smaller coastal ice sheets involved in the shorter oscillations. Would a rise in sea level of a single meter have caused other ice margins to collapse?

The large North American ice sheet is a better candidate for causing sea level rises large enough to trigger reactions in the other ice sheets, particularly during major ice-rafting events. We have seen that these larger events could have raised global sea level as much as 10 meters. But the evidence shows that the first ice-rafted debris deposited in the North Atlantic came from the smaller ice sheets on Iceland and in coastal regions farther north, and only later did the flood of debris arrive on icebergs from North America. This sequence rules out the North American ice sheet as the initial sea level trigger during large ice-rafting events.

Another problem with the hypothesis that ice sheets control millennial oscillations is how oscillations of northern ice sheets could have been transferred to so many other regions. Climate changes in the American West over orbital time scales have been interpreted as a response to splitting of the jet stream caused by changes in the elevation of the North American ice sheet (see Figure 13-17B). Some climate scientists have suggested that this same explanation might work for the shorter millennial-scale changes, but it is difficult to see how changes of just 1% to 10% in ice volume could have altered the height of the North American ice sheet enough to affect the position of the jet stream. Atmospheric circulation would have to have been *extremely* sensitive to small changes in ice elevation to have had significant large-scale effects on climate.

15-8 Interactions Within the Climate System

A second explanation of millennial-scale oscillations focuses on interactions among several components of Earth's climate system, rather than just processes within ice sheets. Oscillations lasting for several thousand years would have to involve interactions among parts of the climate system capable of responding relatively rapidly. At one extreme, this requirement rules out the great masses of ice in the major continental ice sheets, which respond slowly over just a few millennia. At the other extreme, the fastest-responding parts of the climate system (the atmosphere, for example) cannot be the key to the millennial oscillations either, because they cannot on their own sustain climate changes over several centuries.

For millennial-scale oscillations to occur through internal oscillations, several components of the climate system (two or more) must continuously interact with each other as almost-equal partners (Figure 15-19A). Each component must have a response time fast enough to react decisively at millennial scales but slow enough not to become a climate driver that overwhelms and controls the others. Potential partners in this kind of behavior are the deep ocean, the marine margins of the ice sheets, and possibly the carbon (CO_2) system through its links to deep-water flow (Chapter 11). As these components interact, each must do so with a natural lag or delay that keeps its response distinct from that of the others (Figure 15-19B).

This kind of behavior seems most likely to occur in connection with deep-water formation in the North Atlantic. Deep-water formation in this region is heavily dependent on the salt content of the sea surface, and the salinity of the North Atlantic is tied in part to the rate at which freshwater is delivered from melting ice sheets. In this way, deep-water processes are linked to the ice

sheet margins. In addition, deep-water formation in the North Atlantic is tied to heat transport at both local and larger scales. Locally, enormous amounts of heat are handed off to the atmosphere when deep water forms, but not when deep water ceases to form. Also, the conveyor belt circulation may link changes in the North Atlantic to those in other regions (see Figure 15-13).

Initially it seemed likely that the large meltwater influxes during major ice-rafting events would have added enough freshwater to lower the density of the surface ocean enough to slow or stop formation of deep water. But careful investigations have revealed that deep-water formation slowed or stopped at least 1000 years *before* the largest iceberg influxes arrived. With this timing, ice-rafting episodes cannot be the cause of changes in deep-water formation.

Wally Broecker recently hypothesized that a natural **salt oscillator** operates in the North Atlantic when large ice sheets are present. When salty surface water sinks into the deep ocean to form deep water, it removes salt from the sea surface (Figure 15-20A). But at the same time, ocean heat handed off to the atmosphere while deep water is forming helps to melt nearby coastal margins of ice sheets, and this melting gradually increases the delivery of fresh meltwater and reduces the salinity of the sea surface.

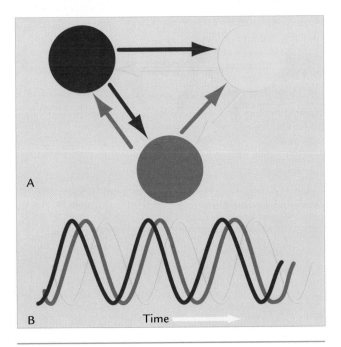

FIGURE 15-19 **Internal climate system oscillations** One explanation offered for millennial oscillations is internal interactions in the climate system. Such interactions would require several components of the climate system to function as nearly equal partners (A), with none dominating the others. Because all components have moderately slow response times, each responds in a pattern distinct from the others (B).

Eventually the sea surface loses so much salt and its density becomes so low that little or no deep water forms (Figure 15-20B). The conveyor-belt flow slows or ceases. With less deep water formed and less salt removed, the salinity of the surface water begins to increase again. At the same time, little or no heat is handed off to the atmosphere, and nearby ice sheets begin to grow again. With less glacial meltwater sent to the ocean, the sea surface becomes saltier. Eventually the rising salinity and density permit more deep water to be formed, completing one full "salt oscillation."

Oscillations in deep-water formation might also alter atmospheric CO_2 values by changing the basic cir-

culation of the deep ocean through mechanisms like those invoked for orbital time scales (Chapter 11). Unfortunately, short-term CO_2 oscillations cannot be detected in ice cores to test this idea.

How would salt-oscillator changes in the North Atlantic propagate elsewhere in the climate system? One mechanism is through the atmosphere. Large temperature changes over the North Atlantic could alter not just temperature and precipitation patterns in Europe and Asia but also the character of several prominent features in the lower-level circulation, such as the low-pressure region now located near Iceland and the high-pressure cell in Siberia.

Significant changes in these basic centers of lower-atmospheric circulation have the potential to disrupt the upper-atmospheric jet stream circulation in the northern hemisphere and propagate into more distant regions. For example, the Santa Barbara Basin lies close to the present-day boundary between ocean water driven from north to south by winds in a subtropical high-pressure cell and water driven from south to north by winds in a low-pressure cell located near Alaska. Disruptions of the jet-stream pattern originating in the North Atlantic have the potential to alter this delicate modern-day balance.

A second mechanism is changes in the surface-ocean conveyor flow. If changes in rates of deep-water formation in the North Atlantic affect the southern hemisphere surface ocean by an opposite-phased seesaw mechanism (Figure 15-13), they may have a nearly global impact. But the only firm evidence to date of changes in rates of deep-water formation comes from the major ice-rafting events spaced at intervals of 5000 years or more. It is still unclear whether or not millennial oscillations lasting less than 4000 years are linked in any way to deep-water changes.

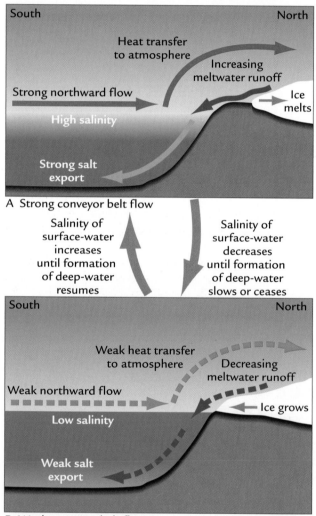

15-9 Causes External to the Climate System: Solar Variability

A third possibility is that the millennial-scale oscillations are driven from outside Earth's climate system. One explanation initially considered was volcanic eruptions. Even though individual eruptions cool climate for only a few years, some climate scientists suggested that many eruptions clustered close together could affect climate over intervals of centuries or even millennia.

Another possibility is changes in the strength of the Sun. As we will discover in Chapter 16, small changes in the Sun's strength have definitely occurred during the last two decades. But what about changes over millennia? Here we will reexamine tree ring evidence that can reveal changes in the solar output over the last several millennia.

As we saw in Chapter 14, scientists have measured the difference between ages derived by counting tree

FIGURE 15-20 **A salt oscillator?** One possible source of millennial climate fluctuations is changes in northward flow of warm, salty surface water along the conveyor belt in the North Atlantic. (A) Strong conveyor flow releases heat that melts ice and lowers the salinity of the North Atlantic, eventually slowing or stopping the formation of deep water. Weak flow (B) then causes salinity to rise, completing the cycle.

rings and those derived by ^{14}C dating of those same rings. The gradually increasing discrepancy between the ages derived from these two methods for the interval before 8000 years ago provided an indication that Earth's magnetic field was weaker in the past, permitting more bombardment by charged cosmic particles (protons) and faster production of ^{14}C atoms.

Within the last 8000 years, however, a second kind of age offset is apparent in these comparisons. Again, differences exist between the ^{14}C ages and the tree ring counts, but these differences persist only for intervals of a few hundred or thousand years (Figure 15-21).

These shorter-term age discrepancies are again interpreted to reflect changes in the rate of production of ^{14}C atoms in Earth's atmosphere, but in this case the main cause of these changes is thought to lie in the strength of the Sun, not in changes of Earth's magnetic shield. The basis of this idea is that particles streaming from the Sun (the "solar wind") today deflect some of the incoming cosmic rays (protons) that would otherwise enter Earth's atmosphere (Figure 15-22). If the amount by which this cosmic bombardment was deflected had changed over intervals of hundreds to thousands of years, it would have altered the rate of ^{14}C production in the atmosphere and might explain the observed short-term differences in the ages derived by the two dating methods.

Evidence confirming the idea that these differences in ^{14}C versus tree ring ages have an origin external to the climate system comes from an isotope of the element beryllium (^{10}Be), which is produced only by collisions with cosmic particles. Over the last 8000 years,

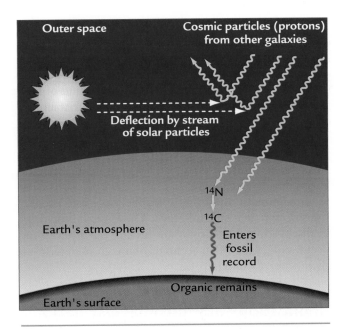

FIGURE 15-22 Changes in strength of solar emissions? Observed changes in rates of ^{14}C production (Figure 15-21) result from changes in rates of cosmic-ray bombardment thought to be caused by changes in solar shielding of Earth's atmosphere.

the history of changes in ^{10}Be recorded in ice cores is very similar to the history of differences between ^{14}C and tree ring ages, an indication that both result from changing deflection of incoming cosmic rays. The best explanation for these changes in deflection is that the Sun's output of energy was not constant (Figure 15-22).

Working on this hypothesis, scientists tested the match between millennial oscillations in climate and changes in rates of ^{14}C production thought to result from changes in the strength of the Sun. Their expectation was that lower rates of production of ^{14}C atoms (Figure 15-21) as a result of greater deflection of cosmic particles by a stronger Sun should correlate with warmer climates.

In fact, little evidence of a relationship between ^{14}C production and climate appears over millennial intervals. The major cycle of ^{14}C production is at 420 years, with much weaker cycles at 2100 (2000–2500) years and (almost negligibly) at 900 years. Even at best, the link to climate is faint. The 2100-year cycle is similar to the timing of changes from one reconstruction of mountain glacier advances over the last 6000 years (Figure 15-17B) and possibly to changes in sea salt and dust in Greenland ice cores over the last 10,000 years (Figure 15-15). The sense of these correlations is that a weaker Sun correlates with a colder climate (mountain glacier advances) and a windier atmosphere (more sea salt in the ice cores). But no significant ^{14}C production cycle is found near the 1500-year period indicated by some

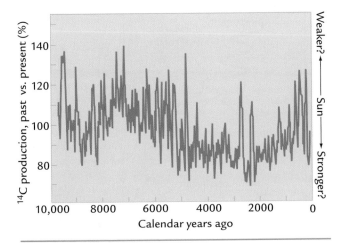

FIGURE 15-21 Millennial-scale changes in the Sun's strength? Ages determined by counting individual tree rings can be compared against ages determined from ^{14}C analyses. The age differences reflect changing ^{14}C production in the atmosphere. (Adapted from M. Stuiver et al., "Climatic, Solar, Oceanic, and Geomagnetic Influences on Late-Glacial and Holocene Atmospheric ^{14}C/^{12}C Change," *Quaternary Research* 35 [1991]: 1–24.)

climate records of the last 8000 years (Figures 15-16, 15-17A).

Another way a 2100-year solar cycle could drive the oscillations apparent in some climate records at periods near 2500 or 1500 years would be by some kind of natural resonant behavior in the climate system that amplified and focused forcing that followed a different rhythm (see Chapter 12). Such amplification would also have to have been enhanced by the presence of ice sheets to account for the much larger millennial fluctuations during glacial intervals.

In summary, the origin of millennial-scale oscillations in climate is still unknown. The biggest difficulties in discovering their origin are finding out where they occur and dating them accurately enough to determine their relative timing.

Implications of Millennial Oscillations for Future Climate

Finding out the origin of the millennial oscillations is important if we are to understand Earth's climate, both past and future. Basic curiosity aside, we need to know how the oscillations are initiated and then transmitted so quickly and so widely through the climate system in order to predict how, when, and where they may occur in the future.

The shorter oscillations could be particularly relevant during the next 500 to 1000 years, an interval when humans will increasingly influence climate. Because millennial oscillations occur faster than orbital-scale changes, they are more likely to have an immediate impact on our future. Will millennial-scale oscillations warm or cool climate during this interval? This relevance motivates climate scientists to examine evidence from the end of the last interglaciation, the last time Earth had as little ice as it does today. What kinds of millennial-scale changes occurred then?

Unfortunately, the answer is not to be found in the Greenland ice cores. The bottom sections of the ice cores contain reliable climate signals only to within 200 meters of the underlying bedrock, about 110,000 years ago. The lowermost 200 meters of ice are folded and squeezed, an indication that interactions with underlying bedrock have compromised these records.

Fortunately, we do have records of the end of the last interglaciation in cores from the North Atlantic Ocean and from nearby Europe. These records tell us there were no ice sheets on North America or Europe during the 10,000 years or more of peak interglacial warmth. They also show that at least one small but significant millennial-scale cooling event occurred in Europe during this interval. Deep-water flow may have been reduced during this brief cooling, and if so, the

loss of ocean heat normally handed off to the atmosphere could have been important.

As the interglaciation came to an end near 115,000 years ago, the ocean began to cool north of Iceland but remained warm south of Iceland, as did the climate in central Europe. At first glance this seems odd: ice had probably begun to grow in northern North America and possibly in northernmost Europe, yet the ocean stayed warm south of Iceland. One interpretation is that deep-water formation remained strong during this interval, and the accompanying hand-off of heat to the atmosphere kept central Europe warm. Only when the ice sheets had grown larger did they begin to cool the sea surface south of Iceland and send large numbers of icebergs into the Atlantic.

In general, the end of the interglaciation 115,000 years ago was defined by slow orbital-scale growth of ice sheets in the northern hemisphere over an interval of 11,000 years or more, but with shorter pulses of cooling punctuating the gradual orbital-scale trend and gaining in intensity as the ice sheets grew (Figure 15-23). The implication for our future, leaving aside for now the next few centuries of greenhouse-gas alteration of climate, is that a slow natural orbital-scale cooling of the northern hemisphere will be interrupted by millennial-scale cool oscillations. Unfortunately, the irregular nature of these oscillations during the last 8000 years makes it hard to predict exactly when such oscillations will occur in the future.

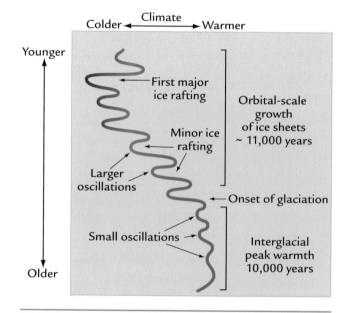

FIGURE 15-23 How did the last interglaciation end? The last interglaciation was brought to an end by two kinds of climate changes. Ice volume slowly grew over 11,000 years or more, but a series of increasingly intense millennial-scale cooling and ice-rafting episodes also punctuated this interval of gradual change.

Key Terms

millennial oscillations
(p. 330)

Dansgaard-Oeschger
cycles (p. 333)

sediment drifts (p. 334)

Heinrich events (p. 335)

bedrock pinning points
(p. 346)

salt oscillator (p. 347)

Review Questions

1. What evidence of millennial oscillations do we find in Greenland ice cores?

2. How do the processes that control $\delta^{18}O$ changes measured in ice sheets differ from those measured in ocean cores?

3. Why is it difficult to correlate millennial oscillations in records from different regions?

4. Describe two kinds of evidence of millennial oscillations from North Atlantic sediments.

5. What other regions show millennial oscillations like those in the North Atlantic and Greenland?

6. What could explain the small size of millennial oscillations during the last 8000 years?

7. What is the evidence for and against millennial oscillations originating from processes internal to ice sheets?

8. What is the evidence that millennial oscillations originate from interactions between deep water and ice sheets?

9. How could ocean flow cause millennial oscillations to have opposite timing north and south of the equator?

10. Do millennial oscillations form true cycles?

11. In your opinion, could changes in the Sun account for millennial oscillations?

Additional Resources

Basic Reading

Bond, G., W. S. Broecker, S. J. Johnsen, J. McManus, L. D. Labeyrie, J. Jouzel, and G. Bonani. 1993. "Correlations Between Climatic Records from North Atlantic Sediments and Greenland Ice." *Nature* 365:143–47.

Crowley, T. J. 1992. "North Atlantic Deep Water Cools the Southern Hemisphere." *Paleoceanography* 7:489–97.

Denton, G. H., and W. Karlen. 1973. "Holocene Climatic Variations: Their Possible Causes." *Quaternary Research* 3:155–205.

Grimm, E. C., G. L. Jacobson, W. A. Watts, B. C. S. Hansen, and K. A. Maasch. 1993. "A 50,000-Year Record of Climate Oscillations from Florida and Its Temporal Correlation with the Heinrich Events." *Science* 261:198–200.

Grootes, P. M., M. Stuiver, J. W. C. White, S. Johnsen, and J. Jouzel. 1993. "Comparison of Oxygen Isotope Records from the GISP2 and GRIP Greenland Ice Cores." *Nature* 366:552–54.

Heinrich, H. 1988. "Origin and Consequences of Cyclic Ice Rafting in the Northeast Atlantic Ocean During the Past 130,000 Years." *Quaternary Research* 29:142–52.

Advanced Reading

Boyle, E. A. 2000. "Is Ocean Thermohaline Circulation Linked to Abrupt Stadial-Interstadial Transitions?" *Quaternary Science Reviews* 19:255–72.

Broecker, W. S., G. Bond, M. Klas, G. Bonani, and W. Wolfli. 1990. "A Salt Oscillator in the Glacial Atlantic? The Concept." *Paleoceanography* 5:469–78.

Clark , P., R. Webb, and L. D. Keigwin. 1999. *Mechanisms of Global Climate Change at Millennial Time Scales.* Geophysical Monograph Series 112. Washington, D.C.: American Geophysical Union.

Hendy, I. L., and J. P. Kennett. 1999. "Latest Quaternary North Pacific Surface-Water Responses Imply Atmospheric-Driven Climatic Instability." *Geology* 27:291–94.

MacAyeal, D. R. 1993. "Binge/Purge Oscillations of the Laurentide Ice Sheet as a Cause of North Atlantic's Heinrich Events." *Paleoceanography* 8:775–84.

McIntyre, A., and Molfino, B. 1996. "Forcing of Atlantic Equatorial and Subpolar Millennial Cycles by Precession." *Science* 274:1867–70.

O'Brien, S. R., P. A. Mayewski, L. D. Meeker, D. A. Meese, M. S. Twickler, and S. I. Whitlow. 1995. "Complexity of Holocene Climate as Reconstructed from a Greenland Ice Core." *Science* 270:1962–64.

Porter, S. C., and Z. S. An. 1995. "Correlation Between Climatic Events in the North Atlantic and China During the Last Glaciation." *Nature* 375:305–8.

Stuiver, M., and T. F. Brazunias. 1993. "Sun, Ocean, Climate, and Atmospheric $^{14}CO_2$: An Evaluation of Causal and Spectral Relationships." *Holocene* 3:289–305.

A major component of Earth's climate system, Arctic sea ice had thinned and retreated significantly by the twenty-first century, perhaps as an early response to a global climate warning that may intensify in the future. (Johnny Johnson/Index Stock Imagery/Picture Quest.)

Climate changes over historical time can be reconstructed from a wide array of archives and proxy indicators. The best archives for intervals of a few thousand years are annually layered ice cores and lake sediments. For recent centuries, annually layered tree rings and corals provide additional coverage, and historical observations are useful for some regions. For the last 100 years, measurements made by instruments are the major source of climatic histories.

PART V

Historical and Future Climate Changes

To this point in the book, records of climate change have been portrayed in the same way as they would be encountered in cores taken from archives of sediment or ice: as vertical columns with the younger ages at the top and the older ages at the bottom. In most of this part, however, time is instead portrayed horizontally, with older ages on the left and younger ages on the right—the way historical events are usually arrayed. Future changes in climate beyond the early twenty-first century are projected out toward the right along the horizontal axis.

Climate changes over the last 1000 years have been smaller than those over tectonic, orbital, and glacial-age millennial time scales, never exceeding 1°C on a global basis. During this relative respite from the larger fluctuations, which are more typical of Earth's history, highly advanced modern civilizations have emerged.

Over the last several hundred years, human activities have begun to alter climate, first at local, then regional, and now at hemispheric and global scales. Human influences on climate are now increasing rapidly and seem destined to continue to do so for decades, or even centuries, to come.

To evaluate the extent of both natural and human causes of climate change, both in the last century and in the future, we explore these issues:

- **Were climate changes during the last 1000 years similar in pattern across the globe, or did they vary from region to region?**
- **Can we see an imprint from millennial-scale and orbital-scale changes during this interval?**
- **Was the warming in the twentieth century caused by human activities or natural climate changes?**
- **What is the sensitivity of Earth's climate system to the by-products of the Industrial Revolution, including greenhouse gases (CO_2 and CH_4) and sulfur dioxide (SO_2)?**
- **What kinds of climate changes lie in Earth's future?**

CHAPTER 16

Historical Changes in Climate

limate changes over the last several thousand years have been small (less than 1°C) and highly variable in pattern from region to region. Over much of this interval, records of climate are based on geological and geochemical indicators stored in annually layered archives: mountain glaciers, tree rings, and corals. Also available in recent centuries are historical observations recorded by humans. Instruments have been used extensively to track the components of the climate system only during the last century, and satellites only in the last few decades.

In this chapter we focus on the contrast between the increasing warmth observed in the twentieth century and the cooler climates that prevailed during the preceding several centuries, known as the Little Ice Age (A.D. 1400–1900). We also investigate the natural causes of climate changes that have been in operation during the last several thousand years, but defer a summary of human causes of climate change until Chapter 17.

The Little Ice Age: Local or Global?

As the last millennium began, scattered evidence from Europe and the high latitudes of the North Atlantic suggests a time of relatively warm climate near 1000 to 1300 called the **Medieval Climatic Optimum**. During this interval, Nordic people settled southwestern Greenland along the fringes of the ice sheet and managed to grow wheat. Sea ice, common today around the Greenland coasts, is rarely mentioned in chronicles from this era.

The subsequent cooling during the **Little Ice Age** (1400–1900) seriously affected the populations of Europe. With colder winters and a shorter growing season, grain and grape crops repeatedly failed in far northern regions where they had been successfully grown during the warmer medieval optimum, causing famines. Lakes, rivers, and ports froze throughout northern Europe during severe winters. The settlements in Greenland were abandoned near the onset of the Little Ice Age, probably because a marginal climate had become inhospitable and perhaps because of conflicts with native Arctic peoples.

Dramatic evidence that Europe's climate was cooler than it is today comes from its mountains. People who lived in the Alps of Switzerland and Austria and the mountains of Norway in the fourteenth and fifteenth centuries experienced large-scale advances of glaciers and chronicled the effects of these major events on their lives. Many of these (and later) ice advances spread over alpine meadows where herds had grazed and destroyed farmhouses. Some advancing glaciers even overran villages.

Working on mountain glacier deposits in the Alps during the last century of the Little Ice Age, the geologist Louis Agassiz first realized that Earth had experienced major ice ages. He saw that large boulders were being eroded, carried down the mountain valleys by ice, and deposited along with finer debris in moraine ridges lower on the mountains. He also found similar ridges almost 1000 meters lower on the mountains than those that were being deposited during the Little Ice Age. Agassiz reasoned that conditions cold enough to allow ice to persist had once been displaced downward by nearly a kilometer during the major ice ages of the past.

Other evidence confirms that temperatures were cooler in Europe and the nearby North Atlantic during the Little Ice Age. One of the longest climate records kept by humans is of the frequency of sea ice along the north and west coasts of Iceland (Figure 16-1). Because fishing is so important to the supply of food on this isolated island, records were routinely kept of times when coastal sea ice made it impossible for ships to go to sea. The record shows the number of weeks per year in which sea ice reached and blocked the northern coast of Iceland.

Sea ice appears to have been infrequent until 1600, although records kept before that date are probably less reliable than later ones. It then increased in frequency and reached a maximum in the nineteenth century before all but disappearing from the coasts of Iceland during the twentieth century, except for occasional years. This sea ice record indicates a major contrast between the cold conditions of the Little Ice Age and the warmth of the twentieth century.

Even though the Little Ice Age was not a true ice age (in the sense that major ice sheets did not actually develop), it may have been a small step in that direction. We have already seen that a persistent long-term cooling trend occurred at high northern latitudes over the last 7000 years. In response to this orbital-scale cooling, a small part of northeastern Canada may have begun moving toward permanent glaciation late in the Little Ice Age.

The Canadian Arctic is a forbidding place of long, cold winters and short, chilly, mosquito-infested summers. In a few locations, small ice caps grow at or near sea level today because summer temperatures are not warm enough to melt them. Rock outcrops in these regions are covered by **lichen**, a primitive mosslike vegetation that can live on bare rock surfaces even under inhospitable conditions (Figure 16-2A). These organisms secrete acids that attack bedrock and break it down into mineral grains. Small lichen growing in the Canadian Arctic today can be dated by their size (Figure 16-2B). They begin life as small specks and then expand into round blobs at predictable rates as they age.

But it is not the living lichens that tell us about the Little Ice Age. Scattered across portions of the Canadian Arctic, and particularly on Baffin Island, between the Canadian mainland and Greenland, are dead relics of a recent ice age that never quite came into being. Broad halos of dead lichens surround the small modern-day ice caps and occur in other regions lacking ice caps (Figure 16-2C). The size of these dead lichens

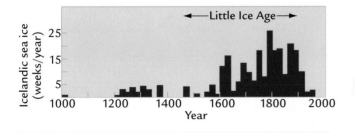

FIGURE 16-1 **Sea ice on the coast of Iceland** The frequency of sea ice along the coast of Iceland increased into the nineteenth century, then declined precipitously in the twentieth century. (Adapted from H. H. Lamb, *Climate–Past, Present, and Future,* vol. 2 [London: Methuen, 1977].)

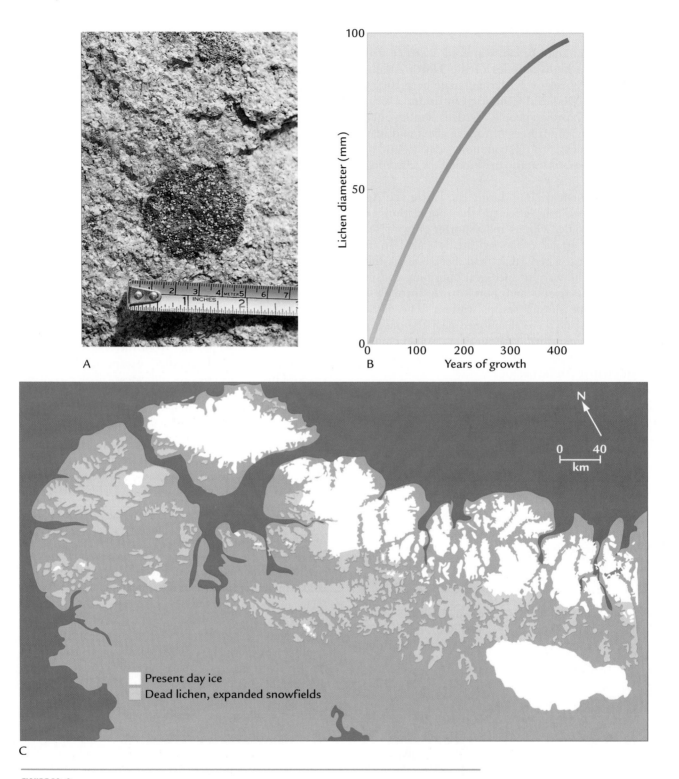

A

B

C

FIGURE 16-2 Lichen halos: Little Ice Age snowfield? Lichens grow on rock surfaces in the Arctic (A) at known rates (B). Halo-like areas of dead lichen in parts of the Canadian Arctic may record an interval of expanded snowfields late in the Little Ice Age (C). (A: courtesy of G. Falconer, Sidney, B.C.; B: adapted from G. H. Denton and W. Karlen, "Holocene Climatic Variations: Their Pattern and Possible Cause," *Quaternary Research* 3 [1973]: 155–205; C: adapted from J. T. Andrews et al., "The Laurentide Ice Sheet: Problems of the Mode and Inception," WMO/IMAP *Symposium Proceedings* 421 [1975].)

indicates they must have developed over intervals of several hundred years in the relatively recent past. Then they all died.

Because lichen are tolerant of extreme cold, it is unlikely that colder temperatures alone killed them. A more likely cause of death is burial beneath snowfields that blocked sunlight through the summer growing season. If summer melting failed to remove the previous winter's snow over an interval of many years, the lichens would have died for lack of photosynthesis. If so, the halos of dead lichens may be an indication that growing snowfields covered large parts of the high terrain on Baffin Island in the fairly recent past.

When did the older lichens die off? The small size of the young lichens now growing on top of the dead ones indicates growth only during the last 100 years—that is, since the end of the Little Ice Age. Expanded snowfields apparently killed the older lichens sometime during the Little Ice Age and kept them from growing back until late in the nineteenth century.

It makes sense that permanent snowfields could have begun to expand across this region. Baffin Island is one of three locations in which the last remnants of the great glacial ice sheets melted near 6000 years ago. And climate model simulations suggest that Baffin Island is one of the regions in which ice sheets probably began to advance during the last 3 million years. Viewed in this context, the growth of snowfields during the Little Ice Age may have been a small first step toward a real ice age, but one that was then ended by the warming of the twentieth century.

But were the Little Ice Age and the medieval optimum hemispheric or even global in scope or were they restricted to a small part of the northern hemisphere? This question is difficult to answer. One problem is that climate changes over the last millennium have been so small ($< 1°C$ globally) that they are difficult to detect amidst the larger changes over longer intervals of Earth's history. A second reason is that records of climate change before the late nineteenth century come from widely scattered locations, with very few from the oceans.

In the next section we will examine several climate records covering the last millennium. As a framework for this search, we will look at each record to see whether or not a Little Ice Age cooling and a twentieth-century warming are evident. For those records that indicate cooler temperatures during the centuries before the twentieth century, we will also check to see which of two possible trends shown in Figure 16-3 occurred: (1) Was the Little Ice Age cooling simply the culmination of a slow orbital-scale cooling? If so, the medieval optimum warm interval may have been just an insignificant blip during an ongoing cooling trend

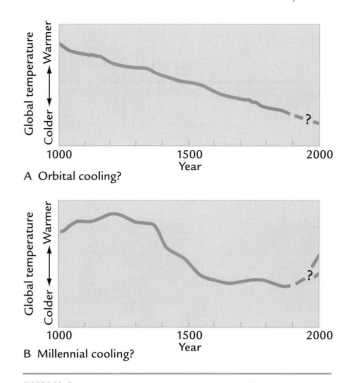

A Orbital cooling?

B Millennial cooling?

FIGURE 16-3 Two interpretations of the Little Ice Age The cool Little Ice Age interval in Europe between 1400 and 1900 could have resulted from (A) slow orbital-scale cooling or (B) a millennial-scale oscillation.

(Figure 16-3A). (2) Or was the Little Ice Age the most recent in a series of distinct millennial oscillations? If so, the medieval optimum and preceding centuries may represent the warm extreme of a millennial oscillation, with a shift toward the cool extreme during the Little Ice Age (Figure 16-3B).

A second issue we will explore is the nature of the warming since 1900. Because the twentieth century was also a time of increasing human impact on the environment, this warming can be interpreted in various ways. If the warming of the twentieth century has been unprecedented in magnitude over the last 1000 years or more of recent climate history, then humans probably played a role in it. But if comparably warm intervals occurred before or even during the Little Ice Age, then the warming of the twentieth century may be largely a result of natural variations in climate. By examining climate records covering the last millennium, we can address this important and contentious issue.

Proxy Measures of Historical Climate

Before the nineteenth century, people rarely kept records of climate. As a result, climate scientists rely mainly on proxy records in archives similar to those

A

B

C

D

FIGURE 16-4 **Coring mountain glaciers** Drilling ice cores from annually layered mountain ice (A) requires hauling equipment and supplies up and down the sides of mountains and ice caps (B) and drilling at elevations approaching 20,000 feet (C), sometimes with towering summer storm clouds rising from lower elevations and looming in the background (D). (Courtesy of L. G. Thompson, Byrd Polar Research Institute, Columbus, Ohio.)

used for earlier intervals of Earth's history, such as lake sediments and ice cores. But new sources of proxy records also come into use for the last millennium from archives such as tree rings and corals. Combined, these various kinds of archives provide limited (but steadily improving) coverage of climate changes across Earth's surface during the last few centuries.

16-1 Ice Cores from Mountain Glaciers

Like the continent-sized masses of ice on Antarctica and Greenland, smaller accumulations of ice on mountains are excellent climate archives. Both glaciers in mountain valleys and small ice caps covering mountain summits are useful (Figure 16-4). Some mountain ice dates back thousands of years into the last glaciation, while other glaciers span only a few hundred years. Layers deposited at the surface contain obvious annual signals (Figure 16-4A), but ice flow may degrade the resolution deeper in the ice to longer-term averages of decades or more.

Retrieving ice cores from mountains is a formidable task. Heavy equipment (including solar-powered ice drills) must be hauled up to the subfreezing mountain summits (Fig. 16-4B). Most mountain ice caps are between 100 and 200 meters thick, and most expeditions drill and sample the entire thickness of ice at several locations on each ice cap (Figure 16-4C and D). Lack of oxygen at these altitudes quickly causes exhaustion and other problems. Once drilled, the ice cores must be lugged down to lower elevations and kept from melting in the warmer air.

Only a few mountain glaciers have been cored to date. By far the most extensive efforts have been those by the geologist Lonnie Thompson in the Peruvian Andes. Several expeditions have retrieved ice cores from various elevations on the Quelccaya ice cap at 5670 meters (about 18,500 feet) above sea level. Based on counts of annual layers and on matching volcanic ash layers with historically documented eruptions, these records extend back to 1500 years ago.

Cores taken in the 1980s show annual-scale changes in δ18O values and dust concentrations that can be averaged over decadal intervals (Figure 16-5A). The variations in δ18O values reflect the same factors that affect δ18O in continent-sized ice sheets: changes in source area, transport paths, and amount of water vapor carried to the glacier, as well as changes in the temperature at which the snow condenses above the ice. Higher dust concentrations indicate some combination of drier source areas and stronger winds.

The Quelccaya glacier record registers a shift toward more positive δ18O values and less dust near 1900, implying that a change toward some combination of warmer temperatures, weaker winds, and different

source areas occurred at that time. The earlier δ18O record also resembles the Little Ice Age pattern shown in Figure 16-3B, with more negative values (a cooling?)

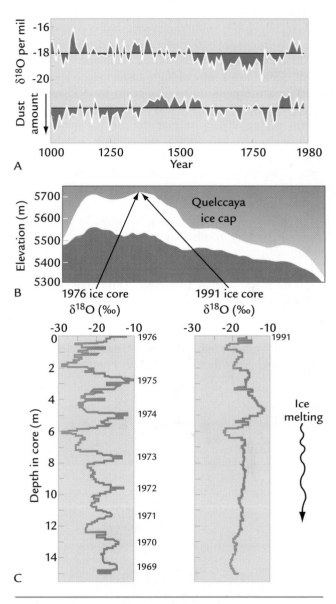

FIGURE 16-5 Quelccaya ice cap in Peru An ice core taken in 1980 from the Quelccaya ice cap in the Peruvian Andes shows more negative δ18O values and higher dust concentrations from 1600 to 1900 than in the twentieth century (A). A return expedition that cored the same location on the ice cap's summit in 1993 (B) found that the annual signal at the surface was being destroyed by melting (C), the first such melting event in the last 1500 years. (A: adapted from L. G. Thompson et al., "The Little Ice Age as Recorded in the Stratigraphy of the Tropical Quelccaya Ice Cap," *Science* 234 [1986]: 361–64; C: adapted from L. G. Thompson et al., "Recent Warming: Ice Core Evidence from Tropical Ice Cores, with Emphasis on Central Asia," *Global and Planetary Change* 7 [1993]: 145–56.)

beginning near 1400. Regardless of the exact combination of factors responsible for these $\delta^{18}O$ trends, this record from a mountaintop in the tropics is remarkably similar to the Little Ice Age pattern in Europe and the nearby Atlantic. In contrast, the dust record shows no such match before 1600: the lowest dust concentrations occur within the early part of the Little Ice Age (1400–1600), and the Medieval Climatic Optimum (1000–1300) was a time of high dust concentrations.

A return expedition to Quelccaya in 1993 encountered something totally unexpected. In the 1970s and 1980s, coring expeditions at the top of the ice cap (Figure 16-5B) had found annual-scale ice layering extending from the most recent ice layers all the way back to the deepest layers deposited 1500 years ago. But in the new cores taken in 1993, meltwater percolating down from the surface had begun to destroy the annual layering in the top of the record (Figure 16-5C). This dramatic finding means that a tropical ice cap that had been continuously recording intact annual layering for 1500 years had suddenly begun to melt. In this location, the warming of the late twentieth century has obviously been unprecedented in the last millennium and a half.

Thompson has also studied ice caps on subtropical mountains in Asia. Measurements covering the last two millennia from a much longer $\delta^{18}O$ record at Dunde ice cap in northern Tibet are shown in Figure 16-6, aver-

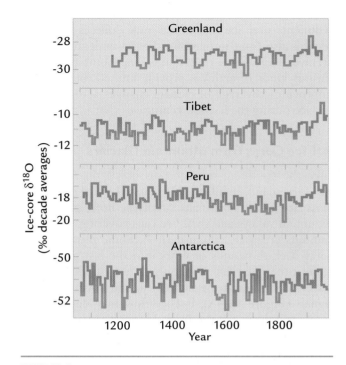

FIGURE 16-7 $\delta^{18}O$ **signals from ice cores in four regions** Cores taken from ice sheets and mountain glaciers in widely separated regions yield widely varying $\delta^{18}O$ signals, and some do not show uniquely positive (warm?) $\delta^{18}O$ values during the twentieth century. (Adapted from L. G. Thompson et al., "Recent Warming: Ice Core Evidence from Tropical Ice Cores, with Emphasis on Central Asia," *Global and Planetary Change* 7 [1993]: 145–56.)

aged over 50-year intervals. Over this interval, the $\delta^{18}O$ record again shows some similarities to Figure 16-3B, with more positive (warmer?) $\delta^{18}O$ values before 1500 and especially around 1000, and more negative (colder?) values during much of the Little Ice Age interval. Values also become more positive (warmer?) toward the twentieth century, although this transition began near 1700, well before the Little Ice Age had ended.

The average $\delta^{18}O$ value during the last 50-year interval measured (1937–87) is more positive than any other 50-year average, not just within the last 2000 years but also across the entire 12,000 years of record at the Dunde site. Either the temperature of snow precipitation was uniquely warm during the mid-twentieth century or some unprecedented change has occurred in the sources or transport paths of the incoming water vapor. Again, the twentieth century looks unique.

In summary, results from the Quelccaya and Dunde ice caps suggest that the climate on low-latitude mountains may have been colder during the Little Ice Age and warmer (perhaps even uniquely so) during the twentieth century. But some ice cores deliver a different message (Figure 16-7). Cores from the Antarctic and

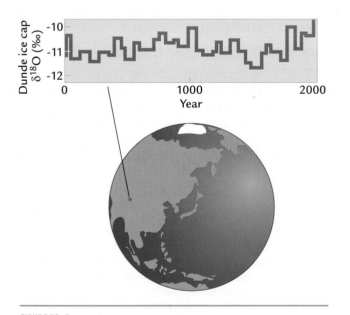

FIGURE 16-6 **Dunde ice cap in Tibet** An ice core from Dunde glacier in eastern Tibet shows lighter $\delta^{18}O$ values during the last 50 years than in any such interval of the previous 2000 years. One interpretation of this trend is that temperatures in the late twentieth century were uniquely warm. (Adapted from L. G. Thompson et al., "Holocene-Late Pleistocene Climatic Ice Core Records from Qinghai-Tibetan Plateau," *Science* 246 [1989]: 361–64.)

Greenland ice sheets do not show similar $\delta^{18}O$ changes, nor do they provide evidence of distinct Little Ice Age cold followed by unique twentieth-century warmth. These site-to-site mismatches indicate that the variations in climate recorded by $\delta^{18}O$ changes in ice cores during the last several centuries have often been regional rather than global. This is not a total surprise: we will see later that similar regional variability is typical of properties measured by instruments in recent decades.

16-2 Tree Rings

The use of tree rings to reconstruct climate change over the last several hundred years or more is called **dendroclimatology**. In regions of large seasonal changes, trees produce annual rings that alternate between lighter-colored, low-density "early wood" during the rapid growth of spring and darker, denser bands of "late wood" at the end of the growing season (Box 16-1).

Warmth and heavy rainfall during the growing season are favorable to tree growth, while cold summers and drought inhibit growth. The basic strategy of dendroclimatologists is to search out regions where trees are most sensitive to climatic stress, usually at the limit of their natural ranges of temperature or precipitation. In such regions, trees often grow alone or in isolated clusters. Years of unfavorable climate (low temperature or precipitation) are stressful to the trees and their growth slows, producing unusually narrow rings. Changes between favorable and unfavorable growth years produce distinctive trends in the width and other properties of tree rings.

Tree ring studies have been carried out in many areas, many of which show some degree of distinctiveness due to local climate responses. Across the Arctic, tree ring studies focus on such trees as spruce and larch (Figure 16-8A). An average signal synthesized from trees spanning the entire circumarctic region (Figure

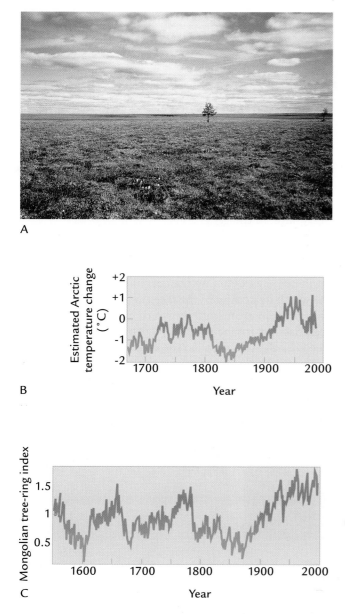

A

B

C

D

FIGURE 16-8 Arctic and Asian tree ring signals Signals from trees on northern continents, such as Siberian larch (A), are combined to create average circumarctic temperature changes over the last several centuries (B). Similar-looking tree ring signals for Central Asia (C) come from studies of larch in the mountains of Mongolia (D). Curve B shows departures from the 1951–80 average. Curve C shows an index of changing tree ring width in Central Asia. (A and D: courtesy of G. C. Jacoby, Lamont-Doherty Earth Observatory of Columbia University; B: adapted from G. C. Jacoby and R. D. D'Arrigo, "Reconstructed Northern Hemisphere Annual Temperature Since 1671 Based on High-Latitude Tree-Ring Data from North America," *Climate Change* 14 [1989]: 39–49; C: adapted from G. C. Jacoby et al., "Mongolian Tree Rings and 20th-Century Warming," *Science* 273 [1996]: 771–73.)

BOX 16-1 TOOLS OF CLIMATE SCIENCE

Analyzing Tree Rings

At chosen sites, a dozen or more trees are sampled by taking radial cores for study. The cores are small (about 0.5 cm in diameter) and do not harm the trees. The investigators date each tree by counting the sequence of tree rings back in time, beginning from the year the cores are taken. Taking multiple cores lets scientists detect annual rings that may be missing from a specific core in a particular tree because of local damage or disease. Because most trees live for a few hundred years, tree ring studies usually produce sequences spanning several centuries. Width and density trends across the annual layers are measured in each core, and records from dozens of cores are averaged to create a single representative signal for each site.

The first step in tree ring analysis is to remove the gradual effects of aging. Trees grow faster when they are young than at maturity, and investigators have to eliminate this growth effect before focusing on the effects of year-to-year changes in climate.

The next step is to relate the sequence of tree ring measurements to nearby instrumental records of climate. In the geographically remote areas typically chosen for tree ring studies, instrumental records typically cover the last 100 years, at most, and overlap only with the later part of the tree ring sequence. By examining the correlations between the width or density of the tree rings and the last half of the instrumental record of climate change, scientists determine what aspects of climate control tree growth. This is called the **calibration interval**

Coring for tree ring studies (A) To study tree rings, scientists drill into trees at sites where trees are stressed by extreme cold or dryness. The cores extracted (B) are so much smaller in diameter than the trees (C) that the trees are not harmed. (Courtesy of G. C. Jacoby, Lamont-Doherty Earth Observatory of Columbia University.)

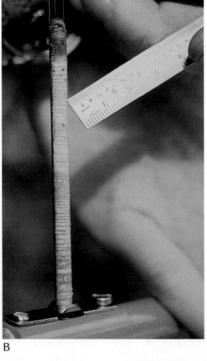

A   B   C

16-8B) shows cool conditions in the late seventeenth and early eighteenth centuries, some warming in the middle and late eighteenth century, a deeper cooling in the mid-nineteenth century, a slow but substantial warming after 1850, a brief cooling between 1950 and 1970, and renewed warming since 1980. Much of this trend is repeated in Alaska, the Canadian Arctic, Scandinavia, and Siberia, with some minor regional variations.

The 320-year trend in figure 16-8B covers the middle and end of the Little Ice Age and the instrumentally recorded warming of the twentieth century. Estimated regional temperature variations larger than 1°C occurred around the Arctic during the Little Ice Age,

BOX 16-1 CONT.   TOOLS OF CLIMATE SCIENCE

because the properties of the tree rings are calibrated against measured changes in climate. The investigators then assess the success of the calibration process by determining how well the earlier half of the known historical climate record can be predicted on the basis of the tree ring measurements from that interval.

In cold regions, temperature early in the growing season is often the strongest control on tree growth. In relatively arid regions, precipitation is more important. In some cases, precipitation in one year affects growth in the next by providing moisture to the soil, which favors growth the following spring. On the other hand, deep winter snows can slow growth in the spring.

The final step is to use correlations defined from the calibration interval to project back to the time span preceding the instrumental record. Such projections assume that the climatic controls on tree ring properties found over the calibration interval were also in operation during the earlier interval. Thus measurements of tree ring properties during the preinstrumental part of the record are converted to estimates of past climate.

For most tree ring studies, the calibration interval coincides with a time of steadily rising atmospheric CO_2 levels. Controlled experiments with vegetation in greenhouses show that growth is enhanced by higher levels of CO_2 used by plants for photosynthesis. As a result, tree growth can speed up even if the temperature or precipitation does not change. Some scientists speculate that correlations between tree ring properties and climate (temperature and precipitation) during the calibration interval are ignoring the fertilizing effect of increasing CO_2 on plants. This CO_2-fertilization effect could masquerade as a climate signal, falsely indicating gradual warming in cold regions or a trend toward wetter climates in dry regions.

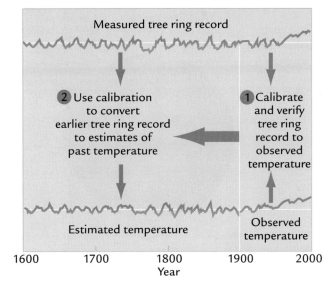

Calibrating tree ring signals Tree ring analysis is based on correlating the width (or density) of individual tree rings with monthly changes in temperature and precipitation recorded in the last half of the instrument record of climate (the last few decades). After these relationships are tested against the first half of the instrument record, they are used to convert older tree ring changes into estimates of past temperature and precipitation.

In arid regions, some evidence suggests that rising CO_2 levels and faster fertilization of leaf pores have reduced the exposure of vegetation to the dry air that normally constrains growth in these regions. If this evidence is valid, then CO_2 fertilization may have been a factor in faster tree growth in dry regions, regardless of other climate changes. But analysis of records from cold regions of both hemispheres suggests that temperature provides a consistent explanation for the long-term trends in tree growth, with no need to call on CO_2 fertilization.

---

which was clearly not a time of unrelieved cold. The warming of the Arctic in the mid-twentieth century reached values unique for the 320 years of record, but temperatures in much of the late twentieth century were similar to those in the eighteenth century.

Another important region for tree ring studies is Central Asia. Because Asia is the largest continent, its

climate is less moderated by the high heat capacity (thermal inertia) of ocean water than other regions. Climate signals derived from tree rings in Mongolia (Figure 16-8C) show a trend similar to that of the circumarctic region. Intervals of warmth occurred during the Little Ice Age, both in the mid-eighteenth century and earlier, and colder temperatures occurred in the late

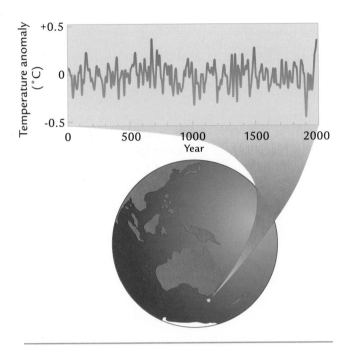

FIGURE 16-9 Tasmanian tree rings Tree ring records from the island of Tasmania, south of Australia, show that temperatures in the twentieth century are nearly unprecedented during the last 2000 years. (Adapted from E. Cook et al., "Interdecadal Climatic Oscillations in the Tasmanian Sector of the Southern Hemisphere: Evidence from Tree Rings over the Past Three Millennia," in *Climatic Variations and Forcing Mechanisms of the Last 2000 Years,* ed. P. D. Jones and R. S. Bradley [Berlin: Springer-Verlag, 1996].)

sixteenth, late seventeenth, and mid- to late nineteenth centuries. The warming of the middle and late twentieth century appears unprecedented within the 450 years of record. These signals come from larch and other trees growing on the high flanks of mountains in Central Asia (Figure 16-8D).

Far fewer tree ring data exist from the high latitudes of the southern hemisphere, in part because most of that region is ocean. Records from pines on the island of Tasmania, south of Australia, extend back more than 2000 years, and the warmth of the last several decades matches any levels reached during that interval (Figure 16-9). The interval of the Little Ice Age after 1500 is cooler than the late twentieth century but does not stand out distinctively against earlier centuries.

The few other long tree ring signals from the southern hemisphere vary widely in character. Tree ring records from New Zealand extending back a few hundred years show unique warmth in the twentieth century, but records from Chile and Argentina show no such trend. Instrumental records from weather stations near the South American sites also do not reveal warming in the late twentieth century, even though the mean trend for the southern hemisphere as a whole is toward greater warmth.

Like ice cores, tree rings tell us that climate has varied from region to region over the last several hundred years, so that no one record fully describes the trends in all areas. Viewed in their entirety, tree ring signals tell us that climate varied significantly within the Little Ice Age, in some regions even warming to levels comparable to those observed during part of the twentieth century, but in others never reaching the warmth observed over the last few decades.

16-3 Corals and Tropical Ocean Temperatures

Observations of climate changes at annual or decadal resolution are not widely available from the oceans because of the slow deposition and mixing of sediments by burrowing organisms. In recent years, climate scientists have begun to exploit corals as climate archives, using annual bands laid down in their $CaCO_3$. Because most corals grow in warm tropical or subtropical oceans, the information they provide covers a large portion of Earth's surface and complements ice core and tree ring studies from higher latitudes and altitudes. Most coral studies come from the tropical Pacific, which is dotted with volcanic islands around which corals grow. The effects of the climatic phenomenon known as El Niño make the Pacific Ocean especially important to study, both for year-to-year and longer-term changes (Box 16-2).

Tropical corals give climate scientists a way to measure the occurrence and intensity of past El Niño events, as well as long-term trends in tropical climate. Corals form annual layers made of $CaCO_3$, and the oxygen incorporated in these layers is derived from seawater. The warming of the eastern Pacific Ocean caused by El Niño events is recorded by changes in the proportions of ^{18}O and ^{16}O isotopes incorporated in the coral skeletons, analogous to the temperature signals recorded in the shells of foraminifera.

For modern corals deposited on the coast of the Galápagos Islands in the eastern Pacific, samples taken from annual layers reveal seasonal temperature changes similar to those measured directly by thermometers in surface waters (Figure 16-10A). The match between $\delta^{18}O$ values and ocean temperatures is not perfect, probably because salinity changes also affect the $\delta^{18}O$ values, but prominent El Niño years can be deciphered by $\delta^{18}O$ minima indicating warm temperatures.

This technique has also been applied to a coral record from the Galápagos spanning almost 400 years (Figure 16-10B). Fluctuations suggesting past El Niño events occur throughout this record. Little long-term

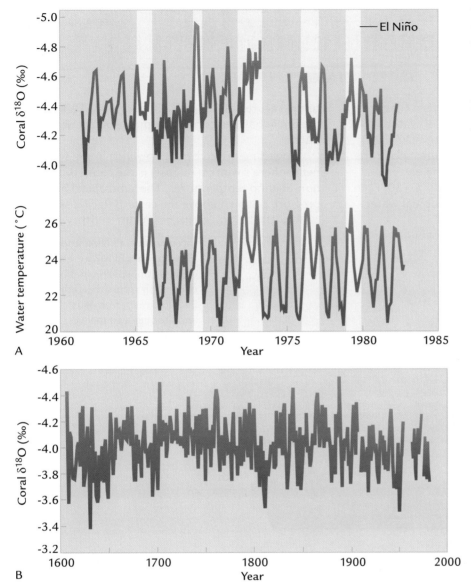

FIGURE 16-10 **Coral δ¹⁸O records: the eastern tropical Pacific** Corals from the Galápagos Islands in the eastern Pacific tend to record low (negative) δ¹⁸O values during warm El Niño years. (Adapted from R. B. Dunbar et al., "Eastern Pacific Sea Surface Temperatures Since 1600 A.D.: The δ¹⁸O Record of Climate Variability," *Paleoceanography* 9 [1994]: 291–315.)

trend is obvious here, other than generally heavier values (cooler temperatures?) near the beginning and end of the record, and just before 1700 and 1800. This tropical Pacific record shows no hint of a Little Ice Age signal or a unique trend in the twentieth century, but a majority of coral records from the tropical Pacific show a long-term decrease in δ¹⁸O consistent with a gradual ocean warming and a trend toward more rainfall toward the present.

Corals spanning 100 years of history are now widely available from the tropical Pacific, and additional records 300 to 400 years in length are being analyzed. This rapidly expanding archive should soon give scientists a clearer picture both of past El Niño variations and of longer-term climatic trends in the tropical Pacific.

In summary, hard work by climate scientists studying a variety of proxy climate indicators (ice cores, tree rings, and tropical corals) has expanded our knowledge of climate changes over the last 1000 years. But despite these efforts, coverage still remains well short of global.

Several attempts have been made to synthesize high-resolution proxy records into a single estimate of northern hemisphere climate during the last millennium. All proxy records used in these reconstructions contain signals that are linked in some way to temperature, but that are also affected by other climatic factors. All of the reconstructions attempted to extract the common temperature trend in the proxy records, with each reconstruction using a somewhat different combination of records. The reconstructed signals all show a gradual

BOX 16-2 CLIMATE INTERACTIONS AND FEEDBACKS

El Niño

E l Niño is an ocean circulation pattern that interrupts the normal circulation of water in the Pacific Ocean every 2 to 7 years. During non-El Niño years, surface temperatures along the coasts of Peru and Ecuador and in the eastern equatorial tropical Pacific are near 18°C (50°F) in winter—far cooler than the typical tropical water

temperatures of 25°C (77°F) or more. This region normally has the coolest tropical surface water on Earth.

Upwelling driven by strong winds in the southern hemisphere's winter (August) is the cause of lower non-El Niño temperatures. The subtropical high pressure cell high above the eastern Pacific sends

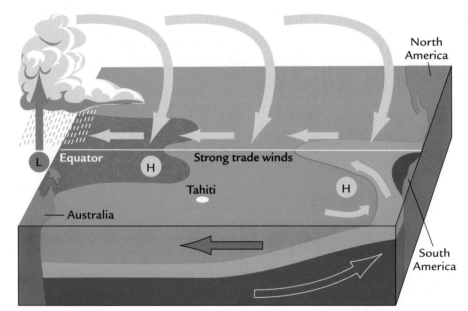

A Non–El Niño year

Circulation in El Niño and non-El Niño years El Niño events change atmospheric and ocean circulation across the entire Pacific Ocean, from Australia and Indonesia to the west coast of South America.

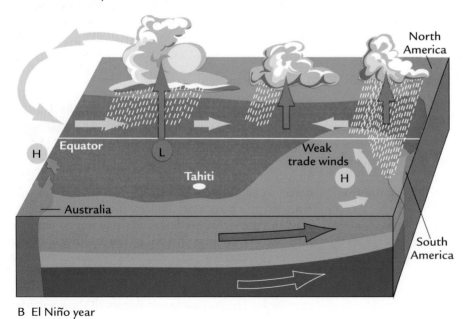

B El Niño year

BOX 16-2 CONT.   CLIMATE INTERACTIONS AND FEEDBACKS

low-level winds northward along the coast of South America in August. These winds drive warm surface waters away from the coast, and cooler water wells up from below. The winds turn westward near the equator in the counterclockwise flow typical around subtropical high-pressure regions in the southern hemisphere. These southern trade winds drive warm surface water away from the equator and toward the southwest, causing cooler water to well up along the equator.

Upwelling waters bring nutrients to the surface, supplying food to an ecosystem ranging from plant plankton up through fish, sea birds, and marine mammals (seals and sea lions). The eastern tropical Pacific is one of the world's richest fisheries, accounting for most of the world's anchovy catch and large amounts of tuna. Another feature of non-El Niño years is lack of rain. Because cool surface waters are a poor source of water vapor for the atmosphere, the coastal deserts of Peru and Chile, in western South America, are normally among the driest regions on Earth.

El Niño years change all this. During El Niño winters, strong winds fail to blow in the eastern and tropical Pacific, upwelling does not occur, and the surface waters along the South American coast warm by between 1° and 5°C. The resulting effects on sea life and human populations can be devastating. Without upwelling, plankton populations crash, and most fish die or migrate away. Without fish, sea birds on tropical islands cannot feed their young and abandon their nests to fly elsewhere in search of food. In severe El Niño

years, a significant fraction of the year's population of young sea birds and mammals dies.

El Niño warming of the surface ocean near the coastal South American deserts also produces a large source of moisture, causing rain to fall in cloudbursts that produce damaging flash floods in normally dry regions with no natural vegetation cover to absorb the water. The warm rains also favor the breeding and spread of tropical diseases such as malaria and cholera among humans.

Even though an El Niño circulation pattern reaches its height during southern hemisphere winter in August, the first hint of unusual warming of the surface ocean is often detected during the previous summer, near Christmas. For this reason, Peruvian fishermen long ago named this phenomenon "El Niño," or "the boy child," referring to the Christ Child.

El Niño events are part of a larger-scale circulation spanning the entire tropical Pacific. In the 1920s the atmospheric scientist Gilbert Walker found matching changes in atmospheric pressure between the western Pacific (northern Australia and Indonesia) and the south-central Pacific island of Tahiti. High pressures at Australia correlate with low pressures in the south-central Pacific, and vice versa.

Low atmospheric pressures are associated with rising air motion and rainfall, while high surface pressure is associated with sinking motion and dry conditions (Chapter 2). This relationship means that the opposing trends in pressure observed through time across the tropical Pacific are part of an enormous circulation cell, with sinking and rising motions occurring at exactly opposite times over

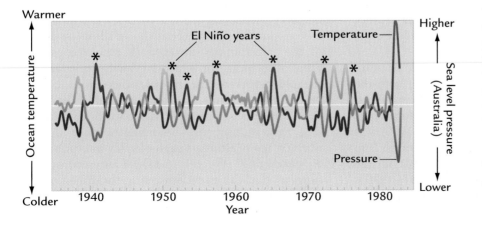

The southern oscillation Long-term changes in atmospheric pressure show that warm El Niño years near South America are times of drier conditions (higher pressures) in northern Australia but wetter conditions (lower pressures) in the south-central Pacific, near Tahiti. (Adapted from E. M. Rasmussen, "El Niño and Variations in Climate," *American Scientist* 73 [1985]: 108–77.)

BOX 16-2 CONT. CLIMATE INTERACTIONS AND FEEDBACKS

northern Australia and Indonesia on the one hand and the south-central Pacific on the other. This persistent trend is called the **southern oscillation.**

El Niño and the southern oscillation are linked. El Niño years of warm ocean temperatures and heavy rains in Peru are also times of unusually high pressures and dry conditions over northern Australia, as well as low pressures and high rainfall in the south-central Pacific. Non-El Niño years, with cool ocean temperatures near South America, are times of low pressure and moist conditions in northern Australia and of higher pressures and reduced rainfall in the south-central Pacific. The linking of these two circulation systems in a larger-scale flow is referred to as **ENSO** (*El Niño–southern oscillation*).

The physical link between these two systems occurs in the lower atmosphere and the upper ocean. The strong east-to-west trade winds that cause upwelling in the eastern Pacific during non-El Niño years drive warm surface water westward across the tropical Pacific. Warm water piles up in the western Pacific at a height several tens of centimeters above the level in the eastern Pacific and forms a natural source of moisture for evaporation and heavy precipitation in northern Australia and Indonesia. Some of the rising air cools, flows eastward at high elevations, and sinks in the east-central Pacific, contributing to the normally cooler and drier conditions near South America.

During El Niño years, trade winds in the eastern Pacific weaken. Without strong winds pushing

water westward, the pool of warm water in the western Pacific cannot be maintained and some water flows back eastward. As this warm eastward flow replaces cool upwelling water in the central and then the eastern Pacific, it becomes a source of latent heat and moisture for local rains. When the flow reaches the Americas, it deflects northward and southward along the coast, bringing warmer and wetter conditions north to California and south to Peru and creating peak El Niño conditions with warm, wet climates and reduced upwelling. Back in the western Pacific, slightly cooler conditions during El Niño years caused by the loss of warm ocean water result in drying in northern Australia and Indonesia.

Eventually El Niño conditions subside and the tropical ocean reverts to its normal state. On occasion it overshoots its normal state and produces abnormally cool sea-surface temperatures in the eastern Pacific. This overshoot is called **La Niña,** or "the girl child."

The unusual temperatures and pressures that develop in tropical regions affected by El Niño can alter circulation patterns outside the tropics. Climate scientists are actively exploring links to other regions, including the tropical Atlantic and North America. For example, El Niño years tend to produce wet, stormy winters along the California coast. Having warnings of such oncoming trends a few months in advance can help protect life and property.

---

but erratic temperature decline for 900 years, ending with a dramatic warming in the twentieth century (Figure 16-11). These trends can be used to address the two questions raised earlier in this chapter.

First, the reconstructed declines in temperature between 1000 and 1900 contain elements of both the orbital pattern shown in Figure 16-3A and the millennial pattern in Figure 16-3B. In a trend similar to the millennial pattern, the Little Ice Age (1400–1900) was cooler than the preceding or following intervals and began with a fairly abrupt cooling near 1450. But the most obvious feature of the signals from 1000 to 1900 is their match with the gradual orbital cooling trend.

Second, the warming of the twentieth century stands out as a unique feature. Both the rate and

amount of warming of the late twentieth century have no precedent during the last 1000 years. The methods used to produce the synthesis shown allows an estimate of the uncertainties involved in each reconstructed value of temperature, which inevitably grow larger for the older part of the record, where map coverage becomes more scarce. The temperatures of the last few decades have just begun to rise above these uncertainties; that is, they are becoming warmer than the warmest temperatures we can possibly interpret history as permitting (Figure 16-11). This reconstruction suggests that the twentieth-century warming was not simply another in a long series of natural climatic oscillations, but something unprecedented for the entire millennium. Such a conclusion would become even

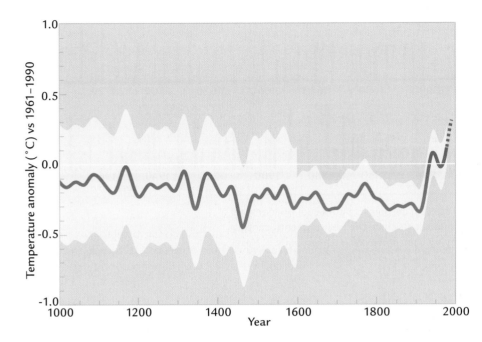

FIGURE 16-11 Temperatures in the northern hemisphere during the last millennium A synthesis of high-resolution climate records spanning all or part of the last millennium shows a gradual cooling in the northern hemisphere for 900 years, followed by an abrupt warming in the twentieth century to temperatures higher than any in the earlier record. Light shading indicates uncertainty in estimated temperature. (Adapted from M. Mann et al., "Northern Hemisphere Temperatures During the Past Millennium: Inferences, Uncertainties, and Limitations," *Geophysical Research Letters* 26 [1999]:759–62.)

more powerful if this kind of synthesis could be extended back another several millennia and into the southern hemisphere.

Historical and Instrumental Observations

Gradually during the course of human history, people began keeping records of climatic phenomena for cultural reasons. Climate scientists who try to use these early historical records to reconstruct past climate must weigh several aspects of their reliability. Were the people recording the phenomena continuously present or relying on secondhand information? Was the task of observing passed to dependable observers through time? Some observers are more easily impressed (and biased) by extreme weather events than by small, less dramatic day-to-day changes.

Historical observational records are also extremely varied. They include the frequency and timing of first and last frosts and of droughts and floods, the timing of autumn lake or river freeze-up and spring ice breakup, the first flowering of shrubs and trees (such as cherry blossoms), and the dates of harvests.

16-4 Historical Records of El Niño

Historical records track past occurrences of El Niño. Immediately after their conquest of the Inca empire in Peru, the Spanish began to make environmental observations. Ships' logs are the major source of information, supplemented by records kept by missionaries and others. These observations include phenomena now understood to result from El Niño events, including sea-surface temperatures warmer than normal, reduced catches of anchovy and other fish, departure of sea birds from coasts and islands, unusually heavy rains and floods, outbreaks of cholera and malaria, and large and sustained rises in sea level. The records start in 1525 and continue through most of the twentieth century (Figure 16-12).

El Niño events recorded during this interval are ranked on a qualitative scale ranging from none to very severe. With 115 events in 465 years, the time between successive El Niño events averages 4 years, but the actual timing varies widely around this number. El Niño events occasionally cluster within certain intervals (the late nineteenth century) and are rare in others (the mid-seventeenth century). As with the $\delta^{18}O$ record in Figure 16-10, no obvious correlation with the Little Ice Age is evident. Nine very severe El Niño events occurred in 465 years, an average of one every 50 years. The last very severe event shown was in 1983; another arrived in 1998, after this record was compiled.

For several reasons, histories of climatic phenomena such as the El Niño record in Figure 16-12 and the Icelandic sea ice record in Figure 16-1 are difficult to synthesize into records of large-scale climate change. For one thing, the records come from widely scattered locations that do not provide even regional coverage, much less global. Also, different indices are sensitive to climate changes during different seasons of the year. The extent of sea ice is sensitive to cold winter temperatures and low ocean salinity, the freezing of lakes to prolonged autumn cold, the blooming of cherry blossoms to warmth in spring, and the length of the growing season to sudden overnight frosts in spring and

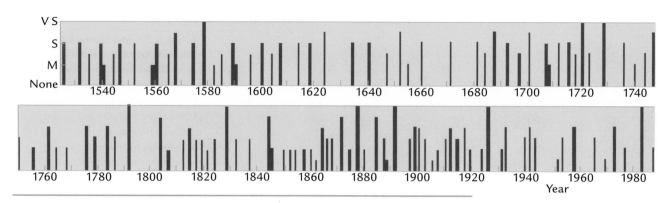

FIGURE 16-12 **Historical records of El Niño** Historical chronicles of unusual phenomena along the South American coast reveal El Niño events since the early sixteenth century. (VS = very strong, S = strong, M = moderate.) (Adapted from W. H. Quinn and V. T. Neal, "The Historical Record of El Niño Events," in *Climate Since A.D. 1500*, ed. R. S. Bradley and P. D. Jones [London: Routledge, 1992].)

autumn. As a result, these indices record changes in parts of the climate system with widely varying response times. For this reason, historical observations recorded before the era of weather instruments give us only anecdotal information about climate changes during preceding centuries.

By the eighteenth and nineteenth centuries, human ingenuity had sparked the creation of instruments for measuring climate. This trend has continued right through the era of satellite observations during the late twentieth century. As technology has developed, coverage of the planet has steadily increased.

16-5 Instrumental Temperature Records

Thermometers have been used to measure temperature changes at some locations in Europe and North America for over 200 years (Figure 16-13). The surface temperature of the ocean has also been measured along heavily traveled shipping routes. But large gaps in coverage have existed at middle and high latitudes of the southern hemisphere, with its stormy oceans, extensive sea ice, and Antarctic ice sheet. Only since the late nineteenth century have enough stations on land (and ocean islands) been recording temperature to enable scientists to make reasonable estimates of the surface temperature of the entire planet (Figure 16-14).

Another problem has been the instruments used to collect temperature data. Ocean temperatures were once measured by scooping up seawater in a canvas bucket and inserting a thermometer. If a few minutes elapsed between collecting the water and measuring its temperature, evaporation could cool the water and reduce its temperature by several tenths of a degree centigrade. Later measurements made with thermometers embedded in the outer parts of the intake valves installed to pull seawater in to cool the ship's engines produced reliable measurements.

Another complicating effect has been population growth around the land stations at which temperature was measured. Near large towns and cities, the spread of asphalt surfaces and loss of vegetation has increased absorption of solar radiation during the day and back radiation of heat from these dark surfaces at night. The

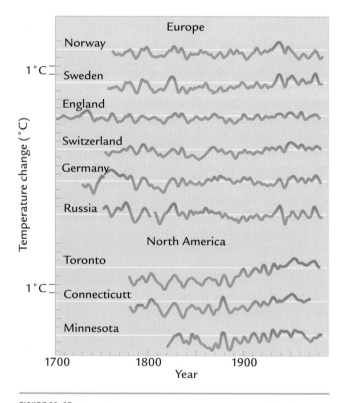

FIGURE 16-13 **Long instrument temperature records** The longest instrument records of regional temperature come from Europe and North America. (Adapted from P. D. Jones and R. S. Bradley, "Climatic Variations in the Longest Instrumental Records," in *Climate Since A.D. 1500*, ed. R. S. Bradley and P. D. Jones [London: Routledge, 1992].)

result has been a significant extra warming effect at stations near population centers. Although this warming reflects real temperature changes at the urban stations, it is not characteristic of changes in rural areas. Care must be taken not to project this bias, called the **urban heat island** effect, to rural regions near growing cities. Adjusting for this bias reduced initial estimates of warming trends by about 30%.

Climate scientists have attempted to overcome these problems and estimate global temperature changes over the last 100 or more years (Figure 16-15). Different temperature reconstructions disagree only to a minor extent, mainly in earlier parts of the century, when station coverage was poor. Brief year-to-year temperature changes aside, the underlying trend over the last 100 years is a warming of about 0.6°C. Temperatures were cooler in the early twentieth century, rose quickly during the 1920s to early 1940s, stabilized or fell slightly from the late 1940s through the late 1970s, and have again risen abruptly since then.

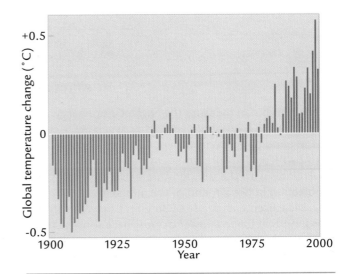

FIGURE 16-15 Change in surface temperature in the twentieth century Reconstructions of global surface temperature based on thermometer measurements show a warming trend of 0.6°C or more during the twentieth century, interrupted by a small cooling from the late 1940s to the mid-1970s. (National Climate Data Center, NOAA, Asheville, N.C.)

Not all climate scientists believe that this instrumental record of climate is valid. They point out that estimates of temperatures in the lower atmosphere based on satellite measurements made since 1979 do not precisely match the surface observations (Box 16-3).

The disagreement between the temperatures measured by satellites and those from surface observations concerns some climate scientists, but other observational evidence leaves no room for doubt that a significant warming occurred during the twentieth century.

16-6 Glaciers and Sea Level

One clear indication of a warming in the twentieth century is the retreat of mountain glaciers. Today, these bodies of ice cover 680 km² of Earth's land surface and represent about 4% of the total surface area of land ice on Earth today, with the rest accounted for by the much larger Antarctic and Greenland ice sheets.

Mountain glaciers respond to local climatic conditions, especially changes in summer temperature but also changes in winter snowfall. Because climate naturally varies from region to region, mountain glaciers in different regions can show varying responses. Individual glaciers also respond to climate changes at different rates, although all react to changes in climate within at most a few decades.

Despite this wide range of possible glacier responses to regionally varying climates, casual historical observations of glacier limits in the early twentieth century and more rigorous scientific studies later show that most

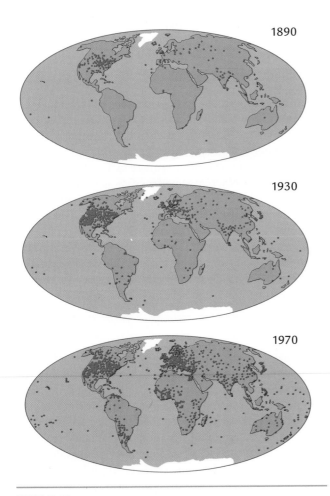

FIGURE 16-14 Temperature stations Coverage of land-based stations that measure surface temperature increased significantly during the twentieth century. (National Climate Data Center, NOAA, Asheville, N.C.)

BOX 16-3   CLIMATE DEBATE

Satellites vs. Surface Temperatures

Satellites can measure the *brightness* or energy emission level of molecules of oxygen (O_2), the second largest constituent (21%) of Earth's atmosphere. This brightness measurement correlates with the temperature of the oxygen molecules in the air, and of the air itself.

Measurements made by satellites between 1979 and 1995 in Earth's lower troposphere show year-to-year changes in estimated temperature similar to those from direct surface measurements. Because the air at Earth's surface is part of the lower troposphere, these two trends should be similar. But on close inspection, the satellite data show no long-term change or even a slight cooling trend, in contrast to the warming trend in the surface data.

Climate scientists have been debating the cause of this discrepancy. One possibility is that the surface instrumental record continues to be biased by inadequate coverage in certain regions or by the effects of urban heat islands not completely removed from that record, but most climate scientists think this explanation is unlikely.

Another possibility is bias in the satellite data. The satellite trend was initially adjusted to correct for small shifts in the timing of satellite orbits and slight differences in the sensitivity of the sensors on successive satellites. More recently, scientists found that the satellites were slowly losing altitude through time, creating a slight artificial cooling effect in the satellite temperature trend. Adjusting for these altitude changes has turned the small cooling trend measured by satellites into a small warming, but still not as large as that in the surface station data.

Another possible explanation for the difference between the two signals is that they simply measure different things. Satellites measure air temperatures averaged across several kilometers, while surface stations measure the temperature at Earth's surface. The surface measurements over the oceans (70% of Earth's surface) are also measurements of *water* temperature, while satellites measure average *air* temperatures above the oceans. Differences often exist between the temperature of seawater and the overlying air, and larger differences could occur between the temperature of the surface ocean and of air averaged through several kilometers.

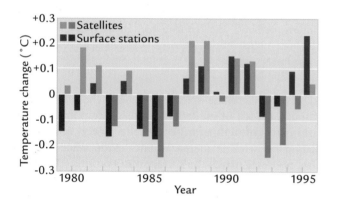

Satellite vs. surface temperature measurements Satellite-based estimates of average temperature in the lower 7 km of Earth's atmosphere since 1979 resemble temperatures recorded by thermometers at Earth's surface in year-to-year variations but disagree slightly in long-term trend. (Adapted from J. W. Hurell and K. E. Trenberth, "Satellite vs. Surface Estimates of Air Temperature Since 1979," *Journal of Climate* 9 [1996]: 2222–32.)

---

glaciers have been in retreat during most of the twentieth century (Figure 16-16). Exceptions to this general rule do occur: for example, some glaciers on the margins of the Atlantic Ocean advanced during the reversal toward cooler climates during the 1960s and 1970s. But mountain glaciers in a wide range of settings have taken part in the prevailing melting trend during the twentieth century. Not only are these ice caps retreating, but the rate of melting has accelerated in recent decades. All tropical mountain glaciers studied are in retreat, and some have disappeared entirely.

The most plausible explanation for the widespread and persistent retreat of mountain glaciers is a warming of climate. The acceleration of melting observed in many regions in the last several decades, including the unprecedented onset of melting high on the Quelccaya ice cap, make sense only if climate is warming, as the surface instrumental record shows (Figure 16-15).

Another source of information on planetary-scale climatic trends is the global average level of the ocean. To compile a history of recent changes in global sea level, climate scientists first have to remove complications produced by lingering vertical recovery of Earth's bedrock surfaces from the melting of glacial maximum ice sheets thousands of years earlier (Chapter 14). Other natural phenomena to avoid (or adjust for) include active tecton-

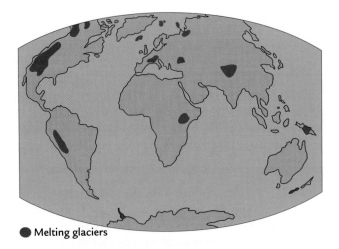

● Melting glaciers

FIGURE 16-16 **Retreat of mountain glaciers in the twentieth century** Mountain glaciers retreated from most regions during the twentieth century, but advanced again in some regions during the 1960s and 1970s. (Adapted from M. F. Meier, "Contribution of Small Glaciers to Global Sea Level," *Science* 226 [1984]: 1418–21.)

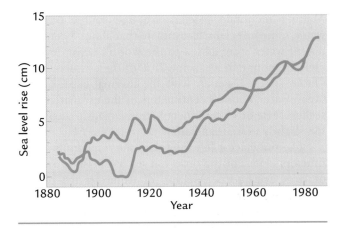

FIGURE 16-17 **Global rise in sea level in the twentieth century** Two reconstructions show that sea level has risen 12 cm or more during the twentieth century, because ice has melted on land and seawater has warmed and expanded. (Adapted from V. Gornitz and S. Lebedeff, "Global Sea Level During the Past Century," in *Sea-Level Fluctuation and Coastal Evolution,* ed. D. Nemmendal, O. H. Pilkey, and J. D. Howard, SEPM Special Publication No. 41 [1987]; and from T. P. Barnett, "Global Sea Level Change," in *Climate Variations and the Greenhouse Effect* [Rockville, Md.: National Climate Program Office, 1988].)

ic uplift caused by mountain building and subsidence resulting from the weight of recently added sediment loads. Human effects also have to be taken into account: pumping of ancient groundwater stored in natural reservoirs below the surface adds water to the ocean, while impoundment of rainfall runoff in reservoirs behind newly constructed dams lowers sea level slightly.

Compilations of sea level history disagree in detail but agree in overall trend (Figure 16-17). They suggest a sea level rise of at least 12 centimeters during the last 100 years. More recent attempts to remove the lingering effects of ice sheets suggest that the true total rise in sea level in the twentieth century may have been larger, at least 15 centimeters and possibly more.

Humans have been slowly adjusting to the rise of the ocean for years. At present about 70% of the world's sandy beaches are retreating. In coastal plain regions with very low slopes, a 15-centimeter rise in sea level can translate into an advance of 1500 centimeters (15 meters, or 50 feet) across the land in a century. Many lighthouses built on the coasts a century or two ago now sit marooned in the ocean, protected for at least the near future by constructed boulder walls. Some, such as Cape Hatteras Light in North Carolina, have been moved inland to more protected positions.

Over a time span as short as the twentieth century, a rise in the average level of the ocean must reflect one of two factors. Either ice on land has melted and added more water to the sea, or the volume of water in the ocean has expanded, or both.

Water expands slightly when heated above 4°C. As a result, warming of the ocean will cause sea level to rise. Scientists estimate that the observed warming of the surface layers of the ocean in the twentieth century has raised sea level by about 5 centimeters, or about 33% of the observed total (Table 16-1). By necessity, this estimate omits the large volume of water lying below the surface layers, because the recent temperature history of these deeper waters is almost completely unknown.

Melting of ice on land is left as the second—and major—cause of the sea level rise of the last century.

Table 16-1 Factors in the Rise of Sea Level in the Twentieth Century [in centimeters]

	Best estimate	Possible range
Thermal expansion		
Surface water	+5	+1.5 to +7
Deep water	?	?
Melting of land ice		
Mountain glaciers	+3	+2 to +4.6
Antarctic ice	?	−10 to +13
Greenland ice	+2.5	+2.3 to +2.5
All factors	+10.5	−4 to +27
Observed sea level rise	+15	+12 to +20

Melting mountain glaciers are one source of extra water. Climate scientists estimate that about 3 centimeters of water were added to the world's ocean level by melting mountain ice, some 20% of the estimated total (Table 16-1). Together with the thermal expansion of ocean water, about 8 centimeters of the estimated total sea level rise of 15 centimeters can be readily accounted for, slightly more than half.

The only factor left to explain the remaining 7 centimeters (45%) of the observed sea level rise is melting of large ice sheets. Melting of Antarctic ice by the atmosphere is unlikely because that ice sheet exists in such a cold, dry environment that any warming could increase the amount of water vapor in the air and feed more snow to it. In fact, some estimates suggest that the Antarctic ice sheet may have grown slightly in the last century. If so, growth of Antarctic ice could have *removed* an amount of water from the ocean equivalent to a lowering of global sea level by 10 centimeters (Table 16-1).

On the other hand, a warming of the ocean around Antarctica could have destabilized the lower margins of the ice sheet, sped up the flow of ice to the ocean, and caused a net loss of volume. Substantial melting of ice shelves on the Antarctic Peninsula (south of South America) has occurred in recent decades, but this region represents only a small part of the ice margin. At present we do not know for certain whether the huge Antarctic ice sheet grew, shrank, or remained at the same size during the last century.

The Greenland ice sheet is the only other possible source of the water needed to account for the remaining sea level rise. Again, scientists do not know for certain from direct observations whether this ice sheet grew or shrank during the twentieth century. Modeling efforts that have simulated the impact of the observed regional warming on this ice sheet suggest that melting of Greenland ice could have raised global sea level by 2 to 3 centimeters during the twentieth century (Table 16-1).

In summary, climate scientists know that global sea level rose by about 15 centimeters during the twentieth century because of some combination of warming-induced expansion of seawater and melting of ice on land. Both explanations require Earth to have warmed in the last century.

16-7 Satellite and Other Observations in Recent Decades

Measurements of Earth's surface and atmosphere increased through the last century, especially with the advent of satellite sensors in the late twentieth century. Although a few ground-based records span much of the century, satellite records cover only a few decades. With such short records, it is difficult to determine whether the trends recorded are the result of Earth's natural climatic variability or a longer-term trend linked to an increasing human impact on climate.

Records of the estimated extent of cloud cover based on surface-station observations extend back to the start of the twentieth century (Figure 16-18). Cloud cover has increased in both hemispheres, especially since 1940. Because these records do not specify what kinds of clouds were increasing, their impact on climate is unknown. The cause of these increases in cloud cover could be warmer surface temperatures, which allow more evaporation of water vapor; cooler high-altitude temperatures, which reduce the capacity of air to hold water vapor; or an increase in the number of particles in the air, which enhance condensation of droplets and form clouds.

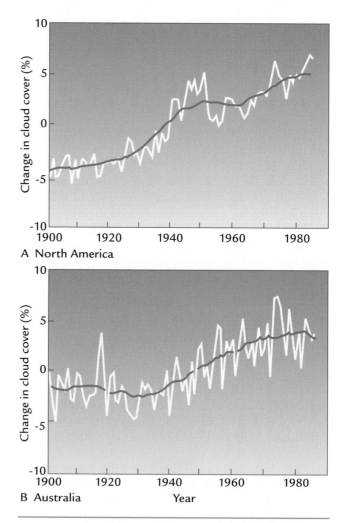

FIGURE 16-18 Increasing cloud cover in the twentieth century Surface observations show increased cloud cover in both hemispheres during the twentieth century. (Adapted from A. Henderson-Sellers, "Continental Cloudiness Changes This Century," *Geojournal* 27 [1992]: 255-62.)

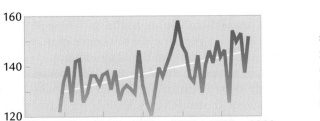

FIGURE 16-19 **The growing season lengthens in Alaska**
Surface temperature measurements indicate the length of
the growing season increased in Alaska during the last half
of the twentieth century. (Adapted from S. W. Running et al.,
"Radar Remote Sensing Proposed for Monitoring Freeze-Thaw
Transitions in Boreal Regions," *EOS* 80 [1999]: 213–21.)

The length of the growing season can be monitored
either at Earth's surface or remotely from satellites.
Measurements in central Alaska indicate an erratic
increase in the length of the growing season by two
weeks over the past 50 years (Figure 16-19). Satellite
sensing of chlorophyll produced by vegetation north of
45°N has also shown that in the mid-1990s the growing
season started a week earlier in spring than in the early
1980s and ended half a week later in autumn for the
northern hemisphere as a whole.

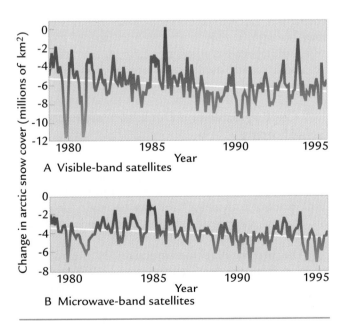

FIGURE 16-20 **Decrease in snow cover over the northern
hemisphere** In the last several decades, two kinds of
satellite measurements show a gradual decrease of snow
cover in the northern hemisphere. (Adapted from *Review of
NASA's Distributed Active Archive Centers,* National Research Council
Report [Washington, D.C.: National Academy Press, 1998].)

Satellite data also indicate a gradual reduction in
snow cover at higher latitudes. Two kinds of satellite
measurements show similar erratic decreases of northern hemisphere snow cover between 1978 and 1995
(Figure 16-20), mainly because of earlier melting of
snow in the spring.

Similarly, the average annual extent of sea ice in the
Arctic Ocean decreased by about 3% in the 1980s and
again by the same amount in the 1990s (Figure 16-21).
An area of sea ice about the size of the state of
Maryland disappeared during each decade, with percentage decreases of 7% in the winter. Measurements
made by soundings from submarines show that sea ice
in the remaining regions thinned by a startlingly large
40% between 1950–76 and the mid-1990s, from an
average thickness of just over 3 meters to less than 2
meters.

All these recent climate records are consistent with
the long-term warming in the twentieth century indicated by surface temperature observations, by glacier
melting, by sea level rise, and by many proxy indicators.
Still, records of just a few decades must be interpreted
with caution. One reason for caution is that northern
hemisphere snow and sea ice increased somewhat in the
1960s, near the end of an interval when surface temperature measurements show that both Arctic and global
mean temperatures cooled slightly. This brief cooling
was a reminder that shorter-term natural oscillations
occur within the climate system, and that records of a
decade or two do not necessarily reveal longer-term
trends. To verify the existence of a longer-term hemispheric or global warming trend with confidence will
require records spanning additional decades.

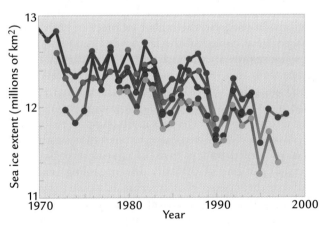

FIGURE 16-21 **Decrease in Arctic sea ice cover** Satellite
measurements show a 6% decrease in the extent of sea ice
cover in the Arctic Ocean since the early 1970s. (Adapted
from K. Vinnikov et al., "Global Warming and Northern
Hemisphere Sea-Ice Extent," *Science* 286 [1999]: 1934–39.)

Natural Causes of Historical Climate Changes

When the last millennium began, climate change was largely the result of natural processes, but by the time it ended, humans had begun to alter climate on a global scale. Here we examine all sources of natural variations in climate during the last 1000 years, with particular attention on the last 100 years. We proceed from longer-term to shorter-term factors.

16-8 Orbital-Scale and Millennial-Scale Controls on Climate

Changes in climate over tectonic time scales are irrelevant to the last 1000 years. Earth's climate has cooled by at least 5°C in the last 50 Myr from greenhouse (ice-free) conditions to its current icehouse state, with ice sheets at both poles. But the rate of this ongoing global cooling has been much too slow (roughly 0.0001°C per 1000 years) to have had any detectable effect on Earth's climate within the last millennium.

Over orbital time scales, climate can change more quickly. The gravitational pull of the Sun, Moon, and other planets cause changes in Earth's tilt and precession that alter the seasonal insolation received at different latitudes in different seasons. Changes at the 23,000-year precession cycle dominate at low and middle latitudes, while changes at the 41,000-year tilt cycle are stronger in polar regions.

Over the last several thousand years, Earth has responded to orbital changes in different ways in different regions. At lower latitudes, the monsoon has greatly diminished in strength, but temperature seems not to have changed significantly in tropical and subtropical regions. In contrast, the higher latitudes of both hemispheres have cooled in the last several thousand years, in some regions by as much as 2°-3°C. A variety of evidence from the Atlantic and Arctic confirms this trend (see Figure 14-20). This cooling in the northern hemisphere has been a response to decreasing summer insolation resulting from changes in tilt and precession.

Averaged across the planet, temperature is estimated to have cooled by less than 1°C during the last 5000 years. If the rate of orbitally driven cooling remained uniform throughout that interval, it would have produced a net global temperature decrease of 0.2°C or less in the last 1000 years. Consistent with this explanation, a slow cooling of about 0.2°C is apparent in the synthesized climate signals for the northern hemisphere shown in Figure 16-11.

Because millennial oscillations can occur over intervals as short as 1500 to 2500 years, their effect on climate during the last 1000 years also needs to be considered. One approach to isolating their effect is to try to track oscillations through the last several thousand years, detect a regular cycle of climate change, and then trace this cycle into the last millennium.

Unfortunately, most available evidence does not clearly show a regular cycle of climate change over the last several thousand years. Various records of mountain glacier advances and retreats can be interpreted as showing either a 2500-year or a 1500-year cycle (see Figure 15-18), and the oscillations seem to have been particularly irregular during the last 4000 years. Other climate records spanning this interval also disagree as to whether a cycle is present, and if so, exactly how long it is. It appears that components of Earth's climate system have gone their separate ways during the last several thousand years, rather than following a unified rhythm. Without regular millennial cycles across most of Earth's surface during this time, we have no easy way to project (reconstruct) the impact of millennial-scale oscillations during the last 1000 years.

The synthesized record of northern hemisphere climate shown in Figure 16-11 provides a different perspective on the possible impact of any millennial oscillations. A record 1000 years in length should cover 60% of a 1500-year cycle and 40% of a 2500-year cycle, enough to give some idea of their true timing and amplitudes (if they exist). The temperature contrast of 0.1° to 0.2°C between the cool Little Ice Age (1400–1900) and the warmth of the preceding centuries (1000–1300) could be attributed to a natural millennial oscillation. But we have already seen that this cooling may be more readily interpreted as part of a longer-term orbital-scale cooling trend. If so, any millennial-scale temperature oscillations were much smaller.

16-9 Century-Scale and Decadal-Scale Factors: Solar Forcing

Millennial-scale changes in the Sun's strength have been hypothesized as a cause of millennial climate oscillations, but the evidence for such a link is not strong. Still, a more convincing case can be made for a solar effect on climate during the last few decades and centuries.

The first satellite measurements of the radiation arriving from the Sun date to late in 1978 and show changes of 2 W/m² occurring in two full cycles covering almost 20 years (Figure 16-22, top). Solar radiation varied by 0.15% (2 W/m² out of 1370) over cycles 11 years in length.

Climate models indicate that a change of 0.15% in the Sun's strength could alter global mean temperature by as much as 0.2°C if it persisted for a long time. For an 11-year cycle, however, the 5.5-year interval between minimum and maximum does not allow the climate system enough time to register a full equilibri-

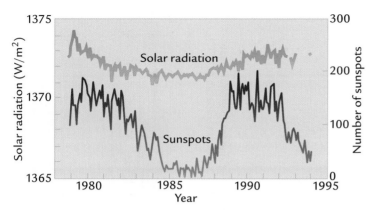

FIGURE 16-22 **Solar radiation and sunspots** Since 1978, satellites have measured changes in the solar radiation arriving at the top of the atmosphere (top), and these changes correlate with the observed numbers of sunspots (bottom). (Adapted from D. V. Hoyt et al., "The Nimbus-7 Solar Total Irradiance: A New Algorithm for Its Derivation," *Journal of Geophysical Research* 97A [1992]: 51–63.)

um response. As a result, Earth's mean temperature in the models warms and cools by less than 0.1°C in response to 11-year variations in the Sun's strength. Temperature changes this small would be difficult to distinguish from the natural variability produced by all the other factors in Earth's climate system.

Fortunately, solar variability can be estimated in a different way. Over the interval of satellite observations, the measured strength of the Sun correlates with the number of **sunspots** (dark circular regions) visible on its surface (Figure 16-22, bottom). Intuitively, it would seem likely that the presence of dark spots on a bright surface would reduce the total amount of emitted solar radiation, but in fact the opposite is observed.

Most of the radiation emitted by the Sun streams out from its polar regions (Figure 16-23) and from bright rings around the sunspots called **faculae**. During years when sunspots are abundant, the amount of radiation emitted in solar flares is at a maximum, because mechanisms within the Sun simultaneously regulate both sunspots and solar emissions. As a result, the amount of solar radiation arriving at Earth during sunspot maxima is at a maximum rather than a minimum.

Although no satellite measurements of the Sun's radiation were made before 1978, we do have a long-term record of sunspots through much of the historical era during which telescopes have been in use. Human observations of cyclic changes in sunspot occurrence through telescopes have been made for almost 500 years, and they confirm that a regular 11-year **sunspot cycle** has existed for at least that long (Figure 16-24).

The relationship between solar radiation and sunspots defined during the satellite era (Figure 16-22) suggests that the strength of the Sun would have correlated with the 11-year sunspot cycle during the much longer presatellite era. If so, scientists have a much greater span of time over which to compare climate change on Earth with solar (sunspot) changes.

Is there evidence from this earlier interval of an 11-year climate cycle that matches the implied 11-year cycle in the Sun's strength? With a few possible exceptions, the answer is no. Some scientists have claimed that they have detected 11-year cycles in various climate records, but the application of rigorous statistical techniques has disproved most of these claims.

Although we now seem to be at another dead end in our efforts to link changes in the Sun's strength to changes in climate, another route remains open. Over at least the last 100 years, Earth's average temperature has followed a trend similar to the sunspot curve

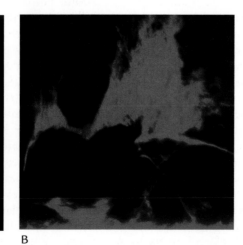

FIGURE 16-23 **Sunspots and solar emissions** Intervals when sunspots are abundant (A) are also times when strong solar emissions from the Sun's polar regions and from the bright margins of the sunspots (B) send increased levels of radiation to Earth. (NOAO.)

A   B

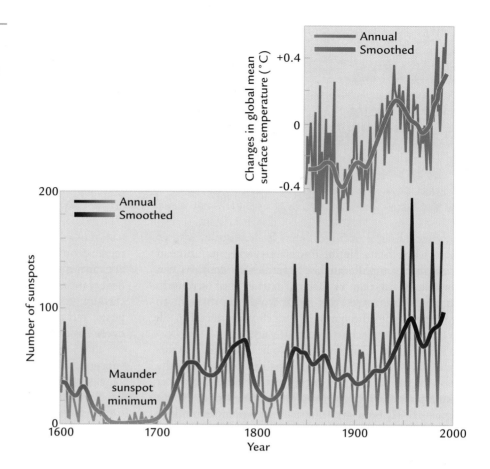

FIGURE 16-24 **Sunspot history from telescopes** Measurements made with telescopes over the last several hundred years show an 11-year sunspot cycle, as well as earlier intervals such as the Maunder minimum, when sunspots were absent for several decades. The longer-term average number of sunspots resembles observed temperature changes during the twentieth century (top). (Adapted from C.-D. Schonwiese et al., "Solar Signals in Global Climatic Change," *Climatic Change* 27 [1994]: 259–81.)

(Figure 16-24). But in this case it is the long-term *average* of sunspot maxima smoothed over several decades that correlates with climate, not the individual 11-year sunspot cycles. The long-term average sunspot trend shows a minimum during the very late nineteenth century, an increase from 1900 to 1960, a decrease during the 1960s and 1970s, and an increase since 1975 (Figure 16-24, bottom). These trends are similar to the cooler temperatures of the late nineteenth century, the warming through 1950, the cooling between 1950 and 1980, and the renewed warming since 1980 (Figure 16-24, top).

It makes good physical sense that solar changes could produce detectable climatic responses over intervals of decades but fail to do so over individual 11-year cycles. The response time of the full climate system is measured in decades rather than years. As a result, relatively small short-term changes in the Sun's strength might not have enough time to affect the system, while longer-term (average) changes would have time to do so.

Some climate scientists have hypothesized that this Sun-climate link existed further back in time, during intervals of larger changes in the Sun's strength. Several intervals almost entirely lacking in sunspots occurred before the twentieth century, including the **Maunder sunspot minimum** from 1645 to 1715 (Figure 16-24) and the **Sporer sunspot minimum** from 1460 to 1550, each named after an early astronomer. Despite the relative crudeness of observational techniques at those times, the existence of these two intervals is certain, implying intervals of decades over which solar emissions may have been unusually weak.

Tree ring studies support the inference of a weaker Sun at these times. Age offsets between ^{14}C dates and counts of annual tree ring layers indicate unusually high production of ^{14}C in Earth's atmosphere resulting from greater bombardment by cosmic particles and consistent with reduced deflection of cosmic particles by weakened solar emissions (see Figures 15-21, 15-22). Other isotopes produced by cosmic particles show the same trend and also imply a weaker Sun.

One estimate is that the Sun was 0.25% weaker than it is now during long sunspot minima, a change almost twice that measured by satellites and persisting over decades rather than just years. These longer sunspot minima correlate with some of the coldest intervals of the Little Ice Age, such as the late fifteenth century and the last few decades of the seventeenth (Figure 16-11). Climate model simulations suggest that a 0.25% decrease in solar radiation during those multidecadal intervals could have caused temperatures to fall as much as 0.2° to 0.5°C below temperatures of the late twentieth century.

But the size of this hypothesized Sun–climate link is uncertain. The trends of estimated northern hemisphere temperature in Figure 16-11 indicate little temperature change between the Maunder sunspot minimum of the late seventeenth century and the interval of more abundant sunspots during the middle of the eighteenth century. This comparison implies a climatic response toward the lower end of the range of model estimates.

In addition, proposed correlations between long-term sunspot averages and Earth's climate that appear convincing over some intervals become weaker, or even reversed in sign, during others. For example, why did the cold of the Little Ice Age persist and even intensify through the middle and late nineteenth century, even though the average number of sunspots (and presumably the Sun's strength) were by then closer to the values typical of the warm twentieth century?

16-10 Annual-Scale Factors: Volcanoes and El Niño

Two factors have affected global climate over time scales of a year or two: large volcanic explosions and major El Niño events. Historically, both have altered Earth's global temperature by less than 1°C, and neither has produced detectable changes lasting more than a few years.

Large volcanic explosions eject sulfur dioxide (SO_2) into the stratosphere, where it turns into sulfate parti-

cles that block a fraction of the incoming solar radiation from reaching Earth's surface. The result is a cooler climate, especially in the year after the eruption, but also for several years afterward, until the sulfate particles settle out (see Figure 7-19). Sulfate particles from volcanic eruptions in the tropics mix into and cool both hemispheres, while those from eruptions outside the tropics stay in their respective hemispheres, as do their cooling effects.

Several large eruptions have occurred in the era of instrumental temperature records, including Katmai (1912), Agung (1963), and El Chichón (1981). Although they may have cooled global climate (Figure 16-25), the amount of cooling is difficult to determine because of uncertainties about the size and sulfur content of each eruption and about the height in the atmosphere to which the SO_2 was injected. In addition, several eruptions coincided with multiyear cooling intervals, and year-to-year variability makes it difficult to isolate their unique contributions.

The first volcanic explosion measured in sufficient detail to assess its effect on climate was the Mount Pinatubo eruption in the Philippines in 1991. Global climate cooled by 0.6°C during the summer after the eruption and produced a net annual cooling of almost 0.3°C (Figure 16-26). Within two years, the cooling effect of Pinatubo had disappeared into the background noise of natural year-to-year temperature variability.

Although major volcanic eruptions can cause such global-scale coolings for a year or two, they cannot affect the longer-term temperature trend over the

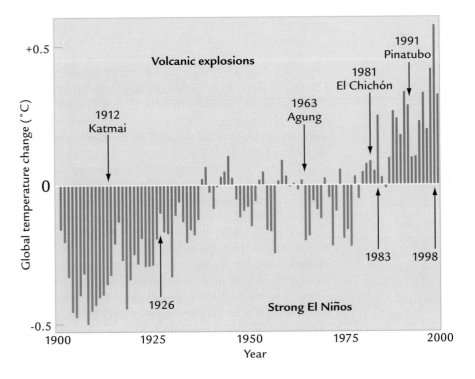

FIGURE 16-25 **Brief episodes of volcanic cooling and El Niño warming** Large volcanic explosions and major El Niño events cause short-term changes that can be large enough to be detected in the instrumental temperature record but leave the long-term baseline trend unaffected. (Adapted from National Climate Data Center, NOAA, Asheville, N.C.)

twentieth century unless they occur in repeated clusters every year or two. The large eruptions of the twentieth century were not spaced closely enough to have such a longer-term effect (Figure 16-25). The history of erup-

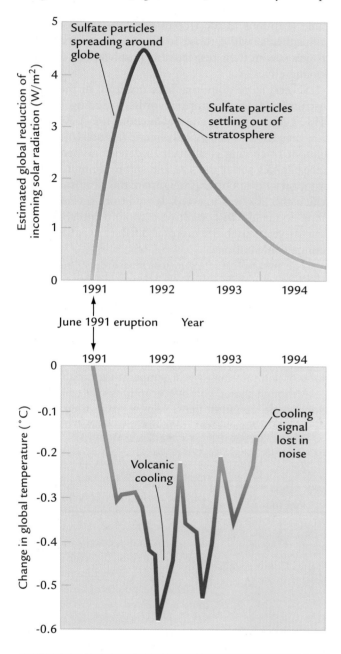

FIGURE 16-26 **Eruption of Mount Pinatubo in 1991 and global cooling** The 1991 eruption of Mount Pinatubo sent sulfur high enough into the stratosphere to cool global climate for the next two years. The cooling effect then disappeared into normal background noise. (Adapted from J. Hansen et al., "Potential Impact of Mount Pinatubo Eruption," *Geophysical Research Letters* 19 [1992]: 215–18, and J. E. Hansen et al., "How Sensitive Is the World's Climate?" *National Geographic Research and Exploration* 9 [1993]: 142–58.)

tions before 1900 is too poorly known to rule out a volcanic contribution to longer-term trends.

The other short-term factor, El Niño events, usually warms the eastern tropical Pacific sea surface for a single year (see Box 16-2). The cause of these events lies in the natural coupled interaction between the tropical Pacific Ocean and the atmosphere. The ocean provides the "memory" that permits the multiyear recurrences of El Niño events. Part of this memory arises from the wind-driven piling up of warm waters in the western Pacific Ocean over a period of years, and the eventual return flow of warm water to the east.

Ocean-atmosphere models constructed in the 1980s successfully predicted several El Niños. They also retroactively "predicted" several prior El Niños, starting from the climatic conditions that had existed in the tropical Pacific before the actual events. These models all incorporate two critical factors: (1) the rapid motion of large volumes of water sloshing back and forth (eastward and westward) along the equator in response to changes in wind strength and other factors, and (2) the great width of the tropical Pacific in comparison with the other oceans.

Over the next several years, more than a dozen groups developed models to predict future El Niño events, but nature soon proved more complicated than they expected. During several years in the early 1990s, a series of weak El Niños that developed but never reached full strength were poorly predicted by the models. Then an unusually strong El Niño developed during a year (1997–98) when none of the models had forecast a major event, and some had forecast none at all. Something is still missing from the models used to predict El Niño events.

Although the ultimate cause of El Niño events remains uncertain, and predicting them is still an imperfect science, we can measure their effects. Unusually large El Niño events like those in 1983 and 1998 can warm the sea surface by 2° to 5°C across a significant fraction of Earth's surface. These regional warmings cause annual global-scale temperature anomalies of as much as +0.1°C.

Although even the larger El Niño events produce relatively small temperature anomalies when averaged at a global scale, some of them (1926, 1983, 1998) can be deciphered in the instrumental temperature record (Figure 16-25). Like volcanic explosions, large El Niño events add to the year-to-year variability in the instrumental temperature record by producing excursions that last for a year or at most two. During the twentieth century, warm El Niño episodes did not alter the baseline level of the longer-term temperature trend, because they occurred too sporadically.

In summary, the gradual increase in global temperature over the last 100 years cannot be explained by short-term volcanic explosions or El Niño events. Of the remaining natural causes of climate change, the long-term orbital response over the last several millennia is a cooling, the reverse of the observed warming. Millennial-scale changes could have contributed to the warming of the last 100 years, but the evidence is inconclusive. Changes in the Sun's strength may explain some of the warming of the twentieth century, but increases in greenhouse gases caused by human activities are also a plausible factor, as we will see in the next chapter.

KEY TERMS

Medieval Climatic Optimum (p. 355)
Little Ice Age (p. 355)
lichen (p. 355)
dendroclimatology (p. 361)
calibration interval (p. 362)
El Niño (p. 366)
southern oscillation (p. 368)
ENSO (p. 368)
La Niña (p. 368)
urban heat island (p. 371)
sunspots (p. 377)
faculae (p. 377)
sunspot cycle (p. 377)
Maunder sunspot minimum (p. 378)
Sporer sunspot minimum (p. 378)

REVIEW QUESTIONS

1. What evidence indicates a cooler climate in Europe and nearby regions during the Little Ice Age?

2. Why are climate changes during the last millennium difficult to detect?

3. What evidence from ice cores suggests that the warming during the twentieth century reached levels unprecedented over the last 1000 years?

4. Why are the rings of environmentally stressed trees ideal for detecting climate signals?

5. How could rising CO_2 levels complicate interpretations of changes in tree rings?

6. What are the major characteristics of El Niño years in comparison with years of normal circulation?

7. What is the urban heat island effect? How does it complicate attempts to synthesize trends of regional, hemispheric, or global temperature change?

8. What other evidence supports the warming shown by surface-station measurements in the twentieth century?

9. Name four kinds of satellite evidence that support a gradual warming of high northern latitudes in the last two decades.

10. What do satellites tell us about the strength of the Sun in the late twentieth century?

11. In what way do sunspot cycles before the twentieth century imply a Sun-climate connection?

12. Did volcanic eruptions and El Niño events affect the gradual trend of global temperature over the last century?

ADDITIONAL RESOURCES

Basic Reading

Barnett, T. P. 1988. "Global Sea Level." In *Climate Variations over the Past Century and the Greenhouse Effect*. Report based on the First Climate Trends Workshop, September 1988. Rockville, Md.: National Climate Program Office, National Oceanic and Atmospheric Administration.

Bradley, R. S., and P. D. Jones. 1992. *Climate Since A.D. 1500*. London: Routledge.

Eddy, J. A. 1976. "The Maunder Minimum." *Science* 192:1189–1201.

Gornitz, V., and S. Lebedeff. 1987. "Global Sea Level Changes During the Last Century." In *Sea level Fluctuations and Coastal Evolution*, ed. D. Nummedal, O. H. Pilkey, and J. Howard. *SEPM* 41:3–16 (special publication).

Grove, J. M. 1988, *The Little Ice Age*. London: Methuen.

Intergovernmental Panel on Climate Change. 1996. *Climate Change, 1995: The Science of Climate Change*, ed. J. T. Houghton et al. Cambridge: Cambridge University Press.

Jacoby, G. C, and R. D'Arrigo. 1993. "Secular Trends in High Northern Latitude Temperature Reconstructions Based on Tree Rings." *Climate Change* 15:163–77.

Lamb, H. H. 1995. *Climate, History and the Modern World*. London: Routledge.

Quinn, W. H., V. T. Neal, and S. E. Antunez de Mayolo. 1987. "El Niño Occurrences over the Past Four and a Half Centuries." *Journal of Geophysical Research* 92:14449–61.

Advanced Reading

Andrews, J. T., R. G. Barry, P. T. Davis, A. S. Dyke, M. Mahaffy, L. D. Williams, and C. Wright. 1975. *The Laurentide Ice Sheet: Problems of the Mode and Speed of Inception*. Proceedings WMO/IAMAP Symposium Publication 421:87–94.

Gagan, M. K., L. K. Ayliffe, J. W. Beck, J. E. Cole, E. R. M. Druffel, R. B. Dunbar, and D. P. Schrag. 2000. "New Views of Tropical Paleoclimate from Corals." *Quaternary Science Reviews* 19:45–64.

Hoyt, D. V., and K. H. Schatten. 1997. *The Role of the Sun in Climate Change*. New York: Oxford University Press.

Lean, J. 1994. "Solar Forcing of Global Climate." In *The Solar Engine and Its Influence on Terrestrial Atmosphere and Climate*, ed. E. Nesme-Ribes, 164–84. New York: Springer-Verlag.

Mann, M. E., R. S. Bradley, and M. K. Hughes. 1999. "Northern Hemisphere Temperatures During the Past Millennium." *Geophysical Research Letters* 26:759–62.

Thompson, L. G., E. Mosley-Thompson, M. E. Davis, P. N. Lin, T. Yao, M. Sdyurgerov, and M. Dai. 1993. "Recent Warming: Ice Core Evidence from Tropical Ice Cores, with Emphasis on Central Asia." *Global and Planetary Change* 7:145–55.

Vinnikov, K., A. Robok, R. J. Stouffer, J. E. Walsh, C. L. Parkinson, D. J. Cavalieri, J. F. B. Mitchell, and V. F. Zakharov. 1999. "Global Warming and Northern Hemisphere Sea Ice Extent." *Science* 286:1934–39.

CHAPTER 17

Humans and Climate Change

Humans arrived late in Earth's long history. The first somewhat human creatures who walked on two legs and used stone tools appeared only within the last 4 Myr, equivalent to less than $\frac{1}{10,000}$ of Earth's age of 4.55 Byr. Subsequent milestones in human history were marked off at intervals that grew shorter by factors of 10: (1) the initial appearance of our species, *Homo sapiens,* in the last 100,000 to 200,000 years; (2) development of the first civilizations in the last 10,000 years; and (3) the arrival and growth of the industrial era within the last few hundred years. These intervals represent ever-smaller fractions of Earth's age: roughly $\frac{1}{100,000}$, one-millionth, and one ten-millionth, respectively.

Over most of their time on Earth, humans have been affected by climate but have not had any measurable impact on the climate system. Within the last two centuries, humans have begun to alter climate, first at regional and then at global scales, although the magnitude of our impacts remains uncertain. In this chapter we trace the progression of humans from passive participants in climate change to active contributors to the process.

The Impact of Climate on Human Evolution

Most anthropologists agree that humans evolved in Africa, because much of the earliest evidence for our ancestry comes from that region, followed somewhat later by similar evidence from southern Asia. All the evidence found in Africa comes from plateau regions along the eastern side of the continent, reaching from Ethiopia and Kenya in the northeast to southernmost South Africa (Figure 17-1). Volcanic activity in this region over many millions of years has deposited basalt layers that can be dated by radiometric (K/Ar) methods. These dated basalts bracket the ages of intervening sediment layers containing the record of human evolution.

Many competing hypotheses have been proposed in attempts to explain human evolution. Some hypotheses focus on social factors, such as the need of early humans to gather food for their infants, who spent several years in a defenseless state. Others focus on the impact of new technology, such as the creation and use of tools to facilitate food gathering and to serve as weapons for defense, hunting, and conquest.

Other scientists propose **climatic hypotheses of human evolution**. Their premise is that evolution was rapid at times when organisms that had become adapted to the habitats produced by one set of climatic conditions were then exposed to new climatic conditions that persisted for thousands of years and formed new habitats. Alteration of their habitat put new demands on our ancestors, but it also created new opportunities for them to exploit. In the end, changes in climate favored those who evolved traits useful for survival in new environments.

Climatic hypotheses focus on two intervals. The first is the time between 6 and 4 Myr ago, when the evolutionary lines that would lead to modern humans and chimpanzees branched off from the line that would lead to the modern great apes (gorillas and baboons). Climatic hypotheses emphasize the importance of the long-term drying trend in Africa during this interval. This drying trend caused tropical rain forest to become interspersed with semiarid grasslands, which gradually spread between groves of trees. The hypothesis is that fragmentation of once continuous forest caused our ancestors to move on the ground for longer distances, requiring more rapid movement to cover longer distances, along with greater resourcefulness.

During the second important interval, between 2.5 and 2 Myr ago, our genus, *Homo*, first appeared. In this case, climatic hypotheses emphasize the onset of northern hemisphere glacial cycles between 2.75 and 2.5 Myr ago as a driving force in human evolution. These hypotheses propose that repeated changes during the glacial cycles between warmer/wetter and cooler/drier climates accelerated the pace of evolution by favoring individuals with greater adaptability.

17-1 Evidence of Human Evolution

Anthropologists focus either on distinctive events that break the continuous process of evolution into separate stages or on quantitative traits that can be measured as they change. Human evolution is marked by five particularly distinctive developments (Figure 17-2): (1) initial branching off from the line of primitive apes between 6 and 4 Myr ago; (2) the onset of bipedalism (a preference for moving upright on two legs) near 4 Myr ago; (3) the use of stone tools beginning near 2.5 Myr ago; (4) the branching of the prehuman line into the genus *Homo* and other forms by 2 Myr ago; and (5) the development of large brains since 2 Myr ago. Each of these changes can be identified and traced from evidence left on the African continent.

Appearance of Human Ancestors The evolution of humans can be traced back to small shrewlike mammals that evolved during the millions of years after the massive extinction of life-forms at the time of the asteroid impact 65 million years ago (Chapter 6). These primitive mammals slowly evolved features now associated with monkeys, such as grasping front paws and long prehensile tails, which led to the lemur family (a

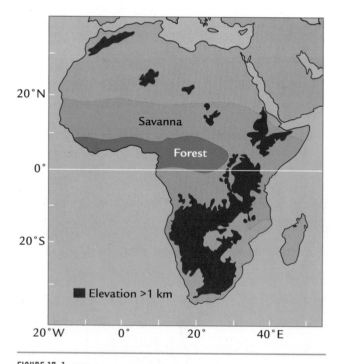

FIGURE 17-1 **African topography and rain forests** Eastern Africa is a region of broad plateaus at elevations not far above 1 km. Most rain forest vegetation occurs today in the wet intertropical convergence zone near the equator, encircled by a broad band of drier savanna.

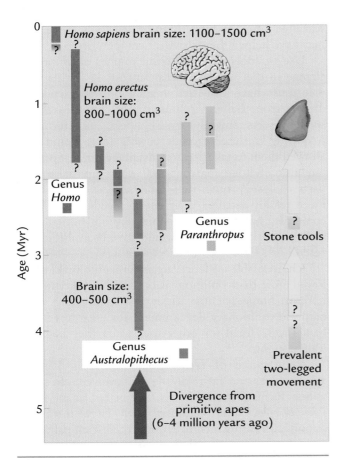

FIGURE 17-2 **Human evolution** The last 5 million years span the evolutionary line from primitive apes to the australopithecines ("southern apes") to our own genus, *Homo,* and finally our own species, *Homo sapiens* ("intelligent man"). The onset of two-legged walking appeared early in this evolutionary progression, followed by the first use of stone tools more than a million years later. (Adapted from P. B. deMenocal, "Plio-Pleistocene African Climate," *Science* 270 [1995]: 53–59.)

tulated that DNA works like an evolutionary clock: the longer the time that has elapsed since two organisms branched off from a common ancestor, the more dissimilar their DNA becomes. If this DNA dissimilarity increases through time at a constant rate, the degree of dissimilarity can be used as a clock to measure elapsed time. Molecular biologists concluded that the line that led to humans diverged from the line that led to the modern great apes between 6 and 4 Myr ago (Figure 17-2).

Walking Upright The evolutionary line that led to modern humans, the *hominids* (from the family *Hominidae,* or humanlike), appeared by 4 Myr ago. Fossil remains of hominid ankle bones with a distinctive structure suggest that walking had become the primary means of movement by 4.3 Myr ago. These hominids, considerably smaller than modern humans, had chimpanzeelike faces but large, strong jaws. They probably spent most of their time in trees, gathering fruits and nuts and avoiding predators, but also moving on the ground when necessary.

FIGURE 17-3 **Early mammals** The line of early mammals from which humans evolved included creatures resembling these modern-day lemurs. (Frans Lanting/Minden Pictures.)

monkeylike tree-climbing animal). Modern lemurs are shown in Figure 17-3.

By 10 Myr ago, one such line had further evolved to primitive apes. Subsequently a group of apes that included both our human ancestors and the chimpanzees branched off from the primitive apes, with modern apes evolving from the other branch. Our prehuman ancestors are thought to have foraged for food in and near woodlands, moving on two legs only occasionally. Radiometric dating of volcanic rocks in East Africa indicated that this branching occurred sometime between 10 and 5 Myr ago, but for a long time the actual timing was difficult to constrain. A new source of evidence then reduced this uncertainty. Molecular biologists began to measure the composition of DNA molecules in the protein of living organisms. They pos-

FIGURE 17-4 Footprints from 3.6 Myr ago Hominid (humanlike) creatures that walked across fresh volcanic ash 3.6 Myr ago in East Africa left their footprints, now fossilized. (Kenneth Garrett, National Geographic Society Image Collection.)

A remarkable deposit dated to 3.6 Myr ago in Tanzania holds footprints of an adult and a child (or a smaller adult) walking across a freshly fallen layer of volcanic ash after it had cooled (Figure 17-4). The tracks show that at one point the child turned, as if to look at something, and then walked on. These human-like apes of the genus *Australopithecus* (known as *australopithecines*) clearly walked upright as their natural means of travel. Scientists argue whether these creatures developed the ability to walk in order to exploit food resources on the grassy savanna lying between stands of trees, or whether they developed an upright posture while they still lived in trees in order to stand on the lower tree limbs (or on the ground) and reach up for fruits and nuts.

Use of Stone Tools The first firm evidence that hominids used stone tools dates to about 2.5 Myr ago. These early tools, used for crude butchering of dead animals, produced marks on the animal bones that indicate crushing and scratching but are sometimes difficult to distinguish from similar marks made by the teeth of carnivores (lions, leopards, cheetahs).

One early hypothesis suggested that humans evolved mainly as "killer apes" because of the aggressive use of tools to kill their prey. But anthropologists today tend to believe that hominids simply made opportunistic use of the remains of animals killed by lions and leopards, constantly contending with hyenas and other

scavengers for the food. Brandishing tools as weapons may have helped them drive off competitors.

Use of tools for butchering implies an important change in diet for our human ancestors. Earlier their diet must have consisted mainly of items collected directly from their environment: fruits, nuts, leaves, and small insects (mainly grasshoppers and termites). Acquiring the use of crude cutting tools would have allowed them to take greater advantage not just of meat but also of bone marrow and other internal animal parts. These protein-rich food sources would have more readily satisfied their energy needs. Stone tools could also have helped hominids dig out buried roots and tubers that otherwise would have been difficult to reach.

The available evidence suggests that toolmaking followed more than 1 million years after the ability to walk upright. Some scientists infer that toolmaking was a natural evolutionary development for creatures whose hands were freed for other uses when they began to walk upright.

Appearance of *Homo* Sometime after 2.5 Myr ago, the ancestral australopithecines evolved into several new forms (Figure 17-2). One line led to the genus *Paranthropus*, stout creatures with large teeth and strong jaws with which to crush protein-rich palm nuts and other hard food in their vegetarian diet. This group became extinct by 1 Myr ago.

The other major group carries our own genus name, *Homo*. These were more graceful (lean-bodied) creatures with larger heads and braincases. The earliest of these creatures is dated (but with some uncertainty) to between 2.4 and 2.3 Myr ago, and our human ancestor *Homo erectus* (upright man) was definitely present by just after 2 Myr ago (Figure 17-2). The wear on the teeth indicates a broader-based diet of meat, fruits, and vegetables.

The first appearance of these humanlike creatures appears to follow closely after the earliest use of stone tools, but the timing is difficult to determine precisely because the fossil record is fragmentary. The use of tools is likely to have favored those prehumans who

TABLE 17-1 Growth in Size (Volume) of Hominid Braincases

Type of hominid	Age (Myr ago)	Braincase (cm³)
Homo sapiens	0.2–0	1100–1500
Homo erectus	2.4–1.8	800–1000
Australopithecus	4.1–3.1	400–500

were able to move easily across the landscape in pursuit of a large variety of seasonally changing food sources. Frequent movement may also have called on a greater use of intellect and imagination.

Brain Size Over time, our human ancestors developed larger brains, shown by the increasing size of preserved skulls that encased and protected the brain. In broad outline, the volume of the braincase tripled over the last 3 or 4 million years (Table 17-1).

The bipedal australopithecines had braincases with volumes of 400 to 500 cm³. The *Homo erectus* ancestors who used stone tools had braincases twice as large, roughly 800 to 1000 cm³. Fully modern humans *(Homo sapiens,* or intelligent man) first appeared between 200,000 and 100,000 years ago, with braincases ranging in size between 1100 and 1500 cm³. A tripling of brain size in 4 million years is unusually rapid; most evolutionary changes are more gradual.

17-2 Did Climate Change Drive Human Evolution?

Does Earth's climate history support the climatic hypotheses of human evolution? To address this question, we will examine existing evidence for climate changes in the region where humans evolved. We will also use results from climate model (GCM) simulations to evaluate the driving forces responsible for the observed climate changes. The causal factors invoked in the climatic hypotheses can be grouped into two categories: (1) gradual tectonic-scale factors that operated over many millions of years and (2) orbital-scale factors linked to the onset of glacial cycles at high latitudes of the northern hemisphere near 2.75 Myr ago.

Tectonic-Scale Changes One critical assumption in climatic hypotheses of human evolution is that a gradual drying trend over the last 20 Myr fragmented earlier forest habitats into areas of trees interspersed with open grasslands, a new habitat to which our human ancestors adapted. What is the evidence in support of such a trend?

Remarkably little evidence has been found across North Africa, because aridity is unfavorable to permanent deposition and preservation of sediment. Many of the longer-term climate records come from the East African plateau (see Figure 17-1) and may reflect changes on a local rather than a regional scale. Newly formed volcanic rocks can alter regional climates by creating new terrain: each newly constructed volcanic feature becomes a potential focal point for monsoonal precipitation but also creates a rain shadow, where less precipitation falls.

Much of the information about long-term climate change in Africa actually comes from ocean sediments that contain material blown out from the continent.

Cores from the tropical eastern Atlantic Ocean show gradual increases in the rate of influx of continental dust (mostly quartz and clay) from North Africa after 4.5 Myr ago (Figure 17-5, left). This trend, also evident in sediments of the western Indian Ocean, has been interpreted as a sign of progressive drying that reduced the vegetation cover and exposed larger areas to erosion by winds. Today over a billion tons of dust are blown from Africa into the Atlantic every year. Dust transport may also have increased because of strengthening of the winds that lift and carry the dust.

Additional evidence for long-term drying in North Africa comes from a sediment sequence taken from the Atlantic Ocean just offshore of the Sahara Desert. These sediments hold a history of pollen changes in northwest Africa spanning discontinuous portions of the last 3.7 Myr. Gradually decreasing amounts of

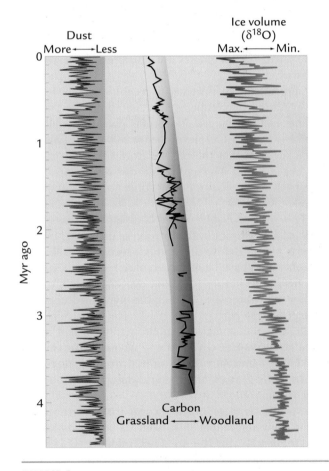

FIGURE 17-5 Long-term changes in African dust and vegetation Over the last 4.5 Myr, increasing amounts of dust were blown from North Africa to the tropical Atlantic (left), and vegetation cover gradually shifted away from trees and shrubs toward warm-season grasses (middle). These changes occurred during a time of global cooling (right). (Adapted from P. B. deMenocal, "Plio-Pleistocene African Climate," *Science* 270 [1995]: 53–59.

forest pollen and increasing amounts of savanna and desert scrub pollen indicate a progressive trend toward dry-adapted vegetation. Other fragmentary pollen records from the East African plateau also reveal a long-term trend toward more open vegetation.

This scattered evidence that forest has given way to more open vegetation is consistent with a long-term drying of climate, but a different interpretation of this trend is also possible. Another cause of the observed vegetation change could be a long-term decrease in atmospheric CO_2.

Near 7 Myr ago a major vegetation shift was under way in warm, low-latitude areas of East Africa and in portions of southern Asia, South America, and North America. In all these regions, woody vegetation (trees and shrubs) began to give way to warm-season grass vegetation. Evidence for this change is found in the type of carbon left behind in soils and in the teeth of grazing animals. Trees and shrubs use one form of photosynthesis, called the **C_3 pathway**, to take carbon from the atmosphere and turn it into organic matter, while warm-season grasses use a different form, the **C_4 pathway**.

Controlled experiments in greenhouses show that C_3 vegetation outcompetes C_4 vegetation when atmospheric CO_2 levels are above 600 ppm, but that the reverse is true for CO_2 levels below 500 ppm. This evidence suggests that the change from C_3 to C_4 carbon that began near 7 Myr ago on all four continents could have been a response to falling concentrations of atmospheric CO_2.

This shift probably altered landscapes across major parts of the tropics and subtropics, including northeastern Africa. Before the shift, C_3 tree and shrub vegetation would have been able to obtain the CO_2 necessary for photosynthesis without exposing leaves to dry air for long intervals. All but the hyperarid cores of deserts of this earlier time may have been relatively green, and the semiarid regions may have had much more extensive tree cover. After 7 Myr ago, the vegetation in the semiarid regions shifted from C_3 trees and shrubs to C_4 grasses, and the arid deserts lost much of their vegetation.

A synthesis of sediment records from northeastern Africa shows a gradual shift from C_3 to C_4 vegetation over the last 4 Myr (Figure 17-5, middle). This trend confirms a shift toward a more open landscape throughout the interval of human evolution, but it does not distinguish between climatic drying and falling CO_2 levels as the cause. In addition, the $\delta^{18}O$ trend toward more positive values throughout the last 4 Myr shows that global climate cooled, along with greater storage of ^{16}O-rich water in ice sheets, since 2.75 Myr ago (Figure 17-5, right).

Climate modeling provides a way to evaluate the factors that drove long-term climate change in Africa. One factor was uplift of the Tibetan plateau. We saw in Chapter 7 that the uplift of Tibet strengthened the Asian summer monsoon and focused heavy rainfall on the southern Himalayas.

The effects of that uplift also reached into northern Africa. The uplifting plateau created a strong counter-clockwise spiral of winds that drove hot, dry air out of the interior of Asia across Arabia and into northeastern Africa (Figure 17-6). These winds produced much drier conditions in summer across a huge arc extending from west-central Asia across the Arabian peninsula and into northeastern Africa. This uplift-induced drying may have continued into the interval of human evolution from 6 to 4 Myr ago, although some scientists doubt that uplift of Tibet continued after about 8 Myr ago.

GCM simulations also indicate that the retreat of a vast interior seaway that once occupied west-central Asia contributed to the long-term drying of this region (Figure 17-6). At maximum size some 30 million years ago, this enormous body of water moderated the climate of western Asia, keeping summers relatively cool

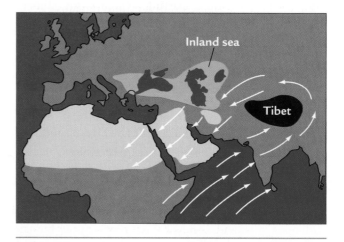

FIGURE 17-6 Large-scale effects of uplift and seaway shrinkage Over the last 20 Myr, tectonic-scale trends gradually made North Africa drier. The uplift of Tibet caused a strong flow of hot, dry northeasterly winds out of the Asian interior and into North Africa in summer. Shrinkage of a large inland sea in west-central Asia further reduced the amount of moisture transported toward Africa. In addition, falling atmospheric CO_2 levels reduced the amount of C_3 (tree and shrub) vegetation in the Sahara and Arabian deserts. (Adapted from W. F. Ruddiman and J. E. Kutzbach, "Plateau Uplift and Climate Change," *Scientific American* 264 [1991]: 66–75, and from G. Ramstein et al., "Effect of Orogeny, Plate Motion, and Land-Sea Distribution on Eurasian Climate over the Past 30 Million Years," *Nature* 386 [1997]: 786–95.)

and winters relatively warm. As the sea shrank (eventually becoming the modern-day Caspian, Black, and Aral seas), western Asia's climate became more continental, with hotter summers and colder winters. Combined with the Tibetan uplift, this trend gradually produced drier climates both in west-central Asia and in the regions of Arabia and northeastern Africa affected by dry air blown out of Asia. But the drying caused by shrinkage of this sea did not extend into the phase of human evolution that began 6 to 4 Myr ago.

Do the climatic data and GCM simulations support a climatic hypothesis of human evolution over the last several million years? The evidence does confirm a gradual trend toward more open vegetation in Africa throughout the time of human evolution, but it does not necessarily confirm increasing aridity as the sole cause of this shift in vegetation. Declining CO_2 levels may have been a more important factor in opening up the forest canopies in the African landscape during the critical interval of human evolution. If so, the climatic hypothesis may still be viable, but in a different (biochemical) form. Taking the next step—determining whether or not this vegetation trend actually did alter the course of human evolution—is difficult because of the incredibly sparse record of hominid remains.

Orbital-Scale Changes (2.75–0 Myr Ago) Other climatic hypotheses of human evolution suggest that both climate and evolution in Africa were influenced by the onset of northern hemisphere glaciation near 2.75 Myr ago. The earliest appearance of the genus *Homo* is dated to between 2.4 and 2.0 Myr ago, not long after the first glacial cycles, near 2.75 Myr ago. Some scientists hypothesize that these early glaciations created cycles of cooling and drying in Africa, caused forest habitat to fragment into grassland, and forced new evolutionary adaptations that produced our early ancestors.

Several kinds of evidence are consistent with the idea that the onset of northern hemisphere glaciation may have affected climate and hominid habitat in Africa. Pollen records from high terrain in East Africa show that the upper tree line retreated down mountainsides and gave way to more open vegetation at or near the onset of the northern hemisphere glacial cycles. Over subsequent intervals of several tens of thousands of years, forest vegetation repeatedly fluctuated up and down the sides of the mountains. These oscillations appear to be connected to the early northern hemisphere glacial cycles.

Records from the eastern and southern plateaus of Africa also show a widespread change from woodland-adapted browsing (grass-eating or leaf-eating) animals to strictly grassland-adapted grazing animals near 2.5 Myr ago. Many browsing animals became extinct, and new grazing animals appeared both by evolution and by immigration from Asia. This evidence indicates a rapid proliferation of open grasslands at the expense of closed forests. This marked change from browsing to grazing animals has been attributed to the cooling and drying produced by the first northern hemisphere glacial cycles, with the drier climates producing greater expanses of open grassy habitat at the expense of reduced areas of dense forest.

Once again, however, the cause of the observed changes in vegetation is still unclear. GCM simulations suggest that the growth of ice sheets at high latitudes of the northern hemisphere would have had relatively small direct effects on temperature and precipitation in Africa. Even the huge ice sheets that existed at the last glacial maximum, 20,000 years ago, caused simulated coolings no larger than 1°–4°C in northern Africa, and smaller effects farther south. The ice sheets were too far from Africa to have had a drastic impact on the climate of its interior. The smaller ice sheets that grew and melted during the first glacial cycles starting 2.75 Myr ago would probably have had smaller direct impacts on African climate.

Still, these early glaciations could have had larger effects on vegetation and habitat in Africa if they were accompanied by drops in atmospheric CO_2 levels, as has occurred during more recent glaciations (see Chapter 11). If lower CO_2 levels occurred during these earlier glaciations, grassy C_4 vegetation might have replaced woody C_3 trees during each glacial cycle, further fragmenting the remaining forest habitat.

What about the more recent stages of human evolution? Unfortunately, the fossil record of human remains is too sparse to determine any correlation between the onset of the much larger glacial cycles near 0.9 Myr ago and later evolutionary changes such as the disappearance of the *Paranthropus* lineage or the first controlled use of fire. Truly modern humans with very large braincases evolved only in the last 200,000 to 100,000 years, within the ongoing sequence of 100,000-year glacial cycles. Some scientists think that the warm, wet interglacial parts of these cycles allowed minor population explosions and that the cool, dry glacial parts of the cycles then winnowed out those less fit to survive under more severe conditions. As with so many aspects of human evolution, the record of human remains is too fragmentary to support or disprove this idea.

17-3 Testing Climatic Hypotheses with Fragmentary Records

Again and again we have bumped up against a persistent problem in trying to test climatic hypotheses of human evolution: the highly fragmentary record of human remains. As we noted earlier, preservation of fossils in

Africa, Arabia, and southern Asia is sparse in part because of prevailing aridity. And for human bones made of easily dissolved calcium phosphate (Ca_3PO_4), preservation is even worse in the acid-rich soils of the rain forests. Because of these problems, the total record of human evolution over 5 million years is based on a few dozen fragments of incomplete skeletons, enough to reveal some of the broad outline of human evolution but not the details.

We have already seen that sampling records of orbital-scale climate change even at intervals as close as a few thousand years can lead to gross misrepresentations of the shape of actual climate signals (Chapter 8). This aliasing problem becomes all the more formidable in records with just one sample every 100,000 years.

For records of human evolution, aliasing can produce an erroneous indication of the time of first acquisition of new physical or technical skills (such as walking or the use of tools and fire) or of the first or last appearance of a new species. With only a few samples, we are unlikely to detect the full record of these events, and instead will see a much reduced range (Figure 17-7). This general rule applies to all fossil remains; the more scarce the fossil type, the less likely we are to obtain a complete record of its existence. And human remains are extremely scarce.

Aliasing affects estimates of the actual time span on Earth of all human ancestors as well as distant cousins in the family tree. The ranges of most species are likely to have been longer than those shown in Figure 17-2 by some unknown amount. Aliasing also complicates attempts to define the relative timing between climate changes and the first use of new skills.

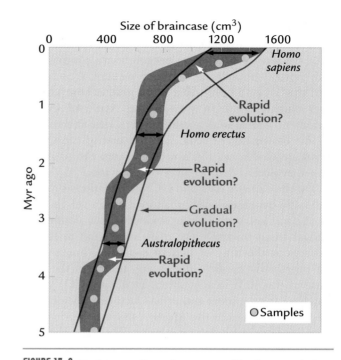

FIGURE 17-8 Undersampling of a measurable characteristic If a slowly evolving characteristic, such as the size of the human brain, varies widely at all points in time (heavy black arrows), scattered sampling (yellow circles) will not reveal this natural variability and widely varying interpretations of the data are possible. In this case, the available samples permit an interpretation of either gradual or rapid evolution.

A second undersampling problem occurs in quantitative measurements of physical traits that evolved over millions of years. In this case, the basic problem is that a broad range of natural variation occurs within human (or prehuman) populations. For example, your classmates have heads that vary widely in size on either side of the average value. The size of a human head is closely tied to the size of the braincase (the part of the skull that cradles the brain), a measurable trait in the fossil record that is important to human evolution.

What effect would this range of natural variation have on interpretations of slow evolutionary changes? The true range of braincase sizes present at any one time is fairly large in comparison with the slow evolution of the mean value (Figure 17-8). If the fossil record provides only one well-preserved specimen every several hundred thousand years, a good chance exists that some of these specimens will not be representative of the mean brain size of the population living at that time, but will fall above or below the mean. Depending on the specific samples collected and analyzed, a highly inaccurate picture of the long-term trend could emerge.

Even if the available fossil record is sparse, we should be able to obtain a general sense of the direction

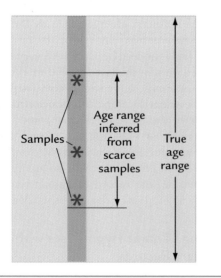

FIGURE 17-7 Undersampling of a fossil record If few samples of the fossil record of an organism are available, the true first and last appearances of that organism will be poorly estimated.

in which the trend is moving, especially if the net amount of evolution far exceeds the natural range of variation present at any one time. But the limitations of the sparse data make it impossible to define the true rates of change. Figure 17-8 shows that sparse data on brain size could be interpreted equally well as either a slow, gradual trend or rapid bursts of change.

The overall outline of human evolution is gradually falling into place, but the sparse fossil record makes it difficult to test hypotheses about the cause or causes of human evolution. As a result, the degree to which climate has affected human evolution is unknown.

The Impact of Climate on Early Civilizations

The first evidence of agriculture dates to between 11,000 and 10,000 calendar years ago in a region of the Middle East called the Fertile Crescent, encompassing present-day Syria, Iraq, Jordan, and Turkey (Figure 17-9). People in this region, the Natufians, abandoned the hunting-and-gathering way of life and began to cultivate wheat, rye, and barley, rather than harvesting grains growing wild. Because agriculture eliminated the need for seasonal migrations to search for food, these people took up residence in permanent dwellings. Within 1000 years, the dwellings began to cluster into permanent village settlements.

Evidence of organized cultivation is based on preserved remains of increasingly diverse types of grains found in regions where they did not naturally grow and where their presence must have been aided by human efforts. Evidence of permanent occupation of villages comes from the dental remains of animals from the settlements. The layering in the teeth of an animal can tell scientists the season in which it died. Such evidence shows that animals were killed in all seasons, because the people stayed in the same place throughout the year. The substantial increases in population in these regions also suggest that a more stable supply of food kept these people in place and well fed. By 10,000 to 9000 years ago, people had begun to domesticate cattle in the Near East and in Egypt.

17-4 Did Climate Affect the Origin of Agriculture?

Before the time agriculture first developed, broad areas of southeastern Europe and the Near East were drier and cooler than they are today, with climates similar to, but less harsh than, those of the preceding glacial maximum (Chapter 13). The Younger Dryas climatic reversal between 13,000 and 11,700 years ago intensified the already dry conditions across the eastern Mediterranean region (Chapter 14). In contrast, in Arabia and Saharan Africa farther south and east, climates wetter than today's prevailed because of the strong summer monsoon.

Somehow, despite the unfavorably arid conditions in the eastern Mediterranean, agriculture developed. People living near dependable water sources and harvesting and eating wild grains may have accidentally scattered some of the grains near their threshing sites, with the discarded grains sprouting in succeeding years. However it happened, primitive farming had begun.

One early hypothesis was that agriculture developed as a response to unfavorably dry climatic conditions that forced humans to cluster together at sites where water was available. Another hypothesis prompted by study of the Younger Dryas event was that humans developed agriculture to avoid negative effects of abrupt climatic changes. Neither hypothesis is easy to test.

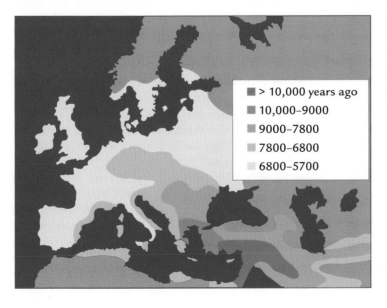

> 10,000 years ago
10,000–9000
9000–7800
7800–6800
6800–5700

FIGURE 17-9 **The spread of agriculture** The practice of agriculture originated in the region north and east of the Mediterranean and gradually spread into Europe, North Africa, and other parts of Asia. (Adapted from N. Roberts, *The Holocene*, 2nd ed. [Oxford: Blackwell, 1998].)

Through succeeding millennia, agriculture spread across much of Europe and into parts of southwest Asia and northernmost Africa (Figure 17-9). Basic agricultural techniques were adapted for local use in regions ranging from the forests of central and northern Europe to the desert margins of Arabia and North Africa.

The first advanced civilizations of the early Egyptian dynasties developed between 6000 and 5000 years ago, when the monsoon was still considerably stronger than it is today. Then and now, Egyptian life centered on the river Nile, fed by monsoon rains in the Ethiopian highlands and flowing northward through hyperarid desert. The Nile ran much stronger 6000 years ago, with larger floods that provided fertile soils and moisture for farming along the floodplain.

Climate in sub-Saharan North Africa turned much drier after 5000 years ago as the summer monsoon weakened. This drying trend affected the civilizations that had come into existence and grown in size during the wetter monsoon climates of the preceding millennia. The weakening of the summer monsoon after 5000 years ago reduced the extent of summer flooding of the Nile and its tributaries (most of them now dry). This change would have put greater stress on populations that had expanded in response to the stable food supply from large crop yields and from the domestication of animals in a more moist monsoonal climate.

At this point, the impact of climate on early civilizations and agriculture is unknown. But these civilizations clearly developed in a region and at a time of enormous fluctuations in climate, especially in the amount of available water, a precious resource in arid regions.

17-5 Sea Level Rise and the Origin of Flood Legends

Many early cultures have a flood legend, a story about a great flood that swept away an earlier civilization. Many flood legends share several features: a deluge sent by a higher authority as punishment for the sins of the people, a warning that a flood was coming, and preflood advice to gather up every kind of animal on a large vessel in order to preserve all life-forms in a postflood world.

The story of Noah in the Hebrew Old Testament is the most widely known of the flood legends, but much the same story is found in the older Babylonian tale of Gilgamesh and in legends of other early cultures in the Old World. These ancient legends were passed down by storytellers over many generations.

In the eighteenth and nineteenth centuries many scientists attempted to reconcile their religious beliefs with their scientific knowledge by advocating a **diluvial hypothesis**. This hypothesis called on a great worldwide flood to explain the widespread deposits of unsorted debris (jumbled mixtures of everything from clay to boulders) strewn across the northern continents. Today we recognize these deposits as moraines left by the retreating ice sheets. Scientists who interpret biblical (and other) flood legends less literally have continued to search for evidence of flooding at a regional scale, focusing on the Near East because of its wealth of early civilizations with flood legends.

In 1998 the geophysicists Bill Ryan and Walter Pitman pulled together evidence collected from the Black Sea region and introduced a dramatic new explanation of ancient flood legends: the **Black Sea flood hypothesis**. During the last glacial maximum, the connection between the Black Sea and the Aegean Sea through the Sea of Marmara in Turkey did not exist. Global sea level stood some 110 meters lower than it does today, and the level of the Aegean, which was linked to the global ocean as part of the Mediterranean Sea, lay well below the threshold needed to connect the two bodies of water (Figure 17-10). A freshwater lake rimmed by reedy swamps and tree-filled grassy plains covered a smaller part of the basin than the modern Black Sea.

As the very last portions of the great northern hemisphere ice sheets melted, the meltwater they returned to the ocean caused the rising Aegean to overflow into the Sea of Marmara and then spill into the freshwater lake (Figure 17-10). Radiocarbon dates in sediment cores from the Black Sea show an abrupt transition from the shells of freshwater molluscs that lived in the lake to the shells of molluscs and plankton that lived in salty ocean water 7600 calendar years ago. Within a short interval, incoming seawater transformed the freshwater lake into a salty inland sea.

But how rapid was that transition? Ryan and Pitman found an enormous gorge lying buried underneath a thin layer of sediments at the bottom of the strait known as the Bosporus, between the Sea of Marmara and the Black Sea. Cut into bedrock, this gorge is clear evidence that an enormous flow of water poured into the Black Sea at some time in the past. The coarse sediments (sand, pebbles, cobbles, and even boulders) along the floor of the gorge are arrayed in enormous dunelike shapes tilted toward the north, indicating a flow from the Aegean into the Black Sea. At the place where the Bosporus now meets the Black Sea, a deep pool is cut into the underlying rock, apparently carved by an immense waterfall created by a torrent of incoming water. Ryan and Pitman calculated that if the deep gorge that is now hidden beneath the Bosporus was filled to capacity, it would have delivered enough water to raise the level of the lake at its end by 15 centimeters (half a foot) a day, fast enough for water to advance across a mile of low-lying landscape every day, day after day, and transform the lake into a sea.

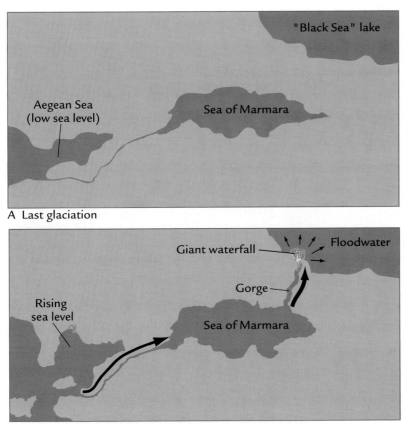

A Last glaciation

B Deglacial flood (7600 years ago)

FIGURE 17-10 The Black Sea flood Slowly melting ice sheets caused the Aegean Sea, an arm of the Mediterranean, to rise until it flooded through 200 kilometers of land into a freshwater lake in the region of the present-day Black Sea. Evidence of the flood includes a deep gorge cut beneath the waters of the Bosporus and a depression carved out by a waterfall in the southern Black Sea region. (Adapted from W. Ryan and W. Pitman, *Noah's Flood* [New York: Simon & Schuster, 1998].)

Ryan and Pitman concluded that the relentless rise of floodwaters caused by the torrent of seawater entering the Black Sea 7600 years ago would have inundated settlements on the shores of the freshwater lake within a single year, displacing thousands of people. They argued further that these displaced people carried with them vivid memories of a true flood disaster, and that their accounts were the origin of the flood legends that proliferated across much of the Old World. They also speculated that these people carried agriculture to many regions as they dispersed (Figure 17-9).

17-6 Possible Impacts of Climate on Other Civilizations

Climate change has been hypothesized as the cause of, or at least a major contributing factor in, the deterioration or collapse of other civilizations. The effects of Little Ice Age cold at high northern latitudes have already been described in Chapter 16. In low-latitude regions where water is scarce, civilizations are more susceptible to drought than to temperature changes.

The general impact of drought is obvious in many cultural records from arid regions: prehistoric cultures tended to expand across the landscape during intervals when rainfall was plentiful, but to move back toward dependable water sources (rivers and lakes) during droughts. Such changes occurred repeatedly along streams and rivers in the southern Colorado Plateau region of the American Southwest between A.D. 300 and 1300. Near 1300, the Anasazi people, who had created and occupied beautiful cave dwellings cut into the sides of cliffs, abruptly abandoned the entire region. Evidence from tree ring studies indicates that their sudden departure occurred during an interval of drought.

Another example is the sudden collapse of the Mayan centers on the Yucatán Peninsula in Mexico just over 1000 years ago (A.D. 860). Evidence from lake sediments indicates that this dislocation occurred during an interval of severe drought. People who had developed a civilization capable of building cities with massive stone temples, carved monumental statues out of stone, and developed a system of writing suddenly abandoned their inland cities and scattered. Some of them moved northward to coastal regions, perhaps because shallow groundwater was more accessible in such regions. Even those surviving coastal populations dwindled.

In most cases, cause-and-effect links between climate change and human cultural dislocations are hard to prove. Wars with neighboring cultures may also have been a factor, along with disease and other factors.

Depletion of resources is a plausible competing explanation for these cultural changes. For example, while the Anasazi did abandon their cliff dwellings in the American Southwest during a time of drought, previous dry intervals of comparable intensity had not driven people from this region. Because the final abandonment of the cliff dwellings also coincided with the near disappearance of tree pollen from climate records, it appears that the Anasazi had cut down most of the juniper and piñon pine trees previously used as fuel for cooking and for winter warmth on the high, cold plateau. If so, the cause of final abandonment may have been depletion of this crucial resource.

Similarly, the Mayas of Central America may have altered their own local environment. They may have contributed to regional drought by cutting trees and reducing the positive moisture feedback effect of transpiration (see Chapter 2). Or their farming methods may have exhausted the available supply of nutrient-rich soils, making agriculture difficult or impossible.

Early Impacts of Humans on Climate

At some point in human history, or more likely over a series of steps spread through time, the relationship between humans and the climate system began to change. In addition to being affected by changes in the climate system, human activities gradually began to affect their environment and Earth's climate.

17-7 Early Impacts of Humans on Large Mammals?

Evidence that early humans may have influenced their environment comes from the extinction of large mammals. In the late stages of the most recent deglaciation, over half of the large mammal species living in North America became extinct within an interval no longer than 1000 years, centered on 12,500 years ago. The list of mammals lost includes giant mammoths and mastodons (larger than modern elephants), horses the size of modern Clydesdales, camels, giant ground sloths, saber-toothed tigers, and beaver as large as modern bears. Two competing explanations for this major extinction pulse have been posed.

One explanation is that the major climate changes at the end of the glacial maximum stressed these populations by creating new environmental combinations to which some mammals were unable to adapt. These conditions included strong summer warming and drying of the land south of the ice sheets by high summer insolation, reduction of habitat in the cooler north by the slow retreat of the ice sheet, and the unusual mixtures of vegetation that developed as forests and grasslands shifted from their glacial positions to their modern locations.

A powerful criticism of this hypothesis is that no such extinction pulse appears in any of the similar ice-age cycles over the preceding hundreds of thousands of years. In fact, the fifty-seven species that went extinct near 12,500 years ago is larger than the number that had disappeared during all of the previous 2.75 million years of glacial cycles. In addition, every major climatic feature of the last deglaciation (high summer insolation, rising CO_2 levels, and rapidly melting ice sheets) had already occurred during several previous deglaciations at the end of earlier 100,000-year cycles of ice buildup without causing pulses of extinction. So critics ask: Why did extinctions occur only during this one deglaciation and not the others?

A second explanation, called the **overkill hypothesis**, put forward by the paleoecologist Paul Martin, is that human hunting caused this major pulse of extinctions. The immediate cause of the extinctions could have been either the first arrival of humans in the Americas or the appearance of a new hunting technology among people already present.

Both the origin and time of arrival of the first humans in the Americas were once thought resolved. They supposedly came by land from Asia near 12,500 years ago, in the late stages of the last deglaciation, crossing into Alaska over a land bridge in the Bering Strait exposed by the lower glacial sea level. Moving down the interior of North America east of the Rockies, these people supposedly passed through the ice-free corridor opened by early melting and separation of the Laurentide ice sheet to the east and the smaller Cordilleran ice sheet over the Canadian Rockies (see Figure 14-2).

Everything about this view is now in dispute. Scattered evidence hints at the arrival of humans 30,000 or 40,000 years ago, although most undisputed ^{14}C dates still support a late-deglacial arrival. Other possibilities suggested recently are that these early people arrived by water, either traveling along northeastern Pacific coastlines or crossing at lower latitudes from southeast Asia. Some scientists infer the presence of early European visitors who did not stay.

A much broader consensus exists in support of evidence that a new hunting technology appeared in the Americas at the same time that the extinctions occurred (at 12,500 years ago). This conclusion holds whether or not humans had first appeared in the Americas earlier. Many of the archeological sites that date near 12,500 calendar years ago contain spears fitted with a new and elegant kind of point fashioned by humans (Figure 17-11). The overkill hypothesis proposes that this new

FIGURE 17-11 **Pulse of mammal extinctions** Woolly mammoths and many other large mammals abruptly became extinct in North America near 12,500 years ago. Distinctive grooved spear points ("Folsom points," named for the site in New Mexico where they were first found, shown here between the bones) suddenly appeared during this interval of widespread extinction. (Denver Museum of Natural History.)

technological development suddenly enabled these people to hunt large mammals more effectively. Smaller mammals did not suffer high extinction rates, perhaps because the hunters concentrated on larger prey.

One criticism of the hunting hypothesis is that some creatures that do not seem likely to have been hunted also went extinct, including some large meat-eating mammals that may have preyed on humans rather than been their prey. A plausible response is that carnivores that depended on the carcasses of large mammals for food may have gone extinct because much of their prey went extinct at the hands of humans. A still-unanswered criticism of the hunting hypothesis is that it does not explain why some large mammals that would seem to have been likely targets for hunters (moose, musk ox, and one species of bison) survived the extinction pulse.

Perhaps climate and humans were both causes. Rapid deglacial climate changes may have put animals

under so much environmental stress that the appearance of a new human technology was sufficient to push many species to extinction. Although these extinctions had a major impact on the mammal population, the ultimate impact on climate was probably minimal, probably just minor changes in the composition of grasslands on which those animals had grazed.

17-8 Impacts of Land Clearance on Climate?

One of the earliest impacts of humans on the landscape at a larger scale was the clearance of forests for farming and fuel. In Europe, widespread forest cutting dates to more than 6000 years ago, while cutting in eastern North America began only in the last few hundred years of European settlement. Forest cutting has also occurred in many regions of the subtropics and tropics.

Climate scientists have hypothesized that these major transformations of the landscape could alter climate on at least a regional scale. Cutting of high-latitude forests could lead to greater albedo (temperature) feedback during winter as bright snow-covered croplands replaced darker conifer and deciduous forests. Cutting of low-latitude forests could also reduce the amount of summer moisture feedback to the atmosphere via transpiration as grasslands or crops replaced trees.

In the 1970s the astronomer Carl Sagan and his colleagues proposed that human cutting of vegetation and the resulting loss of moisture feedback from transpiration might explain the widespread drying recorded across large areas of the subtropics and northern tropics during the last several thousand years. Other scientists cited the resulting drying trend as the reason desert sands buried ancient cities across Arabia and parts of India.

Subsequent advances in climate science have brought many of these earlier claims into doubt or to outright rejection. For example, we now know that drying and large-scale loss of vegetation across much of North Africa, southern Arabia, and India were caused mainly by natural weakening of the orbitally driven summer monsoons, not by human activity.

During recent centuries, clearance of forests has accelerated in many regions as human populations have increased rapidly, especially in the tropics and subtropics. GCM simulations of the impact of these activities on climate have produced contradictory results: some models show significant regional-scale loss of recycled moisture, and others indicate little such effect. Disagreements among models result mainly from differing treatment of the interactions of vegetation with the atmosphere, currently an area of active model development.

The Impacts of Humans on the Atmosphere: The Last 250 Years

Centuries ago, human settlements began to grow into cities where stone and brick, soon joined by asphalt and concrete, replaced natural surfaces, and products from human industry and dwellings poured into rivers, land-fills, and the lower atmosphere. The natural flow of many rivers was blocked by the construction of dams, which created artificial lakes that flooded formerly dry land.

These human activities altered climate to some extent. Replacing soil with asphalt provides a dark surface that absorbs solar radiation and radiates it back to the atmosphere at night. Artificial lakes cool their surroundings and provide additional water vapor for recycling as rain. But these effects have generally been significant only at local or regional scales.

In the last two centuries we have entered a new era. Human activities have now reached the point where they have begun to alter climate on a global scale. We explore here four such factors: (1) increases in the greenhouse gas carbon dioxide (CO_2) caused by land clearance and burning of fossil fuels; (2) increases in the greenhouse gas methane (CH_4) caused by rice farming, tending of animal livestock, and other factors; (3) increases in sulfate aerosols caused by industrial smokestack emissions; and (4) impacts of chlorine-bearing chemicals on ozone in the stratosphere.

17-9 Increases in Carbon Dioxide (CO_2)

The evidence for a human impact on the levels of CO_2 in the atmosphere lies in ice cores. Bubbles of ancient

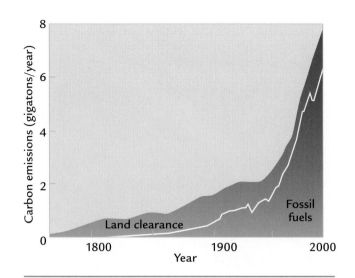

FIGURE 17-13 **Human production of CO_2** Two factors account for the increase in atmospheric CO_2 caused by human activities in the last 250 years: (1) burning of carbon in trees to clear land for agriculture and (2) burning of carbon in fossil fuels—coal, oil, and gas. (Adapted from H. S. Kheshgi et al., "Accounting for the Missing Carbon Sink with the CO_2-Fertilization Effect," *Climate Change* 33 [1996]: 31–62, from data in T. A. Boden et al., *Trends '91: A Compilation of Data on Global Change,* ORNL/CDIAC-46 [Oak Ridge, Tenn.: Oak Ridge National Laboratory, 1991].)

air trapped in the ice show that as the northern hemisphere ice sheets from the last glacial maximum melted away, atmospheric CO_2 values reached a concentration of about 280 ppm and remained near this baseline value for thousands of years. Then, near A.D. 1800, the CO_2 values in ice cores began a gradually accelerating increase.

In 1958 the geochemist Charles Keeling began to measure the CO_2 content of air samples collected in a pristine environment on a mountain in Hawaii. These measurements confirmed a continued acceleration of the CO_2 increase into the late twentieth century (Figure 17-12). By the end of the millennium, atmospheric CO_2 concentrations had passed 365 ppm, an increase of 30% over the 280 ppm baseline value, with almost 70% of the rise occurring since 1950.

The natural baseline CO_2 value of 280 ppm that persisted for thousands of years before the industrial era is called the **preindustrial CO_2 level**. The subsequent increase to values above 365 ppm is the result of human activities that put additional carbon in the atmosphere. This rise is called the **anthropogenic CO_2 increase**.

The additional carbon produced by human activities has come mainly from two sources (Figure 17-13). Throughout the late eighteenth century and most of the nineteenth, the main source of carbon was the clearing of forests to meet the needs of an increasing human population: farmland for agriculture, wood for home

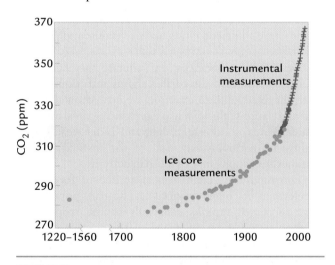

FIGURE 17-12 **Preindustrial and anthropogenic CO_2** The combined atmospheric CO_2 record from bubbles in ice cores and from instrument measurements since 1958 shows an accelerating increase of CO_2 in the last 200 years above the natural baseline of 280 ppm. (Adapted from H. H. Friedli et al., "Ice Core Record of the $^{13}C/^{12}C$ Ratio of Atmospheric CO_2 in the Past Two Centuries," *Nature* 324 [1986]: 237–38.)

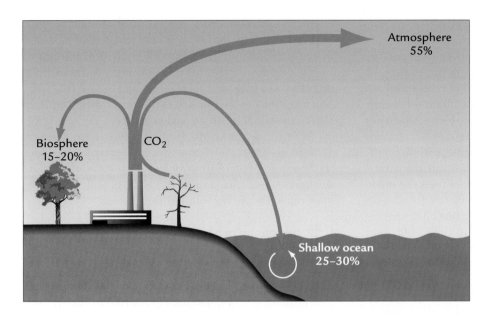

FIGURE 17-14 Where does the CO_2 produced by humans go? Of the carbon added to the climate system by humans, just over half (55%) ends up in the atmosphere, 25–30% enters the surface ocean, and the rest enters the biosphere (vegetation on land, litter, and organic carbon in estuaries).

heating, and charcoal to fuel the furnaces of the Industrial Revolution. Cutting of forests was particularly intensive in forested areas of Europe and eastern North America, which industrialized early.

After 1900, most of the extra carbon added to the atmosphere came from fossil carbon reservoirs buried beneath Earth's surface. At first coal was the main fuel, but later on, oil and natural gas began to be used and became major sources of energy. Gradually the carbon released by fossil fuels came to exceed the amount produced by land clearance. Today industrial carbon emissions (mostly in the northern hemisphere) account for the greatest part of the total. Cutting and burning of tropical rain forest account for most of the present land-clearance contribution.

In recent decades, 55% of the excess carbon produced each year has ended up in the atmosphere and another 25–30% has been added to the ocean (Figure 17-14). Unlike measurements of the well-mixed atmosphere, no single series of measurements can provide a representative average history of changes in CO_2 concentrations in the ocean through time. Because the ocean mixes more slowly than the atmosphere and at rates that vary from region to region, many measurements in many areas are needed to characterize its average change in CO_2.

The upper tens of meters of the ocean mix rapidly with the lower atmosphere, quickly exchanging molecules of gas. As a result, the excess CO_2 from human activities has already been well mixed into this thin layer. But the shallow subsurface ocean below 100 meters is relatively out of touch with the atmosphere. And most of the deeper ocean below 1 kilometer is far more isolated from the surface. As a result, only traces of the excess CO_2 produced in the last two centuries have penetrated much below the surface, and none has

entered the deeper ocean except in the few regions in the North Atlantic where deep water sinks rapidly.

Even in the surface waters, the response of the ocean varies by region and by season. On an annual average, cold high-latitude ocean water acts as a net CO_2 sink and takes CO_2 from the atmosphere, but warm low-latitude ocean surfaces act as a CO_2 source and give some of it back (Figure 17-15). One reason for this pattern is that CO_2 gas is more easily dissolved in cold water than in warm water. Local air-sea exchanges are also governed by the relative concentration of CO_2 in the surface ocean versus the overlying atmosphere,

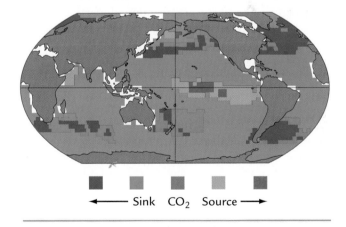

← Sink CO_2 Source →

FIGURE 17-15 Ocean sources and sinks of CO_2 Annually averaged CO_2 concentrations in ocean surface waters are close to those in the overlying atmosphere, but the higher-latitude oceans act as net sinks that absorb carbon from the atmosphere, while the tropical oceans are net sources that give some of it back. (Adapted from T. Takahashi et al., "Global Air-Sea Flux of CO_2: An Estimate Based on Measurements of Sea-Air pCO2 Differences," in *Carbon Dioxide and Climate Change* [Washington, D.C.: National Academy of Sciences, 1997].)

and by other physical and biochemical processes that control carbon exchanges with subsurface waters.

For all these reasons, huge numbers of measurements made over vast regions of the ocean during all seasons (including stormy winters) are required to quantify the slow penetration of CO_2 into and beneath the ocean surface. Despite these problems, growing numbers of measurements confirm that the ocean takes up only about 25–30% of the total human carbon input.

If 55% of the excess carbon ends up in the atmosphere and 25–30% in the oceans, where does the other 15–20% go? The only major reservoir left to take up this carbon is the biosphere, including both live vegetation in the form of trees, grasses, and their roots and dead organic litter in soils and coastal estuaries (Figure 17-14).

Although burning of forests to clear land has been a major source of extra carbon for the atmosphere and the ocean over the last 200 years, some carbon actually moves in the opposite direction, from the atmosphere back to the vegetation. One way this can happen is by regrowth of forests in previously cleared regions. In eastern North America, forests had been almost completely cut by the early twentieth century for farming and for home and industrial fuel. Similar changes occurred in Europe, where clearance had begun much earlier.

A century later, many of these regions look completely different. Young forests or groves of trees now grow in areas where typical photographs from the late nineteenth and early twentieth centuries show landscapes almost completely stripped of trees. As the Midwest opened up to large-scale mechanized farming, farms in the East were abandoned, particularly in New England. Old stone walls that once marked the boundaries of open fields are today found in the middle of young forests. Near eastern cities, rural areas that had been cleared of trees gradually turned into tree-shaded suburbs. In regions that are still rural, coal dug from mines replaced charcoal made from trees and fueled industrial furnaces. This development slowed or stopped forest cutting in the early twentieth century, and forests began to grow back. The widespread regrowth of trees has extracted CO_2 from the atmosphere.

A second way to remove CO_2 from the atmosphere is through **CO_2 fertilization**. Vegetation uses CO_2 during photosynthesis to create the cellulose that forms leaves, blades of grass, tree trunks, and roots. Greenhouse experiments show that most plants obtain carbon more easily from a CO_2-rich atmosphere, so they grow faster. Scattered evidence suggests that the 30% rise in atmospheric CO_2 in the last 200 years has increased this fertilization effect and taken more carbon

from the atmosphere by several possible mechanisms. Vegetation may grow faster; it may become more varied in composition and grow more densely; the amount of woody material in tree branches, trunks, and roots may increase; and trees and shrubs may shed more fresh carbon litter into soils. Some of this litter probably ends up buried in sediments in coastal estuaries.

In any case, atmospheric CO_2 levels have risen by 30% in the last 200 years. Because greenhouse gases trap outgoing radiation from Earth's surface, all climate scientists agree that rising levels of greenhouse gases have caused some warming of the planet.

17-10 Increases in Methane (CH_4)

Another greenhouse gas, methane, has also increased as a result of human activity. Again, the influence of humans on methane concentrations in the atmosphere is evident from air bubbles in ice cores (Figure 17-16). Methane concentrations were close to 700 ppb just before the start of the Industrial Revolution, 250 years ago. Then, like the CO_2 trend, CH_4 concentrations began a slow increase in the nineteenth century, followed by a rapid acceleration after 1900. Instrumental measurements of methane in air samples, which were begun in 1983, show concentrations rising above 1700 ppb, an increase of more than 150% above the preindustrial concentration.

Methane added to the atmosphere comes from sources rich in the remains of vegetation but lacking in oxygen: from swampy bogs where plants decay and

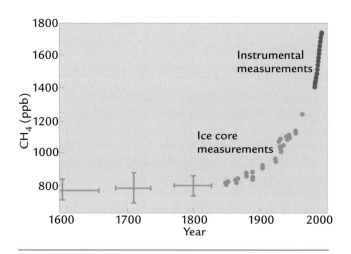

FIGURE 17-16 **Preindustrial and anthropogenic CH_4** The combined atmospheric CH_4 record from bubbles in ice cores and from instrument measurements since the early 1980s shows an accelerating rise of CH_4 above the natural baseline of 700–725 ppb. (Adapted from M. A. K. Khalil and R. A. Rasmussen, "Atmospheric Methane: Trends over the Last 10,000 Years," *Atmospheric Environment* 21 [1987]: 2445–52.)

from cattle and other grazing animals that digest vegetation. In the absence of oxygen, bacteria break down the vegetation and extract its carbon, which combines with hydrogen to form methane gas.

Emissions of CH_4 have increased enormously during the last 250 years, along with the human population that must be fed (Figure 17-17). Today the amount of CH_4 produced by human activity is estimated to be more than twice as large as that from natural sources (Table 17-2). Ever-increasing areas of wet tropical land have been put into cultivation as rice paddies, especially in southeast Asia, and these artificial wetlands produce methane. The growing numbers of cattle and other livestock kept as sources of milk, cheese, and meat have also increased the amount of methane gas expelled into the atmosphere (Figure 17-17). Even earlier human activities could have contributed to the small CH_4 increase that began 5000 years ago, long before the Industrial Revolution (Box 17-1).

Because methane is a greenhouse gas that traps outgoing radiation from Earth's surface, climate scientists

TABLE 17-2 Estimated Present-Day Sources of Atmospheric Methane (gigatons = 10^{15}g per year)

Source	Amount
Natural (30%)	160
Wetlands	115
Termites	20
Oceans	10
Other	15
Anthropogenic sources (70%)	370
Municipal landfills	40
Fossil fuels	100
Rice paddies	80
Livestock	60
Biomass burning	40
Animal waste	25
Domestic sewage	25
All sources	530

Adapted from IPCC Scientific Assessment Working Group, Radiative Forcing of Climate Change, ed. J. T. Houghton et al. (Cambridge: Cambridge University Press, 1994).

agree that the enormous rise in methane in the last two centuries must have caused the planet to warm, adding to the effect of increasing atmospheric CO_2.

17-11 Increases in Sulfate Aerosols

Industrial-era smokestacks emit the gas sulfur dioxide (SO_2) as a by-product of smelting operations in furnaces and from the burning of coal. SO_2 reacts with water vapor and is transformed into sulfate particles, called **sulfate aerosols**. These aerosols stay within the lower several kilometers of the atmosphere, rather than rising to higher elevations like CO_2 and CH_4. Thus these particles could have an impact on regional and even global climate.

Until the 1950s, most smokestacks in Europe and North America were small and most SO_2 emissions stayed close to ground level, producing thick industrial hazes and sulfur-rich acidic air around cites with factories. Later, taller smokestacks were built to disperse emissions to higher levels in the atmosphere (up to 3 km). This effort dramatically improved the air quality in many cities.

The sulfate particles sent higher into the atmosphere during recent decades are picked up by fast-moving winds and dispersed over a broad area. Although sulfates stay in the atmosphere only a few days before rain

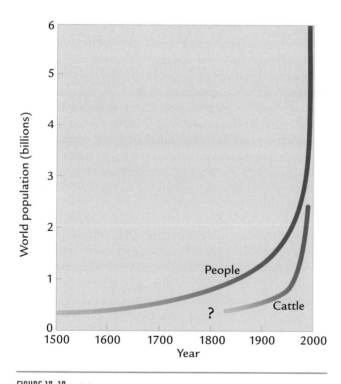

FIGURE 17-17 **Ultimate cause of increasing CH_4** Much of the large increase in atmospheric CH_4 since 1700 is the result of the rapidly rising number of humans, who tend larger herds of cattle and other livestock and also flood growing areas of wetlands to cultivate rice. (D. Merritts et al., *Environmental Geology*, © 1997 by W. H. Freeman and Company, and from T. E. Graedel and P. J. Crutzen, *Atmosphere, Climate, and Change*, Scientific American Library, © 1997 by Lucent Technologies.)

BOX 17-1 CLIMATE DEBATE

Impacts of Humans on CH₄ Levels for 5000 Years?

Near 10,000 years ago, atmospheric methane levels peaked at values of 700–750 ppb. This CH₄ peak coincided with a strong summer monsoon maximum that produced wet climates in the tropics and filled North African lakes to high levels. After this peak, CH₄ levels began a slow decrease toward a minimum value of 625 ppb near 5000 years ago. Because this CH₄ decrease matches the trend of falling lake levels in the tropics, scientists have attributed it to the gradual loss of methane from natural tropical wetlands.

After 5000 years ago, CH₄ levels began a slow rise toward the 700–725 ppb level reached just before the explosive increase of the last two

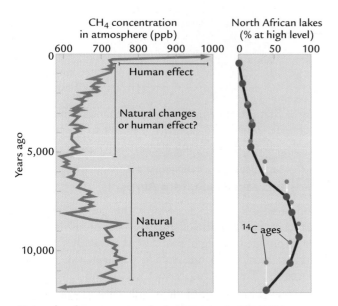

Natural or human control of atmospheric CH₄? CH₄ concentrations peaked at 700–750 ppb near 10,000 years ago and then slowly decreased to 625 ppb near 5000 years ago, a pattern that closely matches the filling and emptying of tropical lakes in North Africa by strong monsoons. The slow rise in CH₄ between 5000 and 300 years ago could have been caused by natural processes or by early agricultural activity. (Left: adapted from T. Blunier et al., "Variations in Atmospheric Methane Mixing Ratio During the Holocene Epoch," *Nature* 374 [1995]: 46–49; right: adapted from J. E. Kutzbach and F. A. Street-Perrott, "Milankovitch Forcing of Fluctuations in the Level of Tropical Lakes," *Nature* 317 [1985]: 130–34.)

centuries. Some climate scientists have proposed that this 100 ppb increase between 5000 years ago and A.D. 1700 was caused by an expansion of natural wetlands at high northern latitudes. In boggy regions at high latitudes, cool, wet climates favor the accumulation of carbon in **peat,** which consists of layers of decaying vegetation. The abundance of decaying vegetation extracts oxygen from the bogs, creating the conditions needed for methane to be produced and expelled to the atmosphere. This is the **peatlands hypothesis.**

An alternative argument is that humans were responsible for at least some and possibly all of the CH₄ increase between 5000 and 300 years ago. The first known use of controlled irrigation for growing rice dates to 5000 years ago. Up to that time, farmers grew or harvested rice in whatever natural wetland areas nature happened to provide. As the wet tropical monsoon climate weakened, farmers developed methods of irrigation to replace this lost resource. At least some part of the CH₄ emissions coming from irrigated rice fields must date back 5000 years. And humans had first begun to tend livestock even earlier, by 10,000 to 9000 years ago. Because CH₄ formed in the stomachs of cattle and other livestock is emitted to the atmosphere, this source of CH₄ would have added to that produced by irrigation.

How do we estimate the quantitative impact of these early human activities on CH₄ emissions so many millennia ago? These sources of methane should have been tied to the number of humans alive at the time, because the only reason for growing rice and tending livestock was to feed people. If CH₄ emissions have always been roughly proportional to the number of living humans, we can use the quantitative relationship between the modern human population and the CH₄ emissions produced by their irrigation efforts and livestock sources as a basis for scaling down to the CH₄ emissions likely to have been emitted by the smaller number of humans living thousands of years ago. This approach suggests that human activities could have been responsible for the 100 ppb increase in methane between 5000 and 300 years ago.

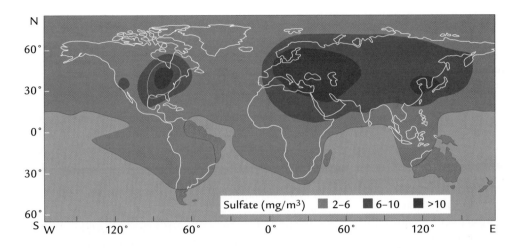

FIGURE 17-18 **Sulfate aerosol plumes** Sulfur dioxide emissions from industrial smokestacks in the North American Midwest, Eastern Europe, and China produce sulfate aerosols that prevailing winds carry eastward in large plumes. (Adapted from R. J. Charlson et al., "Perturbation of the Northern Hemisphere Radiative Balance by Backscattering from Aerosols," *Tellus* 43 [1991]: 152–63.)

removes them, they can be carried 500 or more kilometers downwind from the source regions. Today large plumes of sulfate aerosols are carried far from their sources in eastern Europe, east-central North America, and China (Figure 17-18).

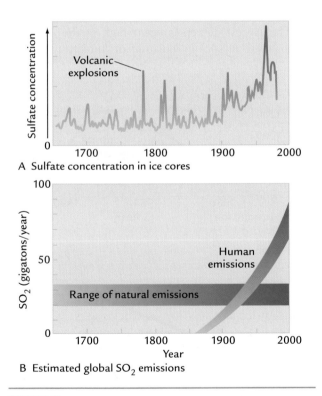

FIGURE 17-19 **Preindustrial and anthropogenic sulfates** (A) Measured sulfate concentrations in Greenland ice reveal a significant regional increase since the nineteenth century. (B) Estimated global SO_2 emissions from smokestacks now exceed natural emissions. (A: adapted from P. A. Mayewski et al., "An Ice-Core Record of Atmospheric Responses to Anthropogenic Sulphate and Nitrate," *Atmospheric Environment* 27 [1990]: 2915–19; B: adapted from R. J. Charlson et al., "Climate Forcing by Anthropogenic Aerosols," *Science* 255 [1992]: 423–30.)

Part of the history of SO_2 emissions can be read in ice cores. The concentration of sulfate in Greenland ice rose sharply in the last century, especially during the industrialization of World War II (Figure 17-19A). Independent estimates of industrial SO_2 production are consistent with this trend (Figure 17-19B). Industrial SO_2 emissions were smaller than those from natural sources through the nineteenth century but became the dominant SO_2 source by the middle of the twentieth. Sulfate concentrations in Greenland ice have dropped sharply since 1980, largely because the United States acted to limit SO_2 emissions from its smokestacks, but this trend is not global. Several other countries have continued to increase SO_2 emissions.

Climate scientists have inferred that large aerosol plumes are capable of causing regional cooling. Like sulfate aerosols created by volcanic explosions, industrial sulfate particles reflect and scatter some of the incoming solar radiation back to space and keep it from reaching Earth's surface. The reduction in radiation cools climate regionally.

A second potential climatic effect of sulfate aerosols is less well understood. The tiny particles form natural centers (nuclei) around which water vapor can condense, forming droplets and then clouds. Clouds can have two opposing effects on climate: their bright surfaces can reflect more incoming solar radiation and cool climate, but high clouds can also absorb more outgoing radiation from Earth's surface and increase the greenhouse effect. As we will see in Chapter 18, the net impact of clouds on climate is poorly known. The net global effect of increased sulfate aerosols is thought to be a cooling, but estimates of the size of the cooling vary widely.

17-12 Increases in Chlorofluorocarbons and Destruction of Ozone

Another gas that occurs naturally in the atmosphere but has declined in abundance in recent decades as a result of human activities is ozone. **Ozone** (O_3) is a triple molecule of oxygen that occurs naturally both in the stratosphere (with the largest amounts from 15 to 30 km) and in the underlying troposphere. Incoming ultraviolet radiation from the Sun liberates individual O atoms from oxygen (O_2) and produces ozone (O_3):

$$O_2 + \underset{(UV)}{\text{Radiation}} \rightarrow O + O$$

$$O + O_2 \rightarrow O_3$$

Ozone is naturally converted back to oxygen (O_2) in the atmosphere by a similar process, but in this case the radiation source can be either ultraviolet light or light in visible wavelengths:

$$O_3 + \underset{(UV \text{ or visible})}{\text{Radiation}} \rightarrow O_2 + O$$

$$O + O_3 \rightarrow 2O_2$$

With far more visible radiation available than ultraviolet, ozone is naturally destroyed much faster than it is produced, so ozone is a short-lived gas. In addition, the rate of conversion back to O_2 increases when certain chemicals are present to speed up the reaction. These chemicals contain chlorine (Cl), fluorine (Fl), and bromine (Br), all of which occur naturally in ocean salt and can be lifted high in the atmosphere by winds. Chlorine reacts with ozone and destroys it, forming chlorine monoxide (ClO):

$$Cl + O_3 \rightarrow ClO + O_2$$

Chlorine then reacts with free oxygen molecules and is liberated from ClO:

$$ClO + O \rightarrow Cl + O_2$$

These liberated chlorine atoms then begin a new cycle of ozone destruction. This cycle of natural ozone destruction is important to humans because ozone in the stratosphere forms a natural protective barrier that shields life-forms from levels of ultraviolet radiation that would otherwise encourage the mutations of cells, with resultant skin cancers.

Ozone also occurs naturally in much greater abundance in the lower troposphere. It originates from both natural and anthropogenic processes, including biomass burning and oil production in refineries. At these lower levels of the atmosphere, ozone generally plays a positive environmental role by cleansing carbon monoxide (CO) and sulfur dioxide (SO_2) from the air. At high concentrations, however, ozone is toxic to plants and an irritant to the eyes and lungs of humans.

In the last century, human activities upset the natural ozone balance in the atmosphere. Several decades ago, James Lovelock, originator of the Gaia hypothesis,

discovered that concentrations of a group of chlorine-bearing gases called **chlorofluorocarbons (CFCs)** were steadily increasing in the atmosphere (Figure 17-20). CFCs are produced entirely by humans for use as refrigerator and air-conditioner coolants, solvents, fire retardants, and foam insulation blown into buildings. Released at ground level, these compounds slowly mix upward through Earth's atmosphere. Because CFCs stay in the atmosphere an average of 100 years, they can reach the stratosphere, where their concentrations have been increasing for decades.

At first scientists thought these CFC increases were completely harmless, but in the early 1970s the atmospheric chemists Sherwood Roland and Mario Molina proposed that rising CFC levels would hasten the natural destruction of ozone by adding extra chlorine to the stratosphere. In the 1980s this proposed connection was confirmed by observations over Antarctica. Measurements showed that the amount of ozone in a column of air over Antarctica during the spring had decreased by about half in just two decades (Figure 17-21A) in a region of unusually abundant chlorine.

The largest decreases occur high in the Antarctic stratosphere in spring. Isolation of Antarctic polar air

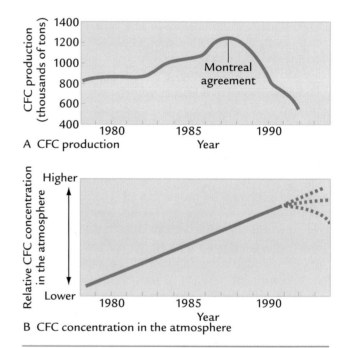

A CFC production

B CFC concentration in the atmosphere

FIGURE 17-20 Anthropogenic CFC increases Because CFCs (chlorofluorocarbons) do not occur naturally in the atmosphere, increasing CFC production by humans in recent decades (A) has caused rising concentrations in the atmosphere (B). (Adapted from T. E. Graedel and P. J. Crutzen, *Atmosphere, Climate, and Change,* Scientific American Library, ©1997 by Lucent Technologies.)

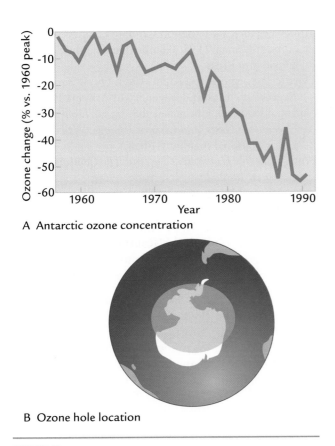

A Antarctic ozone concentration

B Ozone hole location

FIGURE 17-21 **Decline in Antarctic ozone** (A) The total amount of ozone in a column of air over Antarctica in spring decreased by half between the 1960s and 1980s, primarily as a result of rising CFC levels. (B) This ozone depletion has created an "ozone hole" over Antarctica during spring. (A: adapted from T. E. Graedel and P. J. Crutzen, *Atmosphere, Climate, and Change,* Scientific American Library, © 1997 by Lucent Technologies, after J. C. Farman et al., "Large Losses of Total Ozone in Antarctica Reveal Seasonal Interaction," *Nature* 315 [1985]: 201–10; B: adapted from T. E. Graedel and P. J. Crutzen, *Atmosphere, Climate, and Change,* Scientific American Library, ©1997 by Lucent Technologies, after A. J. Krueger, Goddard Space Flight Center, NASA.)

from the rest of Earth's atmosphere in the winter allows CFCs to accumulate to high levels that rapidly destroy ozone when solar radiation increases early the following spring. The region of major ozone depletion in the Antarctic stratosphere is called the **ozone hole** (Figure 17-21B).

This clear connection between CFCs and ozone depletion caused so much alarm that the world's nations signed a treaty in Montreal in 1987 to reduce and ultimately eliminate CFCs. Concentrations of those CFCs that industries have found easiest to replace have since stabilized and begun a slow decline (see Figure 17-20B). Other CFC's still in widespread use have continued to increase, but generally at much slower rates. Stratospheric

ozone levels have stopped falling, but have not yet begun a major recovery toward their natural levels.

Human activities have also contributed to large ozone *increases* in the lower atmosphere, causing periodic smog alerts in many of the world's large cities. When slow-moving air masses allow concentrations of ozone and other pollutants to build to dangerous levels in summer, city dwellers with respiratory problems are urged to remain inside until the air clears.

The increase in CFCs during the late twentieth century added to the total concentration of greenhouse gases. In contrast, the changes in tropospheric and stratospheric ozone are nearly offsetting: increased ozone around smogbound cities in summer represents an increase in the total level of greenhouse gases, while the coincident decrease in stratospheric ozone represents a smaller decrease.

In summary, humans have in recent millennia moved from passive participants in climate change to active agents of change. In the next chapter we will explore a current debate about whether or not humans have become the *primary* agents of global climate change.

KEY TERMS

climatic hypotheses of
  human evolution (p. 384)
C_3 pathway (p. 388)
C_4 pathway (p. 388)
diluvial hypothesis (p. 392)
Black Sea flood hypothesis
  (p. 392)
overkill hypothesis (p. 394)
preindustrial CO_2 level
  (p. 396)
anthropogenic CO_2
  increase (p. 396)

CO_2 fertilization
  (p. 398)
sulfate aerosols
  (p. 399)
peat (p. 400)
peatlands hypothesis
  (p. 400)
ozone (p. 402)
chlorofluorocarbons
  (CFCs) (p. 402)
ozone hole (p. 403)

REVIEW QUESTIONS

1. In what kind of environments and climatic conditions did most of human evolution occur?

2. Why is it difficult to determine whether or not climate affected human evolution?

3. What is the evidence for dramatic flooding in the Black Sea region 7600 years ago?

4. Why are strictly climatic explanations of the extinctions of large mammals 12,500 years ago suspect?

5. What are the main human activities that produce CO_2?

6. Where does the extra CO_2 produced by humans go?

7. What human activities produce methane?

8. How high in the atmosphere do sulfate aerosols from smokestacks reach?

9. Why do chlorofluorocarbons (CFCs) reach much higher in the atmosphere than sulfate aerosols?

10. Why is ozone depletion so distinct over Antarctica?

ADDITIONAL RESOURCES

Basic Reading

Curtis, J. H., D. A. Hodell, and M. Brenner. 1996. "Climatic Variability on the Yucatan Peninsula (Mexico) During the Past 3500 years, and Implications for Maya Cultural Evolution." *Quaternary Research* 48:37–47.

Intergovernmental Panel on Climate Change. 1996. *Climate Change 1995*. Cambridge: Cambridge University Press.

Martin, P. S., and R. G. Klein. 1984. *Quaternary Extinctions*. Tucson: University of Arizona Press.

National Research Council. 1988. *Ozone Depletion, Greenhouse Gases, and Climate Change*. Washington, D.C.: National Academy Press.

Potts, R. 1997. *Humanity's Descent: The Consequences of Ecological Instability*. New York: Avon.

Ryan, W., and W. Pitman. 1998. *Noah's Flood*. New York: Simon & Schuster.

Scientific Assessment of Ozone Depletion. 1994. Geneva: World Meteorological Organization.

Advanced Reading

Charlson, R. J., S. E. Schwartz, J. M. Hales, R. D. Cess, J. A. Coakley, J. E. Hansen, and D. J. Hoffman. 1992. "Climate Forcing by Anthropogenic Aerosols." *Science* 255:423–30.

Lean, J., and D. A. Warrilow. 1989. "Simulations of the Regional Climatic Impact of Amazon Deforestation." *Nature* 342:411–13.

Solomon, S., R. R. Garcia, F. S. Rowland, and D. J. Wuebbles. 1986. "On the Depletion of Antarctic Ozone." *Nature* 321:755–58.

CHAPTER 18

Climate in the Twentieth Century

In this chapter we focus on the **greenhouse debate**, a scientific and political controversy over the cause of climate change during the last century. Nearly every climate scientist agrees with two statements: (1) Earth's average surface temperature has warmed by 0.6°C in the last century, a conclusion based on direct measurements, with additional confirmation from other climatic indicators (Chapter 16); and (2) concentrations of CO_2, CH_4, and other greenhouse gases have been building up in Earth's atmosphere over the last 100 years at an increasing rate because of human activities (Chapter 17). The issue at the heart of the greenhouse debate is not *whether* rising greenhouse-gas levels have contributed to the observed warming but *how much* of the warming has been greenhouse-driven and how much has been caused by Earth's natural climatic variability.

In this chapter we find that uncertainties in the existing evidence preclude a firm assessment of the importance of natural versus human influences on climate in the last century. To counter confusion created by polarized mass media coverage of this issue, we also evaluate arguments used by those holding viewpoints at either extreme of the debate, scrutinizing their claims in the light of insights gained by studying Earth's climate history.

The Impact of Natural Variations in Climate

We can appreciate the important role of natural variations in climate in the greenhouse debate by comparing the three scenarios shown in Figure 18-1. Each case portrays an identical observed climatic warming, but each scenario generates this observed warming from a different combination of natural changes and greenhouse gases.

In scenario A, natural climatic variations are assumed to be small, so that almost all of the observed warming must be caused by greenhouse gases. In scenario B, natural variations are assumed to have pro-

duced a large warming, so that very little of the observed warming is caused by greenhouse gases. In scenario C, natural variability is assumed to have caused a cooling rather than a warming. In this scenario, the warming caused by greenhouse gases must be larger than the warming actually observed.

These widely differing scenarios highlight the central issue of the greenhouse debate: the unknown relative contributions of natural climatic changes and increases in greenhouse gases. Climate scientists can attempt to solve this problem in two ways. First, they can attempt to reconstruct the natural component of climate variation and then subtract it from the observed warming to isolate the warming caused by greenhouse gases. Second, they can attempt to define the impact of greenhouse gases directly.

18.1 Natural Variations on Different Time Scales

The understanding of Earth's climate history gained from the preceding sections of this book gives us some constraints on natural climatic variability during the last century. We have examined variations occurring on four time scales: tectonic, orbital, millennial, and historical. We are now in a position to compare these changes in climate with the temperature changes observed during the twentieth century (Figure 18-2).

Tectonic Time Scale Over the last 50 to 100 million years, Earth's mean temperature has cooled by 5° to 10°C. This extremely slow average rate of cooling would produce a temperature decrease of just 0.00001°C within 100 years, a negligible amount in comparison with the observed 0.6°C warming (Figure 18-2). Slow tectonic cooling can be ignored in the context of the greenhouse debate about the twentieth century.

Orbital Time Scale Between 17,000 and 6000 years ago, a warming caused by orbital changes melted several enormous ice sheets in the northern hemisphere (Chapter 14). Since then, Earth's climate has slowly drifted toward cooler conditions, but not yet into a new glaciation. The total northern hemisphere cooling has not exceeded 1°C in the last 5000 years (Chapters 14 and 16). Within 100 years, such a rate would produce a cooling of 0.02°C, a negligible change in comparison with the 0.6°C warming observed (Figure 18-2).

Orbital-scale changes were more rapid when more ice was present to provide greater positive feedback. Global temperature warmed by 4° to 6°C between 17,000 and 6000 years ago as the large northern hemisphere ice sheets melted, equivalent to an average rate of 0.05°C per 100 years. Rates were probably even faster during the two pulses of rapid deglaciation between 15,000 and 10,000 years ago. But with these ice sheets not in existence during the twentieth century, the slower rate of warming noted above is more likely to have prevailed.

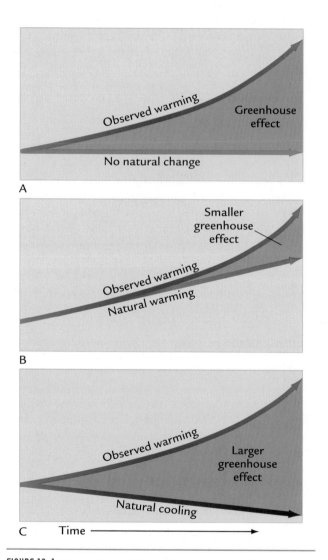

FIGURE 18-1 Natural warming and greenhouse effects
The fraction of the observed warming during the twentieth century that can be attributed to increased concentrations of greenhouse gases depends on the trend in natural climate, variously shown as (A) no change in temperature, (B) a natural warming, and (C) a natural cooling.

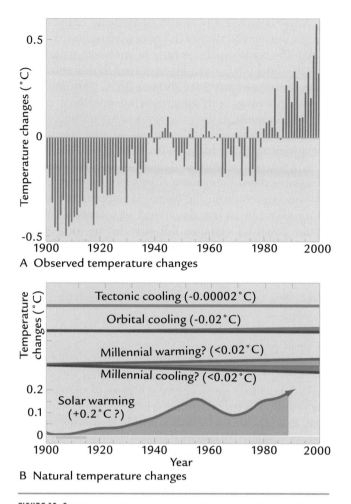

A Observed temperature changes

B Natural temperature changes

FIGURE 18-2 **Observed temperature changes and natural contributions** Compared with the observed change in global temperature during the twentieth century (A), a tectonic-scale cooling trend is undetectable, the orbital-scale cooling is small, and oscillations at millennial time scales are small and uncertain as to direction (B). Decadal-scale changes in solar radiation may have caused significant warming. (A: adapted from National Climate Data Center, NOAA, Asheville, N.C.)

Millennial Time Scales Millennial climatic oscillations could have been significant over the last 100 years. During the last glaciation, millennial-scale temperature oscillations of at least 5° to 6°C occurred in the North Atlantic region, although the size of these changes was probably much smaller when averaged over the entire planet, especially if part of the southern hemisphere warmed as the northern hemisphere cooled (Chapter 15). But large oscillations ceased when the ice sheets melted, and fluctuations have been much smaller during the last 6000 years or more.

The estimated change in northern hemisphere temperature over the 900 years before the twentieth century shows a temperature decrease of 0.2° to 0.25°C

(Chapter 16). This cooling could be either an orbital-scale trend, part of a millennial-scale oscillation, or some combination of the two. The record spans both the Little Ice Age, widely regarded as the most recent millennial-scale cool oscillation, and the four warmer centuries leading up to it.

The net cooling into the Little Ice Age is no larger than 0.1° to 0.2°C, providing one estimate of the size of natural millennial-scale oscillations, but not necessarily a completely valid one. If the short length of this 1000-year record missed the full warmth of the interval just preceding the Little Ice Age, the actual amplitude of this millennial oscillation might be larger. Similarly, if human activities in the industrial era helped to end the Little Ice Age before a full-sized cooling was reached, the estimated amplitude would again be larger.

On the other hand, this 1000-year temperature reconstruction covers only the northern hemisphere. If changes in the southern hemisphere's temperature were opposite in phase from those in the north, or completely unrelated, the total amplitude would be smaller. Given these uncertainties, a good estimate is that natural millennial-scale oscillations over the past 1000 years are unlikely to have altered global climate by more than 0.2°C.

But would the millennial-scale change in temperature during the twentieth century be a warming or a cooling? Again, the timing of these oscillations is not known well enough to permit us to answer this question. Millennial oscillations during the last several thousand years have been hard to correlate from region to region and generally irregular in rhythm, with only scattered indications of true cyclic behavior. As a result, we cannot say for certain whether millennial fluctuations would have caused climate to warm or cool during the twentieth century.

We can attempt to put an upper limit on the likely amplitude of change during the twentieth century. If the shorter millennial-scale cycles average 2000 years in length (a mid-range compromise between the oscillations at 1500 and 2500 years noted in Chapter 15), then a global average temperature change of 0.2°C spread evenly across a 1000-year half-cycle would produce an average rate of warming or cooling during a single century of 0.02°C (Figure 18-2). Such a cooling or warming would again be very small in comparison with the warming observed in the twentieth century.

The only way that natural millennial-scale oscillations might have been significant during the twentieth century would have been if the climate system had naturally flipped from the cold extreme of a millennial oscillation (the Little Ice Age) to the warm extreme (by chance coinciding with the industrial era). Evidence for this kind of behavior does exist during glacial intervals: the large warming at the end of the Younger Dryas oscillation occurred within a century (see Figure 14-7). But the climate record shows no evidence for this kind of behavior at hemispheric or global scales after the last remnants of the great ice sheets disappeared near 6000 years ago. The warming of the twentieth century is unlikely to be the result of such a climatic flip.

In summary, uncertainties in timing, amplitude, and extent of millennial-scale oscillations make it difficult to infer their effect on climate during the twentieth century, and even whether the effect on climate was a warming or a cooling. The evidence in hand suggests that these natural oscillations (whether toward warming or cooling) probably did not exceed 0.2°C on a global basis. If so, millennial-scale oscillations had only a small effect on global climate during the twentieth century.

Decadal Time Scales Natural changes in climate occurring on time scales shorter than decades are generally not relevant to the longer-term temperature trend of the twentieth century. The effects of short-term events disappear too quickly to alter Earth's mean surface temperature. Examples include the 1-year warmings associated with El Niño events and the 2-year or 3-year coolings caused by large volcanic explosions. The only way either of these factors could alter longer-term global climate would be if they occurred so frequently that the temperature effects of individual events overlapped. Such overlaps did not occur during the twentieth century, although some evidence indicates that closely spaced volcanic eruptions could have cooled climate over multiyear intervals in earlier centuries.

Solar variability is the major remaining factor likely to have contributed significantly to global temperature trends during the last 100 years. The correlation between global mean temperature and the long-term average number of sunspots during the last 100 years implies a possible cause-and-effect relationship. In addition, intervals of several decades with few or no sunspots within the last 500 years appear to correlate with times of cooler temperatures.

Some climate scientists argue that solar forcing explains much of the observed temperature changes in the twentieth century, basing their argument on the correlation between global temperature and sunspot trends (shown in Figure 16-24). The problem with this argument is that it demands strong feedbacks within the climate system to make Earth highly sensitive to small changes in the Sun's strength, yet it has to deny the strong action of these same feedbacks in the response of the climate system to changes in concentrations of greenhouse gases. A more consistent interpretation is that the climate system responds with similar sensitivity both to solar forcing and to changes in greenhouse gases, both of which have played roles in temperature changes during the twentieth century.

Various GCM simulations of evolving temperature trends during the twentieth century indicate that solar variability could explain as little as one-tenth to as much as half of the 0.6°C warming observed during the twentieth century. Figure 18-2 shows a 0.2°C warming from a strengthening Sun, equivalent to one-third of the observed warming.

In summary, only a few kinds of natural climatic variations are likely to have affected global temperature during the past century, and these impacts are all somewhat uncertain. Later we will subtract these estimated natural impacts from the observed temperature record in an effort to isolate the effects of increasing greenhouse gases on climate. But first let us look at the greenhouse debate from a different perspective.

Earth's Sensitivity to Greenhouse Gases

In this section we examine two sources of evidence that put constraints on Earth's sensitivity to greenhouse gases: (1) the performance of numerical models of climate (GCMs), which represent one synthesis of the scientific community's accumulated knowledge of the climate system; and (2) climate records available from a few critical intervals of Earth's history during which concentrations of greenhouse gases changed significantly.

18-2 Sensitivity to Greenhouse Gases in Climate Models

All GCMs, along with simpler climate models, have been used to simulate Earth's sensitivity to greenhouse gases. For convenience, this sensitivity can be quantified as the global average change in surface temperature caused by a doubling of CO_2 concentrations from their modern (preindustrial) level of 280 ppm. A CO_2 concentration twice that of the preindustrial era is entered as an initial boundary condition for each model, and the model is run until its simulated temperature comes into equilibrium with this higher CO_2 level. The global average increase in simulated temperature can be called the **2xCO_2 sensitivity** for that model.

Because CO_2 is not the only greenhouse gas that has increased in recent centuries, climate scientists also calculate the combined heat-trapping effects of *all* greenhouse gases by using CO_2 as the standard of reference. For example, the 150% increase in methane concentrations over the last 150 years has had an additional greenhouse effect equivalent to an 8% increase in CO_2. This is called an 8% change in **equivalent CO_2**. The actual 30% increase in CO_2 combined with this equivalent CO_2 increase yields an increase of 38%. Changes in other greenhouse gases (nitrogen oxide, chlorofluoro-carbons, and ozone) have produced additional increases of 12%, for a total equivalent CO_2 increase of close to 50%.

The basic way of quantifying the climatic forcing provided by greenhouse gases is through their impact on Earth's radiative balance, using the same W/m^2 units with which incoming solar radiation and Earth's back radiation are measured (Box 18-1).

GCM simulations over the last several decades have produced a wide range of estimates of Earth's temperature sensitivity to $2xCO_2$ (or equivalent CO_2) levels (Figure 18-3). These estimates average around 2° to 3°C but they vary over a factor of about 10, ranging from a lower limit near 0.5°C to an upper limit of about 5°C.

What accounts for such a wide variation in estimates of the climatic effects of CO_2? One factor that is well constrained and does not vary from model to model is the direct effect of CO_2 and other greenhouse gases in trapping outgoing longwave radiation emitted from Earth's surface (Box 18-1). Climate scientists widely agree that the direct radiative effects of doubling CO_2 (or equivalent amounts of other greenhouse gases) would increase global temperature by 1.25°C.

But this analysis is incomplete: it omits all the feedbacks working within the climate system. Positive feedbacks add to the warming produced by this baseline radiative forcing, while negative feedbacks counter some of the warming. The most prominent feedbacks come from changes in water vapor, the albedo of snow and ice, and clouds.

Water vapor is the major greenhouse gas in Earth's atmosphere today, and it should provide a major positive feedback to any greenhouse warming initiated by increases in CO_2 and other greenhouse gases. This pos-itive feedback specifically applies to the effects of water vapor in a clear, cloudless sky. The maximum possible amount of water vapor that air can hold in different regions or during different seasons today increases greatly at higher temperatures (see Chapter 2). In the modern world, this relationship provides a positive feedback to temperature changes driven by seasonal changes in solar radiation at different latitudes. This feedback helps make the tropics warmer than the poles and summers warmer than winters.

If this modern-day relationship has held up back through time, then water vapor should have produced a predictable (clear-sky) positive feedback to any temperature changes initiated by changes in CO_2. As a result, the initial 1.25°C warming caused by a doubling of CO_2 levels should lead to an increase in water vapor that produces an additional increase in global temperature (Figure 18-4). Most climate models indicate that a doubling of CO_2 would cause an additional temperature increase of 2.5°C from increased water vapor, twice the amount caused by CO_2 alone.

The atmospheric scientist Richard Lindzen disputes this widely held view and proposes that water vapor might actually work as a *negative* feedback. His view is based on the idea that warming would increase the rising motion of moist air only within very narrow areas of the tropics, while it would produce far more widespread downward motion of dry air. He suggests that the larger extent of regions of drier air would act as a thermostat in the tropics by reducing the average amount of tropical water vapor high in the atmosphere. This reduction of water vapor would cause a regional cooling that would oppose most of the warming initiated by increasing CO_2. Other climate scientists have noted that this idea appears inconsistent with several decades

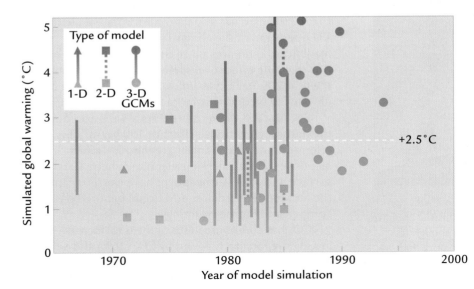

FIGURE 18-3 Model simulations of $2xCO_2$ sensitivity Simulations with several kinds of climate models in recent decades have yielded estimated sensitivities of global mean temperature to doubled concentrations of atmospheric greenhouse gases in the range of +0.5°C to +5°C. (Adapted in part from J. Adem and R. Garduno, "Feedback Effects of Atmospheric CO_2-Induced Warming," *Geofísica Internacional* 37 [1998]: 55–70.)

BOX 18-1 CLIMATE INTERACTIONS AND FEEDBACKS

Direct Radiative Forcing

We can measure the progress of radiation through Earth's climate system as incoming solar radiation, reflection and absorption of solar radiation, trapping and recycling of radiation by greenhouse gases, and outgoing radiation. Each of these processes redistributes radiative energy and each can be measured in units of W/m².

We can simplify these complex transfers of radiation within the climate system to focus on those most relevant to greenhouse gases. Out of 343 W/m² of incoming solar radiation, 240 W/m² penetrate into the climate system. Naturally occurring greenhouse gases, including not only water vapor but also CO_2 and CH_4 at their preindustrial levels, trap roughly 150 W/m² of back radiation from Earth's surface in a natural greenhouse effect. This trapping of energy, along with the internal feedbacks that result, helps to make Earth 31 °C warmer than it would be without greenhouse gases.

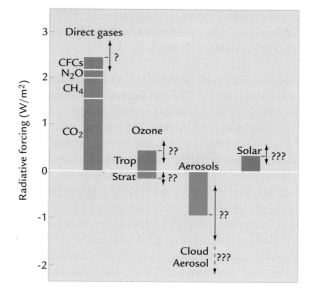

Radiative effects of greenhouse gases Several greenhouse gases have contributed a total of 2.5 W/m² to the greenhouse effect since 1850, with CO_2 accounting for 60% of the total. The contributions of aerosols and solar changes are less certain, as indicated by the increasing numbers of question marks. (Adapted from IPCC Scientific Assessment Working Group, *Radiative Forcing of Climate Change*, ed. J. T. Houghton et al. [Cambridge: Cambridge University Press, 1994].)

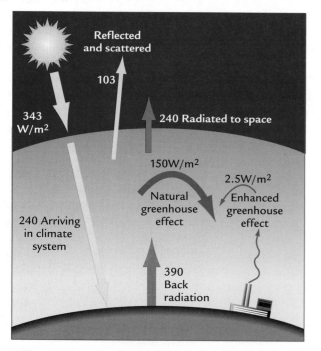

Effects of increases in greenhouse gases on radiation
Human activities since the start of the industrial era have increased greenhouse gas concentrations enough to enhance the natural greenhouse effect by more than 1%. (Adapted from IPCC Scientific Assessment Working Group, *Radiative Forcing of Climate Change*, ed. J. T. Houghton et al. [Cambridge: Cambridge University Press, 1994].)

Greenhouse gases produced by humans add to this natural 150 W/m² greenhouse effect. They do so in part by means of a direct **radiative forcing**; that is, the climatic impact that occurs even without the operation of the climate system's internal feedbacks. Estimates place the cumulative size of this trapping effect over the industrial age through the year 2000 at 2.5 W/m², or about 1% of the total amount of incoming solar radiation (240 W/m²).

Because 2.5 W/m² represents a 1.6% addition to the natural greenhouse effect of 150 W/m², it is referred to as an **enhanced greenhouse effect**. In response to this greenhouse enhancement, Earth's surface and lower atmosphere have warmed, and the atmosphere radiates additional heat to space to compensate for the warming.

CO_2 has contributed 60% of the total increase in radiative forcing, followed by CH_4 (16%) and a combination of other gases (24%). By far the largest increases in greenhouse gases have occurred since 1900, when CO_2 and CH_4 levels began their steepest rise.

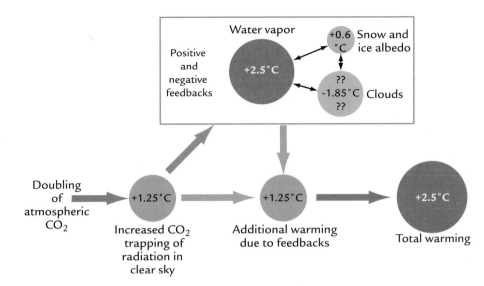

FIGURE 18-4 **Components of 2xCO₂ warming** Higher levels of greenhouse gases alter global temperature both by increasing the amount of heat trapped in a clear (cloud-free) atmosphere and by activating positive and negative feedbacks (water vapor, snow and ice, and clouds) that amplify or reduce the amount of temperature change.

of satellite measurements of seasonal trends. These measurements support the view that water vapor should act as a strong positive feedback.

Another factor in the climate system is positive feedback from ice and snow albedo. If a warming initiated by greenhouse gases causes a retreat of snow and ice toward the poles, the reduction in area of this high-albedo surface will reduce the reflection (and increase the absorption) of solar radiation arriving at Earth's surface at high latitudes. The resulting positive feedback should increase the initial $2xCO_2$ warming by 0.6°C (Figure 18-4).

Among the other feedbacks that operate within the climate system, clouds are by far the largest source of uncertainty in the models. Each type of cloud may have its own effect on climate, and none is well understood. The effects vary with the balance between the amount of solar radiation clouds reflect (thus cooling climate) and the amount of outgoing back radiation from Earth's surface they absorb (thus warming climate).

High, wispy clouds tend to warm Earth's climate slightly, because they are composed of ice crystals that are better at absorbing outgoing radiation than at reflecting incoming radiation. Thicker, lower clouds have just the opposite effect: they cool climate because they are better at reflecting incoming solar radiation than at trapping outgoing radiation from Earth's surface.

A major problem facing scientists is assessing changes in all types of clouds in Earth's atmosphere as the climate warms. Even the best climate models have grid boxes much too large to simulate individual clouds in a realistic way, much less all the critical small-scale physical processes that operate within clouds. These small-scale processes have to be simulated in models on the basis of statistical probability. For example, if the air temperature across a specific region in a model simulation cools to a particular value, the grid boxes within

that region are directed to produce a certain fraction of cloud cover that delivers precipitation during a set fraction of the time interval simulated by the model. This statistical approach is far less satisfying than having the model incorporate the actual processes that produce clouds and precipitation.

It is not even possible at present to predict whether the total amount of cloud cover on Earth would increase or decrease as greenhouse gases increase. A warmer atmosphere will evaporate more water vapor from tropical oceans, thereby increasing the amount of water vapor available to form clouds. But a warmer atmosphere also gains in its capacity to hold water vapor, which reduces the likelihood that vapor will condense into cloud droplets. These two effects compete, and it is unclear which would win in the real world. In addition, human activities produce small particles that may act as nuclei for cloud droplets. Available evidence points to increased cloud cover during the twentieth century, but the types of cloud are poorly known.

The result of these uncertainties is that the treatment of clouds in climate models over the last several decades has yielded net feedback effects ranging from slightly positive (a small warming) to highly negative (a large cooling). In the 1980s, as clouds were first being treated in GCMs as an interactive part of the climate system, several GCM simulations produced unusually high estimated sensitivities of +4° to +5°C in response to a doubling of CO_2 because clouds were treated as a positive feedback (Figure 18-3).

In the 1990s the treatment of clouds in GCMs incorporated the fact that a warming produced by increased CO_2 causes some of the heat-trapping cirrus clouds made of ice crystals to be replaced by Sun-reflecting lower clouds made of water vapor. The resulting increase in solar reflection and decrease in greenhouse trapping then produced an overall cooling feedback in most models.

The current consensus, based on a broad assessment by the **Intergovernmental Panel on Climate Change (IPCC)**, is that the negative feedback from clouds cancels more than half of the positive feedback from water vapor and albedo (Figure 18-4). The combined effect of all positive and negative feedbacks is an estimated 1.25°C warming, equivalent to the initial radiative warming caused by greenhouse gases in the absence of feedbacks. The IPCC estimates the net $2xCO_2$ sensitivity at +2.5°C, close to the longer-term average of climate models over three decades (Figure 18-3), but with the possible range of sensitivity still lying between +1.5°C and +4.5°C.

The changes through time in estimates of Earth's $2xCO_2$ sensitivity in Figure 18-3 show that model development is an ongoing process. Most recent changes in model operations have been based on studies of Earth's climatic behavior in the modern world, and much still remains to be learned, especially about clouds.

18-3 Sensitivity to Greenhouse Gases in Earth's Climate History

The warming of Earth's climate in response to a doubling of greenhouse gases represents only a single point along a continuous curve of varying responses at different CO_2 levels (Figure 18-5). The shape of this curve indicates a larger change in temperature for a given change in the amount of CO_2 in the lower (colder) end of the range than in the higher (warmer) end.

One reason for this varying degree of sensitivity along the trend is that snow and ice cover much more of Earth's surface when CO_2 concentrations are lower and climate is colder. As a result, the greater extent of snow and ice provides larger albedo-temperature feedback to any changes in CO_2. At the higher end of the CO_2 range, snow and ice are so reduced in extent even at the poles that little albedo feedback can occur. A second reason for the varying sensitivity is that the trapping of Earth's back radiation by CO_2 becomes less efficient at higher concentrations as the atmosphere gradually becomes saturated with CO_2.

Although the trend emphasized in Figure 18-5 implies that Earth has a sensitivity to a doubling of CO_2 from 280 to 560 ppm near 2.5°–3°C, we have just seen that climate models indicate a wide range of uncertainty around this estimate (Figure 18-3). In fact, the only firmly known point on this entire plot is the one pairing the preindustrial value of CO_2 with the preindustrial global average temperature. Away from this point, all trends are uncertain.

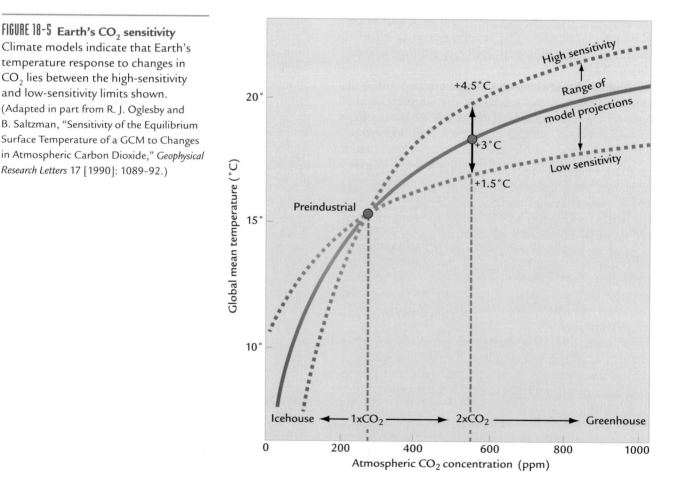

FIGURE 18-5 Earth's CO_2 sensitivity
Climate models indicate that Earth's temperature response to changes in CO_2 lies between the high-sensitivity and low-sensitivity limits shown. (Adapted in part from R. J. Oglesby and B. Saltzman, "Sensitivity of the Equilibrium Surface Temperature of a GCM to Changes in Atmospheric Carbon Dioxide," *Geophysical Research Letters* 17 [1990]: 1089–92.)

The range of uncertainty implied by the models can be represented by two (dashed) curves added to the CO$_2$-temperature relationship in Figure 18-5. These curves represent the higher (+4.5°C) and lower (+1.5°C) ends of the range of model sensitivity to doubled CO$_2$. Together they constrain the possible range of response of the climate system derived from model simulations. Note that the high-sensitivity curve warms faster for a given increase in CO$_2$ concentrations, but also cools faster as CO$_2$ levels decrease.

We can further explore this relationship between CO$_2$ and global temperature by examining key intervals from Earth's climate history. At selected times in the past, CO$_2$ levels different from today's are either precisely known or reasonably well constrained, and estimates of changes in global mean temperature are also constrained within fairly narrow limits. If we can adjust for the impacts of changes caused by forcing other than CO$_2$, the temperature changes at these warmer or colder times in the past should represent the end result of all the interactions and feedbacks caused by changes in CO$_2$ and other greenhouse gases. In effect, we can let Earth's climate system do the math on the greenhouse problem for us.

The Last Glacial Maximum One useful interval is the last glacial maximum, 21,000 years ago. At that time atmospheric CO$_2$ values were near 190 ppm, about 30% lower than the preindustrial level of 280 ppm, and methane values were 350 ppb, almost 50% lower than their preindustrial value. The methane reduction

amounts to a loss of 15% in CO$_2$ equivalent, bringing the net drop in equivalent CO$_2$ concentration to 45%. As discussed in Chapter 13, greenhouse gases must have been the major factor affecting ocean temperatures at tropical and lower subtropical latitudes: solar insolation values were close to those today, and ice sheets were too distant to have had large direct effects through changes in atmospheric circulation.

We also saw in Chapter 13 that the amount of glacial cooling in the tropics is debated, with estimates ranging from as small as −1.5°C (CLIMAP) to as large as −4° or −5°C. If this range of glacial cooling of the tropics resulted only from lower CO$_2$ and CH$_4$ levels, it would indicate that Earth's sensitivity to CO$_2$ largely falls within the range indicated by climate models (Figure 18-6). The smaller (1.5°C) cooling during the glacial maximum indicated by the CLIMAP reconstruction of sea-surface temperature would put Earth's sensitivity at the lower end of the range simulated by the models, while the larger (4° to 5°C) cooling indicated by expanded mountain glaciers and other evidence would push Earth's sensitivity toward or beyond the higher end of the range.

7 Million Years Ago Because no ice cores extend back beyond 400,000 years, a different strategy is needed to estimate CO$_2$ levels for earlier intervals. Sedimentary climate archives from warm, arid regions of four continents record the beginning of a shift from vegetation dominated by a mixture of trees and shrubs to vegetation dominated by warm-season

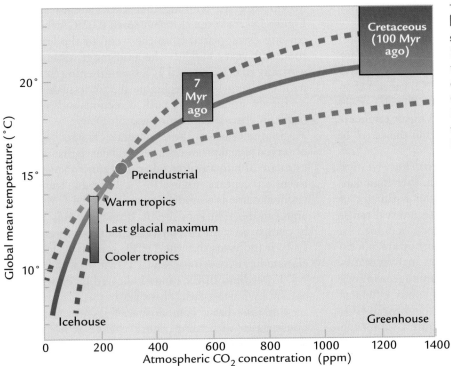

FIGURE 18-6 **Estimates of 2xCO$_2$ sensitivity from Earth history** Estimated changes in global temperature and measured or estimated changes in atmospheric CO$_2$ concentrations from earlier intervals of Earth's history fall within the range of sensitivities indicated by climate models.

grasses near 7 Myr ago (Chapter 17). The fact that similar changes began simultaneously on four continents suggests the need for an explanation that operates at a global scale.

Controlled experiments with plants in greenhouses suggest that a drop in CO_2 through a critical threshold between 600 and 500 ppm could trigger such a change to dry-adapted vegetation in regions of moderate summer drought stress. If this interpretation is correct, Earth may have run its own $2 \times CO_2$ sensitivity experiment near 7 Myr ago, with CO_2 values slowly falling through the $2 \times CO_2$ (560 ppm) level.

No organized attempt has been made to reconstruct global-scale climate near 7 Myr ago, but geologic data point to considerably greater warmth than we have today. The Greenland ice sheet did not yet exist, and the modern belt of tundra and permafrost surrounding the Arctic Ocean was then conifer forest. In the Antarctic, much less ice-rafted debris was being deposited in ocean sediments, indicating less ice in West Antarctica and in coastal regions of East Antarctica where today ice shelves shed icebergs to the ocean. These major differences in Earth's climate at both poles 7 Myr ago clearly require a climate considerably warmer than today's.

Models of ice ablation and accumulation indicate that summer temperatures must have been at least 3°C warmer than in the twentieth century to prevent ice from accumulating on Greenland. Allowing for the likelihood of larger changes in the winter season (based on the absence or near-absence of permafrost and tundra in the circumarctic landscape), annual temperatures in the Arctic were probably at least 5° warmer than they are today.

Allowing for the fact that high latitudes register temperature changes two to three times larger than for the planet as a whole, global mean temperatures 7 Myr ago must have been at least 2° to 2.5°C warmer than they are today. The estimated range of CO_2 values (500 to 600 ppm) and the estimated global mean temperature increase (+2° to +2.5°C) plot within the range of model sensitivities (Figure 18-6).

The Cretaceous (100 Myr Ago) The ice-free Cretaceous world of 100 Myr ago is another candidate for comparison with the modern world. Estimates of greater warmth during this greenhouse interval range from +5° to +11°C, based on methods examined in Chapter 6. CO_2 values for the Cretaceous are poorly constrained: several techniques based on analysis of carbon isotopes in the remains of fossil plants agree in suggesting higher CO_2 values than today's, but estimates range from four to twelve times the preindustrial CO_2 level. As a result, the box plotted in Figure 18-6 for the Cretaceous world of 100 Myr ago has a very large range of uncertainty. The upper limits of possible change in

both temperature and CO_2 lie well off the limits actually plotted.

In summary, three intervals of the past provide evidence that can be used to test the range of CO_2 sensitivity estimated from climate models (Figure 18-6). Estimates of CO_2 and temperature values for all three intervals plot within (or mostly within) the range of model sensitivity, at least crudely validating the basic performance of the GCMs. Unfortunately, these estimates from the past do not help us to narrow the range of model sensitivity. We are still left with a mid-range sensitivity estimate of a 2.5°C warming to a doubling of CO_2, and with an upper limit near +4.5°C and a lower limit near +1.5°C.

Causes of Global Warming in the Twentieth Century

What happens when we try to compare the range of sensitivity estimates from GCMs and from Earth's climate history against the changes in global temperature and in greenhouse gases measured during the last century? Even at first glance, the global warming of 0.6°C since 1900 seems small in comparison with the 30% rise in CO_2, and especially so in comparison with the 50% rise in equivalent CO_2 that results when we combine the effects of *all* greenhouse gases.

To isolate the possible effect of rising greenhouse gases on temperature, we need to subtract the estimated changes in temperature caused by natural variations from the observed changes in the instrumental record. To do this, we add together two changes shown in Figure 18-2: (1) the slow but small 0.02°C orbital-scale cooling extrapolated from the trend of the last several thousand years and (2) the mid-range estimate of a warming of 0.2°C caused by a strengthening Sun. Other natural changes are negligible (tectonic-scale cooling) or highly uncertain in both direction and amplitude (millennial-scale cooling or warming).

Subtraction of these natural changes from the observed signal produces a residual curve that is a rough estimate of human-induced temperature changes in the twentieth century (Figure 18-7). The basic trend of this residual curve is a gradual warming of 0.4°C that accelerates in the 1980s and 1990s. If this warming is directly compared with the observed increase in equivalent CO_2, it indicates that the $2 \times CO_2$ sensitivity for Earth's climate lies between 0.7° and 1°C, a level well below the 1.5° lower limit of the range indicated by climate models and by Earth's earlier history.

But this direct comparison of increases in greenhouse gases with temperature change is not valid. It ignores other factors that may have affected temperatures during the twentieth century.

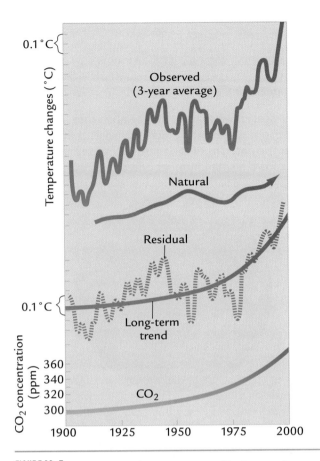

FIGURE 18-7 **Temperature increases caused by humans?**
If estimated natural changes in temperature during the twentieth century are subtracted from the observed temperature changes over that same interval, the residual curve is an estimate of human effects on global temperature.

18-4 Delayed Warming: The Thermal Inertia of the Ocean

One problem with this direct estimate of greenhouse sensitivity is that it fails to allow for the effects of thermal inertia in delaying the full response of the climate system to the higher levels of greenhouse gases. In actuality, the climate system has a natural inertia that slows its overall response to any kind of abrupt change in climate. At times when external factors begin to alter climate, some parts of the climate system react more slowly than others because of their greater thermal inertia. Although the atmosphere responds quickly to imposed changes (including CO_2), ice sheets will not respond fully for many thousands of years (Figure 18-8).

As greenhouse gases have increased over the past century, the most important source of thermal inertia for the climate system as a whole has been the ocean, which covers 70% of Earth's surface. The upper 100 meters of ocean is stirred by lower-atmospheric winds and so is a relatively fast responding part of the climate

system, capable of warming in response to rising levels of greenhouse gas within just a few years.

But most of the ocean below the wind-mixed layer, remote from the surface environment, will only slowly be affected by a change in climate. In a few regions, such as the North Atlantic, rapid formation of deep water has the potential to transmit surface temperature changes into parts of the deeper ocean over time spans of decades. But records of deep-ocean temperature trends over most of the last century do not exist, with a few exceptions (Box 18-2).

Viewed from a global perspective, the slower reaction of the subsurface ocean should delay the ocean's overall response to increasing concentrations of greenhouse gases. This delay results from interactions (mixing and heat exchanges) between the wind-mixed surface layer and the underlying ocean. These exchanges act to slow the otherwise rapid response of the surface ocean.

As a result, the rise in global surface temperature observed at any time in the twentieth century must represent only a part of the warming that is eventually going to occur, because warming is always delayed by the slow response of the system (Figure 18-9). The size of this delayed warming should have been largest at the end of the century because greenhouse gas levels were rising most rapidly late in the century.

The implication of a delayed warming is obvious: estimates of Earth's true 2xCO_2 sensitivity must be larger than the 0.7°–1°C value calculated by comparing the CO_2 increase between 1900 and 2000 with the amount

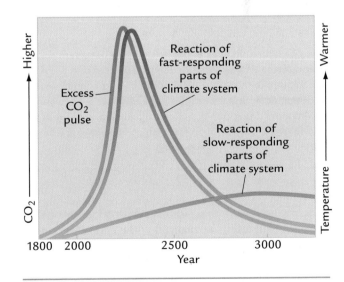

FIGURE 18-8 **Different response times in the climate system**
Parts of the climate system, such as the atmosphere, can respond to imposed changes in days or weeks, while others, such as ice sheets, require thousands of years. The near-surface ocean response occurs over decades.

BOX 18-2 CLIMATE INTERACTIONS AND FEEDBACKS

Deep-Ocean Warming in the Twentieth Century?

Could some of the greenhouse warming be hidden from view in the subsurface ocean? This idea sounds plausible: the ocean has an enormous capacity for storing heat. Temperature increases of just 0.01°C spread worldwide across the deep ocean could mask a greenhouse warming of several degrees Celsius. On the other hand, it is unlikely that greenhouse heat could have penetrated into much of the deep ocean in such a short time.

Unfortunately, deep-ocean temperatures have not been measured consistently over the last century. The two longest records span half a century at the most. Both are from the North Atlantic Ocean and each shows a different trend. One record from the subtropics near Bermuda shows an erratic warming trend of 0.2°C over 40 years. Averaged over the entire water column at this site, the small subsurface warming outweighs the surface cooling and indicates that the entire water column has warmed. If this warming were representative of the entire world ocean, it would indicate an enormous storage of heat.

But a second temperature signal from a higher-latitude site east of Newfoundland shows little net change at similar depths during the same interval, and possibly even a small cooling. The different trends at these two sites, and the presence of

smaller-scale temperature fluctuations that are not understood, suggest it would be unwise to attempt to infer the patterns of long-term ocean temperature change from a small number of records.

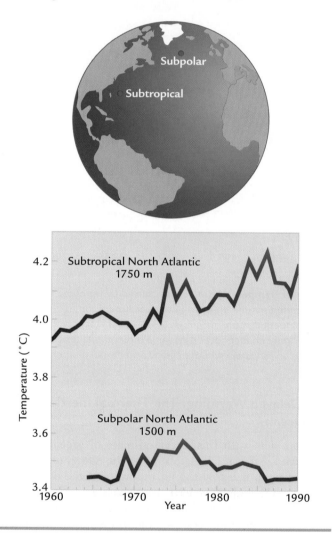

Temperature changes in the deep ocean Long-term monitoring of deep Atlantic temperatures shows warming at 1750 m in the subtropical ocean but little net change at 1500 m at subpolar latitudes. (Adapted from S. Levitus and J. Antonov, "Observational Evidence of Interannual to Decadal-Scale Variability of the Subsurface Temperature-Salinity Structure of the World Ocean," *Climate Change* 31 [1995]: 495–514.)

---

of warming during that century. With some amount of warming still left to be realized at the end of the century, such a direct comparison cannot be valid.

Unfortunately, it is extremely difficult to quantify the delay caused by the slow response of the subsurface ocean and to recalculate Earth's temperature sensitivity to rising levels of greenhouse gases. Ocean models generally indicate that ocean temperatures should take two or more decades to come into equilibrium with rapidly

imposed climate changes. The longer this delay, the larger must be the unrealized warming stored in the climate system. But the mechanisms by which heat penetrates the subsurface ocean and is redistributed through it are complex, and ocean models cannot at this point tell us exactly how large a delayed warming lies in the future.

Adjusting for this delayed warming will increase estimates of Earth's $2xCO_2$ sensitivity well above the

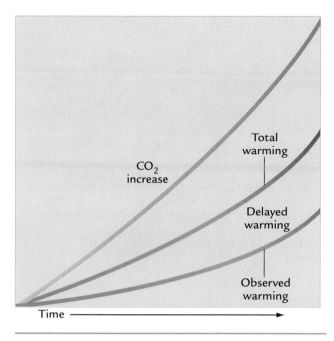

FIGURE 18-9 **Delayed warming in the climate system?** During an interval of rapidly rising CO_2 concentrations, the warming measured at any time is smaller than the warming that will be realized when Earth's delayed response is registered.

$+0.75°$ to $+1°C$ range derived by the direct comparison noted earlier, but probably only into the lower end of the $+1.5°$ to $+4.5°C$ range indicated by GCMs and by Earth's climate history.

18-5 Have Smokestack Sulfates Canceled Part of the Greenhouse Warming?

In the late 1970s the climate scientist Murray Mitchell proposed that a greenhouse gas warming might be significantly reduced by an offsetting cooling effect caused by smokestack emissions of SO_2 and the resulting production of sulfate particles in the atmosphere (Figure 18-10). SO_2 emissions have risen in the same basic exponential trend as CO_2 emissions (Chapter 17), because both are by-products of the era of industrialization, but sulfate particles should push global temperature in the opposite direction. Mitchell suggested that the maximum sulfate cooling effect might conceivably cancel the greenhouse warming, but he also noted that the uncertainties regarding the size of sulfates' impact on climate did not rule out a much smaller (or even a negligible) cooling effect.

Even today the impact of sulfate particles on climate is still somewhat uncertain. One way to try to test Mitchell's idea is to examine regions lying downwind from the three major SO_2 sources, the United States,

Europe, and China. The effects of SO_2 injected into the atmosphere by smokestacks during the last century should be most evident beneath the plumes of sulfate particles spreading downwind from these regions, either as a net cooling or as a reduction in the expected greenhouse warming during recent decades.

If we compare the locations of the three major sulfate plumes (see Figure 17-18) with a map of net temperature changes between 1955–74 and 1975–94 (Figure 18-11), some hint of a relationship emerges, along with mismatches. One region of cooling over Eastern Europe matches the prominent sulfate plume in that area. Another region of cooling over the southeastern United States and the nearby Atlantic only partially overlaps the sulfate plume in the midwestern and eastern United States. A large region of cooling over the central North Pacific lies well east of the sulfate plume in China. In the end, this comparison does not prove a clear climatic response to sulfates, perhaps because it is confounded by other natural regional-scale fluctuations within the climate system.

If sulfate particles emitted from smokestacks actually have caused the climate to cool, this cooling effect would mask part of the true amplitude of the global (and greenhouse gas) warming during the twentieth century and alter estimates of Earth's true sensitivity to changes in the level of CO_2. For example, if sulfates have canceled more than half of the greenhouse gas warming, then the $2xCO_2$ sensitivity of the climate system could lie at the high end of the range simulated by models, between $+2.5°C$ and $+4.5°C$. But if the cooling effects of sulfates have been negligible, Earth's inferred

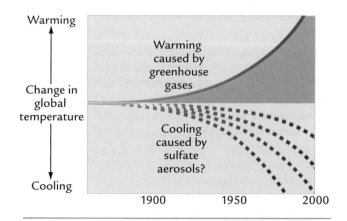

FIGURE 18-10 **Does SO$_2$ cooling counteract greenhouse gas warming?** The warming effect of greenhouse gases may be partly canceled by the cooling effect of sulfates produced by SO_2 emitted from smokestacks. (Adapted from J. M. Mitchell, "The Natural Breakdown of the Present Interglacial and Its Possible Intervention by Human Activities," *Quaternary Research* 2 [1973]: 436–45.)

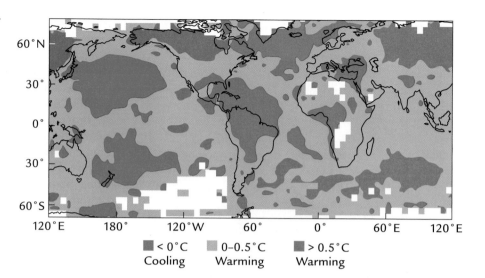

FIGURE 18-11 **Sulfate cooling effect?** Regions of net cooling in the last half of the twentieth century only partly match locations where sulfates have been added to the atmosphere by industrial activity (see Figure 17-18). (Adapted from IPCC Working Group I, *Climate Change 1995: The Science of Climate Change*, ed. J. T. Houghton et al. [Cambridge: Cambridge University Press, 1995].)

$2\times CO_2$ sensitivity would remain much lower. The IPCC estimates that sulfates have canceled about 15–20% of the greenhouse warming, consistent with a $2\times CO_2$ sensitivity of +2.5°C.

The magnitude of this sulfate cooling effect is obviously a major wild card in analyses of the climate system. Knowing its actual effect on climate in the twentieth century would help resolve Earth's true sensitivity to CO_2 and the actual warming effect of greenhouse gases.

If future research were to confirm that sulfate particles produced from smokestack SO_2 do cool climate significantly, humans would face an interesting dilemma suggested by the imaginary sequence in Figure 18-12. If all human emissions of both SO_2 and greenhouse gases were suddenly eliminated entirely, precipitation would wash the excess sulfate particles out of the atmosphere within a few weeks, and their cooling effect would soon disappear. In contrast, an abrupt and total end to CO_2 emissions would not produce a significant reduction in atmospheric CO_2 levels for a long time, because it would take much longer—hundreds to thousands of years—for the excess CO_2 to be mixed into the subsurface ocean, its ultimate fate.

As a result, an abrupt cessation of all industrial and other human emissions to the atmosphere could produce an abrupt climatic warming, caused by the rapid removal of the sulfate cooling effect and the much longer persistence of the CO_2 warming effect (Figure 18-12). This sudden warming would be aggravated for the few decades it takes the climate system to achieve its full delayed warming response to earlier inputs of CO_2. The overall warming would then persist as long as the pulse of excess CO_2 lingered in the atmosphere.

The implication of this simple example is that vigorous actions to clean up smokestack SO_2 emissions could quickly produce some degree of global-scale warming.

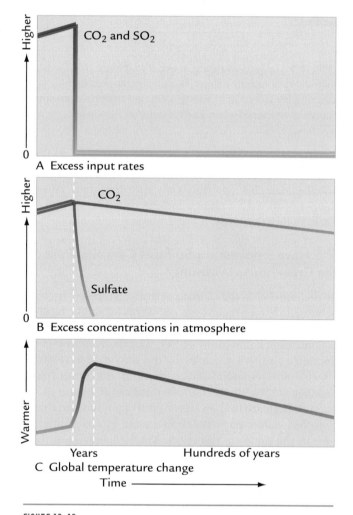

FIGURE 18-12 **What if we abruptly ended CO_2 and SO_2 emissions?** If humans instantly eliminated all industrial and other emissions of greenhouse gases and SO_2, the sulfates and their cooling effect would soon disappear, but the CO_2 and its warming effect would linger for centuries.

One trend to watch in the near future is the effort already under way in the United States to reduce smokestack SO_2 levels. Allowing for the delayed response of the climate system, sulfate reductions in this region could soon produce a warming over the east-central United States, one of the regions on Earth that has not warmed during the twentieth century. The size of this regional warming could be one indication of the strength of the sulfate cooling effect.

18-6 Summary of CO_2 Sensitivity

All climate scientists agree that atmospheric concentrations of CO_2 and other human-produced greenhouse gases have increased markedly in the last century and almost all agree that this increase must have caused climate to warm. But agreement on these issues still allows room for disagreements over whether or not a specific greenhouse warming signal has been detected. These disagreements hinge on the more specific issue of *how much* warming the observed increase in greenhouse gases has caused in the past (and is likely to cause in the future). Scientists are uncertain about this amount, primarily for two reasons.

First, Earth's sensitivity to the known increase in greenhouse gases is uncertain to within a factor of 2 or 3. As a result, even though we can be certain that the observed increase in greenhouse gases must have contributed to the observed warming, scientists cannot yet specify how much it has contributed with any certainty. Second, factors other than greenhouse gases have also affected climate during the last century, and these contributions make it harder to extract the true greenhouse gas signal.

Climate models indicate that Earth's temperature would warm by between 1.5° and 4.5°C for a doubling of CO_2, with uncertainties about clouds the greatest obstacle to narrowing the current range of uncertainty. Climate data from intervals in Earth's earlier history indicate a similar range of sensitivity, with uncertainties about the exact size of past changes in temperature or CO_2 the largest obstacle to narrowing this range.

But CO_2 sensitivities from climate models and from Earth's history do not match the much lower sensitivity implied by direct comparison of the (relatively small) observed warming during the twentieth century with the (relatively large) rise in CO_2 and other greenhouse gases. This mismatch increases if a solar-driven warming is assumed to have occurred during the twentieth century, thus reducing the warming attributed to greenhouse gases.

The low CO_2 sensitivity implied by changes observed during the twentieth century can be reconciled with the sensitivities implied by models and by Earth's climate history only by invoking (1) Earth's delayed response to the rise in greenhouse gases because of the thermal inertia of the ocean and (2) SO_2 emissions as a source of cooling to counteract some of the greenhouse warming. The exact quantitative impact of these factors cannot be determined at present.

It remains to be seen whether or not the low CO_2 sensitivity implied by temperature changes observed in the twentieth century will actually be reconciled with the higher range of sensitivity estimated from the models and from Earth's climate history. Still, *all* estimates of Earth's CO_2 sensitivity agree that rising levels of greenhouse gases have warmed Earth's surface.

The Greenhouse Debate: Proponents and Skeptics

The current greenhouse debate is focused mainly on the warming of the twentieth century. The central issue is how much of the warming observed in the last century was caused by human additions of greenhouse gases to the atmosphere and how much by natural changes in climate.

The full spectrum of scientific opinion is wide, but most climate scientists hold middle-of-the-road opinions about this issue. They are aware of uncertainties in the exact effects of clouds, sulfates, and the strength of the Sun, and they conclude that it is not yet possible to isolate precisely the exact amount of warming caused to date by greenhouse gases. But few doubt that a greenhouse warming has occurred.

Unlike many other issues, the greenhouse debate has not stayed within scientific circles. On the one hand, this issue has drawn the attention of environmentalists, some of whom have called for vigorous changes in economic policy to counteract future increases in emissions of greenhouse gases on behalf of an Earth they see as fragile and easily disturbed by human activities. On the other hand, the environmentalist stance has raised concerns on the part of some industrial leaders, who feel it is premature (or completely unnecessary) to undertake expensive retrofitting to reduce emissions whose climatic impact is of such uncertain magnitude.

A small number of people with relatively extreme views on this issue (scientists and others) at times make inflammatory statements that are picked up and reported by the mass media. The media tend to deal with contentious issues by turning to people with extreme views and ignoring views they consider less newsworthy precisely because they are better balanced. As a result, the public receives a stream of sharply contrasting statements that sound, and often are, in total disagreement.

Politicians near both ends of the political spectrum enter the debate by endorsing the view that suits their position on environmental or economic issues and by criticizing their opponents' views. The natural dynamics of politics further polarizes the debate, confuses the public, and engenders disillusionment about the entire process of scientific investigation of this issue.

This chapter and previous chapters have presented the scientific basis of our understanding of Earth's climate history, and more specifically the greenhouse issue. Now we know why scientists and other citizens have good reason to be uncertain about the exact magnitude of the effect of greenhouse gases on climate. We also know why most scientists neither dismiss greenhouse warming as irrelevant to humanity's future nor at this point see it as an extreme threat to humanity and to the planet. The history of Earth's climate tells us that the truth lies somewhere between these extremes, and no one is sure exactly where.

This section addresses the greenhouse warming debate from a different direction, one specifically intended to counteract some of the negative effects of media coverage and political polarization. The focus here is on extreme statements of the kind often voiced by what can be called greenhouse skeptics and greenhouse proponents, those who hold extreme positions on this issue. These views are compared with what we have learned about the operation of the climate system through the ages.

The skeptic says:

Rising levels of greenhouse gases in the twentieth century have not caused Earth to warm.

As discussed earlier, it is perfectly legitimate to state that scientists have not yet isolated the specific portion of the warming caused by greenhouse gases. But to deny that greenhouse gases have contributed any warming at all is to deny the very validity of the term "greenhouse gases." Perhaps the most persuasive evidence that high CO_2 levels have a greenhouse warming effect comes from the atmosphere of Venus, which is more than 97% CO_2 and has surface temperatures of 460°C, even though less solar radiation reaches its surface than reaches Earth's surface. This greenhouse warming occurs in the absence of any positive feedback from water vapor. If high levels of CO_2 explain why Venus is hot, then rising levels of CO_2 and other greenhouse gases must cause Earth to warm.

Again the skeptic:

The real greenhouse effect on Earth comes from water vapor (H_2O_v), not from CO_2.

This statement is a deceptive half-truth. Although water vapor is indeed the major greenhouse gas in Earth's atmosphere, it comes into play as a positive feedback only in response to changes already under way in response to some other factor (see Figure 18-4). The positive feedback from water vapor operates through the well-defined relationship between air temperature and the amount of water vapor the air can hold (Chapter 2). No one has yet identified a mechanism by which water vapor would change in a completely independent manner and act as a first cause that initiates climate change.

The skeptic:

It is arrogant of humans to think they are capable of altering Earth's climate on a global scale, because Earth's climate system is resilient and its self-healing negative feedbacks will moderate any changes we impose.

The proponent counters:

Positive feedbacks make Earth highly vulnerable to climate changes, and rising levels of greenhouse gases will cause unprecedented changes that will lead to disasters.

As we have seen throughout this book, this debate over vulnerability and resiliency has no one simple answer. Both positive and negative feedbacks operate in the climate system and determine its ultimate response. On very long time scales, chemical weathering apparently moderates imposed climate changes and keeps Earth habitable (Chapter 4). On much shorter time scales, positive feedbacks from water vapor and from snow and ice albedo appear to outweigh negative feedbacks from clouds, but only by a small amount. At present scientists simply do not know the balance of all possible feedbacks working on each time scale and the degree to which Earth is resilient or vulnerable to their combined effects. The recent discovery of large millennial-scale climatic oscillations with uncertain climatic forcing points to potential short-term vulnerability of the climate system, but these changes are still poorly understood (Chapter 15). In addition, much research remains to be done on other feedbacks, particularly those involving vegetation.

The skeptic:

Some areas have warmed very little or even cooled during the last 100 years, so how can you speak of "global" greenhouse gas warming?

Because the climate system is complex and regionally diverse, it is overly simplistic to expect all its components (and regions) to march in lockstep. Regional patterns of climate change have varied earlier, such as the prominent monsoon response in the tropics and the different ice-sheet response in polar regions at orbital time scales (Chapters 9 and 10), and the fact that the American Southwest was much wetter during the last glacial maximum because of a shift in the jet stream, even though most high northern latitudes were drier (Chapter 13). In recent decades, sulfate plumes may cause regional coolings that oppose a more widespread tendency toward greenhouse warming.

The proponent:

The 30% reduction in CO_2 during the last major glaciation 21,000 years ago proves that Earth's climate is highly sensitive to greenhouse gases.

For reasons explained in Chapter 11, the lower CO_2 levels that recur during each glacial maximum are thought to be mainly a response to glaciation rather than a cause of it. On orbital time scales, CO_2 has generally acted as a positive feedback, helping to intensify glaciation trends driven primarily by variations in Earth's orbit.

Again the proponent:

The warming we experienced in the twentieth century is evidence of a powerful greenhouse effect that will grow much larger in the near future.

As we have seen, not all of the warming in the twentieth century can be attributed to greenhouse gases. At this point it is impossible to rule out the claim that the sensitivity of the climate system to changes in CO_2 is relatively small and that significant changes in climate lie far in the future.

The skeptic:

The timing of the observed temperature increase during the twentieth century does not match the observed increase in CO_2: the fast warming between 1920 and 1950 occurred during a time of relatively small input of greenhouse gases, and global temperature actually cooled slightly between 1950 and 1970 as the input of greenhouse gases accelerated.

Because of the presence of positive and negative feedbacks in the climate system, it is simplistic to expect exact one-for-one responses to factors that alter climate. For each time scale examined, more than one process is at work to cause the climate to change. On tectonic time scales, the progressive cooling of the last 55 million years was erratic, possibly because of shorter-term exchanges of CO_2 or CH_4 among various reservoirs (Chapter 7). On orbital time scales (Chapter 14), the

warming that brought the last glacial maximum to an end because of rising summer insolation in the northern hemisphere was interrupted by a millennial-scale reversal toward cooler conditions (the Younger Dryas), after which deglaciation resumed and continued until the major ice sheets had melted (except on Greenland). The presence of shorter-term reversals or pauses in these tectonic-scale or orbital-scale signals do not refute the existence of the underlying trends. Similarly, the cooling in the 1960s and 1970s, perhaps linked to solar forcing, is not an argument against the longer-term warming trend on which it is superimposed.

Despite the extreme statements made by some greenhouse skeptics and proponents, it is clear that human additions of greenhouse gases to the atmosphere during the last century or more constitute a significant—and unintended—experiment run on Earth's climate system at a global scale. As we will see in the final chapter, we are at present only in the early stages of what will become a much larger CO_2 experiment, unless technology intervenes.

KEY TERMS

greenhouse debate (p. 405)
2xCO_2 sensitivity (p. 408)
equivalent CO_2 (p. 408)
radiative forcing (p. 410)
enhanced greenhouse effect (p. 410)
Intergovernmental Panel on Climate Change (IPCC) (p. 412)

REVIEW QUESTIONS

1. Why do uncertainties about natural changes in climate make it difficult to detect the effects of greenhouse gases on temperature during the last century?

2. What relative fractions of the estimated 2.5°C warming for a doubling of CO_2 are caused by the radiative effects of greenhouse gases and by feedbacks within the climate system?

3. What are the strongest positive and negative feedbacks acting to amplify and suppress Earth's temperature response to rising levels of greenhouse gases?

4. How similar are estimates of the sensitivity of Earth's surface temperature to changes in CO_2 derived from climate models and those derived from climate history?

5. What factors complicate attempts to estimate Earth's sensitivity to CO_2 by directly comparing the observed twentieth-century warming with the measured rise in CO_2 and other greenhouse gases?

6. What would happen to Earth's climate if humans abruptly stopped all emissions of gases and particles to the atmosphere?

ADDITIONAL RESOURCES

Basic Reading

Intergovernmental Panel on Climate Change. 1995. *Radiative Forcing of Climate Change*. Cambridge: Cambridge: University Press.

Karl, T. R., and K. E. Trenberth. 1999. "The Human Impact on Climate." *Scientific American* (December), 100–105.

Michaels, P. J. 1992. *Sound and Fury: The Science and Politics of Global Warming*. Washington, D.C.: Cato Institute.

Mitchell, J. M. 1976. "An Overview of Climatic Variability and Its Causal Mechanisms." *Quaternary Research* 6:481–93.

National Research Council. 2000. *Reconciling Observations of Global Temperature Change*. Washington, D.C.: U.S. Government Printing Office.

Office of Science and Technology Policy. 1997. *Climate Change: State of Knowledge*. Washington, D.C.

Schneider, S. H. 1997. *Laboratory Earth: The Planetary Gamble We Can't Afford to Lose*. New York: Basic Books.

Weart, S. R. 1997. "The Discovery of the Risk of Global Warming." *Physics Today* (January), 34–40.

Advanced Reading

Charlson, R. J., S. E. Schwartz, J. M. Hales, R. D. Cess, J. A. Coakley, J. E. Hansen, and D. J. Hoffman. 1992. "Climate Forcing by Anthropogenic Aerosols." *Science* 225:423–30.

Hansen, J. E., A. Lacis, D. Rind, G. Russell, P. Stone, I. Fung, K. Ruedy, and J. Lerner. 1984. "Climate Sensitivity: Analysis of Feedback Mechanisms." *American Geophysical Union Geophysical Monograph Series* 29:49–52.

Hoffert, M. I., and C. Covey. 1992. "Deriving Global Climate Sensitivity from Paleoclimate Reconstructions." *Nature* 360:573–76.

Jones, P. D. 1994. "Recent Warming in Global Temperature Series." *Geophysical Research Letters* 21:1149–52.

Lindzen, R. S. 1990. "Some Coolness to Global Warming." *Bulletin of the American Meteorological Society* 71:288–89.

Manabe, S., and R. J. Stouffer. 1996. "Century-Scale Effect of Increased Atmospheric CO_2 on the Ocean-Atmosphere System." *Nature* 364:215–18.

Mitchell, J. F. B., C. A. Senior, and W. J. Ingram. 1989. "CO_2 and Climate: A Missing Feedback." *Nature* 341:132–34.

Wigley, T. M. L., P. J. Jaunman, B. D. Santer, and K. E. Taylor. 1998. "Relative Detectability of Greenhouse Gas and Aerosol Climate Change Signals." *Climate Dynamics* 14:781–90.

CHAPTER 19

Climate Change in the Next 100 to 1000 Years

The potential effects of CO_2 on climate become more significant as we look into the future. Estimated present-day reserves of fossil fuels (mainly coal) should last for another few hundred years and will add far more CO_2 to the atmosphere than has accumulated so far. Unless human technology or extreme conservation efforts reduces this excess influx, atmospheric CO_2 will increase within two centuries to levels at least two and possibly four or five times higher than those that existed before humans made their influence felt. Levels this high are comparable to those last seen tens of millions of years ago in warmer greenhouse worlds. This warming will overwhelm natural variations in climate and cause climatic and environmental changes unprecedented in human experience. As regional patterns of temperature and precipitation change, impacts on human populations will vary from favorable to unfavorable by region and by season. Atmospheric CO_2 levels will remain high for 1000 years or more, until the ocean absorbs the excess CO_2.

Natural Variations in Climate

How do natural variations in climate affect Earth's various regions? And how sensitive is Earth's climate to the rising levels of greenhouse gases and other constituents of the atmosphere attributable to human activity? The uncertainties inherent in these issues make predictions of climate over the next few decades difficult. Over this interval, natural variability (changes in the Sun's strength) could rival the warming caused by rising concentrations of greenhouse gases.

But half a century from now, as equivalent CO_2 concentrations reach twice the preindustrial value, greenhouse warming will have overwhelmed natural variations in climate. By that time, with the impact of our unintended climatic experiment obvious, the debate over Earth's sensitivity to greenhouse gases will have been settled.

Over the next millennium, natural factors will also continue to drive climate change, sometimes in predictable directions. But these changes will be too small to compete with the effects of greenhouse gases, and these predicted changes will be swamped by the greenhouse warming.

The tectonic-scale tendency toward cooling observed over the last 50 to 100 million years is likely to continue. But at the extremely slow rate of change of 1°C per 10 million years, this trend would cool Earth's climate by only an imperceptible 0.0001°C in the next 1000 years.

Orbitally driven climate changes over the next 1000 years will be mixed in geographic pattern. In the northern hemisphere, a small increase in insolation in summer (and decrease in winter) at lower-middle latitudes caused by precession will be balanced by a small decrease in radiation in summer (and increase in winter) at higher latitudes caused by tilt. In the southern hemisphere, insolation will decrease in summer and increase in winter at all latitudes.

Increased summer insolation over the large landmasses at northern mid-latitudes should slowly strengthen the Asian and African summer monsoons, both currently near their minimum long-term strength. A rise in the strength of monsoons brings dramatic increases in rainfall, but on a global scale its effect on temperature is likely to be small.

Decreased summer insolation at high northern latitudes should cause a natural cooling trend in the future, as it has done for at least the last 1000 years. If this cooling continues at its previous rate, it will amount to no more than 0.2°C within the next 1000 years, far below the projected range of CO_2 warming.

The timing of natural millennial-scale climate change is again difficult to forecast because the oscillations over the last several thousand years have been irregular in duration. The assessment in Chapter 18

indicates that the global-scale amplitude of such changes (whether warming or cooling) over the next 1000 years is unlikely to exceed 0.2°C. Any such changes will be increasingly submerged in a growing greenhouse warming signal.

Over intervals of decades and centuries the Sun's strength will presumably continue to vary, but changes over intervals longer than 11-year cycles cannot be accurately projected. The maximum effect of such changes on global temperature is unlikely to exceed 0.5°C, the upper limit on estimated effects during the last few centuries. Changes in the Sun's strength could contribute significantly to climate changes in the next few decades but will soon be overwhelmed by the growing greenhouse warming.

Natural climatic variability over a few years, such as El Niño events and volcanic eruptions, represent only brief departures from longer-term underlying trends. These short oscillations are irrelevant to projections over the next 1000 years.

Global temperature changes produced by all natural causes combined are unlikely to reach 1°C over the next 1000 years (Figure 19-1). As we shall see, the projected global warming resulting from human additions of

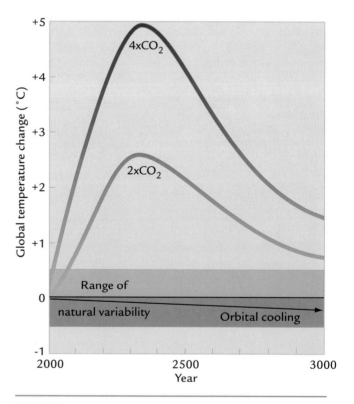

FIGURE 19-1 **Greenhouse and natural changes** The magnitude of the warming caused by increased greenhouse gases should overwhelm the effects of natural climate changes in the next several hundred years. (Adapted from H. S. Kheshgi et al., "Accounting for the Missing Carbon Sink with the CO_2-Fertilization Effect," *Climate Change* 33 [1996]: 31–62.)

greenhouse gases within this same interval is between 2° and 8°C, with a likely value of 4° to 5°C. As a result, natural climate changes will gradually become irrelevant to future projections of climate.

Future Human Impacts on Climate

Humans are likely to have an increasing impact on the environment in general, and on Earth's climate in particular, over the next several hundred years. The largest global-scale impact will be from greenhouse gases, with secondary effects from industrially produced SO_2.

19-1 Projected Carbon Emissions

At the turn of the millennium, atmospheric CO_2 levels continued to rise at a rate of 1.5 ppm (0.4%) per year because of the burning of fossil fuels and the clearing of forests. This rise will continue and probably accelerate in the future, but at unknown rates. Uncertainties in estimating future increases in greenhouse gases center mainly on two issues: (1) How much carbon will human activities emit? (2) How will the climate system distribute this additional CO_2 among its atmospheric, oceanic, and terrestrial reservoirs of carbon?

Projections of future carbon emissions are based on three factors:

$$\begin{matrix} \% \text{ increase} \\ \text{in carbon} \\ \text{emissions} \end{matrix} = \begin{matrix} \% \text{ increase} \\ \text{in population} \end{matrix} \times \begin{matrix} \% \text{ change in} \\ \text{emissions} \\ \text{per person} \end{matrix} \times \begin{matrix} \text{Changes in} \\ \text{efficiency of} \\ \text{carbon use} \end{matrix}$$

Population Increase The number of humans living on Earth is critical to these projections for obvious reasons: expanding numbers of humans require more fuel for industry, transportation, and home heating (burning of fossil fuels) and more land for farming and urban growth (cutting of forests). The number of humans has increased from 1.5 to 6 billion in just the last 100 years as agricultural and medical advances have extended human life expectancy.

Attempts to project increases are complicated by several factors: the tendency of a country's birth rate to fall as per capita income rises, efforts by populous nations such as India and China to slow or stop population growth, and efforts by some organizations to avoid any constraints on reproduction. The most widely cited estimate is that the number of people on Earth will rise rapidly from 6 billion in the year 2000 before leveling out near 11 billion between 2075 and 2100 (Figure 19-2). This projected increase of almost 100% in Earth's population is one multiplier factor in the equation used to predict future atmospheric CO_2 levels.

Emissions per person The second factor, the change in carbon emissions per person on Earth, is linked to the average standard of living. In many nations

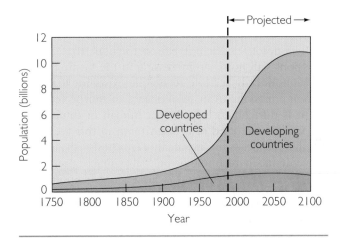

FIGURE 19-2 Future population A United Nations projection of human population during the twenty-first century at first continues the explosive increase of the twentieth century but later levels off at 11 billion people, almost twice the population in the year 2000. (From F. Press and R. Siever, *Understanding Earth*, 2d ed. © 1998 by W. H. Freeman and Company.)

standards of living increase over time, and in the past this process has required more carbon-based fuel for industrialization and day-to-day individual consumption (for cars and home heating or cooling). In the near term, the largest changes will occur in southeast Asian nations moving from semi-industrialized economies to join the industrialized nations. Some developing countries will also move from farm-based economies to semi-industrialized status.

Forecasting living standards is difficult, but the IPCC (Intergovernmental Panel on Climate Change) has estimated a range of possible increases for the next 100 years. These estimates produce a range of possible multipliers for CO_2 emissions in the future.

Efficiency of Use The third factor in the equation, the efficiency with which carbon is used, is the hardest to project. The next few decades will see peaks in annual production of oil and then of natural gas, after which production will begin to decline. Oil and gas are relatively "clean" fuels: they produce abundant energy while adding relatively small amounts of carbon to the atmosphere. When their use begins to decline, coal will gradually take over and become the major carbon-based fuel until the end of the fossil-fuel era.

This shift in the mix of fossil fuels burned has major implications for carbon emissions. Most of the high-grade anthracite coal was burned in the early years of industrialization, before the widespread use of oil and gas. Most of the coal still left to be mined is low-grade bituminous coal, which produces far more CO_2 per unit of usable energy. Greater reliance on coal will cause increased global carbon emissions, even with no increase in the total energy produced. This change adds a third multiplier effect to CO_2 projections.

Technology is the one reason for hope that carbon emissions will not keep increasing in the future. Today no economically feasible technique exists for lowering carbon emissions from smokestacks or motor vehicles, but it is difficult to believe that human ingenuity will not make some kind of breakthrough in this area during the decades—and especially centuries!—ahead. Improvements in technology may also make alternative sources such as solar and wind energy economically competitive with fossil fuels, especially as scarcity begins to make fossil fuels more expensive.

Yet even if technological breakthroughs occur, it is not certain they will be widely adopted. Nations making the difficult transition to industrialization may be reluctant to install expensive technologies simply for the greater good of humanity. They may decide that they should avoid these expenses and pump CO_2 into the atmosphere, just as the industrialized nations have done for two centuries. As for the industrialized nations, political concerns about the burden of new carbon taxes on the general population could slow or stop adoption of expensive technologies.

Estimates of Carbon Emissions At present no tangible basis exists for assuming that technology or conservation will solve the CO_2 problem. Without such solutions, the projected rise in carbon emissions will be enormous. On the basis of projected values of the three terms in the equation given earlier, the IPCC has predicted a range of possible future carbon emissions through the year 2100, and other scientists have projected these trends farther into the future.

The two projections of future CO_2 emissions in Figure 19-3 encompass much of the range of IPCC estimates. Implicit in the upper curve is the assumption of minimal efforts to curb emissions, with economic benefit continuing to be the major basis of decision making by nations and individuals. This trend shows emission rates peaking at a value of three to four times the modern level just before 2200 and then falling back below present-day values near the year 2300.

Yet individuals and nations may take strong action to curb emissions. Under this optimistic assumption, the lower curve shows worldwide carbon emissions peaking in a little more than a decade and then falling below the current level near 2100. This scenario would require either extreme conservation or new technological methods to dispose of carbon.

19-2 Projected CO_2 Concentrations in the Atmosphere

Estimating the future path of CO_2 concentrations in the atmosphere is more difficult than estimating emission

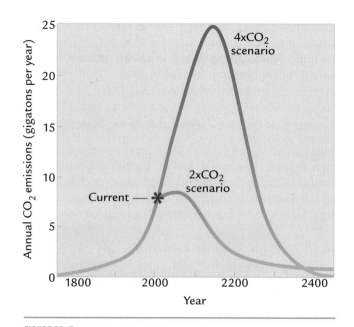

FIGURE 19-3 **Projected carbon emissions** Projections of future carbon emissions vary with uncertainties in future populations, living standards, and conservation efforts and technological innovations. An optimistic projection (the $2xCO_2$ scenario) shows emission rates rising only slightly before declining, but a more direct projection of present trends (the $4xCO_2$ scenario) shows a much larger rise in emissions. Both projections of carbon emissions eventually decline because reserves of fossil fuels will be consumed. (Adapted from H. S. Kheshgi et al., "Accounting for the Missing Carbon Sink with the CO_2-Fertilization Effect," *Climate Change* 33 [1996]: 31–62.)

levels. In addition to uncertainties about emissions, scientists face the question of how the climate system will redistribute the pulse of excess CO_2 among its carbon reservoirs.

The easiest assumption to make is that the atmosphere will continue to receive just over half of the total carbon emissions, as it does today, with the rest entering the ocean and the biosphere. This assumption will remain valid only if the climate system continues to operate as it does today.

Over longer time scales, other considerations come into play. Within a few hundred years, most of the pulse of excess CO_2 added to the atmosphere by humans will be mixed into the subsurface ocean, making ocean water slightly more acidic (Figure 19-4A). Over the course of several thousand years, the acidity produced by this extra CO_2 will cause some of the $CaCO_3$ on the seafloor to dissolve (Figure 19-4B). The ultimate fate of the excess CO_2 pulse will be a slow-acting chemistry experiment.

The processes that will initially mix the excess CO_2 pulse into the deeper ocean are not fully understood.

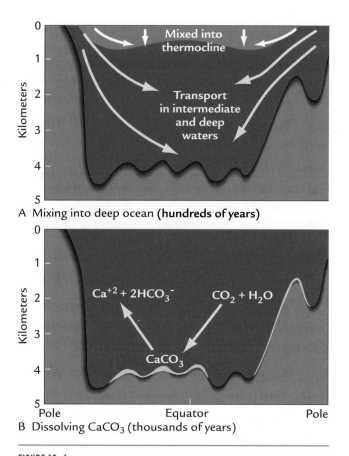

A Mixing into deep ocean **(hundreds of years)**

B Dissolving CaCO₃ **(thousands of years)**

FIGURE 19-4 **The fate of the human CO₂ pulse.** (A) Over the next tens to hundreds of years, the excess CO₂ in the atmosphere will be mixed into the subsurface ocean. (B) Over the next several thousand years, the excess CO₂ will make seawater more acidic and will dissolve CaCO₃ on the seafloor.

Geochemists have gained some insight into this problem by tracking the gradual penetration of tracers produced by recent human activities into the ocean. One such tracer is the pulse of extra ^{14}C produced by nuclear testing in the mid-twentieth century. Almost immediately, winds mixed the bomb-produced ^{14}C into the uppermost 100 meters of the ocean, and some of it has also begun to penetrate to depths as great as 1000 meters in the thermocline. In regions such as the North Atlantic, where deep water forms, ^{14}C from bombs is also slowly moving into the deeper ocean. Using results from these modern tracer studies to predict the mixing of CO₂ into the ocean (Figure 19-4A) is complicated by the possibility that as the surface ocean becomes warmer or lower in salinity in the future, less deep water may form at high latitudes, and less carbon may be absorbed than in the present-day circulation.

Other climate scientists are modeling the role of vegetation in absorbing excess CO₂. On longer time scales, more CO₂ could be taken up by vegetation because of increased fertilization of plants by CO₂, as

well as expansion of trees into the now-frozen north and movement of scrub vegetation into arid regions. But this trend may be countered by two factors. First, humans will probably continue to reduce the area of forests, especially in the tropics. And second, the amount of carbon stored in vegetation may also decline in some areas as some kinds of vegetation are forced out by changes in climate before others can fully replace them.

Specific year-by-year projections of future levels of atmospheric CO₂ depend on the interplay between the continuing input of excess CO₂ to the atmosphere and the slower removal of excess CO₂ to the ocean and the biosphere. Uncertainties in these factors produce a range of possible trends in future atmospheric CO₂ levels.

The two projections of the peak concentration of CO₂ in the atmosphere shown in Figure 19-5 result from the two scenarios in Figure 19-3. These projections differ in both the size and the timing of the CO₂ peak attained. For the lower curve, the 2xCO₂ scenario, CO₂ levels rise to a level nearly twice the preindustrial value near the year 2200 and then begin a gradual decline. For the upper curve, the 4xCO₂ scenario, CO₂ concentrations rise to more than four times the preindustrial value near the year 2250 and then decrease.

These CO₂ projections differ significantly from the emissions trends shown in Figure 19-3. The emissions trends reach their peaks and then fall quickly, while the CO₂ trends in Figure 19-5 reach their peaks later and

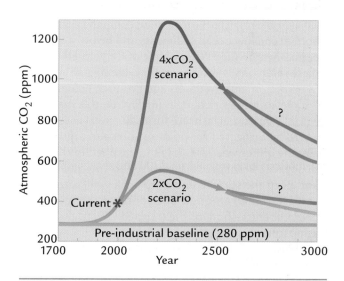

FIGURE 19-5 **Projected CO₂ concentrations** Concentrations of CO₂ in the atmosphere are projected to reach levels somewhere between twice (2xCO₂) and four times (4xCO₂) the preindustrial value of 280 ppm in the next two to three centuries. (Adapted from H. S. Kheshgi et al., "Accounting for the Missing Carbon Sink with the CO₂-Fertilization Effect," *Climate Change* 33 [1996]: 31–62.)

then remain at relatively high levels. The reason for this difference is the time it takes for the ocean to remove the excess CO_2 from the atmosphere. CO_2 molecules can move back and forth between the air and the surface ocean within a few years, but it takes centuries or more for large amounts of excess CO_2 to reach the *subsurface* ocean.

Long after human CO_2 emissions have begun to decrease, the still-sizable yearly increments of CO_2 input will continue to push atmospheric CO_2 concentrations to even higher levels. Not until a century or so after the peak of CO_2 emissions does the rate of human input drop below the rate at which CO_2 is removed into the deep ocean. At this point, CO_2 concentrations in the atmosphere begin a slow decline, but still do not reach the preindustrial level of 280 ppm even a full millennium later.

19-3 Other Human Effects on the Atmosphere

Other emissions by humans may also be important. Methane production is likely to increase, but by a much smaller amount than CO_2. Most of the land that can be used to grow rice is already under irrigation, and future increases in the area of CH_4-emitting wetlands should be negligible. In addition, if the total number of humans on Earth levels out in a half-century or so (Figure 19-2), the increase in CH_4-emitting livestock may also stabilize.

Unlike CO_2, methane stays in the atmosphere only about a decade before being oxidized to other forms. One interesting question is whether a warmer future world will cause large reservoirs of now-frozen methane to melt. If it does, unknown amounts of CH_4 gas could be added to the greenhouse effect (Box 19-1).

BOX 19-1   CLIMATE INTERACTIONS AND FEEDBACKS

Will Frozen Methane Melt?

Methane exists as a gas in the atmosphere, but it also occurs in Earth's colder regions in a frozen form known as **methane clathrate,** a mixture of methane with slushy ice. Clathrates occur in deep-ocean sediments along continental margins, where the pressure produced by overlying water and sediments makes CH_4 stable at temperatures well above freezing (5°C or more). Clathrates also occur in the Arctic, both in shallow ocean sediments and below permafrost on land. The volume of CH_4 stored in these reservoirs is enormous, far exceeding all surface reservoirs (mainly wetlands and livestock) combined.

Without major changes in climate, this methane would remain in clathrate form. But with the warming projected for polar regions and arguably already under way, will CH_4 stay trapped in this slushy ice? Permafrost is expected to begin to melt, and both the surface and deep ocean will warm by at least a small amount. Both of these changes could liberate some of the trapped CH_4 to the atmosphere.

How much CH_4 will be released? Because it will take decades for higher surface temperatures to penetrate far into permafrost and ocean sediments, most scientists doubt that much CH_4 will be released. On the other hand, the pulse of excess CO_2 will be around for several centuries, sufficient to warm polar regions substantially and perhaps

release some of the CH_4. If future greenhouse warming of the poles or the deeper ocean causes the release of even a small fraction of the total mass of frozen methane, it could provide a large positive feedback to the initial greenhouse-gas warming.

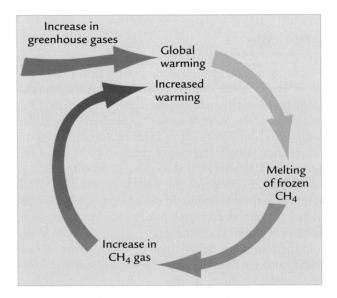

Methane clathrate feedback Future warming of the deep coastal ocean and melting of polar permafrost could release frozen methane and cause additional global warming.

Future additions of SO_2 to the atmosphere by humans are difficult to predict. The relative impact of environmental cleanup efforts in some industrialized nations must be weighed against the increased burning of sulfur-rich coal in nations undergoing industrialization. It is possible that a net increase in SO_2 input to the atmosphere could cause a small future cooling effect that would counter part of the CO_2 warming, particularly during the next several decades.

But in the long run, the impact of the CO_2 buildup is likely to far exceed any cooling effect from sulfates. The short residence time of sulfates in the atmosphere (roughly a week) means that both their concentrations and their impact on climate can be sustained only by continuous input. Once the burning of fossil fuels begins to decrease, the cooling effect of sulfates will diminish, but the warming effect of CO_2 will remain.

Future Climate Change Caused by Increased CO_2

As atmospheric CO_2 levels rise in the future, Earth's climate will continue to warm. Attempts to estimate the amount of future warming are subject to the cumulative uncertainties from three factors: the amount of excess CO_2 emitted as a result of human activities, the levels of atmospheric CO_2 reached as the excess carbon is redistributed among carbon reservoirs, and Earth's sensitivity to higher CO_2 concentrations.

19-4 Projected Temperature Changes

We explore here the two projections of future CO_2 concentrations shown in Figure 19-5, using them as case studies likely to span the full range of possible future CO_2 increases. We assume that Earth's sensitivity to a doubling of CO_2 over preindustrial levels is 2.5°C, the mid-range of sensitivity in most models and in estimates from Earth's earlier history. For each case, we focus on the highest projected CO_2 level reached, in the century between 2200 and 2300. For case 1, the projected peak concentration is $2xCO_2$ and for case 2 it is greater than $4xCO_2$.

For this analysis, we can ignore the fact that Earth's climate system responds to rising CO_2 with a warming delayed by several decades. In both projections, CO_2 levels remain at or near their maximum levels long enough to give the climate system time to reach most of its equilibrium global mean temperature response. But later we will see that some parts of the climate system, specifically the ice sheets, do not have time to respond fully to the higher CO_2 concentrations.

The projected range of future temperatures is plotted in Figure 19-6. The lower projection shows a 2.5°C warming by the year 2350 in response to the $2xCO_2$ peak, more than four times the warming of 0.6°C in the twentieth century. The upper projection shows a 5°C equilibrium warming for the $4xCO_2$ peak, almost ten times the size of the warming in the twentieth century.

The estimated sensitivity of 2.5°C to a doubled CO_2 concentration could be wrong by a factor of 50% or more in either direction. A lower $2xCO_2$ sensitivity of 1.5°C would mean that Earth's temperature would warm more slowly and reach a smaller peak value in the twenty-third century. The relatively small warming in the twentieth century is interpreted by some scientists as a sign that the actual sensitivity is this small. If this lower sensitivity is paired with the projection of a smaller future CO_2 increase to a peak of $2xCO_2$, the average global warming might be as small as 1.5°C. In this case, the impact on Earth's climate would be modest, though not insignificant.

On the other hand, the same uncertainties could work in the opposite direction. Some estimates from Earth's climate history point to a $2xCO_2$ sensitivity of 4° to 5°C; for example, the evidence of a large tropical cooling at the last glacial maximum (Chapter 13). If we combine a high $2xCO_2$ sensitivity of 5°C with the larger projected CO_2 increase to a peak value of $4xCO_2$, global warming might reach values of 8° to 10°C, with

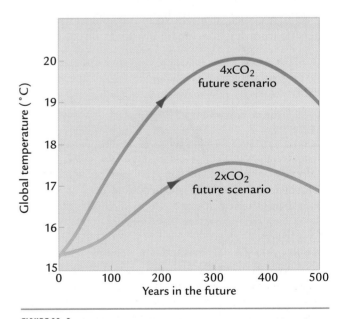

FIGURE 19-6 Projected temperature increases In the future, higher levels of CO_2 and other greenhouse gases are projected to cause global temperature to increase by at least 2.5° (for the $2xCO_2$ scenario) to as much as 5°C (for the $4xCO_2$ scenario).

an unimaginably large impact on Earth's climate. Even in the next century the warming could be as large as 4° to 5°C.

Neither of these extreme cases is considered likely. But future CO_2 levels and Earth's basic sensitivity to changes in CO_2 are so uncertain that neither extreme can be ruled out.

19-5 Partial Analogs from Earth's History: 2xCO_2 and 4xCO_2 Worlds

Atmospheric CO_2 levels of either twice or four times the preindustrial value would be without precedent in the last several million years of Earth history (Figure 19-7). The CO_2 history recorded in the Vostok ice core from Antarctica tells us that CO_2 values have not naturally risen above the 300–325 ppm level for at least the last 400,000 years. The shift from C3 to C4 vegetation suggests that CO_2 levels were last at the 2xCO_2 (560 ppm) value near 7 Myr ago. Concentrations as high as 4xCO_2 (1120 ppm) have probably not existed since at least 40 Myr ago, and possibly since the Cretaceous greenhouse world of 100 Myr ago.

As future CO_2 concentrations reach levels not seen for millions or tens of millions of years, temperatures on Earth will warm toward levels not attained since those times (Figure 19-8). In a sense, CO_2 emissions resulting from human activities will force Earth's climate system to retrace in a few hundred years a journey that took natural forces tens of millions of years to produce.

At first glance, it might seem possible to use these intervals from Earth's climate history as direct analogs

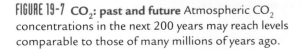

FIGURE 19-7 CO_2: **past and future** Atmospheric CO_2 concentrations in the next 200 years may reach levels comparable to those of many millions of years ago.

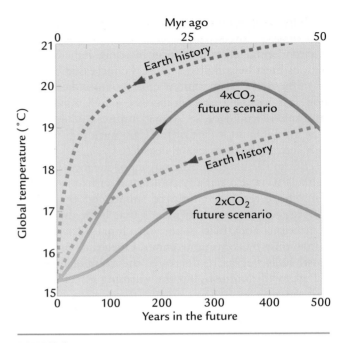

FIGURE 19-8 **Temperature: past and future** Average global temperature in the next 200 years is projected to reach levels comparable to those of many millions of years ago.

for climate in the future. We could simply match the projected future levels of CO_2 to times in the past when CO_2 was at the same estimated level and use the past climatic conditions as analogs for future climate.

But this approach has a serious flaw. The future high-CO_2 pulse will not stay around long enough to bring all parts of the climate system into complete equilibrium with the higher CO_2 levels. The many fast-responding parts of the climate system will quickly adjust to the warming caused by higher CO_2, but the slow-responding ice sheets will not have enough time to adjust fully before most of the pulse of excess CO_2 disappears into the oceans.

The margins of ice sheets can fluctuate rapidly over intervals of a few centuries, but most of the great mass of ice responds only over thousands to tens of thousands of years (Chapter 10). As a result, neither the brief century-long peak in CO_2 nor even the thousand or more years over which CO_2 concentrations will remain above the preindustrial level will be long enough to destroy the ice sheets. The greater warmth will cause melting along the ice margins, particularly in the warmer region around southern Greenland, but the main bulk of the ice will survive. The ice sheets will be like gigantic ice cubes on a summer day, out of place in a warmer world but still unmelted and capable of cooling nearby regions. From the sluggish perspective of the ice sheets, the pulse of excess CO_2 will be a brief episode that ends before it can cause the changes that would occur if it were sustained for 10,000 years or more.

This world of the future will be a strange no-analog combination produced by the slow-responding ice sheets (and deep ocean) and the fast-responding atmosphere, land surface, vegetation, and surface ocean. Near the slow-responding ice sheets, which strongly influence regional climates by their high albedo and their effect on atmospheric winds, the lingering cold caused by the ice sheet will suppress part of the response of the atmosphere and nearby surface ocean to the new warmth caused by high CO_2 levels. Farther from the regional effects of the ice, the fast-responding parts of the climate system will react strongly to the new warmth.

Because this degree of disequilibrium has not occurred in Earth's past, no exact analogs for future climates exist. Still, if we make reasonable allowance for this state of disequilibrium, we can look to the past for general indications of the direction of future climate changes. More specifically, we can use the $2xCO_2$ world of 5 to 10 million years ago and the $4xCO_2$ world of 50 to 100 million years ago as guideposts to future changes.

A $2xCO_2$ World (50–100 Years from Now) CO_2 levels are currently 30% higher than the preindustrial level, with equivalent CO_2 levels 50% higher. Current geopolitical realities suggest it is unlikely that humans will slow the input of excess CO_2 and other greenhouse gases enough to avoid following the projected trend toward a $4xCO_2$ peak. If so, we could reach an equivalent $2xCO_2$ concentration in the next 50 to 100 years, rather than 200 years from now if concentrations rise more slowly. Earth's average temperatures will register a full equilibrium response to this doubled CO_2 value soon afterward, probably by the year 2100.

Climate in 2100 will in some respects be like that 5 to 10 million years ago, the last time atmospheric CO_2 concentrations are thought to have been at twice the preindustrial level. Perhaps the most striking difference in the world of 5 to 10 million years ago was the much-reduced extent of ice in the Arctic Ocean. With less sea ice, the atmosphere was able to extract heat stored in the ocean, and this transfer moderated temperatures, especially in the cold Arctic winter. One consequence of the much warmer winters was that the broad band of permafrost and tundra that now surrounds the Arctic Ocean in Eurasia and North America was absent. In its place was a forest of conifers. The high sensitivity of this region to imposed climate changes is obvious in the northward retreat of tundra and sea ice limits caused by summer insolation values 5% higher than those today near 6000 years ago (Chapter 14).

What will the Arctic be like in the future? Because sea ice and vegetation are relatively fast-responding parts of the climate system, we should expect similarly large transformations of polar sea ice, permafrost, tundra, and northern forests as climate warms (Figure 19-9). Trees should move north of the Arctic circle, and winter sea ice should retreat from the coasts. The

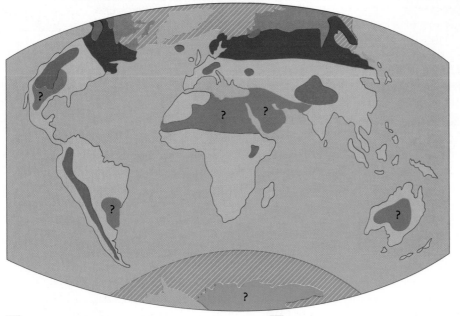

FIGURE 19-9 A $2xCO_2$ world The $2xCO_2$ world likely to exist by the year 2100 will in many ways be similar to the world that existed 5 to 10 million years ago, with less sea ice and permafrost in polar regions, fewer mountain glaciers, and in some regions greener deserts.

- ▇ Mountain glaciers melting
- ▨ Sea ice retreating
- ▇ Permafrost melting
- ▇ Forests moving north
- ▇ Greener deserts
- ▇ Greenland ice sheet melting
- ▇ West Antarctic ice sheet melting
- ▇ East Antarctic ice sheet growing?

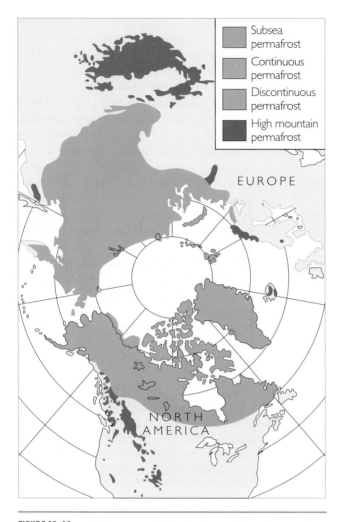

FIGURE 19-10 **Melting permafrost** The large ring of permafrost around the Arctic Ocean today will become vulnerable to gradual melting in the warmth of a 2xCO₂ world. (From F. Press and R. Siever, *Understanding Earth*, 2d ed. © 1998 by W. H. Freeman and Company.)

retreat of sea ice observed over the last few decades may be the beginning of this trend. But it will take much longer, probably hundreds of years, to melt deeper subsurface permafrost (Figure 19-10).

At northern mid-latitudes, forests composed of warm-adapted deciduous trees were also farther north than today 5–10 million years ago, and we should begin to return to this pattern in the next 100 years (Figure 19-9). Coupled models of climate and vegetation simulate a northward shift of cold-adapted hardwood trees such as maple and beech in response to warming, with warm-adapted trees such as oak and hickory moving north to replace them.

At lower latitudes, scrub and tree vegetation were more prevalent 5 to 10 million years ago in several arid regions: the sub-Himalayan region of India and Pakistan, the western American high plains, the South American pampas region of Argentina, and parts of sub-

Saharan Africa and the East African highlands. Higher levels of CO_2 in the atmosphere before 7 million years ago allowed C_3 vegetation (trees and shrubs) to live in arid regions where it cannot survive in the lower-CO_2 world of today. We will pass through this same $2xCO_2$ threshold in about a century, but heading in the other direction, and at a much faster rate. As C_3 shrubs and trees replace C_4 grasses, many semiarid and arid regions should become greener (Figure 19-9).

The warming in the next century will also alter regional patterns of precipitation and evaporation in significant ways. Evaporation will increase worldwide

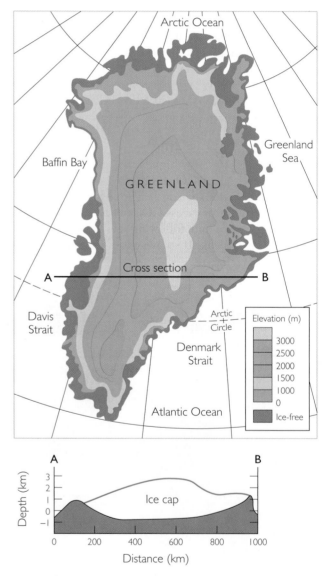

FIGURE 19-11 **The Greenland ice sheet** The lower margins of the Greenland ice sheet should begin to melt in the warmth of a 2xCO₂ world, but most of the higher central surface of the ice sheet will not be affected. (From F. Press and R. Siever, *Understanding Earth*, 2d ed. © 1998 by W. H. Freeman and Company; from R. F. Flint, *Glacial and Quaternary Geology*. © 1971 by Wiley.)

because warmer temperatures will permit air to hold more water vapor. With more water vapor in the air, global average precipitation is also likely to increase, but in patterns that may vary from region to region. With evaporation increasing, those areas that fail to receive more precipitation will become drier, while those that receive more precipitation could become wetter. Unfortunately, because climate model simulations of regional precipitation often disagree, it is difficult to predict moisture trends region by region.

No evidence of mountain glaciers exists in North or South America, Africa, or Asia before about 7 million years ago. The only place where mountain glaciers may have existed was in far northern Scandinavia, where the combined effect of high latitudes and high altitudes made temperatures cold enough for ice to persist. As we have seen, most mountain glaciers on Earth are in retreat today. At a prevailing lapse rate of 6.5°C per kilometer, a 2.5°C warming should cause glaciers to retreat some 330 meters (about 1000 feet) up the sides of mountains. Because mountain glaciers can begin to respond to climate changes within just decades, most should com-

pletely disappear in a $2xCO_2$ world. The melting ice will contribute to a global rise in sea level in the future.

No ice sheet existed on Greenland before 7 Myr ago, apparently because temperatures were too warm to permit a temperate-latitude ice sheet. In the future, of course, we face a different situation: we start off with the Greenland ice sheet already in existence (Figure 19-11), and we need to determine what effect the future warming will have on it. Simulations with ice models indicate widespread melting in response to a 3°C regional warming. Because the central high-elevation parts of the ice sheet will remain too cold to melt appreciably in summer, the vast bulk of ice will be little affected. But the margins of the Greenland ice sheet should be melting rapidly in a $2xCO_2$ world. This melting should increase the flow of ice in streams leading to the margins and cause some loss of ice mass in the ice sheet's interior. This melting will also contribute to a rise in global sea level.

At the opposite pole, the huge ice sheet in the eastern Antarctic (Figure 19-12) not only existed 5 to 10 million years ago but may even have been somewhat

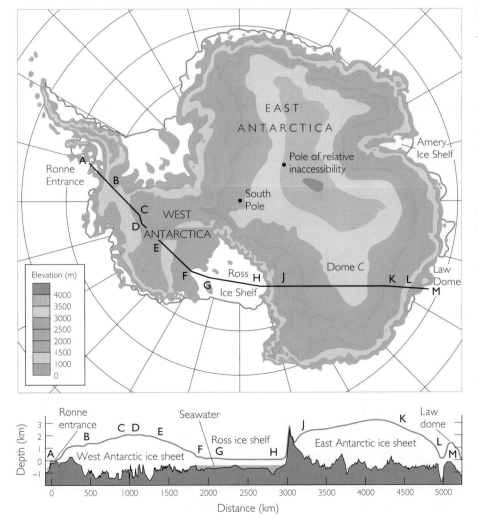

FIGURE 19-12 **The Antarctic ice sheet** The ocean margins of the small, low-altitude Antarctic ice sheet may begin to melt in a $2xCO_2$ world, especially the fringing ice shelves, but the higher, colder East Antarctic ice sheet will probably remain invulnerable to melting and may even grow thicker. (From F. Press and R. Siever, *Understanding Earth*, 2d ed. © 1998 by W. H. Freeman and Company; after Uwe Radok, "The Antarctic Ice," *Scientific American*, August [1985]: 100, based on data from the International Glaciological Project.)

larger than it is now. Today this frigid ice sheet is starved for snow, and the somewhat warmer temperatures of a $2xCO_2$ world might supply more snow for faster and thicker annual accumulations of ice. Unless the future warming causes faster flow in marginal ice streams, that ice sheet could grow, opposing the expected rise in sea level.

The smaller ice sheet in western Antarctica was much smaller still near 10 million years ago, if it existed at all. The extensive ice shelves that now fringe the Antarctic ice sheet may have been vulnerable to destabilization because of extensive contact with a slightly warmer ocean (Figure 19-13). Loss of ice in these regions could then destabilize the ice streams flowing to the ocean and send ice out of the interior of the continent. This evidence suggests that the western Antarctic ice sheet is vulnerable to greater melting in a $2xCO_2$ world, potentially adding to a future rise in sea level.

Expansion of ocean water as it warms should also contribute to a rise in sea level in a $2xCO_2$ world. The IPCC projects that this effect, along with increased melting of mountain glaciers, the Greenland ice sheet, and probably the margins of the western Antarctic ice sheet, is likely to at least double the rate of sea level rise from the 1.5 millimeters per year of the last century to 3 millimeters per year or more. Over a century, such a rate would increase sea level by 30 centimeters (about one foot).

A $4xCO_2$ World (200–300 Years from Now) By one projection, future atmospheric CO_2 concentrations could reach values four to five times the preindustrial level between 2200 and 2300. Projections this far into the future are inherently speculative, in part because of the much greater chance that technological innovations

will avert this outcome. But if we do reach a $4xCO_2$ world in 200–300 years, all the warming trends described for the $2xCO_2$ world will accelerate and new ones will appear. These changes will move Earth's climate toward the world of 50 to 100 million years ago, the last time CO_2 levels were so high.

The world was largely or entirely ice-free 50 million years ago. No evidence has been found of ice sheets or sea ice at either pole. The Arctic margins were surrounded by a mixed forest of hardwoods and evergreens adapted to relatively mild winters. Temperate beech (*Nothofagus*) forests existed on an ice-free Antarctic continent. Although any geologic evidence of mountain glaciers would almost certainly have been destroyed by later erosion, mountain ice probably did not exist anywhere on Earth. A 4° to 5°C warming would cause glacial ice to retreat 660 meters (more than 2000 feet) up the sides of mountains, enough to eliminate today's mountain glaciers.

If CO_2 concentrations do reach $4xCO_2$ in the coming centuries, polar regions will be severely out of equilibrium (Figure 19-14). Both ice sheets should be melting, especially the Greenland ice sheet, but neither should lose much of its vast bulk during the few hundred years of highest CO_2 levels. Sea level should also rise faster as the ice sheets melt at increasing rates and as warming causes thermal expansion of ocean water. Some projections put the rise by this time at 1 to 2 meters, enough for seawater to flood a kilometer or more into flat, low-lying coastal regions.

The East Antarctic ice sheet will continue to hold most of that continent in a deep freeze despite the warmth of the CO_2 pulse, but sea ice around Antarctica should largely disappear. North polar regions far from

FIGURE 19-13 Vulnerable ice shelves? Ice from the interior of Antarctica flows in ice streams to the shelves along the margin. In a warmer world, ice shelves may be vulnerable to destruction by rising ocean temperatures, which may in turn accelerate flow in the ice streams. (Adapted from R. A. Bindschadler et al., "What Is Happening to the West Antarctic Ice Sheet?" *EOS* 79 [1998]: 256–65.)

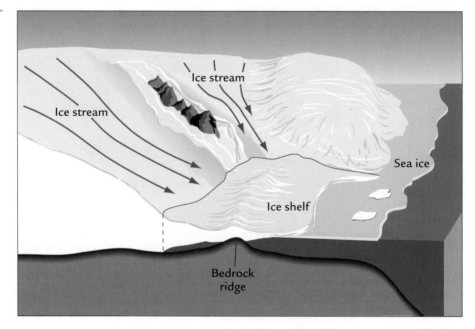

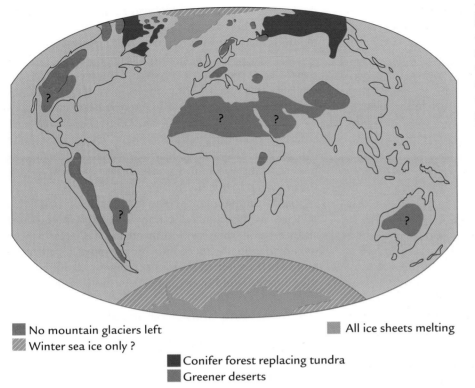

FIGURE 19-14 A 4xCO₂ world The 4xCO₂ world that may come into existence between 2200 and 2300 would be slowly moving toward conditions that existed 50 or more million years ago, when little or no glacial or sea ice was present on Earth, and forests grew in the higher latitudes of the Arctic.

⬛ No mountain glaciers left   ⬛ All ice sheets melting
▨ Winter sea ice only ?

⬛ Conifer forest replacing tundra
⬛ Greener deserts

the Greenland ice sheet will warm dramatically, leaving little or no sea ice, permafrost, or tundra, and a belt of conifer forest slowly moving toward the Arctic Ocean (Figure 19-14). Scrub vegetation should continue to invade arid regions at low latitudes.

19-6 Greenhouse Surprises?

The mixture of slow and fast responses to the large CO_2 pulse of the future will create a climatic disequilibrium unprecedented in Earth history. As a result, entirely new kinds of interactions in the climate system may produce unanticipated phenomena, or **greenhouse surprises**.

One such possibility is that faster melting of Greenland ice could send enough freshwater to the North Atlantic Ocean to lower its salinity and slow or stop the formation of deep water. A relatively small decline in salinity in the Labrador Sea beginning in the 1970s lowered the density of the surface waters enough to prevent them from sinking during the winters of the next two decades. Melting of the margins of the Greenland ice sheet, or altered precipitation patterns caused by greenhouse warming, could add enough low-salinity water to the North Atlantic to stop the formation of deep water for a much longer interval.

One likely consequence would be colder temperatures in northern Europe. Today heat extracted from the North Atlantic during the formation of deep water keeps Europe anomalously warm in winter for its latitude. Without this heat, Europe would become more like northern Canada, perhaps 5°C or more colder than it is now. This dramatic regional cooling would occur in a world experiencing overall warming.

A cut-off of deep-water formation in the North Atlantic might also cause atmospheric CO_2 levels to rise faster than before. Today the North Atlantic is one of the major pathways (sinks) for CO_2 to enter the subsurface ocean. If the rate of uptake of CO_2 in this region slowed or stopped, the extra CO_2 would remain in the atmosphere.

Other greenhouse surprises may lie in the future, many not yet even imagined by climate scientists. In a world with no exact past analog, such surprises may be common.

Monitoring Greenhouse Warming: The Next Few Decades

The central issue of the current greenhouse debate is the portion of the observed warming of the last 100 years caused by greenhouse gases. In the next few decades the debate will fade away as a historical artifact as Earth's climate system yields the answer to this issue.

How will this answer be revealed? Instrumental observations of changes in sensitive parts of Earth's climate system will together reveal Earth's sensitivity to rising levels of greenhouse gases. Many of these measurements already indicate a clear warming trend over the last several decades:

- Ground-based measurements of surface temperature.

- Satellite measurements of temperature in the lower troposphere.

- Tide-gauge measurements of a global rise in sea level.

- Ground-based and satellite observations of shrinking mountain glaciers.

- Ground-based and satellite observations of lengthening growing seasons in northern latitudes.

- Satellite measurements of retreating snow cover at high northern latitudes.

- Satellite measurements of retreating sea ice in the Arctic Ocean.

- Submarine measurements of thinning sea ice in the Arctic.

As the time line of these measurements grows to many decades, it should gradually become possible to distinguish long-term trends caused by greenhouse warming from short-term natural variations of climate. If these indications of warming continue or accelerate in future decades, the judgment of those who now argue that a substantial greenhouse warming is already under way will be confirmed. But if the trends reverse direction or simply slow significantly, the judgment of those scientists who argue for a relatively negligible global warming effect will be vindicated.

The most direct measures of surface temperature changes will continue to come from the array of instruments at surface ground stations and on ships. In addition, satellites will carry increasingly diverse instrumentation to measure Earth's properties from space quickly and efficiently.

In a sense, satellites actually provide proxy measurements of climate because their sensors use indirect means to measure climatic variables such as the extent of ice, snow, vegetation, and changes in temperature. Typically, satellites measure either energy emitted naturally in wave form from Earth's surface or energy initially generated by the satellite, bounced off Earth's surface, and then sensed as it returns. In most cases, these measurements must be filtered to isolate a specific wavelength that responds to some specific feature at or above Earth's surface. In addition, satellite measurements usually have to be adjusted to correct for complicating factors such as slow drift in instrument sensitivity or subtle changes in satellite orbits or intercalibration

between successive generations of instruments. In this sense, satellites are analogous to the indirect proxy methods used to measure Earth's earlier climate history.

Surface-based and satellite observations of the relatively fast-responding components of the climate system (the atmosphere, snow cover, the surface ocean, sea ice, and vegetation) will tell us in the next few decades whether a clear greenhouse warming signal has emerged from the natural climatic variability of the planet. Gradually we will learn how sensitive Earth's climate is to increasing greenhouse gases and how large a warming lies in our future.

Significant changes in the slower-responding components of the climate system—the subsurface ocean and the ice sheets—will occur more slowly and will be more difficult to detect. But recently developed techniques may enable us to detect changes in these parts of the climate system over the next century and beyond.

19-7 Measuring Changes in Ice Sheet Thickness

The size of the Antarctic and Greenland ice sheets will be an important measure of climate change, both as an indication of whether these large masses of ice have begun to respond to the greenhouse warming and because of their potentially large contributions to sea level change. Ice volume is difficult to measure because of the enormous size of ice sheets (kilometers thick, thousands of kilometers wide), the logistical problems inherent in working in such harsh environments, and the difficulty of plumbing the depths of the ice.

Precise and reliable measurements of the volume of ice sheets became possible just before the end of the twentieth century (Figure 19-15). Satellites flying above the ice can now measure the elevation of most of the ice surface to an accuracy of less than 1 meter (less than

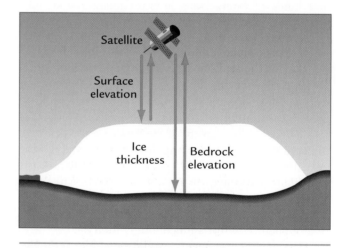

FIGURE 19-15 **Future changes in ice volume** Changes in the thickness of ice sheets in future centuries will be monitored mainly by radar from satellites.

0.1% of the total ice thickness), a level of accuracy suffi-cient to detect changes likely to occur over several decades to a century. Until recently, satellites had great difficulty measuring ice elevation in steeply sloping regions. Multiple passes across the ice sheets, combined with sophisticated computer analysis of the radar images received, have now made it possible to obtain accurate measurements of the entire surface of the ice sheets. But, as with the other climate records listed above, researchers will need to record changes in ice volume many decades into the future to distinguish trends in greenhouse gases from natural variability.

Because the elevation of bedrock under the ice can also change through time, ice thickness (and volume) cannot be determined only from measurements of the surface elevation of the ice. Until recently, the elevation of bedrock beneath the ice was measured by labor-intensive efforts in which stations moved across the ice surface, bounced radar waves off the underlying bedrock, and determined the thickness of the ice by the time the sound waves took to travel back and forth. This technique could be used to measure only selected lines across portions of the ice sheets. More recently, radar measurements of bedrock elevation have been made quickly and inexpensively from satellites, provid-ing full coverage of these continent-sized masses of ice.

19-8 Measuring Ocean Warming and Expansion

Although the course of Earth's future climate will be evaluated mainly by changes in the temperature of the lower atmosphere and the surface ocean, warming could also occur in the subsurface ocean. Trying to detect subsurface changes from measurements made on repeated oceanographic cruises is expensive and ineffi-cient, requiring too many return trips to too many remote locations over too many seasons. Now new tech-nologies may permit monitoring of large-scale changes in ocean temperatures without the need to go to sea.

Satellites have been measuring the altitude of the ocean surface since 1992 (Figure 19-16, top). Because the crust of the seafloor in most regions rises or sinks only slightly over time spans of decades, measurements of the changing height of the sea surface summed over the entire world are mainly measurements of changes in total ocean volume. When satellites have been measur-ing the altitude of the sea surface for many decades, it should be possible to see through shorter-term changes in sea level caused by weather phenomena and El Niño events and quantify sea level rises caused by the melting of glacial ice (adding more water) and the warming of the ocean (expanding the existing water).

Another technological innovation may help scien-tists find out what portion of the future sea level change is caused specifically by warming of the ocean. The velocity of sound waves moving through the ocean is

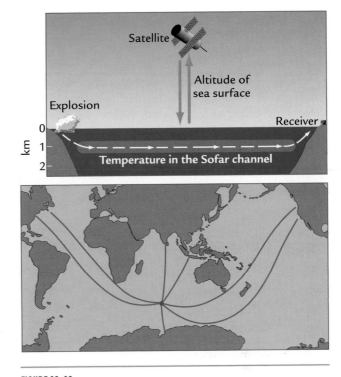

FIGURE 19-16 Future changes in subsurface ocean temperature Future warming of the ocean will be monitored by measuring increases in the height of the ocean caused by thermal expansion and increases in the velocity of sound moving through subsurface ocean layers.

dependent on water temperature: the velocity averages about 1500 meters per second but it increases by 4.6 meters per second per 1°C of warming of the water. A Sofar (Sound fixing and ranging) channel located at a depth near 1 kilometer is particularly favorable to trans-mission of sound waves (Figure 19-16, top). Sound waves moving through the overlying and underlying ocean layers are gradually bent into this channel from above and below because the temperature and density of its water make the waves move slightly faster there than in the surrounding layers. Sound waves traveling through this channel can be transmitted all the way to the opposite side of Earth. Scientists can use these far-traveled waves to measure the average sound velocity, and therefore the average temperature, across large stretches of the subsurface ocean.

An experiment conducted in 1991 at a remote site in the southern Indian Ocean proved that sound generat-ed by explosions detonated in that region could be detected in the distant North Atlantic, Indian, and Pacific Oceans (Figure 19-16, bottom). Encouraged by that initial result, scientists prepared to measure travel times over large stretches of ocean for extended periods of time.

But other scientists had become concerned about the effects these loud sounds (comparable to the full-throttle exhaust from a jet airplane) might have on nearby marine mammals. These concerns delayed further experiments until the volume of sound was reduced and the effects on mammals could be evaluated. Numerous receivers have been set up on ships, ocean islands, and coastlines of continents, and regular measurements are now being taken and evaluated. A few decades from now, when long-term greenhouse trends become distinguishable from short-term variations caused by natural processes, climate scientists should be able to tell whether large areas of the subsurface ocean are warming because greenhouse-gas levels are rising.

The Impacts of Future Increases of Greenhouse Gases on Humans

Another side of the greenhouse debate is the issue of how future changes in climate will be experienced by humans. Because Earth's responses to climate vary from region to region, the range of effects on humans and other life will also vary.

Temperature variations can be two or more times larger in polar regions than in the tropics, especially

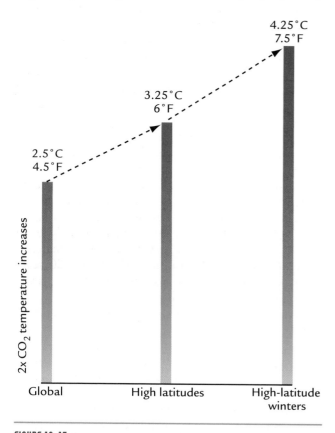

FIGURE 19-17 Larger temperature increases at high latitudes Temperature changes caused by rising CO_2 will be larger than the global average at latitudes above 50°, especially in winter.

over high-latitude landmasses. This difference is apparent in the present-day range of response of landmasses to seasonal changes in the strength of the Sun (Chapter 2) and also in examples from Earth's history, such as the larger high-latitude cooling at the last glacial maximum (Chapter 13) and the larger warming in the Cretaceous greenhouse period (Chapter 6). High latitudes are more responsive because of the positive feedback effects of ice and snow albedo.

This evidence, along with results from climate models, suggests that any imposed climate change could alter temperature at high latitudes twice as much as in the tropics. We focus here on the large changes expected to occur about midway between the poles and the tropics—between 40° and 60°N, the temperate latitudes of major population centers in North America, Europe, and eastern Asia. As CO_2 concentrations reach levels twice the preindustrial value, and as temperatures come into equilibrium with these higher CO_2 levels a few decades later, how will the climate in 50 to 100 years differ from climate today in these regions?

If global mean temperature does warm by 2.5°C in response to a CO_2 doubling, the increase in mean annual temperature at high northern mid-latitudes is likely to be 3° to 3.5°C because of the positive albedo-temperature feedback resulting from the retreat of snow and seasonal ice (Figure 19-17). For the same reason, temperature changes will be larger in winter than in summer, with a warming of perhaps 4° to 5°C during the colder months.

One way to sense what this amount of warming will mean for day-to-day life 50 to 100 years from now is to compare it with present-day seasonal temperature changes. Today, on landmasses at latitudes 40° to 60°N, temperatures cycle back and forth by roughly 25°C (45°F) every six months: mean daily temperatures of 25°C (77°F) in July alternate with temperatures of 0°C (32°F) in January. This 25°C shift over 6 months amounts to an average change of 4.25°C (7.5°F) per month, approximately the same amount as the anticipated cold-season temperature increase at high latitudes in response to a doubling of CO_2.

As a result, the temperature response of northern mid-latitudes to a doubling of CO_2 levels should be roughly equivalent to a one-month shift of the seasons: future Aprils will be like modern Mays, and future Novembers like modern Octobers (Figure 19-18). Summer will last for an extra two months, while winters will be shorter and probably less harsh.

These changes will come on slowly enough and be sufficiently masked by typical year-to-year variability, that the trend will not really be apparent to the average person. But if we could move instantaneously to a time 50 to 100 years in the future, we would find the changes striking.

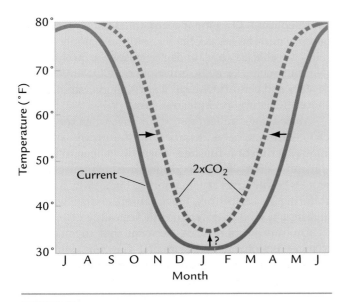

FIGURE 19-18 **Changes in length of seasons** The warming caused by $2xCO_2$ levels will shorten winters and lengthen summers by about one month at middle and high latitudes.

As noted earlier, changes in moisture balance are more difficult to predict than those of temperature. Because warmer air can hold more water vapor than colder air, the increased warmth on Earth will increase the global mean rate of evaporation, and many regions will become more susceptible to drought during the longer summer season if rains are scarce.

But precipitation patterns are highly variable, as shown both by year-to-year differences in the current climate and by disagreements among model simulations. As a result, we have little basis for explicit predictions of regional precipitation trends. We can anticipate that more precipitation will fall as localized summer cloudbursts than as the longer-lasting precipitation typically associated with the passage of frontal systems during cooler seasons.

Value Judgments To this point, future climate changes have been described without value judgments about whether they will be good or bad. In fact, any such single judgment is impossible, because future individuals will experience the changes as good, bad, or neither according to their own likes and dislikes. To those who prefer summer warmth to winter cold, winters shorter by one month at each end may seem wonderful; those who dislike summer heat will suffer.

The way people are affected by future warming will also depend on their livelihoods and locations. Some of the most striking changes will occur in regions where snow and ice retreat northward. At present-day ski resorts located at relatively low latitudes, for example, a loss of two months of subfreezing winter cold may so truncate the snowmaking season that business becomes impossible. On the other hand, in far-northern regions

of Hudson Bay and the Arctic coast of Siberia, the retreat of sea ice may open up new year-round shipping lanes or extend the length of the shipping season in existing lanes.

Water is a scarce and precious commodity. In regions where evaporation increases but precipitation does not, human populations will face serious water shortages. Such changes may be especially acute in the tropics and lower subtropics, where 80% of humans now live.

Changes in rainfall will also affect agriculture, although predicting exactly how is difficult, for reasons noted earlier. In regions where drought increases in frequency because of greater evaporation, agriculture will become more dependent on irrigation. This trend could in turn compound another problem. Heavily irrigated regions such as the North American high plains states (Nebraska, Colorado, Kansas) have for years been pumping deep subsurface water that fell as precipitation during the last glaciation, 21,000 years ago. Because this supply of groundwater is not being replenished rapidly today, increased pumping to replace evaporated surface moisture will draw down the remaining supply even faster.

On the other hand, some of the effects of future warming on agriculture will be positive. Higher levels of CO_2 in the atmosphere should allow some plants to obtain their necessary CO_2 fertilizer for photosynthesis much more quickly and without exposure to the drying effects of evaporation. This fertilizer effect may aid agriculture in some semiarid regions. In addition, warmer winters at high latitudes should extend the northern limit for growing many crops, especially winter wheat across broad regions of Canada and Russia.

One effect on human populations that is likely to be entirely negative is the future rise in sea level caused by melting of land ice and expansion of ocean water. Mountain glaciers in most regions will continue to melt, as will the warmer margins of the Greenland ice sheet. The 15-centimeter rise in sea level that has accompanied the 0.6°C warming of the last 100 years should increase by a factor of about 3 as climate warms by an additional 2°C over the next century.

Although a rise in sea level of 50 centimeters sounds small, in flat lowland regions it could translate into a 500-meter advance of the sea onto the land. In heavily populated regions such as Bangladesh, a sea level rise of half a meter would displace millions of people currently living within 1 meter of present-day sea level.

Especially vulnerable to sea level rise are nations on small coral islands (atolls) scattered across the Pacific Ocean. Many of these atolls rise at most a few meters out of the sea. For them, a half-meter rise in sea level increases the likelihood that storm waves will inundate the entire atoll. Major engineering efforts will also become increasingly necessary in built-up coastal regions such as Florida, Louisiana, and Texas, where sea walls are already required to protect buildings along the coast. These efforts will gradually become more and more expensive.

Although populations along coastlines will become increasingly vulnerable, the rise in sea level will actually be a secondary issue. A far more critical problem is the large number of people moving to coasts already vulnerable to summer hurricanes and typhoons. The half-meter rise in sea level will add only a small increment of vulnerability in comparison with the surge of sea water driven by major hurricanes as they make landfall. Some scientists have suggested that the number of hurricanes may increase with global warming, but this prediction is somewhat speculative.

Some of the largest changes caused by future global warming may have relatively little broad economic impact and yet still be crucial from a local perspective. Large-scale melting of sea ice and permafrost and loss of tundra around the Arctic margins will hardly be a central issue in the lives of people who have never traveled to that barren land. But the loss of those habitats could devastate caribou, polar bears, and the rest of the polar ecosystem, including the cultures of the few native people who still rely on hunting of polar-adapted animals for survival. Loss of vulnerable Arctic habitat would force major changes in their lives.

We have looked only at a small fraction of the changes likely to accompany the warming that could result from large increases in CO_2. As middle and higher latitudes warm, they will develop environmental characteristics more like those of lower latitudes. On the positive side, this change could bring the increased diversity of species typical of regions with warmer climates (assuming that other human impacts don't cause too much fragmentation of habitat), but it could also bring an increase of negative features such as more frequent and more widespread tropical diseases.

Epilogue

Unless technology or extreme conservation efforts greatly reduce CO_2 emissions, Earth is headed toward a warmer future. One of the more promising technological approaches under study today is the idea of piping the excess CO_2 from industrial emissions into the ocean. Because the deeper ocean will be nature's eventual repository of the excess CO_2, it makes sense to consider putting the carbon where it will eventually end up anyway.

As with many technological solutions, cost is a major issue. Many industrial sites lie far from the ocean. How much will it cost, in money and extra fuel expended, to collect CO_2, pipe it many miles across land, and pump it into the ocean? Such a solution is not economically feasible at this time: the costs are prohibitive. Without such a remedy, atmospheric CO_2 level in the decades and centuries ahead will rise higher and higher above the natural preindustrial level.

More than a millennium from now, when the excess CO_2 pulse has been absorbed by the oceans and this huge experiment is finished, what will happen to Earth's climate? Left to function on its own, natural climatic processes will again regain control, but slowly. And just as during its past history, Earth's future climate will vary in different ways at different time scales.

On shorter time scales of centuries to millennia, changes in global temperature of 1°C or less will be driven by changes in the strength of the Sun and by whatever other processes cause millennial-scale oscillations. Over thousands to tens of thousands of years, orbitally driven climate changes will take over, with the likelihood that a cooling of climate at high northern latitudes will gradually initiate the next glaciation of North America and Eurasia. The expansion of snowfields in northern Canada during the Little Ice Age may have been a first step toward the next glaciation, but it was ended by the warming of the twentieth century. Without greenhouse gases to warm climate, this process could resume and intensify. Over still longer tectonic time scales (millions of years), Earth's climate will probably continue its drift toward colder temperatures and more intense glaciations. The relatively pleasant interlude of the last several thousand years, corresponding to most of recorded history, will have come to an end.

KEY TERMS

methane clathrate (p. 428)
greenhouse surprises (p. 435)

REVIEW QUESTIONS

1. What factors will determine how much CO_2 humans add to the atmosphere in the future?

2. Where will all this excess carbon eventually go?

3. In what sense will future CO_2 warming be like past CO_2 warmings? Unlike them?

4. In a $2 \times CO_2$ world, what ice will be melting and what vegetation changes will be under way?

5. How many trends under way in the last few decades can be expected to continue?

6. In a $4 \times CO_2$ world, what ice will be melting and what vegetation changes will be under way?

7. What techniques will allow us to monitor future changes in the thickness of ice sheets?

8. What techniques will allow us to monitor future changes in subsurface ocean temperature?

9. Will future temperature changes be readily apparent to the average person? Why (or why not)?

ADDITIONAL RESOURCES

Graedel, T. E., and P. J. Crutzen. 1997. *Atmosphere, Climate, and Change*. New York: Scientific American Library.

Houghton, J. T., et al., eds. 1995. *Climate Change, 1994: Radiative Forcing of Climate Change and an Evaluation of the IPCC IS92 Emission Scenarios*. Cambridge: Cambridge University Press.

Michaels, P. J. 1992. *Sound and Fury: The Science and Politics of Global Warming*. Washington, D.C.: Cato Institute.

National Research Council. 2000. *Reconciling Observations of Global Temperature Change*. Washington, D.C.: U.S. Government Printing Office.

Office of Science and Technology Policy. 1997. *Climate Change: State of Knowledge*. Washington, D.C.: Executive Office of the President.

Schneider, S. H. 1997. *Laboratory Earth*. New York: Basic Books

Glossary

(Many definitions apply specifically to their use in climate studies.)

ablation The loss of snow or ice from a glacier by melting, calving, and other processes.

accumulation The addition of snow to a glacier.

adiabatic Having to do with an increase in pressure that raises the temperature of a parcel of air (adiabatic warming) or a decrease in pressure that lowers that temperature (adiabatic cooling).

aerosols Extremely small particles or droplets carried in suspension in the air.

albedo The decimal fraction or percentage of incoming solar radiation reflected from a surface.

albedo-temperature feedback A positive feedback that amplifies an initial temperature change by altering the amount of snow cover or sea ice and changing the amount of solar radiation absorbed by Earth's surface.

aliasing The misleading (unrepresentative) signals that result from sampling a record of climate change at too low a resolution.

alkenones Complex organic molecules found in fossil shells of plant plankton and used to reconstruct past temperature.

amplitude Half of the height between peaks and troughs in a regular wave form.

Antarctic bottom water A dense, cold water mass that forms near the Antarctic continent by extreme chilling of surface waters, sinks, and flows along the seafloor below a depth of 4 km.

Antarctic intermediate water A water mass that forms in the Southern Ocean by chilling of seawater exposed in or near sea ice, sinks, and flows northward at depths of 1 to 2 km.

anthropogenic CO_2 increase The steadily increasing concentration of CO_2 in the atmosphere over the last 200 years due to human activities.

anthropogenic forcing All human-related factors that cause climate change.

aphelion The point in Earth's slightly eccentric orbit at which it is farthest from the Sun.

asthenosphere A partially molten layer of rock in the upper mantle that is weak enough to flow and cause movement of the overlying lithospheric plate.

axial precession The wobbling movement of Earth's axis of rotation , which causes it to point in different directions over a cycle of 26,000 years.

back radiation Electromagnetic energy at long (infrared) wavelengths emitted from any material with a temperature above absolute zero (0 K).

Barents ice sheet An ice sheet that covered the present-day Barents Sea, north of Scandinavia, during orbital-scale glaciation cycles.

basal slip Rapid sliding of an ice sheet across its water-lubricated bed, especially in a region where ice streams lie above water-saturated sediment.

bedrock pinning point A high-standing protrusion of bedrock that lies beneath the margin of an ice sheet and slows the flow of ice into the ocean by frictional resistance.

benthic foraminifera Sand-sized organisms (protozoans) that live on and in the seafloor and form shells of $CaCO_3$.

biomass The amount of living matter in a region; also, organic matter used as a source of energy.

biome A region on Earth with a distinctive community of plants.

biome model A vegetation model that simulates the major vegetation type (for example, grassland or desert scrub) that can exist in a region under a given set of climatic conditions.

biosphere The part of the Earth system that supports life, including the oceans, land surfaces, soils, and atmosphere.

biotic proxy An index of past climate change based on measurable variations in the type or abundance of climate-sensitive organisms.

Black Sea flood hypothesis The hypothesis that melting ice sheets caused rising ocean waters to flood into an ancient glacial lake, displacing humans and forming the Black Sea.

BLAG (spreading rate) hypothesis The hypothesis that tectonic-scale climate changes are driven by variations in the global average rate of seafloor spreading, which alter the amount of CO_2 introduced into the atmosphere.

boundary conditions The initial configuration of Earth's properties chosen for a model simulation (such as land-sea distribution, mountain elevation, and atmospheric CO_2 concentration).

burial flux The rate of deposition of a substance in a sedimentary reservoir measured in units of mass per unit of area per unit of time.

C_3 pathway The means by which trees and most shrubs (about 95% of all land plants) obtain CO_2 from the air during the initial step of photosynthesis.

C_4 pathway The means by which grasses that grow during the warm season (about 5% of all land plants) obtain CO_2 from the air during the initial step of photosynthesis.

calibration interval The interval of time (usually 50–100 years) over which the width or density of tree rings can be correlated with historical observations of temperature and precipitation change.

calorie The amount of energy required to raise the temperature of 1 gram of water by 1°C.

calving The process by which a large block of ice breaks off from the margin of a glacier and forms an iceberg that floats in the ocean or in a lake.

carbon isotopes Isotopes of the element carbon with different atomic masses, used to trace the movement of different kinds of carbon through Earth's climate system (^{12}C and ^{13}C) or to measure elapsed time indicated by radioactive decay (^{14}C and ^{12}C).

cardinal points The two equinoxes (spring and autumn) and the two solstices (summer and winter) in Earth's annual revolution around the Sun.

Celsius A temperature scale on which water freezes at 0° and boils at 100°.

channeled scablands A region in Idaho and eastern Washington State in which water impounded in glacial lakes suddenly rushed out and reshaped the landscape, perhaps repeatedly.

chemical weathering Dissolving or other alteration of minerals to a different form by chemical reactions in the presence of water.

chlorofluorocarbons (CFCs) Synthetic chemical compounds generated by human activity and containing chlorine or fluorine that can destroy ozone in the stratosphere.

CLIMAP (Climatic Mapping and Prediction Project) A large cooperative research group during the 1970s and 1980s that first mapped the surface of the ice-age Earth.

climate Fluctuations in Earth's air, water, ice, vegetation, and other properties on time scales longer than one year.

climate data output The climatic properties (such as temperature, precipitation, and winds) that are produced by simulations with climate models.

climate point The point where the equilibrium line that separates net ice melting from net ice accumulation intercepts sea level.

climate proxy A quantifiable indicator of climate change contained in a climate archive and covering an interval that precedes direct instrument measurements of climate.

climate science The study of climate changes and their causes.

climate simulation The use of a numerical model to reproduce climate for a specified set of boundary conditions.

climate system The components of Earth (air, water, ice, vegetation, and land surfaces) that participate in climate change.

climatic hypothesis of human evolution The hypothesis that a trend toward drier climates in Africa changed forest to savanna or grasslands and accelerated human evolution.

clipped responses Climatic responses that are truncated (cut off) in one direction.

closed system A system that does not exchange matter across its boundaries.

coccoliths Tiny disklike plates of $CaCO_3$ produced by algae living in ocean surface waters.

COHMAP (Cooperative Holocene Mapping Project) A cooperative research effort in the 1980s and 1990s that evaluated the causes of climate change during the most recent deglaciation by comparing geologic data with climate model simulations.

conifer forest Forest comprising evergreen, needle-bearing trees.

continental collision The occasional result of plate tectonic processes in which two continents are carried into each other by plate movements, creating high plateaus.

continental crust A layer of rock averaging 30 km thick, having the composition of granite, and comprising the continents.

continental ice sheet A mass of ice kilometers thick covering a continent or a large portion of a continent and moving independently of the underlying bedrock topography.

continental shelf A shallowly submerged extension of a continent beneath the ocean.

continental slope A ramplike structural edge of a continent that slopes into the deep ocean.

control case A model simulation run to reproduce Earth's present-day climate from its present-day boundary conditions.

convection The rising motion of a fluid (air or water) produced when its bottom layer is heated, accompanied by sinking of cooler, denser fluid elsewhere.

convergent margin A boundary between two lithospheric plates that move toward each other and cause a collision of continents or subduction of one plate beneath the other.

coral bands Annual banding in the structure of corals caused by seasonal changes in sunlight, water temperature, and nutrient content.

Cordilleran ice sheet A small ice sheet that covered western Canada and the far northwestern United States during orbital-scale glaciation cycles.

Coriolis effect The apparent deflection of a fluid (air or water) from a straight-line path because of Earth's rotation. The deflection is to the right in the northern hemisphere and to the left in the southern hemisphere.

CO$_2$ fertilization effect The increased growth rate of plants caused by adding CO_2 to the atmosphere.

CO$_2$ saturation The point at which the concentration of CO_2 in the atmosphere reaches so high a level that additional amounts do not increase the greenhouse effect.

Cretaceous An interval of warmer climates and higher sea level between 135 and 65 Myr ago.

Dansgaard-Oeschger cycles Oscillations in various properties (including dust and isotopes of oxygen) recorded in Greenland ice at intervals of 2000 to 7000 years during glacial intervals.

daughter isotope An isotope produced by radiometric decay of another isotope.

deforestation Cutting of forests by humans to clear land for agriculture and other processes.

deglacial two-step The irregular melting of ice sheets during the most recent deglaciation: fast-slow-fast.

δ^{13}C aging The gradual shift toward more negative δ^{13}C values in slow-moving deep water caused by the downward rain of ^{12}C-rich organic matter from overlying surface water.

dendroclimatology The methods used to extract climate signals from changes in the width or density of tree rings, both of which are sensitive to extremes of temperature and precipitation.

dew point The temperature at which cooling air becomes fully saturated with water vapor and permits condensation.

diatoms Silt-sized algae that live in surface waters of lakes, rivers, and oceans and form shells of opal ($SiO_2 \cdot H_2O$).

diffusion The transfer of a property such as heat by random, small-scale movements from a region of higher to lower concentration.

diluvial hypothesis The hypothesis that unsorted sediments found on northern continents resulted from a great flood; these deposits are now recognized as deposits from glaciers.

dissolution A form of chemical weathering in which rocks such as limestone ($CaCO_3$) or rock salt ($NaCl$) are dissolved by water and produce ions that are removed by rivers.

divergent margin A boundary between two lithospheric plates that are moving apart, usually at the crest of an ocean ridge.

Earth system The complex system of Earth's atmosphere, hydrosphere, biosphere, and lithosphere, through which energy and matter circulate.

eccentricity The extent to which Earth's orbit around the Sun departs from a perfect circle.

elastic Capable of deforming rapidly under pressure (as is bedrock under the pressure of a glacier) and rebounding when the pressure is removed.

electromagnetic radiation Self-propogating electric-magnetic waves, which include visible light as well as infrared and ultraviolet waves.

electromagnetic spectrum The complete range of electromagnetic radiation at differing wavelengths.

El Niño A climatic pattern that recurs at intervals of 2 to 7 years and is marked by warm sea surface temperatures in the eastern tropical Pacific, off the west coast of South America.

enhanced greenhouse effect Trapping of Earth's back radiation by greenhouse gases produced by humans, in addition to the warming caused by natural greenhouse gases.

ENSO (*El Niño Southern Oscillation*) The combined oscillations of El Niño (temperature changes in the eastern Pacific) and the southern oscillation (atmospheric pressure changes in the western and south-central Pacific).

eolian sediments Fine sediments deposited by the action of wind.

equilibrium A state of climatic stability toward which the climate system is moving and at which it will eventually remain, unless disturbed.

equilibrium line The level in the atmosphere separating the zones of net addition and loss of ice.

equinoxes The two times during each year (spring and autumn) when the lengths of days and nights are equal.

equivalent CO_2 Changes in all greenhouse gases expressed in terms of an equivalent change in atmospheric CO_2 concentrations.

eustatic Characterized by changes in sea level that are global in scale, rather than the result of local factors such as tectonic uplift or subsidence of the land.

evaporites Minerals or rocks formed by precipitation of crystals from water evaporating in restricted basins in arid climates.

evolution The process by which particular forms of life give rise to other similar forms by gradual genetic changes.

faculae Bright rings that surrounding sunspots and emit large amounts of solar radiation.

Fahrenheit A temperature scale on which water freezes at 32° and boils at 212°.

faint young Sun paradox The paradox in which astronomical models indicate a much weaker Sun through Earth's early history but geologic evidence shows that Earth never froze.

feedback A process internal to Earth's climate system that acts either to amplify changes in climate (positive feedback) or to moderate them (negative feedback).

filtering The technique of extracting and isolating the shape of cycles at specific wavelengths or periods from complex signals.

fluvial sediments Sediments deposited by the action of water.

forcing Any process or disturbance that drives changes in climate.

fractionation A process favoring the transfer of one isotope of an element more than another.

frequency The number of full wave forms (each with one peak and one trough) that occur within a defined interval of time (usually one year). Also, the inverse of the period.

Gaia hypothesis A hypothesis that life regulates climate on Earth.

general circulation model (GCM) A three-dimensional computer model of the global atmosphere (or ocean) that simulates temperature, precipitation, winds, and atmospheric pressure.

geochemical model A model that quantifies the movement of geochemical tracers (minerals, elements, or isotopes) among reservoirs in the climate system.

geochemical tracer A chemical element or isotope whose movement between reservoirs in the climate system can be quantitatively tracked.

geological–geochemical proxy An index of past climate change based on measurable variations in physical or chemical properties of sediments, ice, or other archives.

Gondwana The large continent that existed in the southern hemisphere before the creation of the giant continent Pangaea.

greenhouse debate The controversy over the extent to which rising levels of atmospheric greenhouse gases have warmed Earth's climate during the last 200 years.

greenhouse effect The warming of Earth's surface and lower atmosphere that occurs when its own emitted infrared heat is trapped and reradiated downward by greenhouse gases.

greenhouse era An interval of warm climate on Earth, such as the Cretaceous, with ice sheets absent even in polar regions.

greenhouse gases Gases such as water vapor (H_2O_v), carbon dioxide (CO_2), and methane (CH_4), which trap outgoing infrared radiation emitted by Earth's surface and warm the atmosphere.

greenhouse surprise A climate change in the greenhouse world of the future that cannot be predicted.

grid boxes Geometric units within climate models that have uniform climatic characteristics and exchange heat, energy, and other properties with adjoining grid boxes.

Gulf Stream A narrow current of warm water that emerges from the Gulf of Mexico through the Florida Straits and flows northward along the southeastern coast of the United States.

gyre A spinning cell of water in an ocean basin, particularly at a subtropical latitude.

Hadley cell An atmospheric circulation cell in which air rises in the tropics, flows to the subtropics, sinks near 30° latitude, and flows back toward the tropics as surface trade winds.

half-life The time required for half the number of atoms of a radioactive isotope to decay.

harmonics Secondary cycles related to a wavelike climatic response with a period N, occurring at periods of $N/2$, $N/3$, $N/4$, and so on.

hardwood forest A forest comprising leaf-bearing (deciduous) trees.

heat capacity The amount of heat energy required to raise the temperature of 1 gram of a substance by 1°C.

Heinrich event An interval of rapid flow of icebergs from the margins of ice sheets into the North Atlantic Ocean, causing deposition of sediment layers rich in debris eroded from the land.

historical archives Sources of information on climate based on human observations of natural phenomena made before the era of instrument measurements.

hot spot A point on the surface of a lithospheric plate where magma rising from below causes frequent volcanic activity.

hydrologic cycle The movement of water and water vapor among the atmosphere, land, and ocean through evaporation, precipitation, runoff, and subsurface groundwater flow.

hydrolysis A form of chemical weathering in which water reacts with silicate minerals rich in silicon and oxygen to produce dissolved ions removed in rivers and clays left on the landscape.

hydrothermal Characterized by the circulation of hot fluids through rocks in Earth's outer crust, as at the Mid-Ocean Ridge system.

hypothesis An explanation of observations based on physical principles.

hypsometric curve A graph that summarizes the proportions of Earth's surface that lie at various altitudes above and depths below sea level.

ice dome A high, gently sloping central region of an ice sheet in which snow accumulates and away from which ice flows slowly.

ice-driven response A climate change produced by fluctuations in the size of an ice sheet.

ice-elevation feedback The positive feedback that results when an ice sheet grows to a higher elevation at which accumulation exceeds ablation.

ice flow model A model that simulates ice sheet processes of snow accumulation, internal ice flow, and ablation.

icehouse era An interval of cold climate on Earth, such as the present, with ice sheets present in polar regions.

ice lobe A rounded or arc-shaped outward protrusion of the margin of an ice sheet.

ice-rafted debris Sediments of widely ranging sizes eroded from the land by ice, carried to the ocean, and deposited on the seafloor.

ice saddle A ridge that connects multiple domes of an ice sheet at a slightly lower elevation.

ice shelf A wide body of ice usually hundreds of meters thick that is fed by ice flowing off a continent, partially floats on seawater, and produces icebergs as blocks of ice break off.

ice stream A region of an ice sheet in which the motion of ice is unusually rapid, generally because of water-saturated sediments at the base of the ice.

igneous rock Rock formed by the cooling and solidification of molten magma.

insolation The amount of solar radiation arriving at the top of Earth's atmosphere by latitude and by season.

instrument records Records of climate change measured by devices made by humans, from early thermometers through modern satellite-mounted instruments.

Intergovernmental Panel on Climate Change (IPCC) A large international group of scientists who reflect the current scientific consensus on the impact of greenhouse gases.

intertropical convergence zone (ITCZ) A narrow region within the tropics where warm moist air rises, cools, and loses its water vapor in heavy tropical rainfall.

iron fertilization hypothesis The hypothesis that iron-rich dust blown from the continents during glaciations enhances the productivity of the surface ocean, sends CO_2 into the deep ocean, and reduces the concentration of CO_2 in the atmosphere.

jet stream A narrow meandering stream of air moving rapidly (generally from west to east) at a high latitude and at an altitude averaging 10 km.

Kelvin A scale on which temperature is measured in Celsius degree intervals and on which water freezes at 273 K and boils at 373K, and all motion ceases at 0 K.

La Niña A pattern opposite from that of El Niño, in which sea surface temperatures in the tropical eastern Pacific, off the west coast of South America, become unusually cold.

lapse rate The rate at which temperature falls with elevation in the atmosphere, averaging 6.5°C of cooling for each kilometer of altitude, but more for dry air and less for moist air.

latent heat The quantity of heat gained or lost as a substance changes state (liquid, solid, or gas) at a given temperature and pressure.

latent heat of melting The amount of heat gained as ice melts or released as water freezes.

latent heat of vaporization The amount of heat gained when water turns to water vapor or released when water vapor condenses back to water.

Laurentide ice sheet The largest of the northern hemisphere ice sheets that grow and shrink at orbital cycles, covering east-central Canada and the northern United States east of the Rockies.

lichen Primitive mosslike vegetation that lives on bare rock surfaces, uses sunlight to secrete acids and weather the rock, and slowly grows in a nearly circular shape.

lithosphere The outer rigid shell of Earth (including the upper mantle and oceanic and continental crust), characterized by strong, rocklike properties and divided into plates that move as rigid units during plate tectonic processes.

Little Ice Age An interval between approximately 1400 and 1900 A.D. when temperatures in the northern hemisphere were generally colder than today's, especially in Europe.

loess Silt-sized windblown glacial sediment.

longwave radiation Energy emitted in the infrared part of the electromagnetic spectrum by materials having a temperature above absolute zero (0 K).

macrofossils Larger fragments of vegetation (such as needles, twigs, or cones) that prove the local presence of vegetation on past landscapes, used as a supplement to pollen analysis.

magnetic field The lines of force that are generated by motion in Earth's outer core and cause iron-bearing materials to align in specific directions in relation to the magnetic north pole.

magnetic lineations Long, stripelike regions of ocean crust marked by stronger or weaker magnetism acquired when the crust formed by cooling from molten magma at a ridge crest.

magnetic stratigraphy The correlation and dating of intervals in which Earth's magnetic field was normal (like that of today) or reversed (opposite to that of today).

mantle The middle (and largest) layer of the solid Earth, lying beneath the crust and above the core and composed of silicate minerals.

marine ice sheet An ice sheet whose base lies below sea level, as in western Antarctica.

mass balance The method of tracking the movement of materials within Earth's climate system by applying the law of conservation of mass.

mass balance model A model that tracks the movement of materials within Earth's climate system using conservation of mass to balance inputs and outputs among different reservoirs.

mass spectrometer An instrument that measures different isotopes of the same element (such as ^{16}O vs. ^{18}O and ^{12}C vs. ^{13}C) by separating them by mass.

mass wasting A downhill movement of rock or soil under the force of gravity.

Maunder sunspot minimum An interval between 1645 and 1715 A.D. when astronomers observed very few sunspots on the Sun's surface.

medieval climatic optimum An interval between 1100 and 1300 A.D. in which some northern hemisphere regions were warmer than in the Little Ice Age that followed.

Mediterranean overflow water A water mass that forms in the northern Mediterranean Sea by winter chilling of salty surface waters and flows into the North Atlantic at a depth of 1 km.

methane clathrate A partly frozen slushy mix of methane gas (CH_4) and ice.

Milankovitch theory The theory that orbitally controlled fluctuations in high-latitude solar radiation (insolation) during summer control the size of ice sheets through their effect on melting.

millennial oscillations Fluctuations in climate lasting thousands of years and generally larger during glacial than interglacial intervals.

modulation The tendency for peaks and troughs in a wave form to vary in size in a regular way, such that clusters of large peaks and troughs alternate with clusters of smaller ones.

moraine A pile of unsorted rubble (till) deposited by a glacier at its margin.

monsoon Winds that reverse direction seasonally, blowing onshore in summer and offshore in winter because of different rates of heating and cooling of land and water.

Monterey hypothesis The hypothesis that increased rates of burial of organic carbon on the margins of the Pacific Ocean 17 Myr ago reduced CO_2 levels in the surface ocean and atmosphere.

mountain glacier A body of ice tens to hundred of meters thick and kilometers in length confined to a valley at a high elevation. A mountain ice cap is a similar-sized body of ice lying on the rounded summit of a mountain.

no-analog vegetation A combination of types of vegetation in the past for which no similar (analogous) combination exists today.

nonlinear response A climatic response that occurs on other than a simple one-for-one basis in relation to the forcing.

North Atlantic deep water A water mass that forms in the high-latitude North Atlantic Ocean by winter chilling of salty surface water, sinks, and flows southward at depths of 2 to 4 km.

North Atlantic drift A warm, multipart current flowing northeastward into the high latitudes of the North Atlantic as a continuation of the Gulf Stream.

ocean carbon pump hypothesis The hypothesis that changes in the amount of organic carbon taken up by ocean plankton during photosynthesis and exported to the deep ocean after they die control CO_2 levels in the surface ocean and atmosphere.

ocean crust A layer of rock averaging 7 km thick, having the average composition of basalt, and comprising the ocean floor.

ocean heat transport hypothesis The hypothesis that changes in the amount of heat transported toward polar regions by the ocean cause changes in polar climate.

oceanic gateway A narrow passage between continents that opens or closes and thereby alters ocean circulation.

orbital monsoon hypothesis The hypothesis that orbitally controlled changes in summer insolation at low latitudes drive the strength of the tropical summer monsoon.

orbital tuning The process of constructing a time scale by using the link between astronomically dated changes in solar radiation and the rhythmic climatic responses they cause on Earth.

orographic precipitation Precipitation on the upwind side of a mountain or plateau caused by the forced ascent of warm air to cooler elevations,where the entrained water vapor condenses.

outwash Layered sediments deposited by meltwater streams emerging from a glacier.

overkill hypothesis The hypothesis that the sudden extinction of many mammals 12,500 years ago resulted from human hunting rather than climatic stress.

oxidation A chemical reaction in which electrons are lost from an atom and its charge becomes more positive; also, the addition of oxygen to an element.

ozone A triple molecule of oxygen (O_3) formed by the collision of cosmic particles with normal (O_2) oxygen. Ozone in the stratosphere blocks harmful ultraviolet radiation from the Sun.

ozone hole A region centered over the Antarctic continent in which ozone (O_3) levels in the upper atmosphere (stratosphere) drop to very low values in the spring.

paleomagnetism The study of patterns of ancient magnetism recorded in rocks or sediment.

Pangaea The giant supercontinent that existed between 300 and 175 Myr ago and consisted of all landmasses present on Earth.

parent isotope A radioactive isotope that naturally decays to a daughter isotope.

peat A deposit of decayed carbon-rich plant remains in a wetland environment with little oxygen.

peatlands hypothesis The hypothesis that the increase in methane levels between 5000 and 250 years ago was caused by expanded areas of methane-producing bogs in north polar regions.

perihelion The point in Earth's slightly eccentric orbit at which it is closest to the Sun.

period The time interval between successive peaks or troughs in a series of regular wave forms.

peripheral forebulge A region in which the weight of glacial ice sheets caused bedrock to flow out to the ice margins at great depths and produced a broad upward bulge of the land.

permafrost A permanently frozen mixture of rocks and soil occurring in very cold regions.

phase lag The amount by which one cyclic signal lags behind another signal of the same wavelength (or period).

photosynthesis The process by which plants use nutrients and solar energy to convert water and CO_2 to plant tissue (carbohydrates) and thereby produce oxygen.

physical climate model A numerical model that simulates Earth's climate on the basis of physical principles of fluid motion and transfers of radiative heat energy and momentum.

physical weathering Any mechanical process by which rocks are broken into smaller fragments of the same material.

phytoplankton Small floating organisms (usually algae) that use energy from the Sun and nutrients from the water for the process of photosynthesis.

plane of the ecliptic The plane within which Earth revolves around the Sun.

planktic foraminifera Sand-sized organisms (protozoans) that live in ocean surface waters and form shells made of $CaCO_3$.

plankton Organisms that float in the upper layers of oceans or lakes.

plate tectonics Tectonic interactions resulting from the movement of lithospheric plates.

polar front A sharp boundary zone in a polar ocean between cold, low-salinity waters and warmer, saltier waters; similarly, in the atmosphere, a sharp temperature boundary.

polar position hypothesis The hypothesis that ice sheets exist during interval in Earth's history when landmasses are moved into polar regions by plate tectonic processes.

power spectrum A graphic display of the distribution of power (the square of wave amplitude) against the period (or frequency) of each cycle present in a signal.

precessional index The mathematical product ($\epsilon \sin \omega$) of Earth's sine-wave motion ($\sin \omega$) around the Sun and the eccentricity of its orbit (ϵ)

precession of the ellipse The slow turning of Earth's elliptical orbit in space.

precession of the equinoxes The movement of the solstices and equinoxes around Earth's elliptical orbit over cycles of 23,000 and 19,000 years.

preindustrial CO_2 level The concentration of CO_2 in the atmosphere (280 parts per million) that existed for several thousand years before the Industrial Revolution.

productivity The amount of organic matter synthesized by organisms from inorganic substances per unit of area per unit of time.

proglacial lake A short-lived lake that develops after the retreat of an ice sheet in the bedrock depression left by the weight of the ice.

radiation Electromagnetic energy that drives Earth's climate system. Ultraviolet and visible radiation emitted by the Sun affect climate on Earth, which emits infrared back radiation to space.

radiative forcing The effect of greenhouse gases in trapping (or blocking) solar radiation, expressed in the same units as those of solar energy: watts per square meter (W/m^2).

radiocarbon dating Dating of relatively young carbon-bearing geologic materials by means of ^{14}C, a radioactive isotope that decays with a half-life of 5700 years.

radiolaria Sand-sized organisms that live in surface waters and form shells of opal ($SiO_2 \cdot H_2O$).

radiometric dating Determining the ages of rocks or sediments by measuring the amount of naturally occurring radioactive parent isotopes and their nonradioactive daughter products.

reconstruction A simulation run with a climate model by altering of several boundary conditions in an effort to reproduce a climate that existed at some time in the past.

red beds Sediments or rocks with a red color caused by the oxidation of iron (similar to rust).

regressions Relative motions of the ocean down and off the margins of the land.

reservoir A place of residence for an element or isotope that moves in a cycle.

residence time The average amount of time a tracer of any substance spends in a reservoir.

resolution The degree of detail detected in a climate signal by sampling at a particular interval.

resonant response A strong cyclic response of the climate system to perturbations occurring at the same or other cycles.

response Any change in the climate system caused by a change in climate forcing.

response time The time required for a climatic response to move a defined fraction of the way from its existing value to the value it would hold at its full equilibrium response.

salinity A measure of the salt content of seawater in parts per thousand (‰).

salt oscillator The concept that climatic oscillations in and around the North Atlantic result from changes in surface ocean salinity caused by interactions between the surface and deep ocean, the atmosphere, and the ice sheets.

salt rejection The salt left in seawater when sea ice forms on the ocean.

sapropels Black organic-rich muds deposited on the Mediterranean seafloor as a result of strong inflow from the Nile River, which stifles delivery of oxygen to the deep parts of the basin.

saturation vapor density The maximum amount of water vapor that air can hold at a given temperature.

savanna A semi-arid region of grasses and scattered trees.

Scandinavian ice sheet An ice sheet that covered most of Norway and Sweden as well as the northern part of Germany and France during orbital-scale glaciation cycles.

seafloor spreading The mechanism by which new seafloor is created at an ocean ridge as the adjacent plates move away from the ridge crest at a rate of centimeters per year.

sediment drift A lens-shaped pile of fine sediment (clay and silt) picked up in a region where bottom currents move swiftly and then deposited in a region of the seafloor where currents slow.

sensible heat Heat energy carried by water or air in a form that can be easily felt or sensed, rather than hidden in latent form.

sensitivity test A simulation run with a climate model in which one boundary condition is altered from the (modern) control case to test its effect on climate.

shortwave radiation Electromagnetic energy emitted by the Sun in visible and ultraviolet wavelengths that deliver heat to Earth's climate system.

silicate minerasl Minerals rich in silicon and oxygen and accounting for most of the rocks in Earth's crust and mantle.

sine wave A perfectly regular wave form in which successive peaks and troughs are evenly spaced, with each peak reaching a value of +1 and each trough a value of -1.

sintering The process by which bubbles of air become sealed off and preserved as snow turns to ice at depths of 50 to 100 m within an ice sheet.

snowball Earth hypothesis The hypothesis that Earth was frozen even in the tropics sometime in the interval between 850 and 550 Myr ago.

solstices The times during Earth's yearly revolution around the Sun when the days are longest (summer solstice) and shortest (winter solstice).

southern oscillation Naturally occurring fluctuations in which changes in lower atmospheric surface pressure in the far western Pacific, near northern Australia, are opposite in sense to those in the south-central Pacific, near Tahiti.

specific heat The amount of heat required to raise the temperature of 1 gram of a substance by 1°C.

spectral analysis A numerical technique for detecting and quantifying the distribution of regular (periodic) behavior in a complex signal.

Sporer sunspot minimum An interval between 1460 and 1550 A.D. when very few sunspots were observed on the Sun's surface.

stratosphere The stable layer of the atmosphere lying between 10 and 50 km above Earth's surface and containing most of Earth's ozone.

subduction The slow downward sinking of an ocean plate beneath a continent or an island arc as a result of plate tectonic processes.

sulfate aerosols Fine particles produced in the atmosphere from SO_2 gas emitted by volcanoes or by industrial smokestacks. These particles can block incoming solar radiation.

sunspots Dark areas of temperatures lower than average on the surface of the Sun.

sunspot cycle A natural 11-year cycle in the number of dark spots visible on the face of the Sun, reliably recorded by astronomers for more than four centuries.

surge A sudden and rapid forward movement of the margin of a glacier.

tabular iceberg A large flat-topped slab of ice produced by calving from the seaward margin of an ice shelf.

tectonic plates Divisions of the upper solid Earth (lithosphere) that are 100 km thick and thousands of kilometers in lateral extent and move as rigid units in plate tectonic processes.

termination A 10,000-year interval of rapid melting of ice sheets that brings to an end a longer (90,000-year) interval of slower ice growth.

theory A hypothesis that has survived repeated testing.

thermal expansion coefficient The volumetric expansion of seawater when it warms, and contraction when it cools, by 1 part in 7000 per degree C (for temperatures above 4°C).

thermal inertia The resistance of a component of the climate system to temperature change.

thermocline A layer of water in which temperature changes rapidly in a vertical direction.

thermohaline flow The vertical and lateral movement of subsurface waters in the ocean as a result of contrasts in density caused by differences in temperature and salinity.

thermostat A mechanism that senses changes in temperature and acts to moderate them.

threshold A level at which a sudden change in the basic nature of a climatic response occurs.

tilt The angle between Earth's equatorial plane and the plane of its orbit around the Sun, also equivalent to the angle between Earth's axis of rotation and a line perpendicular to its axis of rotation around the Sun. Also referred to as *obliquity*.

time-dependent model A geochemical model that tracks changes in the rate of movement of tracers among reservoirs within the climate system over time.

time-series analysis A group of techniques for extracting periodic signals (cycles) from complex signals and quantifying their strength, relative timing, and correlation.

transform fault margin A boundary between lithospheric plates in which the plates slide past each other.

transgression Relative movement of the ocean up and across the margins of the land.

transpiration The release of water vapor by plants into the atmosphere.

tree rings Annual bands formed by trees in regions of seasonal climate, with lighter layers formed during rapid growth in the spring and darker layers at the end of growth in the autumn.

troposphere The layer of the atmosphere just above Earth's surface (10 km or more in thickness) in which weather occurs.

tundra A high-latitude or high-altitude environment in which the ground freezes deeply in winter but thaws at the surface in summer, permitting low-growing plants to flourish.

2xCO₂ sensitivity The amount of warming produced by an increase in atmospheric CO_2 levels from the preindustrial level (280 parts per million) to a level of 560 parts per million.

uplift weathering hypothesis The hypothesis that tectonic-scale climate changes are caused when uplift of plateaus and mountains alters the amount of CO_2 removed from the atmosphere by chemical weathering of fragmented rock.

upwelling The rise of cool, nutrient-rich subsurface water to the ocean surface to replace warm nutrient-poor surface water.

urban heat island An urban area where asphalt and other heat-absorbing surfaces absorb solar radiation during the day and radiate it back at night, keeping the area unusually warm.

varves Alternating layers of dark and light (or coarse and fine) sediment that accumulate in annual couplets in lakes or in ocean margin basins with no water turbulence near the bottom.

vegetation-albedo feedback A positive feedback that amplifies an initial temperature change through vegetation changes that alter Earth's surface albedo and the absorption of solar radiation.

vegetation-precipitation feedback A positive feedback that amplifies an initial precipitation change through vegetation changes that alter the transpiration of water vapor and the availability of moisture.

viscous Characterized by a consistency such that the material in question deforms slowly under pressure (as is bedrock under the pressure of a glacier) and only slowly regains its initial shape when the pressure is removed.

warm saline deep water Deep water proposed to have formed in warm salty tropical seas in the past, in contrast to modern sources of deep water in cold polar regions.

water vapor feedback A positive feedback that amplifies an initial change in temperature by altering the amount of water vapor held by the atmosphere and changing the amount of Earth's back radiation trapped by the atmosphere.

wavelength The distance between successive peaks or troughs in a series of regular wave forms.

weather Fluctuations in temperature, precipitation, and winds on time scales of less than a year.

Younger Dryas An interval during the middle of the last deglaciation (near 12,000 years ago) marked by slower melting of ice sheets and a major cooling in the North Atlantic region.

Index